Characterization and Properties of Petroleum Fractions

Characterization and Properties of Petroleum Fractions
First Edition

M. R. Riazi
Professor of Chemical Engineering
Kuwait University

P.O. Box 5969
Safat 13060, Kuwait
riazi@kuc01.kuniv.edu.kw

ASTM Stock Number: MNL50

ASTM
100 Barr Harbor
West Conshohocken, PA 19428-2959

Printed in the U.S.A.

013618106

chem

Library of Congress Cataloging-in-Publication Data

Riazi, M.-R.
    Characterization and properties of petroleum fractions / M.-R. Riazi—1st ed.
      p.   cm.—(ASTM manual series: MNL50)
    ASTM stock number: MNL50
    Includes bibliographical references and index.
    ISBN 0-8031-3361-8
    1. Characterization.   2. Physical property estimation.   3. Petroleum fractions—crude oils.

TP691.R64   2005
666.5—dc22

2004059586

NOTE: This publication does not purport to address all of the safety problems associated with its use. It is the responsibility of the user of this publication to establish appropriate safety and health practices and determine the applicability of regulatory limitations prior to use.

Printed in Philadelphia, PA
January 2005

To
Shiva, Touraj, and Nazly

Contents

Foreword

THIS PUBLICATION, *Characterization and Properties of Petroleum Fractions*, was sponsored by ASTM Committee D02 on Petroleum Fuels and Lubricants. The author is M. R. Riazi, Professor of Chemical Engineering, Kuwait University, Safat, Kuwait. This publication is Manual 50 of ASTM's manual series.

Preface

Scientists do not belong to any particular country, ideology, or religion, they belong to the world community

THE FIELD OF Petroleum Characterization and Physical Properties has received significant attention in recent decades with the expansion of computer simulators and advanced analytical tools and the availability of more accurate experimental data. As a result of globalization, structural changes are taking place in the chemical and petroleum industry. Engineers working in these industries are involved with process simulators to design and operate various units and equipment. Nowadays, a large number of process simulators are being produced that are equipped with a variety of thermodynamic models and choice of predictive methods for the physical properties. A person familiar with development of such methods can make appropriate use of these simulators saving billions of dollars in costs in investment, design, manufacture, and operation of various units in these industries. Petroleum is a complex mixture of thousands of hydrocarbon compounds and it is produced from an oil well in a form of reservoir fluid. A reservoir fluid is converted to a crude oil through surface separation units and then the crude is sent to a refinery to produce various petroleum fractions and hydrocarbon fuels such as kerosene, gasoline, and fuel oil. Some of the refinery products are the feed to petrochemical plants. More than half of world energy sources are from petroleum and probably hydrocarbons will remain the most convenient and important source of energy and as a raw material for the petrochemical plants at least throughout the 21st century. Other fossil type fuels such as coal liquids are also mixtures of hydrocarbons although they differ in type with petroleum oils. From 1970 to 2000, the share of Middle East in the world crude oil reserves raised from 55 to 65%, but this share is expected to rise even further by 2010–2020 when we near the point where half of oil reserves have been produced. The world is not running out of oil yet but the era of cheap oil is perhaps near the end. Therefore, economical use of the remaining oil and treatment of heavy oils become increasingly important. As it is discussed in Chapter 1, use of more accurate physical properties for petroleum fractions has a direct and significant impact on economical operation and design of petroleum processing and production units which in turn would result in a significant saving of existing petroleum reserves.

One of the most important tasks in petroleum refining and related processes is the need for reliable values of the volumetric and thermodynamic properties for pure hydrocarbons and their mixtures. They are important in the design and operation of almost every piece of processing equipment. Reservoir engineers analyze PVT and phase behavior of reservoir fluids to estimate the amount of oil or gas in a reservoir, to determine an optimum operating condition in a separator unit, or to develop a recovery process for an oil or gas field. However, the most advanced design approaches or the most sophisticated simulators cannot guarantee the optimum design or operation of a unit if required input physical properties are not accurate. A process to experimentally determine the volumetric, thermodynamic, and transport properties for all the industrially important materials would be prohibitive in both cost and time; indeed it could probably never be completed. For these reasons accurate estimations of these properties are becoming increasingly important.

Characterization factors of many types permeate the entire field of physical, thermodynamic, and transport property prediction. Average boiling points, specific gravity, molecular weight, critical temperature, critical pressure, acentric factor, refractive index, and certain molecular type analysis are basic parameters necessary to utilize methods of correlation and prediction of the thermophysical properties. For correlating physical and thermodynamic properties, methods of characterizing undefined mixtures are necessary to provide input data. It could be imagined that the best method of characterizing a mixture is a complete analysis. However, because of the complexity of undefined mixtures, complete analyses are usually impossible and, at best, inconvenient. A predictive method to determine the composition or amount of sulfur in a hydrocarbon fuel is vital to see if a product meets specifications set by the government or other authorities to protect the environment.

My first interaction with physical properties of petroleum fluids was at the time that I was a graduate student at Penn State in the late 70s working on a project related to enhanced oil recovery for my M.S. thesis when I was looking for methods of estimation of properties of petroleum fluids. It was such a need and my personal interest that later I joined the ongoing API project on thermodynamic and physical properties of petroleum fractions to work for my doctoral thesis. Since that time, property estimation and characterization of various petroleum fluids has remained one of my main areas of research. Later in the mid-80s I rejoined Penn State as a faculty member and I continued my work with the API which resulted in development of methods for several chapters of the API Technical Data Book. Several years later in late 80s, I continued the work while I was working at the Norwegian Institute of Technology (NTH) at Trondheim where I developed some characterization techniques for heavy petroleum fractions as well as measuring methods for some physical properties. In the 90s while at Kuwait University I got the opportunity to be in direct contact with the oil companies in the region through research, consultation, and conducting special courses for the industry. My association with the University of Illinois at Chicago in early 90s was helpful in the development of equations of state based on velocity of sound. The final revision of the book was completed when I was associated with the University of Texas at Austin and McGill University in Montreal during my leave from Kuwait University.

Characterization methods and estimating techniques presented in this book have been published in various international journals or technical handbooks and included in many commercial softwares and process simulators. They have also been presented as seminars in different oil companies, universities, and research centers worldwide. The major characteristics of these methods are simplicity, generality, accuracy, and availability of input parameters. Many of these methods have been developed by utilizing some scientific fundamentals and combining them with a broad experimental data set to generate semi-theoretical or semi-empirical correlations. Some of these methods have been in use by the petroleum industry and research centers worldwide for the past two decades.

Part of the materials in this book were prepared when I was teaching a graduate course in applied thermodynamics in 1988 while at NTH. The materials, mainly a collection of technical papers, have been continuously updated and rearranged to the present time. These notes have also been used to conduct industrial courses as well as a course on fluid properties in chemical and petroleum engineering. This book is an expansion with complete revision and rewriting of these notes. The main objective of this book is to present the fundamentals and practice of estimating the physical and thermodynamic properties as well as characterization methods for hydrocarbons, petroleum fractions, crude oils, reservoir fluids, and natural gases, as well as coal liquids. However, the emphasis is on the liquid petroleum fractions, as properties of gases are generally calculated more accurately. The book will emphasize manual calculations with practical problems and examples but also will provide good understanding of techniques used in commercial software packages for property estimations. Various methods and correlations developed by different researchers commonly used in the literature are presented with necessary discussions and recommendations.

My original goal and objective in writing this book was to provide a reference for the petroleum industry in both processing and production. It is everyone's experience that in using thermodynamic simulators for process design and equipment, a large number of options is provided to the user for selection of a method to characterize the oil or to get an estimate of a physical property. This is a difficult choice for a user of a simulator, as the results of design calculations significantly rely on the method chosen to estimate the properties. One of my goals in writing this book was to help users of simulators overcome this burden. However, the book is written in a way that it can also be used as a textbook for graduate or senior undergraduate students in chemical, petroleum, or mechanical engineering to understand the significance of characterization, property estimation and methods of their development. For this purpose a set of problems is presented at the end of each chapter. The book covers characterization as well as methods of estimation of thermodynamic and transport properties of various petroleum fluids and products. A great emphasis is given to treatment of heavy fractions throughout the book. An effort was made to write the book in a way that not only would be useful for the professionals in the field, but would also be easily understandable to those non-engineers such as chemists, physicists, or mathematicians who get involved with the petroleum industry. The word *properties* in the title refers to thermodynamic, physical, and transport properties. Properties related to the quality and safety of petroleum products are also discussed. Organization of the book, its uses, and importance of the methods are discussed in detail in Chapter 1. Introduction of similar books and the need for the present book as well as its application in the industry and academia are also discussed in Chapter 1. Each chapter begins with nomenclature and ends with the references used in that chapter. Exercise problems in each chapter contain additional information and methods. More specific information about each chapter and its contents are given in Chapter 1. As Goethe said, "Things which matter most must never be at the mercy of things which matter least."

I am indebted to many people especially teachers, colleagues, friends, students, and, above all, my parents, who have been so helpful throughout my academic life. I am particularly thankful to Thomas E. Daubert of Pennsylvania State University who introduced to me the field of physical properties and petroleum characterization in a very clear and understandable way. Likewise, I am thankful to Farhang Shadman of the University of Arizona who for the first time introduced me to the field of chemical engineering research during my undergraduate studies. These two individuals have been so influential in shaping my academic life and I am so indebted to them for their human characters and their scientific skills. I have been fortunate to meet and talk with many scientists and researchers from both the oil industry and academia from around the world during the last two decades whose thoughts and ideas have in many ways been helpful in shaping the book.

I am also grateful to the institutions, research centers, and oil companies that I have been associated with or that have invited me for lecturing and consultation. Thanks to Kuwait University as well as Kuwait Petroleum Corporation (KPC) and KNPC, many of whose engineers I developed working relations with and have been helpful in evaluation of many of the estimating methods throughout the years. I am thankful to all scientists and researchers whose works have been used in this book and I hope that all have been correctly and appropriately cited. I would be happy to receive their comments and suggestions on the book. Financial support from organizations such as API, NSF, GPA, GRI, SINTEF, Petrofina Exploration Norway, NSERC Canada, Kuwait University, and KFAS that was used for my research work over the past two decades is also appreciated.

I am grateful to ASTM for publishing this work and particularly to Geroge Totten who was the first to encourage me to begin writing this book. His advice, interest, support, and suggestions through the long years of writing the book have been extremely helpful in completing this project. The introductory comments from him as well as those from Philip T. Eubank and José Luis Peña Díez for the back cover are appreciated. I am also grateful to the four unanimous reviewers who tirelessly reviewed the entire and lengthy manuscript with their constructive comments and suggestions which have been reflected in the book. Thanks also to Kathy Dernoga, the publishing manager at ASTM, who was always cooperative and ready to answer my questions and provided me with necessary information and tools during the preparation of this manuscript. Her encouragements and assistance were quite useful in pursuing this work. She also was helpful in the design of the front and back covers of the book as well as providing editorial suggestions. I am thankful to Roberta Storer and Joe Ermigiotti for their excellent job of editing and updating the manuscript. Cooperation of other ASTM staff, especially Monica Siperko, Carla J. Falco, and Marsha Firman is highly appreciated. The art work and most of the graphs and figures were prepared by Khaled Damyar of Kuwait University and his efforts for the book are appreciated. I also sincerely appreciate the publishers and the organizations that gave their permissions to include some published materials, in particular API, ACS, AIChE, GPA, Elsevier (U.K.), editor of Oil & Gas J., McGraw-Hill, Marcel and Dekker, Wiley, SPE, and Taylor and Francis. Thanks to the manager and personnel of KISR for allowing the use of photos of their instruments in the book. Finally and most importantly, I must express my appreciation and thanks to my family who have been helpful and patient during all these years and without whose cooperation, moral support, understanding, and encouragement this project could never have been undertaken. This book is dedicated to my family, parents, teachers, and the world scientific community.

M. R. Riazi
August 2004

Introduction

1

NOMENCLATURE

API	API gravity
A%	Percent of aromatics in a petroleum fraction
D	Diffusion coefficient
CH	Carbon-to-hydrogen weight ratio
d	Liquid density at 20°C and 1 atm
K_W	Watson K factor
k	Thermal conductivity
K_i	Equilibrium ratio of component i in a mixture
$\log_{10}$	Logarithm of base 10
ln	Logarithm of base e
M	Molecular weight
N_{min}	Minimum number of theoretical plates in a distillation column
N%	Percent of naphthenes in a petroleum fraction
n	Sodium D line refractive index of liquid at 20°C and 1 atm, dimensionless
n	Number of moles
P	Pressure
P_c	Critical pressure
P^{sat}	Vapor (saturation) pressure
P%	Percent of paraffins in a petroleum fraction
R	Universal gas constant
R_i	Refractivity intercept
SG	Specific gravity at 15.5°C (60°F)
SUS	Saybolt Universal Seconds (unit of viscosity)
S%	Weight % of sulfur in a petroleum fraction
T	Temperature
T_b	Boiling point
T_c	Critical temperature
T_F	Flash point
T_P	Pour point
T_M	Melting (freezing point) point
V	Volume
x_m	Mole fraction of a component in a mixture
x_v	Volume fraction of a component in a mixture
x_w	Weight fraction of a component in a mixture
y	Mole fraction of a component in a vapor phase

Greek Letters

α	Relative volatility
φ	Fugacity coefficient
ω	Acentric factor
σ	Surface tension
ρ	Density at temperature T and pressure P
μ	Viscosity
ν	Kinematic viscosity

Acronyms

API-TDB	American Petroleum Institute–Technical Data Book
bbl	Barrel
GOR	Gas-to-oil ratio
IUPAC	International Union of Pure and Applied Chemistry
PNA	Paraffin, naphthene, aromatic content of a petroleum fraction
SC	Standard conditions
scf	Standard cubic feet
stb	Stock tank barrel
STO	Stock tank oil
STP	Standard temperature and pressure

IN THIS INTRODUCTORY CHAPTER, first the nature of petroleum fluids, hydrocarbon types, reservoir fluids, crude oils, natural gases, and petroleum fractions are introduced and then types and importance of characterization and physical properties are discussed. Application of materials covered in the book in various parts of the petroleum industry or academia as well as organization of the book are then reviewed followed by specific features of the book and introduction of some other related books. Finally, units and the conversion factors for those parameters used in this book are given at the end of the chapter.

1.1 NATURE OF PETROLEUM FLUIDS

Petroleum is one of the most important substances consumed by man at present time. It is used as a main source of energy for industry, heating, and transportation and it also provides the raw materials for the petrochemical plants to produce polymers, plastics, and many other products. The word *petroleum*, derived from the Latin words *petra* and *oleum*, means literally *rock oil* and a special type of oil called *oleum* [1]. Petroleum is a complex mixture of hydrocarbons that occur in the sedimentary rocks in the form of gases (natural gas), liquids (crude oil), semisolids (bitumen), or solids (wax or asphaltite). Liquid fuels are normally produced from liquid hydrocarbons, although conversion of nonliquid hydrocarbons such as coal, oil shale, and natural gas to liquid fuels is being investigated. In this book, only petroleum hydrocarbons in the form of gas or liquid, simply called *petroleum fluids*, are considered. Liquid petroleum is also simply called *oil*. Hydrocarbon gases in a reservoir are called a *natural gas* or simply a *gas*. An underground reservoir that contains hydrocarbons is called *petroleum reservoir* and its hydrocarbon contents that can be recovered through a producing well is called *reservoir fluid*. Reservoir fluids in the reservoirs are usually in contact with water in porous media conditions and because they are lighter than water, they stay above the water level under natural conditions.

Although petroleum has been known for many centuries, the first oil-producing well was discovered in 1859 by E.L. Drake in the state of Pennsylvania and that marked the birth of modern petroleum technology and refining. The main elements of petroleum are carbon (C) and hydrogen (H) and some small quantities of sulfur (S), nitrogen (N), and oxygen (O). There are several theories on the formation of petroleum. It is generally believed that petroleum is derived from aquatic plants and animals through conversion of organic compounds into hydrocarbons. These animals and plants under aquatic conditions have converted inorganic compounds dissolved in water (such as carbon dioxide) to organic compounds through the energy provided by the sun. An example of such reactions is shown below:

$$(1.1) \qquad 6CO_2 + 6H_2O + energy \rightarrow 6O_2 + C_6H_{12}O_6$$

in which $C_6H_{12}O_6$ is an organic compound called carbohydrate. In some cases organic compounds exist in an aquatic environment. For example, the Nile river in Egypt and the Uruguay river contain considerable amounts of organic materials. This might be the reason that most oil reservoirs are located near the sea. The organic compounds formed may be decomposed into hydrocarbons under certain conditions.

$$(1.2) \qquad (CH_2O)_n \rightarrow xCO_2 + yCH_4$$

in which n, x, y, and z are integer numbers and yCH_z is the closed formula for the produced hydrocarbon compound. Another theory suggests that the inorganic compound calcium carbonate ($CaCO_3$) with alkali metal can be converted to calcium carbide (CaC_2), and then calcium carbide with water (H_2O) can be converted to acetylene (C_2H_2). Finally, acetylene can be converted to petroleum [1]. Conversion of organic matters into petroleum is called *maturation*. The most important factors in the conversion of organic compounds to petroleum hydrocarbons are (1) heat and pressure, (2) radioactive rays, such as gamma rays, and (3) catalytic reactions. Vanadium- and nickel-type catalysts are the most effective catalysts in the formation of petroleum. For this reason some of these metals may be found in small quantities in petroleum fluids. The role of radioactive materials in the formation of hydrocarbons can be best observed through radioactive bombarding of fatty acids (RCOOH) that form paraffin hydrocarbons. Occasionally traces of radioactive materials such as uranium and potassium can also be found in petroleum. In summary, the following steps are required for the formation of hydrocarbons: (1) a source of organic material, (2) a process to convert

organic compounds into petroleum, and (3) a sealed reservoir space to store the hydrocarbons produced. The conditions required for the process of conversion of organic compounds into petroleum (as shown through Eq. (1.2) are (1) geologic time of about 1 million years, (2) maximum pressure of about 17 MPa (2500 psi), and (3) temperature not exceeding 100–120°C (~210–250°F). If a leak occurred sometime in the past, the exploration well will encounter only small amounts of residual hydrocarbons. In some cases bacteria may have biodegraded the oil, destroying light hydrocarbons. An example of such a case would be the large heavy oil accumulations in Venezuela. The hydrocarbons generated gradually migrate from the original beds to more porous rocks, such as sandstone, and form a petroleum *reservoir*. A series of reservoirs within a common rock is called an *oil field*. Petroleum is a mixture of hundreds of different identifiable hydrocarbons, which are discussed in the next section. Once petroleum is accumulated in a reservoir or in various sediments, hydrocarbon compounds may be converted from one form to another with time and varying geological conditions. This process is called *in-situ alteration*, and examples of chemical alteration are thermal maturation and microbial degradation of the reservoir oil. Examples of physical alteration of petroleum are the preferential loss of low-boiling constituents by the diffusion or addition of new materials to the oil in place from a source outside the reservoir [1]. The main difference between various oils from different fields around the world is the difference in their composition of hydrocarbon compounds. Two oils with exactly the same composition have identical physical properties under the same conditions [2].

A good review of statistical data on the amount of oil and gas reservoirs, their production, processing, and consumption is usually reported yearly by the *Oil and Gas Journal* (OGJ). An annual refinery survey by OGJ is usually published in December of each year. OGJ also publishes a forecast and review report in January and a midyear forecast report in July of each year. In 2000 it was reported that total proven oil reserves is estimated at 1016 billion bbl (1.016×10^{12} bbl), which for a typical oil is equivalent to approximately 1.39×10^{11} tons. The rate of oil production was about 64.6 million bbl/d (~3.23 billion ton/year) through more than 900 000 producing wells and some 750 refineries [3, 4]. These numbers vary from one source to another. For example, Energy Information Administration of US Department of Energy reports world oil reserves as of January 1, 2003 as 1213.112 billion bbl according to OGJ and 1034.673 billion bbl according to *World Oil* (www.eia.doe.gov/emeu/iea). According to the OGJ worldwide production reports (*Oil and Gas Journal*, Dec. 22, 2003, p. 44), world oil reserves estimates changed from 999.78 in 1995 to 1265.811 billion bbl on January 1, 2004. For the same period world gas reserves estimates changed from 4.98×10^{15} scf to 6.0683×10^{15} scf. In 2003 oil consumption was about 75 billion bbl/day, and it is expected that it will increase to more than 110 million bbl/day by the year 2020. This means that with existing production rates and reserves, it will take nearly 40 years for the world's oil to end. Oil reserves life (reserves-to-production ratio) in some selected countries is given by OGJ (Dec. 22, 2004, p. 45). According to 2003 production rates, reserves life is 6.1 years in UK, 10.9 years in US, 20 years in Russia, 5.5 years in Canada, 84 years in Saudi Arabia, 143 years in Kuwait, and 247 years in Iraq. As in January 1, 2002, the total number of world oil wells was 830 689, excluding shut or service wells (OGJ, Dec. 22, 2004). Estimates of world oil reserves in 1967 were at 418 billion and in 1987 were at 896 billion bbl, which shows an increase of 114% in this period [5]. Two-thirds of these reserves are in the Middle East, although this portion depends on the type of oil considered. Although some people believe the Middle East has a little more than half of world oil reserves, it is believed that many undiscovered oil reservoirs exist offshore under the sea, and with increase in use of the other sources of energy, such as natural gas or coal, and through energy conservation, oil production may well continue to the end of the century. January 2000, the total amount of gas reserves was about 5.15×10^{15} scf, and its production in 1999 was about 200×10^9 scf/d (5.66×10^9 sm^3/d) through some 1500 gas plants [3]. In January 2004, according to OGJ (Dec. 22, 2004, p. 44), world natural gas reserves stood at 6.068×10^{15} scf (6068.302 trillion scf). This shows that existing gas reserves may last for some 70 years. Estimated natural gas reserves in 1975 were at 2.5×10^{15} scf (7.08×10^{13} sm^3), that is, about 50% of current reserves [6]. In the United States, consumption of oil and gas in 1998 was about 65% of total energy consumption. Crude oil demand in the United State in 1998 was about 15 million bbl/d, that is, about 23% of total world crude production [3]. Worldwide consumption of natural gas as a clean fuel is on the rise, and attempts are underway to expand the transfer of natural gas through pipelines as well as its conversion to liquid fuels such as gasoline. The world energy consumption is distributed as 35% through oil, 31% through coal, and 23% through natural gas. Nearly 11% of total world energy is produced through nuclear and hydroelectric sources [1].

1.1.1 Hydrocarbons

In early days of chemistry science, chemical compounds were divided into two groups: inorganic and organic, depending on their original source. Inorganic compounds were obtained from minerals, while organic compounds were obtained from living organisms and contained carbon. However, now organic compounds can be produced in the laboratory. Those organic compounds that contain only elements of carbon (C) and hydrogen (H) are called *hydrocarbons*, and they form the largest group of organic compounds. There might be as many as several thousand different hydrocarbon compounds in petroleum reservoir fluids. Hydrocarbon compounds have a general closed formula of $C_x H_y$, where x and y are integer numbers. The lightest hydrocarbon is methane (CH_4), which is the main component in a natural gas. Methane is from a group of hydrocarbons called *paraffins*. Generally, hydrocarbons are divided into four groups: (1) paraffins, (2) olefins, (3) naphthenes, and (4) aromatics. Paraffins, olefins, and naphthenes are sometime called *aliphatic* versus aromatic compounds. The International Union of Pure and Applied Chemistry (IUPAC) is a nongovernment organization that provides standard names, nomenclature, and symbols for different chemical compounds that are widely used [7]. The relationship between the various hydrocarbon constituents of crude oils is hydrogen addition or hydrogen loss. Such

interconversion schemes may occur during the formation, maturation, and *in-situ* alteration of petroleum.

Paraffins are also called alkanes and have the general formula of $C_n H_{2n+2}$, where n is the number of carbon atoms. Paraffins are divided into two groups of normal and isoparaffins. Normal paraffins or normal alkanes are simply written as *n*-paraffins or *n*-alkanes and they are open, straight-chain saturated hydrocarbons. Paraffins are the largest series of hydrocarbons and begin with methane (CH_4), which is also represented by C_1. Three *n*-alkanes, methane (C_1), ethane (C_2), and *n*-butane (C_4), are shown below:

Methane   Ethane   *n*-Butane

(CH_4)   (C_2H_6)   (C_4H_{10})

The open formula for *n*-C_4 can also be shown as CH_3—CH_2—CH_2—CH_3 and for simplicity in drawing, usually the CH_3 and CH_2 groups are not written and only the carbon–carbon bonds are drawn. For example, for a *n*-alkane compound of *n*-heptadecane with the formula of $C_{17}H_{36}$, the structure can also be shown as follows:

n-Heptadecane ($C_{17}H_{36}$)

The second group of paraffins is called *isoparaffins*; these are branched-type hydrocarbons and begin with isobutane (methylpropane), which has the same closed formula as *n*-butane (C_4H_{10}). Compounds of different structures but the same closed formula are called *isomers*. Three branched or isoparaffin compounds are shown below:

isobutane   isopentane (methylbutane)   isooctane (2-methylheptane)

(C_4H_{10})   (C_5H_{12})   (C_8H_{18})

In the case of isooctane, if the methyl group (CH_3) is attached to another carbon, then we have another compound (i.e., 3-methylheptane). It is also possible to have more than one branch of CH_3 group, for example, 2,3-dimethylhexane and 2-methylheptane, which are simply shown as following:

2-Methylheptane (C_8H_{18})   2,3-Dimethylhexane (C_8H_{18})

Numbers refer to carbon numbers where the methyl group is attached. For example, 1 refers to the first carbon either

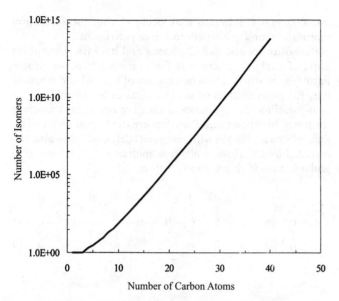

FIG. 1.1—Number of possible alkane isomers.

from the right or from the left. There are 2 isomers for butane and 3 for pentane, but there are 5 isomers for hexane, 9 for heptane, 18 for octane (C_8H_{18}), and 35 for nonane. Similarly, dodecane ($C_{12}H_{26}$) has 355, while octadecane ($C_{18}H_{38}$) has 60 523 and C_{40} has 62×10^{12} isomers [1, 8, 9]. The number of isomers rapidly increases with the number of carbon atoms in a molecule because of the rapidly rising number of their possible structural arrangements as shown in Fig. 1.1. For the paraffins in the range of C_5–C_{12}, the number of isomers is more than 600 although only about 200–400 of them have been identified in petroleum mixtures [10]. Isomers have different physical properties. The same increase in number of isomers with molecular weight applies to other hydrocarbon series. As an example, the total number of hydrocarbons (from different groups) having 20 carbon atoms is more than 300 000 [10]!

Under standard conditions (SC) of 20°C and 1 atm, the first four members of the alkane series (methane, ethane, propane, and butane) are in gaseous form, while from C_5H_{12} (pentane) to *n*-hexadecane ($C_{16}H_{36}$) they are liquids, and from *n*-heptadecane ($C_{17}H_{38}$) the compounds exist as waxlike solids at this standard temperature and pressure. Paraffins from C_1 to C_{40} usually appear in crude oil and represent up to 20% of crude by volume. Since paraffins are fully saturated (no double bond), they are stable and remain unchanged over long periods of geological time.

Olefins are another series of noncyclic hydrocarbons but they are unsaturated and have at least one double bond between carbon–carbon atoms. Compounds with one double bond are called monoolefins or alkenes, such as ethene (also named ethylene: $CH_2=CH_2$) and propene or propylene ($CH_2=CH-CH_3$). Besides *structural isomerism* connected with the location of double bond, there is another type of isomerism called *geometric isomerism*, which indicates the way atoms are oriented in space. The configurations are differentiated in their names by the prefixes *cis-* and *trans-* such as *cis-* and *trans-*2-butene. Monoolefins have a general formula of C_nH_{2n}. If there are two double bonds, the olefin is called *diolefin* (or diene), such as butadiene ($CH_2=CH-CH=CH_2$).

Unsaturated compounds are more reactive than saturated hydrocarbons (without double bond). Olefins are uncommon in crude oils due to their reactivity with hydrogen that makes them saturated; however, they can be produced in refineries through cracking reactions. Olefins are valuable products of refineries and are used as the feed for petrochemical plants to produce polymers such as polyethylene. Similarly compounds with triple bonds such as acetylene ($CH{\equiv}CH$) are not found in crude oils because of their tendency to become saturated [2].

Naphthenes or cycloalkanes are ring or cyclic saturated hydrocarbons with the general formula of C_nH_{2n}. Cyclopentane (C_5H_{10}), cyclohexane (C_6H_{12}), and their derivatives such as *n*-alkylcyclopentanes are normally found in crude oils. Three types of naphthenic compounds are shown below:

Cyclopentane	Methylcyclopentane	Ethylcyclohexane
(C_5H_{10})	(C_6H_{12})	(C_8H_{16})

If there is only one alkyl group from *n*-paraffins (i.e., methyl, ethyl, propyl, *n*-butyl, . . .) attached to a cyclopentane hydrocarbon, the series is called *n*-alkylcyclopentanes, such as the two hydrocarbons shown above where on each junction of the ring there is a CH_2 group except on the alkyl group juncture where there is only a CH group. For simplicity in drawing, these groups are not shown. Similarly there is a homologous naphthenic series of *n*-alkylcyclohexanes with only one saturated ring of cyclohexane, such as ethylcyclohexane shown above. Naphthenic hydrocarbons with only one ring are also called *monocycloparaffins* or *mononaphthenes*. In heavier oils, saturated multirings attached to each other called *polycycloparaffins* or *polynaphthenes* may also be available. Thermodynamic studies show that naphthene rings with five and six carbon atoms are the most stable naphthenic hydrocarbons. The content of cycloparaffins in petroleum may vary up to 60%. Generally, any petroleum mixture that has hydrocarbon compounds with five carbon atoms also contains naphthenic compounds.

Aromatics are an important series of hydrocarbons found in almost every petroleum mixture from any part of the world. Aromatics are cyclic but unsaturated hydrocarbons that begin with benzene molecule (C_6H_6) and contain carbon–carbon double bonds. The name aromatic refers to the fact that such hydrocarbons commonly have fragrant odors. Four different aromatic compounds are shown below:

(C_6H_6)	(C_7H_8)	(C_8H_{10})	($C_{10}H_8$)
Benzene	Toluene	O-xylene	Naphthalene
	(Methylbenzene)	(1,2-Dimethylbenzene)	

In the above structures, on each junction on the benzene ring where there are three bonds, there is only a group of CH, while at the junction with an alkylgroup (i.e., toluene) there is only a C atom. Although benzene has three carbon–carbon double bonds, it has a unique arrangement of electrons that allows benzene to be relatively unreactive. Benzene is, however, known to be a cancer-inducing compound [2]. For this reason, the amount of benzene allowed in petroleum products such as gasoline or fuel oil is limited by government regulations in many countries. Under SC, benzene, toluene, and xylene are in liquid form while naphthalene is in a solid state. Some of the common aromatics found in petroleum and crude oils are benzene and its derivatives with attached methyl, ethyl, propyl, or higher alkyl groups. This series of aromatics is called *alkylbenzenes* and compounds in this homologous group of hydrocarbons have a general formula of C_nH_{2n-6} (where $n \geq 6$). Generally, aromatic series with only one benzene ring are also called *monoaromatics* (MA) or mononuclear aromatics. Naphthalene and its derivatives, which have only two unsaturated rings, are sometime called *diaromatics*. Crude oils and reservoir fluids all contain aromatic compounds. However, heavy petroleum fractions and residues contain multi-unsaturated rings with many benzene and naphthene rings attached to each other. Such aromatics (which under SC are in solid form) are also called *polyaromatics* (PA) or polynuclear aromatics (PNA). In this book terms of mono and polyaromatics are used. Usually, heavy crude oils contain more aromatics than do light crudes. The amount of aromatics in coal liquids is usually high and could reach as high as 98% by volume. It is common to have compounds with napthenic and aromatic rings side by side, especially in heavy fractions. Monoaromatics with one napthenic ring have the formula of C_nH_{2n-8} and with two naphthenic rings the formula is C_nH_{2n-8}. There are many combinations of alkyl-naphthenoaromatics [1, 7].

Normally, high-molecular-weight polyaromatics contain several *heteroatoms* such as sulfur (S), nitrogen (N), or oxygen (O) but the compound is still called an aromatic hydrocarbon. Two types of these compounds are shown below [1]:

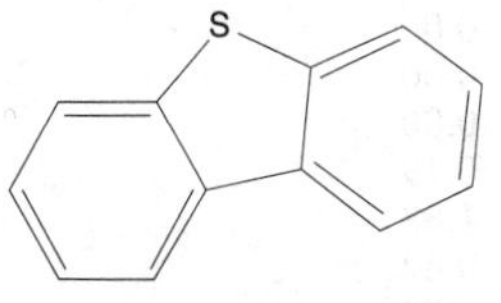

Dibenzothiophene 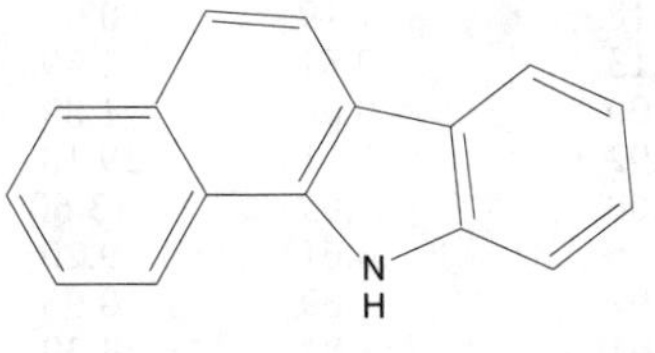   Benzocarbazole ($C_{16}H_{11}N$)

Except for the atoms S and N, which are specified in the above structures, on other junctions on each ring there is either a CH group or a carbon atom. Such heteroatoms in multiring aromatics are commonly found in asphaltene compounds as shown in Fig. 1.2, where for simplicity, C and H atoms are not shown on the rings.

Sulfur is the most important heteroatom in petroleum and it can be found in cyclic as well as noncyclic compounds such as mercaptanes (R—S—H) and sulfides (R—S—R'), where R and R' are alkyl groups. Sulfur in natural gas is usually found in the form of hydrogen sulfide (H_2S). Some natural gases

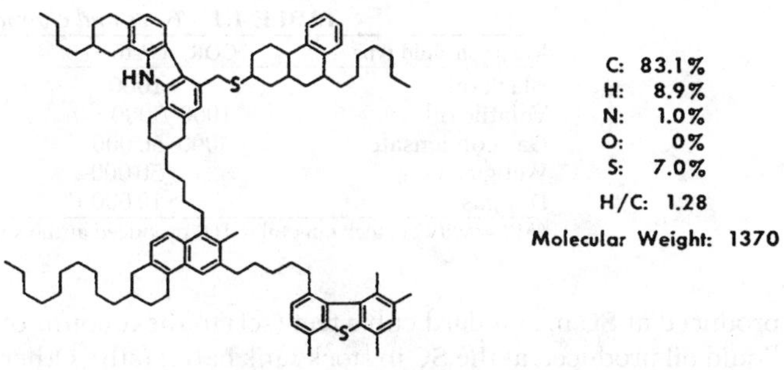

```
C:  83.1%
H:   8.9%
N:   1.0%
O:     0%
S:   7.0%

H/C: 1.28

Molecular Weight: 1370
```

FIG. 1.2—An example of asphaltene molecule. Reprinted from Ref. [1], p. 463, by courtesy of Marcel Dekker, Inc.

contain H_2S as high as 30% by volume. The amount of sulfur in a crude may vary from 0.05 to 6% by weight. In Chapter 3, further discussion on the sulfur contents of petroleum fractions and crude oils will be presented. The presence of sulfur in finished petroleum products is harmful, for example, the presence of sulfur in gasoline can promote corrosion of engine parts. Amounts of nitrogen and oxygen in crude oils are usually less than the amount of sulfur by weight. In general for petroleum oils, it appears that the compositions of elements vary within fairly narrow limits; on a weight basis they are [1]

Carbon (C), 83.0–87.0%
Hydrogen (H), 10.0–14.0%
Nitrogen (N), 0.1–2.0%
Oxygen (O), 0.05–1.5%
Sulfur (S), 0.05–6.0%
Metals (Nickel, Vanadium, and Copper), <1000 ppm (0.1%)

Generally, in heavier oils (lower API gravity, defined by Eq. (2.4)) proportions of carbon, sulfur, nitrogen, and oxygen elements increase but the amount of hydrogen and the overall quality decrease. Further information and discussion about the chemistry of petroleum and the type of compounds found in petroleum fractions are given by Speight [1]. Physical properties of some selected pure hydrocarbons from different homologous groups commonly found in petroleum fluids are given in Chapter 2. Vanadium concentrations of above 2 ppm in fuel oils can lead to severe corrosion in turbine blades and deterioration of refractory in furnaces. Ni, Va, and Cu can also severely affect the activities of catalysts and result in lower products. The metallic content may be reduced by solvent extraction with organic solvents. Organometallic compounds are precipitated with the asphaltenes and residues.

1.1.2 Reservoir Fluids and Crude Oil

The word *fluid* refers to a pure substance or a mixture of compounds that are in the form of gas, liquid, or both a mixture of liquid and gas (vapor). *Reservoir fluid* is a term used for the mixture of hydrocarbons in the reservoir or the stream leaving a producing well. Three factors determine if a reservoir fluid is in the form of gas, liquid, or a mixture of gas and liquid. These factors are (1) composition of reservoir fluid, (2) temperature, and (3) pressure. The most important characteristic of a reservoir fluid in addition to specific gravity (or API gravity) is its gas-to-oil ratio (GOR), which represents the amount of gas

TABLE 1.1—*Types and characteristics of various reservoir fluids.*

Reservoir fluid type	GOR, scf/stb	CH$_4$, mol%	C$_{6+}$, mol%	API gravity of STO[a]
Black oil	<1000	≤50	≥30	<40
Volatile oil	1000–3000	50–70	10–30	40–45
Gas condensate	3000–50 000	70–85	3–10	≥45
Wet gas	≥50 000	≥75	<3	>50
Dry gas	≥10 0000	≥90	<1	No liquid

[a]API gravity of stock tank oil (STO) produced at the surface facilities at standard conditions (289 K and 1 atm).

produced at SC in standard cubic feet (scf) to the amount of liquid oil produced at the SC in stock tank barrel (stb). Other units of GOR are discussed in Section 1.7.23 and its calculation is discussed in Chapter 9. Generally, reservoir fluids are categorized into four or five types (their characteristics are given in Table 1.1). These five fluids in the direction of increasing GOR are black oil, volatile oil, gas condensate, wet gas, and dry gas.

If a gas after surface separator, under SC, does not produce any liquid oil, it is called *dry gas*. A natural gas that after production at the surface facilities can produce a little liquid oil is called *wet gas*. The word wet does not mean that the gas is wet with water, but refers to the hydrocarbon liquids that condense at surface conditions. In dry gases no liquid hydrocarbon is formed at the surface conditions. However, both dry and wet gases are in the category of natural gases. Volatile oils have also been called *high-shrinkage crude oil* and *near-critical oils*, since the reservoir temperature and pressure are very close to the critical point of such oils, but the critical temperature is always greater than the reservoir temperature [11]. Gases and gas condensate fluids have critical temperatures less than the reservoir temperature. Black oils contain heavier compounds and therefore the API gravity of stock tank oil is generally lower than 40 and the GOR is less than 1000 scf/stb. The specifications given in Table 1.1 for various reservoir fluids, especially at the boundaries between different types, are arbitrary and vary from one source to another [9, 11]. It is possible to have a reservoir fluid type that has

properties outside the corresponding limits mentioned earlier. Determination of a type of reservoir fluid by the above rule of thumb based on the GOR, API gravity of stock tank oil, or its color is not possible for all fluids. A more accurate method of determining the type of a reservoir fluid is based on the phase behavior calculations, its critical point, and shape of the phase diagram which will be discussed in Chapters 5 and 9. In general, oils produced from wet gas, gas condensate, volatile oil, and black oil increase in specific gravity (decrease in API gravity and quality) in the same order. Here quality of oil indicates lower carbon, sulfur, nitrogen, and metal contents which correspond to higher heating value. Liquids from black oils are viscous and black in color, while the liquids from gas condensates or wet gases are clear and colorless. Volatile oils produce fluids brown with some red/green color liquid. Wet gas contains less methane than a dry gas does, but a larger fraction of C$_2$–C$_6$ components. Obviously the main difference between these reservoir fluids is their respective composition. An example of composition of different reservoir fluids is given in Table 1.2.

In Table 1.2, C$_{7+}$ refers to all hydrocarbons having seven or higher carbon atoms and is called heptane-plus fraction, while C$_6$ refers to a group of all hydrocarbons with six carbon atoms (hexanes) that exist in the fluid. M_{7+} and SG$_{7+}$ are the molecular weight and specific gravity at 15.5°C (60°F) for the C$_{7+}$ fraction of the mixture, respectively. It should be realized that molecular weight and specific gravity of the whole reservoir fluid are less than the corresponding values for the

TABLE 1.2—*Composition (mol%) and properties of various reservoir fluids and a crude oil.*

Component	Dry gas[a]	Wet gas[b]	Gas condensate[c]	Volatile oil[d]	Black oil[e]	Crude oil[f]
CO$_2$	3.70	0.00	0.18	1.19	0.09	0.00
N$_2$	0.30	0.00	0.13	0.51	2.09	0.00
H$_2$S	0.00	0.00	0.00	0.00	1.89	0.00
C$_1$	96.00	82.28	61.92	45.21	29.18	0.00
C$_2$	0.00	9.52	14.08	7.09	13.60	0.19
C$_3$	0.00	4.64	8.35	4.61	9.20	1.88
iC$_4$	0.00	0.64	0.97	1.69	0.95	0.62
nC$_4$	0.00	0.96	3.41	2.81	4.30	3.92
iC$_5$	0.00	0.35	0.84	1.55	1.38	2.11
nC$_5$	0.00	0.29	1.48	2.01	2.60	4.46
C$_6$	0.00	0.29	1.79	4.42	4.32	8.59
C$_{7+}$	0.00	1.01	6.85	28.91	30.40	78.23
Total	100.00	100.00	100.00	100.00	100.00	100.00
GOR (scf/stb)	...	69917	4428	1011	855	...
M_{7+}	...	113	143	190	209.8	266
SG$_{7+}$ (at 15.5°C)	...	0.794	0.795	0.8142	0.844	0.895
API$_{7+}$	...	46.7	46.5	42.1	36.1	26.6

[a]Gas sample from Salt Lake, Utah [12].
[b]Wet gas data from McCain [11].
[c]Gas condensate sample from Samson County, Texas (M. B. Standing, personal notes, Department of Petroleum Engineering, Norwegian Institute of Technology, Trondheim, Norway, 1974).
[d]Volatile oil sample from Raleigh Field, Smith County, Mississipi (M. B. Standing, personal notes, Department of Petroleum Engineering, Norwegian Institute of Technology, Trondheim, Norway, 1974).
[e]Black oil sample from M. Ghuraiba, M.Sc. Thesis, Kuwait University, Kuwait, 2000.
[f]A crude oil sample produced at stock tank conditions.

heptane-plus fraction. For example, for the crude oil sample in Table 1.2, the specific gravity of the whole crude oil is 0.871 or API gravity of 31. Details of such calculations are discussed in Chapter 4. These compositions have been determined from recombination of the compositions of corresponding separator gas and stock tank liquid, which have been measured through analytical tools (i.e., gas chromatography, mass spectrometry, etc.). Composition of reservoir fluids varies with the reservoir pressure and reservoir depth. Generally in a producing oil field, the sulfur and amount of heavy compounds increase versus production time [10]. However, it is important to note that within an oil field, the concentration of light hydrocarbons and the API gravity of the reservoir fluid increase with the reservoir depth, while its sulfur and C_{7+} contents decrease with the depth [1]. The lumped C_{7+} fraction in fact is a mixture of a very large number of hydrocarbons, up to C_{40} or higher. As an example the number of pure hydrocarbons from C_5 to C_9 detected by chromatography tools in a crude oil from North Sea reservoir fluids was 70 compounds. Detailed composition of various reservoir fluids from the North Sea fields is provided by Pedersen *et al.* [13]. As shown in Chapter 9, using the knowledge of the composition of a reservoir fluid, one can determine a pressure–temperature (*PT*) diagram of the fluid. And on the basis of the temperature and pressure of the reservoir, the exact type of the reservoir fluid can be determined from the *PT* diagram.

Reservoir fluids from a producing well are conducted to two- or three-stage separators which reduce the pressure and temperature of the stream to atmospheric pressure and temperature. The liquid leaving the last stage is called *stock tank oil* (STO) and the gas released in various stages is called *associated gas*. The liquid oil after necessary field processing is called *crude oil*. The main factor in operation and design of an oil–gas separator is to find the optimum operating conditions of temperature and pressure so that the amount of produced liquid (oil) is maximized. Such conditions can be determined through phase behavior calculations, which are discussed in detail in Chapter 9. Reservoir fluids from producing wells are mixed with free water. The water is separated through gravitational separators based on the difference between densities of water and oil. Remaining water from the crude can be removed through dehydration processes. Another surface operation is the desalting process that is necessary to remove the salt content of crude oils. Separation of oil, gas, and water from each other and removal of water and salt from oil and any other process that occurs at the surface are called *surface production operations* [14].

The crude oil produced from the atmospheric separator has a composition different from the reservoir fluid from a producing well. The light gases are separated and usually crude oils have almost no methane and a small C_2–C_3 content while the C_{7+} content is higher than the original reservoir fluid. As an example, the composition of a crude oil produced through a three-stage separator from a reservoir fluid is also given in Table 1.2. Actually this crude is produced from a black oil reservoir fluid (composition given in Table 1.2). Two important characterisitcs of a crude that determine its quality are the API gravity (specific gravity) and the sulfur content. Generally, a crude with the API gravity of less than 20 (SG > 0.934) is called *heavy crude* and with API gravity of greater than 40 (SG < 0.825) is called *light crude* [1, 9]. Similarly, if the sulfur

content of a crude is less than 0.5 wt% it is called a sweet oil. It should be realized that these ranges for the gravity and sulfur content are relative and may vary from one source to another. For example, Favennec [15] classifies heavy crude as those with API less than 22 and light crude having API above 33. Further classification of crude oils will be discussed in Chapter 4.

1.1.3 Petroleum Fractions and Products

A crude oil produced after necessary field processing and surface operations is transferred to a refinery where it is processed and converted into various useful products. The refining process has evolved from simple batch distillation in the late nineteenth century to today's complex processes through modern refineries. Refining processes can be generally divided into three major types: (1) separation, (2) conversion, and (3) finishing. Separation is a physical process where compounds are separated by different techniques. The most important *separation* process is distillation that occurs in a distillation column; compounds are separated based on the difference in their boiling points. Other major physical separation processes are absorption, stripping, and extraction. In a gas plant of a refinery that produces light gases, the heavy hydrocarbons (C_5 and heavier) in the gas mixture are separated through their absorption by a liquid oil solvent. The solvent is then regenerated in a stripping unit. The *conversion* process consists of chemical changes that occur with hydrocarbons in reactors. The purpose of such reactions is to convert hydrocarbon compounds from one type to another. The most important reaction in modern refineries is the cracking in which heavy hydrocarbons are converted to lighter and more valuable hydrocarbons. Catalytic cracking and thermal cracking are commonly used for this purpose. Other types of reactions such as isomerization or *alkylation* are used to produce high octane number gasoline. *Finishing* is the purification of various product streams by processes such as desulfurization or acid treatment of petroleum fractions to remove impurities from the product or to stabilize it.

After the desalting process in a refinery, the crude oil enters the atmospheric distillation column, where compounds are separated according to their boiling points. Hydrocarbons in a crude have boiling points ranging from −160°C (boiling point of methane) to more than 600°C (1100°F), which is the boiling point of heavy compounds in the crude oil. However, the carbon–carbon bond in hydrocarbons breaks down at temperatures around 350°C (660°F). This process is called *cracking* and it is undesirable during the distillation process since it changes the structure of hydrocarbons. For this reason, compounds having boiling points above 350°C (660+°F) called residuum are removed from the bottom of atmospheric distillation column and sent to a vacuum distillation column. The pressure in a vacuum distillation column is about 50–100 mm Hg, where hydrocarbons are boiled at much lower temperatures. Since distillation cannot completely separate the compounds, there is no pure hydrocarbon as a product of a distillation column. A group of hydrocarbons can be separated through distillation according to the boiling point of the lightest and heaviest compounds in the mixtures. The lightest product of an atmospheric column is a mixture of methane and ethane (but mainly ethane) that has the boiling

TABLE 1.3—Some petroleum fractions produced from distillation columns.

Petroleum fraction	Approximate hydrocarbon range	Approximate boiling range °C	Approximate boiling range °F
Light gases	C_2–C_4	−90 to 1	−130–30
Gasoline (light and heavy)	C_4–C_{10}	−1–200	30–390
Naphthas (light and heavy)	C_4–C_{11}	−1–205	30–400
Jet fuel	C_9–C_{14}	150–255	300–490
Kerosene	C_{11}–C_{14}	205–255	400–490
Diesel fuel	C_{11}–C_{16}	205–290	400–550
Light gas oil	C_{14}–C_{18}	255–315	490–600
Heavy gas oil	C_{18}–C_{28}	315–425	600–800
Wax	C_{18}–C_{36}	315–500	600–930
Lubricating oil	>C_{25}	>400	>750
Vacuum gas oil	C_{28}–C_{55}	425–600	800–1100
Residuum	>C_{55}	>600	>1100

Information given in this table is obtained from different sources [1, 18, 19].

range of −180 to −80°C (−260 to −40°F), which corresponds to the boiling point of methane and ethane. This mixture, which is in the form of gas and is known as fuel gas, is actually a *petroleum fraction*. In fact, during distillation a crude is converted into a series of petroleum fractions where each one is a mixture of a limited number of hydrocarbons with a specific range of boiling point. Fractions with a wider range of boiling points contain greater numbers of hydrocarbons. All fractions from a distillation column have a known boiling range, except the residuum for which the upper boiling point is usually not known. The boiling point of the heaviest component in a crude oil is not really known, but it is quite high. The problem of the nature and properties of the heaviest compounds in crude oils and petroleum residuum is still under investigation by researchers [16, 17]. Theoretically, it can be assumed that the boiling point of the heaviest component in a crude oil is infinity. Atmospheric residue has compounds with carbon number greater than 25, while vacuum residue has compounds with carbon number greater than 50 ($M > 800$). Some of the petroleum fractions produced from distillation columns with their boiling point ranges and applications are given in Table 1.3. The boiling point and equivalent carbon number ranges given in this table are approximate and they may vary according to the desired specific product. For example, the light gases fraction is mainly a mixture of ethane, propane, and butane; however, some heavier compounds (C_{5+}) may exist in this fraction. The fraction is further fractionated to obtain ethane (a fuel gas) and propane and butane (petroleum gases). The petroleum gases are liquefied to get liquefied petroleum gas (LPG) used for home cooking purposes. In addition the isobutane may be separated for the gas mixture to be used for improving vapor pressure characteristics (volatility) of gasoline in cold weathers. These fractions may go through further processes to produce desired products. For example, gas oil may go through a cracking process to obtain more gasoline. Since distillation is not a perfect separation process, the initial and final boiling points for each fraction are not exact and especially the end points are approximate values. Fractions may be classified as narrow or wide depending on their boiling point range. As an example, the composition of an Alaska crude oil for various products is given in Table 1.4 and is graphically shown in Fig. 1.3. The weight and volume percentages for the products are near each other. More than 50% of the crude is processed in vacuum distillation unit. The vacuum residuum is mainly resin and asphaltenes-type compounds composed of high

molecular weight multiring aromatics. The vacuum residuum may be mixed with lighter products to produce a more valuable blend.

Distillation of a crude oil can also be performed in the laboratory to divide the mixture into many narrow boiling point range fractions with a boiling range of about 10°C. Such narrow range fractions are sometimes referred to as *petroleum cuts*. When boiling points of all the cuts in a crude are known, then the boiling point distribution (distillation curve) of the

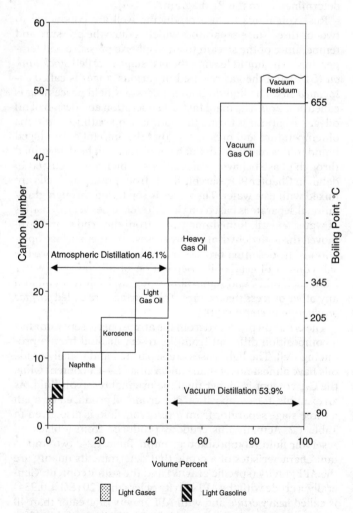

FIG. 1.3—Products and composition of Alaska crude oil.

TABLE 1.4—*Products and composition of alaska crude oil.*

Petroleum fraction	Approximate hydrocarbon range	Approximate boiling range[a] °C	°F	vol%	wt%
Atmospheric distillation					
Light gases	C_2–C_4	−90 to 1	−130–30	1.2	0.7
Light gasoline	C_4–C_7	−1–83	30–180	4.3	3.5
Naphthas	C_7–C_{11}	83–205	180–400	16.0	14.1
Kerosene	C_{11}–C_{16}	205–275	400–525	12.1	11.4
Light gas oil (LGO)	C_{16}–C_{21}	275–345	525–650	12.5	12.2
Sum	C_2–C_{21}	−90–345	−130–650	46.1	41.9
Vacuum distillation (VD)					
Heavy gas oil (HGO)	C_{21}–C_{31}	345–455	650–850	20.4	21.0
Vacuum gas oil (VGO)	C_{31}–C_{48}	455–655	850–1050	15.5	16.8
Residuum	>C_{48}	655+	1050+	18.0	20.3
Sum	C_{21}–C_{48+}	345–655+	650–1050	53.9	58.1
Total Crude	C_2–C_{48+}	−90–655+	−130–650+	100.0	100.0

Information given in this table has been extracted from Ref. [19].
[a]Boiling ranges are interconverted to the nearest 5°C (°F).

whole crude can be obtained. Such distillation data and their uses will be discussed in Chapters 3 and 4. In a petroleum cut, hydrocarbons of various types are lumped together in four groups of paraffins (P), olefins (O), naphthenes (N), and aromatics (A). For olefin-free petroleum cuts the composition is represented by the PNA content. If the composition of a hydrocarbon mixture is known the mixture is called a *defined mixture*, while a petroleum fraction that has an unknown composition is called an *undefined fraction*.

As mentioned earlier, the petroleum fractions presented in Table 1.3 are not the final products of a refinery. They go through further physicochemical and finishing processes to get the characteristics set by the market and government regulations. After these processes, the petroleum fractions presented in Table 1.3 are converted to *petroleum products*. The terms petroleum fraction, petroleum cut, and petroleum product are usually used incorrectly, while one should realize that petroleum fractions are products of distillation columns in a refinery before being converted to final products. Petroleum cuts may have very narrow boiling range which may be produced in a laboratory during distillation of a crude. In general the petroleum products can be divided into two groups: (1) fuel products and (2) nonfuel products. The major *fuel petroleum products* are as follows:

1. Liquefied petroleum gases (LPG) that are mainly used for domestic heating and cooking (50%), industrial fuel (clean fuel requirement) (15%), steam cracking feed stock (25%), and as a motor fuel for spark ignition engines (10%). The world production in 1995 was 160 million ton per year (≅5 million bbl/d) [20]. LPG is basically a mixture of propane and butane.
2. Gasoline is perhaps one of the most important products of a refinery. It contains hydrocarbons from C_4 to C_{11} (molecular weight of about 100–110). It is used as a fuel for cars. Its main characteristics are antiknock (octane number), volatility (distillation data and vapor pressure), stability, and density. The main evolution in gasoline production has been the use of unleaded gasoline in the world and the use of *reformulated gasoline* (RFG) in the United States. The RFG has less butane, less aromatics, and more oxygenates. The sulfur content of gasoline should not exceed 0.03% by weight. Further properties and characteristics of gasoline

will be discussed in Chapter 3. The U.S. gasoline demand in 1964 was 4.4 million bbl/d and has increased from 7.2 to 8.0 million bbl/d in a period of 7 years from 1991 to 1998 [6, 20]. In 1990, gasoline was about a third of refinery products in the United States.
3. Kerosene and jet fuel are mainly used for lighting and jet engines, respectively. The main characteristics are sulfur content, cold resistance (for jet fuel), density, and ignition quality.
4. Diesel and heating oil are used for motor fuel and domestic purposes. The main characteristics are ignition (for diesel oil), volatility, viscosity, cold resistance, density, sulfur content (corrosion effects), and flash point (safety factor).
5. Residual fuel oil is used for industrial fuel, for thermal production of electricity, and as motor fuel (low speed diesel engines). Its main characteristics are viscosity (good atomization for burners), sulfur content (corrosion), stability (no decantation separation), cold resistance, and flash point for safety.

The major *nonfuel petroleum products* are [18] as follows:

1. Solvents are light petroleum cuts in the C_4–C_{14} range and have numerous applications in industry and agriculture. As an example of solvents, white spirits which have boiling points between 135 and 205°C are used as paint thinners. The main characteristics of solvents are volatility, purity, odor, and toxicity. Benzene, toluene, and xylenes are used as solvents for glues and adhesives and as a chemical for petrochemical industries.
2. Naphthas constitute a special category of petroleum solvents whose boiling points correspond to the class of white spirits. They can be classified beside solvents since they are mainly used as raw materials for petrochemicals and as the feeds to steam crackers. Naphthas are thus industrial intermediates and not consumer products. Consequently, naphthas are not subject to government specifications but only to commercial specifications.
3. Lubricants are composed of a main base stock and additives to give proper characteristics. One of the most important characteristics of lubricants is their viscosity and viscosity index (change of viscosity with temperature). Usually aromatics are eliminated from lubricants to improve

their viscosity index. Lubricants have structure similar to isoparaffinic compounds. Additives used for lubricants are viscosity index additives such as polyacrylates and olefin polymers, antiwear additives (i.e., fatty esters), antioxidants (i.e., alkylated aromatic amines), corrosion inhibitors (i.e., fatty acids), and antifoaming agents (i.e., polydimethylsiloxanes). Lubricating greases are another class of lubricants that are semisolid. The properties of lubricants that should be known are viscosity index, aniline point (indication of aromatic content), volatility, and carbon residue.

4. Petroleum waxes are of two types: the paraffin waxes in petroleum distillates and the microcrystalline waxes in petroleum residua. In some countries such as France, paraffin waxes are simply called paraffins. Paraffin waxes are high melting point materials used to improve the oil's pour point and are produced during dewaxing of vacuum distillates. Paraffin waxes are mainly straight chain alkanes (C_{18}–C_{36}) with a very small proportion of isoalkanes and cycloalkanes. Their freezing point is between 30 and 70°C and the average molecular weight is around 350. When present, aromatics appear only in trace quantities. Waxes from petroleum residua (microcrystalline form) are less defined aliphatic mixtures of n-alkanes, isoalkanes, and cycloalkanes in various proportions. Their average molecular weights are between 600 and 800, carbon number range is alkanes C_{30}–C_{60}, and the freezing point range is 60–90°C [13]. Paraffin waxes (when completely dearomatized) have applications in the food industry and food packaging. They are also used in the production of candles, polishes, cosmetics, and coatings [18]. Waxes at ordinary temperature of 25°C are in solid states although they contain some hydrocarbons in liquid form. When melted they have relatively low viscosity.

5. Asphalt is another major petroleum product that is produced from vacuum distillation residues. Asphalts contain nonvolatile high molecular weight polar aromatic compounds, such as asphaltenes (molecular weights of several thousands) and cannot be distilled even under very high vacuum conditions. In some countries asphalt is called bitumen, although some suggest these two are different petroleum products. Liquid asphaltic materials are intended for easy applications to roads. Asphalt and bitumen are from a category of products called hydrocarbon binders. Major properties to determine the quality of asphalt are flash point (for safety), composition (wax content), viscosity and softening point, weathering, density or specific gravity, and stability or chemical resistance.

6. There are some other products such as white oils (used in pharmaceuticals or in the food industry), aromatic extracts (used in the paint industry or the manufacture of plastics), and coke (as a fuel or to produce carbon elecrodes for aluminum refining). Petroleum cokes generally have boiling points above 1100+°C (~2000+°F), molecular weight of above 2500+, and carbon number of above 200+. Aromatic extracts are black materials, composed essentially of condensed polynuclear aromatics and of heterocyclic nitrogen and/or sulfur compounds. Because of this highly aromatic structure, the extracts have good solvent power.

Further information on technology, properties, and testing methods of fuels and lubricants is given in Ref. [21].

In general, more than 2000 petroleum products within some 20 categories are produced in refineries in the United States [1, 19]. Blending techniques are used to produce some of these products or to improve their quality. The product specifications must satisfy customers' requirements for good performance and government regulations for safety and environment protection. To be able to plan refinery operations, the availability of a set of product quality prediction methods is therefore very important.

There are a number of international organizations that are known as standard organizations that recommend specific characteristics or standard measuring techniques for various petroleum products through their regular publications. Some of these organizations in different countries that are known with their abbreviations are as follows:

1. ASTM (American Society for Testing and Materials) in the United States
2. ISO (International Organization for Standardization), which is at the international level
3. IP (Institute of Petroleum) in the United Kingdom
4. API (American Petroleum Institute) in the United States
5. AFNOR (Association Francaise de Normalisation), an official standard organization in France
6. Deutsche Institut fur Normung (DIN) in Germany
7. Japan Institute of Standards (JIS) in Japan

ASTM is composed of several committees in which the *D-02 committee* is responsible for petroleum products and lubricants, and for this reason its test methods for petroleum materials are designated by the prefix D. For example, the test method ASTM D 2267 provides a standard procedure to determine the benzene content of gasoline [22]. In France this test method is designated by EN 238, which are documented in AFNOR information document M 15-023. Most standard test methods in different countries are very similar in practice and follow ASTM methods but they are designated by different codes. For example the international standard ISO 6743/0, accepted as the French standard NF T 60-162, treats all the petroleum lubricants, industrial oils, and related products. The abbreviation NF is used for the French standard, while EN is used for European standard methods [18].

Government regulations to protect the environment or to save energy, in many cases, rely on the recommendations of official standard organizations. For example, in France, AFNOR gives specifications and requirements for various petroleum products. For diesel fuels it recommends (after 1996) that the sulfur content should not exceed 0.05 wt% and the flash point should not be less than 55°C [18].

1.2 TYPES AND IMPORTANCE OF PHYSICAL PROPERTIES

On the basis of the production and refining processes described above it may be said that the petroleum industry is involved with many types of equipment for production, transportation, and storage of intermediate or final petroleum products. Some of the most important units are listed below.

1. Gravity decanter (to separate oil and water)
2. Separators to separate oil and gas
3. Pumps, compressors, pipes, and valves

4. Storage tanks
5. Distillation, absorption, and stripping columns
6. Boilers, evaporators, condensers, and heat exchangers
7. Flashers (to separate light gases from a liquid)
8. Mixers and agitators
9. Reactors (fixed and fluidized beds)
10. Online analyzers (to monitor the composition)
11. Flow and liquid level measurement devices
12. Control units and control valves

The above list shows some, but not all, of the units involved in the petroleum industry. Optimum design and operation of such units as well as manufacture of products to meet market demands and government regulations require a complete knowledge of properties and characteristics for hydrocarbons, petroleum fractions/products, crude oils, and reservoir fluids. Some of the most important characteristics and properties of these fluids are listed below with some examples for their applications. They are divided into two groups of temperature-independent parameters and temperature-dependent properties. The *temperature-independent properties* and parameters are as follows:

1. *Specific gravity* (SG) or *density* (d) at SC. These parameters are temperature-dependent; however, specific gravity at 15.5°C and 1 atm and density at 20°C and 1 atm used in petroleum characterization are included in this category of temperature-independent properties. The specific gravity is also presented in terms of *API gravity*. It is a useful parameter to characterize petroleum fluids, to determine composition (PNA) and the quality of a fuel (i.e., sulfur content), and to estimate other properties such as critical constants, density at various temperatures, viscosity, or thermal conductivity [23, 24]. In addition to its direct use for size calculations (i.e., pumps, valves, tanks, and pipes), it is also needed in design and operation of equipments such as gravity decanters.
2. *Boiling point* (T_b) or distillation curves such as the true boiling point curve of petroleum fractions. It is used to determine volatility and to estimate characterization parameters such as average boiling point, molecular weight, composition, and many physical properties (i.e., critical constants, vapor pressure, thermal properties, transport properties) [23–25].
3. *Molecular weight* (M) is used to convert molar quantities into mass basis needed for practical applications. Thermodynamic relations always produce molar quantities (i.e., molar density), while in practice mass specific values (i.e., absolute density) are needed. Molecular weight is also used to characterize oils, to predict composition and quality of oils, and to predict physical properties such as viscosity [26–30].
4. *Refractive index* (n) at some reference conditions (i.e., 20°C and 1 atm) is another useful characterization parameter to estimate the composition and quality of petroleum fractions. It is also used to estimate other physical properties such as molecular weight, equation of state parameters, the critical constants, or transport properties of hydrocarbon systems [30, 31].
5. Defined characterization parameters such as Watson K, *carbon-to-hydrogen weight ratio*, (CH weight ratio), *refractivity intercept* (R_i), and *viscosity gravity constant* (VGC)

to determine the quality and composition of petroleum fractions [27–29].

6. Composition of petroleum fractions in terms of wt% of paraffins (P%), naphthenes (N%), aromatics (A%), and sulfur content (S%) are important to determine the quality of a petroleum fraction as well as to estimate physical properties through pseudocomponent methods [31–34]. Composition of other constituents such as *asphaltene* and *resin* components are quite important for heavy oils to determine possibility of solid-phase deposition, a major problem in the production, refining, and transportation of oil [35].
7. *Pour point* (T_P), and *melting point* (T_M) have limited uses in wax and paraffinic heavy oils to determine the degree of solidification and the wax content as well as minimum temperature required to ensure fluidity of the oil.
8. *Aniline point* to determine a rough estimate of aromatic content of oils.
9. *Flash point* (T_F) is a very useful property for the safety of handling volatile fuels and petroleum products especially in summer seasons.
10. *Critical temperature* (T_c), *critical pressure* (P_c), and *critical volume* (V_c) known as critical constants or critical properties are used to estimate various physical and thermodynamic properties through equations of state or generalized correlations [36].
11. *Acentric factor* (ω) is another parameter that is needed together with critical properties to estimate physical and thermodynamic properties through equations of state [36].

The above properties are mainly used to characterize the oil or to estimate the physical and thermodynamic properties which are all *temperature-dependent*. Some of the most important properties are listed as follows:

1. *Density* (ρ) as a function of temperature and pressure is perhaps the most important physical property for petroleum fluids (vapor or liquid forms). It has great application in both petroleum production and processing as well as its transportation and storage. It is used in the calculations related to sizing of pipes, valves, and storage tanks, power required by pumps and compressors, and flow-measuring devices. It is also used in reservoir simulation to estimate the amount of oil and gas in a reservoir, as well as the amount of their production at various reservoir conditions. In addition density is used in the calculation of equilibrium ratios (for phase behavior calculations) as well as other properties, such as transport properties.
2. *Vapor pressure* (P^{vap}) is a measure of volatility and it is used in phase equilibrium calculations, such as flash, bubble point, or dew point pressure calculations, in order to determine the state of the fluid in a reservoir or to separate vapor from liquid. It is needed in calculation of equilibrium ratios for operation and design of distillation, absorber, and stripping columns in refineries. It is also needed in determination of the amount of hydrocarbon losses from storage facilities and their presence in air. Vapor pressure is the property that represents ignition characteristics of fuels. For example, the *Reid vapor pressure* (RVP) and boiling range of gasoline govern ease of starting engine, engine warm-up, rate of acceleration, mileage economy, and tendency toward vapor lock [19].

3. *Heat capacity* (C_p) of a fluid is needed in design and operation of heat transfer units such as heat exchangers.
4. *Enthalpy* (H) of a fluid is needed in energy balance calculations, heat requirements needed in design and operation of distillation, absorption, stripping columns, and reactors.
5. *Heat of vaporization* (ΔH_{vap}) is needed in calculation of heat requirements in design and operation of reboilers or condensers.
6. *Heats of formation* (ΔH_f), *combustion* (ΔH_c), and *reaction* (ΔH_r) are used in calculation of heating values of fuels and the heat required/generated in reactors and furnaces in refineries. Such information is essential in design and operations of burners, furnaces, and chemical reactors. These properties together with the *Gibbs free energy* are used in calculation of equilibrium constants in chemical reactions to determine the optimum operating conditions in reactors for best conversion of feed stocks into the products.
7. *Viscosity* (μ) is another useful property in petroleum production, refining, and transportation. It is used in reservoir simulators to estimate the rate of oil or gas flow and their production. It is needed in calculation of power required in mixers or to transfer a fluid, the amount of pressure drop in a pipe or column, flow measurement devices, and design and operation of oil/water separators [37, 38].
8. *Thermal conductivity* (k) is needed for design and operation of heat transfer units such as condensers, heat exchangers, as well as chemical reactors [39].
9. *Diffusivity* or *diffusion coefficient* (D) is used in calculation of mass transfer rates and it is a useful property in design and operation of reactors in refineries where feed and products diffuse in catalyst pores. In petroleum production, a gas injection technique is used in improved oil recovery where a gas diffuses into oil under reservoir conditions; therefore, diffusion coefficient is also required in reservoir simulation and modeling [37, 40–42].
10. *Surface tension* (σ) or *interfacial tension* (IFT) is used mainly by the reservoir engineers in calculation of capillary pressure and rate of oil production and is needed in reservoir simulators [37]. In refineries, IFT is a useful parameter to determine foaming characteristics of oils and the possibility of having such problems in distillation, absorption, or stripping columns [43]. It is also needed in calculation of the rate of oil dispersion on seawater surface polluted by an oil spill [44].
11. *Equilibrium ratios* (K_i) and *fugacity coefficients* (φ_i) are the most important thermodynamic properties in all phase behavior calculations. These calculations include vapor–liquid equilibria, bubble and dew point pressure, pressure–temperature phase diagram, and GOR. Such calculations are important in design and operation of distillation, absorption and stripping units, gas-processing units, gas–oil separators at production fields, and to determine the type of a reservoir fluid [45, 46].

Generally, the first set of properties introduced above (temperature-independent) are the basic parameters that are used to estimate physical and thermodynamic properties given in the second set (temperature-dependent). Properties such as density, boiling point, molecular weight, and refractive index are called *physical properties*. Properties such as enthalpy, heat capacity, heat of vaporization, equilibrium ratios, and fugacity are called *thermodynamic properties*. Viscosity, thermal conductivity, diffusion coefficient, and surface tension are in the category of physical properties but they are also called *transport properties*. In general all the thermodynamic and physical properties are called *thermophysical* properties. But they are commonly referred to as physical properties or simply *properties*, which is used in the title of this book.

A property of a system depends on the thermodynamic state of the system that is determined by its temperature, pressure, and composition. A process to experimentally determine various properties for all the industrially important materials, especially complex mixtures such as crude oils or petroleum products, would be prohibitive in both cost and time, indeed it could probably never be completed. For these reasons accurate methods for the estimation of these properties are becoming increasingly important. In some references the term property prediction is used instead of property estimation; however, in this book as generally adopted by most scientists both terms are used for the same purpose.

1.3 IMPORTANCE OF PETROLEUM FLUIDS CHARACTERIZATION

In the previous section, various basic characteristic parameters for petroleum fractions and crude oils were introduced. These properties are important in design and operation of almost every piece of equipment in the petroleum industry. Thermodynamic and physical properties of fluids are generally calculated through standard methods such as corresponding state correlations or equations of state and other pressure–volume–temperature (PVT) relations. These correlations and methods have a generally acceptable degree of accuracy provided accurate input parameters are used. When using cubic equation of state to estimate a thermodynamic property such as absolute density for a fluid at a known temperature and pressure, the critical temperature (T_c), critical pressure (P_c), acentric factor (ω), and molecular weight (M) of the system are required. For most pure compounds and hydrocarbons these properties are known and reported in various handbooks [36, 47–50]. If the system is a mixture such as a crude oil or a petroleum fraction then the pseudocritical properties are needed for the calculation of physical properties. The pseudocritical properties cannot be measured but have to be calculated through the composition of the mixture. Laboratory reports usually contain certain measured properties such as distillation curve (i.e., ASTM D 2887) and the API gravity or specific gravity of the fraction. However, in some cases viscosity at a certain temperature, the percent of paraffin, olefin, naphthene, and aromatic hydrocarbon groups, and sulfur content of the fraction are measured and reported. Petroleum fractions are mixtures of many compounds in which the specific gravity can be directly measured for the mixture, but the average boiling point cannot be measured. Calculation of average boiling point from distillation data, conversion of various distillation curves from one type to another, estimation of molecular weight, and the PNA composition of fractions are the initial steps in characterization of petroleum fractions [25, 46, 47]. Estimation of other basic parameters introduced in Section 1.2, such as asphaltenes and sulfur contents, CH, flash and pour points, aniline point, refractive index and density at SC, pseudocrtitical properties, and acentric factor, are also considered as parts of characterization of petroleum fractions [24, 28, 29, 51–53]. Some of these properties such as the critical constants and acentric factor are not even known for some heavy pure hydrocarbons and should be estimated from available properties. Therefore characterization methods also apply to pure hydrocarbons [33]. Through characterization, one can estimate the basic parameters needed for the estimation of various physical and thermodynamic properties as well as to determine the composition and quality of petroleum fractions from available properties easily measurable in a laboratory.

For crude oils and reservoir fluids, the basic laboratory data are usually presented in the form of the composition of hydrocarbons up to hexanes and the heptane-plus fraction (C_{7+}), with its molecular weight and specific gravity as shown in Table 1.2. In some cases laboratory data on a reservoir fluid is presented in terms of the composition of single carbon numbers or simulated distillation data where weight fraction of cuts with known boiling point ranges are given. Certainly because of the wide range of compounds existing in a crude oil or a reservoir fluid (i.e., black oil), an average value for a physical property such as boiling point for the whole mixture has little significant application and meaning. Characterization of a crude oil deals with use of such laboratory data to present the mixture in terms of a defined or a continuous mixture. One commonly used characterization technique for the crudes or reservoir fluids is to represent the hydrocarbon-plus fraction (C_{7+}) in terms of several narrow-boiling-range cuts called *psuedocomponents* (or *pseudofractions*) with known composition and characterization parameters such as, boiling point, molecular weight, and specific gravity [45, 54, 55]. Each pseudocomponent is treated as a petroleum fraction. Therefore, characterization of crude oils and reservoir fluids require characterization of petroleum fractions, which in turn require pure hydrocarbon characterization and properties [56]. It is for this reason that properties of pure hydrocarbon compounds and hydrocarbon characterization methods are first presented in Chapter 2, the characterization of petroleum fractions is discussed in Chapter 3, and finally methods of characterization of crude oils are presented in Chapter 4. Once characterization of a petroleum fraction or a crude oil is done, then a physical property of the fluid can be estimated through an appropriate procedure. In summary, characterization of a petroleum fraction or a crude oil is a technique that through available laboratory data one can calculate basic parameters necessary to determine the quality and properties of the fluid.

Characterization of petroleum fractions, crude oils, and reservoir fluids is a state-of-the-art calculation and plays an important role in accurate estimation of physical properties of these complex mixtures. Watson, Nelson, and Murphy of Universal Oil Products (UOP) in the mid 1930s proposed initial characterization methods for petroleum fractions [57]. They introduced a characterization parameter known as Watson or UOP characterization factor, K_W, which has been used extensively in characterization methods developed in the following years. There are many characterization methods

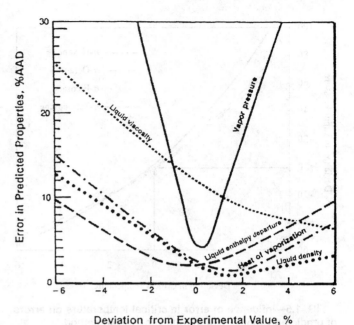

FIG. 1.4—Influence of error in critical temperature on errors in predicted physical properties of toluene. Taken from Ref. [58] with permission.

suggested in the literature or in process simulators and each method generates different characterization parameters that in turn would result different estimated final physical property with subsequent impact in design and operation of related units. To decide which method of characterization and what input parameters (where there is a choice) should be chosen depends very much on the user's knowledge and experience in this important area.

To show how important the role of characterization is in the design and operation of units, errors in the prediction of various physical properties of toluene through a modified BWR equation of state versus errors introduced to actual critical temperature (T_c) are shown in Fig. 1.4 [58]. In this figure, errors in the prediction of vapor pressure, liquid viscosity, vapor viscosity, enthalpy, heat of vaporization, and liquid density are calculated versus different values of critical temperature while other input parameters (i.e., critical pressure, acentric factor, etc.) were kept constant. In the use of the equation of state if the actual (experimental) value of the critical temperature is used, errors in values of predicted properties are generally within 1–3% of experimental values; however, as higher error is introduced to the critical temperature the error in the calculated property increases to a much higher magnitude. For example, when the error in the value of the critical temperature is zero (actual value of T_c), predicted vapor pressure has about 3% error from the experimental value, but when the error in T_c increases to 1, 3, or 5%, error in the predicted vapor pressure increases approximately to 8, 20, and 40%, respectively. Therefore, one can realize that 5% error in an input property for an equation of state does not necessarily reflect the same error in a calculated physical property but can be propagated into much higher errors, while the predictive equation is relatively accurate if actual input parameters are used. Similar results are observed for other physical

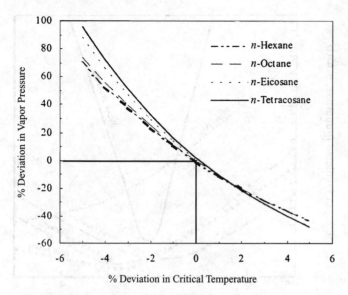

FIG. 1.5—Influence of error in critical temperature on errors of predicted vapor pressure from Lee–Kesler method.

properties and with other correlations for the estimation of physical properties [59]. Effect of the error in the critical temperature on the vapor pressure of different compounds predicted from the Lee–Kesler method (see Section 7.3.2) is shown in Fig. 1.5. When the actual critical temperature is used, the error in the predicted vapor pressure is almost negligible; however, if the critical temperature is under-predicted by 5%, the error in the vapor pressure increases by 60–80% for the various compounds evaluated.

As shown in Chapter 6, vapor pressure is one of the key parameters in the calculation of equilibrium ratios (K_i) and subsequent relative volatility (α_{12}), which is defined in a binary system of components 1 and 2 as follows:

$$(1.3) \qquad K_1 = \frac{y_1}{x_1}$$

$$(1.4) \qquad \alpha_{12} = \frac{K_1}{K_2} = \frac{y_1}{x_1} \times \frac{x_2}{y_2}$$

where x_1 and x_2 are the mole fractions of components 1 and 2 in the liquid phase, respectively. Similarly y_1 and y_2 are the mole fractions in the vapor phase for components 1 and 2, respectively. For an ideal binary system at low pressure, the equilibrium ratio K_i is directly proportional to the vapor pressure as will be seen in Chapter 6.

The most important aspect in the design and operation of distillation columns is the number or trays needed to make a specific separation for specific feed and products. It has been shown that a small error in the value of relative volatility could lead to a much greater error in the calculation of number of trays and the length of a distillation column [60]. The minimum number of trays required in a distillation column can be calculated from the knowledge of relative volatility through the *Fenske Equation* given below [61].

$$(1.5) \qquad N_{\min} = \frac{\ln[x_D(1 - x_B)/x_B(1 - x_D)]}{\ln(\alpha_{12})} - 1$$

where $N_{\min}$ is the minimum number of plates, and x_D and x_B are the mole fraction of the light component in the distillate (top) and bottom products, respectively. Equation (1.5) is developed for a binary mixture; however, a similar equation has been developed for multicomponent mixtures [61]. For different values of α, errors calculated for the minimum number of trays versus errors introduced in the value of α through Eq. (1.5) are shown in Fig. 1.6. As is shown in this figure, a −5% error in the value of α when its value is 1.1 can generate an error of more than 100% in the calculation of minimum number of trays. It can be imagined that the error in the actual number of trays would be even higher than 100%. In addition, the calculated numbers of trays are theoretical and when converted to real number of trays through overall column efficiency, the error may increase to several hundred percent. The approach of building the column higher to have a safe design is quite expensive.

As an example, a distillation column of diameter 4.5 m and height 85 m has an investment cost of approximately $4 million (€4.5 million) as stated by Dohrn and Pfohl [60]. Error in the calculation of relative volatility, α, could have been caused by the error in calculation of vapor pressure, which itself could have been caused by a small error in an input parameter such as critical temperature [58, 59]. Therefore, from this simple analysis one can realize the extreme cost and loss in the investment that can be caused by a small error in the estimation of critical temperature. Similar other examples have been given in the literature [62]. Nowadays, investment in refineries or their upgrading costs billions of dollars. For example, for a typical refinery of 160 000 bbl/d (8 million tons/year) capacity, the cost of construction in Europe is about $2 billion [18]. This is equivalent to refining cost of $7.5/bbl while this number for refineries of 1980s was about $2/bbl. In addition to the extra cost of investment, inappropriate design of units can cause extra operating costs and shorten the plant life as well as produce products that do not match the original design specifications. The use of a proper characterization method to calculate more accurate

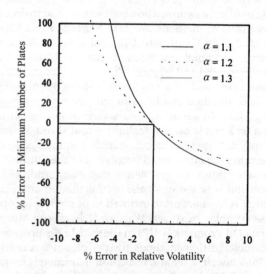

FIG. 1.6—Effect of error in the relative volatility on the error of minimum number of plates of a distillation column.

properties of petroleum fractions can save a large portion of such huge additional investment and operating costs.

1.4 ORGANIZATION OF THE BOOK

As the title of the book portrays and was discussed in Sections 1.2 and 1.3, the book presents methods of characterization and estimation of thermophysical properties of hydrocarbons, defined mixtures, undefined petroleum fractions, crude oils, and reservoir fluids. The entire book is written in nine chapters in a way such that in general every chapter requires materials presented in previous chapters. In addition there is an appendix and an index. Chapter 1 gives a general introduction to the subject from basic definition of various terms, the nature of petroleum, its formation and composition, types of petroleum mixtures, and the importance of characterization and property prediction to specific features of the book and its application in the petroleum industry and academia. Because of the importance of units in property calculations, the last section of Chapter 1 deals with unit conversion factors especially between SI and English units for the parameters used in the book. Chapter 2 is devoted to properties and characterization of pure hydrocarbons from C_1 to C_{22} from different hydrocarbon groups, especially from homologous groups commonly found in petroleum fluids. Properties of some nonhydrocarbons found with petroleum fluids such as H_2O, H_2S, CO_2, and N_2 are also given. Basic parameters are defined at the beginning of the chapter, followed by characterization of pure hydrocarbons. Predictive methods for various properties of pure hydrocarbons are presented and compared with each other. A discussion is given on the state-of-the-arts development of predictive methods. The procedures presented in this chapter are essential for characterization of petroleum fractions and crude oils discussed in Chapters 3 and 4.

Chapter 3 discusses various characterization methods for petroleum fractions and petroleum products. Characterization parameters are introduced and analytical instruments in laboratory are discussed. In this chapter one can use minimum laboratory data to characterize petroleum fractions and to determine the quality of petroleum products. Estimation of some basic properties such as molecular weight, molecular-type composition, sulfur content, flash, pour point and freezing points, critical constants, and acentric factor for petroleum fractions are presented in this chapter. A theoretical discussion on development of characterization methods and generation of predictive correlations from experimental data is also presented. Methods of Chapter 3 are extended to Chapter 4 for the characterization of various reservoir fluids and crude oils. Chapters 2–4 are perhaps the most important chapters in this book, as the methods presented in these chapters influence the entire field of physical properties in the remaining chapters.

In Chapter 5, PVT relations, equations of state, and corresponding state correlations are presented [31, 63–65]. The use of the velocity of light and sound in developing equations of state is also presented [31, 66–68]. Equations of state and corresponding state correlations are powerful tools in the estimation of volumetric, physical, transport, and thermodynamic properties [64, 65, 69]. Procedures outlined in Chapter 5 will be used in the prediction of physical properties discussed in the follow-up chapters. Fundamental thermodynamic relations for calculation of thermodynamic properties are presented in Chapter 6. The last three chapters of the book show applications of methods presented in Chapters 2–6 for calculation of various physical, thermodynamic, and transport properties. Methods of calculation and estimation of density and vapor pressure are given in Chapter 7. Thermal properties such as heat capacity, enthalpy, heat of vaporization, heats of combustion and reaction, and the heating value of fuels are also discussed in Chapter 7. Predictive methods for transport properties namely viscosity, thermal conductivity, diffusivity, and surface tension are given in Chapter 8 [30, 31, 42, 43, 69, 70]. Finally, phase equilibrium calculations, estimation of equilibrium ratios, GOR, calculation of pressure–temperature (*PT*) diagrams, solid formations, the conditions at which asphaltene, wax, and hydrate are formed, as well as their preventive methods are discussed in Chapter 9.

The book is written according to the standards set by ASTM for its publication. Every chapter begins with a general introduction to the chapter. Since in the following chapters for most properties several predictive methods are presented, a section on conclusion and recommendations is added at the end of the chapter. Practical problems as examples are presented and solved for each property discussed in each chapter. Finally, the chapter ends by a set of exercise problems followed by a citation section for the references used in the chapter.

The Appendix gives a summary of definitions of terms and properties used in this manual according to the ASTM dictionary as well as the Greek letters used in this manual. Finally the book ends with an index to provide a quick guide to find specific subjects.

1.5 SPECIFIC FEATURES OF THIS MANUAL

In this part several existing books in the area of characterization and physical properties of petroleum fractions are introduced and their differences with the current book are discussed. Then some special features of this book are presented.

1.5.1 Introduction of Some Existing Books

There are several books available that deal with physical properties of petroleum fractions and hydrocarbon systems. The most comprehensive and widely used book is the *API Technical Data Book—Petroleum Refining* [47]. It is a book with 15 chapters in three volumes, and the first edition appeared in mid 1960s. Every 5 years since, some chapters of the book have been revised and updated. The project has been conducted at the Pennsylvania State University and the sixth edition was published in 1997. It contains a data bank on properties of pure hydrocarbons, chapters on characterization of petroleum fractions, thermodynamic and transport properties of liquid and gaseous hydrocarbons, their mixtures, and undefined petroleum fractions. For each property, one predictive method that has been approved and selected by the API-TDB committees as the best available method is presented. This book will be referred as *API-TDB* throughout this book.

Another important book in this area is *The Properties of Gases and Liquids* that was originally written by Reid and Sherwood in 1950s and it has been revised and updated nearly every decade. The fifth and latest edition was published in November 2000 [36] by three authors different from the original two authors. The book has been an excellent reference for students and practical engineers in the industry over the past five decades. It discusses various methods for prediction of properties of pure hydrocarbons as well as nonhydrocarbons and their defined mixtures. However, it does not treat undefined petroleum fractions, crude oils, and reservoir fluids. Most of the methods for properties of pure compounds require the chemical structure of compounds (i.e., group contribution techniques). The book compares various methods and gives its recommendations for each method.

There are several other books in the area of properties of oils that document empirically developed predictive methods, among them is the book *Properties of Oils and Natural Gases,* by Pedersen *et al.* [13]. The book mainly treats reservoir fluids, especially gas condensates from North Sea, and it is mainly a useful reference for reservoir engineers. Books by McCain [11], Ahmed [71], Whitson [45], and Danesh [72] are all written by reservoir engineers and contain information mainly for phase behavior calculations needed in petroleum production and reservoir simulators. However, they contain some useful information on methods of prediction of some physical properties of petroleum fractions. Another good reference book was written by Tsonopoulos *et al.* [73] on thermodynamic and transport properties of coal liquids in the mid 1980s. Although there are many similarities between coal liquids and petroleum fractions, the book does not consider crude oils and reservoir fluids. But it provides some useful correlations for properties of coal liquids. The book by Wauquier [18] on petroleum refining has several useful chapters on characterization and physical properties of petroleum fractions and finished products. It also provides the test methods according to European standards. Some organizations' Web sites also provide information on fluid physical properties. A good example of such online information is provided by National Institute of Standards (http://webbook.nist.gov) which gives molecular weight, names, formulas, structure, and some data on various compounds [74].

1.5.2 Special Features of the Book

This book has objectives and aims that are different from the books mentioned in Section 1.5.1. The main objective of this book has been to provide a quick reference in the area of petroleum characterization and properties of various petroleum fluids for the people who work in the petroleum industry and research centers, especially in petroleum processing (*downstream*), petroleum production (*upstream*), and related industries. One special characteristic of the book is its discussion on development of various methods which would help the users of process/reservoir simulators to become familiar with the nature of characterization and property estimation methods for petroleum fractions. This would in turn help them to choose the proper predictive method among the

many methods available in a process simulator. However, the book has been written in a language that is understandable to undergraduate and graduate students in all areas of engineering and science. It contains practical solved problems as well as exercise problems so that the book would be suitable as a text for educational purposes.

Special features of this book are Chapters 2, 3, and 4 that deal with the characterization of hydrocarbons, petroleum fractions, and crude oils and their impact on the entire field of property prediction methods. It discusses both light as well as heavy fractions and presents methods of prediction of the important characteristics of petroleum products from minimum laboratory data and easily measurable parameters. It presents several characterization methods developed in recent years and not documented in existing references. The book also presents various predictive methods, including the most accurate and widely used method for each property and discusses points of strength, weaknesses, and limitations. Recommended methods are based on the generality, simplicity, accuracy, and availability of input parameters. This is another special feature of the book. In Chapters 5 and 6 it discusses equations of state based on the velocity of sound and light and how these two measurable properties can be used to predict thermodynamic and volumetric properties of fluids, especially heavy compounds and their mixtures [31, 63, 66–68]. Significant attention is given throughout the book on how to estimate properties of heavy hydrocarbons, petroleum fractions, crude oils, and reservoir fluids. Most of the methods developed by Riazi and coworkers [23, 24, 26–33, 51–56, 63, 65–70], which have been in use by the petroleum industry [47, 75–82], are documented in this book. In addition, a new experimental technique to measure diffusion coefficients in reservoir fluids under reservoir conditions is presented in Chapter 8 [42]. In Chapter 9 some new methods for determination of onset of solid formation are introduced. Reported experimental data on characteristics and properties of various oils from different parts of the world are included in various chapters for direct evaluations and testing of methods. Although both gases and liquids are treated in the book, emphasis is on the liquid fractions. Generally, the methods of estimation of properties of gases are more accurate than those for liquid systems. Most of the methods presented in the book are supported by some scientific basis and they are not simply empirical correlations derived from a certain group of data. This widens the application of the methods presented in the book to different types of oils. However, all basic parameters and necessary engineering concepts are defined in a way that is understandable for those nonengineer scientists who are working in the petroleum or related industry. Nearly all methods are expressed through mathematical relations so they are convenient for computer applications; however, most of them are simple such that the properties can be calculated by hand calculators for a quick estimate whenever applicable special methods are given for coal liquid fractions. This is another unique feature of this book.

1.6 APPLICATIONS OF THE BOOK

The information that is presented in the book may be applied and used in all areas of the petroleum industries: production, processing, and transportation. It can also be used as a textbook for educational purposes. Some of the applications of the materials covered in the book were discussed in Sections 1.2 and 1.3. The applications and uses of the book may be summarized as follows.

1.6.1 Applications in Petroleum Processing (Downstream)

Engineers, scientists, and operators working in various sectors of petroleum processing and refining or related industries can use the entire material discussed in the book. It helps laboratory people in refineries to measure useful properties and to test the reliability of their measurements. The book should be useful for engineers and researchers to analyze experimental data and develop their own predictive methods. It is also intended to help people who are involved with development of computer softwares and process simulators for design and operation of units and equipments in petroleum refineries. Another objective was to help users of such simulators to be able to select an appropriate predictive method for a particular application based on available data on the fraction.

1.6.2 Applications in Petroleum Production (Upstream)

Reservoir, chemical, and mechanical engineers may use the book in reservoir simulators, design and operation of surface separators in production fields, and feasibility studies for enhanced oil recovery projects, such as gas injection projects. Another application of the book by reservoir engineers is to simulate laboratory data on PVT experiments for the reservoir fluids, determination of the nature and type of reservoir fluids, and calculation of the initial amounts of oil and gas in the reservoir. Reservoir engineers may also use Chapter 9 to determine the conditions that a solid may form, amount of solid formation, and method of its prevention during production. Practically all chapters of the book should be useful for reservoir engineers.

1.6.3 Applications in Academia

Although the original goal and aim in writing this book was to prepare a reference manual for the industry, laboratories, and research institutions in the area of petroleum, it has been written in a way such that it can also be used as a textbook for educational purposes. It can be used as a text for an elective course for either undergraduate (senior level) or graduate level. Students from chemical, petroleum, and mechanical engineering fields as well as from chemistry and physics can take the course and understand the contents of the book. However, it should not be hard for students from other fields of engineering and science to use this book. The book may also be used to conduct short courses in the petroleum industry.

1.6.4 Other Applications

There are several other areas in which the book can be used. One may use this book to determine the quality of crude oils, petroleum fuels, and products for marketing and government

organizations that set the standards for such materials. As an example, the amount of sulfur or aromatic contents of a fuel can be estimated through minimum laboratory data to check if they meet the market demand or government regulations for environmental protection. This book can be used to determine properties of crude oil, its products, and natural gases that are needed for transportation and storage. Examples of such properties are density, boiling point, flash and pour points, sulfur content, vapor pressure, and viscosity.

The book can also be used to determine the properties of oils for clean-up operations where there is an oil spill on seawater. To simulate the fate of an oil spill and the rate of its disappearance at least the following properties are needed in order to use appropriate simulators [44, 83–85]:

- Characterization of petroleum fractions (Chapter 3)
- Pour point (Chapter 3)
- Characterization of crude oil (Chapter 4)
- Solubility parameter (Chapters 4, 6, and 9)
- Density (Chapters 5 and 7)
- Vapor pressure (Chapter 7)
- Viscosity, diffusion coefficient, and surface tension (Chapter 8)

Accurate prediction of the fate of a crude oil spill depends on the characterization technique used to estimate the physical properties. For example, to estimate how much of the initial oil would be vaporized after a certain time, accurate values of the diffusion coefficient, vapor pressure, and molecular weight are needed in addition to an appropriate characterization method to split the crude into several pseudocomponents [83].

1.7 DEFINITION OF UNITS AND THE CONVERSION FACTORS

An estimated physical property is valuable only if it is expressed in an appropriate unit. The most advanced process simulators and the most sophisticated design approaches fail to perform properly if appropriate units are not used. This is particularly important for the case of estimation of physical properties through various correlations or reporting the experimental data. Much of the confusion with reported experimental data arises from ambiguity in their units. If a density is reported without indicating the temperature at which the density has been measured, this value has no use. In this part basic units for properties used in the book are defined and conversion factors between the most commonly used units are given for each property. Finally some units specifically used in the petroleum industry are introduced. Interested readers may also find other information on units from online sources (for example, http://physics.nist.gov/cuu/contents/index.html).

1.7.1 Importance and Types of Units

The petroleum industry and its research began and grew mainly in the United States during the last century. The relations developed in the 1930s, 1940s, and 1950s were mainly graphical. The best example of such methods is the Winn nomogram developed in the late 1950s [86]. However, with the birth of the computer and its expansion, more analytical methods in the form of equations were developed in the 1960s and mainly in the 1970s and 1980s. Nearly all correlations and graphical methods that were developed until the early 1980s are in English units. However, starting from the 1980s many books and handbooks appeared in the SI units (from Le Systeme International d'Unites). The general trend is to unify all engineering books and documents in SI units to be used by the international community. However, many books, reports, handbooks, and equations and figures in various publications are still in English units. The United States and United Kingdom both officially use the English system of units. Therefore, it is essential that engineers be familiar with both unit systems of English and SI. The other unit system that is sometimes used for some properties is the cgs (centimeter, gram, second) unit, which is derived from the SI unit.

Since the book is prepared for an international audience, the primary unit system used for equations, tables, and figures is the SI; however, it has been tried to present equivalent of numbers and values of properties in both SI and English units. There are some figures that are taken from other references in the literature and are in English units and they have been presented in their original form. There are some special units that are commonly used to express some special properties. For example, viscosity is usually expressed in centipoise (cp), kinematic viscosity in centistoke (cSt), density in g/cm^3, specific gravity (SG) at standard temperature of $60°F$, or the GOR in scf/stb. For such properties, these primary units have been used throughout the book, while their respective equivalent values in SI are also presented.

1.7.2 Fundamental Units and Prefixes

Generally there are four fundamental quantities of length (L), mass (M), time (t), and temperature (T) and when their units are known, units of all other derived quantities can be determined. In the SI system, units of length, mass, and temperature are meter (m), kilogram (kg), and Kelvin (K), respectively. In English units these dimensions have the units of foot (ft), pound mass (lb_m), and degrees Rankine ($°R$), respectively. The unit of time in all unit systems is the second (s), although in English unit, hour (h) is also used for the unit of time. From these units, unit of any other quantity in SI is known. For example the unit of force is SI is $kg·m/s^2$ which is called *Newton* (N) and as a result the unit of pressure must be N/m^2 or *Pascal* (Pa). Since 1 Pa is a very small quantity, larger units such as kPa (1000 Pa) or mega Pascal (MPa) are commonly used. The standard prifixes in SI units are as follows:

Giga (G) = 10^9
Mega (M) = 10^6
Kilo (k) = 10^3
Hecto (h) = 10^2
Deka (da) = 10^1
Deci (d) = 10^{-1}
Centi (c) = 10^{-2}
Milli (m) = 10^{-3}
Micro (μ) = 10^{-6}
Nano (n) = 10^{-9}

As an example 1 000 000 Pa can be expressed as 1 MPa. These prefixes are not used in conjunction with the English units. However, in the English system of units when volumetric quantities of gases are presented in large numbers, usually every 1000 units is expressed by one prefix of M. For example, 2000 scf of gas is expressed as 2 Mscf and similarly 2 000 000 scf is written as 2 MMscf. Other symbols usually used to express large quantities are b for billion (1000 million or 10^9) and tr for trillion (one million millions or 10^{12}).

1.7.3 Units of Mass

The mass is shown by *m* and its unit in SI is kg (kilogram), in cgs is g (gram), and in the English unit system is lb_m (pound-mass). On many occasions the subscript m is dropped for lb when it is referred to mass. In the English unit system, units of ounce (oz) and grains are also used for mass units smaller than a pound. For larger values of mass, unit of *ton* is used, which is defined in three forms of long, short, and metric. Generally the term ton is applied to the metric ton (1000 kg). The conversion factors are as follows:

$$1\,kg = 1000\,g = 2.204634\,lb = 35.27392\,oz$$
$$1\,lb = 0.45359\,kg = 453.59\,g = 16\,oz = 7000\,grain$$
$$1\,g = 0.001\,kg = 0.002204634\,lb = 15.4324\,grain$$
$$1\,ton\,(metric) = 1000\,kg = 2204.634\,lb$$
$$1\,ton\,(short) = 2000\,lb = 907.18\,kg$$
$$1\,ton\,(long) = 2240\,lb = 1016\,kg = 1.12\,ton\,(short)$$
$$= 1.1016\,ton\,(metric)$$

1.7.4 Units of Length

The unit of length in SI is meter (m), in cgs is centimeter (cm), and in English unit system is foot (ft). Smaller values of length in English system are presented in inch (in.). The conversion factors are as follows:

$$1\,m = 100\,cm = 10^{-3}\,km = 1000\,mm = 10^6\,microns\,(\mu m)$$
$$= 10^{10}\,angstroms\,(Å) = 3.28084\,ft = 39.37008\,in.$$
$$= 1.0936\,yd\,(yard)$$
$$1\,ft = 12\,in. = 0.3048\,m = 30.48\,cm = 304.8\,mm$$
$$= 3.048 \times 10^{-4}\,km = 1/3\,yd$$
$$1\,cm = 10^{-2}\,m = 10^{-5}\,km = 10\,mm = 0.0328084\,ft$$
$$= 0.393701\,in.$$
$$1\,km = 1000\,m = 3280.48\,ft = 3.93658 \times 10^4\,in.$$
$$1\,in. = 2.54\,cm = 0.0833333\,ft = 0.0254\,m = 2.54 \times 10^{-5}\,km$$
$$1\,mile = 1609.3\,m = 1.609\,km = 5279.8\,ft$$

1.7.5 Units of Time

The unit of time in all major systems is the second (s); however, for large values of time other units such as minute (min), hour (h), day (d), and sometimes even year (year) are used appropriately. The conversion factors among these units are as follows:

$$1\,\text{year} = 365\,\text{d} = 8760\,\text{h} = 5.256 \times 10^5\,\text{min} = 3.1536 \times 10^7\,\text{s}$$

$$1\,\text{d} = 2.743973 \times 10^{-3}\,\text{year} = 24\,\text{h} = 1440\,\text{min} = 8.64 \times 10^4\,\text{s}$$

$$1\,\text{h} = 1.14155 \times 10^{-4}\,\text{year} = 4.16667 \times 10^{-2}\,\text{d}$$
$$= 60\,\text{min} = 3600\,\text{s}$$

$$1\,\text{min} = 1.89934 \times 10^{-6}\,\text{year} = 6.94444 \times 10^{-4}\,\text{d}$$
$$= 1.66667 \times 10^{-2}\,\text{h} = 60\,\text{s}$$

$$1\,\text{s} = 3.17098 \times 10^{-8}\,\text{year} = 1.157407 \times 10^{-5}\,\text{d}$$
$$= 2.77777 \times 10^{-4}\,\text{h} = 1.66667 \times 10^{-2}\,\text{min}$$

1.7.6 Units of Force

As mentioned above, the unit of force in the SI system is Newton (N) and in the English unit system is pound-force (lb_f). 1 lb_f is equivalent to the weight of a mass of 1 lb_m at the sea level where the acceleration of gravity is 32.174 ft/s^2 (9.807 m/s^2). In the cgs system, the unit of force is dyne (dyn). Another unit for the force in the metric system is kg_f, which is equivalent to the weight of a mass of 1 kg at the sea level. The conversion factors are as follows:

$$1\,\text{N} = 1\,\text{kg} \cdot \text{m/s}^2 = 10^5\,\text{dyn} = 0.2248\,lb_f = 1.01968 \times 10^{-1}\,kg_f$$

$$1\,lb_f = 4.4482\,\text{N} = 0.45359\,kg_f$$

$$1\,kg_f = 9.807\,\text{N} = 2.204634\,lb_f$$

$$1\,\text{dyn} = 10^{-5}\,\text{N} = 2.248 \times 10^{-6}\,lb_f$$

1.7.7 Units of Moles

Another unit to present amount of matter especially in engineering calculations is *mole* (mol), which is defined as the ratio of mass (m) to molecular weight (M).

$$(1.6) \qquad n = \frac{m}{M}$$

In SI system the unit of mole is kmol, where m in the above equation is in kg. In the English system, the unit of mol is lbmol. In the cgs system, the unit of mol is gmol, which is usually written as mol. For example, for methane (molecular weight 16.04) 1 mol of the gas has mass of 16.04 g. One mole of any substance contains 6.02×10^{23} number of molecules (Avogadro's number). The conversion factors between various units of moles are the same as given for the mass in Section 1.7.3.

$$1\,\text{kmol} = 1000\,\text{mol} = 2.204634\,\text{lbmol}$$

$$1\,\text{lbmol} = 0.45359\,\text{kmol} = 453.59\,\text{mol}$$

$$1\,\text{mol} = 0.001\,\text{kmol} = 0.002204634\,\text{lbmol}$$

1.7.8 Units of Molecular Weight

Molecular weight or molar mass shown by M is a number that 1 mol of any substance has equivalent mass of M g. In the SI system the unit of M is kg/kmol and in the English system the unit is lb/lbmol, while in the cgs system the unit of M is g/mol. Molecular weight is represented by the same number in all unit systems regardless of the system used. As an example, methane has the molecular weight of 16 g/mol, 16 lb/lbmol, and 16 kg/kmol in the unit systems of cgs, SI, and English, respectively. For this reason, in many cases the

unit for the molecular weight is not mentioned; however, one must realize that it is not a dimensionless parameter. Most recent compilations of molar masses are provided by Coplen [87].

1.7.9 Units of Pressure

Pressure is the force exerted by a fluid per unit area; therefore, in the SI system it has the unit of N/m^2, which is called Pascal (Pa), and in the English system has the unit of lb_f/ft^2 (psf) or lb_f/in.^{2} (psi). Other units commonly used for the pressure are the *bar* (bar) and *standard atmosphere* (atm). Pressure may also be expressed in terms of mm Hg. In this book units of MPa, kPa, bar, atm, or psi are commonly used for pressure. The conversion factors are given as follows:

$$1\,\text{atm} = 1.01325\,\text{bar} = 101\,325\,\text{Pa} = 101.325\,\text{kPa}$$
$$= 0.101325\,\text{MPa} = 14.696\,\text{psi}$$

$$1\,\text{atm} = 1.0322\,kg_f/\text{cm}^2 = 760\,\text{mm Hg (torr)} = 29.921\,\text{in. Hg}$$
$$= 10.333\,\text{m H}_2\text{O (4°C)}$$

$$1\,\text{bar} = 0.98692\,\text{atm} = 1 \times 10^5\,\text{Pa} = 100\,\text{kPa}$$
$$= 0.1\,\text{MPa} = 14.5038\,\text{psi}$$

$$1\,\text{Pa} = 1 \times 10^{-3}\,\text{kPa} = 1 \times 10^{-6}\,\text{MPa} = 9.8692 \times 10^{-6}\,\text{atm}$$
$$= 1 \times 10^{-5}\,\text{bar} = 1.45037 \times 10^{-4}\,\text{psi}$$

$$1\,\text{psi} = 6.804573 \times 10^{-2}\,\text{atm} = 6.89474 \times 10^{-2}\,\text{bar}$$
$$= 6.89474 \times 10^{-3}\,\text{MPa}$$

$$1\,\text{psf} = 144\,\text{psi} = 9.79858\,\text{atm} = 9.92843\,\text{bar} = 0.99285\,\text{MPa}$$

$$1\,kg_f/\text{cm}^2 = 0.96784\,\text{atm} = 0.98067\,\text{bar} = 14.223\,\text{psi}$$

The actual pressure of a fluid is the *absolute pressure*, which is measured relative to vacuum. However, some pressure measurement devices are calibrated to read zero in the atmosphere and they show the difference between the absolute and atmospheric pressure. This difference is called *gage pressure*. Normally "a" is used to indicate the absolute value (i.e., psia, bara) and "g" is used to show the gage pressure (i.e., psig). However, for absolute pressure very often "a" is dropped from the unit (i.e., psi, atm, bar). Another unit for the pressure is *vacuum pressure* that is defined for pressure below atmospheric pressure. Relations between these units are as follows:

$$(1.7) \qquad P_{\text{gage}} = P_{\text{abs}} - P_{\text{atm}}$$

$$(1.8) \qquad P_{\text{abs}} = P_{\text{atm}} - P_{\text{vac}}$$

Generally gage pressure unit is used to express pressures above the atmospheric pressures and vacuum pressure unit is used for pressures below atmospheric and may be expressed in various units (i.e., mm Hg, psi).

1.7.10 Units of Temperature

Temperature (T) is the most important parameter affecting properties of fluids and it is represented in Centigrade (°C) and Kelvin (K) in the SI system and in Fahrenheit (°F) and degrees Rankine (°R) in the English unit system. Temperature in most equations is in absolute degrees of Kelvin or Rankine. However, according to the definition of Kelvin and degrees Rankine where there is a temperature difference (ΔT), unit of °C is the same as K and °F is the same as °R. These temperature units are related through the following relations:

$$(1.9) \qquad T(\text{K}) = T(^\circ\text{C}) + 273.15$$

$$(1.10) \qquad T(^\circ\text{R}) = T(^\circ\text{F}) + 459.67$$

$$(1.11) \qquad \Delta T(\text{K}) = \Delta T(^\circ\text{C})$$

$$(1.12) \qquad \Delta T(^\circ\text{R}) = \Delta T(^\circ\text{F})$$

$$(1.13) \qquad T(^\circ\text{R}) = 1.8 T(\text{K})$$

$$(1.14) \qquad T(^\circ\text{F}) = 1.8 T(^\circ\text{C}) + 32$$

As an example, absolute temperature of 100 K is equivalent to 1.8×100 or 180°R. Therefore, the conversion factors between K and $^\circ$R are as follows:

1 K = 1.8°R (for absolute temperature T and the temperature difference, ΔT)

1°C = 1.8°F (only for the temperature difference, ΔT)

1.7.11 Units of Volume, Specific Volume, and Molar Volume—The Standard Conditions

Volume (V) has the dimension of cubic length (L^3) and thus in SI has the unit of m^3 and in English its unit is cubic feet (cf or ft^3). Some units particularly used for liquids in the SI system are liter (L), cm^3 (cc), or milliliter (mL) and in English units are gallon (in U.S. or Imperial) and barrel (bbl). Volume of one unit mass of a fluid is called specific volume and the volume of 1 mol of a fluid is called molar volume. Some of the conversion factors are as follows.

$$1\,\text{m}^3 = 10^6\,\text{cm}^3 = 1000\,\text{L} = 35.315\,\text{ft}^3 = 264.18\,\text{gallon (U.S.)}$$
$$= 35.316\,\text{ft}^3 = 6.29\,\text{bbl}$$
$$1\,\text{ft}^3 = 2.8316 \times 10^{-2}\,\text{m}^3 = 28.316\,\text{L} = 7.4805\,\text{gallon (U.S.)}$$
$$1\,\text{bbl} = 42\,\text{gallon(U.S.)} = 158.98\,\text{L} = 34.973\,\text{gallon (Imperial)}$$
$$1\,\text{gallon (U.S.)} = 0.8327\,\text{gallon (Imperial)}$$
$$= 0.023809\,\text{bbl} = 3.7853\,\text{L}$$
$$1\,\text{mL} = 1\,\text{cm}^3 = 10^{-3}\,\text{L} = 10^{-6}\,\text{m}^3 = 0.061024\,\text{in.}^3$$

For the molar volumes some of the conversion factors are given as follows:

$$1\,\text{m}^3/\text{kmol} = 1\,\text{L/mol} = 0.001\,\text{m}^3/\text{mol} = 1000\,\text{cm}^3/\text{mol}$$
$$= 16.019\,\text{ft}^3/\text{lbmol}$$
$$1\,\text{ft}^3/\text{lbmol} = 6.24259 \times 10^{-2}\,\text{m}^3/\text{kmol}$$
$$= 6.24259 \times 10^{-5}\,\text{m}^3/\text{mol} = 62.4259\,\text{cm}^3/\text{mol}$$
$$1\,\text{cm}^3/\text{mol} = 1\,\text{mL/mol} = 1\,\text{L/kmol} = 0.001\,\text{m}^3/\text{kmol}$$
$$= 1.6019 \times 10^{-2}\,\text{ft}^3/\text{lbmol}$$

It should be noted that the same conversion factors apply to specific volumes. For example,

$$1\,\text{ft}^3/\text{lb} = 6.24259 \times 10^{-2}\,\text{m}^3/\text{kg} = 62.4259\,\text{cm}^3/\text{g}$$

Since volume and specific or molar volumes depend on temperature and pressure of the system, values of volume in any unit system are meaningless if the conditions are not specified. This is particularly important for gases in which both temperature and pressure strongly influence the volume. For this reason, to express amount of gases in terms of volume, normally some SC are defined. The SC in the metric SI units are 0°C and 1 atm and in the English system are 60°F and 1 atm. Under these conditions molar volume of any gas is equivalent to 22.4 L/mol (in SI) and 379 scf/lbmol (in English units). In reservoir engineering calculations and petroleum industry in general, the SC in the SI units are also set at 60°F (15.5°C or 289 K) and 1 atm. The choice of standard temperature and pressure (STP) varies from one source to another. In this book when the standard T and P are not specified the STP refers to 289 K and 1 atm, which is equivalent to the STP in English unit system rather than SI system (273 K and 1 atm). However, for liquid systems the volume is less affected by pressure and for this reason specification of temperature alone is sufficient.

1.7.12 Units of Volumetric and Mass Flow Rates

Most processes in the petroleum industry are continuous and usually the volume or mass quantities are expressed in the form of rate defined as volume or mass per unit time. One particular volumetric flow rate used for liquids in the English system is gallon (U.S.) per minute and is known as GPM. Some of the conversion factors for these quantities are

$$1\,\text{m}^3/\text{s} = 1 \times 10^3\,\text{L/s} = 1.5851 \times 10^4\,\text{GPM}$$
$$= 5.4345 \times 10^5\,\text{bbl/d} = 1.27133 \times 10^5\,\text{ft}^3/\text{h}$$
$$1\,\text{ft}^3/\text{h} = 7.86558 \times 10^{-4}\,\text{m}^3/\text{s} = 0.12468\,\text{GPM}$$
$$= 4.27466\,\text{bbl/d}$$
$$1\,\text{GPM} = 2.228 \times 10^{-3}\,\text{ft}^3/\text{s} = 8.0205\,\text{m}^3/\text{h} = 34.285\,\text{bbl/d}$$
$$1\,\text{bbl/d} = 2.9167 \times 10^{-2}\,\text{GPM} = 1.8401 \times 10^{-4}\,\text{m}^3/\text{s}$$
$$= 0.23394\,\text{ft}^3/\text{h}$$

The conversion factors for the mass rates are as follows:

$$1\,\text{kg/s} = 7.93656 \times 10^3\,\text{lb/h} = 3.5136 \times 10^7\,\text{ton/year}$$
$$1\,\text{lb/s} = 1.63295 \times 10^3\,\text{kg/h} = 39.1908\,\text{ton/d}$$

The same conversion factors apply to molar rates.

1.7.13 Units of Density and Molar Density

Density shown by d or ρ is defined as mass per unit volume and it is reciprocal of specific volume. The conversion factors can be obtained from reversing those of specific volume in Section 1.7.11.

$$1\,\text{kg/m}^3 = 6.24259 \times 10^{-2}\,\text{lb/ft}^3 = 1 \times 10^{-3}\,\text{g/cm}^3$$
$$= 8.3455 \times 10^{-3}\,\text{lb/gal}$$
$$1\,\text{lb/ft}^3 = 16.019\,\text{kg/m}^3 = 1.6019 \times 10^{-2}\,\text{g/cm}^3$$
$$= 0.13368\,\text{lb/gal}$$
$$1\,\text{g/cm}^3 = 1\,\text{kg/L} = 10^3\,\text{kg/m}^3 = 62.4259\,\text{lb/ft}^3$$
$$= 8.3455\,\text{lb/gal}$$
$$1\,\text{lb/gal} = 1.19825 \times 10^2\,\text{kg/m}^3 = 7.4803\,\text{lb/ft}^3$$
$$= 0.119825\,\text{g/cm}^3$$

Density may also be presented in terms of number of moles per unit volume, which is called *molar density* and is reciprocal of molar volume. It can be obtained by dividing absolute density to molecular weight. The conversion factors for molar density are exactly the same as those for the absolute density (i.e., 1 mol/cm^3 = 62.4259 lbmol/ft^3). In practical calculations the conversion factors may be simplified without major error in the calculations. For example, 62.4 instead of 62.4259 or 7.48 instead of 7.4803 are used in practical calculations. In expressing values of densities, similar to specific volumes, the SC must be specified. Generally densities of liquid hydrocarbons are reported either in the form of specific gravity at 15.5°C (60°F) or the absolute density at 20°C and 1 atm in g/cm³.

1.7.14 Units of Specific Gravity

For liquid systems, the specific gravity (SG) is defined as the ratio of density of a liquid to that of water, and therefore, it is a dimensionless quantity. However, the temperature at which specific gravity is reported should be specified. The specific gravity is also called *relative density* versus absolute density. For liquid petroleum fractions and crude oils, densities of both the oil and water are expressed at the SC of 60°F (15.5°C) and 1 atm, and they are usually indicated as SG at 60°F/60°F or simply SG at 60°F. Another unit for the specific gravity of liquid hydrocarbons is defined by the American Petroleum Institute (API) and is called *API degree* and is defined in terms of SG at 60°F (API = 141.5/SG − 131.5). For gases, the specific gravity is defined as the ratio of density of the gas to that of the air at the SC, which is equivalent to the ratio of molecular weights. Further discussion on specific gravity, definitions, and methods of calculation are given in Chapter 2 (Section 2.1.3).

1.7.15 Units of Composition

Composition is the most important characteristic of homogenous mixtures in which two or more components are uniformly mixed in a single phase. Because of the nature of petroleum fluids, accurate knowledge of composition is important. Generally composition is expressed as percentage (%) or as fraction (percent/100) in terms of weight, mole, and volume. Density of the components (or pseudocomponents) constituting a mixture is required to convert composition from weight basis to volume basis or vice versa. Similarly conversion of composition from mole basis to weight basis or vice versa requires molecular weight of the constituting components (or pseudocomponents). Mole, weight, and volume fractions are shown by x_m, x_w, and x_v, respectively. Mole, weight, and volume percentages are shown by mol%, wt%, and vol%, respectively. Some references use mol/mol, wt/wt, and vol/vol to express fractional compositions. For normalized compositions, the sum of fractions for all components in a mixture is 1 ($\sum x_i = 1$) and the sum of all percentages is 100. If the molecular weights of all components in a mixture are the same, then the mole fraction and weight fraction are identical. Similarly, if the density (or specific gravity) of all components is the same, the weight and volume fractions are identical. The formula to calculate weight fraction from mole fraction is given as

$$(1.15) \qquad x_{wi} = \frac{x_{mi} M_i}{\sum_{i=1}^{N} x_{mi} M_i}$$

where N is the total number of components, M_i is the molecular weight, and x_{wi} and x_{mi} are the weight and mole fractions of component i, respectively. The conversion from weight to volume fraction can be obtained from the following equation:

$$(1.16) \qquad x_{vi} = \frac{x_{wi}/SG_i}{\sum_{i=1}^{N} x_{wi}/SG_i}$$

in which x_{vi} is the volume fraction and SG_i is the specific gravity of component i. In Eq. (1.16) density (d) can also be used instead of specific gravity. If mole and weight fractions are multiplied by 100, then composition is calculated on the percentage basis. In a similar way the conversion of composition from volume to weight and then to mole fraction can be obtained by reversing the above equations. The composition of a component in a liquid mixture may also be presented by its molar density, units of which were discussed in Section 1.7.13. Generally, a solution with solute molarity of 1 has 1 mol of solute per 1 L of solution (1 mol/L). Through use of both molecular weight of solute and density of solution one can obtain weight fraction from molarity. Another unit to express concentration of a solute in a liquid solution is *molality*. A solution with molality of 1 has 1 mol of solute per 1 kg of liquid solvent.

Another unit for the composition in small quantities is the ppm (*part per million*), which is defined as the ratio of unit weight (or volume) of a component to 10^6 units of weight or volume for the whole mixture. Therefore, ppm can be presented in terms of both volume or weight. Usually in gases the ppm is presented in terms of volume and in liquids it is expressed in terms of weight. When ppm is presented in terms of weight, its relation with wt% is 1 ppm = 10^{-4} wt%. For example, the maximum allowable concentration of H_2S in air for prolonged exposure is 10 ppm or 0.001 wt%. There is another smaller unit defined as *part per billion* known as ppb (1 ppm = 1000 ppb). In the United States a gas is considered "sweet" if the amount of its H_2S content is no more than one quarter grain per 100 scf of gas. This is almost equivalent to 4×10^{-4} mol fraction [88]. This is in turn equivalent to 4 ppm on the gas volume basis. Gas composition may also be represented in terms of partial pressure where sum of all partial pressures is equivalent to the total pressure.

In general, the composition of gases is presented in volume or mole fractions, while the liquid composition may be presented in any form of weight, mole, or volume. For gases at low pressures (≤1 atm where a gas may be considered an ideal gas) mole fraction and volume fractions are the same. However, generally under any conditions, volume and mole fractions are considered the same for gases and vapor mixtures. For narrow boiling range petroleum fractions with compositions presented in terms of PNA percentages, it is assumed that densities and molecular weights for all three representative pseudocompoents are nearly the same. Therefore, with a good degree of approximation, it is assumed that the PNA composition in all three unit systems are the same and for this reason on many occasions the PNA composition is represented only in terms of percentage (%) or fraction without indicating their weight or volume basis. However, this is not the case for the crude or reservoir fluid compositions where the composition is presented in terms of boiling point (or carbon number) and not in the form of molecular type. The following example shows conversion of composition from one type to another for a crude sample.

TABLE 1.5—*Conversion of composition of a crude oil sample from mole to weight and volume percent.*

Component	mol%	Molecular weight (M)	Specific gravity (SG)	wt%	vol%
C_2	0.19	30.07	0.356	0.03	0.06
C_3	1.88	44.10	0.508	0.37	0.64
iC_4	0.62	58.12	0.563	0.16	0.25
nC_4	3.92	58.12	0.584	1.02	1.52
iC_5	2.11	72.15	0.625	0.68	0.95
nC_5	4.46	72.15	0.631	1.44	1.98
C_6 (fraction)	8.59	82.00^a	0.690	3.15	3.97
C_{7+} (fraction)	78.23	266.00	0.895	93.15	90.63
Sum	100.00			100.00	100.00

^{a}This is molecular weight of C_6 hydrocarbon group and should not be mistaken with M of nC_6 which is 86.2.

Example 1.1—The composition of a Middle East crude oil is given in Table 1.5 in terms of mol% with known molecular weight and specific gravity for each component/pseudocomponent. Calculate the composition of the crude in both wt% and vol%.

Solution—In this table values of molecular weight and specific gravity for pure compounds are obtained from Chapter 2 (Table 2.1), while for the C_6 group, values are taken from Chapter 4 and for the C_{7+} fraction, values are given by the laboratory. Conversion calculations are based on Eqs. (1.15) and (1.16) on the percentage basis and the results are also given in Table 1.5. In this calculation it is seen that in terms of wt% and vol%, heavier compounds (i.e., C_{7+}) have higher values than in terms of mol%. ♦

1.7.16 Units of Energy and Specific Energy

Energy in various forms (i.e., heat, work) has the unit of Joule ($1 \, J = 1 \, N \cdot m$) in the SI and $ft \cdot lb_f$ in the English system. Values of heat are also presented in terms of calorie (in SI) and BTU (British Thermal Unit) in the English system. There are two types of joules: absolute joules and international joules, where 1 Joule (int.) = 1.0002 Joule (abs.). In this book only absolute joules is used and it is designated by J. There are also two types of calories: thermochemical and Internationational Steam Tables, where 1 cal (international steam tables) = 1.0007 cal (thermochemical) as defined in the API-TDB [47]. In this book cal refers to the international steam tables unless otherwise is specified. In the cgs system the unit of energy is dyn·cm, which is also called erg. The unit of power in the SI system is J/s or watt (W). Therefore, kW·h equivalent to 3600 kJ is also a unit for the energy. The product of pressure and volume (PV) may also present the unit of energy. Some of the conversion factors for the units of energy are given as follows:

$$1 \, J = 1 \, N \cdot m = 10^{-3} \, kJ = 10^7 \, erg = 0.23885 \, cal$$
$$= 9.4783 \times 10^{-4} \, Btu = 2.778 \times 10^{-7} \, kW \cdot h$$

$$1 \, J = 3.725 \times 10^{-7} \, hp \cdot h = 0.73756 \, ft \cdot lb_f = 9.869 \, L \cdot atm$$

$$1 \, cal \, (\text{International Tables}) = 3.9683 \times 10^{-3} \, Btu = 4.187 \, J$$
$$= 3.088 \, ft \cdot lb_f = 1.1630 \times 10^{-6} \, kW \cdot h$$

$$1 \, cal \, (\text{thermochemical}) = 1 \, cal = 3.9657 \times 10^{-3} \, Btu$$
$$= 4.184 \, J = 3.086 \, ft \cdot lb_f = 1.1622 \times 10^{-6} \, kW \cdot h$$

$$1 Btu = 1055 \, J = 251.99 \, cal = 778.16 \, ft \cdot lb_f$$
$$= 2.9307 \times 10^{-4} \, kW \cdot h$$

$$1 \, ft \cdot lb_f = 1.3558 \, J = 0.32384 \, cal = 1.2851 \times 10^{-3} \, Btu$$
$$= 3.766 \times 10^{-7} \, kW \cdot h$$

$$1 \, kW \cdot h = 3600 \, kJ = 3412.2 \, Btu = 2.655 \times 10^6 \, ft \cdot lb_f$$

Energy per unit mass is called *specific energy* that may be used to present properties such as specific enthalpy, specific internal energy, specific heats of reaction, and combustion or the heating values of fuels. Some of the conversion factors are given below.

$$1 \, J/g = 10^3 \, J/kg = 1 \, kJ/kg = 0.42993 \, Btu/lb$$
$$1 \, Btu/lb = 2.326 \, J/g = 0.55556 \, cal/g$$

The same conversion factors apply to the units of molar energy such as molar enthalpy.

1.7.17 Units of Specific Energy per Degrees

Properties such as heat capacity have the unit of specific energy per degrees. The conversion factors are as follows:

$$1 \, \frac{J}{g \cdot °C} = 1 \times 10^{-3} \, \frac{J}{kg \cdot °C} = 1 \, \frac{kJ}{kg \cdot °C} = 0.23885 \, \frac{Btu}{lb \cdot °F}$$

$$1 \, \frac{cal}{g \cdot °C} = 1 \, \frac{Btu}{lb \cdot °F} = 4.1867 \, \frac{J}{g \cdot °C}$$

As mentioned in Section 1.7.13, for the difference in temperature (ΔT), units of °C and K are the same. Therefore, the units of heat capacity may also be represented in terms of specific energy per Kelvin or degrees Rankine (i.e., $1 \frac{Btu}{lb \cdot °F} = 1 \frac{Btu}{lb \cdot °R} = 1 \frac{cal}{g \cdot °C} = 1 \frac{cal}{g \cdot K}$). The same conversion factors apply to units of molar energy per degrees such as molar heat capacity.

Another parameter which has the unit of molar energy per degrees is the universal *gas constant* (R) used in thermodynamic relations and equations of state. However, the unit of temperature for this parameter is the absolute temperature (K or °R), and °C or °F may never be used in this case. Similar conversion factors as those used for the heat capacity given above also apply to the units of gas constants in terms of molar energy per absolute degrees.

$$1 \, \frac{Btu}{lbmol \cdot °R} = 1 \, \frac{cal}{mol \cdot K} = 1.0007 \, \frac{cal(\text{thermochemical})}{mol \cdot K}$$
$$= 4.1867 \times 10^3 \, \frac{J}{kmol \cdot K}$$

Numerical values of the gas constant are given in Section 1.7.24.

1.7.18 Units of Viscosity and Kinematic Viscosity

Viscosity (absolute viscosity) shown by μ is a property that characterizes the fluidity of fluids and it has the dimension of mass per length per time ($M/L \cdot t$). If the relation between dimensions of force (F) and mass (M) is used ($F = M \cdot L \cdot t^{-2}$), then absolute viscosity finds the dimension of $F \cdot t \cdot L^{-2}$ which is the same as dimension for the product of pressure and time. Therefore, in the SI system the unit of viscosity is $Pa \cdot s$ ($N \cdot m^{-2} \cdot s$). In the cgs system the unit of viscosity is in $g/cm \cdot s$ that is called *poise* (p) and its hundredth is called centipoise (cp), which is equivalent to milli-$Pa \cdot s$ ($mPa \cdot s$). The conversion factors in various units are given below.

$$1\,cp = 1.02 \times 10^{-4}\,kg_f \cdot s/m^2 = 1 \times 10^{-3}\,Pa \cdot s = 1\,mPa \cdot s$$
$$= 10^{-2}\,p = 2.089 \times 10^{-5}\,lb_f \cdot s/ft^2 = 2.419\,lb/h \cdot ft$$
$$= 3.6\,kg/h \cdot m$$

$$1\,Pa \cdot s = 1\,kg/m \cdot s = 1000\,cp = 0.67194\,lb/ft \cdot s$$

$$1\,lb/h \cdot ft = 8.634 \times 10^{-6}\,lb_f \cdot s/ft^2 = 0.4134\,cp = 1.488\,kg/h \cdot m$$

$$1\,kg_f \cdot s/m^2 = 9.804 \times 10^3\,cp = 9.804\,Pa \cdot s = 0.20476\,lb_f \cdot s/ft^2$$

$$1\,lb_f \cdot s/ft^2 = 4.788 \times 10^4\,cp = 4.884\,kg_f \cdot s/m^2$$

The ratio of viscosity to density is known as *kinematic viscosity* (ν) and has the dimension of L/t^2. In the cgs system, the unit of kinematic vsicosity is cm^2/s also called *stoke* (St) and its hundredth is centistoke (cSt). The conversion factors are given below.

$$1\,ft^2/h = 2.778 \times 10^{-4}\,ft^2/s = 0.0929\,m^2/h = 25.81\,cSt$$

$$1\,ft^2/s = 9.29 \times 10^4\,cSt = 334.5\,m^2/h$$

$$1\,cSt = 10^{-2}\,St = 10^{-6}\,m^2/s = 1\,mm^2/s = 3.875 \times 10^{-2}\,ft^2/h$$
$$= 1.076 \times 10^{-5}\,ft^2/s$$

$$1\,m^2/s = 10^4\,St = 10^6\,cSt = 3.875 \times 10^4\,ft^2/h$$

Another unit to express kinematic viscosity of liquids is *Saybolt universal seconds* (SUS), which is the unit for the Saybolt universal viscosity (ASTM D 88). Definition of viscosity gravity constant (VGC) is based on SUS unit for the viscosity at two reference temperatures of 100 and 210°F (37.8 and 98.9 °C). The VGC is used in Chapter 3 to estimate the composition of heavy petroleum fractions. The relation between SUS and cSt is a function of temperature and it is given in the API TDB [47]. The analytical relations to convert cSt to SUS are given below [47].

$$SUS_{eq} = 4.6324\nu_T$$
$$+ \frac{[1.0 + 0.03264\nu_T]}{[(3930.2 + 262.7\nu_T + 23.97\nu_T^2 + 1.646\nu_T^3) \times 10^{-5}]}$$

(1.17)

where ν_T is the kinematic viscosity at temperature T in cSt. The SUS_{eq} calculated from this relation is converted to the SUS_T at the desired temperature of T through the following relation.

(1.18) $$SUS_T = [1 + 1.098 \times 10^{-4}(T - 311)]SUS_{eq}$$

where T is the temperature in kelvin (K). For conversion of cST to SUS at the reference temperature of 311 K (100°F), only Eq. (1.17) is needed. Equation (1.18) is the correction term for temperatures other than 100°F. For kinematic viscosities greater than 70 cSt, Eqs. (1.17) and (1.18) can be simplified to the following form at the temperatures of 311 (100°F) and 372 K, (210°F) respectively [1].

(1.19) $$SUS_{100F} = 4.632\nu_{100F} \quad \nu_{100F} \geq 75\,cSt$$

(1.20) $$SUS_{210F} = 4.664\nu_{210F} \quad \nu_{210F} \geq 75\,cSt$$

where ν_{100F} is the kinematic viscosity at 100°F (311 K) in cSt. As an example, a petroleum fraction with kinematic viscosity of 5 cSt at 311 K has an equivalent Saybolt Universal Viscosity of 42.4 SUS as calculated from Eq. (1.17).

Another unit for the viscosity is SFS (Saybolt foural seconds) expressed for Saybolt foural viscosity, which is measured in a way similar to Saybolt universal viscosity but measured by a larger orifice (ASTM D 88). The conversion from cSt to SFS is expressed through the following equations at two reference temperatures of 122°F (323 K) and 210°F (372 K) [47].

(1.21) $$SFS_{122F} = 0.4717\nu_{122F} + \frac{13924}{\nu_{122F}^2 - 72.59\nu_{122F} + 6816}$$

(1.22) $$SFS_{210F} = 0.4792\nu_{210F} + \frac{5610}{\nu_{210F}^2 + 2130}$$

For conversion of Saybolt foural viscosity (SFS) to kinematic viscosity (cSt.), the above equations should be used in reverse or to use tabulated values given by API-TDB [47]. As an example, an oil with Saybolt foural viscosity of 450 SFS at 210°F has a kinematic viscosity of 940 cSt. Generally, viscosity of highly viscous oils is presented by SUS or SFS units.

1.7.19 Units of Thermal Conductivity

Thermal conductivity (k) as discussed in Chapter 8 represents amount of heat passing through a unit area of a medium for one unit of temperature gradient (temperature difference per unit length). Therefore, it has the dimension of energy per time per area per temperature gradient. In the SI units it is expressed in $J/s \cdot m \cdot K$. Since thermal conductivity is defined based on a temperature difference (ΔT), the unit of °C may also be used instead of K. Because J/s is defined as watt (W), the unit of thermal conductivity in the SI system is usually written as $W/m \cdot K$. In the English system, the unit of thermal conductivity is $\frac{Btu}{ft \cdot h \cdot °F}$ and in some references is written as $\frac{Btu}{h \cdot ft^2 \cdot °F/ft}$, which is the ratio of heat flux to the temperature gradient. The conversion factors between various units are given below.

$$1\,W/m \cdot K\ (J/s \cdot m \cdot °C) = 0.5778\,Btu/ft \cdot h \cdot °F$$
$$= 1.605 \times 10^{-4}\,Btu/ft \cdot s \cdot °F$$
$$= 0.8593\,kcal/h \cdot m \cdot °C$$

$$1\,Btu/ft \cdot h \cdot °F = 1.7307\,W/m \cdot K$$

$$1\,cal/cm \cdot s \cdot °C = 242.07\,Btu/ft \cdot h \cdot °F = 418.95\,W/m \cdot K$$

1.7.20 Units of Diffusion Coefficients

Diffusion coefficient or diffusivity represents the amount of mass diffused in a medium per unit area per unit time per unit concentration gradient. As shown in Chapter 8, it has the same dimension as the kinematic viscosity, which is squared length per time (L^2/t). Usually it is expressed in cm^2/s.

$$1\,cm^2/s = 10^{-4}\,m^2/s = 9.29 \times 10^{-6}\,ft^2/s = 3.3445 \times 10^{-4}\,ft^2/h$$

1.7.21 Units of Surface Tension

Surface tension or interfacial tension (σ) as described in Section 8.6 (Chapter 8) has the unit of energy (work) per unit area and the SI unit of surface tension is J/m^2 = N/m. Since N/m is a large unit the values of surface tension are expressed in milli-N/m (mN/m) which is the same as the cgs unit of surface tension (dyn/cm). The conversion factors for this property are as follows:

$$1\,dyn/cm = 1\,erg/cm^2 = 10^{-3}\,J/m^2 = 1\,mJ/m^2$$
$$= 10^{-3}\,N/m = 1\,mN/m$$

1.7.22 Units of Solubility Parameter

Prediction of solubility parameter (δ) for petroleum fractions and crude oil is discussed in Chapters 4 and 10 and it has the unit of (energy/volume)$^{0.5}$. The traditional unit of δ is in (cal/cm^3)$^{0.5}$. Another form of the unit for the solubility parameter is (pressure)$^{0.5}$. Some conversion factors are given below.

$$1\,(cal_{th}/cm^3)^{0.5} = 2.0455\,(J/cm^3)^{0.5} = 2.0455\,(MPa)^{0.5}$$
$$= 2.0455 \times 10^3\,(J/m^3)^{0.5}$$
$$= 2.0455 \times 10^3\,(Pa)^{0.5} = 10.6004\,(Btu/ft^3)^{0.5}$$
$$= 31.6228\,(kcal_{th}/m^3)^{0.5}$$
$$= 6.4259\,(atm)^{0.5} = 2.05283\,(ft \cdot lb_f/ft^3)^{0.5}$$
$$1\,(MPa)^{0.5} = 0.4889\,(cal_{th}/cm^3)^{0.5} = 1\,(J/cm^3)^{0.5} = 10^3\,(Pa)^{0.5}$$

Values of surface tension in the literature are usually expressed in (cal/cm^3)$^{0.5}$ where cal represents thermochemical unit of calories.

1.7.23 Units of Gas-to-Oil Ratio

Gas-to-oil ratio is an important parameter in determining the type of a reservoir fluid and in setting the optimum operating conditions in the surface separators at the production field (Chapter 9, Section 9.2.1). In some references such as the API-TDB [47], this parameter is called *gas-to-liquid ratio* and is shown by GLR. GOR represents the ratio of volume of gas to the volume of liquid oil from a separator under the SC of 289 K and 101.3 kPa (60°F and 14.7 psia) for both the gas and liquid. Units of volume were discussed in Section 1.7.13. Three types of units are commonly used: the oilfield, the metric, and the English units.

• Oilfield units: standard cubic feet (scf) is used for the volume of gas, and *stock tank barrels* (stb) is used for the volume of oil. Therefore, GOR has the unit of scf/stb.
• Metric units: standard cubic meters (sm^3) is used for the gas, and stock tank cubic meters (stm^3) unit is used for the oil. The volume of liquid oil produced is usually presented under the stock tank conditions, which are 60°F (15.5°C) and 1 atm. Therefore, GOR unit in this system is sm^3/stm^3.
• English unit: scf is used for the gas, and *sock tank cubic feet* (stft^3) is used for the liquid volume. Thus the GOR has the

units of scf /stft^3. This unit is exactly the same as sm^3/stm^3 in the SI unit.

The conversion factors between these three units for the GOR (GLR) are given as follows:

$$1\,scf/stb = 0.1781\,scf/stft^3 = 0.1781\,sm^3/stm^3$$
$$1\,sm^3/stm^3 = 1\,scf/stft^3 = 5.615\,scf/stb$$

1.7.24 Values of Universal Constants

1.7.24.1 Gas Constant

The universal gas constant shown by R is used in equations of state and thermodynamic relations in Chapters 5, 6, 8, and 10. It has the unit of energy per mole per absolute degrees. As discussed in Section 1.7.17, its dimension is similar to that of molar heat capacity. The value of R in the SI unit is 8314 J/ kmol · K. The energy dimension may also be expressed as the product of pressure and volume (PV), which is useful for application in the equations of state. Value of R in terms of energy unit is more useful in the calculation of thermodynamic properties such as heat capacity or enthalpy. Values of this parameter in several other units are given as follows.

$$R = 8.314\,J/mol \cdot K = 8314\,J/kmol \cdot K = 8.314\,m^3 Pa/mol \cdot K$$
$$= 83.14\,cm^3 bar/mol \cdot K$$
$$= 82.06\,cm^3 \cdot atm/mol \cdot K = 1.987\,cal_{th}/mol \cdot K$$
$$= 1.986\,cal/mol \cdot K = 1.986\,Btu/lbmol \cdot R$$
$$= 0.7302\,ft^3 \cdot atm/lbmol \cdot R = 10.73\,ft^3 \cdot psia\,/lbmol \cdot R$$
$$= 1545\,ft \cdot lb_f/lbmol \cdot R$$

1.7.24.2 Other Numerical Constants

The Avogadro number is the number of molecules in 1 mol of a substance.

$$N_A = Avogadro\ number = 6.022 \times 10^{23}\,mol^{-1}$$

For example 1 mol of methane (16 g) consists of 6.022×10^{23} molecules. Other constants are

Boltzman constant $= k_B = R/N_A = 1.381 \times 10^{-23}$ J/K.

Planck constant $= h = 6.626 \times 10^{-34}$ J · s.

Speed of light in vacuum $= c = 2.998 \times 10^8$ m/s.

Numerical constants

$$\pi = 3.14159265$$
$$e = 2.718\,281\,828$$
$$\ln x = \log_{10} x/\log_{10} e = 2.30258509\,\log_{10} x.$$

1.7.25 Special Units for the Rates and Amounts of Oil and Gas

Amounts of oil and gas are usually expressed in volumetric quantities. In the petroleum industry the common unit for volume of oil is barrel (bbl) and for the gas is standard cubic feet (scf) both at the conditions of 60°F (15.5°C) and 1 atm. The production rate for the crude is expressed in bbl/d and for the gas in scf/d.

In some cases, amount of crude oil is expressed in the metric ton. Conversion from volume to weight or vice versa requires density or specific gravity (API) of the oil. For a light

Saudi Arabian crude of 35.5 API (SG = 0.847), the following conversion factors apply between weight and volume of crudes and the rates:

$$1 \text{ ton} \cong 7.33 \text{ bbl} = 308 \text{ gallon (U.S.)} \quad 1 \text{ bbl} \cong 0.136 \text{ ton}$$
$$1 \text{ bbl/d} \cong 50 \text{ ton/year}$$

For a Middle East crude of API 30, 1 ton $\cong$ 7.19 bbl (1 bbl $\cong$ 0.139 ton).

Another way of expressing quantities of various sources of energy is through their heating values. For example, by burning 1×10^6 tons of a crude oil, the same amount of energy can be produced that is produced through burning 1.5×10^9 tons of coal. Of course this value very much depends on the type of crude and the coal. Therefore, such evaluations and comparisons are approximate. In summary, 1 million tons of a typical crude oil is equivalent to other forms of energy:

$$1 \times 10^6 \text{ tons of crude oil} \cong 1.111 \times 10^9 \text{sm}^3 \ (39.2 \times 10^9 \text{scf})$$
$$\text{of natural gas}$$
$$\cong 1.5 \times 10^9 \text{ tons of coal}$$
$$\cong 12 \times 10^9 \text{ kW} \cdot \text{h of electricity}$$

The $\cong$ sign indicates the approximate values, as they depend on the type of oil or gas. For a typical crude, the heating value is approximately 10 500 cal/g (18 900 Btu/lb) and for the natural gas is about 1000 Btu/scf (37.235×10^3 kJ/sm³). Approximately 1 million tons of a typical crude oil can produce an energy equivalent to 4×10^9 kW $\cdot$ h of electricity through a typical power plant. In 1987 the total nuclear energy produced in the world was equivalent to 404×10^6 tons of crude oil based on the energy produced [5]. In the same year the total hydroelectric energy was equivalent to 523.9×10^6 tons of crude oil. In 1987 the total coal reserves in the world were estimated at 1026×10^9 tons, while the total oil reserves were about 122×10^9 tons. However, from the energy point of view the total coal reserves are equivalent to only 0.68×10^9 tons of crude oil. The subject of heating values will be discussed further in Chapter 7 (see Section 7.4.4).

Unit conversion is an important art in engineering calculations and as was stated before with the knowledge of the definition of some basic units for only a few fundamental quantities (energy, length, mass, time, and temperature), the unit for every other property can be obtained. The basic idea in the unit conversion is that a value of a parameter remains the same when it is multiplied by a factor of unity in a way that the initial units are eliminated and the desired units are kept. The following examples demonstrate how a unit can be converted to another unit system without the use of tabulated conversion factors.

Example 1.2—The molar heating value of methane is 802 kJ/mol. Calculate the heating value of methane in the units of cal/g and Btu/lb. The molecular weight of methane is 16.0.

Solution—In this calculation a practicing engineer has to remember the following basic conversion factors: 1 lb = 453.6 g, 1 cal = 4.187 J, and 1 Btu = 252 cal. The value of molecular weight indicates that 1 mol = 16 g. In the conversion process the initial unit is multiplied by a series of known conversion

factors with ratios of unity as follows:

$$802 \frac{\text{kJ}}{\text{mol}} = \left(802 \frac{\text{kJ}}{\text{mol}}\right) \times \left|\frac{\text{mol}}{16 \text{ g}}\right| \times \left|\frac{1000 \text{ J}}{\text{kJ}}\right| \times \left|\frac{\text{cal}}{4.187 \text{ J}}\right|$$
$$= \left(\frac{802 \times 1000}{16 \times 4.187}\right) [\text{cal/g}] = 11971.58 \text{ cal/g}$$

The conversion to the English unit is performed in a similar way:

$$11971.58 \text{ cal/g} = (11971.58 \text{ cal/g}) \times \left|\frac{453.6 \text{ g}}{\text{lb}}\right| \times \left|\frac{\text{Btu}}{252 \text{ cal}}\right|$$
$$= \left(\frac{11971.58 \times 453.6}{252}\right) [\text{Btu/lb}]$$
$$= 21549.2 \text{ Btu/lb}$$

In the above calculations all the ratio of terms inside the $||$ sign have values of unity. ◆

Example 1.3—Thermal conductivity of a kerosene sample at 60°C is 0.07 Btu/h $\cdot$ ft$\cdot$°F. What is the value of thermal conductivity in mW/mK from the following procedures:

1. Use of appropriate conversion factor in Section 1.7.19.
2. Direct calculation with use of conversion factors for fundamental dimensions.

Solution—

1. In Section 1.7.19 the conversion factor between SI and English units is given as:
   1 W/mK = 0.5778 Btu/ft $\cdot$ h $\cdot$ °F. With the knowledge that W = 1000 mW, the conversion is carried as:

$$0.07 \text{ Btu/h} \cdot \text{ft} \cdot °F = \left(0.07 \left|\frac{\text{Btu}}{\text{h} \cdot \text{ft} \cdot °F}\right|\right) \times \left|\frac{1000 \text{ mW}}{\text{W}}\right|$$
$$\times \left|\frac{\text{W/mK}}{0.5778 \text{ Btu/ h} \cdot \text{ft} \cdot °F}\right| = 121.1 \text{ mW/mK}$$

2. The conversion can be carried out without use of the conversion tables if a practicing engineer is familiar with the basic definitions and conversion factors. These are 1 W = 1 J/s, 1 W = 1000 mW, 1 cal = 4.187 J, I Btu = 251.99 cal, 1 h = 3600 s, 1 ft = 0.3048 m, 1 K = 1°C = 1.8°F (for the temperature difference). It should be noted that thermal conductivity is defined based on temperature difference.

$$0.07 \text{ Btu/h} \cdot \text{ft} \cdot °F$$
$$= \left[0.07 \left|\frac{\text{Btu}}{\text{h} \cdot \text{ft} \cdot °F}\right|\right] \times \left|\frac{251.99 \text{ cal}}{\text{Btu}}\right| \times \left|\frac{4.187 \text{ J}}{\text{cal}}\right| \times \left|\frac{\text{h}}{3600 \text{ s}}\right|$$
$$\times \left|\frac{\text{W}}{\text{J/s}}\right| \times \left|\frac{1000 \text{ mW}}{\text{W}}\right| \times \left|\frac{\text{ft}}{0.3048 \text{ m}}\right| \times \left|\frac{1.8°F}{°C}\right| \times \left|\frac{°C}{\text{K}}\right|$$
$$= \left|\frac{0.07 \times 251.99 \times 4.187 \times 1000 \times 1.8}{3600 \times 0.3048}\right| \times \left|\frac{\text{mW}}{\text{m} \cdot \text{K}}\right|$$
$$= 121.18 \text{ mW/mK} \qquad ◆$$

Examples 1.2 and 1.3 show that with the knowledge of only very few conversion factors and basic definitions of fundamental units, one can obtain the conversion factor between any two unit systems for any property without use of a reference conversion table.

1.8 PROBLEMS

1.1. State one theory for the formation of petroleum and give names of the hydrocarbon groups in a crude oil. What are the most important heteroatoms and their concentration level in a crude oil?

1.2. The following compounds are generally found in the analysis of a crude oil: ethane, propane, isobutane, n-butane, isopentane, n-pentane, 2,2-dimethylbutane, cyclopentane, cyclohexane, n-hexane, 2-methylpentane, 3-methylpentane, benzene, methylcyclopentane, 1,1-dimethylcyclopentane, and hydrocarbons from C_7 and heavier grouped as C_{7+}.
   a. For each compound, draw the chemical structure and give the formula. Also indicate the name of hydrocarbon group that each compound belongs to.
   b. From the above list give the compounds that possibly exist in a gasoline fraction.

1.3. Give the names of $n\text{-}C_{20}$, $n\text{-}C_{30}$, $n\text{-}C_{40}$, and three isomers of n-heptane according to the IUPAC system.

1.4. List the 10 most important physical properties of crude and its products that are required in both the design and operation of an atmospheric distillation column.

1.5. What thermodynamic and physical properties of gas and/or liquid fluids are required for the following two cases?
   a. Design and operation of an absorption column with chemical reaction [40, 89].
   b. Reservoir simulation [37].

1.6. What is the characterization of petroleum fractions, crude oils, and reservoir fluids? Explain their differences.

1.7. Give the names of the following compounds according to the IUPAC system.

a.

(b)   (c)   (d)

e. $CH_2{=}CH{-}CH{=}CH{-}CH_2{-}CH_3$

1.8. From an appropriate reference find the following data in recent years.
   a. What is the distribution of refineries in different parts of the world (North America, South America, Western Europe, Africa, Middle East, Eastern Europe and Former Soviet Union, and Asia Pacific)?
   b. Where is the location of the biggest refinery in the world and what is its capacity in bbl/d?
   c. What is the history of the rate of production of gasoline, distillate, and residual from refineries in the world and the United States for the last decade?

1.9. Characteristics of three reservoir fluids are given below. For each case determine the type of the reservoir fluid using the rule of thumb.
   a. GOR = 20 scf/stb
   b. GOR = 150 000 scf/stb
   c. CH_4 mol% = 70, API gravity of STO = 40

1.10. GOR of a reservoir fluid is 800 scf/stb. Assume the molecular weight of the stock tank oil is 260 and its specific gravity is 0.87.
   a. Calculate the GOR in sm^3/stm^3 and the mole fraction of gases in the fluid.
   b. Derive a general mathematical relation to calculate GOR from mole fraction of dissolved gas (x_A) through STO gravity (SG) and oil molecular weight (M). Calculate x_A using the developed relation.

1.11. The total LPG production in 1995 was 160 million tons/year. If the specific gravity of the liquid is assumed to be 0.55, what is the production rate in bbl/d?

1.12. A C_{7+} fraction of a crude oil has the following composition in wt%. The molecular weight and specific gravity of each pseudocomponent are also given below. Calculate the composition of crude in terms of vol% and mol%.

Pseudocomponent	wt%	M	SG
C_{7+} (1)	17.3	110	0.750
C_{7+} (2)	23.6	168	0.810
C_{7+} (3)	31.8	263	0.862
C_{7+} (4)	16.0	402	0.903
C_{7+} (5)	11.3	608	0.949
Total C_{7+}	100		

1.13. It is assumed that a practicing engineer remembers the following fundamental unit conversion factors without a reference.

1 ft = 0.3048 m = 12 in.
1 atm = 101.3 kPa = 14.7 psi
1 K = 1.8°R
1 Btu = 252 cal
1 cal = 4.18 J
1 kg = 2.2 lb
g = 9.8 m/s²
1 lbmol = 379 scf
Molecular weight of methane = 16 g/mol

Calculate the following conversion factors using the above fundamental units.

   a. The value of gas constant is 1.987 cal/mol·K. What is its value in psi·ft³/lbmol·R?
   b. Pressure of 5000 psig to atm
   c. 1 kg$_f$/cm² to kPa
   d. 1 Btu/lb·°F to J/kg·K
   e. 1 Btu/lbmol to cal/g
   f. 1000 scf of methane gas to lbmol
   g. 1 MMM scf of methane to kg

h. 1 cp to lb/ft.h
    i. 1 Pa · s to cp
    j. 1 g/cm^3 to lb/ft^3
1.14. A crude oil has API gravity of 24. What is its density in g/cm^3, lb/ft^3, kg/L, kg/m^3?
1.15. Convert the following units for the viscosity.
    a. Crude viscosity of 45 SUS (or SSU) at 60°C (140°F) to cSt.
    b. Viscosity of 50 SFS at 99°C (210°F) to cSt.
    c. Viscosity of 100 cp at 38°C (100°F) to SUS
    d. Viscosity of 10 cp at 99°C (210°F) to SUS
1.16. For each ton of a typical crude oil give the equivalent estimates in the following terms:
    a. Volume of crude in bbl.
    b. Tons of equivalent coal.
    c. Standard cubic feet (scf) and sm^3 of natural gas.
1.17. In terms of equivalent energy values, compare existing reserves for three major fossil types and nonrenewable sources of energy: oil, natural gas, and coal by calculating
    a. the ratio of existing world total gas reserves to the world total oil reserves.
    b. the ratio of existing world total coal reserves to the world total oil reserves.
    c. the percent share of amount of each energy source in total reserves of all three sources.

REFERENCES

[1] Speight, J. G., *The Chemistry and Technology of Petroleum*, 3rd ed., Marcel Dekker, New York, 1998.
[2] Jahn, F., Cook, M., and Graham, M., *Hydrocarbon Exploration and Production*, Developments in Petroleum Science 46, Elsevier, Amsterdam, 1998.
[3] *Oil and Gas Journal Data Book*, 2000 edition, PennWell Publishing, Tulsa, OK, 2000.
[4] *International Petroleum Encyclopedia, Volume 32*, PennWell Publishing, Tulsa, OK, 1999.
[5] *BP Statistical Review of World Energy*, British Petroleum, United Kingdom, June 1988.
[6] *Basic Petroleum Data Book*, Petroleum Industry Statistics, Volume XVIII, Number 1, American Petroleum Institute (API), Washington, DC, 1988.
[7] The International Union of Pure and Applied Chemistry (IUPAC), 2003, Research Triangle Park, NC, URL: http://www.iupac.org.
[8] Nelson, W. L., *Petroleum Refinery Engineering*, McGraw-Hill, New York, 1958.
[9] Manning, F. S. and Thompson, R. R., *Oilfield Processing*, PennWell Publishing, Tulsa, OK, 1995.
[10] Altagelt, K. H. and Boduszynski, M. M., *Composition and Analysis of Heavy Petroleum Fractions*, Marcel Dekker, New York, 1994.
[11] McCain, Jr., W. D., *The Properties of Petroleum Fluids*, 2nd ed., PennWell Publishing, Tulsa, OK, 1990.
[12] Speight, J. G., *Gas Processing—Environmental Aspects and Methods*, Butterworth Heinemann, Oxford, England, 1993.
[13] Pedersen, K. S., Fredenslund, Aa., and Thomassen, P., *Properties of Oils and Natural Gases*, Gulf Publishing, Houston, TX, 1989.
[14] Arnold, K. and Stewart, M., *Surface Production Operations, Design of Oil-Handling Systems and Facilities, Volume 1*, 2nd ed., Gulf Publishing, Houston, TX, 1998.
[15] Favennec, J. P., *Petroleum Refining, Volume 5: Refinery Operation and Management*, Editions Technip, Paris, 1998.
[16] Mansoori, G. A., "Modeling of Asphaltene and Other Heavy Organic Depositions," *Journal of Petroleum Science and Engineering*, Vol. 17, 1997, pp. 101–111.
[17] Goual, L. and Firoozabadi, A., "Measuring Asphaltenes and Resins, and Dipole Moment in Petroleum Fluids," *American Institute of Chemical Engineers Journal*, Vol. 48, No. 11, 2002, pp. 2646–2662.
[18] Wauquier, J.-P., *Petroleum Refining, Volume 1: Crude Oil, Petroleum Products, Process Flowsheets*, Editions Technip, Paris, 1995.
[19] Gary, J. H. and Handwerk, G. E., *Petroleum Refining—Technology and Economics*, 3rd ed., Marcel & Dekker, New York, 1994.
[20] Beck, R. J., "Demand Growth to Continue for Oil, Resume for Gas this Year in the U.S.," *Oil and Gas Journal*, January 26, 1998, p. 57.
[21] Totten, G. E., Shah, R. J., and Westbrook, S. R., *Fuels and Lubricants Handbook: Technology, Properties, Performance, and Testing*, Eds., ASTM MNL 37, ASTM International, West Conshohocken, PA, 2003, pp. 1075.
[22] ASTM, *Annual Book of Standards*, Section Five, Petroleum Products, Lubricants, and Fossil Fuels (in Five Volumes), ASTM International, West Conshohocken, PA, 2002.
[23] Riazi, M. R. and Daubert, T. E., "Simplify Property Predictions," *Hydrocarbon Processing*, Vol. 59, No. 3, 1980, pp. 115–116.
[24] Riazi, M. R. and Daubert, T. E., "Characterization Parameters for Petroleum Fractions," *Industrial and Engineering Chemistry Research*, Vol. 26, 1987, pp. 755–759.
[25] Riazi, M. R. and Al-Otaibi, G. N., "Estimation of Viscosity of Petroleum Fractions," *Fuel*, Vol. 80, 2001, pp. 27–32.
[26] Riazi, M. R. and Daubert, T. E., "Molecular Weight of Heavy Fractions from Viscosity," *Oil and Gas Journal*, Vol. 58, No. 52, 1987, pp. 110–113.
[27] Riazi, M. R. and Daubert, T. E., "Prediction of the Composition of Petroleum Fractions," *Industrial and Engineering Chemistry, Process Design and Development*, Vol. 19, No. 2, 1980, pp. 289–294.
[28] Riazi, M. R. and Daubert, T. E., "Prediction of Molecular Type Analysis of Petroleum Fractions and Coal Liquids," *Industrial and Engineering Chemistry, Process Design and Development*, Vol. 25, No. 4, 1986, pp. 1009–1015.
[29] Riazi, M. R., Nasimi, N., and Roomi, Y., "Estimating Sulfur Content of Petroleum Products and Crude Oils," *Industrial and Engineering Chemistry Research*, Vol. 38, No. 11, 1999, pp. 4507–4512.
[30] Riazi, M. R., Al-Enezi, G., and Soleimani, S., "Estimation of Transport Properties of Liquids," *Chemical Engineering Communications*, Vol. 176, 1999, pp. 175–193.
[31] Riazi, M. R. and Roomi, Y., "Use of the Refractive Index in the Estimation of Thermophysical Properties of Hydrocarbons and Their Mixtures," *Industrial and Engineering Chemistry Research*, Vol. 40, No. 8, 2001, pp. 1975–1984.
[32] Riazi, M. R., *Prediction of Thermophysical Properties of Petroleum Fractions*, Doctoral Dissertation, Department of Chemical Engineering, Pennsylvania State University, University Park, PA, 1979.
[33] Riazi, M. R. and Al-Sahhaf, T., "Physical Properties of n-Alkanes and n-Alkyl Hydrocarbons: Application to Petroleum Mixtures," *Industrial and Engineering Chemistry Research*, Vol. 34, 1995, pp. 4145–4148.
[34] Daubert, T. E., "Property Predictions," *Hydrocarbon Processing*, Vol. 59, No. 3, 1980, pp. 107–112.
[35] Riazi, M. R. and Roomi, Y. A., "Compositional Analysis of Petroleum Fractions," *Preprints of Division of Petroleum*

Chemistry Symposia, American Chemical Society (ACS), Vol. 40, No. 8, 2001, pp. 1975–1980.

[36] Poling, B. E., Prausnitz, J. M., and O'Connell, J. P., *Properties of Gases and Liquids*, 5th ed., McGraw-Hill, New York, 2000.

[37] Mattax, C. C. and Dalton, R. L., *Reservoir Simulation*, SPE Nomograph 13, Society of Petroleum Engineers, Richardson, TX, 1990.

[38] Riazi, M. R. and Faghri, A., "Effect of the Interfacial Drag on Gas Absorption with Chemical Reaction in a Vertical Tube," *American Institute of Chemical Engineers Journal*, Vol. 32, No. 4, 1986, pp. 696–699.

[39] Dabir, B., Riazi, M. R., and Davoudirad, H. R., "Modeling of Falling Film Reactors," *Chemical Engineering Science*, Vol. 51, No. 11, 1996, pp. 2553–2558.

[40] Riazi, M. R., Whitson, C. H., and da Silva, F., "Modelling of Diffusional Mass Transfer in Naturally Fractured Reservoirs," *Journal of Petroleum Science and Engineering*, Vol. 10, 1994, pp. 239–253.

[41] Riazi, M. R. and Whitson, C. H., "Estimating Diffusion Coefficients of Dense Fluids," *Industrial and Engineering Chemistry Research*, Vol. 32, No. 12, 1993, pp. 3081–3088.

[42] Riazi, M. R., "A New Method for Experimental Measurement of Diffusion Coefficient in Reservoir Fluids," *Journal of Petroleum Science and Engineering*, Vol. 14, 1996, pp. 235–250.

[43] Kister, H. Z., *Distillation Operations*, McGraw-Hill, New York, 1990.

[44] Villoria, C. M., Anselmi, A. E., Intevep, S. A., and Garcia, F. R., "An Oil Spill Fate Model," SPE23371, Society of Petroleum Engineers, 1991, pp. 445–454.

[45] Whitson, C. H. and Brule, M. R., *Phase Behavior*, Monograph Volume 20, Society of Petroleum Engineers, Richardson, TX, 2000.

[46] Firoozabadi, A., *Thermodynamics of Hydrocarbon Reservoirs*, McGraw-Hill, New York, 1999.

[47] Daubert, T. E. and Danner, R. P., Eds., *API Technical Data Book—Petroleum Refining*, 6th ed., American Petroleum Institute (API), Washington, DC, 1997.

[48] AIChE DIPPR®Database, Design Institute for Physical Property Data (DIPPR), EPCON International, Houston, TX, 1996.

[49] Rowley, R. L., Wilding, W. V., Oscarson, J. L., Zundel, N. A., Marshall, T. L., Daubert, T. E., and Danner, R. P., *DIPPR Data Compilation of Pure Compound Properties*, Design Institute for Physical Properties (DIPPR), Taylor & Francis, New York, 2002 (http://dippr.byu.edu).

[50] Thermodynamic Research Center, National Institute of Standards and Technology (NIST), Boulder, CO, URL: http://www.nist.gov/.

[51] Riazi, M. R. and Daubert, T. E., "Analytical Correlations Interconvert Distillation Curve Types," *Oil and Gas Journal*, August 25, 1986, pp. 50–57.

[52] Riazi, M. R. and Daubert, T. E., "Improved Characterization of Wide Boiling Range Undefined Petroleum Fractions," *Industrial and Engineering Chemistry Research*, Vol. 26, 1987, pp. 629–632.

[53] Riazi, M. R. and Daubert, T. E., "Predicting Flash Points and Pour Points of Petroleum Fractions," *Hydrocarbon Processing*, September 1987, pp. 81–84.

[54] Riazi, M. R., "A Distribution Model for C_{7+} Fractions Characterization of Petroleum Fluids," *Industrial and Engineering Chemistry Research*, Vol. 36, 1997, pp. 4299–4307.

[55] Riazi, M. R., "Distribution Model for Properties of Hydrocarbon-Plus Fractions," *Industrial and Engineering Chemistry Research*, Vol. 28, 1989, pp. 1731–1735.

[56] Riazi, M. R. and Al-Sahhaf, T. A., "Physical Properties of Heavy Petroleum Fractions and Crude Oils," *Fluid Phase Equilibria*, Vol. 117, 1996, pp. 217–224.

[57] Watson, K. M., Nelson, E. F., and Murphy, G. B., "Characterization of Petroleum Fractions," *Industrial and Engineering Chemistry*, Vol. 27, 1935, pp. 1460–1464.

[58] Brule, M. R., Kumar, K. H., and Watanasiri, S., "Characterization Methods Improve Phase Behavior Predictions," *Oil & Gas Journal*, February 11, 1985, pp. 87–93.

[59] Riazi, M. R., Al-Sahhaf, T. A., and Al-Shammari, M. A., "A Generalized Method for Estimation of Critical Constants," *Fluid Phase Equilibria*, Vol. 147, 1998, pp. 1–6.

[60] Dohrn, R. and Pfohl, O. "Thermophysical Properties—Industrial Directions," Paper presented at the *Ninth International Conference on Properties and Phase Equilibria for Product and Process Design* (PPEPPD 2001), Kurashiki, Japan, May 20–25, 2001.

[61] McCabe, W. L., Smith, J. C., and Harriot, P., *Unit Operations of Chemical Engineering*, 4th ed., McGraw-Hill, New York, 1985.

[62] Peridis, S., Magoulas, K., and Tassios, D., "Sensitivity of Distillation Column Design to Uncertainties in Vapor–Liquid Equilibrium Information," *Separation Science and Technology*, Vol. 28, No. 9, 1993, pp. 1753–1767.

[63] Riazi, M. R. and Mansoori, G. A., "Simple Equation of State Accurately Predicts Hydrocarbon Densities," *Oil and Gas Journal*, July 12, 1993, pp. 108–111.

[64] Lee, B. I. and Kesler, M. G., "A Generalized Thermodynamic Correlation Based on Three- Parameter Corresponding States," *American Institute of Chemical Engineers Journal*, Vol. 21, No. 5, 1975, pp. 510–527.

[65] Riazi, M. R. and Daubert, T. E., "Application of Corresponding States Principles for Prediction of Self-Diffusion Coefficients in Liquids," *American Institute of Chemical Engineers Journal*, Vol. 26, No. 3, 1980, pp. 386–391.

[66] Riazi, M. R. and Mansoori, G. A., "Use of the Velocity of Sound in Predicting the *PVT* Relations," *Fluid Phase Equilibria*, Vol. 90, 1993, pp. 251–264.

[67] Shabani, M. R., Riazi, M. R., and Shaban, H. I., "Use of Velocity of Sound in Predicting Thermodynamic Properties from Cubic Equations of State," *Canadian Journal of Chemical Engineering*, Vol. 76, 1998, pp. 281–289.

[68] Riazi, M. R. and Roomi, Y. A., "Use of Velocity of Sound in Estimating Thermodynamic Properties of Petroleum Fractions," *Preprints of Division of Petroleum Chemistry Symposia*, American Chemical Society (ACS), August 2000, Vol. 45, No. 4, pp. 661–664.

[69] Riazi, M. R. and Faghri, A., "Prediction of Thermal Conductivity of Gases at High Pressures," *American Institute of Chemical Engineers Journal*, Vol.31, No.1, 1985, pp. 164–167.

[70] Riazi, M. R. and Faghri, A., "Thermal Conductivity of Liquid and Vapor Hydrocarbon Systems: Pentanes and Heavier at Low Pressures," *Industrial and Engineering Chemistry, Process Design and Development*, Vol. 24, No. 2, 1985, pp. 398–401.

[71] Ahmed, T., *Hydrocarbon Phase Behavior*, Gulf Publishing, Houston, TX, 1989.

[72] Danesh, A., *PVT and Phase Behavior of Petroleum Reservoir Fluids*, Elsevier, Amsterdam, 1998.

[73] Tsonopoulos, C., Heidman, J. L., and Hwang, S.-C., *Thermodynamic and Transport Properties of Coal Liquids*, An Exxon Monograph, Wiley, New York, 1986.

[74] National Institute of Standards and Technology (NIST), Boulder, CO, 2003 (http://webbook.nist.gov/chemistry/).

[75] Ahmed, T., *Reservoir Engineering Handbook*, Gulf Publishing, Houston, TX, 2000.

[76] Edmister, W. C. and Lee, B. I., *Applied Hydrocarbon Thermodynamics*, 2nd ed., Gulf Publishing, Houston, TX, 1985.

[77] HYSYS, "Reference Volume 1., Version 1.1", HYSYS Reference Manual for Computer Software, HYSYS Conceptual Design, Hyprotech Ltd., Calgary, Alberta, Canada, 1996.

[78] EPCON, "API Tech Database Software," EPCON International, Houston, TX, 2000 (www.epcon.com).

[79] Stange, E. and Johannesen, S. O., "HYPO*S, A Program for Heavy End Characterization," An Internal Program from the Norsk Hydro Oil and Gas Group (Norway), Paper presented at the 65th *Annual Convection*, Gas Processors Association, San Antonio, TX, March 10, 1986.

[80] Aspen Plus, "Introductory Manual Software Version," Aspen Technology, Inc., Cambridge, MA, December 1986.

[81] PRO/ II, "Keyword Manual," Simulation Sciences Inc., Fullerton, CA, October 1992.

[82] Riazi, M. R., "Estimation of Physical Properties and Composition of Hydrocarbon Mixures," Analytical Advances for Hydrocarbon Research, edited by Hsu, C. Samuel, Kluwer Academic/Plenum Publishers, New York, NY, 2003, pp. 1–26.

[83] Riazi, M. R. and Al-Enezi, G., "A Mathematical Model for the Rate of Oil Spill Disappearance from Seawater for Kuwaiti Crude and Its Products," *Chemical Engineering Journal*, Vol. 73, 1999, pp. 161–172.

[84] Riazi, M. R. and Edalat, M., "Prediction of the Rate of Oil Removal from Seawater by Evaporation and Dissolution," *Journal of Petroleum Science and Engineering*, Vol. 16, 1996, pp. 291–300.

[85] Fingas, M. F., "A Literature Review of the Physics and Predictive Modeling of Oil Spill Evaporation," *Journal of Hazardous Materials*, Vol. 42, 1995, pp. 157–175.

[86] Winn, F. W., "Physical Properties by Nomogram," *Petroleum Refiners*, Vol. 36, No. 21, 1957, pp. 157.

[87] Coplen, T. B., "Atomic Weights of the Elements 1999," *Pure and Applied Chemistry*, Vol. 73, No. 4, 2001, pp. 667–683.

[88] Maddox, R. N., *Gas Conditioning and Processing, Volume 4,* Campbell Petroleum Series, John M. Campbell, Norman, OK, 1985.

[89] Riazi, M. R., "Estimation of Rates and Enhancement Factors in Gas Absorption with Zero-Order Reaction and Gas Phase Mass Transfer Resistances," *Chemical Engineering Science*, Vol. 41, No. 11, 1986, pp. 2925–2929.

Characterization and Properties of Pure Hydrocarbons

2

NOMENCLATURE

API   API gravity defined in Eq. (2.4)

A, B Parameters in a potential energy relation

A, B Parameters in a two-parameter cubic equation of state

$a, b, \ldots i$ Correlation constants in various equations

CH   Carbon-to-hydrogen weight ratio

d_{20} Liquid density at 20°C and 1 atm, g/cm³

d_c Critical density defined by Eq. (2.9), g/cm³

F Intermolecular force

I Refractive index parameter defined in Eq. (2.36)

K_W Watson (UOP) K factor defined by Eq. (2.13)

ln   Natural logarithm (base e)

$\log_{10}$ Logarithm to the base 10

M Molecular weight, g/mol (kg/kmol)

n Sodium D line refractive index of liquid at 20°C and 1atm, dimensionless

N_A Avogadro's number

N_C Carbon number (number of carbon atoms in a hydrocarbon molecule)

P_c Critical pressure, bar

P^{vap} Vapor (saturation) pressure, bar

R Universal gas constant, 8.314 J/mol·K

R_i Refractivity intercept defined in Eq. (2.14)

R_m Molar refraction defined in Eq. (2.34), cm³/g

R^2 R squared, defined in Eq. (2.136)

r Distance between molecules

SG   Specific gravity of liquid substance at 15.5°C (60°F) defined by Eq. (2.2), dimensionless

SG_g Specific gravity of gas substance at 15.5°C (60°F) defined by Eq. (2.6), dimensionless

T_b Boiling point, K

T_c Critical temperature, K

T_F Flash point, K

T_M Melting (freezing point) point, K

V Molar volume, cm³/gmol

V Saybolt universal viscosity, SUS

V_c Critical volume (molar), cm³/mol (or critical specific volume, cm³/g)

VGC   Viscosity gravity constant defined by Eq. (2.15)

Z_c Critical compressibility factor defined by Eq. (2.8), dimensionless

Greek Letters

Γ Potential energy defined in Eq. (2.19)

α Polarizability defined by Eq. (2.33), cm³/mol

ε Energy parameter in a potential energy relation

μ Absolute (dynamic) viscosity, cp [mPa·s]. Also used for dipole moment

ν Kinematic viscosity defined by Eq. (2.12), cSt [mm²/s]

θ A property of hydrocarbon such as $M, T_c, P_c, V_c, I, d, T_b, \ldots$

ρ Density at a given temperature and pressure, g/cm³

σ Surface tension, dyn/cm (= mN/m)

σ Size parameter in potential energy relation

ω Acentric factor defined by Eq. (2.10), dimensionless

Superscript

°   Properties of n-alkanes from Twu correlations

Subscripts

A   Aromatic

N   Naphthenic

P   Paraffinic

T   Value of a property at temperature T

°   A reference state for T and P

∞   Value of a property at $M \to \infty$

20   Value of a property at 20°C

38(100)   Value of kinematic viscosity at 38°C (100°F)

99(210)   Value of kinematic viscosity at 99°C (210°F)

Acronyms

%AAD   Average absolute deviation percentage defined by Eq. (2.135)

API-TDB   American Petroleum Institute—Technical Data Book

%D   Absolute deviation percentage defined by Eq. (2.134)

EOS   Equation of state

IUPAC   International Union of Pure and Applied Chemistry

%MAD   Maximum absolute deviation percentage

NIST   National Institute of Standards and Technology

RK   Redlich–Kwong

vdW   van der Waals

R^2 R squared, Defined in Eq. (2.136)

As discussed in chapter 1, the characterization of petroleum fractions and crude oils depends on the characterization and properties of pure hydrocarbons. Calculation of the properties of a mixture depends on the properties of its constituents. In this chapter, first basic parameters and properties of pure compounds are defined. These properties are either temperature-independent or values of some basic properties at a fixed temperature. These parameters are the basis for calculation of various physical properties discussed in this book. Reported values of these parameters for more than 100 selected pure compounds are given in Section 2.2. These values will be used extensively in the following chapters, especially in Chapters 3 to determine the quality and properties of petroleum fractions. In Section 2.3, the characterization of hydrocarbons is introduced, followed by the development of a generalized correlation for property estimation that is a unique feature of this chapter. Various correlations and methods for the estimation of these basic parameters for pure hydrocarbons and narrow boiling range petroleum fractions are presented in different sections. Finally, necessary discussion and recommendations for the selection of appropriate predictive methods for various properties are presented.

2.1 DEFINITION OF BASIC PROPERTIES

In this section, all properties of pure hydrocarbons presented in Section 2.2 are defined. Some specific characteristics of petroleum products, such as cetane index and pour point, are defined in Chapter 3. Definitions of general physical properties such as thermal and transport properties are discussed in corresponding chapters where their estimation methods are presented.

2.1.1 Molecular Weight

The units and definition of molecular weight or molar mass, M, was discussed in Section 1.7.8. The molecular weight of a pure compound is determined from its chemical formula and the atomic weights of its elements. The atomic weights of the elements found in a petroleum fluid are C = 12.011, H = 1.008, S = 32.065, O = 16.0, and N = 14.01, as given by the IUPAC standard [1]. As an example, the molecular weight of methane (CH_4) is calculated as $12.011 + 4 \times 1.008 = 16.043$ kg/kmol or 16.043 g/mol (0.01604 kg/mol) or 16.043 lb/lbmol. Molecular weight is one of the characterization parameters for hydrocarbons.

2.1.2 Boiling Point

The boiling point of a pure compound at a given pressure is the temperature at which vapor and liquid exist together at equilibrium. If the pressure is 1 atm, the boiling point is called the normal boiling point. However, usually the term *boiling point*, T_b, is used instead of normal boiling point and for other pressures the term *saturation temperature* is used. In some cases, especially for heavy hydrocarbons in which thermal cracking may occur at high temperatures, boiling points at pressures other than atmospheric is specified. Boiling points of heavy hydrocarbons are usually measured at 1, 10, or 50 mm Hg. The conversion of boiling point from low pressure to normal boiling point requires a vapor pressure relation and methods for its calculation for petroleum fractions are discussed in Chapter 3. The boiling point, when available, is one of the most important characterization parameters for hydrocarbons and is frequently used in property estimation methods.

2.1.3 Density, Specific Gravity, and API Gravity

Density is defined as mass per unit volume of a fluid. Density is a state function and for a pure compound depends on both temperature and pressure and is shown by ρ. Liquid densities decrease as temperature increases but the effect of pressure on liquid densities at moderate pressures is usually negligible. At low and moderate pressures (less than a few bars), saturated liquid density is nearly the same as actual density at the same temperature. Methods of the estimation of densities of fluids at various conditions are discussed in Chapters 5 and 7. However, liquid density at the reference conditions of 20°C (293 K) and 1 atm is shown by d and it is used as a characterization parameter in this chapter as well as Chapter 3. Parameter d is also called absolute density to distinguish from relative density. Other parameters that represent density are specific volume ($1/d$), molar volume (M/d), and molar density (d/M). Generally, absolute density is used in this book as the characteristic parameter to classify properties of hydrocarbons.

Liquid density for hydrocarbons is usually reported in terms of specific gravity (SG) or relative density defined as

$$(2.1) \qquad SG = \frac{\text{density of liquid at temperature } T}{\text{density of water at temperature } T}$$

Since the standard conditions adopted by the petroleum industry are 60°F (15.5°C) and 1 atm, specific gravities of liquid hydrocarbons are normally reported at these conditions. At a reference temperature of 60°F (15.5°C) the density of liquid water is 0.999 g/cm³ (999 kg/m³) or 8.337 lb/gal(U.S.). Therefore, for a hydrocarbon or a petroleum fraction, the specific gravity is defined as

$$(2.2) \quad SG\,(60°F/60°F) = \frac{\text{density of liquid at 60°F in g/cm}^3}{0.999 \text{ g/cm}^3}$$

Water density at 60°F is 0.999 or almost 1 g/cm³; therefore, values of specific gravities are nearly the same as the density of liquid at 15.5°C (289 K) in g/cm³. The Society of Petroleum Engineers usually uses γ for the specific gravity and in some references it is designated by S. However, in this book SG denotes the specific gravity. Since most of hydrocarbons found in reservoir fluids have densities less than that of water, specific gravities of hydrocarbons are generally less than 1. Specific gravity defined by Eq. (2.2) is slightly different from the specific gravity defined in the SI system as the ratio of the density of hydrocarbon at 15°C to that of water at 4°C designated by d_4^{15}. Note that density of water at 4°C is exactly 1 g/cm³ and therefore d_4^{15} is equal to the density of hydrocarbon at 15°C in g/cm³. The relation between these two specific gravities is approximately given as follows:

$$(2.3) \qquad\qquad SG = 1.001 d_4^{15}$$

In this book specific gravity refers to SG at 60°F/60°F (15.5°C). In the early years of the petroleum industry, the American

Petroleum Institute (API) defined the API gravity (degrees API) to quantify the quality of petroleum products and crude oils. The API gravity is defined as [2]

$$(2.4) \qquad \text{API gravity} = \frac{141.5}{\text{SG (at } 60°\text{F)}} - 131.5$$

Degrees API was derived from the degrees Baumé in which it is defined in terms of specific gravity similar to Eq. (2.4) except numerical values of 140 and 130 were used instead of 141.5 and 131.5, respectively. Liquid hydrocarbons with lower specific gravities have higher API gravity. Aromatic hydrocarbons have higher specific gravity (lower API gravity) than do paraffinic hydrocarbons. For example, benzene has SG of 0.8832 (API of 28.72) while *n*-hexane with the same carbon number has SG of 0.6651 (API gravity of 81.25). A liquid with SG of 1 has API gravity of 10. Once Eq. (2.4) is reversed it can be used to calculate specific gravity from the API gravity.

$$(2.5) \qquad \text{SG} = \frac{141.5}{\text{API gravity} + 131.5}$$

The definition of specific gravity for gases is somewhat different. It is defined as relative density of gas to density of air at standard conditions. In addition, density of gases is a strong function of pressure. Since at the standard conditions (15.5°C and 1 atm) the density of gases are estimated from the ideal gas law (see Chapter 5), the specific gravity of a gas is proportional to the ratio of molecular weight of gas (M_g) to the molecular weight of air (28.97).

$$(2.6) \qquad \text{SG}_g = \frac{M_g}{28.97}$$

Therefore, to obtain the specific gravity of a gas, only its molecular weight is needed. For a mixture, M_g can be determined from the gas composition, as discussed in Chapter 3.

2.1.4 Refractive Index

Refractive index or refractivity for a substance is defined as the ratio of velocity of light in a vacuum to the velocity of light in the substance (fluid) and is a dimensionless quantity shown by n:

$$(2.7) \qquad n = \frac{\text{velocity of light in the vacuum}}{\text{velocity of light in the substance}}$$

In other words, when a light beam passes from one substance (air) to another (a liquid), it is bent or refracted because of the difference in speed between the two substances. In fact, refractive index indicates the degree of this refraction. Refractive index is a state function and depends on the temperature and pressure of a fluid. Since the velocity of light in a fluid is less than the velocity of light in a vacuum, its value for a fluid is greater than unity. Liquids have higher values of refractive index than that of gases. For gases the values of refractive index are very close to unity.

All frequencies of electromagnetic radiation (light) travel at the same speed in vacuum (2.998×10^8 m/s); however, in a substance the velocity of light depends on the nature of the substance (molecular structure) as well as the frequency of the light. For this reason, standard values of refractive index must be measured at a standard frequency. Usually the refractive index of hydrocarbons is measured by the sodium D line

at 20°C and 1 atm. The instrument to measure the refractive index is called a *refractometer* and is discussed in Chapter 3. In some references the values of refractive index are reported at 25°C; however, in this book the refractive index at 20°C and 1 atm is used as a characterization parameter for hydrocarbons and petroleum fractions. As is shown in this chapter and Chapter 3, refractive index is a very useful characterization parameter for pure hydrocarbons and petroleum fractions, especially in relation with molecular type composition. Values of n vary from about 1.3 for propane to 1.6 for some aromatics. Aromatic hydrocarbons have generally higher n values than paraffinic compounds as shown in Table 2.1.

2.1.5 Critical Constants (T_c, P_c, V_c, Z_c)

The critical point is a point on the pressure–volume–temperature diagram where the saturated liquid and saturated vapor are identical and indistinguishable. The temperature, pressure, and volume of a pure substance at the critical point are called *critical temperature* (T_c), *critical pressure* (P_c), and *critical volume* (V_c), respectively. In other words, the critical temperature and pressure for a pure compound are the highest temperature and pressure at which the vapor and liquid phase can coexist at equilibrium. In fact, for a pure compound at temperatures above the critical temperature, it is impossible to liquefy a vapor no matter how high the pressure is. A fluid whose temperature and pressure are above the critical point is called supercritical fluid. For pure compounds, critical temperature and pressure are also called true critical temperature and true critical pressure. However, as will be discussed in Chapter 3, *pseudocritical properties* are defined for mixtures and petroleum fractions, which are different from true critical properties. Pseudocritical properties are important in process calculations for the estimation of thermophysical properties of mixtures.

The critical compressibility factor, Z_c, is defined from T_c, P_c, and V_c according to the general definition of compressibility factor.

$$(2.8) \qquad Z_c = \frac{P_c V_c}{RT_c}$$

where R is the universal gas constant. According to Eq. (2.8), Z_c is dimensionless and V_c must be in terms of molar volume (i.e., cm³/mol) to be consistent with R values given in Section 1.7.24. Critical temperature, pressure, and volume (T_c, P_c, V_c) are called the *critical constants* or *critical properties*. Critical constants are important characteristics of pure compounds and mixtures and are used in corresponding states correlations and equations of state (EOS) to calculate *PVT* and many other thermodynamic, physical, and transport properties. Further discussion on the critical point of a substance is given in Chapter 5. As was discussed in Section 1.3, the results of EOS calculations very much depend on the values of critical properties used. Critical volume may be expressed in terms of specific critical volume (i.e., m³/kg), molar critical volume (i.e., m³/kmol), or critical density d_c (i.e., kg/m³) or critical molar density (i.e., kmol/m³). Critical density is related to the critical molar volume as

$$(2.9) \qquad d_c = \frac{M}{V_c}$$

Experimental values of critical properties have been reported for a large number of pure substances. However, for hydrocarbon compounds, because of thermal cracking that occurs at higher temperatures, critical properties have been measured up to C_{18} [2]. Recently some data on critical properties of *n*-alkanes from C_{19} to C_{36} have been reported [3]. However, such data have not yet been universally confirmed and they are not included in major data sources. Reported data on critical properties of such heavy compounds are generally predicted values and vary from one source to another. For example, the API-TDB [2] reports values of 768 K and 11.6 bar for the critical temperature and pressure of *n*-eicosane, while these values are reported as 767 K and 11.1 bar by Poling *et al.* [4]. Generally, as boiling point increases (toward heavier compounds), critical temperature increases while critical pressure decreases. As shown in Section 2.2, aromatics have higher T_c and P_c relative to those of paraffinic compounds with the same carbon atoms.

2.1.6 Acentric Factor

Acentric factor is a parameter that was originally defined by Pitzer to improve accuracy of corresponding state correlations for heavier and more complex compounds [5, 6]. Acentric factor is a defined parameter and not a measurable quantity. It is a dimensionless parameter represented by ω and is defined as

$$(2.10) \qquad \omega = -\log_{10}\left(P_r^{vap}\right) - 1.0$$

where
P_r^{vap} = reduced vapor pressure, P^{vap}/P_c, dimensionless
P^{vap} = vapor pressure at $T = 0.7\ T_c$ (reduced temperature of 0.7), bar
P_c = critical pressure, bar
T = absolute temperature, K
T_c = critical temperature, K

Acentric factor is defined in a way that for simple fluids such as argon and xenon it is zero and its value increases as the size and shape of molecule changes. For methane $\omega = 0.001$ and for decane it is 0.489. Values reported for acentric factor of pure compounds are calculated based on Eq. (2.10), which depends on the values of vapor pressure. For this reason values reported for the acentric factor of a compound may slightly vary from one source to another depending on the relation used to estimate the vapor pressure. In addition, since calculation of the acentric factor requires values of critical temperature and pressure, reported values for ω also depend on the values of T_c and P_c used.

2.1.7 Vapor Pressure

In a closed container, the vapor pressure of a pure compound is the force exerted per unit area of walls by the vaporized portion of the liquid. Vapor pressure, P^{vap}, can also be defined as a pressure at which vapor and liquid phases of a pure substance are in equilibrium with each other. The vapor pressure is also called saturation pressure, P^{sat}, and the corresponding temperature is called saturation temperature. In an open air under atmospheric pressure, a liquid at any temperature below its boiling point has its own vapor pressure that is less than 1 atm. When vapor pressure reaches 1 atm,

the saturation temperature becomes the normal boiling point. Vapor pressure increases with temperature and the highest value of vapor pressure for a substance is its critical pressure (P_c) in which the corresponding temperature is the critical temperature (T_c). When a liquid is open to the atmosphere at a temperature T in which the vapor pressure of liquid is P^{vap}, vol% of the compound vapors in the air is

$$(2.11) \qquad \text{vol\%} = 100 \times \left(\frac{P^{vap}}{P_a}\right)$$

where P_a is the atmospheric pressure. Derivation of Eq. (2.11) is based on the fact that vapor pressure is equivalent to partial pressure (mole fraction × total pressure) and in gases under low-pressure conditions, mole fraction and volume fraction are the same. At sea level, where $P_a = 1$ atm, calculation of vol% of hydrocarbon vapor in the air from Eq. (2.11) is simply 100 P^{vap}, if P^{vap} is in atm.

Vapor pressure is a very important thermodynamic property of any substance and it is a measure of the volatility of a fluid. Compounds with a higher tendency to vaporize have higher vapor pressures. More volatile compounds are those that have lower boiling points and are called light compounds. For example, propane (C_3) has boiling point less than that of *n*-butane (nC_4) and as a result it is more volatile. At a fixed temperature, vapor pressure of propane is higher than that of butane. In this case, propane is called the light compound (more volatile) and butane the heavy compound. Generally, more volatile compounds have higher critical pressure and lower critical temperature, and lower density and lower boiling point than those of less volatile (heavier) compounds, although this is not true for the case of some isomeric compounds. Vapor pressure is a useful parameter in calculations related to hydrocarbon losses and flammability of hydrocarbon vapor in the air (through Eq. 2.11). More volatile compounds are more ignitable than heavier compounds. For example, *n*-butane is added to gasoline to improve its ignition characteristics. Low-vapor-pressure compounds reduce evaporation losses and chance of vapor lock. Therefore, for a fuel there should be a compromise between low and high vapor pressure. However, as will be seen in Chapter 6, one of the major applications of vapor pressure is in calculation of equilibrium ratios (K_i values) for phase equilibrium calculations. Methods of calculation of vapor pressure are given in detail in Chapter 7. For pure hydrocarbons, values of vapor pressure at the reference temperature of 100°F (38°C) are provided by the API [2] and are given in Section 2.2. For petroleum fractions, as will be discussed in Chapter 3, method of Reid is used to measure vapor pressure at 100°F. *Reid vapor pressure* (RVP) is measured by the ASTM test method D 323 and it is approximately equivalent to vapor pressure at 100°F (38°C). RVP is a major characteristic of gasoline fuel and its prediction is discussed in Chapter 3.

2.1.8 Kinematic Viscosity

Kinematic viscosity is defined as the ratio of absolute (dynamic) viscosity μ to absolute density ρ at the same temperature in the following form:

$$(2.12) \qquad \nu = \frac{\mu}{\rho}$$

As discussed in Section 1.7.18, kinematic viscosity is expressed in cSt, SUS, and SFS units. Values of kinematic viscosity for pure liquid hydrocarbons are usually measured and reported at two reference temperatures of 38°C (100°F) and 99°C (210°F) in cSt. However, other reference temperatures of 40°C (104°F), 50°C (122°F), and 60°C (140°F) are also used to report kinematic viscosities of petroleum fractions. Liquid viscosity decreases with an increase in temperature (see Section 2.7). Kinematic viscosity, as it is shown in Chapter 3, is a useful characterization parameter, especially for heavy fractions in which the boiling point may not be available.

2.1.9 Freezing and Melting Points

Petroleum and most petroleum products are in the form of a liquid or gas at ambient temperatures. However, for oils containing heavy compounds such as waxes or asphaltinic oils, problems may arise from solidification, which cause the oil to lose its fluidity characteristics. For this reason knowledge of the freezing point is important and it is one of the major specifications of jet fuels and kerosenes. For a pure compound the *freezing point* is the temperature at which liquid solidifies at 1 atm pressure. Similarly the *melting point*, T_M, is the temperature that a solid substance liquefies at 1 atm. A pure substance has the same freezing and melting points; however, for petroleum mixtures, there are ranges of melting and freezing points versus percent of the mixture melted or frozen. For a mixture, the initial melting point is close to the melting point of the lightest compound in the mixture, while the initial freezing point is close to the freezing point (or melting point) of the heaviest compound in the mixture. Since the melting point increases with molecular weight, for petroleum mixtures the initial freezing point is greater than the initial melting point. For petroleum mixtures an equivalent term of pour point instead of initial melting point is defined, which will be discussed in Chapter 3. Melting point is an important characteristic parameter for petroleum and paraffinic waxes.

2.1.10 Flash Point

Flash point, T_F, for a hydrocarbon or a fuel is the minimum temperature at which vapor pressure of the hydrocarbon is sufficient to produce the vapor needed for spontaneous ignition of the hydrocarbon with the air with the presence of an external source, i.e., spark or flame. From this definition, it is clear that hydrocarbons with higher vapor pressures (lighter compounds) have lower flash points. Generally flash point increases with an increase in boiling point. Flash point is an important parameter for safety considerations, especially during storage and transportation of volatile petroleum products (i.e., LPG, light naphtha, gasoline) in a high-temperature environment. The surrounding temperature around a storage tank should always be less than the flash point of the fuel to avoid possibility of ignition. Flash point is used as an indication of the fire and explosion potential of a petroleum product. Estimation of the flash point of petroleum fractions is discussed in Chapter 3, and data for flash points of some pure hydrocarbons are given in Table 2.2. These data were obtained using the *closed cup* apparatus as described in ASTM D 93 (ISO 2719) test method. There is another method of measuring flash point known as *open cup* for those oils with

flash point greater than 80°C (ASTM D 92 or ISO 2592 test methods). Flash point should not be mistaken with *fire point*, which is defined as the minimum temperature at which the hydrocarbon will continue to burn for at least 5 s after being ignited by a flame.

2.1.11 Autoignition Temperature

This is the minimum temperature at which hydrocarbon vapor when mixed with air can spontaneously ignite without the presence of any external source. Values of autoignition temperature are generally higher than flash point, as given in Table 2.2 for some pure hydrocarbons. Values of autoignition temperature for oils obtained from mineral sources are in the range of 150–320°C (300–500°F), for gasoline it is about 350°C (660°F), and for alcohol is about 500°C (930°F) [7]. With an increase in pressure the autoignition temperature decreases. This is particularly important from a safety point of view when hydrocarbons are compressed.

2.1.12 Flammability Range

To have a combustion, three elements are required: fuel (hydrocarbon vapor), oxygen (i.e., air), and a spark to initiate the combustion. One important parameter to have a good combustion is the ratio of air to hydrocarbon fuel. The combustion does not occur if there is too much air (little fuel) or too little air (too much fuel). This suggests that combustion occurs when hydrocarbon concentration in the air is within a certain range. This range is called *flammability range* and is usually expressed in terms of lower and upper volume percent in the mixture of hydrocarbon vapor and air. The actual volume percent of hydrocarbon vapor in the air may be calculated from Eq. (2.11) using vapor pressure of the hydrocarbon. If the calculated vol% of hydrocarbon in the air is within the flammability range then the mixture is flammable by a spark or flame.

2.1.13 Octane Number

Octane number is a parameter defined to characterize antiknock characteristic of a fuel (gasoline) for spark ignition engines. Octane number is a measure of fuel's ability to resist auto-ignition during compression and prior to ignition. Higher octane number fuels have better engine performance. The octane number of a fuel is measured based on two reference hydrocarbons of *n*-heptane with an assigned octane number of zero and isooctane (2,2,4-trimethylpentane) with assigned octane number of 100. A mixture of 70 vol% isooctane and 30 vol% *n*-heptane has an octane number of 70. There are two methods of measuring octane number of a fuel in the laboratory. The methods are known as *motor octane number* (MON) and *research octane number* (RON). The MON is indicative of high-speed performance (900 rpm) and is measured under heavy road conditions (ASTM D 357). The RON is indicative of normal road performance under low engine speed (600 rpm) city driving conditions (ASTM D 908). The third type of octane number is defined as *posted octane number* (PON), which is the arithmetic average of the MON and RON [PON = (MON + RON)/2]. Generally isoparaffins have higher octane number than do normal paraffins. Naphthenes have relatively higher octane number than do corresponding paraffins and aromatics have very high octane numbers. The octane number of a fuel can be improved by adding tetra-ethyl-lead (TEL) or methyl-tertiary-butyl-ether (MTBE). Use of lead (Pb) to improve octane number of fuels is limited in many industrial countries. In these countries MTBE is used for octane number improvement. However, there are problems of groundwater contamination with MTBE. MTBE has MON and RON of 99 and 115, respectively [8]. Lead generally improves octane number of fuels better than MTBE. The addition of 0.15 g Pb/L to a fuel of RON around 92 can improve its octane number by 2–3 points. With 0.6 g Pb/L one may improve the octane number by 10 points [8]. However, as mentioned above, because of environmental hazards use of lead is restricted in many North American and West European countries. Values of the octane number measured without any additives are called clear octane number. For pure hydrocarbons values of clear MON and RON are given in Section 2.2. Estimation of the octane number of fuels is discussed in Chapter 3.

2.1.14 Aniline Point

The aniline point for a hydrocarbon or a petroleum fraction is defined as the minimum temperature at which equal volumes of liquid hydrocarbon and aniline are miscible. Aniline is an aromatic compound with a structure of a benzene molecule where one atom of hydrogen is replaced by the $-NH_2$ group $(C_6H_5-NH_2)$. The aniline point is important in characterization of petroleum fractions and analysis of molecular type. As discussed in Chapter 3, the aniline point is also used as a characterization parameter for the ignition quality of diesel fuels. It is measured by the ASTM D 611 test method. Within a hydrocarbon group, aniline point increases with molecular weight or carbon number, but for the same carbon number it increases from aromatics to paraffinic hydrocarbons. Aromatics have very low aniline points in comparison with paraffins, since aniline itself is an aromatic compound and it has better miscibility with aromatic hydrocarbons. Generally, oils with higher aniline points have lower aromatic content. Values of the aniline point for pure hydrocarbons are given in Table 2.2, and its prediction for petroleum fractions is discussed in Chapter 3.

2.1.15 Watson K

Since the early years of the petroleum industry it was desired to define a characterization parameter based on measurable parameters to classify petroleum and identify hydrocarbon molecular types. The Watson characterization factor denoted by K_W is one of the oldest characterization factors originally defined by Watson *et al.* of the Universal Oil Products (UOP) in mid 1930s [9]. For this reason the parameter is sometimes called UOP characterization factor and is defined as

$$(2.13) \qquad K_W = \frac{(1.8T_b)^{1/3}}{SG}$$

where

T_b = normal boiling point K
SG = specific gravity at 15.5°C

In the original definition of K_W, boiling point is in degrees Rankine and for this reason the conversion factor of 1.8 is used to have T_b in the SI unit. For petroleum fractions

T_b is the mean average boiling point (also see Chapter 3). The purpose of definition of this factor was to classify the type of hydrocarbons in petroleum mixtures. The naphthenic hydrocarbons have K_W values between paraffinic and aromatic compounds. In general, aromatics have low K_W values while paraffins have high values. However, as will be discussed in Chapter 3 there is an overlap between values of K_W from different hydrocarbon groups. The Watson K was developed in 1930s by using data for the crude and products available in that time. Now the base petroleum stocks in general vary significantly from those of 1930s [10, 11]. However, because it combines two characterization parameters of boiling point and specific gravity it has been used extensively in the development of many physical properties for hydrocarbons and petroleum fractions [2, 11, 12].

2.1.16 Refractivity Intercept

Kurtz and Ward [13] showed that a plot of refractive index against density for any homologous hydrocarbon group is linear. For example, plot of refractive index of *n*-paraffins versus density (d_{20}) in the carbon number range of C_5–C_{45} is a straight line represented by equation $n = 1.0335 + 0.516d_{20}$, with R^2 value of 0.9998 ($R^2 = 1$, for an exact linear relation). Other hydrocarbon groups show similar performance with an exact linear relation between n and d. However, the intercept for various groups varies and based on this observation they defined a characterization parameter called *refractivity intercept*, R_i, in the following form:

$$(2.14) \qquad R_i = n - \frac{d}{2}$$

where n and d are refractive index and density of liquid hydrocarbon at the reference state of 20°C and 1 atm in which density must be in g/cm³. R_i is high for aromatics and low for naphthenic compounds, while paraffins have intermediate R_i values.

2.1.17 Viscosity Gravity Constant

Another parameter defined in the early years of petroleum characterization is the *viscosity gravity constant* (VGC). This parameter is defined based on an empirical relation developed between Saybolt viscosity (SUS) and specific gravity through a constant. VGC is defined at two reference temperatures of 38°C (100°F) and 99°C (210°F) as [14]

$$(2.15) \qquad VGC = \frac{10SG - 1.0752 \log_{10}(V_{38} - 38)}{10 - \log_{10}(V_{38} - 38)}$$

$$(2.16) \qquad VGC = \frac{SG - 0.24 - 0.022 \log_{10}(V_{99} - 35.5)}{0.755}$$

where

V_{38} = viscosity at 38°C (100°F) in SUS (Saybolt Universal Seconds)
V_{99} = Saylbolt viscosity (SUS) at 99°C (210°F)

Conversion factors between cSt and SUS are given in Section 1.7.18. Equations (2.15) and (2.16) do not give identical values for a given compound but calculated values are close to each other, except for very low viscosity oils. Equation (2.16) is recommended only when viscosity at 38°C (100°F) is not available. VGC varies for paraffinic hydrocarbons from 0.74 to 0.75, for naphthenic from 0.89 to 0.94, and for aromatics from 0.95 to 1.13 [15]. In Chapter 3, VGC along with other parameters has been used to estimate the composition of petroleum fractions. Values of VGC for some hydrocarbons are given in Table 2.3. The main limitation in use of VGC is that it cannot be defined for compounds or fractions with viscosities less than 38 SUS (~3.6 cSt) at 38°C. A graphical method to estimate VGC of petroleum fractions is presented in Chapter 3.

ASTM D 2501 suggests calculation of VGC using specific gravity and viscosity in mm²/s (cSt) at 40°C, v_{40} in the following form:

$$(2.17) \quad VGC = \frac{SG - 0.0664 - 0.1154 \log_{10}(v_{40} - 5.5)}{0.94 - 0.109 \log_{10}(v_{40} - 5.5)}$$

Values of VGC calculated from Eq. (2.17) are usually very close to values obtained from Eq. (2.15). If viscosity at 40°C is available, use of Eq. (2.17) is recommended for calculation of VGC. Another relation to calculate VGC in metric units was proposed by Kurtz *et al.* [16] in terms of kinematic viscosity and density at 20°C, which is also reported in other sources [17].

$$(2.18) \quad VGC = \frac{d - 0.1384 \log_{10}(v_{20} - 20)}{0.1526[7.14 - \log_{10}(v_{20} - 20)]} + 0.0579$$

in which d is density at 20°C and 1 atm in g/cm³ and v_{20} is the kinematic viscosity at 20°C in cSt. In this method viscosity of oil at 20°C must be greater than 20 cSt. However, when there is a choice Eq. (2.15) should be used for the procedures described in Chapter 3.

Example 2.1—API RP-42 [18] reports viscosity of some heavy hydrocarbons. 1,1-Di-(alphadecalyl)hendecane ($C_{31}H_{56}$) is a naphthenic compound with molecular weight of 428.8 and specific gravity of 0.9451. The kinematic viscosity at 38°C (100°F) is 20.25 cSt. Calculate the viscosity gravity constant for this compound.

Solution—Using Eq. (1.17), the viscosity is converted from cSt to SUS: $V_{38} = 99.5$ SUS. Substituting values of V_{38} and SG = 0.9451 into Eq. (2.15) gives VGC = 0.917. The VGC may be calculated from Eq. (2.17) with direct substitution of viscosity in the cSt unit. Assuming there is a slight change in viscosity from 38 to 40°C, the same value of viscosity at 38°C is used for v_{40}. Thus $v_{40} \cong 20.25$ cSt (mm²/s) and Eq. (2.17) gives VGC = 0.915. The small difference between calculated values of VGC because in Eq. (2.17) viscosity at 40°C must be used, which is less than the viscosity at 38°C. Calculated VGC is within the range of 0.89–0.94 and thus the hydrocarbon must be a naphthenic compound (also see Fig. 3.22 in Chapter 3). ◆

2.1.18 Carbon-to-Hydrogen Weight Ratio

Carbon-to-hydrogen weight ratio, CH weight ratio, is defined as the ratio of total weight of carbon atoms to the total weight of hydrogen in a compound or a mixture and is used to characterize a hydrocarbon compound. As was discussed in Section 1.1.1, hydrocarbons from different groups have different formulas. For example, alkanes (paraffins) have the general formula of C_nH_{2n+2}, alkylcyclopentanes or alkylcyclohexanes (naphthenes) have formula of C_nH_{2n}, and alkylbenzenes (aromatics) have formula of C_nH_{2n-6} ($n \geq 6$). This shows that at the same carbon number, the atomic ratio of number of carbon (C) atoms to number of hydrogen (H) atoms increases from paraffins to naphthenes and aromatics. For example, *n*-hexane (C_6H_{14}), cyclohexane (C_6H_{12}), and benzene (C_6H_6) from three different hydrocarbon groups all have six carbon atoms, but have different CH atomic ratios of 6/14, 6/12, and 6/6, respectively. If CH atomic ratio is multiplied by the ratio of atomic weights of carbon (12.011) to hydrogen (1.008), then CH weight ratio is obtained. For example, for *n*-hexane, the CH weight ratio is calculated as (6/14)×(12.011/1.008) = 5.107. This number for benzene is 11.92. Therefore, CH weight ratio is a parameter that is capable of characterizing the hydrocarbon type. In addition, within the same hydrocarbon group, the CH value changes from low to high carbon number. For example, methane has CH value of 2.98, while pentane has CH value of 4.96. For extremely large molecules ($M \to \infty$), the CH value of all hydrocarbons regardless of their molecular type approaches the limiting value of 5.96. This parameter is used in Section 2.3 to estimate hydrocarbon properties, and in Chapter 3 it is used to estimate the composition of petroleum fractions. In some references HC atomic ratio is used as the characterizing parameter. According to the definition, the CH weight ratio and HC atomic ratio are inversely proportional. The limiting value of HC atomic ratio for all hydrocarbon types is 2.

Another use of CH weight ratio is to determine the quality of a fossil-type fuel. Quality and the value of a fuel is determined from its heat of combustion and heating value. Heating value of a fuel is the amount of heat generated by complete combustion of 1 unit mass of the fuel. For example, *n*-hexane has the heating value of 44734 kJ/kg (19232 Btu/lb) and benzene has the heating value of 40142 kJ/kg (17258 Btu/lb). Calculation of heating values are discussed in Chapter 7. From this analysis it is clear that as CH value increases the heating value decreases. Hydrogen (H_2), which has a CH value of zero, has a heating value more than that of methane (CH_4) and methane has a heating value more than that of any other hydrocarbon. Heavy aromatic hydrocarbons that have high CH values have lower heating values. In general, by moving toward lower CH value fuel, not only do we have better heating value but also better and cleaner combustion of the fuel. It is for this reason that the use of natural gas is preferable to any other type of fuel, and hydrogen is an example of a perfect fuel with zero CH weight ratio (CH = 0), while black carbon is an example of the worst fuel with a CH value of infinity. Values of CH for pure hydrocarbons are given in Section 2.2 and its estimation methods are given in Section 2.6.3.

2.2 DATA ON BASIC PROPERTIES OF SELECTED PURE HYDROCARBONS

2.2.1 Sources of Data

There are several sources that provide data for physical properties of pure compounds. Some of these sources are listed below.

1. API: *Technical Data Book—Petroleum Refining* [2]. The first chapter of API-TDB compiles basic properties of more than 400 pure hydrocarbons and some nonhydrocarbons that are important in petroleum refining. For some compounds where experimental data are not available, predicted values from the methods recommended by the API are given.

2. DIPPR: Design Institute for Physical and Property Data [19]. The project initially supported by the AIChE began in early 1980s and gives various physical properties for both hydrocarbon and nonhydrocarbon compounds important in the industry. A computerized version of this data bank is provided by EPCON [20].

3. TRC Thermodynamic Tables—Hydrocarbons [21]. The Thermodynamic Research Center (formerly at the Texas A&M University) currently at the National Institute of Standards and Technology (NIST) at Boulder, CO, (http://www.trc.nist.gov/) in conjunction with the API Research Project 44 [22] has regularly published physical and basic properties of large number of pure hydrocarbons.

4. API Research Project 44 [22]. This project sponsored by the API was conducted at Texas A&M University and provides physical properties of selected hydrocarbons.

5. API Research Project 42 [18]. This data compilation completed in the 1960s provides experimental data on density, refractive index, viscosity, and vapor pressure for more than 300 hydrocarbons with carbon number greater than C_{11}.

6. Dortmund Data Bank (DDB) [23]. This project on physical properties has been conducted at the University of Oldenburg in Germany. DDB contains experimental data from open literature on various thermodynamic properties of pure compounds and some defined mixtures. Data have been programmed in a computer software convenient for extracting data. Majority of data are on thermodynamic properties, such as vapor–liquid equilibrium (VLE), activity coefficients, and excess properties. However, data on viscosity, density, vapor pressure, thermal conductivity, and surface tension have also been complied as mentioned in their Web site. Unfortunately they have not compiled characteristic data on hydrocarbons and petroleum fractions important in the petroleum industry. Also the data on transport properties are mainly for pure compounds at atmospheric pressures.

7. The fourth and fifth editions of *The Properties of Gases and Liquids* [4] also provide various properties for more than 400 pure compounds (hydrocarbons and nonhydrocarbons). However, data in this book have been mainly taken from the TRC Tables [21].

8. There are also some free online sources that one may use to obtain some physical property data. The best example is the one provided by NIST (http://webbook.nist.gov). Various universities and researchers have also developed special online sources for free access to some data on physical properties. For example the Center for Research in Computational Thermochemistry (CRCT) of Ecole Ploytechnique Montreal provides online calculational software at http://www.crct.polymtl.ca/fact/index.php/. The Center for Applied Thermodynamic Studies (CATS) at the University of Idaho also provides softwares for property calculations at its website (http://www.webpages.uidaho.edu/~cats/). G. A. Mansoori in his personal Web site also provides some online

sources for physical property data (http://tigger.uic.edu/~mansoori/Thermodynamic.Data.and.Property-html/).

In many occasions different sources provide different values for a particular property depending on the original source of data. Calculated properties such as critical constants and acentric factor for compounds heavier than C_{18} should be taken with care as in different sources different methods have been used to predict these parameters.

2.2.2 Properties of Selected Pure Compounds

The basic properties of pure hydrocarbons from different groups that will be used in the predictive methods presented in the following chapters are tabulated in Tables 2.1 and 2.2. The basic properties of M, T_M, T_b, SG, d_{20}, n_{20}, T_c, P_c, V_c, Z_c, and ω are presented in Table 2.1. Secondary properties of kinematic viscosity, API gravity, K_W, vapor pressure, aniline point, flash and autoignition points, flammability range, and octane number are given in Table 2.2. Compounds selected are mainly hydrocarbons from paraffins, naphthenes, and aromatics that constitute crude oil and its products. However, some olefinic and nonhydrocarbons found with petroleum fluids are also included. Most of the compounds are from homologous hydrocarbon groups that are used as model compounds for characterization of petroleum fractions discussed in Chapter 3. The properties tabulated are the basic properties needed in characterization techniques and thermodynamic correlations for physical properties of petroleum fractions. Although there are separate chapters for estimation of density, viscosity, or vapor pressure, these properties at some reference temperatures are provided because of their use in the characterization methods given in Chapter 3 and 4. Other physical properties such as heat capacity or transport properties are given in corresponding chapters where the predictive methods are discussed. Data for more than 100 selected compounds are presented in this section and are limited to C_{22} mainly due to the lack of sufficient experimental data for heavier compounds. Data presented in Tables 2.1 and 2.2 are taken from the API-TDB [2, 22]. Standard methods of measurement of these properties are presented in Chapter 3.

Example 2.2—Assume large amount of toluene is poured on the ground in an open environment at which the temperature is 38°C (100°F). Determine if the area surrounding the liquid surface is within the flammability range.

Solution—From Table 2.2, the flammability range is 1.2–7.1 vol% of toluene vapor. From this table, the vapor pressure of toluene at 38°C is 0.071 bar (0.07 atm). Substituting this vapor pressure value in Eq. (2.11) gives the value of vol% = 100 × 0.07/1.0 = 7% of toluene in the air mixture. This number is within the flammability range (1.2 < 7 < 7.1) and therefore the surrounding air is combustible. ◆

2.2.3 Additional Data on Properties of Heavy Hydrocarbons

Some data on density, refractive index, and viscosity of some heavy hydrocarbons are given in Table 2.3. These data are taken from API RP 42 [18]. Values of R_i and VGC in the table

TABLE 2.1—Basic physical properties of some selected compounds.

Compound	Formula	N_C	M	T_M °C	T_b °C	SG, at 60°F	d_{20}, g/cm³	n_{20}	T_C, °C	P_C, bar	V_C, cm³/mol	Z_C	ω
Paraffins													
Methane	CH₄	1	16.0	−182.5	−161.5	0.2999	...	...	−82.59	45.99	98.65	0.2864	0.0115
Ethane	C₂H₆	2	30.1	−182.8	−88.6	0.3554	0.3386	...	32.17	48.72	145.48	0.2792	0.0995
Propane	C₃H₈	3	44.1	−187.7	−42.0	0.5063	0.4989	...	96.68	42.48	200.14	0.2765	0.1523
n-Butane	C₄H₁₀	4	58.1	−138.3	−0.5	0.5849	0.5791	1.3326	151.97	37.96	255.09	0.2740	0.2002
n-Pentane	C₅H₁₂	5	72.2	−129.7	36.1	0.6317	0.6260	1.3575	196.55	33.70	313.05	0.2702	0.2515
n-Hexane	C₆H₁₄	6	86.2	−95.3	68.7	0.6651	0.6605	1.3749	234.45	30.25	371.22	0.2661	0.3013
n-Heptane	C₇H₁₆	7	100.2	−90.6	98.4	0.6902	0.6857	1.3876	267.05	27.40	427.88	0.2610	0.3495
n-Octane	C₈H₁₈	8	114.2	−56.8	125.7	0.7073	0.7031	1.3974	295.55	24.90	486.35	0.2561	0.3996
n-Nonane	C₉H₂₀	9	128.3	−53.5	150.8	0.7220	0.7180	1.4054	321.45	22.90	543.67	0.2519	0.4435
n-Decane	C₁₀H₂₂	10	142.3	−29.6	174.2	0.7342	0.7302	1.4119	344.55	21.10	599.58	0.2463	0.4923
n-Undecane	C₁₁H₂₄	11	156.3	−25.6	195.9	0.7439	0.7400	1.4151	365.85	19.50	658.69	0.2418	0.5303
n-Dodecane	C₁₂H₂₆	12	170.3	−9.6	216.3	0.7524	0.7485	1.4195	384.85	18.20	715.67	0.2381	0.5764
n-Tridecane	C₁₃H₂₈	13	184.4	−5.4	235.5	0.7611	0.7571	1.4235	401.85	16.80	774.60	0.2319	0.6174
n-Tetradecane	C₁₄H₃₀	14	198.4	5.9	253.6	0.7665	0.7627	1.4269	419.85	15.70	829.82	0.2261	0.6430
n-Pentadecane	C₁₅H₃₂	15	212.4	9.9	270.7	0.7717	0.7680	1.4298	434.85	14.80	888.49	0.2234	0.6863
n-Hexadecane	C₁₆H₃₄	16	226.4	18.2	286.9	0.7730	0.7729	1.4325	449.85	14.00	944.33	0.2199	0.7174
n-Heptadecane	C₁₇H₃₆	17	240.5	22.0	302.2	0.7752	0.7765	1.4348	462.85	13.40	999.83	0.2189	0.7697
n-Octadecane	C₁₈H₃₈	18	254.5	28.2	316.7	0.7841	0.7805	1.4369	473.85	12.70	1059.74	0.2167	0.8114
n-Nonadecane	C₁₉H₄₀	19	268.5	31.9	329.9	0.7880	0.7844	1.4388	484.85	12.10	1119.82	0.2150	0.8522
n-Eicosane	C₂₀H₄₂	20	282.6	36.4	343.8	0.7890	0.7871	1.4405	494.85	11.60	1169.50	0.2125	0.9069
n-Heneicosane	C₂₁H₄₄	21	296.6	40.2	356.5	0.7954	0.7906	1.4440	504.85	11.10	1229.41	0.2110	0.9420
n-Docosane	C₂₂H₄₆	22	310.6	44.0	368.6	0.7981	0.7929	1.4454	513.85	10.60	1289.49	0.2089	0.9722
isobutane (2-Methylpropane)	C₄H₁₀	4	58.1	−159.6	−11.7	0.5644	0.5584	...	134.99	36.48	262.71	0.2824	0.1808
isopentane (2-Methylbutane)	C₅H₁₂	5	72.2	−159.9	27.8	0.6265	0.6213	1.3537	187.28	33.81	305.84	0.2701	0.2275
2-Methylpentane	C₆H₁₄	6	86.2	−95.3	60.3	0.6577	0.6529	1.3715	224.35	30.05	371.22	0.2661	0.2774
2-Methylhexane	C₇H₁₆	7	100.2	−118.3	90.1	0.6822	0.6778	1.3849	257.22	27.34	421.00	0.2610	0.3277
2-Methylheptane	C₈H₁₈	8	114.2	−109.0	117.7	0.7029	0.6987	1.3949	286.49	24.84	487.78	0.2604	0.3772
2,2,4-Trimethylpentane (isooctane)	C₈H₁₈	8	114.2	−107.4	99.2	0.6988	0.6945	1.3915	270.81	25.68	467.81	0.2656	0.3022
2-Methyloctane	C₉H₂₀	9	128.3	−80.4	143.3	0.7176	0.7134	1.4031	313.60	22.90	541.27	0.2541	0.4212
2-Methylnonane	C₁₀H₂₂	10	142.3	−74.7	167.0	0.7307	0.7266	1.4100	336.85	21.20	582.70	0.2436	0.4723
Olefins													
Ethene (Ethylene)	C₂H₄	2	28.1	−169.2	−103.7	0.1388	...	...	9.19	50.40	131.00	0.2813	0.0865
Propene (Propylene)	C₃H₆	3	42.1	−185.3	−47.7	0.5192	0.5111	...	92.42	46.65	188.36	0.2891	0.1398
1-Butene	C₄H₈	4	56.1	−185.4	−6.3	0.6001	0.5938	...	146.80	40.43	239.24	0.2770	0.1905
1-Pentene	C₅H₁₀	5	70.1	−165.2	30.0	0.6456	0.6402	1.3715	191.63	35.13	295.10	0.2683	0.2312
1-Hexene	C₆H₁₂	6	84.2	−139.8	63.5	0.6790	0.6740	1.3879	230.88	31.40	354.13	0.2654	0.2804
1-Heptene	C₇H₁₄	7	98.2	−118.9	93.6	0.7015	0.6971	1.3998	264.14	28.30	413.15	0.2617	0.3310
1-Octene	C₈H₁₆	8	112.2	−101.7	121.3	0.7181	0.7140	1.4087	293.50	25.68	460.26	0.2509	0.3764
1-Nonene	C₉H₁₈	9	126.2	−81.4	146.9	0.7330	0.7290	...	320.10	23.30	528.04	0.2494	0.4171
1-Decene	C₁₀H₂₀	10	140.3	−66.3	170.6	0.7450	0.7410	...	343.25	22.18	584.08	0.2528	0.4800
Acetylene	C₂H₂	2	26.0	−80.8	−83.8	...	0.4001	...	35.17	61.39	112.97	0.2706	0.1873
1,3-Butadiene	C₄H₆	4	54.1	−108.9	−4.4	...	0.6219	...	152.02	42.77	220.49	0.2668	...

Naphthenes													
Cyclopentane	C₅H₁₀	5	70.1	−93.8	49.3	0.7502	0.7456	1.4065	238.61	45.02	257.89	0.2729	0.1959
Methylcyclopentane	C₆H₁₂	6	84.2	−142.4	71.8	0.7540	0.7491	1.4097	259.64	37.85	318.92	0.2725	0.2302
Ethylcyclopentane	C₇H₁₄	7	98.2	−138.4	103.5	0.7712	0.7667	1.4198	296.37	33.98	375.14	0.2692	0.2716
Propylcyclopentane	C₈H₁₆	8	112.2	−117.3	131.0	0.7811	0.7768	1.4263	322.85	30.20	428.03	0.2609	0.3266
n-Butylcyclopentane	C₉H₁₈	9	126.2	−108.0	156.6	0.7893	0.7851	1.4316	347.85	27.20	483.11	0.2545	0.3719
n-Pentylcyclopentane	C₁₀H₂₀	10	140.3	−83.0	180.5	0.7954	...	1.4352	370.65	24.50	536.79	0.2457	0.4184
n-Hexylcyclopentane	C₁₁H₂₂	11	154.3	−73.0	202.9	0.8006	...	1.4386	390.95	22.20	592.40	0.2382	0.4646
n-Heptylcyclopentane	C₁₂H₂₄	12	168.3	−53.0	223.9	0.8051	...	1.4416	409.45	20.10	647.30	0.2293	0.5100
n-Octylcyclopentane	C₁₃H₂₆	13	182.4	−44.0	243.5	0.8088	...	1.4446	426.35	18.30	702.38	0.2210	0.5525
n-Nonylcyclopentane	C₁₄H₂₈	14	196.4	−29.0	262.0	0.8121	...	1.4467	441.75	16.70	757.64	0.2129	0.5956
n-Decylcyclopentane	C₁₅H₃₀	15	210.4	−22.1	279.4	0.8149	...	1.4486	455.95	15.30	811.76	0.2049	0.6314
n-Undecylcyclopentane	C₁₆H₃₂	16	224.4	−10.0	295.8	0.8175	...	1.4503	469.05	14.00	867.27	0.1968	0.6741
n-Dodecylcyclopentane	C₁₇H₃₄	17	238.5	−5.0	311.2	0.8197	...	1.4518	481.25	12.80	921.48	0.1881	0.7163
n-Tridecylcyclopentane	C₁₈H₃₆	18	252.5	5.0	325.9	0.8217	...	1.4531	492.45	11.80	977.26	0.1812	0.7582
n-Tetradecylcyclopentane	C₁₉H₃₈	19	266.5	9.0	340.0	0.8235	...	1.4543	502.85	10.90	1031.55	0.1743	0.7949
n-Pentadecylcyclopentane	C₂₀H₄₀	20	280.5	17.0	353.0	0.8252	...	1.4554	512.55	10.00	1087.59	0.1665	0.8395
n-Hexadecylcyclopentane	C₂₁H₄₂	21	294.6	21.0	366.0	0.8267	...	1.4564	521.55	9.20	1141.97	0.1590	0.8755
n-Heptadecylcyclopentane	C₂₂H₄₄	22	308.6	27.0	377.0	0.8280	...	1.4573	538.01	11.91	1198.28	0.2115	0.9060
Cyclohexane	C₆H₁₂	6	84.2	6.5	80.7	0.7823	0.8021	1.4262	280.43	40.73	307.89	0.2725	0.2096
Methylcyclohexane	C₇H₁₄	7	98.2	−126.6	100.9	0.7748	0.7702	1.4231	299.04	34.71	367.79	0.2684	0.2350
Ethylcyclohexane	C₈H₁₆	8	112.2	−111.3	131.8	0.7926	0.7884	1.4330	336.00	30.40	430.13	0.2582	0.2455
Propylcyclohexane	C₉H₁₈	9	126.2	−94.9	156.8	0.7981	0.7940	1.4371	366.00	28.07	476.81	0.2519	0.2595
n-Butylcyclohexane	C₁₀H₂₀	10	140.3	−74.7	181.0	0.8033	0.7993	1.4408	393.85	25.70	534.16	0.2476	0.2743
Aromatics													
Benzene	C₆H₆	6	78.1	5.5	80.1	0.8832	0.8780	1.5011	289.01	48.98	258.94	0.2714	0.2100
Methylbenzene (Toluene)	C₇H₈	7	92.1	−95.0	110.6	0.8741	0.8685	1.4969	318.65	41.06	315.80	0.2635	0.2621
Ethylbenzene	C₈H₁₀	8	106.2	−95.0	136.2	0.8737	0.8678	1.4959	344.05	36.06	373.81	0.2627	0.3026
Propylbenzene	C₉H₁₂	9	120.2	−99.6	159.2	0.8683	0.8630	1.4920	365.23	32.00	439.71	0.2651	0.3447
n-Butylbenzene	C₁₀H₁₄	10	134.2	−87.9	183.3	0.8660	0.8610	1.4898	387.40	28.87	496.89	0.2612	0.3938
n-Pentylbenzene	C₁₁H₁₆	11	148.2	−75.0	205.5	0.8624	0.8583	1.4878	406.75	26.04	549.74	0.2532	0.4378
n-Hexylbenzene	C₁₂H₁₈	12	162.3	−61.2	226.1	0.8622	0.8581	1.4864	424.85	23.80	592.64	0.2431	0.4790
n-Heptylbenzene	C₁₃H₂₀	13	176.3	−48.0	246.1	0.8617	0.8576	1.4854	440.85	21.80	648.27	0.2381	0.5272
n-Octylbenzene	C₁₄H₂₂	14	190.3	−36.0	264.4	0.8602	0.8562	1.4845	455.85	20.20	703.41	0.2344	0.5670
n-Nonylbenzene	C₁₅H₂₄	15	204.4	−24.2	282.1	0.8596	0.8557	1.4838	467.85	18.95	752.70	0.2315	0.6331
n-Decylbenzene	C₁₆H₂₆	16	218.4	−14.4	297.9	0.8590	0.8551	1.4832	479.85	17.70	812.55	0.2297	0.6797
n-Undecylbenzene	C₁₇H₂₈	17	232.4	−5.2	313.3	0.8587	0.8548	1.4828	490.85	16.72	867.64	0.2284	0.7333
n-Dodecylbenzene	C₁₈H₃₀	18	246.4	2.8	327.6	0.8595	0.8556	1.4821	506.85	15.60	923.08	0.2221	0.7333
n-Tridecylbenzene	C₁₉H₃₂	19	260.5	10.0	341.3	0.8584	0.8545	1.4821	516.85	14.80	977.25	0.2202	0.7799
n-Tetradecylbenzene	C₂₀H₃₄	20	274.5	16.0	354.0	0.8587	0.8553	1.4818	526.85	14.00	1029.88	0.2168	0.8130
n-Pentadecylbenzene	C₂₁H₃₆	21	288.5	22.0	366.0	0.8587	0.8540	1.4815	535.85	13.30	1089.71	0.2155	0.8567
n-Hexadecylbenzene	C₂₂H₃₈	22	302.5	27.0	378.0	0.8586	0.8541	1.4813	544.85	12.70	1140.80	0.2130	0.8996
o-xylene (o-dimethylbenzene)	C₈H₁₀	8	106.2	−25.2	144.4	0.8849	0.8799	...	357.18	37.34	369.17	0.2630	0.3104
m-xylene	C₈H₁₀	8	106.2	−47.9	139.1	0.8691	0.8643	1.4972	343.90	35.36	375.80	0.2590	0.3259

(Continued)

TABLE 2.1—(*Continued*)

Compound	Formula	N_C	M	T_M, °C	T_b, °C	SG at 60°F	d_{20}, g/cm³	n_{20}	T_C, °C	P_C, bar	V_C, cm³/mol	Z_C	ω
p-xylene	C_8H_{10}	8	106.2	13.3	138.4	0.8654	0.8608	1.4958	343.08	35.11	379.11	0.2598	0.3215
isopropylbenzene	C_9H_{12}	9	120.2	−96.0	152.4	0.8682	0.8632	1.4915	357.95	32.09	426.95	0.2611	0.3258
1,2,3-Trimethylbenezene	C_9H_{12}	9	120.2	−25.4	176.1	0.8985	0.8942	1.5130	391.38	34.54	414.20	0.2590	0.3664
isobutylbenzene	$C_{10}H_{14}$	10	134.2	−51.5	172.8	0.8577	0.8532	1.4865	377.00	30.40	455.83	0.2564	0.3797
o-Cymene	$C_{10}H_{14}$	10	134.2	−71.5	178.2	0.8812	0.8765	1.5005	388.85	29.30	489.35	0.2605	0.3372
Cyclohexylbenzene	$C_{12}H_{16}$	12	160.3	7.0	240.1	0.8617	0.9430	...	470.85	28.80	529.25	0.2464	0.3783
Naphthalene	$C_{10}H_8$	10	128.2	80.3	218.0	1.0281	1.0238	...	475.20	40.51	412.89	0.2688	0.3022
1-Methylnaphthalene	$C_{11}H_{10}$	11	142.2	−30.5	244.7	1.0242	1.1020	...	498.89	36.60	465.17	0.2652	0.3478
Styrene	C_8H_8	8	104.2	−30.6	145.2	0.9097	0.9049	...	362.85	38.40	351.76	0.2555	0.2971
Diphenyl methane	$C_{13}H_{12}$	13	168.2	25.2	264.3	1.0101	1.0244	...	494.85	29.20	547.20	0.2502	0.4615
Nonhydrocarbons													
Water	H_2O	...	18.0	0.0	100.0	1.0000	0.9963	...	373.98	220.55	55.91	0.2292	0.3449
Carbon dioxide	CO_2	1	44.0	−56.6	−78.5	0.8172	0.7745	...	31.06	73.83	93.96	0.2743	0.2236
Hydrogen Sulfide	H_2S	...	34.1	−85.5	−60.4	0.8012	0.7901	...	100.38	89.63	98.51	0.2843	0.0942
Nitrogen	N_2	...	28.0	−210.0	−195.8	0.8094	...	...	−146.95	34.00	89.21	0.2891	0.0377
Oxygen	O_2	...	32.0	−218.8	−183.0	1.1420	...	...	−118.57	50.43	73.32	0.2877	0.0222
Ammonia	NH_3	...	17.0	−77.7	−33.4	0.6165	0.6093	...	132.50	112.80	72.51	0.2425	0.2526
Carbon monoxide	CO	1	28.0	−205.0	−191.5	...	...	...	−140.23	34.99	94.43	0.2990	0.0482
Hydrogen	H_2	...	2.0	−259.2	−252.8	...	...	...	−239.96	13.13	64.28	0.3059	−0.2160
Air	N_2+O_2	...	29.0	−214.0	−194.5	...	...	...	−140.70	37.74	91.45	0.3134	0.0074
Methanol	CH_3OH	1	32.04	−97.7	64.7	0.7993	0.7944	1.3265*	239.49	80.97	118.00	0.2240	0.5640
Ethanol	C_2H_5OH	2	46.07	−114.1	78.29	0.7950	0.7904	1.3594*	240.77	61.48	167.00	0.2400	0.6452
Isobutanol	C_4H_9OH	4	74.12	−108.0	107.66	0.8063	0.8015	1.3938*	274.63	43.02	272.97	0.2580	0.5848

N_C: carbon number; d_{20}: liquid density at 20°C; V_C: critical volume; Z_C: critical compressibility factor; ω: acentric factor. Data source: API-TDB [2].
n_{20}: refractive index at 20°C; P_C: critical pressure; T_b: normal boiling point; T_M: freezing point; T_C: critical temperature;
*SG of light gases are at low temperatures where the gas is in liquid form.

40

TABLE 2.2—*Secondary properties of some selected compounds.*

Compound	Formula	N_C	API gravity	K_W	P^{sat} at 100°F, bar	Viscosity, cSt $\nu_{38(100)}$	$\nu_{99(210)}$	CH	T_F, °C	AI, °C	AP, °C	Flammability range, vol.% Min	Max	RON clear
Paraffins														
Methane	CH_4	1	340.3	19.53	344.737	...	...	2.98	...	...	536.9	5.00	15.00	...
Ethane	C_2H_6	2	266.6	19.49	55.1579	...	...	3.97	...	...	471.9	2.90	13.00	1.6
Propane	C_3H_8	3	148.0	14.74	12.9621	0.1775	...	4.47	...	83.2	449.9	2.10	9.50	1.8
n-Butane	C_4H_{10}	4	110.4	13.49	3.5608	0.2533	0.1686	4.77	...	70.8	287.9	1.80	8.40	93.8
n-Pentane	C_5H_{12}	5	92.5	13.02	1.0745	0.3394	0.2643	4.96	-40.0	...	242.9	1.40	8.30	61.7
n-Hexane	C_6H_{14}	6	81.2	12.79	0.3435	0.4152	0.3515	5.11	-21.7	68.7	224.9	1.20	7.70	24.8
n-Heptane	C_7H_{16}	7	73.5	12.67	0.1112	0.5046	0.3997	5.21	-4.1	...	203.9	1.00	7.00	0.0
n-Octane	C_8H_{18}	8	68.6	12.66	0.0370	0.6364	0.4694	5.30	12.9	...	205.9	0.96	...	...
n-Nonane	C_9H_{20}	9	64.5	12.66	0.0125	0.8078	0.5537	5.36	30.9	...	204.9	0.87	2.90	...
n-Decane	$C_{10}H_{22}$	10	61.2	12.67	0.0042	1.0154	0.6397	5.42	45.9	...	200.9	0.78	2.60	...
n-Undecane	$C_{11}H_{24}$	11	58.7	12.71	0.0014	1.2588	0.7469	5.46	65.0	...	201.9	...	...	...
n-Dodecane	$C_{12}H_{26}$	12	56.6	12.74	0.0005	1.5452	0.8624	5.50	73.9	...	202.9	...	...	...
n-Tridecane	$C_{13}H_{28}$	13	54.4	12.76	0.0002	1.8634	0.9885	5.53	78.9	...	201.9	...	...	...
n-Tetradecane	$C_{14}H_{30}$	14	53.1	12.82	0.0001	2.2294	1.1328	5.56	100.0	...	199.9	...	...	...
n-Pentadecane	$C_{15}H_{32}$	15	51.9	12.87	0.0000	2.6415	1.2859	5.59	113.9	...	201.9	...	...	...
n-Hexadecane	$C_{16}H_{34}$	16	51.6	12.97	0.0000	3.1229	1.4413	5.61	135.1	...	201.9	...	...	...
n-Heptadecane	$C_{17}H_{36}$	17	51.0	13.05	0.0000	3.6045	1.5815	5.63	147.8	...	201.9	...	...	...
n-Octadecane	$C_{18}H_{38}$	18	49.0	13.01	0.0000	4.1620	1.7940	5.64	165.1	...	201.9	...	...	...
n-Nonadecane	$C_{19}H_{40}$	19	48.1	13.04	0.0000	4.6090	1.9889	5.66	167.9	...	201.9	...	...	...
n-Eicosane	$C_{20}H_{42}$	20	47.8	13.12	0.0000	5.3165	2.1703	5.67	166.9	...	201.9	...	...	...
n-Heneicosane	$C_{21}H_{44}$	21	46.4	13.11	...	...	2.4099	5.69	176.9	...	201.9	...	...	...
n-Docosane	$C_{22}H_{46}$	22	45.8	13.15	...	...	...	5.70	184.9	...	201.9	...	...	...
isobutane (2-Methylpropane)	C_4H_{10}	4	119.2	13.78	5.0199	0.2773	0.1873	4.77	-57.2	107.7	460.1	1.80	8.40	0.1
isopentane (2-Methylbutane)	C_5H_{12}	5	94.4	13.01	1.4110	0.3066	...	4.96	-35.2	70.8	420.1	1.40	8.30	92.3
2-Methylpentane	C_6H_{14}	6	83.6	12.82	0.4666	0.3862	...	5.11	-23.2	73.8	...	1.20	7.70	73.4
2-Methylhexane	C_7H_{16}	7	75.9	12.72	0.1562	0.4730	0.3635	5.21	4.1	74.1	...	1.00	7.00	42.4
2-Methylheptane	C_8H_{18}	8	69.8	12.65	0.0528	0.5908	0.3738	5.30	...	73.9	...	0.98	...	20.6
2,2,4-Trimethylpentane (isooctane)	C_8H_{18}	8	71.0	12.52	0.1181	0.6077	...	5.30	-12.2	79.5	...	1.00	...	100.0
2-Methyloctane	C_9H_{20}	9	65.7	12.66	0.0177	0.7382	0.4329	5.36	22.9	...	...	...	...	...
2-Methylnonane	$C_{10}H_{22}$	10	62.1	12.66	0.0058	0.9636	0.5401	5.42	40.9	...	...	...	...	...

(Continued)

TABLE 2.2—(Continued)

Compound	Formula	N_C	API gravity	K_W	p^{sat} at 100°F, bar	Viscosity, cSt $\nu_{38(100)}$	Viscosity, cSt $\nu_{99(210)}$	CH	T_F, °C	AI, °C	AP, °C	Flammability range, vol.% Min	Flammability range, vol.% Max	RON clear
Olefins														
Ethene (Ethylene)	C_2H_4	2	888.0	48.49	15.7812	…	…	5.96	−108.2	…	…	2.30	32.30	…
Propene (Propylene)	C_3H_6	3	141.0	14.26	4.2953	0.1801	…	5.96	…	…	455.1	2.00	11.00	100.2
1-Butene	C_4H_8	4	104.3	13.05	1.3203	…	…	5.96	…	…	383.9	1.60	9.30	97.4
1-Pentene	C_5H_{10}	5	87.7	12.66	0.4143	…	…	5.96	−18.2	19.1	272.9	1.50	8.70	90.9
1-Hexene	C_6H_{12}	6	76.9	12.46	0.1353	0.3415	0.3043	5.96	−31.2	22.8	253.1	1.00	7.50	76.4
1-Heptene	C_7H_{14}	7	70.2	12.41	0.0453	0.4317	0.3590	5.96	0.1	27.3	263.1	0.80	6.90	54.5
1-Octene	C_8H_{16}	8	65.5	12.42	0.0152	0.5657	0.4294	5.96	20.9	32.6	230.1	0.80	6.80	28.7
1-Nonene	C_9H_{18}	9	61.5	12.43	0.0051	0.7004	0.5027	5.96	26.9	38.1	236.9	0.60	6.00	…
1-Decene	$C_{10}H_{20}$	10	58.4	12.45	…	0.8848	…	5.96	47.1	44.2	235.1	0.55	5.70	…
Acetylene	C_2H_2	2	…	…	4.0906	…	…	11.92	…	…	305.1	2.50	80.00	…
1,3 Butadiene	C_4H_6	4	…	…	…	0.2030	0.2248	7.94	−18.2	…	428.9	…	…	…
Naphthenes														
Cyclopentane	C_5H_{10}	5	57.1	11.12	0.6839	0.4973	…	5.96	−39.2	16.8	361.1	1.40	9.40	100.1
Methylcyclopentane	C_6H_{12}	6	56.2	11.31	0.3106	0.5646	…	5.96	−27.2	33.1	328.9	1.20	8.35	91.4
Ethylcyclopentane	C_7H_{14}	7	52.0	11.39	0.0970	0.6199	0.4613	5.96	−4.1	36.8	260.1	1.10	6.70	67.2
Propylcyclopentane	C_8H_{16}	8	49.7	11.51	0.0325	0.7257	0.5148	5.96	15.9	44.5	269.1	0.95	6.40	31.2
n-Butylcyclopentane	C_9H_{18}	9	47.8	11.63	0.0108	0.9118	0.6200	5.96	31.9	48.8	250.1	0.80	5.90	97.0
n-Pentylcyclopentane	$C_{10}H_{20}$	10	46.4	11.75	…	1.1280	0.7300	5.96	…	…	…	0.74	5.47	…
n-Hexylcyclopentane	$C_{11}H_{22}$	11	45.2	11.86	…	1.4150	0.8500	5.96	…	…	…	0.68	5.20	…
n-Heptylcyclopentane	$C_{12}H_{24}$	12	44.3	11.97	…	1.7480	0.9800	5.96	…	…	…	0.62	5.06	…
n-Octylcyclopentane	$C_{13}H_{26}$	13	43.5	12.07	…	2.1300	1.1200	5.96	…	…	…	0.57	5.01	…
n-Nonylcyclopentane	$C_{14}H_{28}$	14	42.7	12.16	…	2.5700	1.2700	5.96	…	…	…	0.53	5.07	…
n-Decylcyclopentane	$C_{15}H_{30}$	15	42.1	12.25	…	3.0500	1.4400	5.96	…	…	…	0.50	5.24	…
n-Undecylcyclopentane	$C_{16}H_{32}$	16	41.6	12.33	…	3.6300	1.6100	5.96	…	…	…	0.47	5.53	…
n-Dodecylcyclopentane	$C_{17}H_{34}$	17	41.1	12.41	…	4.2500	1.7800	5.96	…	…	…	0.44	5.95	…
n-Tridecylcyclopentane	$C_{18}H_{36}$	18	40.7	12.48	…	4.9500	1.9800	5.96	…	…	…	0.41	6.53	…
n-Tetradecylcyclopentane	$C_{19}H_{38}$	19	40.3	12.55	…	5.7100	2.1900	5.96	…	…	…	0.39	7.33	…
n-Pentadecylcyclopentane	$C_{20}H_{40}$	20	40.0	12.61	…	6.5600	2.4000	5.96	…	…	…	0.37	8.38	…
n-Hexadecylcyclopentane	$C_{21}H_{42}$	21	39.7	12.67	…	7.4900	…	5.96	…	…	…	0.35	9.79	…
n-Heptadecylcyclopentane	$C_{22}H_{44}$	22	39.4	12.73	…	…	…	5.96	…	…	…	0.34	11.67	…
Cyclohexane	C_6H_{12}	6	49.4	11.00	0.2274	0.9419	…	5.96	−20.0	31.1	260.1	1.30	8.00	83.0
Methylcyclohexane	C_7H_{14}	7	51.1	11.31	0.1106	0.7640	0.4757	5.96	−5.9	41.1	285.1	1.15	7.20	74.8
Ethylcyclohexane	C_8H_{16}	8	47.0	11.35	0.0333	0.8629	0.5122	5.96	22.1	43.8	261.9	0.90	6.60	45.6
Propylcyclohexane	C_9H_{18}	9	45.8	11.50	0.0117	1.0010	0.5759	5.96	30.9	49.8	248.1	0.95	5.90	17.8
n-Butylcyclohexane	$C_{10}H_{20}$	10	44.6	11.64	0.0040	1.2539	0.6100	5.96	47.9	54.4	246.1	0.85	5.50	…
Aromatics														
Benzene	C_6H_6	6	28.7	9.74	0.2216	0.5927	0.3306	11.92	−11.2	−30.0	560.1	1.40	7.10	…
Methylbenzene (Toluene)	C_7H_8	7	30.4	10.11	0.0710	0.5604	0.3433	10.43	4.9	−30.0	480.1	1.20	7.10	105.8
Ethylbenzene	C_8H_{10}	8	30.5	10.34	0.0257	0.6540	0.3970	9.53	15.1	−30.0	430.1	1.00	6.70	100.8
Propylbenzene	C_9H_{12}	9	31.5	10.59	0.0100	0.7977	0.4534	8.94	30.1	−30.0	456.1	0.88	6.00	101.5
n-Butylbenzene	$C_{10}H_{14}$	10	31.9	10.82	0.0033	0.9483	0.5186	8.51	50.1	−30.0	410.1	0.80	5.80	100.4

Compound	Formula	N_C	API		p^{sat}	v	CH			AI			RON
n-Pentylbenzene	C₁₁H₁₆	11	32.6	11.03	0.0011	1.1824	8.19	65.1	…	…	0.80	5.50	…
n-Hexylbenzene	C₁₂H₁₈	12	32.6	11.19	0.0004	1.4419	7.94	80.1	…	…	0.70	5.30	…
n-Heptylbenzene	C₁₃H₂₀	13	32.7	11.35	0.0001	1.7546	7.75	95.1	…	…	0.70	5.10	…
n-Octylbenzene	C₁₄H₂₂	14	33.0	11.50	0.0000	2.0974	7.58	107.1	…	…	0.70	4.90	…
n-Nonylbenzene	C₁₅H₂₄	15	33.1	11.63	0.0000	2.5329	7.45	98.9	…	…	0.60	4.70	…
n-Decylbenzene	C₁₆H₂₆	16	33.2	11.75	0.0000	3.0119	7.33	106.9	…	…	0.60	4.60	…
n-Undecylbenzene	C₁₇H₂₈	17	33.3	11.86	0.0000	3.5672	7.23	143.9	…	…	0.60	4.40	…
n-Dodecylbenzene	C₁₈H₃₀	18	33.1	11.94	0.0000	4.1857	7.15	140.9	…	…	0.60	4.30	…
n-Tridecylbenzene	C₁₉H₃₂	19	33.3	12.05	0.0000	4.9566	7.07	164.9	…	…	0.50	4.20	…
n-Tetradecylbenzene	C₂₀H₃₄	20	33.3	12.13	0.0000	5.7107	7.01	173.9	…	…	0.50	4.10	…
n-Pentadecylbenzene	C₂₁H₃₆	21	33.3	12.20	0.0000	6.5716	6.95	182.9	…	…	0.50	4.00	…
n-Hexadecylbenzene	C₂₂H₃₈	22	33.3	12.28	0.0000	7.4138	6.90	192.9	…	…	0.50	3.90	…
o-xylene (o-dimethylbenzene)	C₈H₁₀	8	28.4	10.27	0.0182	…	9.53	16.9	…	…	1.00	6.00	…
m-xylene	C₈H₁₀	8	31.3	10.42	0.0226	0.5936	9.53	25.1	−30.0	463.1	1.10	7.00	104.0
p-xylene	C₈H₁₀	8	32.0	10.46	0.0237	0.6167	9.53	25.1	−30.0	465.1	1.10	7.00	103.4
isopropylbenzene	C₉H₁₂	9	31.5	10.54	0.0129	0.7474	8.94	43.9	−15.0	528.1	0.88	6.50	102.1
1,2,3-Trimethylbenzene	C₉H₁₂	9	26.0	10.37	0.0049	0.8317	8.94	51.2	…	…	0.88	5.20	101.4
isobutylbenzene	C₁₀H₁₄	10	33.5	10.84	0.0057	0.9895	8.51	55.1	…	…	0.80	6.00	101.6
o-Cymene	C₁₀H₁₄	10	29.1	10.59	0.0045	0.9739	8.51	53.1	…	…	0.80	5.20	100.6
Cyclohexylbenzene	C₁₂H₁₆	12	32.7	11.30	0.0002	2.0086	8.94	98.9	…	…	0.70	5.40	…
Naphthalene	C₁₀H₈	10	6.1	9.34	0.0002	…	14.89	80.1	…	526.1	0.88	5.90	…
1-Methylnaphthalene	C₁₁H₁₀	11	6.7	9.54	…	2.1837	13.11	82.1	…	…	0.80	5.30	…
Styrene	C₈H₈	8	24.0	10.00	0.0170	0.6637	11.92	31.9	…	…	1.10	6.10	103.0
Diphenyl methane	C₁₃H₁₂	13	8.6	9.79	0.0001	2.1883	12.91	130.1	…	490.1	0.70	5.20	…
Nonhydrocarbons													
Water	H₂O	…	10.0	8.76	0.0655	0.7063	0.0003	…	…	…	…	…	…
Carbon dioxide	CO₂	1	…	8.63	27.2295	…	0.1447	…	…	…	…	…	…
Hydrogen Sulfide	H₂S	…	…	9.06	14.5632	…	…	…	…	260.1	4.30	45.50	…
Nitrogen	N₂	…	…	…	…	…	…	…	…	…	…	…	…
Oxygen	O₂	…	…	4.78	…	…	…	…	…	…	…	…	…
Ammonia	NH₃	…	…	12.26	…	0.1990	0.1373	…	…	…	16.00	25.00	…
Carbon monoxide	CO	1	…	…	…	…	…	…	…	650.9	12.50	74.00	…
Hydrogen	H₂	…	…	…	…	…	…	…	…	608.9	4.00	75.00	…
Air	N₂+O₂	…	…	…	…	…	…	…	…	400.1	…	…	…
Methanol	CH₃OH	1	45.52	10.60	0.3189	0.5900	2.98	10.85	…	463.85	2.3	36.0	130
Ethanol	C₂H₅OH	2	46.48	10.80	0.1603	1.0995	3.97	12.85	…	422.85	4.3	19.0	…
Isobutanol	C₄H₉OH	4	44.0	10.94	0.0335	2.8700	4.77	27.85	…	…	1.7	10.9	125

N_C: carbon number; p^{sat}: true vapor pressure at 100 °F (311 K); v: kinematic viscosity; CH: carbon-to-hydrogen weight ratio; T_M: freezing point; AP: aniline point; AI: autoignition temperature; RON: research octane number. Data sources: API-TDB [2]; for autoignition temperature: DIPPR [20].

TABLE 2.3—*Additional data on refractive index and viscosity of heavy hydrocarbons.*

Components	Formula	M	SG	d	v	V	n	R_i	VGC
Paraffins									
n-Tetracosane	$C_{24}H_{50}$	338.6	0.8028	0.7980	2.79	35.5	1.4504	1.051	0.745
11-n-Butyldocosane	$C_{26}H_{54}$	366.7	0.8081	0.8041	2.78	35.5	1.4461	1.044	0.749
5-n-Bytryldocosane	$C_{26}H_{54}$	366.7	0.8099	0.8059	3.02	36.3	1.4466	1.044	0.750
9-Ethyl-9-n-Heptyloctadecane	$C_{27}H_{56}$	380.7	0.8117	0.8076	3.48	37.8	1.4513	1.048	0.746
7-n-Hexyldocosane	$C_{28}H_{58}$	394.7	0.8121	0.8080	3.32	38.2	1.4517	1.048	0.749
11-n-Decylheneicosane	$C_{31}H_{64}$	436.8	0.8158	0.8117	3.66	38.4	1.454	1.084	0.7481
n-Dotriacontane	$C_{32}H_{66}$	450.8	0.8160	0.8119	5.01	42.7	1.455	1.049	0.730
11-n-Decyldocosane	$C_{32}H_{66}$	450.8	0.8170	0.8129	3.91	39.2	1.4543	1.048	0.747
11-n-Decyltetracosane	$C_{34}H_{70}$	478.9	0.8194	0.8153	4.45	40.9	1.4556	1.048	0.746
9-n-Octylhexacosane	$C_{38}H_{78}$	478.9	0.8187	0.8156	4.70	14.7	1.456	1.049	0.743
13-n-Dodeecylhexacosane	$C_{38}H_{78}$	507.0	0.8209	0.8168	4.90	42.4	1.4567	1.040	0.745
13-n-Undecylpentacosane	$C_{38}H_{78}$	535.0	0.8230	0.8189	5.54	4.44	1.4577	1.048	0.744
Naphthenes									
1,3-Dicyclohexylcyclohexane	$C_{18}H_{32}$	248.4	0.9490	0.940	4.61	41.4	1.5065[a]	1.036	0.917
1,2-Dicyclohexylcyclohexane	$C_{18}H_{32}$	248.4	0.949	0.9404	4.86	42.2	1.5062	1.036	0.917
2(ar)-n-Butyl-3(ar)-n-hexyl-tetralin	$C_{20}H_{32}$	272.5	0.9182	0.9086	2.92	35.9	1.5075	1.048	0.909
Perhydridibenzo-(a,i) fluorene	$C_{21}H_{34}$	286.5	1.0081	0.9981	10.66	61.6	1.5277	1.0287	0.976
Di(alpha-decalyl)-methane	$C_{21}H_{36}$	288.5	0.9807	0.9709	10.96	62.6	1.518	1.031	0.940
1,2-Di(alpha-decalyl)ethane	$C_{22}H_{38}$	302.5	0.9748	0.9661	12.2	67.2	1.5176	1.034	0.930
1,1-Di(alpha-decalyl)ethane	$C_{22}H_{38}$	302.5	0.9854	0.9765	16.5	83.9	1.5217	1.033	0.938
1,5-Dicyclopentyl-3(2-cyclopentylethyl) pentane	$C_{22}H_{38}$	304.5	0.9040	0.895	3.96	39.5	1.4853	1.0378	0.866
1,7-Dicyclopentyl-4(2-phenylethy) heptane	$C_{25}H_{40}$	340.6	0.9298	0.9205	3.92	39.2	1.507	1.046	0.899
1-Phenyl-3(2-cyclohexylethyl)-6-cyclopentylhexane	$C_{25}H_{40}$	340.6	0.9318	0.9230	4.80	42	1.5086	1.0465	0.893
6-n-Ocylperhydrobenz(de) anthracene	$C_{25}H_{44}$	344.6	0.9530	0.9432	7.91	52.1	1.5081	1.036	0.909
Cholestane	$C_{27}H_{48}$	372.8	0.9578	0.9482	21.15	103.4	1.52[a]	1.050	0.897
1,1-Di(alphadecalyl)hendecane	$C_{31}H_{56}$	428.8	0.9451	0.9356	20.25	99.5	1.5062	1.038	0.885
Aromatics									
1,1-Diphenylethylene	$C_{14}H_{12}$	180.2	1.0330	1.0235	3.96[b]	39.1[b]	1.6078	1.097	1.033
4,5-Dimethyl-9,10-dihydropenanthrene	$C_{16}H_{14}$	206.3	1.1090	1.0998	4.93	42.2	1.6433[a]	1.094	1.127
4,5-Dimethyl-9,10-dihydropenanthrene	$C_{16}H_{16}$	208.3	1.0738	1.0684	4.43	40.8	1.6286	1.095	1.089
1,2-Diphenylbenze	$C_{18}H_{18}$	230.3	1.0923	1.0814	4.53	41.2	1.6447	1.105	1.107
9-n-Butylanthracene	$C_{16}H_{16}$	239.3	1.0600	1.0530	4.61	41.4	1.6006	1.074	1.064
1,2,3,4,5,6,8,13,14,15,15-Dodecahydrochrysene	$C_{18}H_{24}$	240.4	1.0572	1.0492	8.16	52.9	1.5752	1.051	1.049
1,5-Di-phenyl-3(2-phenylethyl)2-pentene	$C_{25}H_{28}$	326.5	1.0210	1.0159	3.42	37.6	1.5816	1.074	1.030
1,5-Di-phenyl-3(2-phenylethyl)pentane	$C_{25}H_{40}$	328.5	1.0175	1.0076	3.81	38.8	1.5725	1.070	1.015
1,7-Dicyclopentyl-4-(2-phenylethyl)heptane	$C_{25}H_{28}$	340.6	0.9251	0.9205	3.92	39.2	1.5070	1.047	0.919
1,2,3,4,5,6,8,9,10,17,18-Dodecahydro-9(n-octyl)naphthecene	$C_{26}H_{40}$	352.6	0.9845	0.9796	13.09	70.5	1.5467	1.060	0.950
1,10-Di-(alphanaphyl)decane	$C_{30}H_{34}$	394.6	1.0400	1.0296	12.27	67.4	1.6094	1.095	1.016

M: molecular weight;
SG: specific gravity at 15.56°C (60/60°F);
d: density at 20°C, g/cm³;
v: kinematic viscosity at 210°F (99°C), cSt;
V: Saybolt viscosity at 210°F (99°C), SUS;
n: refractive index at 20°C;
R_i: refractivity intercept;
VGC: viscosity gravity constant. Data source: API-RP 42 [18].
[a] Calculated values of n.
[b] Kinematic or Saybolt viscosities at 100°F (38°C).

are calculated by Eqs. (2.14) and (2.15) and will be used to develop predictive methods for the composition of heavy fractions discussed in Chapter 3 (Section 3.5).

2.3 CHARACTERIZATION OF HYDROCARBONS

The work on characterization of pure hydrocarbons began in 1933 when Watson and Nelson for the first time developed two empirical charts relating molecular weight to either boiling point and K_W or boiling point and API gravity [24]. In these charts boiling point and specific gravity (or API gravity) are used as the two independent input parameters. Since then the work on characterization and methods of estimation of basic properties of pure hydrocarbons and petroleum fractions has continued to the present time. Methods developed in 1930s till 1960s were mainly graphical, while with the use of computer, methods developed in 1970s till present time are in the forms of analytical correlations. The best example of chart-type correlations, which has been in use by the industry, is the Winn nomogram that relates molecular weight, critical pressure, aniline point, and CH weight ratio to boiling point and specific gravity [25]. A version of Winn nomograph as used by the API [2] is presented in Fig. 2.12. Some of the analytical correlations that are used in the industry are Cavett [26], Kesler–Lee [12], Lee–Kesler [27], Riazi–Daubert [28, 29], Twu [30], and Riazi–Sahhaf [31]. Most of these correlations use boiling point and specific gravity as the input parameters to estimate parameters such as molecular weight, critical constants, and acentric factor. Most recently Korsten [32] has developed a characterization scheme that uses boiling point and a parameter called double-bond equivalent (DBE) as the input parameters. DBE can be estimated from H/C atomic ratio. In another paper Korsten [33] lists and evaluates various correlations for estimation of critical properties of pure hydrocarbons. Tsonopolus et al. [34] give the list of correlations developed for characterization of coal liquids in terms of boiling point and specific gravity. There are some methods of estimation of properties of pure compounds that are based on various group contribution techniques. The most accurate methods of group contributon for various properties with necessary recommendations are given in the fifth edition of *Properties of Gases and Liquids* [4]. Even some of these group contribution methods require properties such as molecular weight or boiling point. Examples of such procedures are the Lydersen and Ambrose methods [4]. The problem with group contribution methods is that the structure of the compound must be known. For this reason they are not appropriate for undefined petroleum fractions. However, they can be used to predict properties of pure compounds when experimental data are not available (i.e., critical properties of heavy pure hydrocarbons). In fact, on this basis the properties of hydrocarbons heavier than C_{18} have been predicted and reported by the API [2].

As discussed, during the past 70 years many methods in the forms of charts and equations were proposed to estimate the basic properties of hydrocarbons from the knowledge of the boiling point and specific gravity or the molecular weight. Nearly all of these correlations are empirical in nature

without any theoretical explanation. Boiling point and specific gravity were used in most correlations based on experience and their availability. However, the characterization methods proposed by Riazi and Daubert [28, 29, 35] are based on the theory of intermolecular forces and EOS parameters [36]. Although EOS are discussed in Chapter 5, their application in the development of analytical correlations to characterize hydrocarbons are discussed here. In the following parts in this section several characterization schemes developed by Riazi et al. [28, 29, 31, 35, 37] are presented along with other methods.

2.3.1 Development of a Generalized Correlation for Hydrocarbon Properties

Properties of a fluid depend on the intermolecular forces that exist between molecules of that fluid [38, 39]. As summarized by Prausnitz et al. [39] these forces are grouped into four categories. (1) Electrostatic forces between charged molecules (ions) and between permanent dipoles or higher multipoles. These forces result from the chemical structure of molecules and are important in polar compounds (i.e., water, methanol, ethanol, etc.). (2) Induction forces on molecules that are polarizable when subjected to an electric field from polar compounds. These forces are also called *dipole forces* and are determined by dipole moment of molecules (μ), which is proportional to polarizability factor, α, and the field strength. These forces are proportional to $\mu^2 \times \alpha$. (3) The third type of forces are attraction (dispersion forces) and repulsion between nonpolar molecules. These forces, also called *London forces*, are static in nature and are proportional to α^2. (4) The last are special (chemical) forces leading to association or complex formation such as chemical bonds. According to London these forces are additive and except for very polar compounds, the strongest forces are of the London type. For light and medium hydrocarbon compounds, London forces are the dominant force between the molecules.

The intermolecular force, F, is related to the potential energy, Γ, according to the following relation:

$$(2.19) \qquad F = -\frac{d\Gamma}{dr}$$

where r is the distance between molecules. The negative of the potential energy, $-\Gamma(r)$, is the work required to separate two molecules from the intermolecular distance r to infinite separation. Equation of state parameters can be estimated from the knowledge of the potential energy relation [39]. Most hydrocarbon compounds, especially the light and medium molecular weight hydrocarbons, are considered as nonpolar substances. There are two forces of attraction (dispersion forces) and repulsion between nonpolar molecules. The common convention is that the force of attraction is negative and that of repulsion is positive. As an example, when molecules of methane are 1 nm apart, the force of attraction between them is 2×10^{-8} dyne [39]. The following relation was first proposed by Mie for the potential energy of nonpolar molecules [39]:

$$(2.20) \qquad \Gamma = \frac{A_\circ}{r^n} - \frac{B_\circ}{r^m}$$

where

$$\Gamma_{rep} = \frac{A_\circ}{r^n}$$

$$\Gamma_{att} = -\frac{B_\circ}{r^m}$$

Γ_{rep} = repulsive potential
Γ_{att} = attractive potential
$A_\circ$ = parameter characterizing the repulsive force (>0)
$B_\circ$ = parameter characterizing the attractive force (>0)
r = distance between molecules
n, m = positive numbers, $n > m$

The main characteristics of a two-parameter potential energy function is the minimum value of potential energy, Γ_{min}, designated by $\varepsilon = -\Gamma_{min}$ and the distance between molecules where the potential energy is zero ($\Gamma = 0$) which is designated by σ. London studied the theory of dispersion (attraction) forces and has shown that $m = 6$ and it is frequently convenient for mathematical calculations to let $n = 12$. It can then be shown that Eq. (2.20) reduces to the following relation known as Lennard–Jones potential [39]:

$$(2.21) \qquad \Gamma = 4\varepsilon\left[\left(\frac{\sigma}{r}\right)^{12} - \left(\frac{\sigma}{r}\right)^6\right]$$

In the above relation, ε is a parameter representing molecular energy and σ is a parameter representing molecular size. Further discussion on intermolecular forces is given in Section 5.3.

According to the principle of statistical thermodynamics there exists a universal EOS that is valid for all fluids that follow a two-parameter potential energy relation such as Eq. (2.21) [40].

$$(2.22) \qquad Z = f_1(\varepsilon, \sigma, T, P)$$

$$(2.23) \qquad Z = \frac{PV_{T,P}}{RT}$$

where
Z = dimensionless compressibility factor
$V_{T,P}$ = molar volume at absolute temperature, T, and pressure, P
f_1 = universal function same for all fluids that follow Eq. (2.21).

By combining Eqs. (2.20)–(2.23) we obtain

$$(2.24) \qquad V_{T,P} = f_2(A_\circ, B_\circ, T, P)$$

where $A_\circ$ and $B_\circ$ are the two parameters in the potential energy relation, which differ from one fluid to another. Equation (2.24) is called a two-parameter EOS. Earlier EOS such as van der Waals (vdW) and Redlich–Kwong (RK) developed for simple fluids all have two parameters A and B [4] as discussed in Chapter 5. Therefore, Eq. (2.24) can also be written in terms of these two parameters:

$$(2.25) \qquad V_{T,P} = f_3(A, B, T, P)$$

The three functions f_1, f_2, and f_3 in the above equations vary in the form and style. The conditions at the critical point for any PVT relation are [41]

$$(2.26) \qquad \left(\frac{\partial P}{\partial V}\right)\bigg|_{T_c,P_c} = 0$$

$$(2.27) \qquad \left(\frac{\partial^2 P}{\partial V^2}\right)\bigg|_{T_c,P_c} = 0$$

Application of Eqs. (2.26) and (2.27) to any two-parameter EOS would result in relations for calculation of parameters A and B in terms of T_c and P_c, as shown in Chapter 5. It should be noted that EOS parameters are generally designated by lower case a and b, but here they are shown by A and B. Notation $a, b, c, \ldots$ are used for correlation parameters in various equations in this chapter. Applying Eqs. (2.26) and (2.27) to Eq. (2.25) results into the following three relations for T_c, P_c, and V_c:

$$(2.28) \qquad T_c = f_4(A, B)$$
$$(2.29) \qquad P_c = f_5(A, B)$$
$$(2.30) \qquad V_c = f_6(A, B)$$

Functions f_4, f_5, and f_6 are universal functions and are the same for all fluids that obey the potential energy relation expressed by Eq. (2.20) or Eq. (2.21). In fact, if parameters A and B in a two-parameter EOS in terms of T_c and P_c are rearranged one can obtain relations for T_c and P_c in terms of these two parameters. For example, for van der Waals and Redlich–Kwong EOS the two parameters A and B are given in terms of T_c and P_c [21] as shown in Chapter 5. By rearrangement of the vdW EOS parameters we get

$$T_c = \left(\frac{8}{27R}\right)AB^{-1} \quad P_c = \left(\frac{1}{27}\right)AB^{-2} \quad V_c = 3B$$

and for the Redlich–Kwong EOS we have

$$T_c = \left[\frac{(0.0867)^5 R}{0.4278R}\right]^{2/3} A^{2/3}B^{-2/3}$$

$$P_c = \left[\frac{(0.0867)^5 R}{(0.4278)^2}\right]^{1/3} A^{2/3}B^{-5/3} \quad V_c = 3.847B$$

Similar relations can be obtained for the parameters of other two-parameter EOS. A generalization can be made for the relations between T_c, P_c, and V_c in terms of EOS parameters A and B in the following form:

$$(2.31) \qquad [T_c, P_c, V_c] = aA^bB^c$$

where parameters a, b, and c are the constants which differ for relations for T_c, P_c, and V_c. However, these constants are the same for each critical property for all fluids that follow the same two-parameter potential energy relation. In a two-parameter EOS such as vdW or RK, V_c is related to only one parameter B so that V_c/B is a constant for all compounds. However, formulation of V_c through Eq. (2.30) shows that V_c must be a function of two parameters A and B. This is one of the reasons that two-parameters EOS are not accurate near the critical region. Further discussion on EOS is given in Chapter 5.

To find the nature of these two characterizing parameters one should realize that A and B in Eq. (2.31) represent the two parameters in the potential energy relation, such as ε and σ in Eq. (2.22). These parameters represent energy and size characteristics of molecules. The two parameters that are readily measurable for hydrocarbon systems are the boiling point, T_b, and specific gravity, SG; in fact, T_b represents the energy parameter and SG represents the size parameter. Therefore, in Eq. (2.31) one can replace parameters A and B by T_b and SG. However, it should be noted that T_b is not the same as parameter A and SG is not the same as parameter B, but it is their combination that can be replaced. There are many other parameters that may represent A and B in Eq. (2.31). For example, if Eq. (2.25) is applied at a reference state of T_0 and P_0, it can be written as

$$(2.32) \qquad V_{T_0, P_0} = f_3(A, B, T_0, P_0)$$

where V_{T_0, P_0} is the molar volume of the fluid at the reference state. The most convenient reference conditions are temperature of 20°C and pressure of 1 atm. By rearranging Eq. (2.32) one can easily see that one of the parameters A or B can be molar volume at 20°C and pressure of 1 atm [28, 36, 42].

To find another characterization parameter we may consider that for nonpolar compounds the only attractive force is the London dispersion force and it is characterized by factor polarizability, α, defined as [38, 39]

$$(2.33) \qquad \alpha = \left(\frac{3}{4\pi N_A}\right) \times \left(\frac{M}{\rho}\right) \times \left(\frac{n^2 - 1}{n^2 + 2}\right)$$

where
N_A = Avogadro's number
M = molecular weight
ρ = absolute density
n = refractive index

In fact, polarizability is proportional to molar refraction, R_m, defined as

$$(2.34) \qquad R_m = \left(\frac{M}{\rho}\right) \times \left(\frac{n^2 - 1}{n^2 + 2}\right)$$

$$(2.35) \qquad V = \frac{M}{\rho}$$

$$(2.36) \qquad I = \frac{n^2 - 1}{n^2 + 2}$$

in which V is the molar volume and I is a characterization parameter that was first used by Huang to correlate hydrocarbon properties in this way [10, 42]. By combining Eqs. (2.34)–(2.36) we get

$$(2.37) \qquad I = \frac{R_m}{V} = \frac{\text{actual molar volume of molecules}}{\text{apparent molar volume of molecules}}$$

R_m, the molar refraction, represents the actual molar volume of molecules, V represents the apparent molar volume and their ratio, and parameter I represents the fraction of total volume occupied by molecules. R_m has the unit of molar volume and I is a dimensionless parameter. R_m/M is the specific refraction and has the same unit as specific volume. Parameter I is proportional to the volume occupied by the molecules and it is close to unity for gases ($I_g \cong 0$), while for liquids it is greater than zero but less than 1 ($0 < I_{liq} < 1$). Parameter I can represent molecular size, but the molar volume, V, is a parameter that characterizes the energy associated with the molecules. In fact as the molecular energy increases so does the molar volume. Therefore, both V and I can be used as two independent parameters to characterize hydrocarbon properties. Further use of molar refraction and its relation with EOS

parameters and transport properties are discussed in Chapters 5, 6, and 8. It is shown by various investigators that the ratio of T_b/T_c is a characteristic of each substance, which is related to either T_c or T_b [36, 43]. This ratio will be used to correlate properties of pure hydrocarbons in Section 2.3.3. Equation (2.31) can be written once for T_c in terms of V and I and once for parameter T_b/T_c. Upon elimination of parameter V between these two relations, a correlation can be obtained to estimate T_c from T_b and I. Similarly through elimination of T_c between the two relations, a correlation can be derived to estimate V in terms of T_b and I [42].

It should be noted that although both density and refractive index are functions of temperature, both theory and experiment have shown that the molar refraction ($R_m = VI$) is nearly independent of temperature, especially over a narrow range of temperature [38]. Since V at the reference temperature of 20°C and pressure of 1 atm is one of the characterization parameters, I at 20°C and 1 atm must be the other characterization parameter. We chose the reference state of 20°C and pressure of 1 atm because of availability of data. Similarly, any reference temperature, e.g. 25°C, at which data are available can be used for this purpose. Liquid density and refractive index of hydrocarbons at 20°C and 1 atm are indicated by d_{20} and n_{20}, respectively, where for simplicity the subscript 20 is dropped in most cases. Further discussion on refractive index and its methods of estimation are given elsewhere [35].

From this analysis it is clear that parameter I can be used as one of the parameters A or B in Eq. (2.31) to represent the size parameter, while T_b may be used to represent the energy parameter. Other characterization parameters are discussed in Section 2.3.2. In terms of boiling point and specific gravity, Eq. (2.31) can be generalized as following:

$$(2.38) \qquad \theta = a T_b^b SG^c$$

where T_b is the normal boiling point in absolute degrees (kelvin or rankine) and SG is the specific gravity at 60°F(15.5°C). Parameter θ is a characteristic property such as molecular weight, M, critical temperature, T_c, critical pressure, P_c, critical molar volume, V_c, liquid density at 20°C, d_{20}, liquid molar volume at 20°C and 1 atm, V_{20}, or refractive index parameter, I, at 20°C. It should be noted that θ must be a temperature-independent property. As mentioned before, I at 20°C and 1 atm is considered as a characteristic parameter and not a temperature-dependent property. Based on reported data in the 1977 edition of API-TDB, constants a, b, and c were determined for different properties and have been reported by Riazi and Daubert [28]. The constants were obtained through linear regression of the logarithmic form of Eq. (2.38). Equation (2.38) in its numerical form is presented in Sections 2.4–2.6 for basic characterization parameters. In other chapters, the form of Eq. (2.38) will be used to estimate the heat of vaporization and transport properties as well as interconversion of various distillation curves. The form of Eq. (2.38) for T_c is the same as the form Nokay [44] and Spencer and Daubert [45] used to correlate the critical temperature of some hydrocarbon compounds. Equation (2.38) or its modified versions (Eq. 2.42), especially for the critical properties and molecular weight, have been in use by industry for many years [2, 8, 34, 46–56]. Further application of this equation will be discussed in Section 2.9.

One should realize that Eq. (2.38) was developed based on a two-parameter potential energy relation applicable to non-polar compounds. For this reason, this equation cannot be used for systems containing polar compounds such as alcohol, water, or even some complex aromatic compounds that are considered polar. In fact constants a, b, and c given in Eq. (2.38) were obtained based on properties of hydrocarbons with carbon number ranging from C_5 to C_{20}. This is almost equivalent to the molecular weight range of 70–300. In fact molecular weight of n-C_{20} is 282, but considering the extrapolation power of Eq. (2.38) one can use this equation up to molecular weight of 300, which is roughly equivalent to boiling point of 370°C (700°F). Moreover, experimental data on the critical properties of hydrocarbons above C_{18} were not available at the time of development of Eq. (2.38). For heavier hydrocarbons additional parameters are required as will be shown in Section 2.3.3 and Chapter 3. The lower limit for the hydrocarbon range is C_5, because lighter compounds of C_1–C_4 are mainly paraffinic and in the gaseous phase at normal conditions. Equation (2.38) is mainly applied for undefined petroleum fractions that have average boiling points higher than the boiling point of C_5 as will be seen in Chapter 3. Methods of calculation of properties of natural gases are discussed in Chapter 4.

Example 2.3—Show that the molecular weight of hydrocarbons, M, can be correlated with the boiling point, T_b, and specific gravity, SG, in the form of Eq. (2.38).

Solution—It was already shown that molar volume at 20°C and 1 atm, V_{20}, can be correlated to parameters A and B of a potential energy function through Eq. (2.32) as follows: $V_{20} = g_1(A, B)$. In fact, parameter V_{20} is similar in nature to the critical molar volume, V_c, and can be correlated with T_b and SG as Eq. (2.38): $V_{20} = aT_b^b SG^c$. But $V_{20} = M/d_{20}$, where d_{20} is the liquid density at 20°C and 1 atm and is considered a size parameter. Since T_b is chosen as an energy parameter and SG is selected as a size parameter, then both d_{20} and SG represent the same parameter and can be combined with the energy parameter as $M = aT_b^b SG^c$, which has the same form of Eq. (2.38) for $\theta = M$. ♦

2.3.2 Various Characterization Parameters for Hydrocarbon Systems

Riazi and Daubert [29] did further study on the expansion of the application of Eq. (2.38) by considering various input parameters. In fact instead of T_b and SG we may consider any pair of parameters θ_1 and θ_2, which are capable of characterizing molecular energy and the molecular size. This means that Eq. (2.31) can be expressed in terms of two parameters θ_1 and θ_2:

$$(2.39) \qquad \theta = a\theta_1^b \theta_2^c$$

However, one should realize that while these two parameters are independent, they should represent molecular energy and molecular size. For example, the pairs such as (T_b, M) or (SG, I) cannot be used as the pair of input parameters (θ_1, θ_2). Both SG and I represent size characteristics of molecules and they are not a suitable characterization pair. In the development of a correlation to estimate the properties of hydrocarbons, all compounds from various hydrocarbon groups are considered. Properties of hydrocarbons vary from one hydrocarbon type to another and from one carbon number to another. Hydrocarbons and their properties can be tabulated as a matrix of four columns with many rows. Columns represent hydrocarbon families (paraffins, olefins, naphthenes, aromatics) while rows represent carbon number. Parameters such as T_b, M, or kinematic viscosity at 38°C (100°F), $\nu_{38(100)}$, vary in the vertical direction with carbon number, while parameters such as SG, I, and CH vary significantly with hydrocarbon type. This analysis is clearly shown in Table 2.4 for C_8 in paraffin and aromatic groups and properties of C_7 and C_8 within the same group of paraffin family. As is clearly shown by relative changes in various properties, parameters SG, I, and CH clearly characterize hydrocarbon type, while T_b, M, and $\nu_{38(100)}$ are good parameters to characterize the carbon number within the same family. Therefore, a correlating pair should be selected in a way that characterizes both the hydrocarbon group and the compound carbon number. A list of properties that may be used as pairs of correlating parameters (θ_1, θ_2) in Eq. (2.39) are given below [29].

θ:	T_c (K), P_c (bar), V_c (cm³/g), M, T_b (K), SG, I (20°C), CH
(θ_1, θ_2) Pairs:	(T_b, SG), (T_b, I), (T_b, CH), (M, SG), (M, I), (M, CH), $(\nu_{38(100)}, SG)$, $(\nu_{38(100)}, I)$, $(\nu_{38(100)}, CH)$

The accuracy of Eq. (2.39) was improved by modification of its a parameter in the following form:

$$(2.40) \qquad \theta = a\, \exp[b\theta_1 + c\theta_2 + d\theta_1\theta_2]\theta_1^e \theta_2^f$$

Values of constants a–f in Eq. (2.40) for various parameters of θ and pairs of (θ_1, θ_2) listed above are given in Table 2.5. It should be noted that the constants reported by Riazi–Daubert have a follow-up correction that was reported later in the same volume [29]. These constants are obtained from properties of hydrocarbons in the range of C_5–C_{20} in

TABLE 2.4—*Comparison of properties of adjacent members of paraffin family and two families of C_8 hydrocarbons.*

Hydrocarbon group	T_b, K	M	$\nu_{38(100)}$, cSt	SG	I	CH
Paraffin family						
C_7H_{16}(n-heptane)	371.6	100.2	0.5214	0.6882	0.236	5.21
C_8H_{18}(n-octane)	398.8	114.2	0.6476	0.7070	0.241	5.30
% Difference in property	+7.3	+14.2	+24.2	+2.7	+2.1	+1.7
Two Families (C_8)						
Paraffin (n-octane)	398.8	114.2	0.6476	0.7070	0.241	5.30
Aromatic (ethylbenzene)	409.4	106.2	0.6828	0.8744	0.292	9.53
% Difference in property	+2.7	−7.0	5.4	+23.7	+21.2	+79.8

TABLE 2.5—Coefficients of Eq. (2.40) $\theta = a\,\exp(b\theta_1 + c\theta_2 + d\theta_1\theta_2)\theta_1^e\,\theta_2^f$ for various properties.

θ	θ_1	θ_2	a	b	c	d	e	f	AAD%
T_c	T_b	SG	9.5232	-9.314×10^{-4}	-0.54444	6.48×10^{-4}	0.81067	0.53691	0.5
T_c	T_b	I	4.487601×10^5	-1.3171×10^{-3}	-16.9097	4.5236×10^{-3}	0.6154	4.3469	0.6
T_c	T_b	CH	8.6649	0	0	0	0.67221	0.10199	0.7
T_c	M	SG	308	-1.3478×10^{-4}	-0.61641	0	0.2998	1.0555	0.7
T_c	M	I	1.347444×10^6	2.001×10^{-4}	-13.049	0	0.2383	4.0642	0.8
T_c	M	CH	20.74	1.385×10^{-3}	-0.1379	-2.7×10^{-4}	0.3526	1.4191	1.0
T_c	$\nu_{38(100)}$	I	2.4522×10^3	-0.0291	-1.2664		0.1884	0.7492	1.0
P_c	T_b	SG	3.195846×10^5	-8.505×10^{-3}	-4.8014	5.749×10^{-3}	-0.4844	4.0846	2.7
P_c	T_b	I	8.4027×10^{23}	-0.012067	-74.5612	0.0342	-1.0303	18.4302	2.6
P_c	T_b	CH	9.858968	-3.8443×10^{-3}	-0.3454		-0.1801	3.3223	3.5
P_c	M	SG	3116.632	-1.8078×10^{-3}	-0.3084	0	-0.8063	1.6015	2.7
P_c	M	I	2.025×10^{16}	-0.01415	-48.5809	0.0451	-0.8097	12.9148	2.3
P_c	M	CH	56.26043	-2.139×10^{-3}	-0.265	0	-0.6616	2.4004	3.1
P_c	$\nu_{38(100)}$	I	393.306	0	0	0	-0.4974	2.052	4.2
V_c	T_b	SG	6.049×10^{-2}	-2.6422×10^{-3}	-0.26404	1.971×10^{-3}	0.7506	-1.2028	1.8
V_c	T_b	I	6.712×10^{-3}	-2.720×10^{-3}	0.91548	7.920×10^{-3}	0.5775	-2.1548	1.8
V_c	T_b	CH	1.409×10^{1}	-1.6594×10^{-3}	0.05345	2.6649×10^{-4}	0.1657	-1.4439	2.4
V_c	M	SG	7.529×10^{-1}	-2.657×10^{-3}	0.5287	2.6012×10^{-3}	0.20378	-1.3036	1.9
V_c	M	I	6.3429×10^{-5}	-2.0208×10^{-3}	14.1853	4.5318×10^{-3}	0.2556	-4.60413	2.1
V_c	M	CH	1.597×10^{1}	-2.3533×10^{-3}	0.1082	3.826×10^{-4}	0.0706	-1.3362	2.6
V_c	$\nu_{38(100)}$	I	2.01×10^{-10}	-0.16318	36.09011	0.4608	0.1417	-10.65067	1.7
M	$\nu_{38(100)}$	SG	1032.1	9.78×10^{-4}	-9.53384	2.0×10^{-3}	0.97476	6.51274	2.1
M	T_b	SG	8.9205×10^{-6}	1.55833×10^{-5}	4.2376	0	2.0935	-1.9985	2.3
M	T_b	I	1.81456×10^{-3}	0	0	0	1.9273	-0.2727	2.3
M	$\nu_{38(100)}$	CH	1.51723×10^6	-0.195411	-9.638970	0.162470	0.56370	6.89383	3.4
M	$\nu_{38(100)}$	I	4×10^{-9}	-8.9854×10^{-2}	38.106	0	0.6675	-10.6	3.5
T_b	M	SG	3.76587	3.77409×10^{-3}	2.984036	-4.25288×10^{-3}	4.0167×10^{-1}	-1.58262	1.0
T_b	M	I	75.775	0	0	0	0.4748	0.4283	1.1
T_b	M	CH	20.25347	-1.57415×10^{-4}	-4.5707×10^{-2}	9.22926×10^{-6}	0.512976	0.472372	1.1
T_b	$\nu_{38(100)}$	SG	2379.86	-1.3051×10^{-2}	-1.68759	-2.1247×10^{-2}	0.262914	1.3489	1.7
T_b	$\nu_{38(100)}$	I	0.050629	-6.5236×10^{-2}	14.9371	6.029×10^{-2}	0.3228	-3.8798	1.6
SG	T_b	I	2.43810×10^7	-4.19400×10^{-4}	-23.553500	3.98736×10^{-4}	-0.341800	6.9195	0.5
SG	T_b	CH	2.86706×10^{-3}	-1.83321×10^{-3}	-0.081635	6.49168×10^{-5}	0.890041	0.73238	1.0
SG	M	I	1.1284×10^6	-1.58800×10^{-3}	-20.594000	7.33400×10^{-5}	-0.07710	6.30280	0.5
SG	M	CH	6.84403×10^{-2}	-1.48844×10^{-3}	-0.079250	4.92112×10^{-5}	0.289844	0.919255	1.3
SG	$\nu_{38(100)}$	I	3.80830×10^7	-6.14060×10^{-2}	-26.393400	0.253300	-0.235300	8.04224	0.5
I	T_b	SG	0.02343	7.0294×10^{-4}	2.46832	-1.0268×10^{-3}	0.05721	-0.71990	0.5
I	T_b	CH	5.60121×10^{-3}	-1.7774×10^{-4}	-6.0737×10^{-2}	-7.9452×10^{-5}	0.44700	0.98960	0.5
I	M	SG	0.42238	3.1886×10^{-4}	-0.200996	-4.2451×10^{-4}	-0.00843	1.11782	0.5
I	M	CH	4.23900×10^{-2}	-5.6946×10^{-4}	-6.836×10^{-2}	0.0	0.16560	0.82910	0.9
I	$\nu_{38(100)}$	SG	0.263760	1.7458×10^{-2}	0.231043	-1.8441×10^{-2}	-0.01128	0.77078	0.6
I	$\nu_{38(100)}$	CH	8.71600×10^{-2}	6.1396×10^{-2}	-7.019×10^{-2}	-2.5935×10^{-3}	0.05166	0.84599	0.4
CH	T_b	SG	3.47028	1.4850×10^{-2}	16.94020	-0.012491	-2.72522	-6.79769	2.5
CH	T_b	I	8.39640×10^{-13}	7.7171×10^{-3}	71.65310	-0.02088	-1.37730	-13.61390	1.7
CH	M	SG	2.35475	9.3485×10^{-3}	4.74695	-8.0172×10^{-3}	-0.68418	-0.76820	2.6
CH	M	I	2.9004×10^{-13}	7.8276×10^{-3}	60.34840	-2.4450×10^{-2}	-0.37884	-12.34051	1.7
CH	$\nu_{38(100)}$	SG	2.52300×10^{-12}	0.48281	29.98797	-0.55768	-0.14657	-20.31303	2.5
CH	$\nu_{38(100)}$	I	2.1430×10^{-12}	0.28320	53.73160	-0.91085	-0.17158	-10.88065	1.6

Coefficients for Eq. (2.40) have been obtained from data on approximately 140 pure hydrocarbons in the molecular weight range of 70–300 and boiling point range of 300–620 K. Units: V_c in cm³/g, T_b and T_c in K, P_c in bar, and $\nu_{38(100)}$ in cSt. With permission from Ref. [29].

a similar approach as the constants of Eq. (2.38) were obtained. Similarly, Eq. (2.40) can be applied to hydrocarbon systems in the molecular weight range of 70–280, which is approximately equivalent to the boiling range of 30–350°C (~80–650°F). However, they may be used up to C_{22} or molecular weight of 300 (~boiling point 370°C) with good accuracy. In obtaining the constants, Eq. (2.40) was first converted into linear form by taking logarithm from both side of the equation and then using a linear regression program in a spreadsheet the constants were determined. The value of R^2 (index of correlation) is generally above 0.99 and in some cases near 0.999. However, when viscosity or CH parameters are used the R^2 values are lower. For this reason use of kinematic viscosity or CH weight ratio should be used as a last option when other parameters are not available. Properties of heavy hydrocarbons are discussed in the next section. When Eq. (2.40) is applied to petroleum fractions, the choice of input parameters is determined by the availability of experimental data; however, when a choice exists the following trends determine the characterizing power of input parameters used in Eq. (2.39) or (2.40): The first choice for θ_1 is T_b, followed by M, and then $\nu_{38(100)}$, while for the parameter θ_2 the first choice is SG, followed by parameters I and CH. Therefore the pair of (M, SG) is preferable to (M, CH) when the choice exists.

2.3.3 Prediction of Properties of Heavy Pure Hydrocarbons

One of the major problems in characterization of heavy petroleum fractions is the lack of sufficient methods to predict basic characteristics of heavy hydrocarbons. As mentioned in the previous section, Eqs. (2.38) or (2.40) can be applied to hydrocarbons up to molecular weight of about 300. Crude oils and reservoir fluids contain fractions with molecular weights higher than this limit. For example, products from vacuum distillation have molecular weight above this range. For such fractions application of either Eq. (2.38) or (2.40) leads to some errors that will affect the overall property of the whole crude or fluid. While similar correlations may be developed for higher molecular weight systems, experimental data are limited and most data (especially for critical properties for such compounds) are predicted values. As mentioned in the previous section, the heavy hydrocarbons are more complex and two parameters may not be sufficient to correlate properties of these compounds.

One way to characterize heavy fractions, as is discussed in the next chapter, is to model the fraction as a mixture of pseudocompounds from various homologous hydrocarbon families. In fact, within a single homologous hydrocarbon group, such as n-alkanes, only one characterization parameter is sufficient to correlate the properties. This single characterization parameter should be one of those parameters that best characterizes properties in the vertical direction such as carbon number (N_C), T_b, or M. As shown in Table 2.4, parameters SG, I_{20}, and CH weight ratio are not suitable for this purpose. Kreglewiski and Zwolinski [57] used the following relation to correlate critical temperature of n-alkanes:

$$(2.41) \qquad \ln(\theta_\infty - \theta) = a - bN_C^{2/3}$$

where θ_∞ represents value of a property such as T_c at $N_C \rightarrow \infty$, and θ is the value of T_c for the n-alkane with carbon number of

N_C. Later, this type of correlation was used by other investigators to correlate T_c and P_c for n-alkanes and alkanols [58–60]. Based on the above discussion, M or T_b may also be used instead of N_C. Equation (2.41) suggests that for extremely high molecular weight hydrocarbons ($M \rightarrow \infty$), critical temperature or pressure approaches a finite value ($T_{c\infty}$, $P_{c\infty}$). While there is no proof of the validity of this claim, the above equation shows a good capability for correlating properties of n-alkanes for the molecular weight range of interest in practical applications.

Based on Eq. (2.41), the following generalized correlation was used to characterize hydrocarbons within each homologous hydrocarbon group:

$$(2.42) \qquad \ln(\theta_\infty - \theta) = a - bM^c$$

The reason for the use of molecular weight was its availability for heavy fractions in which boiling point data may not be available due to thermal cracking. For four groups of n-alkanes, n-alkylcycopentanes, n-alkylcyclohexanes, and n-alkylbenzenes, constants in Eq. (2.42) were determined using experimental data reported in the 1988 edition of API-TDB [2] and 1986 edition of TRC [21]. The constants for T_M, T_b, SG, d_{20}, I, T_{br} (T_b/T_c), P_c, d_c, ω, and σ are given in Table 2.6 [31]. Carbon number range and absolute and average absolute deviations (AAD) for each property are also given in Table 2.6. Errors are generally low and within the accuracy of the experimental data. Equation (2.42) can be easily reversed to estimate M from T_b for different families if T_b is chosen as the characterizing parameter. Then estimated M from T_b can be used to predict other properties within the same group (family), as is shown later in this chapter. Similarly if N_C is chosen as the characterization parameter, M for each family can be estimated from N_C before using Eq. (2.42) to estimate various properties. Application and definition of surface tension are discussed in Chapter 8 (Sec 8.6).

Constants given in Table 2.6 have been obtained from the properties of pure hydrocarbons in the carbon number ranges specified. For T_M, T_b, SG, d, and I, properties of compounds up to C_{40} were available, but for the critical properties values up to C_{20} were used to obtain the numerical constants. One condition imposed in obtaining the constants of Eq. (2.42) for the critical properties was the criteria of internal consistency at atmospheric pressure. For light compounds critical temperature is greater than the boiling point ($T_{br} < 1$) and the critical pressure is greater than 1 atm ($P_c > 1.01325$ bar). However, this trend changes for very heavy compounds where the critical pressure approaches 1 atm. Actual data for the critical properties of such compounds are not available. However, theory suggests that when $P_c \rightarrow 1.01325$ bar, $T_c \rightarrow T_b$ or $T_{br} \rightarrow 1$. And for infinitely large hydrocarbons when $N_C \rightarrow \infty$ ($M \rightarrow \infty$), $P_c \rightarrow 0$. Some methods developed for prediction of critical properties of hydrocarbons lead to $T_{br} = 1$ as $N_C \rightarrow \infty$[43]. This can be true only if both T_c and T_b approach infinity as $N_C \rightarrow \infty$. The value of carbon number for the compound whose $P_c = 1$ atm is designated by N_c^*. Equation (2.42) predicts values of $T_{br} = 1$ at N_c^* for different homologous hydrocarbon groups. Values of N_c^* for different hydrocarbon groups are given in Table 2.7. In practical applications, usually values of critical properties of hydrocarbons and fractions up to C_{45} or C_{50} are needed. However, accurate prediction of critical properties at N_c^* ensures that

TABLE 2.6—*Constants of Eq. (2.42) for various parameters.*

θ	C No. Range	θ_∞	a	b	c	AAD^b	$\%AAD^b$
				Constants in Eq. (2.42)			
Constants for physical properties of *n*-alkanes [31][a]							
T_M	C_5–C_{40}	397	6.5096	0.14187	0.470	1.5	0.71
T_b	C_5–C_{40}	1070	6.98291	0.02013	2/3	0.23	0.04
SG	C_5–C_{19}	0.85	92.22793	89.82301	0.01	0.0009	0.12
d_{20}	C_5–C_{40}	0.859	88.01379	85.7446	0.01	0.0003	0.04
I	C_5–C_{40}	0.2833	87.6593	86.62167	0.01	0.00003	0.002
$T_{br} = T_b/T_c$	C_5–C_{20}	1.15	−0.41966	0.02436	0.58	0.14	0.027
$-P_c$	C_5–C_{20}	0	4.65757	0.13423	0.5	0.14	0.78
d_c	C_5–C_{20}	0.26	−3.50532	1.5×10^{-6}	2.38	0.002	0.83
$-\omega$	C_5–C_{20}	0.3	−3.06826	−1.04987	0.2	0.008	1.2
σ	C_5–C_{20}	33.2	5.29577	0.61653	0.32	0.05	0.25
Constants for physical properties of *n*-alkylcyclopentanes							
T_M	C_7–C_{41}	370	6.52504	0.04945	2/3	1.2	0.5
T_b	C_6–C_{41}	1028	6.95649	0.02239	2/3	0.3	0.05
SG	C_7–C_{25}	0.853	97.72532	95.73589	0.01	0.0001	0.02
d_{20}	C_5–C_{41}	0.857	85.1824	83.65758	0.01	0.0003	0.04
I	C_5–C_{41}	0.283	87.55238	86.97556	0.01	0.00004	0.003
$T_{br} = T_b/T_c$	C_5–C_{18}	1.2	0.06765	0.13763	0.35	1.7	0.25
$-P_c$	C_6–C_{18}	0	7.25857	1.13139	0.26	0.4	0.9
$-d_c$	C_6–C_{20}	−0.255	−3.18846	0.1658	0.5	0.0004	0.11
$-\omega$	C_6–C_{20}	0.3	−8.25682	−5.33934	0.08	0.002	0.54
σ	C_6–C_{25}	30.6	14.17595	7.02549	0.12	0.08	0.3
Constants for physical properties of *n*-alkylcyclohexane							
T_M	C_7–C_{20}	360	6.55942	0.04681	0.7	1.3	0.7
T_b	C_6–C_{20}	1100	7.00275	0.01977	2/3	1.2	0.29
SG	C_6–C_{20}	0.845	−1.51518	0.05182	0.7	0.0014	0.07
d_{20}	C_6–C_{21}	0.84	−1.58489	0.05096	0.7	0.0005	0.07
I	C_6–C_{20}	0.277	−2.45512	0.05636	0.7	0.0008	0.06
$T_{br} = T_b/T_c$	C_6–C_{20}	1.032	−0.11095	0.1363	0.4	2	0.3
$-P_c$	C_6–C_{20}	0	12.3107	5.53366	0.1	0.15	0.5
$-d_c$	C_6–C_{20}	−0.15	−1.86106	0.00662	0.8	0.0018	0.7
$-\omega$	C_7–C_{20}	0.6	−5.00861	−3.04868	0.1	0.005	1.4
σ	C_6–C_{20}	31	2.54826	0.00759	1.0	0.17	0.6
Constants for physical properties of *n*-alkylbenzenes							
T_M	C_9–C_{42}	375	6.53599	0.04912	2/3	0.88	0.38
T_b	C_6–C_{42}	1015	6.91062	0.02247	2/3	0.69	0.14
$-$SG	C_6–C_{20}	−0.8562	224.7257	218.518	0.01	0.0008	0.1
$-d_{20}$	C_6–C_{42}	−0.854	238.791	232.315	0.01	0.0003	0.037
$-I$	C_6–C_{42}	−0.2829	137.0918	135.433	0.01	0.0001	0.008
$T_{br} = T_b/T_c$	C_6–C_{20}	1.03	−0.29875	0.06814	0.5	0.83	0.12
$-P_c$	C_6–C_{20}	0	9.77968	3.07555	0.15	0.22	0.7
$-d_c$	C_6–C_{20}	−0.22	−1.43083	0.12744	0.5	0.002	0.8
$-\omega$	C_6–C_{20}	0	−14.97	−9.48345	0.08	0.003	0.68
σ	C_6–C_{20}	30.4	1.98292	−0.0142	1.0	0.4	1.7

With permission from Ref. [31].
[a] Data sources: T_M T_b, and d are taken from TRC [21]. All other properties are taken from API-TDB-1988 [2]. Units: T_M, T_b, and T_c are in K; d_{20} and d_c are in g/cm^3; P_c is in bar; σ is in dyn/cm.
[b] AD and AAD% given by Eqs. (2.134) and (2.135).

the estimated critical properties by Eq. (2.42) are realistic for hydrocarbons beyond C_{18}. This analysis is called *internal consistency* for correlations of critical properties.

In the characterization method proposed by Korsten [32, 33] it is assumed that for extremely large hydrocarbons ($N_C \to \infty$), the boiling point and critical temperature also approach infinity. However, according to Eq. (2.42) as $N_C \to \infty$ or ($M \to \infty$), properties such as T_b, SG, d, I, T_{br}, P_c, d_c,

ω, and σ all have finite values. From a physical point of view this may be true for most of these properties. However, Korsten [33] suggests that as $N_C \to \infty$, P_c and d_c approach zero while T_b, T_c, and most other properties approach infinity. Goossen [61] developed a correlation for molecular weight of heavy fractions that suggests boiling point for extremely large molecules approches a finite value of $T_{b\infty} = 1078$. In another paper [62] he shows that for infinite paraffinic chain length,

TABLE 2.7—*Prediction of atmospheric critical pressure from Eq. (2.42).*

Hydrocarbon type	N_c^* calculated at $T_b = T_c$	N_c^* calculated at $P_c = 1.01325$	Predicted P_c (bar) at $T_b = T_c$
n-Alkanes	84.4	85	1.036
n-Alkylcyclopentanes	90.1	90.1	1.01
n-Alkylcyclohexanes	210.5	209.5	1.007
n-Alkylbenzenes	158.4	158.4	1.013

With permission from Ref. [31].

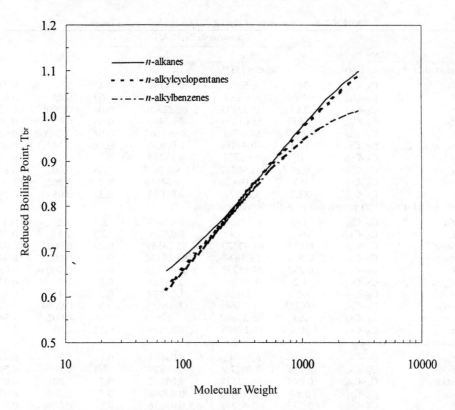

FIG. 2.1—Reduced boiling point of homologous hydrocarbon groups from Eq. (2.42).

$d_\infty = 0.8541$ and $n_\infty = 1.478$ ($I_\infty = 0.283$), while the values obtained through Eq. (2.42) (see Table 2.6) are $T_{b\infty} = 1070$, $d_\infty = 0.859$, and $I_\infty = 0.2833$. One can see how close the values are although they have been derived by two different methods. However, these values are of little practical application as long as a proposed correlation satisfies the condition of $T_{br} = 1$ at $P_c = 1.0133$ bar. Equation (2.42) will be used later in Chapter 4 to develop physical properties of single carbon number (SCN) cuts up to C_{50} for the estimation of properties of heavy crude oils and reservoir oils. Graphical presentation of Eq. (2.42) for T_{br} and T_c versus molecular weight of different hydrocarbon families is shown in Figs. 2.1 and 2.2 for molecular weights up to 3000 ($N_C \cong 214$).

One direct application of critical properties of homologous hydrocarbons is to calculate phase equilibrium calculations for wax precipitation and cloud point of reservoir fluids and crude oils as shown by Pan *et al.* [63, 64]. These investigators evaluated properties calculated through Eq. (2.42) and modified this equation for the critical pressure of PNA hydrocarbons with molecular weight above 300 through the following relation:

$$(2.43) \qquad P_c = a - b \exp(-cM)$$

where a, b, and c are given for the three hydrocarbon groups in Table 2.8 [64]. However, Eq. (2.43) does not hold the internal consistency at P_c of 1 atm, which was imposed in deriving the constants of Eq. (2.42). But this may not affect results for practical calculations as critical pressures of even the heaviest compounds do not reach to atmospheric pressure. A comparison between Eq. (2.42) and (2.43) for the critical pressure of paraffins, naphthenes, and aromatics is shown in Fig. 2.3.

Pan *et al.* [63, 64] also recommend use of the following relation for the acentric factor of aromatics for hydrocarbons with $M < 800$:

$$(2.44) \qquad \ln \omega = -36.1544 + 30.94 M^{0.026261}$$

and when $M > 800$, $\omega = 2.0$. Equation (2.42) is recommended for calculation of other thermodynamic properties based on the evaluation made on thermodynamic properties of waxes and asphaltenes [63, 64].

For homologous hydrocarbon groups, various correlations may be found suitable for the critical properties. For example, another relation that was found to be applicable to critical pressure of *n*-alkyl families is in the following form:

$$(2.45) \qquad P_c = (a + bM)^{-n}$$

where P_c is in bar and M is the molecular weight of pure hydrocarbon from a homologous group. Constant n is greater than unity and as a result as $M \to \infty$ we have $P_c \to 0$, which satisfies the general criteria for a P_c correlation. Based on data on P_c of *n*-alkanes from C_2 to C_{22}, as given in Table 2.1, it was found that $n = 1.25$, $a = 0.032688$, and $b = 0.000385$, which gives $R^2 = 0.9995$ with average deviation of 0.75% for 21 compounds. To show the degree of extrapolation of this equation, if data from C_2 to C_{10} (only nine compounds) are used to

TABLE 2.8—*Coefficients of Eq. (2.43).*

Coefficient	Paraffins	Naphthenes	Aromatics
a	0.679091	2.58854	4.85196
b	−22.1796	−27.6292	−42.9311
c	0.00284174	0.00449506	0.00561927

Taken from Pan *et al.* [63, 64].

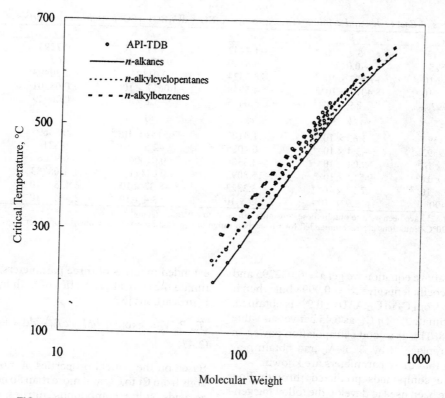

FIG. 2.2—Critical temperature of homologous hydrocarbon groups from Eq. (2.42).

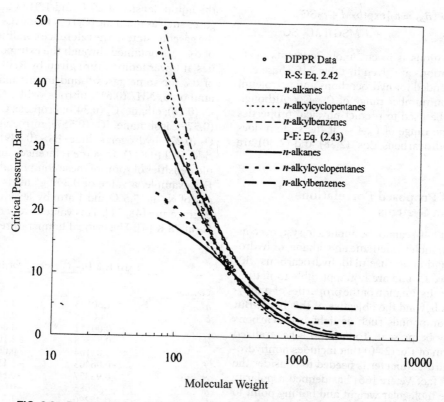

FIG. 2.3—Prediction of critical pressure of homologous hydrocarbon groups from Eqs. (2.42) and (2.43).

TABLE 2.9—*Constants in Eqs. (2.46a and 2.46b)* $\theta = a_1 \exp(b_1\theta_1 + c_1SG + d_1\theta_1SG)\theta_1^e SG^f$ *for various properties of heavy hydrocarbons.*

θ	θ_1	a_1	b_1	c_1	d_1	e_1	f_1	AAD%
T_c	T_b	35.9413	-6.9×10^{-4}	-1.4442	4.91×10^{-4}	0.7293	1.2771	0.3
P_c	T_b	6.9575	-0.0135	-0.3129	9.174×10^{-3}	0.6791	-0.6807	5.7
V_c	T_b	6.1677×10^{10}	-7.583×10^{-3}	-28.5524	0.01172	1.20493	17.2074	2.5
I	T_b	3.2709×10^{-3}	8.4377×10^{-4}	4.59487	-1.0617×10^{-3}	0.03201	-2.34887	0.1
d_{20}	T_b	0.997	2.9×10^{-4}	5.0425	-3.1×10^{-4}	-0.00929	1.01772	0.07

θ	θ_1	a_2	b_2	c_2	d_2	e_2	f_2	AAD%
T_b	M	9.3369	1.65×10^{-4}	1.4103	-7.5152×10^{-4}	0.5369	-0.7276	0.3
T_c	M	218.9592	-3.4×10^{-4}	-0.40852	-2.5×10^{-5}	0.331	0.8136	0.2
P_c	M	8.2365×10^4	-9.04×10^{-3}	-3.3304	0.01006	-0.9366	3.1353	6.2
V_c	M	9.703×10^6	-9.512×10^{-3}	-15.8092	0.01111	1.08283	10.5118	1.6
I	M	1.2419×10^{-2}	7.27×10^{-4}	3.3323	-8.87×10^{-4}	6.438×10^{-3}	-1.61166	0.2
d_{20}	M	1.04908	2.9×10^{-4}	-7.339×10^{-2}	-3.4×10^{-4}	3.484×10^{-3}	1.05015	0.09

Data generated from Eq. (2.42) have been used to obtain these constants. Units: V_c in cm^3/mol; T_c, and T_b in K; P_c in bar; d_{20} in g/cm^3 at 20°C. Equations are recommended for the carbon range of C_{20}–C_{50}; however, they may be used for the C_5–C_{20} with lesser degree of accuracy.

obtain a and b in the above equation we get $a = 0.032795$ and $b = 0.000381$. These coefficients give $R^2 = 0.9998$ but when it is used to estimate P_c from C_2 to C_{22} AAD of 0.9% is obtained. These coefficients estimate P_c of n-C_{36} as 6.45 bar versus value given in DIPPR as 6.8. This is a good extrapolation power. In Eq. (2.45) one may replace M by T_b or N_c and obtain new coefficients for cases that these parameters are known.

Properties of pure compounds predicted through Eqs. (2.42) and (2.43) have been used to develop the following generalized correlations in terms of (T_b, SG) or (M, SG) for the basic properties of heavy hydrocarbons from all hydrocarbon groups in the C_6–C_{50} range [65].

$$T_c, \ P_c, \ V_c, \ I, \ d_{20} = a_1 \left[\exp(b_1 T_b + c_1 SG + d_1 T_b SG)\right] T_b^{e_1} SG^{f_1}$$
(2.46a)

$$T_b, \ T_c, \ P_c, \ V_c, \ I, \ d_{20} = a_2[\exp(b_2 M + c_2 SG + d_2 M SG)] M^{e_2} SG^{f_2}$$
(2.46b)

where V_c in these relations is in cm^3/mol. Constants a_1–f_1 and a_2–f_2 in these relations are given in Table 2.9. These correlations are recommended for hydrocarbons and petroleum fractions in the carbon number range of C_{20}–C_{50}. Although these equations may be used to predict physical properties of hydrocarbons in the range of C_6–C_{20}, if the system does not contain heavy hydrocarbons Eqs. (2.38) and (2.40) are recommended.

2.3.4 Extension of Proposed Correlations to Nonhydrocarbon Systems

Equations (2.38) and (2.40) cannot be applied to systems containing hydrocarbons, such as methane and ethane, or hydrogen sulfide. These equations are useful for hydrocarbons with carbon numbers above C_5 and are not applicable to natural gases or refinery gases. Estimation of the properties of nonhydrocarbon systems is beyond the objective of this book. But in reservoir fluids, compounds such as light hydrocarbons or H_2S and CO_2 may be present. To develop a generalized correlation in the form of Eq. (2.40) that includes nonhydrocarbons, usually a third parameter is needed to consider the effects of polarity. In fact Vetere [66] has defined a polarity factor in terms of the molecular weight and boiling point to predict properties of polar compounds. Equation (2.40) was

extended in terms of three parameters, T_b, d_{20}, and M, to estimate the critical properties of both hydrocarbons and nonhydrocarbons [37].

$$T_c, P_c, V_c = \exp[a + bM + cT_b + dd_{20} + eT_b d_{20}] M^f T_b^{g+hM} d_{20}^i$$
(2.47)

Based on the critical properties of more than 170 hydrocarbons from C_1 to C_{18} and more than 80 nonhydrocarbons, such as acids, sulfur compounds, nitriles, oxide gases, alcohols, halogenated compounds, ethers, amines, and water, the nine parameters in Eq. (2.47) were determined and are given in Table 2.10. In using Eq. (2.47), the constant d should not be mistaken with parameter d_{20} used for liquid density at 20°C. As in the other equations in this chapter, values of T_b and T_c are in kelvin, P_c is in bar, and V_c is in cm^3/g. Parameter d_{20} is the liquid density at 20°C and 1.0133 bar in g/cm^3. For light gases such as methane (C_1) or ethane (C_2) in which they are in the gaseous state at the reference conditions, a fictitious value of d_{20} was obtained through the extrapolation of density values at lower temperature given by Reid *et al.* [4]. The values of d_{20} for some gases found in this manner are as follows: ammonia, NH_3 (0.61); nitrous oxide, N_2O (0.79); methane, C_1 (0.18); ethane, C_2 (0.343); propane, C_3 (0.5); n-butane, nC_4 (0.579); isobutane, iC_4 (0.557); nitrogen, N_2 (0.135); oxygen, O_2 (0.22); hydrogen sulfide, H_2S (0.829); and hydrogen chloride, HCl (0.837). In some references different values for liquid densities of some of these compounds have been reported. For example, a value of 0.809 g/cm^3 is reported as the density of N_2 at 15.5°C and 1 atm by several authors in reservoir engineering [48, 51]. This value is very close to the density of N_2 at 78 K [4]. The critical temperature of N_2 is 126.1 K and

TABLE 2.10—*Constants for Eq. (2.47).*

$\theta \rightarrow$ Constants	T_c, K	P_c, MPa	V_c, cm^3/g
a	1.60193	10.74145	-8.84800
b	0.00558	0.07434	-0.03632
c	-0.00112	-0.00047	-0.00547
d	-0.52398	-2.10482	0.16629
e	0.00104	0.00508	-0.00028
f	-0.06403	-1.18869	0.04660
g	0.93857	-0.66773	2.00241
h	-0.00085	-0.01154	0.00587
i	0.28290	1.53161	-0.96608

therefore at temperature 288 or 293 K it cannot be a liquid and values reported for density at these temperatures are fictitious. In any case the values given here for density of N_2, CO_2, C_1, C_2, and H_2S should not be taken as real values and they are only recommended for use in Eq. (2.47). It should be noted that d_{20} is the same as the specific gravity at 20°C in the SI system (d_4^{20}). This equation was developed based on the fact that nonhydrocarbons are mainly polar compounds and a two-parameter potential energy relation cannot represent the intermolecular forces between molecules, therefore a third parameter is needed to characterize the system. This method would be particularly useful to estimate the bulk properties of petroleum fluids containing light hydrocarbons as well as nonhydrocarbon gases. Evaluation of this method is presented in Section 2.9.

2.4 PREDICTION OF MOLECULAR WEIGHT, BOILING POINT, AND SPECIFIC GRAVITY

Molecular weight, M, boiling point, T_b, and specific gravity, SG, are perhaps the most important characterization parameters for petroleum fractions and many physical properties may be calculated from these parameters. Various methods commonly used to calculate these properties are presented here. As mentioned before, the main application of these correlations is for petroleum fractions when experimental data are not available. For pure hydrocarbons either experimental data are available or group contribution methods are used to estimate these parameters [4]. However, methods suggested in Chapter 3 to estimate properties of petroleum fractions are based on the method developed from the properties of pure hydrocarbons in this chapter.

2.4.1 Prediction of Molecular Weight

For pure hydrocarbons from homologous groups, Eq. (2.42) can be reversed to obtain the molecular weight from other properties. For example, if T_b is available, M can be estimated from the following equation:

$$(2.48) \qquad M = \left\{ \frac{1}{b}[a - \ln(T_{b\infty} - T_b)] \right\}^{1/c}$$

where values of a, b, c, and $T_{b\infty}$ are the same constants as those given in Table 2.6 for the boiling point. For example, for *n*-alkanes, M can be estimated as follows:

$$(2.49) \quad M_p = \left\{ \frac{1}{0.02013}[6.98291 - \ln(1070 - T_b)] \right\}^{3/2}$$

in which M_p is molecular weight of *n*-alkane (*n*-paraffins) whose normal boiling point is T_b. Values obtained from Eq. (2.49) are very close to molecular weight of *n*-alkanes. Similar equations can be obtained for other hydrocarbon groups by use of values given in Table 2.6. Once M is determined from T_b, then it can be used with Eq. (2.42) to obtain other properties such as specific gravity and critical constants.

2.4.1.1 Riazi–Daubert Methods

The methods developed in the previous section are commonly used to calculate molecular weight from boiling point and

specific gravity. Equation (2.38) for molecular weight is [28]

$$(2.50) \qquad M = 1.6607 \times 10^{-4} T_b^{2.1962} SG^{-1.0164}$$

This equation fails to properly predict properties for hydrocarbons above C_{25}. This equation was extensively evaluated for various coal liquid samples along with other correlations by Tsonopoulos *et al.* [34]. They recommended this equation for the estimation of the molecular weight of coal liquid fractions. Constants in Eq. (2.40) for molecular weight, as given in Table 2.5, were modified to include heavy hydrocarbons up to molecular weight of 700. The equation in terms of T_b and SG becomes

$$(2.51) \quad \begin{aligned} M = 42.965[&\exp(2.097 \times 10^{-4} T_b - 7.78712 SG \\ &+ 2.08476 \times 10^{-3} T_b SG)]T_b^{1.26007} SG^{4.98308} \end{aligned}$$

This equation can be applied to hydrocarbons with molecular weight ranging from 70 to 700, which is nearly equivalent to boiling point range of 300–850 K (90–1050°F), and the API gravity range of 14.4–93. These equations can be easily converted in terms of Watson K factor (K_W) and API degrees using their definitions through Eqs. (2.13) and (2.4). A graphical presentation of Eq. (2.51) is shown in Fig. 2.4. (Equation (2.51) has been recommended by the API as it will be discussed later.) Equation (2.51) is more accurate for light fractions ($M < 300$) with an %AAD of about 3.5, but for heavier fractions the %AAD is about 4.7. This equation is included in the API-TDB [2] and is recognized as the standard method of estimating molecular weight of petroleum fractions in the industry.

For heavy petroleum fractions boiling point may not be available. For this reason Riazi and Daubert [67] developed a three-parameter correlation in terms of kinematic viscosity based on the molecular weight of heavy fractions in the range of 200–800:

$$(2.52) \quad M = 223.56 \left[\nu_{38(100)}^{(-1.2435+1.1228 SG)} \nu_{99(210)}^{(3.4758-3.038 SG)} \right] SG^{-0.6665}$$

The three input parameters are kinematic viscosities (in cSt) at 38 and 98.9°C (100 and 210°F) shown by $\nu_{38(100)}$ and $\nu_{99(210)}$, respectively, and the specific gravity, SG, at 15.5°C. It should be noted that viscosities at two different temperatures represent two independent parameters: one the value of viscosity and the other the effect of temperature on viscosity, which is another characteristic of a compound as discussed in Chapter 3. The use of a third parameter is needed to characterize complexity of heavy hydrocarbons that follow a three-parameter potential energy relation. Equation (2.52) is only recommended when the boiling point is not available. In a case where specific gravity is not available, a method is proposed in Section 2.4.3 to estimate it from viscosity data. Graphical presentation of Eq. (2.52) is shown in Fig. 2.5 in terms of API gravity. To use this figure, based on the value of $\nu_{38(100)}$ a point is determined on the vertical line, then from values of $\nu_{99(210)}$ and SG, another point on the chart is specified. A line that connects these two points intersects with the line of molecular weight where it may be read as the estimated value.

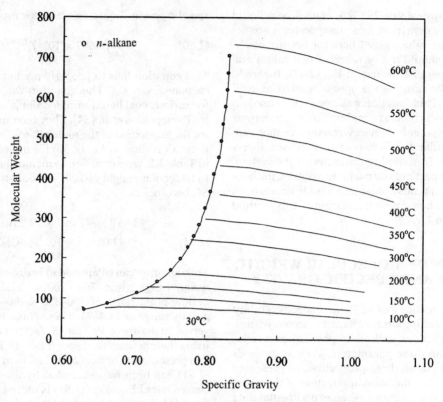

FIG. 2.4—Estimation of molecular weight from Eq. (2.51).

2.4.1.2 ASTM Method

ASTM D 2502 method [68] provides a chart to calculate the molecular weight of viscous oils using the kinematic viscosities measured at 100°F (38°C) and 210°F (99°C). The method was empirically developed by Hirschler in 1946 [69] and is presented by the following equation.

$$(2.53) \qquad M = 180 + K(H_{38(100)} + 60)$$

where

$$K = 4.145 - 1.733 \log_{10}(\text{VSF} - 145)$$
$$\text{VSF} = H_{38(100)} - H_{99(210)}$$
$$H = 870 \log_{10}[\log_{10}(\nu + 0.6)] + 154$$

in which ν is the kinematic viscosity in cSt. This equation was developed some 60 years ago and requires kinematic viscosities at 38 and 99°C in cSt as the only input parameters. The Hirschler method was included in the API-TDB in 1964 [2], but in the 1987 revision of API-TDB it was replaced by Eq. (2.52). Riazi and Daubert [67] extensively compared Eq. (2.52) with the Hirschler method and they found that for some 160 fractions in the molecular weight range of 200–800 the percent average absolute deviation (%AAD) for these methods were 2.7% and 6.9%, respectively. Even if the constants of the Hirschler correlation were reobtained from the data bank used for the evaluations, the accuracy of the method improved only slightly from 6.9 to 6.1% [67].

Example 2.4—The viscosity and other properties of 5-*n*-butyldocosane, $C_{26}H_{54}$, as given in API RP-42 [18] are $M = 366.7$, $SG = 0.8099$, $\nu_{38(100)} = 11.44$, and $\nu_{99(210)} = 3.02$ cSt. Calculate the molecular weight with %AD from the API method, Eq. (2.52), and the Hirschler method (ASTM 2502), Eq. (2.53).

Solution—In using Eq. (2.52) three parameters of $\nu_{38(100)}$, $\nu_{99(210)}$, and SG are needed.
$M = 223.56 \times [11.44^{(-1.2435+1.1228\times0.8099)} \times 3.02^{(3.4758-3.038\times0.8099)}]$
$\times (0.8099^{-0.6665}) = 350.2$, %AD = 4.5%. From Eq. (2.53):
$H_{38} = 183.3$, $H_{99} = -65.87$, VSF = 249.17, $K = 0.6483$, $M = 337.7$, %AD = 7.9%. ♦

2.4.1.3 API Methods

The API-TDB [2] adopted methods developed by Riazi and Daubert for the estimation of the molecular weight of hydrocarbon systems. In the 1982 edition of API-TDB, a modified version of Eq. (2.38) was included, but in its latest editions (from 1987 to 1997) Eqs. (2.51) and (2.52) are included after recommendations made by the API-TDB Committee.

2.4.1.4 Lee—Kesler Method

The molecular weight is related to boiling point and specific gravity through an empirical relation as follows [13]:

$$
\begin{aligned}
M = {}& -12272.6 + 9486.4\text{SG} + (8.3741 - 5.9917\text{SG})T_b \\
& + (1 - 0.77084\text{SG} - 0.02058\text{SG}^2) \\
& \times (0.7465 - 222.466/T_b)10^7/T_b \\
& + (1 - 0.80882\text{SG} + 0.02226\text{SG}^2) \\
& \times (0.3228 - 17.335/T_b)10^{12}/T_b^3
\end{aligned}
$$

(2.54)

High-molecular-weight data were also used in obtaining the constants. The correlation is recommended for use up to a boiling point of about 750 K (~850°F). Its evaluation is shown in Section 2.9.

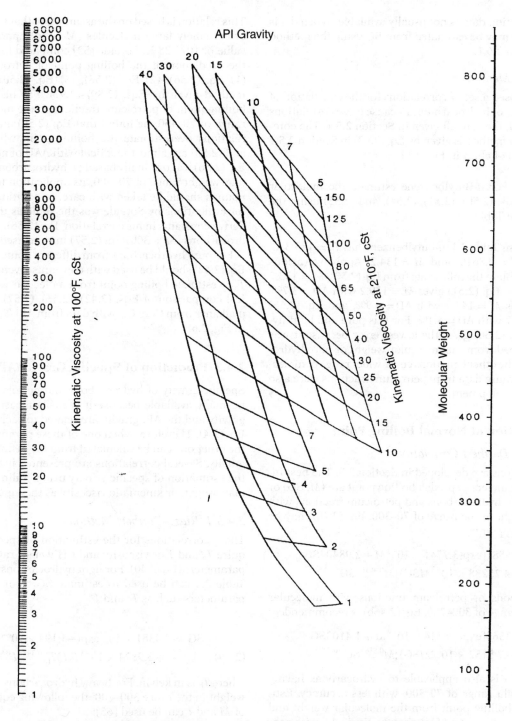

FIG. 2.5—Estimation of molecular weight from Eq. (2.52). Taken from Ref. [67] with permission.

2.4.1.5 Goossens Correlation

Most recently Goossens [61] correlated M to T_b and d_{20} in the following form using the data on 40 pure hydrocarbons and 23 petroleum fractions:

$$(2.55) \qquad M = 0.01077 T_b^\beta / d_4^{20}$$

where $\beta = 1.52869 + 0.06486 \ln[T_b/(1078 - T_b)]$. Inspection of this equation shows that it has the same structure as

Eq. (2.38) but with a variable b and $c = -1$. Parameter b is considered as a function of T_b, while SG in Eq. (2.38) is replaced by d_4^{20} the specific gravity at 20/4°C (d_4^{20} is the same as d_{20} in g/cm³). The data bank used to develop this equation covers the carbon range of C_5–C_{120} ($M \sim 70$–1700, $T_b \sim$ 300–1000 K, and $d \sim 0.63$–1.08). For the same 63 data points used to obtain the constants of Eq. (2.55), the average error was 2.1%. However, practical application of Eq. (2.55) is limited to much lower molecular weight fractions because heavy fraction distillation data is not usually available. When d_{20} is not available it may be estimated from SG using the method given in Section 2.6.1.

2.4.1.6 Other Methods

Twu [30] proposed a set of correlations for the calculation of M, T_c, P_c, and V_c of hydrocarbons. Because these correlations are interrelated, they are all given in Section 2.5.1. The computerized Winn method is given by Eq. (2.93) in Section 2.5.1 and in the form of chart in Fig. 2.12.

Example 2.5—For *n*-Butylbenzene estimate the molecular weight from Eqs. (2.50), (2.51), (2.54), and (2.55) using the input data from Table 2.1

Solution—From Table 2.1, *n*-butylbenzene has $T_b = 183.3°C$, SG = 0.8660, $d = 0.8610$, and $M = 134.2$. Applying various equations we obtain the following: from Eq. (2.50), $M = 133.2$ with AD = 0.8%, Eq. (2.51) gives $M = 139.2$ with AD = 3.7%, Eq. (2.54) gives $M = 143.4$ with AD = 6.9%, and Eq. (2.55) gives $M = 128.7$ with AD = 4.1%. For this pure and light hydrocarbon, Eq. (2.50) gives the lowest error because it was mainly developed from the molecular weight of pure hydrocarbons while the other equations cover wider range of molecular weight because data from petroleum fractions were also used in their development. ♦

2.4.2 Prediction of Normal Boiling Point

2.4.2.1 Riazi–Daubert Correlations

These correlations are developed in Section 2.3. The best input pair of parameters to predict boiling point are (M, SG) or (M, I). For light hydrocarbons and petroleum fractions with molecular weight in the range of 70–300, Eq. (2.40) may be used for boiling point:

$$T_b = 3.76587[\exp(3.7741 \times 10^{-3}M + 2.98404SG$$
$$(2.56) \quad - 4.25288 \times 10^{-3}MSG)]M^{0.40167}SG^{-1.58262}$$

For hydrocarbons or petroleum fractions with molecular weight in the range of 300–700, Eq. (2.46b) is recommended:

$$T_b = 9.3369[\exp(1.6514 \times 10^{-4}M + 1.4103SG$$
$$(2.57) \quad - 7.5152 \times 10^{-4}MSG)]M^{0.5369}SG^{-0.7276}$$

Equation (2.57) is also applicable to hydrocarbons having molecular weight range of 70–300, with less accuracy. Estimation of the boiling point from the molecular weight and refractive index parameter (I) is given by Eq. (2.40) with constants in Table 2.5. The boiling point may also be calculated through K_W and API gravity by using definitions of these parameters given in Eqs. (2.13) and (2.4).

2.4.2.2 Soreide Correlation

Based on extension of Eq. (2.56) and data on the boiling point of some C_{7+} fractions, Soreide [51, 52] developed the following correlation for the normal boiling point of fractions in the range of 90–560°C.

$$T_b = 1071.28 - 9.417 \times 10^4 \exp(-4.922 \times 10^{-3}M$$
$$(2.58) \quad - 4.7685SG + 3.462 \times 10^{-3}MSG) M^{-0.03522}SG^{3.266}$$

This relation is based on the assumption that the boiling point of extremely large molecules ($M \to \infty$) approaches a finite value of 1071.28 K. Soreide [52] compared four methods for the prediction of the boiling point of petroleum fractions: (1) Eq. (2.56), (2) Eq. (2.58), (3) Eq. (2.50), and (4) Twu method given by Eqs. (2.89)–(2.92). For his data bank on boiling point of petroleum fractions in the molecular weight range of 70–450, he found that Eq. (2.50) and the Twu correlations overestimate the boiling point while Eqs. (2.56) and (2.58) are almost identical with AAD of about 1%. Since Eq. (2.56) was originally based on hydrocarbons with a molecular weight range of 70–300, its application to heavier compounds should be taken with care. In addition, the database for evaluations by Soreide was the same as the data used to derive constants in his correlation, Eq. (2.58). For heavier hydrocarbons ($M > 300$) Eq. (2.57) may be used.

For pure hydrocarbons from different homologous families Eq. (2.42) should be used with constants given in Table 2.6 for T_b to estimate boiling point from molecular weight. A graphical comparison of Eqs. (2.42), (2.56), (2.57), and (2.58) for *n*-alkanes from C_5 to C_{36} with data from API-TDB [2] is shown in Fig. 2.6.

2.4.3 Prediction of Specific Gravity/API Gravity

Specific gravity of hydrocarbons and petroleum fractions is normally available because it is easily measurable. Specific gravity and the API gravity are related to each other through Eq. (2.4). Therefore, when one of these parameters is known the other one can be calculated from the definition of the API gravity. Several correlations are presented in this section for the estimation of specific gravity using boiling point, molecular weight, or kinematic viscosity as the input parameters.

2.4.3.1 Riazi–Daubert Methods

These correlations for the estimation of specific gravity require T_b and I or viscosity and CH weight ratio as the input parameters (Eq. 2.40). For light hydrocarbons, Eq. (2.40) and Table 2.5 can be used to estimate SG from different input parameters such as T_b and I.

$$SG = 2.4381 \times 10^7 \exp(-4.194 \times 10^{-4}T_b - 23.5535I$$
$$(2.59) \quad + 3.9874 \times 10^{-3}T_bI)T_b^{-0.3418}I^{6.9195}$$

where T_b is in kelvin. For heavy hydrocarbons with molecular weight in the range 300–700, the following equation in terms of M and I can be used [65]:

$$SG = 3.3131 \times 10^4 \exp(-8.77 \times 10^{-4}M - 15.0496I$$
$$(2.60) \quad + 3.247 \times 10^{-3}MI)M^{-0.01153}I^{4.9557}$$

Usually for heavy fractions, T_b is not available and for this reason, M and I are used as the input parameters. This equation also may be used for hydrocarbons below molecular weight of 300, if necessary. The accuracy of this equation is about 0.4 %AAD for 130 hydrocarbons in the carbon number range of C_7–C_{50} ($M \sim 70$–700).

For heavier fractions (molecular weight from 200 to 800) and especially when the boiling point is not available the following relation in terms of kinematic viscosities developed by

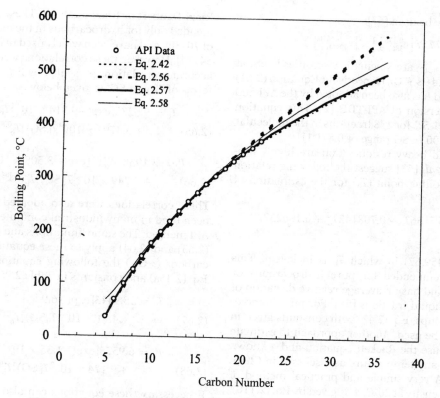

FIG. 2.6—Estimation of boiling point of *n*-alkanes from various methods.

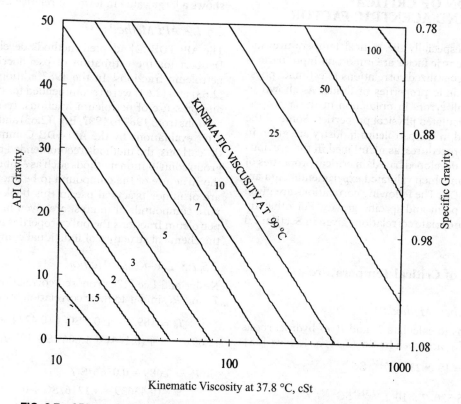

FIG. 2.7—API gravity and viscosity of heavy hydrocarbon fractions by Eq. (2.61).

Riazi and Daubert may be used [67]:

$$(2.61) \qquad SG = 0.7717 \left[\nu_{38(100)}^{0.1157} \right] \times \left[\nu_{99(210)}^{-0.1616} \right]$$

in which $\nu_{38(100)}$ and $\nu_{99(210)}$ are kinematic viscosities in cSt at 100 and 210°F (37.8 and 98.9°C), respectively. Equation (2.61) is shown in Fig. 2.7 and has also been adopted by the API and is included in 1987 version of API-TDB [2]. This equation gives an AAD of about 1.5% for 158 fractions in the molecular weight range of 200–500 (~SG range of 0.8–1.1).

For coal liquids and heavy residues that are highly aromatic, Tsonopoulos *et al.* [58] suggest the following relation in terms of normal boiling point (T_b) for the estimation of specific gravity.

$$SG = 0.553461 + 1.15156 T_\circ - 0.708142 T_\circ^2 + 0.196237 T_\circ^3$$
$$(2.62)$$

where $T_\circ = (1.8 T_b - 459.67)$ in which T_b is in kelvin. This equation is not recommended for pure hydrocarbons or petroleum fractions and has an average relative deviation of about 2.5% for coal liquid fractions [58]. For pure homologous hydrocarbon groups, Eq. (2.42) with constants given in Table 2.6 for SG can be used. Another approach to estimate specific gravity is to use the Rackett equation and a known density data point at any temperature as discussed in Chapter 5 (Section 5.8). A very simple and practical method of estimating SG from density at 20°C, d, is given by Eq. (2.110), which will be discussed in Section 2.6.1. Once SG is estimated the API gravity can be calculated from its definition, i.e., Eq. (2.4).

2.5 PREDICTION OF CRITICAL PROPERTIES AND ACENTRIC FACTOR

Critical properties, especially the critical temperature and pressure, and the acentric factor are important input parameters for EOS and generalized correlations to estimate physical and thermodynamic properties of fluids. As shown in Chapter 1 even small errors in prediction of these properties greatly affect calculated physical properties. Some of the methods widely used in the petroleum industry are given in this section. These procedures, as mentioned in the previous sections, are mainly developed based on critical properties of pure hydrocarbons in which validated experimental data are available only up to C_{18}. The following correlations are given in terms of boiling point and specific gravity. For other input parameters, appropriate correlations given in Section 2.3 should be used.

2.5.1 Prediction of Critical Temperature and Pressure

2.5.1.1 Riazi–Daubert Methods

Simplified equations to calculate T_c and P_c of hydrocarbons in the range of C_5–C_{20} are given by Eq. (2.38) as follows [28].

$$(2.63) \qquad T_c = 19.06232 T_b^{0.58848} SG^{0.3596}$$

$$(2.64) \qquad P_c = 5.53027 \times 10^7 T_b^{-2.3125} SG^{2.3201}$$

where T_c and T_b are in kelvin and P_c is in bar. In the literature, Eqs. (2.50), (2.63), and (2.64) are usually referred to as

Riazi-Daubert or Riazi methods. These equations are recommended only for hydrocarbons in the molecular weight range of 70–300 and have been widely used in industry [2, 47, 49, 51, 54, 70]. However, these correlations were replaced with more accurate correlations presented by Eq. (2.40) and Table 2.5 in terms of T_b and SG as given below:

$$T_c = 9.5233[\exp(-9.314 \times 10^{-4} T_b - 0.544442 SG$$
$$(2.65) \qquad + 6.4791 \times 10^{-4} T_b SG)] T_b^{0.81067} SG^{0.53691}$$

$$P_c = 3.1958 \times 10^5 [\exp(-8.505 \times 10^{-3} T_b - 4.8014 SG$$
$$(2.66) \qquad + 5.749 \times 10^{-3} T_b SG)] T_b^{-0.4844} SG^{4.0846}$$

These correlations were also adopted by the API and have been used in many industrial computer softwares under the API method. The same limitations and units as those for Eqs. (2.63) and (2.64) apply to these equations. For heavy hydrocarbons (>C_{20}) the following equations are obtained from Eq. (2.46a) and constants in Table 2.9:

$$T_c = 35.9413[\exp(-6.9 \times 10^{-4} T_b - 1.4442 SG$$
$$(2.67) \qquad + 4.91 \times 10^{-4} T_b SG)] T_b^{0.7293} SG^{1.2771}$$

$$P_c = 6.9575[\exp(-1.35 \times 10^{-2} T_b - 0.3129 SG$$
$$(2.68) \qquad + 9.174 \times 10^{-3} T_b SG)] T_b^{0.6791} SG^{-0.6807}$$

If necessary these equations can also be used for hydrocarbons in the range of C_5–C_{20} with good accuracy. Equation (2.67) predicts values of T_c from C_5 to C_{50} with %AAD of 0.4%, but Eq. (2.68) predicts P_c with AAD of 5.8%. The reason for this high average error is low values of P_c (i.e, a few bars) at higher carbon numbers which even a small absolute deviation shows a large value in terms of relative deviation.

2.5.1.2 API Methods

The API-TDB [2] adopted methods developed by Riazi and Daubert for the estimation of pseudocritical properties of petroleum fractions. In the 1982 edition of API-TDB, Eqs. (2.63) and (2.64) were recommended for critical temperature and pressure of petroleum fractions, respectively, but in its editions from 1987 to 1997, Eqs. (2.65) and (2.66) are included after evaluations by the API-TDB Committee. For pure hydrocarbons, the methods recommended by API are based on group contribution methods such as Ambrose, which requires the structure of the compound to be known. These methods are of minor practical use in this book since properties of pure compounds of interest are given in Section 2.3 and for petroleum fractions the bulk properties are used rather than the chemical structure of individual compounds.

2.5.1.3 Lee–Kesler Method

Kesler and Lee [12] proposed correlations for estimation of T_c and P_c similar to their correlation for molecular weight.

$$T_c = 189.8 + 450.6\,SG + (0.4244 + 0.1174\,SG)T_b$$
$$(2.69) \qquad + (0.1441 - 1.0069\,SG)10^5/T_b$$

$$\ln P_c = 5.689 - 0.0566/SG$$
$$\qquad - (0.43639 + 4.1216/SG + 0.21343/SG^2) \times 10^{-3} T_b$$
$$(2.70) \qquad + (0.47579 + 1.182/SG + 0.15302/SG^2) \times 10^{-6} \times T_b^2$$
$$\qquad - (2.4505 + 9.9099/SG^2) \times 10^{-10} \times T_b^3$$

where T_b and T_c are in kelvin and P_c is in bar. In these equations attempts were made to keep internal consistency among T_c and P_c that at P_c equal to 1 atm, T_c is coincided with normal boiling point, T_b. The correlations were recommended by the authors for the molecular range of 70–700 ($\sim C_5$–C_{50}). However, the values of T_c and P_c for compounds with carbon numbers greater than C_{18} used to develop the above correlations were not based on experimental evidence.

2.5.1.4 Cavett Method

Cavett [26] developed empirical correlations for T_c and P_c in terms of boiling point and API gravity, which are still available in some process simulators as an option and in some cases give good estimates of T_c and P_c for light to middle distillate petroleum fractions.

$$
\begin{aligned}
T_c = {} & 426.7062278 + (9.5187183 \times 10^{-1})(1.8T_b - 459.67) \\
& - (6.01889 \times 10^{-4})(1.8T_b - 459.67)^2 \\
& - (4.95625 \times 10^{-3})(\text{API})(1.8T_b - 459.67) \\
& + (2.160588 \times 10^{-7})(1.8T_b - 459.67)^3 \\
& + (2.949718 \times 10^{-6})(\text{API})(1.8T_b - 459.67)^2 \\
& + (1.817311 \times 10^{-8})(\text{API}^2)(1.8T_b - 459.67)^2
\end{aligned}
$$

(2.71)

$$
\begin{aligned}
\log(P_c) = {} & 1.6675956 + (9.412011 \times 10^{-4})(1.8T_b - 459.67) \\
& - (3.047475 \times 10^{-6})(1.8T_b - 459.67)^2 \\
& - (2.087611 \times 10^{-5})(\text{API})(1.8T_b - 459.67) \\
& + (1.5184103 \times 10^{-9})(1.8T_b - 459.67)^3 \\
& + (1.1047899 \times 10^{-8})(\text{API})(1.8T_b - 459.67)^2 \\
& - (4.8271599 \times 10^{-8})(\text{API}^2)(1.8T_b - 459.67) \\
& + (1.3949619 \times 10^{-10})(\text{API}^2)(1.8T_b - 459.67)^2
\end{aligned}
$$

(2.72)

In these relations P_c is in bar while T_c and T_b are in kelvin and the API gravity is defined in terms of specific gravity through Eq. (2.4). Terms $(1.8T_b - 459.67)$ come from the fact that the unit of T_b in the original relations was in degrees fahrenheit.

2.5.1.5 Twu Method for T_c, P_c, V_c, and M

Twu [30] initially correlated critical properties (T_c, P_c, V_c), specific gravity (SG), and molecular weight (M) of n-alkanes to the boiling point (T_b). Then the difference between specific gravity of a hydrocarbon from other groups (SG) and specific gravity of n-alkane (SG°) was used as the second parameter to correlate properties of hydrocarbons from different groups. This type of correlation, known as a perturbation expansion, was first introduced by Kesler–Lee–Sandler (KLS) [71] and later used by Lin and Chao [72] to correlate critical properties of hydrocarbons using n-alkane as a reference fluid and the specific gravity difference as the correlating parameter. However, KLS correlations did not find practical application because they defined a new third parameter similar to the acentric factor which is not available for petroleum mixtures. Lin and Chao (LC) correlated T_c, $\ln(P_c)$, ω, SG, and T_b of n-alkanes from C_1 to C_{20} to molecular weight, M. These properties for all other hydrocarbons in the same molecular weight were correlated to the difference in T_b and SG of the substance of interest with that of n-alkane. Therefore, LC correlations require three input parameters of T_b, SG, and M for each property. Each correlation for each property contained as many as 33 numerical constants. These correlations are

included in some references [49]. However, the Twu correlations although based on the same format as the KLS or LC require input parameters of T_b and SG and are applicable to hydrocarbons beyond C_{20}. For heavy hydrocarbons similar to the approach of Lee–Kesler [12], Twu [30] used the critical properties back calculated from vapor pressure data to expand his data bank on the critical constants of pure hydrocarbon compounds. For this reason the Twu correlations have found a wider range of application. The Twu correlations for the critical properties, specific gravity, and molecular weight of n-alkanes are as follows:

$$
\begin{aligned}
T_c° = {} & T_b(0.533272 + 0.34383 \times 10^{-3} \times T_b \\
& + 2.52617 \times 10^{-7} \times T_b^2 - 1.658481 \times 10^{-10} \times T_b^3 \\
& + 4.60773 \times 10^{24} \times T_b^{-13})^{-1}
\end{aligned}
$$

(2.73)

$$
\alpha = 1 - T_b/T_c°
$$

(2.74)

$$
\begin{aligned}
P_c° = {} & (1.00661 + 0.31412\alpha^{1/2} + 9.16106\alpha \\
& + 9.5041\alpha^2 + 27.35886\alpha^4)^2
\end{aligned}
$$

(2.75)

$$
V_c° = (0.34602 + 0.30171\alpha + 0.93307\alpha^3 + 5655.414\alpha^{14})^{-8}
$$

(2.76)

$$
\text{SG}° = 0.843593 - 0.128624\alpha - 3.36159\alpha^3 - 13749.5\alpha^{12}
$$

(2.77)

$$
\begin{aligned}
T_b = {} & \exp(5.12640 + 2.71579\beta - 0.286590\beta^2 - 39.8544/\beta \\
& - 0.122488/\beta^2) - 13.7512\beta + 19.6197\beta^2
\end{aligned}
$$

(2.78)

where T_b is the boiling point of hydrocarbons in kelvin and $\beta = \ln(M°)$ in which $M°$ is the molecular weight n-alkane reference compound. Critical pressure is in bar and critical volume is in cm³/mol. Data on the properties of n-alkanes from C_1 to C_{100} were used to obtain the constants in the above relations. For heavy hydrocarbons beyond C_{20}, the values of the critical properties obtained from vapor pressure data were used to obtain the constants. The author of these correlations also indicates that there is internal consistency between T_c and P_c as the critical temperature approaches the boiling point. Equation (2.78) is implicit in calculating $M°$ from T_b. To solve this equation by iteration a starting value can be found from the following relation:

(2.79)
$$
M° = T_b/(5.8 - 0.0052T_b)
$$

For other hydrocarbons and petroleum fractions the relation for the estimation of T_c, P_c, V_c, and M are as follows:

Critical temperature

(2.80)
$$
T_c = T_c°[(1 + 2f_T)/(1 - 2f_T)]^2
$$

$$
\begin{aligned}
f_T = {} & \Delta \text{SG}_T[-0.27016/T_b^{1/2} \\
& + (0.0398285 - 0.706691/T_b^{1/2})\Delta \text{SG}_T]
\end{aligned}
$$

(2.81)

(2.82)
$$
\Delta \text{SG}_T = \exp[5(\text{SG}° - \text{SG})] - 1
$$

Critical volume

(2.83)
$$
V_c = V_c°[(1 + 2f_V)/(1 - 2f_V)]^2
$$

$$f_V = \Delta SG_V [0.347776/T_b^{1/2}$$
$$(2.84) \qquad + (-0.182421 + 2.248896/T_b^{1/2})\Delta SG_V]$$

$$(2.85) \qquad \Delta SG_V = \exp[4(SG^{\circ 2} - SG^2)] - 1$$

Critical pressure

$$(2.86) \quad P_c = P_c^{\circ}(T_c/T_c^{\circ}) \times (V_c^{\circ}/V_c)[(1 + 2f_P)/(1 - 2f_P)]^2$$

$$f_P = \Delta SG_P[(2.53262 - 34.4321/T_b^{1/2} - 2.30193T_b/1000)$$
$$(2.87) \qquad + (-11.4277 + 187.934/T_b^{1/2} + 4.11963T_b/1000)\Delta SG_P]$$

$$(2.88) \qquad \Delta SG_P = \exp[0.5(SG^{\circ} - SG)] - 1$$

Molecular weight

$$(2.89) \qquad \ln(M) = (\ln M^{\circ})[(1 + 2f_M)/(1 - 2f_M)]^2$$

$$f_M = \Delta SG_M[\chi + (-0.0175691 + 0.143979/T_b^{1/2})\Delta SG_M]$$
$$(2.90)$$

$$(2.91) \qquad \chi = |0.012342 - 0.244541/T_b^{1/2}|$$

$$(2.92) \qquad \Delta SG_M = \exp[5(SG^{\circ} - SG)] - 1$$

In the above relations T_b and T_c are in kelvin, V_c is in cm^3/mol, and P_c is in bar. One can see that these correlations should be solved simultaneously because they are highly interrelated to each other and for this reason relations for estimation of M and V_c based on this method are also presented in this part.

Example 2.6—Estimate the molecular weight of *n*-eicosane ($C_{20}H_{42}$) from its normal boiling point using Eq. (2.49) and the Twu correlations.

Solution—*n*-Eicosane is a normal paraffin whose molecular weight and boiling point are given in Table 2.1 as $M = 282.55$ and $T_b = 616.93$ K. Substituting T_b in Eq. (2.49) gives $M = 282.59$ (%AD = 0.01%). Using the Twu method, first an initial guess is calculated through Eq. (2.79) as $M^{\circ} = 238$ and from iteration the final value of M° calculated from Eq. (2.78) is 281.2 (%AD = 0.48%). Twu method for estimation of properties of hydrocarbons from other groups is shown later in the next example. ♦

2.5.1.6 Winn–Mobil Method

Winn [25] developed a convenient nomograph to estimate various physical properties including molecular weight and the pseudocritical pressure for petroleum fractions. Mobil [73] proposed a similar nomograph for the estimation of pseudocritical temperature. The input data in both nomographs are boiling point (or K_W) and the specific gravity (or API gravity). As part of the API project to computerize the graphical methods for estimation of physical properties, these nomographs were reduced to equation forms for computer applications by Riazi [36] and were later reported by Sim and Daubert [74]. These empirically developed correlations have forms similar to Eq. (2.38) and for M, T_c, and P_c are as follows.

$$(2.93) \qquad M = 2.70579 \times 10^{-5} T_b^{2.4966} SG^{-1.174}$$

$$(2.94) \qquad \ln T_c = -0.58779 + 4.2009 T_b^{0.08615} SG^{0.04614}$$

$$(2.95) \qquad P_c = 6.148341 \times 10^7 T_b^{-2.3177} SG^{2.4853}$$

where T_b and T_c are in kelvin and P_c is in bar. Comparing values estimated from these correlations with the values from the original figures gives AAD of 2, 1, and 1.5% for M, T_c, and P_c, respectively, as reported in Ref. [36]. In the literature these equations are usually referred as Winn or Sim–Daubert and are included in some process simulators. The original Winn nomograph for molecular weight and some other properties is given in Section 2.8.

2.5.1.7 Tsonopoulos Correlations

Based on the critical properties of aromatic compounds, Tsonopoulos *et al.* [34] proposed the following correlations for estimation of T_c and P_c for coal liquids and aromatic-rich fractions.

$$\log_{10} T_c = 1.20016 + 0.61954(\log_{10} T_b)$$
$$(2.96) \qquad + 0.48262(\log_{10} SG) + 0.67365(\log_{10} SG)^2$$

$$\log_{10} P_c = 7.37498 - 2.15833(\log_{10} T_b)$$
$$(2.97) \qquad + 3.35417(\log_{10} SG) + 5.64019(\log_{10} SG)^2$$

where T_b and T_c are in kelvin and P_c is in bar. These correlations are mainly recommended for coal liquid fractions and they give average errors of 0.7 and 3.5% for the estimation of critical temperature and pressure of aromatic hydrocarbons.

2.5.2 Prediction of Critical Volume

Critical volume, V_c, is the third critical property that is not directly used in EOS calculations, but is indirectly used to estimate interaction parameters (k_{ij}) needed for calculation of mixture pseudocritical properties or EOS parameters as will be discussed in Chapter 5. In some corresponding state correlations developed to estimate transport properties of fluids at elevated pressure, reduced density (V_c/V) is used as the correlating parameter and values of V_c are required as shown in Chapter 8. Critical volume is also used to calculate critical compressibility factor, Z_c, as shown by Eq. (2.8).

2.5.2.1 Riazi–Daubert Methods

A simplified equation to calculate V_c of hydrocarbons in the range of C_5–C_{20} is given by Eq. (2.38) as follows.

$$(2.98) \qquad V_c = 1.7842 \times 10^{-4} T_b^{2.3829} SG^{-1.683}$$

in which V_c is in cm^3/mol and T_b is in kelvin. When evaluated against more than 100 pure hydrocarbons in the carbon range of C_5–C_{20} an average error of 2.9% was observed. This equation may be used up to C_{35} with reasonable accuracy. For heavier hydrocarbons, V_c is given by Eq. (2.46a) and in terms of T_b and SG is given as

$$V_c = 6.2 \times 10^{10}[\exp(-7.58 \times 10^{-3}T_b - 28.5524SG$$
$$(2.99) \qquad + 1.172 \times 10^{-2}T_b SG)]T_b^{1.20493} SG^{17.2074}$$

where V_c is in cm^3/mol. Although this equation is recommended for hydrocarbons heavier than C_{20} it may be used, if necessary, for the range of C_5–C_{50} in which the AAD is about 2.5%. To calculate V_c from other input parameters, Eqs. (2.40) and (2.46b) with Tables 2.5 and 2.9 may be used.

2.5.2.2 Hall–Yarborough Method

This method for estimation of critical volume follows the general form of Eq. (2.39) in terms of M and SG and is given as [75]:

$$(2.100) \qquad V_c = 1.56 \, M^{1.15} \text{SG}^{-0.7935}$$

Predictive methods in terms of M and SG are usually useful for heavy fractions where distillation data may not be available.

2.5.2.3 API Method

In the most recent API-TDB [2], the Reidel method is recommended to be used for the critical volume of pure hydrocarbons given in terms of T_c, P_c, and the acentric factor as follows:

$$(2.101) \qquad V_c = \frac{RT_c}{P_c[3.72 + 0.26(\alpha_R - 7.00)]}$$

in which R is the gas constant and α_R is the Riedel factor given in terms of acentric factor, ω.

$$(2.102) \qquad \alpha_R = 5.811 + 4.919\omega$$

In Eq. (2.101), the unit of V_c mainly depends on the units of T_c, P_c, and R used as the input parameters. Values of R in different unit systems are given in Section 1.7.24. To have V_c in the unit of cm³/mol, T_c must be in kelvin and if P_c is in bar, then the value of R must be 83.14. The API method for calculation of critical volume of mixtures is based on a mixing rule and properties of pure compounds, as will be discussed in Chapter 5. Twu's method for estimation of critical volume is given in Section 2.5.1.

2.5.3 Prediction of Critical Compressibility Factor

Critical compressibility factor, Z_c, is defined by Eq. (2.8) and is a dimensionless parameter. Values of Z_c given in Table 2.1 show that this parameter is a characteristic of each compound, which varies from 0.2 to 0.3 for hydrocarbons in the range of C_1–C_{20}. Generally it decreases with increasing carbon number within a homologous hydrocarbon group. Z_c is in fact value of compressibility factor, Z, at the critical point and therefore it can be estimated from an EOS. As it will be seen in Chapter 5, two-parameter EOS such as van der Waals or Peng–Robinson give a single value of Z_c for all compounds and for this reason they are not accurate at the critical region. Three-parameter EOS or generalized correlations generally give more accurate values for Z_c. On this basis some researchers correlated Z_c to the acentric factor. An example of such correlations is given by Lee–Kesler [27]:

$$(2.103) \qquad Z_c = 0.2905 - 0.085\omega$$

Other references give various versions of Eq. (2.103) with slight differences in the numerical constants [6]. Another version of this equation is given in Chapter 5. However, such equations are only approximate and no single parameter is capable of predicting Z_c as its nature is different from that of acentric factor.

Another method to estimate Z_c is to combine Eqs. (2.101) and (2.102) and using the definition of Z_c through Eq. (2.8) to develop the following relation for Z_c in terms of acentric

factor, ω:

$$(2.104) \qquad Z_c = \frac{1.1088}{\omega + 3.883}$$

Usually for light hydrocarbons Eq. (2.103) is more accurate than is Eq. (2.104), while for heavy compounds it is the opposite; however, no comprehensive evaluation has been made on the accuracy of these correlations.

Based on the methods presented in this chapter, the most appropriate method to estimate Z_c is first to estimate T_c, P_c, and V_c through methods given in Sections 2.5.1 and 2.5.2 and then to calculate Z_c through its definition given in Eq. (2.8). However, for consistency in estimating T_c, P_c, and V_c, one method should be chosen for calculation of all these three parameters. Figure 2.8 shows prediction of Z_c from various correlations for n-alkanes from C_5 to C_{36} and comparing with data reported by API-TDB [2].

Example 2.7—The critical properties and acentric factor of n-hexatriacontane ($C_{36}H_{74}$) are given as follows [20]: $T_b = 770.2$ K, SG $= 0.8172$, $M = 506.98$, $T_c = 874.0$ K, $P_c = 6.8$ bar, $V_c = 2090$ cm³/mol, $Z_c = 0.196$, and $\omega = 1.52596$. Calculate M, T_c, P_c, V_c, and Z_c from the following methods and for each property calculate the percentage relative deviation (%D) between estimated value and other actual value.

a. Riazi–Daubert method: Eq. (2.38)
b. API methods
c. Riazi–Daubert extended method: Eq. (2.46a)
d. Riazi–Sahhaf method for homologous groups, Eq. (2.42), P_c from Eq. (2.43)
e. Lee–Kesler methods
f. Cavett method (only T_c and P_c), Z_c from Eq. (2.104)
g. Twu method
h. Winn method (M, T_c, P_c) and Hall–Yarborough for V_c
i. Tabulate %D for various properties and methods.

Solution—(a) Riazi–Daubert method by Eq. (2.38) for M, T_c, P_c, and V_c are given by Eqs. (2.50), (2.63), (2.64), and (2.98). (b) The API methods for prediction of M, T_c, P_c, V_c, and Z_c are expressed by Eqs. (2.51), (2.65), (2.66), (2.101), and (2.104), respectively. (c) The extended Riazi–Daubert method expressed by Eq. (2.46a) for hydrocarbons heavier than C_{20} and constants for the critical properties are given in Table 9. For T_c, P_c, and V_c this method is presented by Eqs. (2.67), (2.68), and (2.99), respectively. The relation for molecular weight is the same as the API method, Eq. (2.51). (d) Riazi–Sahhaf method is given by Eq. (42) in which the constants for n-alkanes given in Table 2.6 should be used. In using this method, if the given value is boiling point, Eq. (2.49) should be used to calculate M from T_b. Then the predicted M will be used to estimate other properties. In this method P_c is calculated from Eq. (2.43). For parts a, b, c, g, and h, Z_c is calculated from its definition by Eq. (2.8). (e) Lee–Kesler method for M, T_c, P_c, and Z_c are given in Eqs. (2.54), (2.69), (2.70), and (2.103), respectively. V_c should be back calculated through Eq. (2.8) using T_c, P_c, and Z_c. (f) Similarly for the Cavett method, T_c and P_c are calculated from Eqs. (2.71) and (2.72), while V_c is back calculated from Eq. (2.8) with Z_c calculated from Eq. (2.104). (g) The Twu methods are expressed by Eqs. (2.73)–(2.92) for M, T_c, P_c, and V_c. Z_c is calculated from Eq. (2.8). (h) The Winn

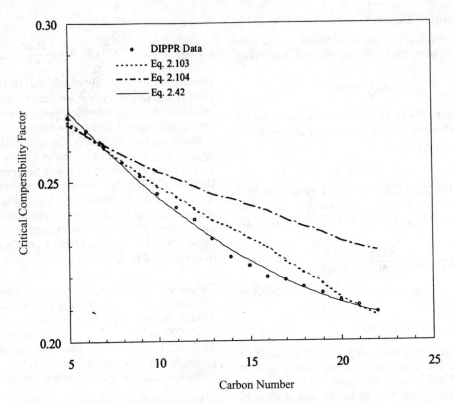

FIG. 2.8—Estimation of critical compressibility factor of *n*-alkanes from various methods.

method for M, T_c, and P_c are given by Eqs. (2.93)–(2.95). In part h, V_c is calculated from the Hall–Yarborough through Eq. (2.100) and Z_c is calculated through Eq. (2.8). Summary of results is given in Table 2.11. No judgement can be made on accuracy of these different methods through this single-point evaluation. However, methods of Riazi–Sahhaf (Part d) and Twu (Part g) give the most accurate results for this particular case. The reason is that these methods have specific relations for *n*-alkanes family and *n*-hexatriacontane is hydrocarbon from this family. In addition, the values for the critical properties from DIPPR [20] are estimated values rather than true experimental values. ♦

2.5.4 Prediction of Acentric Factor

Acentric factor, ω, is a defined parameter that is not directly measurable. Accurate values of the acentric factor can be obtained through accurate values of T_c, P_c, and vapor pressure

with use of Eq. (2.10). Attempts to correlate ω with parameters such as T_b and SG all have failed. However, for homologous hydrocarbon groups the acentric factor can be related to molecular weight as given by Eqs. (2.42) or (2.44). For other compounds the acentric factor should be calculated through its definition, i.e., Eq. (2.10), with the use of a correlation to estimate vapor pressure. Use of an accurate correlation for vapor pressure would result in a more accurate correlation for the acentric factor. Methods of the calculation of the vapor pressure are discussed in Chapter 7. There are three simple correlations for the estimation of vapor pressure that can be used in Eq. (2.10) to derive corresponding correlations for the acentric factor. These three methods are presented here.

2.5.4.1 Lee–Kesler Method

They proposed the following relations for the estimation of acentric factor based on their proposed correlation for vapor pressure [27].

TABLE 2.11—*Prediction of critical properties of n-hexatriacontane from different methods^a (Example 2.7).*

Part	Method(s)	M Est.**	M %D	T_c, K Est.	T_c, K %D	P_c, bar Est.	P_c, bar %D	V_c, cm³/mol Est.	V_c, cm³/mol %D	Z_c Est.	Z_c %D
	Data from DIPPR [20]	507.0	...	874.0	...	6.8	...	2090.0	...	0.196	...
a	R-D: Eq. (2.38)	445.6	−12.1	885.8	1.3	7.3	7.4	1894.4	−9.3	0.188	−4.2
b	API Methods	512.7	1.1	879.3	0.6	7.37	8.4	1849.7	−11.5	0.205	4.6
c	R–D (ext.): Eq. (2.46a)	...	...	870.3	−0.4	5.54	−18.5	1964.7	−6.0	0.150	−23.3
d	R–S: Eqs. 2.42 &2.43	506.9	0	871.8	−0.3	5.93	−12.8	1952.5	−6.6	0.16	−18.4
e	L–K Methods	508.1	0.2	935.1	7.0	5.15	−24.3	2425.9	16.0	0.161	−18.0
f	Cavett & Eq. (2.104)	...	...	915.5	4.7	7.84	15.3	...	...	...	...
g	Twu	513.8	1.3	882.1	0.9	6.02	−11.4	2010.0	−3.8	0.165	−15.8
h	Winn and H−Y	552.0	8.9	889.5	1.77	7.6	11.8	2362.9	13.1	0.243	24.0

^{a}The references for the methods are (a) R-D: Riazi–Daubert [28]; (b) API: Methods in the API-TDB [2]; (c) Extended Riazi–Dubert [65]; (d) Riazi–Sahhaf [31]; (e) Kesler–Lee [12] and Lee–Kesler [27]; (f) Cavett [26]; Twu [31]; (h) Winn [25] and Hall–Yarborough [75]. Est.: Estimated value. %D: % relative deviation defined in Eq. (2.134).

For $T_{br} \leq 0.8$ ($\leq C_{20} \sim M \leq 280$)

$$(2.105) \quad \omega = \frac{-\ln P_c/1.01325 - 5.92714 + 6.09648/T_{br} + 1.28862 \ln T_{br} - 0.169347 T_{br}^6}{15.2518 - 15.6875/T_{br} - 13.4721 \ln T_{br} + 0.43577 T_{br}^6}$$

where P_c is in bar and T_{br} is the reduced boiling point which is defined as

$$(2.106) \quad T_{br} = T_b/T_c$$

and Kesler–Lee [12] proposed the following relation for $T_{br} > 0.8$ ($\sim > C_{20} \sim M > 280$):

$$(2.107) \quad \begin{aligned} \omega = &-7.904 + 0.1352 K_W - 0.007465 K_W^2 + 8.359 T_{br} \\ &+ (1.408 - 0.01063 K_W)/T_{br} \end{aligned}$$

in which K_W is the Watson characterization factor defined by Eq. (2.13). Equation (2.105) may also be used for compounds heavier than C_{20} ($T_{br} > 0.8$) without major error as shown in the example below

2.5.4.2 Edmister Method

The Edmister correlation [76] is developed on the same basis as Eq. (2.105) but using a simpler two-parameter equation for the vapor pressure derived from Clapeyron equation (see Eq. 7.15 in Chapter 7).

$$(2.108) \quad \omega = \left(\frac{3}{7}\right) \times \left(\frac{T_{br}}{1 - T_{br}}\right) \times \left[\log_{10}\left(\frac{P_c}{1.01325}\right)\right] - 1$$

where $\log_{10}$ is the logarithm base 10, T_{br} is the reduced boiling point, and P_c is the critical pressure in bar. As is clear from Eqs. (2.105) and (2.108), these two methods require the same three input parameters, namely, boiling point, critical temperature, and critical pressure. Equations (2.105) and (2.108) are directly derived from vapor pressure correlations discussed in Chapter 7.

2.5.4.3 Korsten Method

The Edmister method underestimates acentric factor for heavy compounds and the error tends to increase with increasing molecular weight of compounds because the vapor pressure rapidly decreases. Most recently Korsten [77] modified the Clapeyron equation for vapor pressure of hydrocarbon systems and derived an equation very similar to the

Edmister method:

$$(2.109) \quad \omega = 0.5899 \left(\frac{T_{br}^{1.3}}{1 - T_{br}^{1.3}}\right) \times \left[\log\left(\frac{P_c}{1.01325}\right)\right] - 1$$

To compare this equation with the Edmister equation, the factor (3/7), which is equivalent to 0.42857 in Eq. (2.108), has been replaced by 0.58990 and the exponent of T_{br} has been changed from 1 to 1.3 in Eq. (2.109).

One can realize that accuracy of these methods mainly depends on the accuracy of the input parameters. However, for pure compounds in which experimental data on pure hydrocarbons are available the Lee–Kesler method, Eq. (2.105), gives an AAD of 1–1.3%, while the Edmister method gives higher error of about 3–3.5%. The Korsten method is new and it has not been extensively evaluated for petroleum fractions, but for pure hydrocarbons it seems that it is more accurate than the Edmister method but less accurate than the Lee–Kesler method. Generally, the Edmister method is not recommended for pure hydrocarbons and is used to calculate acentric factors of undefined petroleum fractions. For petroleum fractions, the pseudocritical temperature and pressure needed in Eqs. (2.105) and (2.108) must be estimated from methods discussed in this section. Usually, when the Cavett or Winn methods are used to estimate T_c and P_c, the acentric factor is calculated by the Edmister method. All other methods for the estimation of critical properties use Eq. (2.105) for calculation of the acentric factor. Equation (2.107) is applicable for heavy fractions and a detailed evaluation of its accuracy is not available in the literature. Further evaluation of these methods is given in Section 2.9. The methods of calculation of the acentric factor for petroleum fractions are discussed in the next chapter.

Example 2.8—Critical properties and acentric factor of *n*-hexatriacontane ($C_{36}H_{74}$) are given as by DIPPR [20] as $T_b = 770.2K$, SG = 0.8172, $T_c = 874.0$ K, $P_c = 6.8$ bar, and $\omega = 1.52596$. Estimate the acentric factor of *n*-hexatriacontane using the following methods:

a. Kesler–Lee method with T_c, P_c from DIPPR
b. Lee–Kesler method with T_c, P_c from DIPPR
c. Edmister method with T_c, P_c from DIPPR
d. Korsten method with T_c, P_c from DIPPR
e. Riazi–Sahhaf correlation, Eq. (2.42)

TABLE 2.12—*Prediction of acentric factor of n-hexatriacontane from different methods (Example 2.8).*

Part	Method for ω	Method for T_c & P_c^a	T_c, K	P_c, bar	Calc. ω	% Rel. dev.
a	Kesler–Lee	DIPPR	874.0	6.8	1.351	−11.5
b	Lee–Kesler	DIPPR	874.0	6.8	1.869	22.4
c	Edmister	DIPPR	874.0	6.8	1.63	6.8
d	Korsten	DIPPR	874.0	6.8	1.731	13.5
e	Riazi–Sahhaf	not needed	$\cdots$	$\cdots$	1.487	−2.6
f	Korsten	R-D-80	885.8	7.3	1.539	0.9
g	Lee–Kesler	API	879.3	7.4	1.846	21.0
h	Korsten	Ext. RD	870.3	5.54	1.529	0.2
i	Lee–Kesler	R-S	871.8	5.93	1.487	−2.6
j	Edmister	Winn	889.5	7.6	1.422	−6.8
k	Kesler–Lee	L-K	935.1	5.15	0.970	−36.4
l	Lee–Kesler	Twu	882.1	6.03	1.475	−3.3

^{a}R-D-80: Eqs. (2.63) and (2.64); API: Eqs. (2.65) and (2.66); Ext. RD: Eqs. (2.67) and (2.68);
R-S: Eqs. (2.42) and (2.43); Winn: Eqs. (2.94) and (2.95); L-K: Eqs. (2.69) and (2.70);
Twu: Eqs. (2.80) and (2.86).

f. Lee–Kesler method with T_c, P_c obtained from Part a in Example 2.6

g. Lee-Kesler method with T_c, P_c obtained from Part b in Example 2.6

h. Lee–Kesler method with T_c, P_c obtained from Part c in Example 2.6

i. Lee–Kesler method with T_c, P_c obtained from Part d in Example 2.6

j. Edmister method with T_c, P_c obtained from Part h in Example 2.6

k. Lee–Kesler method with T_c, P_c obtained from Part e in Example 2.6

l. Lee–Kesler method with T_c, P_c obtained from Part g in Example 2.6

m. Tabulate %D for estimated value of acentric factor in each method.

Lee–Kesler method refers to Eq. (2.105) and Kesler–Lee to Eq. (2.107).

Solution—All three methods of Lee–Kesler, Edmister, and Korsten require T_b, T_c, and P_c as input parameters. The method of Kesler–Lee requires K_W in addition to T_{br}. From definition of Watson K, we get $K_W = 13.64$. Substituting these values from various methods one calculates the acentric factor. A summary of the results is given in Table 2.12. The least accurate method is the Kesler–Lee correlations while the most accurate method is Korsten combined with Eqs. (2.67) and (2.68) for the critical constants. ♦

2.6 PREDICTION OF DENSITY, REFRACTIVE INDEX, CH WEIGHT RATIO, AND FREEZING POINT

Estimation of density at different conditions of temperature, pressure, and composition (ρ) is discussed in detail in Chapter 5. However, liquid density at 20°C and 1 atm designated by d in the unit of g/cm^3 is a useful characterization parameter which will be used in Chapter 3 for the compositional analysis of petroleum fractions especially in conjunction with the definition of refractivity intercept by Eq. (2.14). The sodium D line refractive index of liquid petroleum fractions at 20°C and 1 atm, n, is another useful characterization parameter. Refractive index is needed in calculation of refractivity intercept and is used in Eq. (2.40) for the estimation of various properties through parameter I defined by Eq. (2.36). Moreover refractive index is useful in the calculation of density and transport properties as discussed in Chapters 5 and 8. Carbon-to-hydrogen weight ratio is needed in Chapter 3 for the estimation of the composition of petroleum fractions. Freezing point, T_F, is useful for analyzing solidification of heavy components in petroleum oils and to determine the cloud point temperature of crude oils and reservoir fluids as discussed in Chapter 9 (Section 9.3.3).

2.6.1 Prediction of Density at 20°C

Numerical values of d_{20} for a given compound is very close to the value of SG, which represents density at 15.5°C in the unit of g/cm^3 as can be seen from Tables 2.1 and 2.3. Liquid density generally decreases with temperature. Variation of density with temperature is discussed in Chapter 6. However, in this section methods of estimation of density at 20°C, d_{20}, are presented to be used for the characterization methods discussed in Chapter 3. The most convenient way to estimate d_{20} is through specific gravity. As a rule of thumb $d_{20} = 0.995$ SG. However, a better approximation is provided through calculation of change of density with temperature ($\Delta d/\Delta T$), which is negative and for hydrocarbon systems is given as [7]

$$（2.110）\qquad \Delta d/\Delta T = -10^{-3} \times (2.34 - 1.9 d_T)$$

where d_T is density at temperature T in g/cm^3. This equation may be used to obtain density at any temperature once a value of density at one temperature is known. This equation is quite accurate within a narrow temperature range limit. One can use the above equation to obtain a value of density, d_{20}, at 20°C (g/cm^3) from the specific gravity at 15.5°C as

$$（2.111）\qquad d_{20} = SG - 4.5 \times 10^{-3}(2.34 - 1.9SG)$$

Equation (2.111) may also be used to obtain SG from density at 20 or 25°C.

$$（2.112）\qquad \begin{aligned} SG &= 0.9915 d_{20} + 0.01044 \\ SG &= 0.9823 d_{25} + 0.02184 \end{aligned}$$

Similarly density at any other temperature may be calculated through Eq. (2.110). Finally, Eq. (2.38) may also be used to estimate d_{20} from T_b and SG in the following form:

$$（2.113）\qquad d_{20} = 0.983719 T_b^{0.002016} SG^{1.0055}$$

This equation was developed for hydrocarbons from C_5 to C_{20}; however, it can be safely used up to C_{40} with AAD of less than 0.1%. A comparison is made between the above three methods of estimating d for some n-paraffins with actual data taken from the API-TDB [2]. Results of evaluations are given in Table 2.13. This summary evaluation shows that Eqs. (2.111) and (2.113) are almost equivalent, while as expected the rule of thumb is less accurate. Equation (2.111) is recommended for practical calculations.

2.6.2 Prediction of Refractive Index

The refractive index of liquid hydrocarbons at 20°C is correlated through parameter I defined by Eq. (2.14). If parameter I is known, by rearranging Eq. (2.14), the refractive index, n, can be calculated as follows:

$$（2.114）\qquad n = \left(\frac{1 + 2I}{1 - I} \right)^{1/2}$$

For pure and four different homologous hydrocarbon compounds, parameter I is predicted from Eq. (2.42) using molecular weight, M, with constants in Table 2.6. If boiling point is available, M is first calculated by Eq. (2.48) and then I is calculated. Prediction of I through Eq. (2.42) for various hydrocarbon groups is shown in Fig. 2.9. Actual values of refractive index from API-TDB [2] are also shown in this figure.

For all types of hydrocarbons and narrow-boiling range petroleum fractions the simplest method to estimate parameter I is given by Riazi and Daubert [28] in the form of Eq. (2.38) for the molecular weight range of 70–300 as follows:

$$（2.115）\qquad I = 0.3773 T_b^{-0.02269} SG^{0.9182}$$

TABLE 2.13—*Prediction of density (at 20°C) of pure hydrocarbons.*

n-Paraffin	T_b, K	SG	d, g/cm³	Estimated density, g/cm³					
				Eq. (2.113)	%AD	Eq. (2.111)	%AD	0.995SG	%AD
n-C$_5$	309.2	0.6317	0.6267	0.6271	0.06	0.6266	0.02	0.6285	0.29
n-C$_{10}$	447.3	0.7342	0.7303	0.7299	0.05	0.7299	0.05	0.7305	0.03
n-C$_{15}$	543.8	0.7717	0.768	0.7677	0.03	0.7678	0.03	0.7678	0.02
n-C$_{20}$	616.9	0.7890	0.7871	0.7852	0.24	0.7852	0.24	0.7851	0.26
n-C$_{25}$	683.2	0.8048	0.7996	0.8012	0.20	0.8012	0.19	0.8008	0.15
n-C$_{30}$	729.3	0.8123	0.8086	0.8088	0.03	0.8087	0.01	0.8082	0.04
n-C$_{36}$	770.1	0.8172	0.8146	0.8138	0.09	0.8137	0.12	0.8131	0.18
Overall					0.10		0.10		0.14

where T_b is in Kelvin. This equation predicts n with an average error of about 1% for pure hydrocarbons from C$_5$ to C$_{20}$. More accurate relations are given by Eq. (2.40) and Table 2.5 in terms of various input parameters. The following method developed by Riazi and Daubert [29] and included in the API-TDB [2] have accuracy of about 0.5% on n in the molecular weight range of 70–300.

$$I = 2.34348 \times 10^{-2} \left[\exp\left(7.029 \times 10^{-4} T_b + 2.468 \text{SG}\right.\right.$$
$$(2.116) \qquad \left.\left.- 1.0267 \times 10^{-3} T_b \text{SG}\right)\right] T_b^{0.0572} \text{SG}^{-0.720}$$

where T_b is in kelvin. For heavier hydrocarbons (>C$_{20}$) the following equation derived from Eq. (2.46b) in terms of M and SG can be used.

$$I = 1.2419 \times 10^{-2} \left[\exp\left(7.272 \times 10^{-4} M + 3.3223 \text{SG}\right.\right.$$
$$(2.117) \qquad \left.\left.- 8.867 \times 10^{-4} M\text{SG}\right)\right] M^{0.006438} \text{SG}^{-1.6117}$$

Equation (2.117) is generally applicable to hydrocarbons with a molecular weight range of 70–700 with an accuracy of less than 0.5%; however, it is mainly recommended for carbon

numbers greater than C$_{20}$. If other parameters are available Eqs. (2.40) may be used with constants given in Tables 2.5 and 2.9. The API method to estimate I for hydrocarbons with $M > 300$ is similar to Eq. (2.116) with different numerical constants. Since for heavy fractions the boiling point is usually not available, Eq. (2.117) is presented here. Another relation for estimation of I for heavy hydrocarbons in terms of T_b and SG is given by Eq. (2.46a) with parameters in Table 2.9, which can be used for heavy hydrocarbons if distillation data is available.

Once refractive index at 20°C is estimated, the refractive index at other temperatures may be predicted from the following empirical relation [37].

$$(2.118) \qquad n_T = n_{20} - 0.0004(T - 293.15)$$

where n_{20} is refractive index at 20°C (293 K) and n_T is the refractive index at the temperature T in which T is in kelvin. Although this equation is simple, but it gives sufficient accuracy for practical applications. A more accurate relation can be developed by considering the slope of dn_T/dT (value

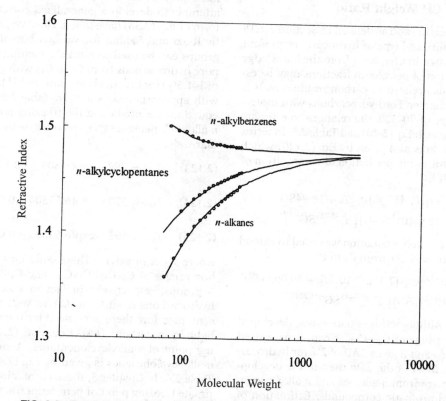

FIG. 2.9—**Prediction of refractive indices of pure hydrocarbons from Eq. (2.42).**

of −0.0004 in Eq. (2.118)) as a function of n_{20} rather than a constant. Another approach to estimate refractive index at temperatures other than 20°C is to assume that specific refraction is constant for a given hydrocarbon:

$$(2.119) \quad \text{Specific refraction} = \frac{I_T}{d_T} = \frac{I_{20}}{d_{20}} = \text{constant}$$

where I_{20} is the refractive index parameter at 20°C and I_T is its value at temperature T. Similarly d_T is density at temperature T. In fact the value of specific refraction is the same at all temperatures [38]. If I_{20}, d_{20}, and d_T are known, then I_T can be estimated from the above equation. Value of n_T can be calculated from I_T and Eq. (2.114). Equation (2.119) has the same accuracy as Eq. (2.118), but at the temperatures far from the reference temperature of 20°C accuracy of both methods decrease. Because of simplicity, Eq. (2.118) is recommended for calculation of refractive index at different temperatures. It is obvious that the reference temperature in both Eqs. (2.118) and (2.119) can be changed to any desired temperature in which refractive index is available. Refractive index is also related to another property called *dielectric constant*, ε, which for non-polar compounds at any temperature is $\varepsilon = n^2$. For example, at temperature of 20°C, a paraffinic oil has dielectric constant of 2.195 and refractive index of 1.481 ($n^2 = 2.193$). Dielectric constants of petroleum products may be used to indicate the presence of various constituents such as asphaltenes, resins, etc. [11]. However, for more complex and polar molecules such as multiring aromatics, this simple relation between ε and n^2 is not valid and they are related through dipole moment. Further discussion on the methods of estimation of refractive index is given by Riazi and Roomi [37].

2.6.3 Prediction of CH Weight Ratio

Carbon-to-hydrogen weight ratio as defined in Section 2.1.18 is indicative of the quality and type of hydrocarbons present in a fuel. As will be shown in Chapter 3 from the knowledge of CH value, composition of petroleum fractions may be estimated. CH value is also related to carbon residues as it is discussed in the next chapter. For hydrocarbons with molecular weight in the range of 70–300, the relations to estimate CH values are given through Eq. (2.40) and Table 2.5. In terms of T_b and SG the relation is also given by Eq. (2.120) which is also recommended for use in prediction of composition of petroleum fractions [78].

$$\text{CH} = 3.4707 \left[\exp\left(1.485 \times 10^{-2} T_b + 16.94 \text{SG} \right. \right.$$
$$(2.120) \qquad \left. \left. -1.2492 \times 10^{-2} T_b \text{SG}\right) \right] T_b^{-2.725} \text{SG}^{-6.798}$$

where T_b is in kelvin. The above equation was used to extend its application for hydrocarbons from C_6 to C_{50}.

$$\text{CH} = 8.7743 \times 10^{-10} \left[\exp\left(7.176 \times 10^{-3} T_b + 30.06242 \text{SG} \right. \right.$$
$$(2.121) \qquad \left. \left. -7.35 \times 10^{-3} T_b \text{SG}\right) \right] T_b^{-0.98445} \text{SG}^{-18.2753}$$

where T_b is in kelvin. Although this equation was developed based on data in the range of C_{20}–C_{50}, it can also be used for lower hydrocarbons and it gives AAD of 2% for hydrocarbons from C_{20} to C_{50}. Most of the data used in the development of this equation are from n-alkanes and n-alkyl monocyclic naphthenic and aromatic compounds. Estimation of CH weight ratio from other input parameters is possible through Eq. (2.40) and Table 2.5. Once CH weight ratio is

determined the atomic HC ratio can be calculated from their definitions as described in Section 2.1.18:

$$(2.122) \quad \text{HC (atomic ratio)} = \frac{11.9147}{\text{CH(weight ratio)}}$$

Example 2.9—Estimate the values of CH (weight) and HC (atomic) ratios for n-tetradecylbenzene ($C_{20}H_{34}$) from Eqs. (2.120) and (2.121) and compare with the actual value. Also draw a graph of CH values from C_6 to C_{50} for the three homologous hydrocarbon groups from paraffins, naphthenes, and aromatics based on Eq. (2.121) and actual values.

Solution—The actual values of CH weight and HC atomic ratios are calculated from the chemical formula and Eq. (2.122) as CH = $(20 \times 12.011)/(34 \times 1.008) = 7.01$, $\text{HC}_{(\text{atomic})} = 34/20 = 1.7$. From Table 2.1, for n-tetradecylbenzene ($C_{20}H_{34}$), $T_b = 627$ K and SG = 0.8587. Substituting these values into Eq. (2.120) gives CH = 7.000, and from Eq. (2.122) atomic HC ratio = 1.702. The error from Eq. (2.134) is %D = 0.12%. Equation (2.121) gives CH = 6.998, which is nearly the same as the value obtained from Eq. (2.120) with the same error. Similarly CH values are calculated by Eq. (2.121) for hydrocarbons ranging from C_6 to C_{50} in three homologous hydrocarbon groups and are shown with actual values in Fig. 2.10. ♦

2.6.4 Prediction of Freezing/Melting Point

For pure compounds, the normal freezing point is the same as the melting point, T_M. Melting point is mainly a parameter that is needed for predicting solid–liquid phase behavior, especially for the waxy oils as shown in Chapter 9. All attempts to develop a generalized correlation for T_M in the form of Eq. (2.38) have failed. However, Eq. (2.42) developed by Riazi and Sahhaf for various homologous hydrocarbon groups can be used to estimate melting or freezing point of pure hydrocarbons from C_7 to C_{40} with good accuracy (error of 1–1.5%) for practical calculations [31]. Using this equation with appropriate constants in Table 2.6 gives the following equations for predicting the freezing point of n-alkanes (P), n-alkycyclopentanes (N), and n-alkybenzenes (A) from molecular weight.

$$(2.123) \quad T_{MP} = 397 - \exp(6.5096 - 0.14187M^{0.47})$$

$$(2.124) \quad T_{MN} = 370 - \exp(6.52504 - 0.04945M^{2/3})$$

$$(2.125) \quad T_{MA} = 395 - \exp(6.53599 - 0.04912M^{2/3})$$

where T_M is in kelvin. These equations are valid in the carbon ranges of C_5–C_{40}, C_7–C_{40}, and C_9–C_{40} for the P, N, and A groups, respectively. In fact in wax precipitation linear hydrocarbons from C_1 to C_{15} as well as aromatics are absent, therefore there is no need for the melting point of very light hydrocarbons [64]. Equation (2.124) is for the melting point of n-alkylcyclopentanes. A similar correlation for n-alkylcyclohexanes is given by Eq. (2.42) with constants in Table 2.6. In Chapter 3, these correlations will be used to estimate freezing point of petroleum fractions.

Won [79] and Pan *et al.* [63] also proposed correlations for the freezing points of hydrocarbon groups. The Won

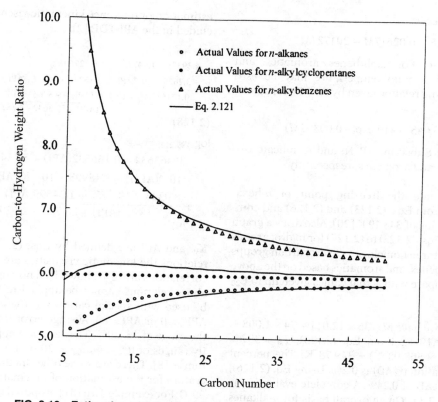

FIG. 2.10—Estimation of CH weight Ratio from Eq. (2.121) for various families.

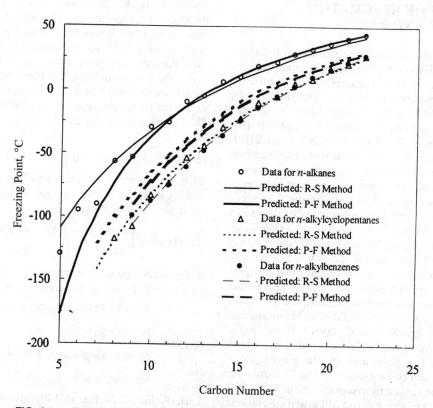

FIG. 2.11—Estimation of freezing point of pure hydrocarbons for various families. [R–S refers to Eqs. (2.123)–(2.125); P-F refers to Eqs. (2.126) and (2.127).

correlation for *n*-alkanes is

$$(2.126) \qquad T_{MP} = 374.5 + 0.02617M - 20172/M$$

where T_{MP} is in kelvin. For naphthenes, aromatics, and isoparaffins the melting point temperature may be estimated from the following relation given by Pan–Firrozabadi–Fotland [63].

$$(2.127) \qquad T_{M(iP,N,A)} = 333.45 - 419 \exp(-0.00855M)$$

where T_M is in kelvin. Subscripts iP, N, and A indicate isoparaffins, naphthenes, and aromatics, respectively.

Example 2.10—Estimate the freezing point of *n*-hexatriacontane ($C_{36}H_{74}$) from Eqs. (2.123) and (2.126) and compare with the actual value of 348.19 K [20]. Also draw a graph of predicted T_M from Eqs. (2.123) to (2.127) for hydrocarbons from C_7 to C_{40} for the three homologous hydrocarbon groups from paraffins, naphthenes, and aromatics based on the above two methods and compare with actual values given up to C_{20} given in Table 2.2.

Solution—For *n*-C_{36}, we have $M = 36 \times 12.011 + 74 \times 1.008 = 508.98$ and $T_M = 348.19$ K. From Eq. (2.123), $T_M = 397 - \exp(6.5096 - 0.14187 \times 508.98^{0.47}) = 349.78$ K. The percent absolute relative deviation (%AD) is 0.2%. Using Eq. (2.126), $T_M = 348.19$ K with %AD of 0.24%. A complete evaluation is demonstrated in Fig. 2.11. On an overall basis for *n*-alkanes Eq. (2.126) is more accurate than Eq. (2.123) while for naphthenes and aromatics, Eqs. (2.124) and (2.125) are more accurate than Eq. (2.127). ♦

2.7 PREDICTION OF KINEMATIC VISCOSITY AT 38 AND 99°C

Detailed prediction of the viscosities of petroleum fractions will be discussed in Chapter 8. However, kinematic viscosity defined by Eq. (2.12) is a characterization parameter needed to calculate parameters such as VGC (Section 2.1.17), which will be used in Chapter 3 to determine the composition of petroleum fractions. Kinematic viscosity at two reference temperatures of 100°F ($37.78 \approx 38$°C) and 210°F ($98.89 \approx 99$°C) are generally used as basic characterization parameters and are designated by $v_{38(100)}$ and $v_{99(210)}$, respectively. For simplicity in writing, the reference temperatures of 100 and 210°F are presented as 38 and 99°C rather than accurate values of 37.78 and 98.89. Kinematic viscosity decreases with temperature and for highly viscous oils values of $v_{99(210)}$ are reported rather than $v_{38(100)}$. The temperature dependency of viscosity is discussed in Chapter 8 and as will be seen, the viscosity of petroleum fractions is one of the most complex physical properties to predict, especially for very heavy fractions and multiring aromatic/naphthenic compounds. Heavy oils with API gravities less than 10 could have kinematic viscosities of several millions cSt at 99°C (210°F). These viscosity values would be almost impossible to predict from bulk properties such as boiling point and specific gravity. However, there are some relations proposed in the literature for the estimation of these kinematic viscosities from T_b and SG or their equivalent parameters K_W and API gravity. Relations developed by Abbott *et al.* [80] are commonly used for the

estimation of reference kinematic viscosities and are also included in the API-TDB [2]:

$$
\begin{aligned}
\log v_{38(100)} = {}& \\
& 4.39371 - 1.94733K_W + 0.12769K_W^2 \\
& + 3.2629 \times 10^{-4}API^2 - 1.18246 \times 10^{-2}K_WAPI \\
& + \frac{0.171617K_W^2 + 10.9943(API) + 9.50663 \times 10^{-2}(API)^2 - 0.860218K_W(API)}{(API) + 50.3642 - 4.78231K_W}
\end{aligned}
$$

$$(2.128)$$

$$
\begin{aligned}
\log v_{99(210)} = {}& \\
& -0.463634 - 0.166532(API) + 5.13447 \\
& \times 10^{-4}(API)^2 - 8.48995 \times 10^{-3}K_WAPI \\
& + \frac{8.0325 \times 10^{-2}K_W + 1.24899(API) + 0.19768(API)^2}{(API) + 26.786 - 2.6296K_W}
\end{aligned}
$$

$$(2.129)$$

K_W and API are defined by Eqs. (2.13) and (2.4). In these relations the kinematic viscosities are in cSt (mm²/s). These correlations are also shown by a nomograph in Fig. 2.12. The above relations cannot be applied to heavy oils and should be used with special care when $K_W < 10$ or $K_W > 12.5$ and API < 0 or API > 80. Average error for these equations is in the range of 15–20%. They are best applicable for the viscosity ranges of $0.5 < v_{38(100)} < 20$ mm²/s and $0.3 < v_{99(210)} < 40$ mm²/s [8]. There are some other methods available in the literature for the estimation of kinematic viscosities at 38 and 99°C. For example Twu [81] proposed two correlations for the kinematic viscosities of *n*-alkanes from C_1 to C_{100} in a similar fashion as his correlations for the critical properties discussed in Section 2.5.1. Errors of ±100% are common for prediction of viscosities of typical oils through this method [17].

Once kinematic viscosities at two temperatures are known, ASTM charts (ASTM D 341-93) may be used to obtain viscosity at other temperatures. The ASTM chart is an empirical relation between kinematic viscosity and temperature and it is given in Fig. 2.13 [68]. In using this chart two points whose their viscosity and temperature are known are located and a straight line should connect these two points. At any other temperature viscosity can be read from the chart. Estimated values are more accurate within a smaller temperature range. This graph can be represented by the following correlation [8]:

$$(2.130) \qquad \log[\log(v_T + 0.7 + c_T)] = A_1 + B_1 \log T$$

where v_T is in cSt, T is the absolute temperature in kelvin, and log is the logarithm to base 10. Parameter c_T varies with value of v_T as follows [8]:

$$(2.131) \quad c_T = \begin{cases} 0.085(v_T - 1.5)^2 & \text{if } v_T < 1.5 \text{ cSt } [\text{mm}^2/\text{s}] \\ 0.0 & \text{if } v_T \geq 1.5 \text{ cSt } [\text{mm}^2/\text{s}] \end{cases}$$

If the reference temperatures are 100 and 210°F (38 and 99°C), then A_1 and B_1 are given by the following relations:

$$
\begin{aligned}
& A_1 = 12.8356 \times (2.57059D_1 - 2.49268D_2) \\
& B_1 = 12.8356(D_2 - D_1) \\
(2.132) \quad & D_1 = \log[\log(v_{38(100)} + 0.7 + c_{38(100)})] \\
& D_2 = \log[\log(v_{99(210)} + 0.7 + c_{99(210)})]
\end{aligned}
$$

Various forms of Eq. (2.130) are given in other sources [2, 11, 17]. Errors arising from use of Eq. (2.130) are better or

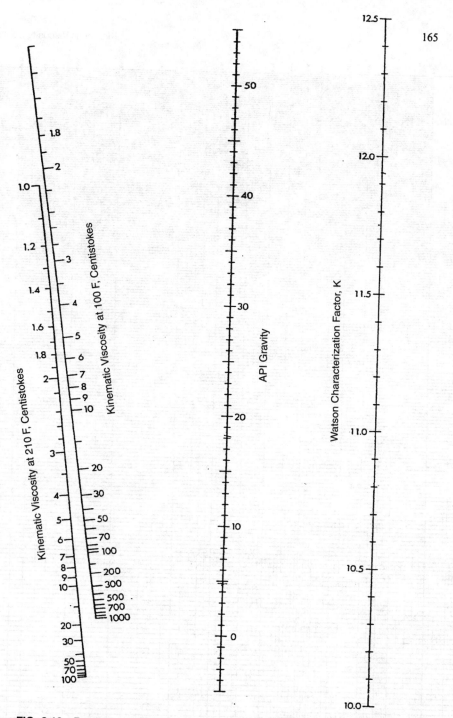

FIG. 2.12—Prediction of kinematic viscosity from K_W and the API gravity. With permission from Ref. [2].

at least in the same range of errors for the prediction of viscosity from Eqs. (2.128) and (2.129). Similarly constants A_1 and B_1 in Eq. (2.130) can be determined when values of viscosity at two temperatures other than 100 and 210°F are known. When ν_T is being calculated from Eq. (2.130) at temperature T, a trial and error procedure is required to determine parameter c_T. The first estimate is calculated by assuming $\nu_T > 1.5$ cSt and thus $c_T = 0$. If calculated value is less than 1.5 cSt, then c_T is calculated from Eq. (2.131). Extrapolated values

from Fig. 2.12 or Eq. (2.130) should be taken with caution. An application of this method to estimate kinematic viscosity of petroleum fractions is demonstrated in Chapter 3. Further discussion on the estimation of viscosity is given in Chapter 8.

Consistency Text—One way to check reliability of a predicted physical property is to perform a consistency test through different procedures. For example, laboratory reports may consist of viscosity data at a temperature other than 38 or

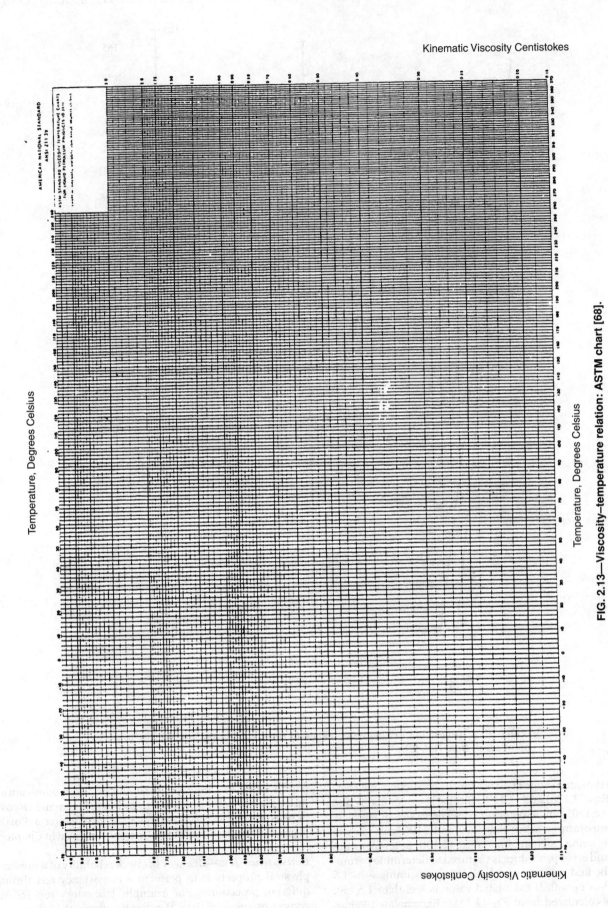

Temperature, Degrees Celsius

Temperature, Degrees Celsius

Kinematic Viscosity Centistokes

FIG. 2.13—Viscosity–temperature relation: ASTM chart [68].

99°C. In many cases kinematic viscosity at 40°C (122°F) or 60°C (140°F) is reported. One may estimate the kinematic viscosities at 38 and 99°C through Eqs. (2.128) and (2.129) and then use the ASTM chart (or Eq. 2.128) to obtain the value of viscosity at 40 or 60°C. If the interpolated value is far from laboratory data then the estimation method cannot be trusted and other methods should be considered.

Another consistency test can be made through estimation of the molecular weight by Eq. (2.52) using estimated viscosities by Eqs. (2.128) and (2.129). If value of M calculated through Eq. (2.52) is near the value of M estimated from T_b and SG by Eqs. (2.51) or (2.50), then all estimated values can be trusted. Such consistency tests can be extended to all other physical properties. The following example demonstrates the test method.

Example 2.11—The viscosity of a pure multiring hydrocarbon from an aromatic group (naphthecene type compound) with formula $C_{26}H_{40}$ has been reported in the API RP-42 [18]. Data available are $M = 352.6$, SG $= 0.9845$, and $\nu_{99(210)} = 13.09$ cSt. Estimate the kinematic viscosity of this hydrocarbon at 38 and 99°C (100 and 210°F) by Eqs. (2.128) and (2.129). How can you assess the validity of your estimated kinematic viscosity at 38°C?

Solution—To estimate the viscosity through the Abbott correlations, K_W and API gravity are needed. However, T_b is not available and should be estimated from M and SG. Since $M > 300$, we use Eq. (2.57) in terms of M and SG to estimate T_b as follows: $T_b = 720.7$ K. Using Eqs. (2.13) and (2.4), K_W and API are calculated as $K_W = 11.08$ and API $= 12.23$. Using Eqs. (2.128) and (2.129) the viscosities are calculated as $\nu_{38(100)} = 299.76$ cSt. $\nu_{99(210)} = 11.35$ cSt. At 99°C the estimated value can be directly evaluated against the experimental data: $\%D = [(11.08 - 13.09)/13.09)] \times 100 = -15.4\%$. To evaluate accuracy of estimated viscosity at 38°C a consistency test is required. Since the actual value of molecular weight, M, is given, one can estimate M through Eq. (2.52) using estimated values of $\nu_{38(100)}$, $\nu_{99(210)}$, and SG as the input parameter. The estimated M is 333.8 which in comparison with actual value of 352.6 gives %AD of 11.36%. This error is acceptable considering that Eq. (2.52) has been developed based on data of petroleum fractions and the fact that input parameters are estimated rather than actual values. Therefore, we can conclude that the consistency test has been successful and the value of 299.8 cSt as viscosity of this hydrocarbon at 38°C is acceptable. The error on estimated viscosity at 99°C is 15.4%, which is within the range of errors reported for the method. It should be realized that the equations for prediction of kinematic viscosity and estimation of molecular weight by Eq. (2.52) were originally recommended for petroleum fractions rather than pure compounds. ♦

2.8 THE WINN NOMOGRAM

Development of estimation techniques through graphical methods was quite common in the 1930s through the 1950s when computational tools were not available. *Nomogram* or *nomograph* usually refers to a graphical correlation between different input parameters and desired property when more than two input parameters are involved. By drawing a straight line between values of input parameters, a reading can be made where the straight line intersects with the line (or curve) of the desired property. The best example and widely used nomogram is the one developed by Winn in 1957 [25]. This nomogram, which is also included in the API-TDB [2], relates molecular weight (M), CH weight ratio, aniline point, and Watson K to boiling point and specific gravity (or API gravity) on a single chart and is shown in Fig. 2.14.

Application of this figure is mainly for petroleum fractions and the mean average boiling point defined in Chapter 3 is used as the boiling point, T_b. If any two parameters are available, all other characterization parameters can be determined. However, on the figure, the best two input parameters are T_b and SG that are on the opposite side of the figure. Obviously use of only M and T_b as input parameters is not suitable since they are near each other on the figure and an accurate reading for other parameters would not be possible. Similarly CH and SG are not suitable as the only two input parameters. Previously the computerized form of the Winn nomogram for molecular weight was given by Eq. (2.95). Use of the nomogram is not common at the present time especially with availability of personal computers (PCs) and simulators, but still some process engineers prefer to use a nomogram to have a quick estimate of a property or to check their calculations from analytical correlations and computer programs.

If the boiling point is not available, methods discussed in Section 2.4.2 may be used to estimate the boiling point before using the figure. Equation (2.50) for molecular weight may be combined with Eq. (2.13) to obtain a relation for the estimation of K_W from M and SG [51].

$$(2.133) \qquad K_W = 4.5579 M^{0.15178} SG^{-0.84573}$$

This equation gives an approximate value for K_W and should be used with care for hydrocarbons heavier than C_{30}. A more accurate correlation for estimation of K_W can be obtained if the boiling point is calculated from Eq. (2.56) or (2.57) and used in Eq. (2.13) to calculate K_W.

Example 2.12—Basic properties of *n*-tridecylcyclohexane ($C_{19}H_{38}$) are given in Table 2.1. Use M and SG as available input parameters to calculate

a. K_W from Eq. (2.133).
b. K_W from most accurate method.
c. K_W from Winn Nomogram.
d. CH weight ratio from M and SG.
e. CH weight ratio from Winn method.
f. %D for each method in comparison with the actual values.

Solution—From Table 2.1, $M = 266.5$, $T_b = 614.7$ K, and SG $= 0.8277$. HC atomic ratio $= 38/19 = 2.0$. Using Eq. (2.122), CH weight ratio $= 11.9147/2.0 = 5.957$. From definition of K_W, i.e., Eq. (2.13), the actual value of K_W is calculated as $K_W = (1.8 \times 614.7)^{1/3}/0.8277 = 12.496$.

a. From Eq. (2.133), $K_W = 4.5579 \times [(266.5)^{0.15178}] \times [(0.8277)^{-0.84573}] = 12.485$. %D $= -0.09\%$.
b. The compound is from the *n*-alkylcyclohexane family and the most accurate way of predicting its boiling point is through Eq. (2.42) with constants given in Table 2.6 which

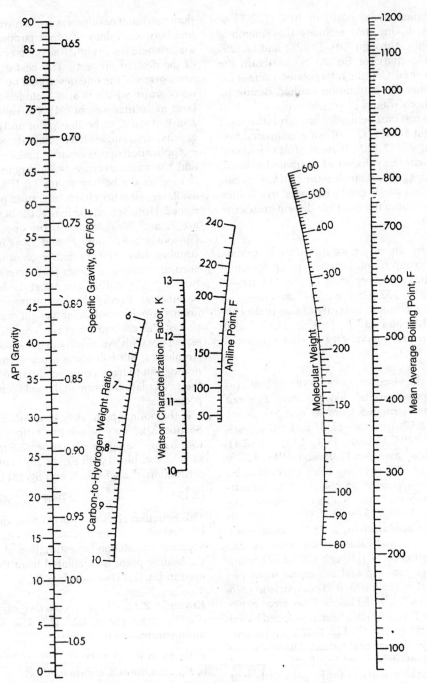

FIG. 2.14—Winn nomogram for characterization of petroleum fractions. With permission from Ref. [2].

gives $T_b = 1100 - \exp[7.00275 - 0.01977 \times (266.5)^{2/3}] = 615.08$ K. Using $T_b = 615.08$ and SG = 0.8277 in Eq. (2.13) gives $K_W = 12.498$. The %D is +0.016%.

c. When using Winn nomogram (Fig. 2.14) it is easier to convert SG to API gravity, which through Eq. (2.4) is 39.46. A straight line between points 266.5 on the M line and 39.5 on the API gravity line intersects the Watson K line at $K_W = 12.27$ and the intersection with the CH line is at CH = 6.1. The %D for K_W is −1.8%.

d. CH weight ratio can be estimated from Eq. (2.40) using M and SG as input parameters with constants in Table 2.5,

which result in CH value of 6.2 with %D = +4% (CH$_{actual}$ = 5.96).

$$CH = 2.35475 \times [\exp(9.3485 \times 10^{-3} \times 266.5 + 4.74695 \times 0.8277 - 8.01719 \times 10^{-3} \times 266.5 \times 0.8277)] \times [(266.5)^{-0.68418}] \times [(0.8277)^{-0.7682}] = 6.2$$

e. CH weight ratio from the Winn method was obtained in Part c as CH = 6.1, which gives %D = +2.3%.

f. For all parts %D is calculated from Eq. (2.134). Part b gives the most accurate K_W value because T_b was calculated accurately. However, Part b in this case is also accurate. Estimation of the CH value is less accurate than prediction of boiling point and gives errors higher than K_W. ♦

2.9 ANALYSIS AND COMPARISON OF VARIOUS CHARACTERIZATION METHODS

Generally there are a large number of pure hydrocarbons and their properties can be used for evaluation purposes. However, hydrocarbons from certain groups (i.e., paraffins, naphthenes, and aromatics) are more abundant in petroleum fractions and can be used as a database for evaluation purposes. Molecular weight, critical properties, and acentric factor are important properties and their predictive methods are presented in this chapter. Errors in any of these properties greatly influence the accuracy of the estimated physical property. Methods of estimation of these properties from bulk properties such as boiling point and specific gravity that are presented in this chapter have been in use in the petroleum industry for many years. In some process simulators a user should select a characterization method out of more than a dozen methods included in the simulator [56]. In each application, the choice of characterization method by the user strongly influences the simulation results. Although there has not been a general and comprehensive evaluation of various characterization methods, a conclusion can be made from individual's experiences reported in the literature. In this section first we discuss criteria for evaluation of various methods and then different predictive methods for molecular weight, critical constants and acentric factor are compared and evaluated.

2.9.1 Criteria for Evaluation of a Characterization Method

Methods of characterization and correlations presented in this chapter are mainly based on properties of pure hydrocarbons. However, some of these correlations such as Eq. (2.52) for estimation of the molecular weight of heavy fractions or the correlations presented for prediction of the kinematic viscosity are based mainly on the properties of fractions rather than pure compounds. The main application of these correlations is for basic properties of undefined petroleum fractions in which bulk properties of a fraction are used to estimate a desired parameter. Therefore, the true evaluation of these characterization methods should be made through properties of petroleum fractions as will be discussed in upcoming chapters. However, evaluation of these methods with properties of pure hydrocarbons can be used as a preliminary criteria to judge the accuracy of various correlations. A method, which is more accurate than other methods for pure hydrocarbons, is not necessarily the best method for petroleum mixtures. A database for pure hydrocarbons consists of many compounds from different families. However, evaluations made by some researchers are based primarily on properties of limited pure hydrocarbons (e.g., n-alkanes). The conclusions through such evaluations cannot be generalized to all hydrocarbons and petroleum fractions. Perhaps it is not a fair comparison if a data set used to develop a method is also used to evaluate the other methods that have used other databases. Type

of compounds selected, the source of data, number of data points, and the basis for the evaluation all affect evaluation outcome. The number of numerical constants in a correlation and number of input parameters also affect the accuracy. Usually older methods are based on a fewer and less accurate data than newer methods. It would be always useful to test different methods on a set of data that have not been used in obtaining the correlation coefficients. The most appropriate procedure would be to compare various methods with an independent data set not used in the development of any methods considered in the evaluation process. Another fair comparison of two different correlation would be to use the same database and reobtain the numerical constants in each correlation from a single database. This was done when Eqs. (2.52) and (2.53) were compared, as discussed in Section 2.4.1. These are the bases that have been used to compare some of the correlations presented in this chapter.

Basically there are two parameters for the evaluation of a correlation. One parameter is the percent average absolute deviation (%AAD). Average errors reported in this chapter and throughout the book are based on percent relative deviation (%D). These errors are defined as following:

$$(2.134) \quad \%D = \left(\frac{\text{estimated value} - \text{actual value}}{\text{actual value}} \right) \times 100$$

$$(2.135) \quad \%AAD = \left(\frac{1}{N} \right) \sum |\%D|$$

where N is the total number of data points and summation is made on all the points. $|\%D|$ is called percent absolute deviation and it is shown by %AD. The maximum value of $|\%D|$ in a data set is referred as %MAD. The second parameter is called R squared (R^2) that is considered as an index of the correlation when parameters of a correlation are obtained from a data set. A value of 1 means perfect fit while values above 0.99 generally give good correlation. For a set of data with X column (independent variable) and Y column (dependent variable) the parameter is defined as

$$(2.136) \quad R^2 = \frac{[N(\sum XY) - (\sum X)(\sum Y)]^2}{[N\sum X^2 - (\sum X)^2] \times [N\sum Y^2 - (\sum Y)^2]}$$

where X and Y are values of the independent and corresponding dependent variables and N is the number of data points. The $\sum$ is the summation over all N values of X, X^2, Y, Y^2, and XY as indicated in the above equation. The R^2 value can be interpreted as the proportion of the variance in y attributable to the variance in x and it varies from 0 to maximum value of 1.

For most of the correlations presented in this chapter such as Eqs. (2.40), (2.42), or (2.46a) the %AAD for various properties is usually given in the corresponding tables where the constants are shown. Most of these properties have been correlated with an R^2 value of minimum 0.99. Some of these properties such as kinematic viscosity or CH weight ratio showed lower values for R^2. Evaluation of some of the other correlations is made through various examples presented in this chapter.

Nowadays with access to sophisticated mathematical tools, it is possible to obtain a very accurate correlation from any data set. For example, when the method of *neural network* is used to obtain correlations for estimation of critical properties, a very accurate correlation can be obtained for a large number of compounds [82]. However, such correlations contain as many as 30 numerical values, which limit their power of extrapolatability. It is our experience that when a correlation is based on some theoretical foundation, it has fewer constants with a wider range of application and better extrapolatability. This is particularly evident for the case of Eq. (2.38) developed based on the theory of intermolecular forces and EOS parameters. Equation (2.38) has only three parameters that are obtained from data on properties of pure hydrocarbons from C_5 to C_{20}. This equation for various properties can be safely used up to C_{30}. Tsonopoulos *et al.* [34] and Lin *et al.* [83] have extensively evaluated Eq. (2.50) for estimation of the molecular weight of different samples of coal liquids, which are mainly aromatics, and compared with other sophisticated multiparameter correlations specifically developed for the molecular weight of coal liquids. Their conclusion was that Eq. (2.50) gave the lowest error even though only pure component data were used to develop this equation. Further evaluation of characterization methods for molecular weight and critical properties are given in the following parts.

2.9.2 Evaluation of Methods of Estimation of Molecular Weight

As mentioned above most of the evaluations made on Eq. (2.50) for the molecular weight of petroleum fractions below 300 suggest that it predicts quite well for various

TABLE 2.14—*Evaluation of methods for estimation of molecular weight of petroleum fractions.*[a]

Method	Equation(s)	Abs Dev %**	
		AAD%	MAD%
API (Riazi–Daubert)	(2.51)	3.9	18.7
Twu	(2.89)–(2.92)	5.0	16.1
Kesler–Lee	(2.54)	8.2	28.2
Winn	(2.93)	5.4	25.9

[a]Number of data points: 625; Ranges of data: $M \sim 70$–700, $T_b \sim 300$–850, SG ~ 0.63–0.97
[b]Defined by Eqs. (2.134) and (135). Reference [29].

fractions. This equation has been included in most process simulators [54–56]. Whitson [51, 53] has used this equation and its conversion to K_W (Eq. 2.133) for fractions up to C_{25} in his characterization methods of reservoir fluids. A more general form of this equation is given by Eq. (2.51) for the molecular weight range of 70–700. This equation gives an average error of 3.4% for fractions with $M < 300$ and 4.7% for fractions with $M > 300$ for 625 fractions from Penn State database on petroleum fractions. An advantage of Eq. (2.51) over Eq. (2.50) is that it is applicable to both light and heavy fractions. A comparative evaluation of various correlations for estimation of molecular weight is given in Table 2.14 [29]. Process simulators [55] usually have referred to Eq. (2.50) as Riazi–Daubert method and Eq. (2.51) as the API method. The Winn method, Eq. (2.93), has been also referred as Sim–Daubert method in some sources [55, 84].

For pure hydrocarbons the molecular weight of three homologous hydrocarbon groups predicted from Eq. (2.51) is drawn versus carbon number in Fig. 2.15. For a given carbon number the difference between molecular weights of

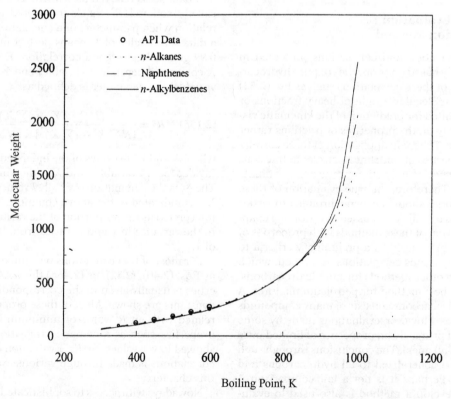

FIG. 2.15—Evaluation of Eq. (2.51) for molecular weight of pure compounds.

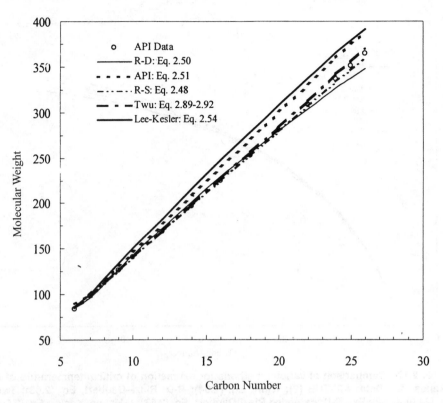

FIG. 2.16—Evaluation of various methods for prediction of molecular weight of *n*-alkylcycohexanes. Riazi–Daubert: Eq. (2.50); API: Eq. (2.51); Riazi–Sahhaf: Eq. (2.48); Lee–Kesler: Eq. (2.54); Twu: Eqs. (2.89)–(2.92).

hydrocarbons from different groups is small. Actual values of molecular weight of *n*-alkylbenzenes up to C_{20} as reported by API-TDB [2] are also shown on the figure. Equation (2.51) is not the best method for the prediction of molecular weight of pure compounds as it was primarily developed for petroleum fractions. Various methods for the estimation of molecular weight for *n*-alkylcyclohexanes with the API data (up to C_{26}) are shown in Fig. 2.16 for the range of C_6–C_{50}. At higher carbon numbers the deviation between the methods increases. The Twu method accurately estimates molecular weight of low-molecular-weight pure hydrocarbons; however, at higher molecular weights it deviates from actual data. A comparison between evaluations presented in Fig. 2.16 and Table 2.14 shows that a method that is accurate for prediction of properties of pure hydrocarbons is not necessarily the best method for petroleum fractions. Evaluation of method of prediction of molecular weight from viscosity (Eqs. (2.52) and (2.53)) has been discussed in Section 2.4.1.

2.9.3 Evaluation of Methods of Estimation of Critical Properties

Evaluation of correlations for estimation of critical properties of pure compounds can be made directly with the actual values for hydrocarbons up to C_{18}. However, when they are applied to petroleum fractions, pseudocritical properties are calculated which are not directly measurable. These values should be evaluated through other properties that are measurable but require critical properties for their calculations. For example, enthalpies of petroleum fractions are calcu-

lated through generalized correlations which require critical properties as shown in Chapters 6 and 7. The phase behavior prediction of reservoir fluids also requires critical properties of petroleum cuts that make up the fluid as discussed in Chapter 9. These two indirect methods are the basis for the evaluation of correlations for estimation of critical properties. These evaluations very much depend on the type of fractions evaluated. For example, Eqs. (2.63)–(2.66) for estimation of T_c and P_c have been developed based on the critical data from C_5 to C_{18}; therefore, their application to heavy fractions is not reliable although they can be safely extrapolated to C_{25}–C_{30} hydrocarbons. In the development of these equations, the internal consistency between T_c and P_c was not imposed as the correlations were developed for fractions with $M < 300$. These correlations were primarily developed for light fractions and medium distillates that are produced from atmospheric distillation columns.

For pure hydrocarbons from homologous families, Eq. (2.42) with constants in Table (2.6) provide accurate values for T_c, P_c, and V_c. Prediction of T_c and P_c from this equation and comparison with the API-TDB data are shown in Figs. 2.2 and 2.3, respectively. Evaluation of various methods for critical temperature, pressure, and volume of different hydrocarbon families is demonstrated in Figs. 2.17–2.19 respectively. A summary of evaluations for T_c and P_c of hydrocarbons from different groups of all types is presented in Table 2.15 [29]. Discontinuity of API data on P_c of *n*-alkylcyclopentanes, as seen in Fig. 2.18, is due to prediction of P_c for heavier hydrocarbons ($>C_{20}$) through a group contribution method.

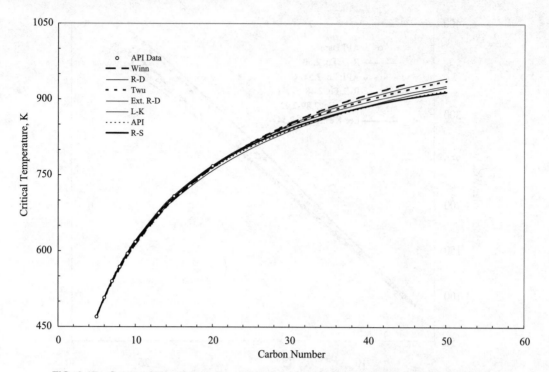

FIG. 2.17—Comparison of various methods for estimation of critical temperature of *n*-alkanes. API Data: API-TDB [2]; Winn: Eq. (2.94); R-D: Riazi–Daubert, Eq. (2.63); Twu: Eq. (2.80)–(2.82); Ext. R-D: Extended Riazi-Daubert, Eq. (2.67); L-K: Lee–Kesler: Eq. (2.69); API: Eq. (2.55); R-S: Riazi–Sahhaf, Eq. (2.42); and Table 2.6.

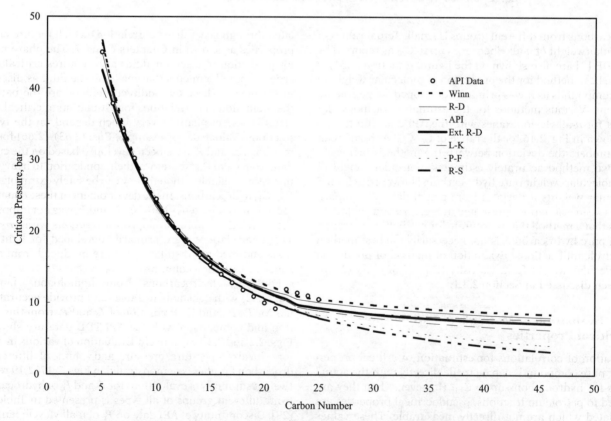

FIG. 2.18—Comparison of various methods for estimation of critical pressure of *n*-alkylcyclopentanes. API Data: API-TDB [2]; Winn: Eq. (2.95); R-D: Riazi–Daubert, Eq. (2.64); API: Eq. (2.56); Ext. R-D: Extended Riazi–Daubert, Eq. (2.68); L-K: Lee–Kesler, Eq. (2.70); P-F: Plan–Firoozabadi, Eq. (2.43); and Table 2.8; R-S: Riazi–Sahhaf, Eq. (2.42); and Table 2.6.

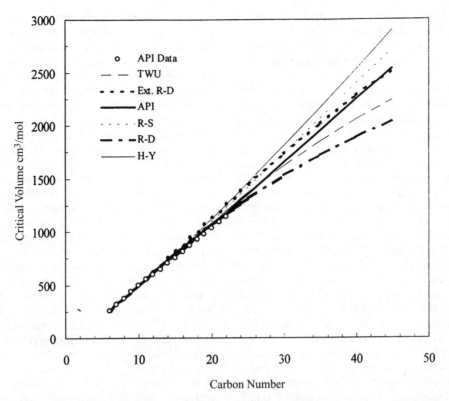

FIG. 2.19—Comparison of various methods for estimation of critical volume of *n*-alkylbenzenes. API Data: API-TDB [2]; Twu: Eqs. (2.83)–(2.85); API: Eq. (2.101); R-S: Riazi–Sahhaf, Eq. (2.42) and Table 2.6; R-D: Riazi–Daubert, Eq. (2.98); H-Y: Hall–Yarborough, Eq. (2.100).

Evaluation of these methods for critical properties of hydrocarbons heavier than C_{20} was not possible due to the lack of confirmed experimental data. Application of these methods for critical properties of petroleum fractions and reservoir fluids is based on the accuracy of predicted physical property. These evaluations are discussed in Chapter 3, where the method of pseudocomponent is introduced for the estimation of properties of petroleum fractions. Generally, a more accurate correlation for properties of pure hydrocarbons does not necessarily give better prediction for petroleum fractions especially those containing heavy compounds.

Evaluation of methods of estimation of critical properties for petroleum fractions is a difficult task as the results depend on the type of petroleum fraction used for the evaluation. The Riazi and Daubert correlations presented by Eq. (2.63) and (2.64) or the API methods presented by Eqs. (2.65) and (2.66) were developed based on critical property data from C_5 to C_{18}; therefore, their application to petroleum fractions containing very heavy compounds would be less accurate. The Kesler–Lee and the Twu method were originally developed based on some calculated data for critical properties of heavy hydrocarbons and the consistency of T_c and P_c were observed at $P_c = 1$ atm at which T_b was set equal to T_c. Twu used some values of T_c and P_c back-calculated from vapor pressure data for hydrocarbons heavier than C_{20} to extend application of his correlations to heavy hydrocarbons. Therefore, it is expected that for heavy fractions or reservoir fluids containing heavy compounds these methods perform better than Eqs. (2.63)–(2.66)

TABLE 2.15—*Evaluation of various methods for prediction of critical temperature and pressure of pure hydrocarbons from C_5 to C_{20}.*

		Abs Dev %**			
		T_c		P_c	
Method	Equation(s)	AD%	MAD%	AD%	MAD%
API	(2.65)–(2.66)	0.5	2.2	2.7	13.2
Twu	(2.73)–(2.88)	0.6	2.4	3.9	16.5
Kesler–Lee	(2.69)–(2.70)	0.7	3.2	4	12.4
Cavett	(2.71)–(2.73)	3.0	5.9	5.5	31.2
Winn (Sim–Daubert)	(2.94)–(2.95)	1.0	3.8	4.5	22.8
Riazi–Daubert	(2.63)–(2.64)	1.1	8.6	3.1	9.3
Lin & Chao	Reference [72]	1.0	3.8	4.5	22.8

[a]Data on T_c and P_c of 138 hydrocarbons from different families reported in API-TDB were used for the evaluation process [29].
[b]Defined by Eqs. (2.134) and (2.135).

for T_c and P_c as observed by some researchers [51, 85]. However, Eq. (2.42) and subsequently derived Eqs. (2.67) and (2.68) have the internal consistency and can be used from C_5 to C_{50} although they are developed for hydrocarbons from C_{20} to C_{50}

The 1980 Riazi–Daubert correlations for T_c and P_c were generally used and recommended by many researchers for light fractions ($M < 300$, carbon number $< C_{22}$). Yu *et al.* [84] used 12 different correlations to characterize the C_{7+} plus fraction of several samples of heavy reservoir fluids and bitumens. Based on the results presented on gas-phase composition, GOR, and saturation pressure, Eqs. (2.63) and (2.64) showed better or equivalent predictions to other methods. Whitson [53] made a good analysis of correlations for the critical properties and their effects on characterization of reservoir fluids and suggested the use of Eqs. (2.63) and (2.64) for petroleum cuts up to C_{25}. But later [51] based on his observation for phase behavior prediction of heavy reservoir fluids, he recommended the use of Kesler–Lee or Twu for estimation of T_c and P_c of such fluids, while for estimation of critical volume he uses Eq. (2.98). Soreide [52] in an extensive evaluation of various correlations for the estimation of critical properties recommends use of the API-TDB [2] method for estimation T_c and P_c (Eqs. (2.65) and (2.66)) but he recommends Twu method for the critical volume. His recommendations are based on phase behavior calculations for 68 samples of North Sea reservoir fluids. In a recently published *Handbook of Reservoir Engineering* [48], and calculations made on phase behavior of reservoir fluids [86], Eqs. (2.65), (2.66) have been selected for the estimation of critical properties of undefined petroleum fractions. Another possibility to reduce the

error associated with critical properties of heavy fractions is to back-calculate the critical properties of the heaviest end of the reservoir fluid from an EOS based on a measured physical property such as density or saturation pressure [51, 52, 70]. Firoozabadi *et al.* [63, 64] have studied extensively the wax and asphaltene precipitation in reservoir fluids. They analyzed various methods of calculating critical properties of heavy petroleum fractions and used Eq. (2.42) for the critical properties and acentric factor of paraffins, naphthenes, and aromatics, but they used Eq. (2.43) for the critical pressure of various hydrocarbon groups with $M > 300$. Their evaluation was based on the calculation of the cloud point of different oils. It is believed that fractions with molecular weight greater than 800 ($\sim C_{57}$) mainly contain aromatic hydrocarbons [63] and therefore Eq. (2.42) with constants given in Table 2.6 for aromatics is an appropriate correlation to estimate the properties of such fractions.

More recently Jianzhong *et al.* [87] reviewed and evaluated various methods of estimation of critical properties of petroleum and coal liquid fractions. Their work followed the work of Voulgaris *et al.* [88], who recommended use of Eq. (2.38) for estimation of critical properties for the purpose of prediction of physical properties of petroleum fractions and coal liquids. They correctly concluded that complexity of correlations does not necessarily increase their accuracy. They evaluated Lee–Kesler, Riazi–Daubert, and Twu methods with more than 318 compounds ($>C_5$) including those found in coal liquids with boiling point up to 418°C (785°F) and specific gravity up to 1.175 [87]. They suggested that Eq. (2.38) is the most suitable and accurate relation especially when the coefficients are modified. Based on their database, they

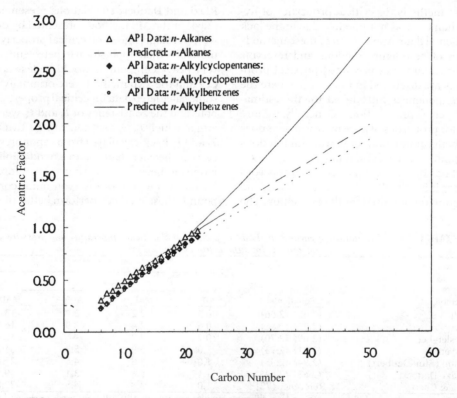

FIG. 2.20—Prediction of acentric factor of pure hydrocarbons from Eq. (2.42).

obtained the coefficients for T_c, P_c, and V_c in Eq. (2.38) with use of d_{20} (liquid density at 20°C and 1 atm in g/cm³) instead of SG (T_c, P_c, $V_c = aT_b^b d_{20}^c$). They reported the coefficients as [87] T_c/K ($a = 18.2394$, $b = 0.595251$, $c = 0.347420$), P_c/bar ($a = 2.95152, b = -2.2082, c = 2.22086$), and V_c/cm³/mol ($a = 8.22382 \times 10^{-5}$, $b = 2.51217$, $c = -1.62214$). Equation (2.38) with these coefficients have not been extensively tested against data on properties of petroleum fractions as yet but for more than 300 pure hydrocarbons gives average errors of 0.7, 3.8, and 2.9% for T_c, P_c, and V_c, respectively [87].

2.9.4 Evaluation of Methods of Estimation of Acentric Factor and Other Properties

For the calculation of the acentric factor of pure hydrocarbons Eq. (2.42) is quite accurate and will be used in Chapter 3 for the pseudocomponent method. Firoozabadi suggests that for aromatics with $M > 800$, $\omega = 2$. Generally there are three methods for the estimation of the acentric factor of undefined petroleum fractions. Perhaps the most accurate method is to estimate the acentric factor through its definition, Eq. (2.10), and vapor pressure estimated from a reliable method [86]. This method will be further discussed in Chapter 7 along with methods of calculation of vapor pressure. For pure hydrocarbons the Lee–Kesler method is more accurate than the Edmister method [36]. The Korsten method for estimating acentric factor is new and has not yet been evaluated extensively. For three different homologous hydrocarbon families from C₆ to C₅₀, values of acentric factor calculated from Eq. (2.42) are compared with values reported in the API-TDB [2] and they are shown in Fig. 2.20. Prediction of acentric factors from different methods for *n*-alkylcyclopentanes and *n*-alkylbenzenes

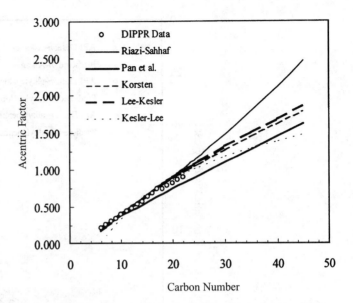

FIG. 2.22—Prediction of acentric factor of *n*-alkylbenzenes from various methods. DIPPR Data: DIPPR [20]; Riazi–Sahhaf: Eq. (2.42) and Table 2.6; Pan *et al.*: Ref. [63, 64], Eq. (2.44); Korsten: Eq. (2.109); Lee–Kesler: Eq. (2.105); Kesler–Lee: Eq. (2.107).

are presented in Figs. 2.21 and 2.22, respectively. The Riazi–Sahhaf method refers to Eq. (2.42) and coefficients given in Table 2.6 for different hydrocarbon families. In Fig. 2.22 the Pan *et al.* [63, 64] method refers to Eq. (2.44), which has been recommended for *n*-alkylbenzenes (aromatics). The Lee–Kesler method, Eq. (2.105), has been generally used for the estimation of accentric factor of undefined petroleum fractions [27]. The Kesler–Lee method refers to Eq. (2.107), which was recommended by Kesler–Lee [12] for estimation of the acentric factor of hydrocarbons with $T_{br} > 0.8$, which is nearly equivalent to hydrocarbons with molecular weights greater than 300. However, our experience shows that this equation is accurate for pure compounds when true critical temperatures are used and high errors can occur when the predicted critical temperature is used in the equation. For heavy hydrocarbons and petroleum fractions ($M > 300$) with estimated critical properties, either the method of pseudo-component discussed in Chapter 3 or the Lee–Kesler may be the most appropriate method. The accuracy of a method to estimate acentric factor also depends on the values of T_c and P_c used to calculate ω as was shown in Example 2.7. Usually the Cavett correlations for T_c and P_c are used together with the Edmister method. Evaluation of these methods for the prediction of properties of undefined petroleum fractions is discussed in Chapter 3.

The accuracy of correlations presented for estimation of other properties such as density, refractive index, boiling point, and CH has been discussed in the previous section where these methods are presented. Prediction of the refractive index for pure hydrocarbons is shown in Fig. 2.9. Prediction of viscosity at 38°C (100°F), ν_{38}, through Eq. (2.128) for pure hydrocarbons from three hydrocarbon groups is shown in Fig. 2.23. Further assessment of accuracy of these methods

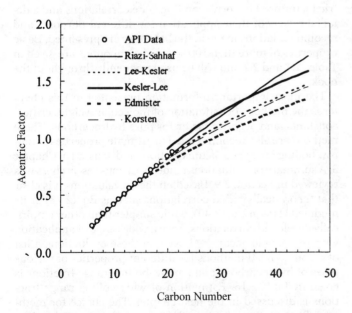

FIG. 2.21—Prediction of acentric factor of *n*-alkylcyclopentanes from various methods. API Data: API-TDB [2]; R-S: Riazi–Sahhaf, Eq. (2.42) and Table 2.6; L-K: Lee–Kesler, Eq. (2.105); K-L: Kesler–Lee, Eq. (2.107); Edmister: Eq. (2.108); Korsten: Eq. (2.109).

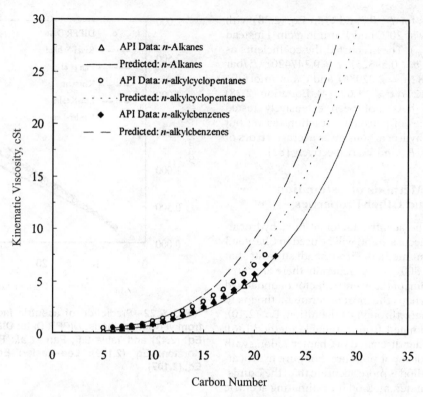

FIG. 2.23—Prediction of kinematic viscosity of pure hydrocarbons at 37.8°C from Eq. (2.128).

is discussed in the following chapters where properties of petroleum fractions are calculated.

2.10 CONCLUSIONS AND RECOMMENDATIONS

In this chapter methods of characterization of pure hydrocarbons have been presented. These methods will be used in Chapters 3 and 4 for the characterization of petroleum fractions and crude oils, respectively. This chapter is an important chapter in this book as the method selected for the characterization of hydrocarbons affects the accuracy of estimation of every physical property throughout the book. In this chapter the basic characterization parameters have been introduced and based on the theory of intermolecular forces, a generalized correlation for the characterization of hydrocarbon systems has been developed. It is shown that fundamentally developed correlations are simpler with a wider field of application and accuracy. For light fractions ($M < 300$), generally two-parameter correlations are sufficient for practical calculations, while for heavier hydrocarbons or nonhydrocarbons the use of a third parameter is needed. The two characterization parameters should represent the energy and size characteristics of molecules. Characterization parameters such as T_b, M, and $\nu_{38(100)}$ may be used as energy, parameters while SG, I, and CH could be used as size parameters. Boiling point and specific gravity are the most easily measurable and appropriate characterization parameters followed by molecular weight and refractive index. Viscosity and CH

parameters may be used as the last option for prediction of properties of hydrocarbons. Various methods of estimation of these parameters as well as critical properties and acentric factor used in corresponding state correlations and a detailed review of their application for different purposes and recommendations made in the literature are presented. Basic properties of more than 100 selected compounds are given in Tables 2.1 and 2.2 and will be used frequently throughout the book.

The most important information presented in this chapter is the methods of estimation of molecular weight, critical constants, and acentric factor for pure hydrocarbons. These methods are also recommended to estimate properties of narrow boiling range petroleum fractions as discussed in Chapter 3. A summary of evaluations made by various researchers was reviewed in Section 2.9. Based on these evaluations it is clear that theoretically based correlations such as Eq. (2.38) or its modified version Eq. (2.40), while simpler than other empirically developed correlations, have a wide range of application with reasonable accuracy. Based on these evaluations a list of recommended methods for different properties of various types of hydrocarbons and narrow boiling range fractions is given in Table 2.16. Estimation of wide boiling range fractions is discussed in the next chapter. The choice for methods of calculation of properties not presented in Table 2.16 is generally narrow and comments have been made where the methods are introduced in each section. The information presented in this chapter should help practical engineers to develop new correlations or to select an appropriate characterization scheme when using a process simulator.

TABLE 2.16—*Recommended methods for the prediction of the basic properties of pure hydrocarbons and narrow boiling range petroleum fractions[a].*

Property	Range of M	Method	Equation
M	70–700	API [2]	2.51
	70–300	Riazi–Dabubert [28]	2.50
	200–800	API [2]	2.52(b)
	70–700	Twu [30]	2.89–2.92(c)
T_c	70–300	API [2]	2.65
	70–700	Lee-Kesler [12]	2.69
	70–700	Extended API [65]	2.67
	70–800	Riazi–Sahhaf [31]	2.42[d]
	<70[e]	Riazi et al. [37]	2.47[e]
P_c	70–300	API [2]	2.66
	70–700	Lee–Kesler [12] or Twu [30]	2.70
	70–700	Extended API [65]	2.68
	70–300	Riazi–Sahhaf [31]	2.42[d]
	300–800	Pan–Firoozabadi–Fotland [63]	2.43[d]
	<70[e]	Riazi et al. [37]	2.47[e]
V_c	70–350	Riazi–Daubert [28]	2.98
	300–700	Extended R–D [65]	2.99
	70–700	Riazi–Sahhaf [31]	2.42[d]
	<70[e]	Riazi et al. [37]	2.47[e]
Z_c	70–700	By definition of Z_c	2.8
ω	70–300	Lee–Kesler [27]	2.105
	300–700	Korsten [77]	2.109
	70–700	Riazi–Sahhaf [31]	2.42[f]
	300–700	Pan–Firoozabadi–Fotland [63]	2.44[g]
T_b	70–300	Riazi–Daubert [29]	2.56
	300–700	Extended R–D [65]	2.57
	70–700	Riazi–Sahhaf [31]	2.42[d]
SG	All range	Denis et al. [8]	2.112
	70–300	Riazi–Daubert [29]	2.59
	70–700	Extended R–D [65]	2.60
	200–800	API [2]	2.61[d]
I	70–300	Riazi–Daubert [29]	2.116
	300–700	Extended R–D [65]	2.117
	70–700	Riazi–Sahhaf [31]	2.42[d]
d	All range	Denis et al. [8]	2.111
	70–350	Riazi–Daubert [28]	2.113
T_M	70–700	Pan–Firoozabadi–Fotland [63]	2.126[h]
		Riazi–Sahhaf [31]	2.124 and 2.125[i]

Methods recommended for pure homologous hydrocarbons (designated by c–i) are also recommended for the pseudocomponent method discussed in Chapter 3 for petroleum fractions. The 300 boundary is approximate and methods recommended for the range of 70–300 may be used safely up to molecular weight of 350 and similarly methods recommended for the range 300–700 may be used for molecular weight as low as 250.
[a] For narrow boiling range fractions a midpoint distillation temperature can be used as T_b.
[b] Only when T_b is not available.
[c] Recommended for pure hydrocarbons from all types.
[d] Recommended for pure homologous hydrocarbon groups.
[e] For compounds and fractions with molecular weight less than 70 and those containing nonhydrocarbon compounds (H_2S, CO_2, N_2, etc.) Eq. (2.47) is recommended.
[f] Equation (2.42) is applicable to acentric factor of n-alkylbenzenes up to molecular weight of 300.
[g] Equation (2.44) is applicable to acentric factor of aromatics for $300 < M < 800$ and for $M > 800$, $\omega = 2$ should be used.
[h] For pure hydrocarbons from n-alkanes family.
[i] For pure hydrocarbons from n-alkylcylopentanes (naphthenes) and n-alkylbenzenes (aromatics) families.

2.11 PROBLEMS

2.1. For light hydrocarbons and narrow boiling range fractions usually a few measured parameters are available. For each one of the following cases determine the best two parameters from the set of available data that are suitable to be used for property predictions:
a. T_b, M, SG
b. CH, $\nu_{38(100)}$, n_{20}
c. CH, n_{20}, SG
d. T_b, M, n_{20}, CH
e. $\nu_{38(100)}$, T_b, CH, M

2.2. For heavy and complex hydrocarbons or petroleum fractions, basic properties can be best determined from three parameters. Determine the best three parameters for each of the following cases:
a. T_b, M, SG, $\nu_{38(100)}$
b. CH, $\nu_{38(100)}$, n_{20}, $\nu_{99(210)}$, API Gravity
c. CH, n_{20}, SG, M, $\nu_{99(210)}$
d. T_b, M, n_{20}, CH, K_W

2.3. You wish to develop a predictive correlation for prediction of molar volume, V_T, in terms of $\nu_{38(100)}$, SG, and temperature T. How do you propose a simple relation with temperature dependent parameters for estimation of molar volume?

2.4. A tank contains pure hydrocarbon liquid from the n-paraffin group. Determine the lightest hydrocarbon from the *n*-alkane family that can exist in an open vessel at the environment of 38°C (100°F) and 1 atm without danger of flammability in the vapor phase near the vessel.

2.5. Develop three relations for estimation of CH weight ratio of *n*-paraffins, *n*-alkylcyclopentanes, and *n*-alkylbenzene in terms of their respective molecular weight. For each group calculate CH_∞ (and HC_∞). Show graphical presentation of the predicted values versus actual values of CH for the three families on a single graph.

2.6. Predict the refractive index of *n*-paraffins, *n*-alkylcyclopentanes, and *n*-alkylbenzene versus carbon number from C_6 to C_{50} using Eq. (2.46a) and compare graphically with values from Eq. (2.42). In using Eq. (2.46a) it is necessary to obtain M from N_C in each family, and then from Eq. (2.42) T_b and SG may be estimated for each carbon number in each family.

2.7. A pure hydrocarbon has molecular weight of 338.6 and specific gravity of 0.8028. Using appropriate methods calculate
   a. boiling point, T_b.
   b. refractivity intercept, R_i.
   c. kinematic viscosity at 38 and 99°C.
   d. VGC from three different methods.

2.8. For *n*-butylcyclohexane, critical properties and molecular weight are give in Table 2.1. Use T_b and SG as the input partameters and calculate
   a. M, T_c, P_c, d_c, and Z_c from the API-TDB-87 methods.
   b. M, T_c, P_c, d_c, and Z_c the Lee–Kesler correlations.
   c. M, T_c, P_c, d_c, and Z_c from the Riazi–Daubert correlations (Eq. 2.38).
   d. M, T_c, P_c, d_c, and Z_c from the Twu correlations.
   e. Compare values from each method with actual values and tabulate the %D.

2.9. Use calculated values of T_c and P_c in Problem 2.8 to calculate acentric factor from the Lee–Kesler and Korsten correlations for each part, then obtain the errors (%D) for each method.

2.10. Estimate the acentric factor of isooctane from Lee–Kesler, Edmister, and Korsten correlations using input data from Table 2.1. Calculate the %D for each method.

2.11. Estimate the kinematic viscosity of *n*-heptane at 38 and 99°C and compare with the experimental values reported by the API-TDB [2]. Also estimate viscosity of n-heptane at 50°C from Eq. (2.130) and the ASTM viscosity–temperature chart.

2.12. For *n*-alkylcylopentanes from C_5 to C_{10}, estimate d_{20} from SG using the rule of thumbs and a more accurate method. Compare the results with actual values from Table 2.1. For these compounds also estimate refractive index at 25°C using M as the only input data available. Use both methods for the effect of temperature on refractive index as discussed in Section 2.6.2 and compare your results with the values reported by the API-TDB [2].

REFERENCES

[1] Coplen, T. B., "Atomic Weights of the Elements 1999," *Pure and Applied Chemistry*, Vol. 73, No. 4, 2001, pp. 667–683.

[2] Daubert, T. E. and Danner, R. P., Eds., *API Technical Data Book—Petroleum Refining*, 6th ed., American Petroleum Institute (API), Washington, DC, 1997.

[3] Nikitin, E. D., Pavlov, P. A., and Popov, A. P., "Vapor–Liquid Critical Temperatures and Pressures of Normal Alkanes from 19 to 36 Carbon Atoms, Naphthalene and *m*-Terphenyl Determined by the Pulse-Heating Technique," *Fluid Phase Equilibria*, Vol. 141, 1997, pp. 155–164.

[4] Reid, R. C., Prausnitz, J. M., and Poling, B. E., *Properties of Gases and Liquids*, 4th ed., Mc-Graw Hill, New York, 1987. Poling, B. E., Prausnitz, J. M., O'Connell, J. P., Properties of Gases and Liquids, 5th ed., Mc-Graw Hill, New York, 2001.

[5] Pitzer, K. S., "The Volumetric and Thermodynamic Properties of Fluids, I: Theoretical Basis and Virial Coefficients," *Journal of American Chemical Society*, Vol. 77, 1955, pp. 3427–3433.

[6] Pitzer, K. S., Lippmann, D. Z., Curl, Jr., R. F., Huggins, C. M., and Petersen, D. E., "The Volumetric and Thermodynamic Properties of Fluids, II: Compressibility Factor, Vapor Pressure, and Entropy of Vaporization," *Journal of American Chemical Society*, Vol. 77, 1955, pp. 3433–3440.

[7] Denis, J., Briant, J., and Hipeaux, J. C., *Lubricant Properties Analysis and Testing*, Translated to English by G. Dobson, Editions Technip, Paris, 1997.

[8] Wauquier, J.-P., *Petroleum Refining, Vol. 1: Crude Oil, Petroleum Products, Process Flowsheets*, Translated from French by David H. Smith, Editions Technip, Paris, 1995.

[9] Watson, K. M., Nelson, E. F., and Murphy, G. B., "Characterization of Petroleum Fractions," *Industrial and Engineering Chemistry*, Vol. 27, 1935, pp. 1460–1464.

[10] Daubert, T. E., "Property Predictions," *Hydrocarbon Processing*, Vol. 59, No. 3, 1980, pp. 107–112.

[11] Speight, J. G., *The Chemistry and Technology of Petroleum*, 3rd ed., Marcel Dekker, New York, 1998.

[12] Kesler, M. G. and Lee, B. I., "Improve Prediction of Enthalpy of Fractions," *Hydrocarbon Processing*, Vol. 55, 1976, pp. 153–158.

[13] Kurtz, Jr., S. S. and Ward, A. L., "The Refractivity Intercept and the Specific Refraction Equation of Newton, I: Development of the Refractivity Intercept and Composition with Specific Refraction Equations," *Journal of Franklin Institute*, Vol. 222, 1936, pp. 563–592.

[14] Hill, J. B. and Coats, H. B., "The Viscosity Gravity Cosntant of Petroleum Lubricating Oils," *Industrial and Engineering Chemistry*, Vol. 20, 1928, pp. 641–644.

[15] Riazi, M. R. and Daubert, T. E., "Prediction of the Composition of Petroleum Fractions," *Industrial and Engineering Chemistry, Process Design and Development*, Vol. 19, No. 2, 1980, pp. 289–294.

[16] Kurtz, S. S., King, R. W., Stout, W. J., Partikian, D. G., and Skrabek, E. A., "Relationsship Between Carbon-Type Composition, Viscosity Gravity Constant, and Refractivity Intercept of Viscous Fractions of Petroleum," *Analytical Chemistry*, Vol. 28, 1956, pp. 1928–1936.

[17] Altgelt, K. H. and Boduszynski, M. M., *Composition and Analysis of Heavy Petroleum Fractions*, Marcel Dekker, New York, 1994.

[18] *API Research Project 42: Properties of Hydrocarbons of High Molecular Weight*, American Petroleum Institute, New York, 1966.

[19] Daubert, T. E., Danner, R. P., Sibul, H. M., and Stebbins, C. C., *Physical and Thermodynamic Properties of Pure Compounds: Data Compilation*, DIPPR-AIChE, Taylor & Francis, Bristol, PA, 1994 (extant) (www.aiche.org/dippr). Updated reference: Rowley, R. L., Wilding, W. V., Oscarson, J. L., Zundel, N. A., Marshall, T. L., Daubert, T. E., and Danner, R. P., *DIPPR Data Compilation of Pure Compound Properties*, Design Institute for Physical Properties (DIPPR), Taylor & Francis, New York, 2002 (http://dippr.byu.edu).

[20] AIChE DIPPR®Database, Design Institute for Physical Property Data (DIPPR), EPCON International, Houston, 1996.

[21] Hall, K. R., Ed., *TRC Thermodynamic Tables—Hydrocarbons*, Thermodynamic Research Center, The Texas A&M University System, College Station, TX, 1993. Updated reference: *TRC Thermodynamic Tables—Hydrocarbons*, Thermodynamic Research Center, National Institute of Standards and Technology, Boulder, CO, 2001. (http://www.trc.nist.gov/).

[22] *API Research Project 44: Selected Values of Properties of Hydrocarbons and Related Compounds*, Tables of Physcial and Thermodynamic Properties of Hydrocarbons, A&M Press, College Station, TX, 1978 (extant).

[23] Dortmund Data Bank (DDB), Laboratory for Thermophysical Properties, LTP GmbH, Institute at the University of Oldenburg (www.ddbst.de).

[24] Watson, K. M. and Nelson, E. F., "Improved Methods for Approximating Critical and Thermal Properties of Petroleum Fractions," *Industrial and Engineering Chemistry*, Vol. 25, 1933, pp. 880–887.

[25] Winn, F. W., "Physical Properties by Nomogram," *Petroleum Refiners*, Vol. 36, No. 21, 1957, pp. 157–159.

[26] Cavett, R. H., "Physical Data for Distillation Calculations, Vapor–Liquid Equilibria," *Proceeding of 27th API Meeting*, API Division of Refining, Vol. 42, No. 3, 1962, pp. 351–366.

[27] Lee, B. I. and Kesler, M. G., "A Generalized Thermodynamic Correlation Based on Three- Parameter Corresponding States," *American Institute of Chemical Engineers Journal*, Vol. 21, 1975, pp. 510–527.

[28] Riazi, M. R. and Daubert, T. E., "Simplify Property Predictions," *Hydrocarbon Processing*, Vol. 59, No. 3, 1980, pp. 115–116.

[29] Riazi, M. R. and Daubert, T. E., "Characterization Parameters for Petroleum Fractions," *Industrial and Engineering Chemistry Research*, Vol. 26, 1987, pp. 755–759. (Corrections, p. 1268.)

[30] Twu, C. H., "An Internally Consistent Correlation for Predicting the Critical Properties and Molecular Weights of Petroleum and Coal-Tar Liquids," *Fluid Phase Equilbria*, Vol. 16, 1984, pp. 137–150.

[31] Riazi, M. R. and Al-Sahhaf, T., "Physical Properties of *n*-Alkanes and *n*-Alkyl Hydrocarbons: Application to Petroleum Mixtures," *Industrial and Engineering Chemistry Research*, Vol. 34, 1995, pp. 4145–4148.

[32] Korsten, H., "Characterization of Hydrocarbon Systems by DBE Concept," *American Institute of Chemical Engineers Journal*, Vol. 43, No. 6, 1997, pp. 1559–1568.

[33] Korsten, H., "Critical Properties of Hydrocarbon Systems," *Chemical Engineering and Technology*, Vol. 21, No. 3, 1998, pp. 229–244.

[34] Tsonopoulos, C., Heidman, J. L., and Hwang, S.-C., *Thermodynamic and Transport Properties of Coal Liquids*, An Exxon Monograph, Wiley, New York, 1986.

[35] Riazi, M. R. and Roomi, Y., "Use of the Refractive Index in the Estimation of Thermophysical Properties of Hydrocarbons and Their Mixtures," *Industrial and Engineering Chemistry Research*, Vol. 40, No. 8, 2001, pp. 1975–1984.

[36] Riazi, M. R., *Prediction of Thermophysical Properties of Petroleum Fractions*, Doctoral Dissertation, Department of Chemical Engineering, Pennsylvania State University, University Park, PA, 1979.

[37] Riazi, M. R., Al-Sahhaf, T. A., and Al-Shammari, M. A., "A Generalized Method for Estimation of Critical Constants," *Fluid Phase Equilibria*, Vol. 147, 1998, pp. 1–6.

[38] Hirschfelder, O. J., Curtiss, C. F., and Bird, R. B., *Molecular Theory of Gases and Liquids*, Wiley, New York, 1964.

[39] Prausnitz, J. M., Lichtenthaler, R. N., and de Azevedo, E. G., *Molecular Thermodynamics of Fluid-Phase Equilibria*, Prentice-Hall, New Jersey, 1986.

[40] Hill, T. L., *An Introduction to Statistical Thermodynmaics*, Addison Wesley, Boston, MA, 1960.

[41] Smith, J. M., Van Ness, H. C., and Abbott, M. M., *Introduction to Chemical Engineering Thermodynamics*, 5th ed., McGraw-Hill, New York, 1996.

[42] Huang, P. K., *Characterization and Thermodynamic Correlations for Undefined Hydrocarbon Mixtures*, Ph.D. Dissertation, Pennsylvania State University, University Park, PA, 1977.

[43] Ambrose, D., "Correlation and Estimation of Vapor–Liquid Critical Properties, I: Critical Temperatures of Organic Compounds," National Physical Laboratory, Teddington, Middlesex, UK, TW11 0LW, NPL Report, Vol. 92, 1978, NPL Report 98, 1979.

[44] Nokay, R., "Estimate Petrochemical Properties," *Chemical Engineering*, Vol. 6, No. 4, 1959, pp. 147–148.

[45] Spencer, C. F. and Daubert, T. E., "A Critical Evaluation of Methods for the Prediction of Critical Propertis of Hydrocarbons," *American Institute of Chemical Engineers Journal*, Vol. 19, No. 3, 1973, pp. 482–486.

[46] Edmister, W. C. and Lee, B. I., *Applied Hydrocarbon Thermodynamics*, 2nd ed., Gulf Publishing, Houston, TX, 1985.

[47] Stange, E. and Johannesen, S. O., "HYPO*S, A Program for Heavy End Characterization," An Internal Program from the Norsk Hydro Oil and Gas Group (Norway), *Paper presented at the 65th Annual Convection*, Gas Processors Association, San Antonio, TX, March 10, 1986.

[48] Ahmed, T., *Reservoir Engineering Handbook*, Gulf Publishing, Houston, TX, 2000.

[49] Ahmed, T., *Hydrocarbon Phase Behavior*, Gulf Publishing, Houston, TX, 1989.

[50] "API Tech Database Software," EPCON International, Houston, TX, 2000 (www.epcon.com).

[51] Whitson, C. H. and Brule, M. R., *Phase Behavior*, Monograph Vol. 20, Society of Petroleum Engineers, Richardson, TX, 2000.

[52] Soreide, I., *Improved Phase Behavior Predictions of Petroleum Reservoir Fluids from a Cubic Equation of State*, Doctor of Engineering Dissertation, Norwegian Institute of Technology, Trondheim, Norway, 1989.

[53] Whitson, C. H., "Effect of C_{7+} Properties on Equation-of-State Predictions," *Society of Petroleum Engineers Journal*, December 1984, pp. 685–696.

[54] Aspen Plus, "Introductory Manual Software Version," Aspen Technology, Inc., Cambridge, MA, December 1986.

[55] PRO/ II, "Keyword Manual," Simulation Sciences Inc., Fullerton, CA, October 1992.

[56] HYSYS, "Reference Volume 1, Version 1.1," HYSYS Reference Manual for Computer Software, HYSYS Conceptual Design, Hyprotech Ltd., Calgary, Alberta, Canada, 1996.

[57] Kreglewski, A. and Zwolinski, B. J., "A New Relation for Physical Properties of *n*-Alkanes and *n*-Alkyl Compounds," *Journal of Physical Chemistry*, Vol. 65, 1961, pp. 1050–1052.

[58] Tsonopoulos, C., "Critical Constants of *n*-Alkanes from Methane to Polyethylene," *American Institute of Chemical Engineers Journal*, Vo. 33, No. 12, 1987, pp. 2080–2083.

[59] Gray, Jr., R. D., Heidman, J. L., Springer, R. D., and Tsonopolous, C., "Characterization and Property Prediction for Heavy Petroleum and Sysnthetic Liquids," *Fluid Phase Equilibria*, Vol. 53, 1989, pp. 355–376.

[60] Teja, A. S., Lee, R. J., Rosenthal, D. J., and Anselme, M., "Correlation of the Critical Properties of *n*-Alkanes and Alkanols," *Fluid Phase Equilibria*, Vol. 56, 1990, pp. 153–169.

[61] Goossens, A. G., "Prediction of Molecular Weight of Petroleum Fractions," *Industrial and Engineering Chemistry Research*, Vol. 35, 1996, pp. 985–988.

[62] Goossens, A. G., "Prediction of the Hydrogen Content of Petroleum Fractions," *Industrial and Engineering Chemistry Research*, Vol. 36, 1997, pp. 2500–2504.

[63] Pan, H., Firoozabadi, A., and Fotland, P., "Pressure and Composition Effect on Wax Precipitation: Experimental Data and Model Results," *Society of Petroleum Engineers Production and Facilities*, Vol. 12, No. 4, 1997, pp. 250–258.

[64] Firoozabadi, A., *Thermodynamics of Hydrocarbon Reservoirs*, McGraw Hill, New York, 1999.

[65] Riazi, M. R. and Adwani, H. A., "Some Guidelines for Choosing a Characterization Method for Petroleum Fractions in Process Simulators," *Chemical Engineering Research and Design*, IchemE, Vol. 83, 2005.

[66] Vetere, A., "Methods to predict the critical constants of organic compounds," *Fluid Phase Equilibria*, Vol. 109, 1995, pp. 17–27.

[67] Riazi, M. R., Daubert, T. E., "Molecular Weight of Heavy Fractions from Viscosity," *Oil and Gas Journal*, Vol. 58, No. 52, 1987, pp. 110–113.

[68] ASTM, *Annual Book of Standards*, Section Five: Petroleum Products, Lubricants, and Fossil Fuels (in Five Volumes), ASTM International, West Conshohocken, PA, 2002.

[69] Hirschler, A. E., "Molecular Mass of Viscous Hydrocarbon Oils: Correlation of Density with Viscosities," *Journal of the Institute of Petroleum*, Vol. 32, No. 1946, 267, pp. 133–161.

[70] Chorn, L. G. and Mansoori, G. A., C_{7+} *Characterization*, Taylor & Francis, New York, 1989.

[71] Kesler, M. G., Lee, B. I., and Sandler, S. I., "A Third Parameter for Use in Generalized Thermodynamic Correlations," *Industrial and Engineering Chemistry Fundamentals*, Vol. 18, 1979, pp. 49–54.

[72] Lin, H. M. and Chao, K. C., "Correlation of Critical Properties and Acentric Factor of Hydrocarbons and Derivatives," *American Institute of Chemical Engineers Journal*, Vol. 30, No. 6, 1984, pp. 981–983.

[73] Mobil Oil Company, "Engineering Data Book," New York, 1961.

[74] Sim, W. J. and Daubert, T. E., "Predcition of Vapor-Liquid Equilibria of Undefined Mixtures," *Industrial and Engineering Chemistry, Process Design and Development*, Vol. 19, 1980, pp. 386–393.

[75] Hall, K. R. and Yarborough, L., "New Simple Correlation for Predicting Critical Volume," *Chemical Engineering*, Vol. 78, No. 25, 1971, p. 76–77.

[76] Edmister, W. C., "Applied Hydrocarbon Thermodynamics, Part 4: Compressibility Factors and Equations of State," *Petroleum Refiner*, Vol. 37, 1958, pp. 173–179.

[77] Korsten, H., "Internally Consistent Prediction of Vapor Pressure and Related Properties," *Industrial and Engineering Chemistry Research*, 2000, Vol. 39, pp. 813–820.

[78] Riazi, M. R. and Daubert, T. E., "Prediction of Molecular Type Analysis of Petroleum Fractions and Coal Liquids," *Industrial and Engineering Chemistry, Process Design and Development*, Vol. 25, No. 4, 1986, pp. 1009–1015.

[79] Won, K. W., Thermodynamics for Solid Solution–Liquid–Vapor Equilibria: Wax Formation from Heavy Hydrocarbon Mixtures," *Fluid Phase Equilibria*, Vol. 30, 1986, pp. 265–279.

[80] Abbott, M. M., Kaufmann, T. G. and Domash, L., "A Correlation for Predicting Liquid Viscosities of Petroleum Fractions," *Canadian Journal of Chemical Engineering*, Vol. 49, No. 3, 1971, pp. 379–384.

[81] Twu, C. H., "Internally Consistent Correlation for Predicting Liquid Viscosities of Petroleum Fractions," *Industrial and Engineering Chemistry, Process Design and Development*, Vol. 24, 1985, 1287–1293.

[82] Riazi, M. R. and Elkamel, A. "Using Neural Network Models to Estimate Critical Constants", *Simulators International*, International Society for Computer Simulation, Vol. 29, No. 3, 1997, pp. 172–177.

[83] Lin, H. M., Kim, H., Guo, T., and Chao, K. C., "Equilibrium Vaporization of a Coal Liquid from a Kentucky No. 9 Coal," *Industrial and Engineering Chemistry, Process Design and Development*, Vol. 24, No. 4, 1985, pp. 1049–1055.

[84] Yu, J. M., Wu, R. S., and Batycky, J. P., "A C_{7+} Characterization Method for Bitumen Mixture Phase Behavior Prediction," C_{7+} *Characterization*, L. G. Chorn and G. A. Mansoori Eds., Taylor & Francis, New York, 1989, pp. 169–196.

[85] Danesh, A., *PVT and Phase Behavior of Petroleum Reservoir Fluids*, Elsevier, Amsterdam, 1998.

[86] Manafi, H., Mansoori, G. A. and Ghotbi, S., "Phase Behavior Prediction of Petroleum Fluids with Minimum Characterization Data," *Journal of Petroleum Science and Engineering*, Vol. 22, 1999, pp. 67–93.

[87] Jianzhong, Zh., Biao, Zh., Suoqi, Zh., Renan, W., and Guanghua, Y., "Simplified Prediction of Critical Properties of Nonpolar Compounds, Petroleum, and Coal Liquid Fractions," *Industrial and Engineering Chemistry Research*, Vol 37, No. 5, 1998, pp. 2059–2060.

[88] Voulgaris, M., Stamatakis, S., Magoulas, K., and Tassios, D., "Prediction of Physical Properties for nonpolar Compounds, Petroleum, and Coal Liquid Fractions," *Fluid Phase Equilibria*, Vol. 64, 1991, pp. 73–106.

Characterization of Petroleum Fractions

3

NOMENCLATURE

AP	Aniline point, °C (unless specified otherwise)
API	API gravity defined in Eq. (2.4)
$A, B, \ldots, F$	Correlation coefficients in various equations
$a, b..i$	Correlation coefficients in various equations
CABP	Cubic average boiling point, K
CH	Carbon-to-hydrogen weight ratio
d	Liquid density at 20°C and 1 atm, g/cm^3
K_W	Watson (UOP) K factor defined by Eq. (2.13)
I	Refractive index parameter defined in Eq. (2.36)
M	Molecular weight, g/mol [kg/kmol]
MABP	Molal average boiling point, K
MeABP	Mean average boiling point, K
n	Sodium D line refractive index of liquid at 20°C and 1 atm, dimensionless
N_C	Carbon number (number of carbon atoms in a hydrocarbon molecule)
P_c	Critical pressure, bar
R_i	Refractivity intercept in Eq. (2.14)
SG	Specific gravity of liquid substance at 15.5°C (60°F) defined by Eq. (2.2), dimensionless
SG_g	Specific gravity of gas substance at 15.5°C (60°F) defined by Eq. (2.6), dimensionless
SL	ASTM D 86 slope between 10 and 90% points, °C (K)/vol%
S%	Weight percent of sulfur in a fraction
T_b	Boiling point, K
T_c	Critical temperature, K
T_F	Flash point, K
T_M	Melting (freezing point) point, K
T_{10}	Temperature on distillation curve at 10% volume vaporized, K
T_{50}	Temperature on distillation curve at 50% volume vaporized, K
V	Molar volume, cm^3/gmol
V	Saybolt universal viscosity, SUS
V_c	Critical volume (molar), cm^3/gmol
VGC	Viscosity gravity constant defined by Eqs. (2.15)–(2.18)
VABP	Volume average boiling point, K
WABP	Weight average boiling point, K

Greek Letters

Γ	Gamma function
μ	Absolute (dynamic) viscosity, cp [mPa·s]. Also used for dipole moment
ν	Kinematic viscosity defined by Eq. (2.12), cSt [mm^2/s]

θ	A property of hydrocarbon such as: M, T_c, P_c, V_c, I, d, T_b, ...
ρ	Density at a given temperature and pressure, g/cm^3
σ	Surface tension, dyn/cm [=mN/m]
ω	Acentric factor defined by Eq. (2.10), dimensionless

Superscript

$\circ$	Properties of n-alkanes from Twu correlations

Subscripts

A	Aromatic
N	Naphthenic
P	Paraffinic
T	Value of a property at temperature T
$\circ$	A reference state for T and P
∞	Value of a property at $M \to \infty$
20	Value of a property at 20°C
39(100)	Value of kinematic viscosity at 37.8°C (100°F)
99(210)	Value of kinematic viscosity at 98.9°C (210°F)

Acronyms

API-TDB	American Petroleum Institute—Technical Data Book
ASTM D	ASTM International (test methods by D committee)
%AD	Absolute deviation percentage defined by Eq. (2.134)
%AAD	Average absolute deviation percentage defined by Eq. (2.135)
EFV	Equilibrium flash vaporization
EOS	Equation of state
FBP	Final boiling point (end point)
GC	Gas chromatography
GPC	Gel permeation chromatography
HPLC	High performance liquid chromatography
KISR	Kuwait Institute for Scientific Research
IBP	Initial boiling point
IR	Infrared
MA	Monoaromatic
MS	Mass spectroscometry
PA	Poly (di- tri-, and higher) aromatic
PIONA	Paraffin, isoparaffin, olefin, naphthene, and aromatic
RVP	Reid vapor pressure
RS	R squared (R^2), defined by Eq. (2.136)
SD	Simulated distillation
TBP	True boiling point
UV	Ultraviolet

IN THIS CHAPTER methods of characterization of petroleum fractions and products are discussed. Petroleum fractions are mixtures of hydrocarbon compounds with a limited boiling point range. Experimental methods on measurement of basic properties that can be obtained from laboratory testing are first presented and then methods of prediction of properties that are not available will be discussed. Two general methods are presented: one for defined mixtures and another for undefined mixtures in which the composition is not known but some bulk properties are available. Petroleum fractions are also divided into light and heavy as well as narrow and wide boiling range mixtures in which different characterization methods are proposed. In addition to methods of estimation of characterization parameters discussed in Chapter 2 for pure hydrocarbons, predictive methods for some characteristics specifically applicable to petroleum fractions are presented in this chapter. These characteristic parameters include distillation curve types and their interconversions, hydrocarbon type composition, sulfur content, carbon residue, octane number, pour, cloud, aniline, and smoke points that affect the quality of a fuel. Standard test methods recommended by ASTM are given for various properties. Finally, minimum laboratory data needed to characterize various fractions as well as analysis of laboratory data and criteria for development of a predictive method are discussed at the end of this chapter. Most of methods presented in this chapter will also be used in Chapter 4 to characterize crude oils and reservoir fluids.

3.1 EXPERIMENTAL DATA ON BASIC PROPERTIES OF PETROLEUM FRACTIONS

In this section characterization parameters that are usually measured in the laboratory as well as methods of their measurements are discussed. Generally not all of these parameters are reported in a laboratory report, but at least from the knowledge of some of these properties, all other basic properties for the fraction can be determined from the methods presented in this chapter.

3.1.1 Boiling Point and Distillation Curves

Pure compounds have a single value for the boiling point; however, for mixtures the temperature at which vaporization occurs varies from the boiling point of the most volatile component to the boiling point of the least volatile component. Therefore, boiling point of a defined mixture can be represented by a number of boiling points for the components existing in the mixture with respect to their composition. For a petroleum fraction of unknown composition, the boiling point may be presented by a curve of temperature versus vol% (or fraction) of mixture vaporized. Different mixtures have different boiling point curves as shown in Fig. 3.1 for a gas oil petroleum product [1]. The curves indicate the vaporization temperature after a certain amount of liquid mixture vaporized based on 100 units of volume. The boiling point of the lightest component in a petroleum mixture is called *initial boiling point* (IBP) and the boiling point of the heaviest compound is called the *final boiling point* (FBP). In some references the FBP is also called the *end point*. The difference

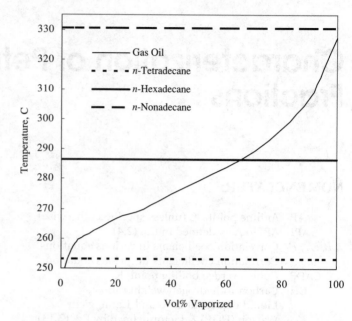

FIG. 3.1—Distillation curve for a gas oil and three pure hydrocarbons.

between FBP and IBP is called *boiling point range* or simply boiling range. For petroleum fractions derived from a crude oil, those with wider boiling range contain more compounds than fractions with narrower boiling range. This is due to the continuity of hydrocarbon compounds in a fraction. Obviously, in general, for defined mixtures this is not the case. For a pure component the boiling range is zero and it has a horizontal distillation curve as shown in Fig. 3.1 for three *n*-alkane compounds of C_{14}, C_{16}, and C_{19}. For the gas oil sample shown in Fig. 3.1 the IBP is 248°C (477°F) and the FBP is 328°C (621°F). Therefore its boiling range is 80°C (144°F) and compounds in the mixture have approximate carbon number range of C_{14}–C_{19}. Crude oils have boiling ranges of more than 550°C (~1000°F), but the FBPs are not accurate. For heavy residues and crude oils the FBPs may be very large or even infinite as the heaviest components may never vaporize at all. Generally, values reported as the IBP and FBP are less reliable than other points. FBP is in fact the maximum temperature during the test and its measurement is especially difficult and inaccurate. For heavy fractions it is possible that some heavy compounds do not vaporize and the highest temperature measured does not correspond to the boiling point of heaviest component present in the mixture. If the temperature is measured until, i.e. 60% vaporized, then the remaining 40% of the fraction is called residue. The boiling point curve of petroleum fractions provides an insight into the composition of feedstocks and products related to petroleum refining processes. There are several methods of measuring and reporting boiling points of petroleum fractions that are described below.

3.1.1.1 ASTM D 86

ASTM D 86 is one of the simplest and oldest methods of measuring and reporting boiling points of petroleum fractions and is conducted mainly for products such as naphthas, gasolines, kerosenes, gas oils, fuel oils, and other similar petroleum products. However, this test cannot be conducted

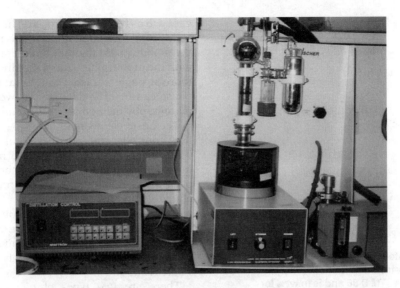

FIG. 3.2—Experimental apparatus for measurement of boiling point of petroleum fractions by ASTM D 86 method (courtesy of Kuwait Institute for Scientific Research).

for mixtures containing very light gases or very heavy compounds that cannot be vaporized. The test is conducted at atmospheric pressure with 100 mL of sample and the result is shown as a distillation curve with temperatures at 0, 5, 10, 20, 30, 40, 50, 60, 70, 80, 90, 95, and 100% volume vaporized. The final boiling point (at 100%) is the least accurate value and it is usually less than the true final boiling point. In many cases only a few temperatures are reported. An exposed thermometer is used and temperatures are reported without stem corrections. For heavy products, temperatures are reported at maximum of 90, 70, or even 50% volume vaporized. This is due to the cracking of heavy hydrocarbons at high temperatures in which vaporization temperatures do not represent boiling points of the original compounds in the mixture. The cracking effect is significant at temperatures above 350°C (660°F); however, ASTM D 86 temperatures reported above 250°C (480°F) should be used with caution. Corrections applied to consider the effects of cracking are applicable from 250 to 500°C; however, these procedures have not been widely used and generally have not been confirmed. In the new revisions of API-TDB-97 no correction for cracking in ASTM D 86 temperatures has been recommended [2]. An apparatus to measure distillation of petroleum fractions by ASTM D 86 method is shown in Fig. 3.2.

3.1.1.2 True Boiling Point

ASTM D 86 distillation data do not represent actual boiling point of components in a petroleum fraction. Process engineers are more interested in actual or true boiling point (TBP) of cuts in a petroleum mixture. Atmospheric TBP data are obtained through distillation of a petroleum mixture using a distillation column with 15–100 theoretical plates at relatively high reflux ratios (i.e., 1–5 or greater). The high degree of fractionation in these distillations gives accurate component distributions for mixtures. The lack of standardized apparatus and operational procedure is a disadvantage, but variations between TBP data reported by different laboratories for the same sample are small because a close approach to

complete component separation is usually achieved. Measurement of TBP data is more difficult than ASTM D 86 data in terms of both time and cost. TBP and ASTM D 86 curves for a kerosene sample are shown in Fig. 3.3 based on data provided by Lenoir and Hipkin [1]. As shown in this figure the IBP from TBP curve is less than the IBP from ASTM D 86 curve, while the FBP of TBP curve is higher than that of ASTM curve. Therefore, the boiling range based on ASTM D 86 is less than the actual true boiling range. In TBP, the IBP is the vapor temperature that is observed at the instant that the first drop of condensate falls from the condenser.

3.1.1.3 Simulated Distillation by Gas Chromatography

Although ASTM D 86 test method is very simple and convenient, it is not a consistent and reproducible method. For this reason another method by gas chromatography (GC) is being recommended to present distillation data. A distillation curve produced by GC is called a *simulated distillation* (SD) and the method is described in ASTM D 2887 test method. Simulated distillation method is simple, consistent, and

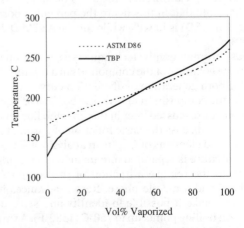

FIG. 3.3—ASTM D 86 and TBP curves for a kerosene sample.

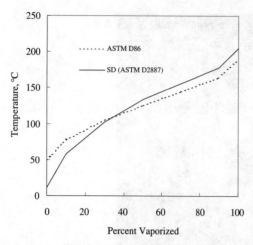

FIG. 3.4—Simulated and ASTM D 86 distillation curves for a petroleum fraction. (The percent is in vol% for ASTM D 86 and is in wt% for ASTM D 2887.)

reproducible and can represent the boiling range of a petroleum mixture without any ambiguity. This method is applicable to petroleum fractions with a FBP of less than 538°C (1000°F) and a boiling range of greater than 55°C (100°F) and having a vapor pressure sufficiently low to permit sampling at ambient temperature. The ASTM D 2887 method is not applicable to gasoline samples and the ASTM D 3710 test method is recommended for such fractions. Distillation curves by SD are presented in terms of boiling point versus wt% of mixture vaporized because as described below in gas chromatography composition is measured in terms of wt% or weight fraction. Simulated distillation curves represent boiling points of compounds in a petroleum mixture at atmospheric pressure; however, as will be shown later SD curves are very close to actual boiling points shown by TBP curves. But these two types of distillation data are not identical and conversion methods should be used to convert SD to TBP curves. In comparison with ASTM D 86, the IBP from a SD curve of a petroleum mixture is less than IBP from ASTM D 86 curve, while the FBP from SD curve is higher than the FBP from ASTM D 86 of the same mixture (see Fig. 3.4). This is the same trend as that of TBP curves in comparison with ASTM curves as was shown in Fig. 3.3. A typical SD curve for a gas oil sample is shown in Fig. 3.4. Note that in this figure the percent vaporized for ASTM D 2887 (SD) is in wt% while for the ASTM D 86 curve is in vol%.

The gas chromatography technique is a separation method based on the volatility of the compounds in a mixture. The GC is used for both generation of distillation curves as well as to determine the composition of hydrocarbon gas or liquid mixtures, as will be discussed later in this chapter. For this reason in this part we discuss the basic function of chromatography techniques and elements of GC. In an analysis of a mixture by a GC, the mixture is separated into its individual compounds according to the relative attraction of the components for a stationary and a mobile phase. Recent advances in chromatography make it possible to identify and separate compounds with boiling points up to 750°C (1380°F). A small fluid sample (few microliters for liquid and 5 mL for gas samples) is injected by a needle injector into a heated zone in which the sample is vaporized and carried by a high-purity carrier

gas such as helium or nitrogen. The stationary phase is either solid or liquid. A component that is more strongly attracted to the mobile phase than to the stationary phase is swept along with the mobile phase more rapidly than a component that is more strongly attracted to the stationary phase. The mobile phase can be a liquid phase as well; in this case the chromatography method is called *liquid chromatography* (LC).

The basic elements of a GC are a cylinder of carrier gas, flow controller and pressure regulator, sample injector, column, detector, recorder, and thermostats for cylinder, column, and detector. The sample after injection enters a heated oven where it enters the GC column (stationary phase). The eluted components by the carrier gas called *effluents* enter a detector where the concentration of each component may be determined. The presence of a component in the carrier gas leaving the column is continuously monitored through an electric signal, which eventually is fed to a recorder for a visual readout.

There are two types of columns, packed or capillary columns, and two types of detectors, flame ionization detector or thermal conductivity detector. Packed columns have inner diameters of 5–8 mm and length of 1–5 m. Column and detector types depend on the nature of samples being analyzed by the GC. The capillary columns are equivalent to hundreds of theoretical equilibrium stages and can be used in preference to packed columns. The inner diameter of capillary columns is about 0.25–0.53 mm and their length is about 10–150 m. The stationary phase is coated on the inside wall of columns. The *flame ionization detector* (FID) is highly sensitive to all organic compounds (10^{-12} g) but is not sensitive to inorganic compounds and gases such as H_2O, CO_2, N_2, and O_2. The FID response is almost proportional to the mass concentration of the ionized compound. Hydrogen of high purity is used as the fuel for the FID. The *thermal conductivity detector* (TCD) is sensitive to almost all the compounds but its sensitivity is less than that of FID. TCD is often used for analysis of hydrocarbon gas mixtures containing nonhydrocarbon gases. The *retention time* is the amount of time required for a given component spent inside the column from its entrance until its emergence from the column in the effluent. Each component has a certain retention time depending on the structure of compound, type of column and stationary phase, flow rate of mobile phase, length, and temperature of column. More volatile compounds with lower boiling points have lower retention times. Detector response is measured in millivolts by electric devices. The written record obtained from a chromatographic analysis is called a *chromatograph*. Usually the time is the abscissa (x axis) and mV is the ordinate (y axis). A typical chromatograph obtained to analyze a naphtha sample from a Kuwaiti crude is shown in Fig. 3.5. Each peak corresponds to a specific compound. Qualitative analysis with GC is done by comparing retention times of sample components with retention times of reference compounds (standard sample) developed under identical experimental conditions. With proper flow rate and temperature, the retention time can be reproduced within 1%. Every component has only one retention time; however, components having the same boiling point or volatility but different molecular structure cannot be identified through GC analysis. In Fig. 3.5, compounds with higher retention time (x coordinate) have higher boiling points and the actual boiling point or the compound can be determined by comparing the peak

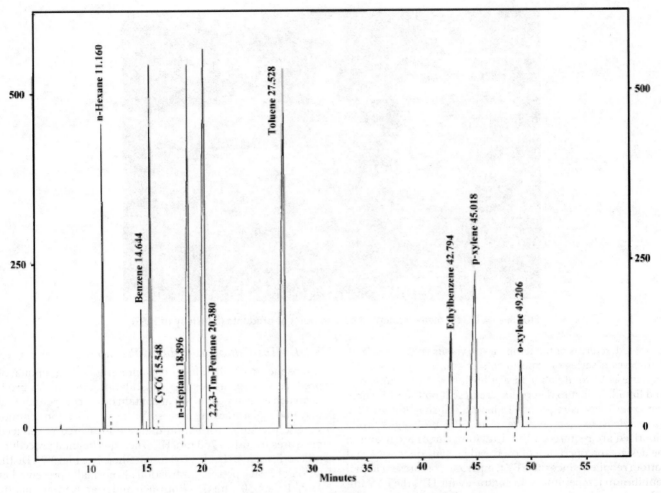

FIG. 3.5—A typical chromatograph for a Kuwaiti naphtha sample.

with the similar peak of a known compound with a known boiling point. In the quantitative analysis of a mixture, it can be shown that the area under a particular component peak (as shown in Fig. 3.5) is directly proportional to the total amount (mass) of the component reaching the detector. The amount reaching the detector is also proportional to the concentration (weight percent or weight fraction) of the component in the sample injected. The proportionality constant is determined with the aid of standards containing a known amount of the sample component. Modern GCs are equipped with a computer that directly measures the areas under each peak and composition can be directly determined from the computer printout. A printout for the chromatograph of Fig. 3.5 is shown in Table 3.1 for the naphtha sample. The area per-

cent is the same as composition in wt% with boiling points of corresponding components. In analysis of samples by a GC, the composition is always determined in wt% and not in vol% or fraction. For this reason the output of a GC analysis for a simulated distillation is a curve of temperature (boiling point) versus wt% of mixture vaporized, as can be seen in Fig. 3.4. Further information for use of GC for simulated distillation up to 750°C is provided by Curvers and van den Engel [3]. A typical GC for measurement of boiling point of petroleum products is shown in Fig. 3.6.

3.1.1.4 Equilibrium Flash Vaporization

Equilibrium flash vaporization (EFV) is the least important type of distillation curve and is very difficult to measure. It is

TABLE 3.1—*Calculation of composition of a naphtha sample with GC chromatograph shown in Fig. 3.4.*

No.	Name	Time, min	Area	Area %	T_b,° C
1	*n*-Hexane	11.16	1442160	3.92	68.7
2	Benzene	14.64	675785	1.84	80.1
3	Cyclohexane	15.55	3827351	10.40	80.7
4	*n*-Heptane	18.90	5936159	16.14	98.4
5	2,2,3-Trimethyl-pentane	20.38	8160051	22.18	109.8
6	Toluene	27.53	8955969	24.34	110.6
7	Ethylbenzene	42.79	1678625	4.56	136.2
8	*p*-Xylene	45.02	4714426	12.82	138.4
9	*o*-Xylene	49.21	1397253	3.8	144.4
	Total		36787780	100	

FIG. 3.6—A GC for measurement of boiling point of products (courtesy of KISR).

presented in terms of the temperature versus vol% vaporized. It involves a series of experiments at constant atmospheric pressure with total vapor in equilibrium with the unvaporized liquid. In fact to determine each point on the EFV curve one experiment is required. To have a full shape of an EFV curve at least five temperatures at 10, 30, 50, 70, and 90 vol% vaporized are required. EFV distillation curves are useful in the design and operation of overhead partial condensers and bottom reboilers since the EFV temperatures represent actual equilibrium temperatures. In contrast with TBP, the EFV initial temperature of a mixture is greater than the IBP of ASTM D 86 curve, while the FBP from a EFV curve is lower than the FBP from the TBP curve for the same mixture. EFV curves at pressures above atmospheric up to pressures of 15 bar may also be useful for design and operation of vaporizing or condensing vessels under pressure.

3.1.1.5 Distillation at Reduced Pressures

Atmospheric distillation curves present boiling points of products from an atmospheric distillation column. For products such as heavy gas oils that contain heavy compounds and may undergo a cracking process during vaporization at atmospheric pressure, distillation data are measured at reduced pressures from 1 to 760 mm Hg. The experimental procedure is described in ASTM D 1160 test method (see Fig. 3.7). Distillation of heavy petroleum fractions is normally presented at 1, 2, 10, or 50 mm Hg. Both a manual and an automatic method are specified. The temperature of the vapor should not exceed 400°C (750°F). ASTM D 1160 distillation data are measured more accurately than ASTM D 86 since it is conducted at low pressure. For this reason ASTM D 1160 curves are closer to TBP curves at the same pressure base. Conversion of distillation data from low pressure to equivalent

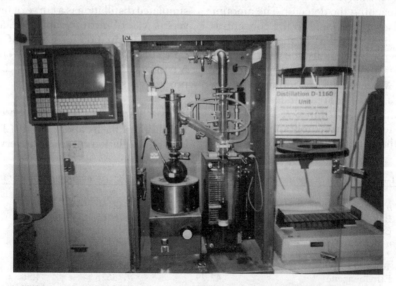

FIG. 3.7—An apparatus for experimental measurement of boiling point at reduced pressures by ASTM D 1160 test method (courtesy of KISR).

atmospheric boiling points are given in the ASTM Manual [4] and will be discussed later in this chapter.

3.1.2 Density, Specific Gravity, and API Gravity

Specific gravity (SG) or relative density and the API gravity are defined in Section 2.1.3 and for pure hydrocarbons are given in Tables 2.1 and 2.2. Aromatic oils are denser than paraffinic oils. Once specific gravity is known, the API gravity can be determined by Eq. (2.4), which corresponds to the ASTM D 287 method. The standard temperature to measure the specific gravity is 15.56°C (60°F); however, absolute density is usually reported at 20°C. Specific gravity or density for a petroleum mixture is a bulk property that can be directly measured for the mixture. Specific gravity is a property that indicates the quality of a petroleum product and, as was shown in Chapter 2, is a useful property to estimate various physical properties of petroleum fluids. A standard test method for density and specific gravity of liquid petroleum products and distillates in the range of 15–35°C through use of a digital density meter is described in ASTM D 4052 method [4]. The apparatus must be calibrated at each temperature and this test method is equivalent to ISO 12185 and IP 365 methods. Another method using a hydrometer is described under ASTM D 1298 test method. Hydrometer is a glass float with lead ballast that is floated in the liquid. The level at which hydrometer is floating in the liquid is proportional to the specific gravity of the liquid. Through graduation of the hydrometer the specific gravity can be read directly from the stalk of hydrometer. This method is simpler than the ASTM 4052 method but is less accurate. The French standard procedure for measuring density by hydrometer is described under NFT 60-101 test method. With some hydrometers densities with accuracy of 0.0005 g/mL can be measured. A digital density meter model DMA 48 from PARA (Austria) is shown in Fig. 3.8.

3.1.3 Molecular Weight

Molecular weight is another bulk property that is indicative of molecular size and structure. This is an important property that usually laboratories do not measure and fail to report when reporting various properties of petroleum fractions. This is perhaps due to the low accuracy in the measurement of the molecular weight of petroleum fractions, especially for heavy fractions. However, it should be realized that experimental uncertainty in reported values of molecular weight is less than the errors associated with predictive methods for this very useful parameter. Since petroleum fractions are mixtures of hydrocarbon compounds, mixture molecular weight is defined as an average value called number average molecular weight or simply molecular weight of the mixture and it is calculated as follows:

$$M = \sum_i x_i M_i \tag{3.1}$$

where x_i and M_i are the mole fraction and molecular weight of component i, respectively. Molecular weight of the mixture, M, represents the ratio of total mass of the mixture to the total moles in the mixture. Exact knowledge of molecular weight of a mixture requires exact composition of all compounds in the mixture. For petroleum fractions such exact knowledge is not available due to the large number of components present in the mixture. Therefore, experimental measurement of mixture molecular weight is needed in lieu of exact composition of all compounds in the mixture.

There are three methods that are widely used to measure the molecular weight of various petroleum fractions. These are *cryoscopy*, the *vapor pressure* method, and the *size exclusion chromatography* (SEC) method. For heavy petroleum fractions and asphaltenic compounds the SEC method is commonly used to measure distribution of molecular weight

FIG. 3.8—PARA model DMA 48 digital density meter (courtesy of Chemical Engineering Department at Kuwait University).

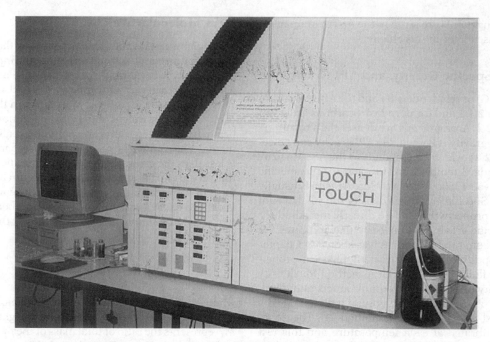

FIG. 3.9—A GPC from waters model 150-C plus (courtesy of Chemical Engineering Department at Kuwait University).

in the fraction. The SEC method is mainly used to determine molecular weights of polymers in the range of 2000 to 2×10^6. This method is also called gel permeation chromatography (GPC) and is described in the ASTM D 5296 test method. In the GPC method, by comparing the elution time of a sample with that of a reference solution the molecular weight of the sample can be determined. A GPC instrument is shown in Fig. 3.9. The SEC experiment is usually performed for heavy residues and asphaltenes in crude oils and gives the wt% of various constituents versus molecular weight as will be discussed in Chapter 4.

The vapor pressure method is based on the measurement of the difference between vapor pressure of sample and that of a known reference solvent with a vapor pressure greater than that of the sample. A solution of about 1 g of sample in 25 mL of the reference solvent is prepared. This solution, which has vapor pressure less than that of the solvent, tends to condense the vapors of solvent on the sample thus creating a temperature difference which is measured by two thermistors. The molarity of the solution is calculated using calibrated curves. This method is described by the ASTM D 2503 test method and is applicable to oils with an initial boiling point greater than 220°C [5]. A typical experimental error and uncertainty in measuring the molecular weight is about 5%.

The third and most widely used method of determining the molecular weight of an unknown petroleum mixture is by the cryoscopy method, which is based on freezing point depression. The freezing point of a solution is a measure of the solution's concentration. As the concentration of the solute increases, the freezing point of the solution will be lower. The relation between freezing point depression and concentration is linear. For organic hydrocarbons, benzene is usually used as the solvent. Special care should be taken when working with benzene [6]. Calibration curves can be prepared by measuring the freezing points of different solute concentrations with a known solute and a known solvent. A cryoscope can measure the freezing point depression with an accuracy

of about 0.001°C. The relation to obtain molecular weight of a sample is [6]

$$(3.2) \qquad M = \frac{1000 \times K_f \times m_1}{\Delta T \times m_2}$$

where K_f is molal freezing point depression constant of the solvent and is about 5.12°C/mol. ΔT is the freezing point depression and the reading from the cryoscope. m_1 is the mass of solute and m_2 is the mass of solvent both in grams. It generally consists of refrigerator, thermometer and the apparatus to hold the sample. A cryoscope is shown in Fig. 3.10.

3.1.4 Refractive Index

Refractive index or *refractivity* is defined in Section 2.1.4 and its values at 20°C for pure hydrocarbons are given in Table 2.1. Refractive indexes of hydrocarbons vary from 1.35 to 1.6; however, aromatics have refractive index values greater than naphthenes, which in turn have refractive indexes greater than paraffins. Paraffinic oils have lower refractive index values. It was shown in Chapter 2 that refractive index is a useful parameter to characterize hydrocarbon systems and, as will be seen later in this chapter, it is needed to estimate the composition of undefined petroleum fractions. Refractive index is the ratio of the speed of light in a vacuum to that of a medium. In a medium, the speed of light depends on the wavelength and temperature. For this reason refractive index is usually measured and reported at 20°C with the D line sodium light. For mixtures, refractive index is a bulk property that can be easily and accurately measured by an instrument called a *refractometer*. Refractive index can be measured by digital refractometers with a precision of ±0.0001 and temperature precision of ±0.1°C. The amount of sample required to measure refractive index is very small and ASTM D 1218 provides a test method for clear hydrocarbons with values of refractive indexes in the range of 1.33–1.5 and the temperature range of 20–30°C. In the ASTM D 1218 test method the Bausch and

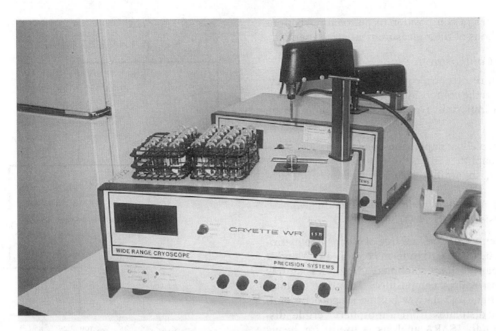

FIG. 3.10—Model 5009 wide range cryoscope to measure molecular weight (courtesy of Chemical Engineering Department at Kuwait University).

Lomb refractometer is used. Refractive index of viscous oils with values up to 1.6 can be measured by the ASTM D 1747 test method. Samples must have clear color to measure their refractive index; however, for darker and more viscous samples in which the actual refractive index value is outside the range of application of refractometer, samples can be diluted by a light solvent and refractive index of the solution should be measured. From the composition of the solution and refractive index of pure solvent and that of the solution, refractive index of viscous samples can be determined. A Model Abbe refractometer (Leica) is shown in Fig. 3.11. This refractometer measures refractive index of liquids within the temperature range of −20 to 100°C with temperature accuracy of ±0.01°C. Because of simplicity and importance of refractive index it would be extremely useful if laboratories measure and report its value at 20°C for a petroleum product, especially if the composition of the mixture is not reported.

3.1.5 Compositional Analysis

Petroleum fractions are mixtures of many different types of hydrocarbon compounds. A petroleum mixture is well defined if the composition and structure of all compounds present in the mixture are known. Because of the diversity and number of constituents of a petroleum mixture, the determination of such exact composition is nearly impossible. Generally, hydrocarbons can be identified by their carbon number or by their molecular type. Carbon number distribution may be determined from fractionation by distillation or by molecular weight distribution as discussed earlier in this section. However, for narrow boiling range petroleum products and petroleum cuts in which the carbon number range is quite limited, knowledge of molecular type of compounds is very important. As will be seen later, properties of petroleum fractions with detailed compositional analysis can be estimated with a higher degree of accuracy than for undefined fractions. After distillation data, molecular type composition is the most important characteristic of petroleum fractions. In this

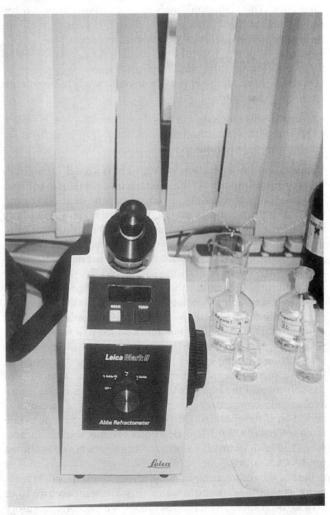

FIG. 3.11—Leica made Abbe refractometer (courtesy of Chemical Engineering Department at Kuwait University).

section various types of composition of petroleum fractions and different methods of their measurement are presented.

3.1.5.1 Types of Composition

Based on the nature of petroleum mixture, there are several ways to express the composition of a petroleum mixture. Some of the most important types of composition are given below:

- PONA (paraffins, olefins, naphthenes, and aromatics)
- PNA (paraffins, naphthenes, and aromatics)
- PIONA (paraffins, isoparaffins, olefins, naphthenes, and aromatics)
- SARA (saturates, aromatics, resins, and asphalthenes)
- Elemental analysis (C, H, S, N, O)

Since most petroleum fractions are free of olefins, the hydrocarbon types can be expressed in terms of only PINA and if paraffins and isoparaffins are combined a fraction is simply expressed in terms of PNA composition. This type of analysis is useful for light and narrow boiling range petroleum products such as distillates from atmospheric crude distillation units. But the SARA analysis is useful for heavy petroleum fractions, residues, and fossil fuels (i.e., coal liquids), which have high contents of aromatics, resins, and asphaltenes. The elemental analysis gives information on hydrogen and sulfur contents as well as C/H ratio, which are indicative of the quality of petroleum products.

3.1.5.2 Analytical Instruments

Generally three methods may be used to analyze petroleum fractions. These are

- separation by solvents
- chromatography methods
- spectroscopic methods

The method of separation by solvents is based on solubility of some compounds in a mixture in a particular solvent. The remaining insoluble compounds may be in a solid or another immiscible liquid phase. This method is particularly useful for heavy petroleum fractions and residues containing asphaltenes, resins, and saturate hydrocarbons. The degree of solubility of a compound in a solvent depends on the chemical structure of both the solute and the solvent. If the two structures are similar there is a greater degree of solubility. For example, high-molecular-weight asphaltenes are not soluble in a low-molecular-weight paraffinic solvent such as *n*-heptane. Therefore, if *n*-heptane is added to a heavy oil, asphaltenes precipitate while the other constituents form a soluble solution with the solvent. If solvent is changed to propane, because of the greater difference between the structure of the solvent and the high-molecular-weight asphaltenes, more asphaltenic compounds precipitate. Similarly if acetone is added to a deasphalted oil (DAO), resins precipitate while low-molecular-weight hydrocarbons remain soluble in acetone. In Fig. 3.12 an all-solvent fractions procedure is shown for SARA analysis [7].

One of the disadvantages of the all-solvent separation technique is that in some instances a very low temperature (0 to −10°C) is required, which causes inconvenience in laboratory operation. Another difficulty is that in many cases large volumes of solvent may be required and solvents must have sufficiently low boiling point so that the solvent can be completely

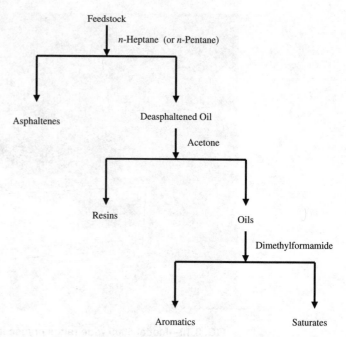

FIG. 3.12—An all-solvent fractionation procedure. Reprinted from Ref. [7], p. 267, by courtesy of Marcel Dekker, Inc.

removed from the product [7]. ASTM [4] provides several methods based on solvent separation to determine amounts of asphaltenes. In ASTM D 2007 test method *n*-pentane is used as the solvent, while in ASTM D 4124 asphaltene is separated by *n*-heptane. Schematics of these test methods are shown in Figs. 3.13 and 3.14, respectively, as given by Speight [7]. Asphaltenes are soluble in liquids with a surface tension above 25 dyne/cm such as pyridine, carbon disulfide, carbon tetrachloride, and benzene [7].

The principle of separation by chromatography technique was described in Section 3.1.1.3. If the mobile phase is gas the instrument is called a gas chromatograph (GC), while for

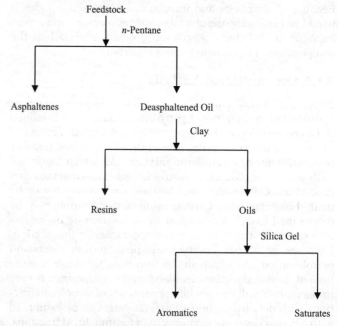

FIG. 3.13—The ASTM D 2007 procedure. Reprinted from Ref. [7], p. 280, by courtesy of Marcel Dekker, Inc.

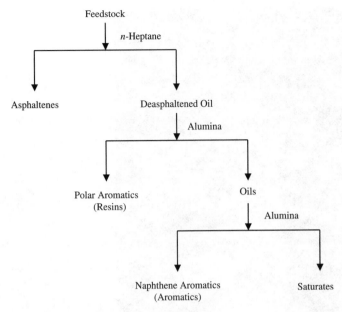

FIG. 3.14—Separation of asphaltenes and resins from petroleum fractions. Reprinted from Ref. [7], p. 281, by courtesy of Marcel Dekker, Inc.

liquid mobile phase it is called a *liquid chromatograph* (LC). As discussed earlier, components can be separated by their boiling points through GC analysis. In advanced petroleum refineries automatic online GCs are used for continuous analysis of various streams to control the quality of products. A

stream may be analyzed every 20 min and automatic adjustment can be made to the refinery unit. In crude assay analysis distillation is being replaced by chromatography techniques. The LC method is used for less volatile mixtures such as heavy petroleum fractions and residues. Use of LC for separation of saturated and aromatic hydrocarbons is described in ASTM D 2549 test method. Various forms of chromatography techniques have been applied to a wide range of petroleum products for analysis, such as PONA, PIONA, PNA, and SARA. One of the most useful types of liquid chromatography is *high performance liquid chromatography* (HPLC), which can be used to identify different types of hydrocarbon groups. One particular application of HPLC is to identify asphaltene and resin type constituents in nonvolatile feedstocks such as residua. Total time required to analyze a sample by HPLC is just a few minutes. One of the main advantages of HPLC is that the boiling range of sample is immaterial. A HPLC analyzer is shown in Fig. 3.15.

The accuracy of chromatography techniques mainly depends on the type of detector used [7]. In Section 3.1.1.3, flame ionization (FID) and thermal conductivity (TCD) detectors are described, which are widely used in GC. For LC the most common detectors are refractive index detector (RID) and wavelength UV (ultraviolet) detector. UV spectroscopy is particularly useful to identify the types of aromatics in asphaltenic fractions. Another spectroscopy method is conventional *infrared* (IR) spectroscopy, which yields information about the functional features of various petroleum constituents. For example, IR spectroscopy will aid in the identification of N—H and O—H functions and the nature of polymethylene

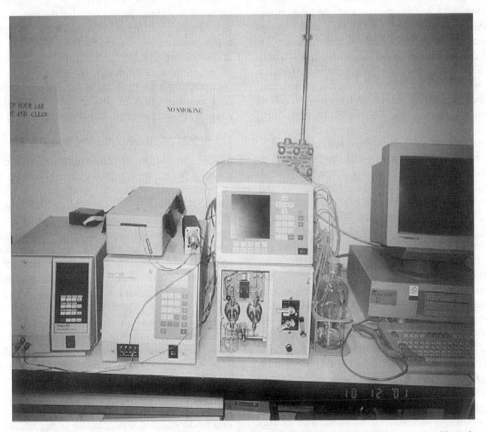

FIG. 3.15—A HPLC instrument (courtesy of Chemical Engineering Department at Kuwait University).

FIG. 3.16—HP made GC–MS model 5890 Series II. (courtesy of Chemical Engineering Department at Kuwait University).

chains (C—H) and the nature of any polynuclear aromatic systems [7].

Another type of analysis of petroleum fractions to identify molecular groups is by spectrometric methods such as *mass spectrometry* (MS). In general, there is a difference between spectroscopy and spectrometry methods although in some references this difference is not acknowledged. Spectroscopy refers to the techniques where the molecules are excited by various sources, such as UV and IR, to return to their normal state. Spectrometry refers to the techniques where the molecules are actually ionized and fragmented. Evolution of spectroscopic methods comes after chromatography techniques, nonetheless, and in recent decades they have received considerable attention. While volatile and light petroleum products can be analyzed by gas chromatography, heavier and nonvolatile compounds can be analyzed and identified by spectrometric methods. One of the most important types of spectrometry techniques in analysis of petroleum fractions is mass spectrometry (MS). In this method, masses of molecular and atomic components that are ionized in the gaseous state by collision with electrons are measured. The advantage of MS over other spectrometric methods is its high reproducibility of quantitative analysis and information on molecular type in complex mixtures. Mass spectrometry can provide the most detailed quantitative and qualitative information about the atomic and molecular composition of organic and inorganic compounds. However, use of MS is limited to organic compounds that are stable up to 300°C (570°F). At higher temperatures thermal decomposition may occur and the analysis will be biased [7]. Through MS analysis, hydrocarbons of similar boiling points can be identified. In the MS analysis, molecular weight, chemical formula of hydrocarbons, and their amounts can be determined. The most powerful instrument to analyze petroleum distillates is the combination of a GC and an MS called GC–MS instrument, which separates compounds both through boiling point and molecular weight. For heavy petroleum fractions containing high-boiling-point compounds an integrated LC–MS unit may be suitable for analysis of mixtures; however, use of LC–MS is more difficult than GC–MS because in LC–MS solvent must be removed from the elute before it can be analyzed by MS. A GC–MS instrument from Hewlett Packard (HP) is shown in Fig. 3.16

Another type of separation is by SEC or GPC, which can be used to determine molecular weight distribution of heavy petroleum fractions as discussed in Section 3.1.3. Fractions are separated according to their size and molecular weight and the method is particularly useful to determine the amount of asphaltenes. Asphaltenes are polar multiring aromatic compounds with molecular weight above 1000 (see Fig. 1.2). It is assumed that in this molecular weight range only aromatics are present in a petroleum fraction [8].

3.1.5.3 PNA Analysis

As determination of the exact composition of a petroleum fraction is nearly an impossible task, for narrow boiling range petroleum fractions and products a useful type of compositional analysis is to determine the amounts of paraffins (P), naphthenes (N), and aromatics (A). As mentioned before, most petroleum products are olefin free and PNA analysis provides a good knowledge of molecular type of mixture constituents. However, some analyzers give the amount of isoparaffins and olefins as well. These analyzers are called PIONA analyzer, and a Chrompack Model 940 PIONA analyzer is shown in Fig. 3.17. An output of this type of analyzer is similar to the GC output; however, it directly gives wt% of *n*-paraffins, isoparaffins, olefins, naphthenes, and aromatics. The composition is expressed in wt%, which can be converted to mole, weight, and volume fractions as will be shown later in this chapter.

3.1.5.4 Elemental Analysis

The main elements present in a petroleum fraction are carbon (C), hydrogen (H), nitrogen (N), oxygen (O), and sulfur (S). The most valuable information from elemental analysis

FIG. 3.17—A chrompack model 940 PIONA analyzer (courtesy of Chemical Engineering Department at Kuwait University).

that can be obtained is on the C/H ratio and sulfur content of a petroleum mixture from which one can determine the quality of oil. As boiling points of fractions increase or their API gravity decrease the amount of C/H ratio, sulfur content, nitrogen content, and the metallic constituents increase, signifying a reduction in the quality of an oil. Sulfur content of very heavy fractions can reach 6–8% and the nitrogen content can reach 2.0–2.5 wt%. There are specific methods to measure these elements individually. However, instruments do exist that measure these elements all together; these are called elemental analyzers. One of these apparatuses is CHN analyzers in which there is a simultaneous combustion in pure oxygen at 1000°C. Carbon is reduced to CO_2, H is reduced to H_2O, and N is converted to nitrogen oxides. Nitrogen oxides are then reduced over copper at 650°C to nitrogen by eliminating oxygen. A mixture of CO_2, H_2O, and N_2 is separated by gas chromatography with TCD. In a similar fashion, sulfur is oxidized to SO_2 and is detected by TCD after detection of CO_2, N_2, and H_2O. Oxygen is determined by passing it over carbon at high temperature and converted to CO, which is measured by a GC [5]. ASTM test methods for elemental analysis of petroleum products and crude oils include hydrogen content (ASTM D 1018, D 3178, D 3343), nitrogen content (ASTM D 3179, D 3228, D 3431), and sulfur content (ASTM D 129/IP 61, D 1266/IP 107, D 1552, D 4045). An elemental analyzer Model CHNS-932 (Leco Corp., St. Joseph, MI, USA) is shown in Fig. 3.18. In this analyzer, the CO_2, H_2O, and SO_2 are detected by infrared detector (IRD) while N_2 is determined by the TCD method.

Another group of heteroatoms found in petroleum mixtures are metallic constituents. The amount of these metals are in the range of few hundreds to thousand ppm and their amounts increase with increase in boiling points or decrease in the API gravity of oil. Even a small amount of these metals, particularly nickel, vanadium, iron, and copper, in the feedstocks for catalytic cracking have negative effects on the activity of catalysts and result in increased coke forma-

tion. Metallic constituents are associated with heavy compounds and mainly appear in residues. There is no general method to determine the composition of all metals at once but ASTM [4] provides test methods for determination of various metallic constituents (i.e., ASTM D 1026, D 1262, D 1318, D 1368, D 1548). Another method is to burn the oil sample in which metallic compounds appear in inorganic ashes. The ash should be digested by an acid and the solution is examined for metal species by atomic absorption spectroscopy [7].

3.1.6 Viscosity

Absolute and kinematic viscosities are defined in Section 2.1.8 and experimental data for the kinematic viscosity of some pure hydrocarbons are given in Table 2.2. Viscosity of petroleum fractions increase with a decrease in the API gravity and for residues and heavy oils with the API gravity of less than 10 (specific gravity of above 1), viscosity varies from several thousands to several million poises. Viscosity is a bulk property that can be measured for all types of petroleum fractions in liquid form. Kinematic viscosity is a useful characterization parameter for heavy fractions in which boiling point data are not available due to thermal decomposition during distillation. Not only is viscosity an important physical property, but it is a parameter that can be used to estimate other physical properties as well as the composition and quality of undefined petroleum fractions as shown later in this chapter. Since viscosity varies with temperature, values of viscosity must be reported with specified temperature. Generally, kinematic viscosity of petroleum fractions are measured at standard temperatures of 37.8°C (100°F) and 98.9°C (210°F). However, for very heavy fractions viscosity is reported at temperatures above 38°C, i.e., 50°C (122°F) or 60°C (140°F). When viscosity at two temperatures are reported from the method of Section 2.7 one can obtain the viscosity at other temperatures. Measurement of viscosity is easy but the method and

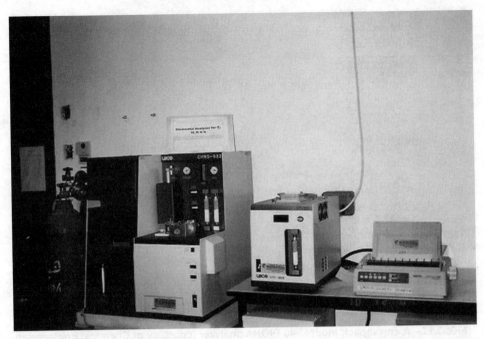

FIG. 3.18—Leco made CHNS-932 model elemental analyzer (courtesy of Chemical Engineering Department at Kuwait University).

the instrument depend on the type of sample. For Newtonian and high-shear fluids such as engine oils, viscosity can be measured by a capillary U-type viscometer. An example of such viscometers is the Cannon–Fenske viscometer. The test method is described in ASTM D 445, which is equivalent to ISO 3104 method, and kinematic viscosity is measured at temperatures from 15 to 100°C (~60 to 210°F). In this method, repeatability and reproducibility are 0.35 and 0.7%, respectively [5]. Another type of viscometer is a rotary viscometer, which is used for a wide range of shear rates, especially for low shear rate and viscous fluids such as lubricants and heavy petroleum fractions. In these viscometers, fluid is placed between two surfaces, one is fixed and the other one is rotating. In these viscometers absolute viscosity can be measured and an example of such viscometers is the Brookfield viscometer. Details of measurement and prediction of viscosity of petroleum fractions are given in Chapter 8. As the viscosity of petroleum fractions, especially the heavy oils, is one of the most difficult properties to estimate, its experimental value is highly useful and desirable.

3.2 PREDICTION AND CONVERSION OF DISTILLATION DATA

Various distillation curves are introduced in Section 3.1.1. For simplicity ASTM is used to refer to ASTM D 86 distillation curve, similarly TBP, SD, and EFV refer to true boiling point, simulated distillation (ASTM D 2887), and equilibrium flash vaporization, respectively. Petroleum fractions have a range of boiling points. To use the correlations introduced in Chapter 2, a single value for boiling point is required. For this reason there is a need for the definition of an average boiling point or a characteristic boiling point based on a distillation curve. Availability of one type of distillation curve

for simplicity in experimental measurement and the need for another type for its application requires conversion methods between various distillation curves. The tedious procedures necessary to obtain experimental EFV data have given impetus to the development of correlations for predicting EFV data from the analytical ASTM and TBP distillations. Simulated distillation by gas chromatography appears to be the most simple, reproducible, and consistent method to describe the boiling range of a hydrocarbon fraction unambiguously. TBP is the most useful distillation curve, while available data might be ASTM D 86, ASTM D 2887, or ASTM D 1160 distillation curves. ASTM [4] has accepted this technique as a tentative method for the "Determination of Boiling Range Distribution of Petroleum Fractions by Gas Chromatography" (ASTM D 2887). In most cases distillation data are reported in terms of ASTM D 86 or SD. In this section methods of calculation of average boiling points, interconversion of various distillation curves, and prediction of complete distillation curves from a limited data are presented.

3.2.1 Average Boiling Points

Boiling points of petroleum fractions are presented by distillation curves such as ASTM or TBP. However, in prediction of physical properties and characterization of hydrocarbon mixtures a single characteristic boiling point is required. Generally an average boiling point for a fraction is defined to determine the single characterizing boiling point. There are five average boiling points defined by the following equations [9]. Three of these average boiling points are VABP (volume average boiling point), MABP (molal average boiling point) and WABP (weight average boiling point), defined for a mixture of n components as

$$(3.3) \qquad \text{ABP} = \sum_{i=1}^{n} x_i T_{bi}$$

where ABP is the VABP, MABP, or WABP and x_i is the corresponding volume, mole, or weight fraction of component i. T_{bi} is the normal boiling point of component i in kelvin. Two other average boiling points are CABP (cubic average boiling point) and MeABP (mean average boiling point) defined as

$$(3.4) \quad CABP = \left(\frac{1}{1.8}\right)\left[\sum_{i=1}^{n} x_{vi}\,(1.8T_{bi} - 459.67)^{1/3}\right]^3 + 255.37$$

$$(3.5) \quad MeABP = \frac{MABP + CABP}{2}$$

where T_{bi} in Eq. (3.4) is in kelvin. The conversion factors in Eq. (3.4) come from the fact that the original definition of CABP is in degrees Fahrenheit. For petroleum fractions in which volume, weight, or mole fractions of components are not known, the average boiling points are calculated through ASTM D 86 distillation curve as

$$(3.6) \quad VABP = \frac{T_{10} + T_{30} + T_{50} + T_{70} + T_{90}}{5}$$

where T_{10}, T_{30}, T_{50}, T_{70}, and T_{90} are ASTM temperatures at 10, 30, 50, 70, and 90 vol% distilled. ASTM distillation curves can be characterized by the magnitudes of temperatures and overall slope of the curve. A parameter that approximately characterizes slope of a distillation curve is the slope of a linear line between 10 and 90% points. This slope shown by SL is defined as

$$(3.7) \quad SL = \frac{T_{90} - T_{10}}{80}$$

where T_{10} and T_{90} are the ASTM D 86 temperatures at 10 and 90% of volume vaporized. The 10–90 slope, SL, in some references is referred to as the Engler slope and is indicative of a variety of compounds in a petroleum fraction. When the boiling points of compounds are near each other the value of SL and the boiling range of the fraction are low. For petroleum fractions, WABP, MABP, CABP, and MeABP are correlated through an empirical plot to VAPB and SL in Chapter 2 of the API-TDB [2]. Analytical correlations based on the API plot were developed by Zhou [10] for use in a digital computer. For heavy fractions and vacuum distillates in which distillation data by ASTM D 1160 are available, they should first be converted to ASTM D 86 and then average boiling points are calculated. Analytical correlations for estimation of average boiling points are given by the following equations in terms of VABP and SL [2, 10].

$$(3.8) \quad ABP = VABP - \Delta T$$

$$\ln(-\Delta T_W) = -3.64991 - 0.02706(VABP - 273.15)^{0.6667}$$
$$(3.9) \quad\quad\quad + 5.163875 SL^{0.25}$$

$$\ln(\Delta T_M) = -1.15158 - 0.01181(VABP - 273.15)^{0.6667}$$
$$(3.10) \quad\quad\quad + 3.70612 SL^{0.333}$$

$$\ln(\Delta T_C) = -0.82368 - 0.08997(VABP - 273.15)^{0.45}$$
$$(3.11) \quad\quad\quad + 2.456791 SL^{0.45}$$

$$\ln(\Delta T_{Me}) = -1.53181 - 0.0128(VABP - 273.15)^{0.6667}$$
$$(3.12) \quad\quad\quad + 3.646064 SL^{0.333}$$

where ABP is an average boiling point such as WABP, MABP, CABP, or MeABP and ΔT is the corresponding correction temperature for each ABP. All temperatures are in kelvin.

VABP and SL are defined in Eqs. (3.6) and (3.7). Once ΔT is calculated for each case, corresponding ABP is calculated from Eq. (3.8). Equations (3.8)–(3.12) calculate values of various ABP very close to those obtained from empirical plot in the API-TDB [2]. The following example shows application of these equations in calculation of various ABP.

Example 3.1—A low boiling naphtha has the ASTM D 86 temperatures of 77.8, 107.8, 126.7, 155, and 184.4°C at 10, 30, 50, 70, and 90 vol% distilled [11]. Calculate VABP, WABP, MABP, CABP, and MeABP for this fraction.

Solution—From Eqs. (3.6) and (3.7) VABP and SL are calculated as follows: VABP = $(77.8 + 107.8 + 126.7 + 155 + 184.4)/5 = 130.3°C = 403.5$ K, and SL = $(184.4 - 77.8)/80 = 1.333°C$ (K)/%. From Eqs. (3.9)–(3.12) various correction temperatures are calculated: $\Delta T_W = -3.3°C$, $\Delta T_M = 13.8°C$, $\Delta T_C = 3.2°C$, and $\Delta T_{Me} = 8.6°C$. From Eq. (3.8) various average boiling points are calculated: WABP = 133.7, MABP = 116.5, CABP = 127.1, and MeABP = 121.7°C. ♦

Application and estimation of various boiling points are discussed by Van Winkle [12]. Since the materials boil over a range of temperature, any one average boiling point fails to be useful for correlation of all properties. The most useful type of ABP is MeABP, which is recommended for correlation of most physical properties as well as calculation of Watson K as will be discussed later in this chapter. However, for calculation of specific heat, VABP is recommended [12]. In Example 3.1, MeABP is 121.7°C, which varies from 126.7 for the ASTM D 86 temperature at 50 vol% distilled (T_{50}). However, based on our experience, for narrow boiling range fractions with SL < 0.8°C/% the MeABP is very close to 50% ASTM temperature. As an example, for a gas oil sample [11] with ASTM temperatures of 261.7, 270, 279.4, 289.4, and 307.2°C at 10, 30, 50, 70, and 90 vol%, the MeABP is calculated as 279, which is very close to 50% ASTM temperature of 279.4°C. For this fraction the value of SL is 0.57°C/%, which indicates the boiling range is quite narrow. Since none of the average boiling points defined here represent the true boiling point of a fraction, the 50% ASTM temperature may be used as a characteristic boiling point instead of average boiling point. In this case it is assumed that the difference between these temperatures is within the range of experimental uncertainty for the reported distillation data as well as the correlation used to estimate a physical property.

3.2.2 Interconversion of Various Distillation Data

Work to develop empirical methods for converting ASTM distillations to TBP and EFV distillations began in the late 1920s and continued through the 1950s and 1960s by a large number of researchers [13–18]. All of the correlations were based on discordant experimental data from the literature. Experimental ASTM, TBP, and EFV data on which the empirical correlations are based suffer a lack of reproducibility because there were no standardized procedures or apparatus available. All of these correlations were evaluated and compared to each other by House et al. [19] to select most appropriate methods for inclusion in the API-TDB. As a result of their evaluations, the following methods were adopted in the API Data Book as the best method: Edmister–Pollock [14] for ASTM to TBP, Edmister–Okamoto [15–17] for ASTM to EFV, and Maxwell for conversion of TBP to EFV [19]. Most of these correlations were in graphical forms and inconvenient for computer applications. Later, Arnold computerized these graphical methods through a set of nth order polynomials [20]. Correlation to convert ASTM D 2887 (SD) to ASTM D 86 were first developed by Ford using multiplier regression analysis [21]. In the mid 1980s Riazi and Daubert [22] developed analytical methods for the conversion of distillation curves based on the generalized correlation for hydrocarbon properties given by Eq. (2.2). These methods were adopted by the API in the fifth edition of API-TDB-88 [2] to replace the previous methods. Continued interests from the petroleum industry for these conversion methods led to development of further methods. The latest methods for the conversion of distillation curves were developed by Daubert in mid 1990s [23] through modifying Riazi–Daubert correlations. In this section the API methods (Riazi–Daubert and Daubert) for conversion of distillation data are presented, which are also recommended and used in other references and industrial software [24, 25].

3.2.2.1 Riazi–Daubert Method

Riazi and Daubert methods for the interconversion of various distillation data are based on the generalized correlation for property estimation of hydrocarbons in the form of Eq. (2.38). Available distillation temperature and specific gravity of the fraction are used as the input parameters to estimate the desired distillation data in the following form [22]:

$$(3.13) \qquad T_i(\text{desired}) = a\left[T_i(\text{available})\right]^b \text{SG}^c$$

where T_i (available) is the available distillation temperature at a specific vol% distilled and T_i(desired) is the desired distillation data for the same vol% distilled, both are in kelvin. SG is the specific gravity of fraction at 15.5°C and $a, b,$ and c are correlation parameters specific for each conversion type and each vol% point on the distillation curve. For example, if this equation is used to convert ASTM to EFV at 10%, T_i (available) is ASTM temperature at 10% and T_i(desired) is the EFV temperature at 10% and constants $a, b,$ and c are specific for this conversion type at 10% of volume vaporized.

3.2.2.1.1 ASTM D 86 and TBP Conversion—If distillation data available are in the form of ASTM D 86 and desired distillation is TBP, Eq. (3.13) can be used, but for this particular type of conversion value of constant c for all points is zero and the equation reduces to

$$(3.14) \qquad \text{TBP} = a(\text{ASTM D } 86)^b$$

where both TBP and ASTM temperatures are for the same vol% distilled and are in kelvin. Constants a and b at various points along the distillation curve with the range of application are given in Table 3.2.

For a total of 559 data points for 80 different samples, Eq. (3.14) gives an average absolute deviation (AAD) of about 5°C, while the Edmister–Pollock method [14] gives an AAD of about 7°C. Generally predictions at 0% give higher errors and are less reliable. Details of evaluations are given in our previous publications [22, 26]. Equation (3.14) can be easily reversed to predict ASTM from TBP data, but this is a rare application as usually ASTM data are available. If TBP distil-

TABLE 3.2—*Correlation constants for Eq. (3.14).*

Vol%	a	b	ASTM D 86 range,[a] °C
0	0.9177	1.0019	20–320
10	0.5564	1.0900	35–305
30	0.7617	1.0425	50–315
50	0.9013	1.0176	55–320
70	0.8821	1.0226	65–330
90	0.9552	1.0110	75–345
95	0.8177	1.0355	75–400

Source: Ref. [22].
[a]Temperatures are approximated to nearest 5.

lation curve is available then ASTM curve can be estimated as

$$(3.15) \qquad \text{ASTM D } 86 = \left(\frac{1}{a}\right)^{1/b}(\text{TBP})^{1/b}$$

where constants a and b are given in Table 3.2 as for Eq. (3.14).

3.2.2.1.2 ASTM D 86 and EFV Conversions—Application of Eq. (2.13) to this type of conversion gives

$$(3.16) \qquad \text{EFV} = a(\text{ASTM D } 86)^b(\text{SG})^c$$

where constants $a, b,$ and c were obtained from more than 300 data points and are given in Table 3.3. Equation (3.16) was evaluated with more than 300 data points from 43 different samples and gave AAD of 6°C, while the method of Edmister–Okamoto [15] gave an AAD of 10°C [22, 26].

In using these equations if specific gravity of a fraction is not available, it may be estimated from available distillation curves at 10 and 50% points as given by the following equation:

$$(3.17) \qquad \text{SG} = aT_{10}^b T_{50}^c$$

where constants $a, b,$ and c for the three types of distillation data, namely, ASTM D 86, TBP, and EFV, are given in Table 3.4. Temperatures at 10 and 50% are both in kelvin.

3.2.2.1.3 SD to ASTM D 86 Conversions—The equation derived from Eq. (3.13) for the conversion of simulated distillation (ASTM D 2887) to ASTM D 86 distillation curve has the following form:

$$(3.18) \qquad \text{ASTM D } 86 = a(\text{SD})^b(F)^c$$

where constant F is a parameter specifically used for this type of conversion and is given by the following equation:

$$(3.19) \qquad F = 0.01411(\text{SD }10\%)^{0.05434}(\text{SD }50\%)^{0.6147}$$

in which SD 10% and SD 50% are the SD temperatures in kelvin at 10 and 50 wt% distilled, respectively. Parameter F calculated from Eq. (3.19) must be substituted in Eq. (3.18) to estimate ASTM D 86 temperature at corresponding percent point expressed in volume basis. Equation (3.18) cannot be

TABLE 3.3—*Correlation constants for Eq. (3.16).*

Vol%	a	b	c	ASTM D 86 range,[a] °C
0	2.9747	0.8466	0.4209	10–265
10	1.4459	0.9511	0.1287	60–320
30	0.8506	1.0315	0.0817	90–340
50	3.2680	0.8274	0.6214	110–355
70	8.2873	0.6871	0.9340	130–400
90	10.6266	0.6529	1.1025	160–520
100	7.9952	0.6949	1.0737	190–430

Source: Ref. [22].
[a]Temperatures are approximated to nearest 5.

TABLE 3.4—Correlation constants for Eq. (3.17).

Distillation type	T_{10} range,^{a}°C	T_{50} range,^{a}°C	SG range	a	b	c	No. of data points	AAD %
ASTM D 86	35–295	60–365	0.70–1.00	0.08342	0.10731	0.26288	120	2.2
TBP	10–295	55–320	0.67–0.97	0.10431	0.12550	0.20862	83	2.6
EFV	79–350	105–365	0.74–0.91	0.09138	−0.0153	0.36844	57	57

Source: Ref. [22].
^{a}Temperatures are approximated to nearest 5.

used in a reverse form to predict SD from ASTM D 86, but this type of conversion is usually not desired as most predictive methods use ASTM D 86 data while laboratories report SD data. Constants a, b, and c in Eq. (3.18) were obtained from 81 different samples and 567 data points and are given in Table 3.5 with the range of SD data at each percentage along the distillation curve.

Equation (3.18) and the method of Ford published by ASTM, included in the earlier editions of API-TDB [21], were evaluated by some 570 data points and gave AAD of 5 and 5.5°C, respectively [22, 26]. Larger errors were observed at the initial and final boiling points (0 and 100%) but excluding these points the AAD reduces to about 3°C for conversions within the range of 10–90% distilled.

The procedures given in this section should be used with the range of data specified in Tables 3.1–3.4. Use of these equations outside the specified ranges could cause large errors. Graphical forms of these equations for conversion of various distillation curves are given in Reference [22] as well as in the fourth edition of the API-TDB-88 [2]. One of the advantages of these equations is that they can be used in reversed form. This means one may estimate EFV from TBP data through conversion of TBP to ASTM by Eq. (3.15) and then using Eq. (3.16) to estimate EFV from calculated ASTM curve. The example below shows this conversion process.

Example 3.2—For a blend of naphtha–kerosene sample, ASTM, TBP, and EFV distillation curves are given in the API-TDB [2]. These data are represented in Table 3.6. Use the Riazi–Daubert methods to predict EFV curve from TBP curve.

Solution—TBP data are used as available input data. Equation (3.15) should be used to estimate ASTM D 86 from TBP. For the initial point at 0%, the calculations are as follows. ASTM D 86 = $(1/0.9177)^{1/1.0019}$ $(10 + 273)^{1/1.0019}$ = 305 K = 305 − 273 = 32°C. The actual data for the initial ASTM temperature is 35°C, which is close to the calculated value. Now to estimate EFV from Eq. (3.16), specific gravity, is required which is not given by the problem. SG can be estimated from Eq. (3.17) and constants given in Table 3.3 for the TBP. From Table 3.6, T_{10}(TBP) = 71.1 and T_{50}(TBP) = 204.4°C. Using these values in Eq. (3.17) gives SG = 0.10431 $(71.1 + 273)^{0.1255}$ $(204.4 + 273)^{0.20862}$ = 0.7862. Now from

calculated ASTM and SG, the EFV temperatures can be estimated from Eq. (3.16) with constants given in Table 3.2. EFV = 2.9747 $(32 + 273)^{0.8466}$ $(0.7862)^{0.4209}$ = 340.9 K = 340.9 − 273 = 67.9°C. The calculated value is very close to the actual value of 68.3°C (see Table 3.5). Similarly EFV values at other points are calculated and results are shown in Fig. 3.19. Predicted EFV curve from TBP are very close to the actual EFV curve. The AAD between predicted EFV and experimental data is 2.6 K. It should be noted that if experimental ASTM data and specific gravity were used, the predicted values of EFV would be even closer to the experimental values. ◆

3.2.2.2 Daubert's Method

Daubert and his group developed a different set of equations to convert ASTM to TBP, SD to ASTM, and SD to TBP [23]. These methods have been included in the sixth edition of API-TDB [2] and are given in this section. In these methods, first conversion should be made at 50% point and then the difference between two cut points are correlated in a form similar to Eq. (3.14). In this method SD data can be converted directly to TBP without calculating ASTM as was needed in the Riazi–Daubert method.

3.2.2.2.1 ASTM and TBP Conversion—The following equation is used to convert an ASTM D 86 distillation at 50% point temperature to a TBP distillation 50% point temperature.

$$TBP(50\%) = 255.4 + 0.8851[ASTM\ D\ 86(50\%) − 255.4]^{1.0258}$$
(3.20)

where ASTM (50%) and TBP (50%) are temperatures at 50% volume distilled in kelvin. Equation (3.20) can also be used in a reverse form to estimate ASTM from TBP. The following equation is used to determine the difference between two cut points:

$$Y_i = AX_i^B$$
(3.21)

where

Y_i = difference in TBP temperature between two cut points, K (or °C)

X_i = observed difference in ASTM D 86 temperature between two cut points, K (or °C)

A, B = constants varying for each cut point and are given in Table 3.7

TABLE 3.5—Correlation constants for Eq. (3.18).

Vol%	a	b	c	SD range,^{a}°C
0	5.1764	0.7445	0.2879	−20–200
10	3.7452	0.7944	0.2671	25–230
30	4.2749	0.7719	0.3450	35–255
50	1.8445	0.5425	0.7132	55–285
70	1.0751	0.9867	0.0486	65–305
90	1.0849	0.9834	0.0354	80–345
100	1.7991	0.9007	0.0625	95–405

Source: Ref. [22].
^{a}Temperatures are approximated to nearest 5.

TABLE 3.6—Data on various distillation curves for a naphtha–kerosene blend [2].

Vol% distilled	ASTM D 86, °C	TBP, °C	EFV, °C
0	35.0	10.0	68.3
10	79.4	71.1	107.2
30	145.6	143.3	151.1
50	201.7	204.4	182.2
70	235.6	250.6	207.2
90	270.6	291.7	228.3

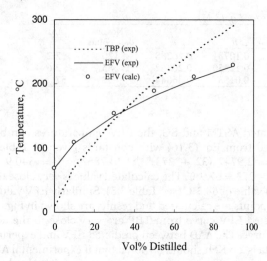

FIG. 3.19—Prediction of EFV from TBP curve for a naphtha–kerosene blend (Example 3.2).

To determine the true boiling point temperature at any percent distilled, calculation should begin with 50% TBP temperature and addition or subtraction of the proper temperature difference Y_i.

$$(3.22)\quad\begin{aligned}
\text{TBP}(0\%) &= \text{TBP}(50\%) - Y_4 - Y_5 - Y_6 \\
\text{TBP}(10\%) &= \text{TBP}(50\%) - Y_4 - Y_5 \\
\text{TBP}(30\%) &= \text{TBP}(50\%) - Y_4 \\
\text{TBP}(70\%) &= \text{TBP}(50\%) + Y_3 \\
\text{TBP}(90\%) &= \text{TBP}(50\%) + Y_3 + Y_2 \\
\text{TBP}(100\%) &= \text{TBP}(50\%) + Y_3 + Y_2 + Y_1
\end{aligned}$$

This method was developed based on samples with ASTM 50% point temperature of less than 250°C (480°F), but it is recommended for extrapolation up to fractions with ASTM 50% temperature of 315°C (600°F) as suggested by the API [2]. Average absolute deviation for this method as reported by the API-TDB [2] is about 4.6°C for some 70 samples. Predicted TBP at 0 and 100% are the least accurate values followed by values at 10 and 90% points as it is shown in the following example.

Example 3.3—ASTM D 86 and TBP distillation data for a kerosene sample [1] are given in Table 3.7. Predict the TBP curve from ASTM data using Riazi–Daubert and Daubert's methods and calculate AAD for each method.

Solution—The Riazi–Daubert method for conversion of ASTM to TBP data is presented by Eq. (3.14) and constants in Table 3.2. The Daubert's method is expressed by Eqs. (30.20)–

TABLE 3.7—*Correlation constants for Eq. (3.21).*

i	Cut point range, %	A	B	Maximum allowable $X_i,^a$°C
1	100–90	0.1403	1.6606	...
2	90–70	2.6339	0.7550	55
3	70–50	2.2744	0.8200	85
4	50–30	2.6956	0.8008	140
5	30–10	4.1481	0.7164	140
6	10–0	5.8589	0.6024	55

Source: Refs. [2, 23].
^{a}Temperatures are approximated to nearest 5.

(3.22). The summary of results is given in Table 3.8. The overall average absolute deviations (AAD) for Eqs. (3.14) and (3.20) are calculated as 2.2 and 3.8°C, respectively. As it is seen in Table 3.8, Eqs. (3.20)–(3.22) are more accurate at 30, 50, and 70% points than at the lower or higher ends of the distillation curve. ◆

3.2.2.2.2 SD to TBP Conversion—As described before, simulated distillation by gas chromatography (ASTM D 2887) is now commonly used as a means of measuring boiling points of light petroleum fractions. SD curves are expressed in terms of temperature versus wt% distilled, while TBP curves are expressed in terms of temperature versus vol% distilled. In the Daubert's method of conversion of SD to TBP it is assumed that TBP at 50 vol% distilled is equal to SD temperature at 50 wt% distilled. Equations for conversion of SD to TBP are similar to equations developed for conversion of ASTM to TBP.

$$(3.23)\quad\text{TBP}(50\,\text{vol}\%) = \text{SD}(50\,\text{wt}\%)$$

where SD (50 wt%) and TBP (50 vol%) are temperatures at 50% distilled in kelvin (or °C). The difference between adjacent cut points is calculated from the following equation as given by the API-TDB [2].

$$(3.24)\quad V_i = C W_i^D$$

where

V_i = difference in TBP temperature between two cut points, K (or °C)

W_i = observed difference in SD temperature between two cut points, K (or °C)

C, D = constants varying for each cut point and are given in Table 3.9

To determine the true boiling point temperature at any percent distilled, calculation should begin with 50% TBP temperature and addition or subtraction of the proper temperature difference V_i.

$$(3.25)\quad\begin{aligned}
\text{TBP}(5\%) &= \text{TBP}(50\%) - V_5 - V_6 - V_7 \\
\text{TBP}(10\%) &= \text{TBP}(50\%) - V_5 - V_6 \\
\text{TBP}(30\%) &= \text{TBP}(50\%) - V_5 \\
\text{TBP}(70\%) &= \text{TBP}(50\%) + V_4 \\
\text{TBP}(90\%) &= \text{TBP}(50\%) + V_4 + V_3 \\
\text{TBP}(95\%) &= \text{TBP}(50\%) + V_4 + V_3 + V_2 \\
\text{TBP}(100\%) &= \text{TBP}(50\%) + V_4 + V_3 + V_2 + V_1
\end{aligned}$$

This method is applicable to fractions with TBP 50% points in the range of 120–370°C (250–700°F). Average absolute deviation for this method as reported by the API-TDB [2] is about 7.5°C for about 21 samples. Based on 19 datasets it was observed that errors in direct conversion of SD to TBP is slightly higher than if SD is converted first to ASTM and then estimated ASTM is converted to TBP by Eqs. (3.20)–(3.22). Details of these evaluations are given by the API [2]. Predicted TBP at 5, 95, and 100% are the least accurate values followed by values at 10 and 90% points as is shown in the following example.

Example 3.4—Experimental ASTM D 2887 (SD) and TBP distillation data for a petroleum fraction are given in Table 3.9 as taken from API [2]. Predict the TBP curve from SD data using

TABLE 3.8—*Prediction of TBP from ASTM for a kerosene sample of Example 3.3.*

Vol% distilled	ASTM D 86 exp,°C	TBP exp,°C	Eq. (3.14)		Eqs. (3.20)–(3.22)	
			TBP calc,°C	AD,°C	TBP calc,°C	AD,°C
0	165.6	146.1	134.1	12.0	133.1	13.0
10	176.7	160.6	160.6	0.0	158.1	2.5
30	193.3	188.3	188.2	0.1	189.2	0.9
50	206.7	209.4	208.9	0.5	210.6	1.2
70	222.8	230.6	230.2	0.4	232.9	2.3
90	242.8	255.0	254.7	0.3	258.1	3.1
Overall AAD, °C				2.2		3.8

Riazi–Daubert and Daubert's methods and calculate AAD for each method.

Solution—The Riazi–Daubert methods do not provide a direct conversion from SD to TBP, but one can use Eqs. (3.18) and (3.19) to convert SD to ASTM D 86 and then Eq. (3.14) should be used to convert ASTM to TBP data. From Eq. (3.19) and use of SD at 10% and 50% points, the value of parameter F is calculated as 0.8287. Value of SD temperature at 50 wt% is 168.9°C, from Eq. (3.18) with appropriate constants in Table 3.5 one can obtain ASTM D 86 (50%) = 166.3°C. Substituting this value for ASTM into Eq. (3.14) gives TBP (50 vol%) = 167.7°C, while the experimental value as given in Table 3.10 is 166.7°C. The AD is then calculated as $167.7 - 166.7 = 1$°C. Daubert's method for conversion of SD to TBP is direct and is presented through Eqs. (3.23)–(3.25). According to Eq. (3.23), TBP (50%) = SD (50%) = 168.9°C, which gives an AD of 2.2°C for this point. A summary of complete calculation results is given in Table 3.10. The overall AAD for Eqs. (3.14) and (3.18) is 4.8, while for Eqs. (3.23)–(3.25) is 2.2°C.

Results presented in Example 3.4 show that Eqs. (3.23)–(3.25) are more accurate than Eqs. (3.14) and (3.18) for the conversion of SD to TBP. One of the reasons for such a result is that the sample presented in Table 3.10 to evaluate these methods is taken from the same data bank used to develop correlations of Eqs. (3.23)–(3.25). In addition these equations provide a direct conversion of SD to TBP. However, one should realize that Eqs. (3.23)–(3.25) are based on only 19 datasets and this limits the application of these equations. While Eqs. (3.14), (3.18), and (3.19) are based on much larger data banks with wider range of application. As available data on both SD and TBP are very limited, a concrete recommendation on superiority of these two approaches cannot be made at this time.

3.2.2.2.3 SD to ASTM D 86 Conversion—Equations to convert SD (ASTM D 2887) distillation data to ASTM D 86

TABLE 3.9—*Correlation constants for Eq. (3.24).*

i	Cut point range, %	C	D	Maximum allowable W_i,°C
1	100–95	0.03849	1.9733	15
2	95–90	0.90427	0.8723	20
3	90–70	0.37475	1.2938	40
4	70–50	0.25088	1.3975	40
5	50–30	0.08055	1.6988	40
6	30–10	0.02175	2.0253	40
7	10–0	0.20312	1.4296	20

Source: Taken with permission from Refs. [2, 23].
[a]Temperatures are approximated to nearest 5.

data are similar to the equations developed by Daubert [2, 23] to convert ASTM or SD to TBP as given in this section. The equations are summarized as following:

$$ASTM\ D\ 86(50\,vol\%) = 255.4 + 0.79424$$
$$(3.26) \quad [SD(50\,wt\%) - 255.4]^{1.0395}$$

where SD (50 wt%) and ASTM D 86 (50 vol%) are temperatures at 50% distilled in kelvin. The difference between adjacent cut points is calculated from the following equation as given by the API-TDB [2].

$$(3.27) \qquad U_i = ET_i^F$$

where

U_i = difference in ASTM D 86 temperatures between two cut points, K (or °C)

T_i = observed difference in SD temperatures between two cut points, K (or °C)

E, F = constants varying for each cut point and are given in Table 3.11

To determine the ASTM D 86 temperature at any percent distilled, calculations should begin with 50% ASTM D 86 temperature and addition or subtraction of the proper temperature difference U_i.

$$ASTM\ D\ 86(0\%) = ASTM\ D\ 86(50\%)$$
$$- U_4 - U_5 - U_6$$
$$ASTM\ D\ 86(10\%) = ASTM\ D\ 86(50\%)$$
$$- U_4 - U_5$$
$$(3.28) \quad ASTM\ D\ 86(30\%) = ASTM\ D\ 86(50\%) - U_4$$
$$ASTM\ D\ 86(70\%) = ASTM\ D\ 86(50\%) + U_3$$
$$ASTM\ D\ 86(90\%) = ASTM\ D\ 86(50\%)$$
$$+ U_3 + U_2$$
$$ASTM\ D\ 86(100\%) = ASTM\ D\ 86(50\%)$$
$$+ U_3 + U_2 + U_1$$

This method is applicable to fractions with ASTM D 86 50% points in the range of 65–315°C (150–600°F). The average absolute deviation for this method as reported by the API-TDB [2] is about 6°C for some 125 samples and approximately 850 data points. Predicted ASTM temperatures at 0 and 100% are the least accurate values followed by values at 10 and 90% points as is shown in the following example.

Example 3.5—Experimental ASTM D 2887 (SD) and ASTM D 86 distillation data for a petroleum fraction are given in Table 3.11 as taken from the API-TDB [2]. Predict the ASTM

TABLE 3.10—*Prediction of TBP from SD for a petroleum fraction of Example 3.4.*

Wt% or vol% distilled	ASTM D 2887 (SD) exp.,°C	TBP exp.,°C	Eqs. (3.18) and (3.14)		Eq. (3.23)–(3.25)	
			TBP calc,°C	AD,°C	TBP calc,°C	AD,°C
10	151.7	161.1	146.1	15.0	164.3	3.2
30	162.2	163.3	157.1	6.2	166.9	3.5
50	168.9	166.7	167.7	1.0	168.9	2.2
70	173.3	169.4	170.7	1.3	170.9	1.5
90	181.7	173.9	179.1	5.3	176.7	2.8
Overall AAD, °C				4.8		2.2

D 86 curve from SD data using Riazi–Daubert and Daubert's methods and calculate AAD for each method.

Solution—Both methods provide direct methods for conversion of SD to ASTM D 86 and calculations are similar to those presented in Examples 3.3. and 3.4. Equations (3.26)–(3.28) are used for Daubert's method, while Eqs. (3.18) and (3.19) are used for Riazi–Daubert method. A summary of complete calculation results is given in Table 3.12. The overall AAD for Eq. (3.18) is 1.5, while for Eqs. (3.26)–(3.28) is 2.0°C. ♦

Results presented in Example 3.5 show that Eq. (3.18) is slightly more accurate than Eqs. (3.26)–(3.28) for the conversion of SD to ASTM D 86. This is consistent with AAD reported for these methods. However, Eqs. (3.26)–(3.28) are based on a larger data set than is Eq. (3.18). In general, Riazi–Daubert methods are simpler and easily reversible, while the existing API methods are slightly more complex. The advantage of Daubert's methods is that the predicted curve is smooth and uniform, while in the Riazi–Daubert methods every point is predicted independent of adjacent point and lack of smoothness in predicted curve is possible, although this is rarely observed in our experience. Since in the Daubert's methods temperatures at 0 and 10% points are calculated from predicted values at 30 and 50% points, larger errors are observed at the lower (0, 5, and 10% distilled) or upper ends (90, 95, and 100% distilled) of predicted distillation curves. In general the accuracy of both methods in the prediction of distillation curves at 0 and 100% points are limited. This is mainly due to the experimental uncertainty in measurement of temperatures at the end points.

3.2.2.3 Interconverion of Distillation Curves at Reduced Pressures

Normal boiling points of heavy petroleum fractions such as products of a vacuum distillation column cannot be measured due to the thermal decomposition of heavy hydrocarbons at high temperatures. For this reason distillation data are reported at reduced pressures of 1–50 mm Hg, as described

earlier in this chapter under ASTM D 1160 test method. For prediction of physical and thermodynamic properties normal boiling points are required. For this reason methods of calculation of equivalent atmospheric boiling point (EABP) are important. One has to recognize that EABP is not a real boiling point as for such heavy fractions there is no actual and real experimental value for the normal boiling point. This parameter can be obtained from conversion of distillation curves at low pressures to equivalent distillation curves at atmospheric pressures and it is just an apparent normal boiling point. The basis of such conversion is vapor pressure correlation for the fraction of interest, which will be discussed in Chapter 6. In this part we present calculation methods for the conversion of ASTM D 1160 to atmospheric distillation curve and for the prediction of atmospheric TBP curves from ASTM D 1160. It should be noted that ASTM D 1160 does not refer to any specific pressure. The pressure may vary from 1 to 50 mm Hg. When D 1160 curve is converted to a distillation curve at atmospheric pressure through a vapor pressure correlation the resulting distillation curve is not equivalent to ASTM D 86 or to TBP curve. The resulting distillation curve is referred to as equivalent atmospheric ASTM D 1160. Another low pressure distillation data is TBP distillation curve at 1, 10, or 50 mm Hg. Through vapor pressure correlations TBP at reduced pressures can be converted to atmospheric TBP. There is a procedure for the conversion of ASTM D 1160 to TBP at 10 mm Hg which is presented in this section. Therefore, to convert ASTM D 1160 to TBP at atmospheric pressure one has to convert D 1160 at any pressure to D 1160 at 10 mm Hg and then to convert resulting D 1160 to TBP at 10 mm Hg. This means if ASTM D 1160 at 1 mm Hg is available, it must be first converted to D 1160 at 760 mm Hg, then to D 1160 at 10 mm Hg followed by conversion to TBP at 10 mm Hg and finally to TBP at 760 mm Hg. A summary chart for various conversions is presented at the end of this section.

3.2.2.3.1 Conversion of a Boiling Point at Sub- or Super-Atmospheric Pressures to the Normal Boiling Point or Vice Versa—The conversion of boiling point or saturation temperature at subatmospheric ($P < 760$ mm Hg) or super-atmospheric ($P > 760$ mm Hg) conditions to normal boiling point is based on a vapor pressure correlation. The method widely used in the industry is the correlation developed for petroleum fractions by Maxwell and Bonnell [27], which is also used by the API-TDB [2] and other sources [24] and is presented here. This correlation is given for several pressure ranges as follows:

$$(3.29) \qquad T_b' = \frac{748.1QT}{1 + T(0.3861Q - 0.00051606)}$$

TABLE 3.11—*Correlation constants for Eq. (3.27).*

i	Cut point range, %	E	F	Maximum allowable T_i,°C
1	100–90	2.13092	0.6596	55
2	90–70	0.35326	1.2341	55
3	70–50	0.19121	1.4287	55
4	50–30	0.10949	1.5386	55
5	30–10	0.08227	1.5176	85
6	10–0	0.32810	1.1259	85

Source: Taken with permission from Ref. [2].
[a] Temperatures are approximated to nearest 5.

TABLE 3.12—*Prediction of ASTM D 86 from SD for a petroleum fraction of Example 3.5.*

Vol% distilled	ASTM D 2887 (SD) exp.,°C	ASTM D 86 exp.,°C	Eqs. (3.18) and (3.19) ASTM D 86 calc.,°C	AD,°C	Eqs. (3.25)–(3.28) ASTM D 86 calc.,°C	AD,°C
10	33.9	56.7	53.2	3.4	53.5	3.2
30	64.4	72.8	70.9	1.9	68.2	4.5
50	101.7	97.8	96.0	1.8	96.8	1.0
70	140.6	131.7	131.3	0.4	132.5	0.9
90	182.2	168.3	168.3	0.0	167.8	0.6
Overall AAD, °C				1.5		2.0

$$Q = \frac{6.761560 - 0.987672 \log_{10} P}{3000.538 - 43 \log_{10} P} \quad (P < 2 \text{ mm Hg})$$

$$Q = \frac{5.994296 - 0.972546 \log_{10} P}{2663.129 - 95.76 \log_{10} P} \quad (2 \le P \le 760 \text{ mm Hg})$$

$$Q = \frac{6.412631 - 0.989679 \log_{10} P}{2770.085 - 36 \log_{10} P} \quad (P > 760 \text{ mm Hg})$$

$$T_b = T_b' + 1.3889 F (K_W - 12) \log_{10} \frac{P}{760}$$

$$F = 0 \qquad (T_b < 367 \text{ K}) \text{ or when } K_W$$
$$\text{is not available}$$

$$F = -3.2985 + 0.009 T_b \qquad (367 \text{ K} \le T_b \le 478 \text{ K})$$

$$F = -3.2985 + 0.009 T_b \qquad (T_b > 478 \text{ K})$$

where

P = pressure at which boiling point or distillation data is available, mm Hg

T = boiling point originally available at pressure P, in kelvin

T_b' = normal boiling point corrected to $K_W = 12$, in kelvin

T_b = normal boiling point, in kelvin

K_W = Watson (UOP) characterization factor [$= (1.8 T_b)^{1/3}$ /SG]

F = correction factor for the fractions with K_W different from 12

$\log_{10}$ = common logarithm (base 10)

The original evaluation of this equation is on prediction of vapor pressure of pure hydrocarbons. Reliability of this method for normal boiling point of petroleum fractions is unknown. When this equation is applied to petroleum fractions, generally K_W is not known. For these situations, T_b' is calculated with the assumption that K_W is 12 and $T_b = T_b'$. This is to equivalent to the assumption of $F = 0$ for low-boiling-point compounds or fractions. To improve the result a second round of calculations can be made with K_W calculated from estimated value of T_b'. When this equation is applied to distillation curves of crude oils it should be realized that value of K_W may change along the distillation curve as both T_b and specific gravity change.

Equation (3.29) can be easily used in its reverse form to calculate boiling points (T) at low or elevated pressures from normal boiling point (T_b) as follows:

$$(3.30) \qquad T = \frac{T_b'}{748.1 Q - T_b'(0.3861 Q - 0.00051606)}$$

where

$$T_b' = T_b - 1.3889 F (K_W - 12) \log_{10} \frac{P}{760}$$

where all the parameters are defined in Eq. (3.29). The main application of this equation is to estimate boiling points at 10 mm Hg from atmospheric boiling points. At $P = 10$ mm Hg, $Q = 0.001956$ and as a result Eq. (3.30) reduces to the following simple form:

$$(3.31) \qquad T(10 \text{ mm Hg}) = \frac{0.683398 T_b'}{1 - 1.63434 \times 10^{-4} T_b'}$$

in which T_b' is calculated from T_b as given in Eq. (3.30) and both are in kelvin. Temperature T (10 mm Hg) is the boiling point at reduced pressure of 10 mm Hg in kelvin. By assuming $K_W = 12$ (or $F = 0$) and for low-boiling fractions value of normal boiling point, T_b, can be used instead of T_b' in Eq. (3.31). To use these equations for the conversion of boiling point from one low pressure to another low pressure (i.e., from 1 to 10 mm Hg), two steps are required. In the first step, normal boiling point or T (760 mm Hg) is calculated from T (1 mm Hg) by Eq. (3.29) and in the second step T (10 mm Hg) is calculated from T (760 mm Hg) or T_b through Eqs. (3.30) and (3.31).

In the mid 1950s, another graphical correlations for the estimation of vapor pressure of high boiling hydrocarbons were proposed by Myers and Fenske [28]. Later two simple linear relations were derived from these charts to estimate T (10 mm Hg) from the normal boiling point (T_b) or boiling point at 1 mm Hg as follows [29]:

$$T(10 \text{ mm Hg}) = 0.8547 T(760 \text{ mm Hg}) - 57.7 \quad 500 \text{ K} < T(760 \text{ mm}) < 800 \text{ K}$$
$$T(10 \text{ mm Hg}) = 1.07 T(1 \text{ mm Hg}) + 19 \qquad 300 \text{ K} < T(1 \text{ mm}) < 600 \text{ K}$$
$$(3.32)$$

where all temperatures are in kelvin. These equations reproduce the original figures within 1%; however, they should be used within the temperatures ranges specified. Equations (3.30) and (3.31) are more accurate than Eq. (3.32) but for quick hand estimates the latter is more convenient. Another simple relation for quick conversion of boiling point at various pressures is through the following correction, which was proposed by Van Kranen and Van Nes, as given by Van Nes and Van Westen [30].

$$\log_{10} P_T = 3.2041 \left(1 - 0.998 \times \frac{T_b - 41}{T - 41} \times \frac{1393 - T}{1393 - T_b} \right)$$
$$(3.33)$$

where T is the boiling point at pressure P_T and T_b is the normal boiling point. P_T is in bar and T and T_b are in K. Accuracy of this equation is about 1%.

3.2.2.3.2 Conversion of a Distillation Curve from Sub- or Super- Atmospheric Pressures to a Distillation Curve at Atmospheric Pressure—The method of conversion of boiling points through Eqs. (3.29)–(3.32) can be used to every point on a distillation curve under either sub- or superatmospheric pressure conditions. In these equations T_b or T (760 mm Hg) represent a point along the distillation curve at atmospheric pressure. It can be applied to any of TBP, EFV, or ASTM D 1160 distillation curves. However, it should be noted that these equations convert distillation curves from one pressure to another within the same type. For example, it is not possible to use these equations to directly convert ASTM D 1160 at 10 mm Hg to TBP at 760 mm Hg. Such conversions require two steps that are discussed in the following section. The only distillation curve type that might be reported under superatmospheric pressure ($P > 1.01325$ bar) condition is the EFV distillation curve. TBP curve may be at 1, 10, 100, or 760 mm Hg pressure. Experimental data on ASTM D 1160 are usually reported at 1, 10, or 50 mm Hg. ASTM D 86 distillation is always reported at atmospheric pressure. It should be noted that when ASTM D 1160 distillation curve is converted to or reported at atmospheric pressure (760 mm Hg) it is not equivalent to or the same as ASTM D 86 distillation data. They are different types of distillation curves and there is no direct conversion between these two curves.

3.2.2.3.3 Conversion of ASTM D 1160 at 10 mm Hg to TBP Distillation Curve at 10 mm Hg—The only method widely used under subatmospheric pressure condition for conversion of distillation curves is the one developed by Edmister–Okamoto [17], which is used to convert ASTM D 1160 to TBP, both at 10 mm Hg. This method is graphical and it is also recommended by the API-DTB [2]. In this method it is assumed the at 50% points ASTM D 1160 and TBP temperatures are equal. The Edmister–Okamoto chart is converted into equation form through regression of values read from the figure in the following form [2]:

$$TBP(100\%) = ASTM\ D\ 1160(100\%)$$
$$TBP(90\%) = ASTM\ D\ 1160(90\%)$$
$$TBP(70\%) = ASTM\ D\ 1160(70\%)$$
(3.34) $$TBP(50\%) = ASTM\ D\ 1160(50\%)$$
$$TBP(30\%) = ASTM\ D\ 1160(50\%) - F_1$$
$$TBP(10\%) = ASTM\ D\ 1160(30\%) - F_2$$
$$TBP(0\%) = ASTM\ D\ 1160(10\%) - F_3$$

where functions F_1, F_2, and F_3 are given in terms of temperature difference in the ASTM D 1160:

$$F_1 = 0.3 + 1.2775(\Delta T_1) - 5.539 \times 10^{-3}(\Delta T_1)^2 + 2.7486 \times 10^{-5}(\Delta T_1)^3$$

$$F_2 = 0.3 + 1.2775(\Delta T_2) - 5.539 \times 10^{-3}(\Delta T_2)^2 + 2.7486 \times 10^{-5}(\Delta T_2)^3$$

$$F_3 = 2.2566(\Delta T_3) - 266.2 \times 10^{-4}(\Delta T_3)^2 + 1.4093 \times 10^{-4}(\Delta T_3)^3$$

$$\Delta T_1 = ASTM\ D\ 1160(50\%) - ASTM\ D\ 1160(30\%)$$
$$\Delta T_2 = ASTM\ D\ 1160(30\%) - ASTM\ D\ 1160(10\%)$$
$$\Delta T_3 = ASTM\ D\ 1160(10\%) - ASTM\ D\ 1160(0\%)$$

in the above relations all temperatures are either in °C or in kelvin.

3.2.2.4 Summary Chart for Interconverion of Various Distillation Curves

A summary of all conversion methods is shown in Fig. 3.20. It should be noted that any distillation curve at low pressure (i.e., ASTM D 1160 or EFV at 1, 10, 50, mm Hg or TBP at 1 mm Hg) should be first converted to TBP distillation curve at 10 mm Hg before they are converted to TBP at atmospheric pressure.

Example 3.6—For a petroleum fraction the ASTM D 1160 distillation data at 10 mm Hg are given in Table 3.13. Predict the TBP curve at atmospheric pressure.

Solution—ASTM D 1160 data have been converted to TBP at 10 mm Hg by Eq. (3.34). Then Eq. (3.29) with $P = 10$ mm Hg and $Q = 0.001956$ is used to convert TBP from 10 to 760 mm Hg. A summary of results is given in Table 3.13. The second and less accurate method to convert TBP from 10 to 760 mm Hg is through Eq. (3.32), which in its reverse form becomes T (760 mm Hg) = 1.17T (10 mm Hg) + 67.51. Estimated TBP at 760 mm Hg through this relation is presented in the last column of Table 3.13. ♦

3.2.3 Prediction of Complete Distillation Curves

In many cases distillation data for the entire range of percent distilled are not available. This is particularly the case when a fraction contains heavy compounds toward the end of distillation curve. For such fractions distillation can be performed to a certain temperature. For example, in a TBP or ASTM curve, distillation data may be available at 10, 30, 50, and 70% points but not at 90 or 95% points, which are important for process engineers and are characteristics of a petroleum product. For heavier fractions the distillation curves may even end at 50% point. For such fractions it is important that values of temperatures at these high percentage points to be estimated from available data. In this section a distribution function for both boiling point and density of petroleum fractions is presented so that its parameters can be determined from as few as three data points on the curve. The function can predict the boiling point for the entire range from initial point to 95% point. This function was proposed by Riazi [31] based on a probability distribution model for the properties of heptane plus fractions in crude oils and reservoir fluids and its detailed characteristics are discussed in Section 4.5.4. The distribution model is presented by the following equation (see Eq. 4.56):

$$(3.35) \qquad \frac{T - T_\circ}{T_\circ} = \left[\frac{A}{B} \ln \left(\frac{1}{1 - x} \right) \right]^{1/B}$$

in which T is the temperature on the distillation curve in kelvin and x is the volume or weight fraction of the mixture distilled. A, B, and $T_\circ$ are the three parameters to be determined from available data on the distillation curve through a linear regression. $T_\circ$ is in fact the initial boiling point (T at $x = 0$) but has to be determined from actual data with $x > 0$. The experimental value of $T_\circ$ should not be included in the regression process since it is not a reliable point. Equation (3.35)

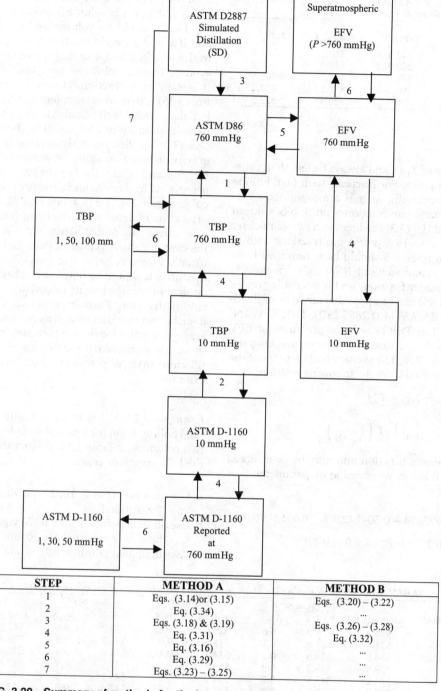

FIG. 3.20—Summary of methods for the interconversion of various distillation curves.

STEP	METHOD A	METHOD B
1	Eqs. (3.14)or (3.15)	Eqs. (3.20) – (3.22)
2	Eq. (3.34)	...
3	Eqs. (3.18) & (3.19)	Eqs. (3.26) – (3.28)
4	Eq. (3.31)	Eq. (3.32)
5	Eq. (3.16)	...
6	Eq. (3.29)	...
7	Eqs. (3.23) – (3.25)	...

does not give a finite value for T at $x = 1$ (end point at 100% distilled). According to this model the final boiling point is infinite (∞), which is true for heavy residues. Theoretically, even for light products with a limited boiling range there is a very small amount of heavy compound since all compounds in a mixture cannot be completely separated by distillation. For this reason predicted values from Eq. (3.35) are reliable up to $x = 0.99$, but not at the end point. Parameters A, B, and $T_\circ$ in Eq. (3.35) can be directly determined by using Solver (in Tools) in Excel spreadsheets. Another way to determine the constants in Eq. (3.35) is through its conversion into the

following linear form:

$$(3.36) \qquad Y = C_1 + C_2 X$$

where $Y = \ln [(T - T_\circ) / T_\circ]$ and $X = \ln \ln [1 / (1 - x)]$. Constants C_1 and C_2 are determined from linear regression of Y versus X with an initial guess for $T_\circ$. Constants A and B are determined from C_1 and C_2 as $B = 1/C_2$ and $A = B \exp (C_1 B)$. Parameter $T_\circ$ can be determined by several estimates to maximize the R squared (RS) value for Eq. (3.36) and minimize the AAD for prediction of T form Eq. (3.35). If the initial boiling point in a distillation curve is available it can be used as the

TABLE 3.13—*Conversion of ASTM D 1160 to TBP at 760 mm Hg for the petroleum fraction of Example 3.6.*

Vol% distilled	ASTM D 1160, °C	TBP$_{10}^{a}$, °C	TBP$_{760}^{b}$, °C	TBP$_{760}^{b}$, °C
10	150	142.5	280.1	280.7
30	205	200.9	349.9	349.0
50	250	250	407.2	406.5
70	290	290	453.1	453.3
90	350	350	520.4	523.5

^{a}Eq. (3.34).
^{b}Eq. (3.29).
^{c}Eq. (3.32).

initial guess, but value of T_o should always be less than value of T for the first data point. For fractions with final boiling point very high and uncertain, such as atmospheric or vacuum residues and heptane-plus fraction of crude oils, value of B can be set as 1.5 and Eq. (3.35) reduces to a two-parameter equation. However, for various petroleum fractions with finite boiling range parameter B should be determined from the regression analysis and value of B for light fractions is higher than that of heavier fractions and is normally greater than 1.5. Equation (3.35) can be applied to any type of distillation data, ASTM D 86, ASTM D 2887 (SD), TBP, EFV, and ASTM D 1160 as well as TBP at reduced pressures or EFV at elevated pressures. In the case of SD curve, x is cumulative weight fraction distilled. The average boiling point of the fraction can be determined from the following relation:

$$T_{av} = T_o(1 + T_{av}^*)$$

(3.37)
$$T_{av}^* = \left(\frac{A}{B}\right)^{\frac{1}{B}} \Gamma\left(1 + \frac{1}{B}\right)$$

in which Γ is the gamma function and may be determined from the following relation when value of parameter B is greater than 0.5.

$$\Gamma\left(1 + \frac{1}{B}\right) = 0.992814 - 0.504242B^{-1} + 0.696215B^{-2}$$
$$- 0.272936B^{-3} + 0.088362B^{-4}$$

(3.38)

Development of these relations is discussed in Chapter 4. In Eq. (3.35), if x is volume fraction, then T_{av} calculated from Eq. (3.37) would be volume average boiling point (VABP) and if x is the weight fraction then T_{av} would be equivalent to the weight average boiling point (WABP). Similarly mole average boiling point can be estimated from this equation if x is in mole fraction. However, the main application of Eq. (3.35) is to predict complete distillation curve from a limited data available. It can also be used to predict boiling point of residues in a crude oil as will be shown in Chapter 4. Equation (3.35) is also perfectly applicable to density or specific gravity distribution along a distillation curve for a petroleum fraction and crude oils. For the case of density, parameter T is replaced by d or SG and density of the mixture may be calculated from Eq. (3.37). When Eqs. (3.35)–(3.38) are used for prediction of density of petroleum fractions, the value of RS is less than that of distillation data. While the value of B for the case of density is greater than that of boiling point and is usually 3 for very heavy fractions (C_{7+}) and higher for lighter mixtures. It should be noted that when Eqs. (3.35)–(3.38) are applied to specific gravity or density, x should be cumulative volume fraction. Further properties and application of this distribution function as well as methods of calculation of average properties for the mixture are given in Chapter 4. Herein we demonstrate use for this method for prediction of distillation curves of petroleum fractions through the following example.

Example 3.7—ASTM D 86 distillation data from initial to final boiling point for a gas oil sample [1] are given in the first two columns of Table 3.14. Predict the distillation curve for the following four cases:

a. Use data points at 5, 10, 20, 30, 40, 50, 60, 70, 80, 90, and 95 vol% distilled.
b. Use all data points from 5 to 70 vol% distilled.
c. Use three data points at 10, 30, and 50%.
d. Use three data points at 30, 50, and 70%.

TABLE 3.14—*Prediction of ASTM D 86 distillation curve for gas oil sample of Example 3.7.*

Vol% distilled	Temp. exp., K	Data Set A Pred, K	Data Set A AD, K	Data Set B Pred, K	Data Set B AD, K	Data Set C Pred, K	Data Set C AD, K	Data Set D Pred, K	Data Set D AD, K
0	520.4	526.0	5.6	525.0	4.6	530.0	9.6	512.0	8.4
5	531.5	531.6	0.1	531.5	0.0	532.7	1.2	526.4	5.1
10	534.8	534.6	0.2	534.7	0.1	534.8	0.0	531.2	3.6
20	539.8	539.5	0.4	539.6	0.2	538.9	0.9	537.8	2.0
30	543.2	543.8	0.7	544.0	0.8	543.1	0.0	543.1	0.0
40	548.2	548.1	0.1	548.1	0.0	547.6	0.5	547.9	0.3
50	552.6	552.5	0.1	552.4	0.2	552.6	0.0	552.6	0.0
60	557.0	557.3	0.3	557.0	0.1	558.5	1.4	557.4	0.4
70	562.6	562.9	0.3	562.2	0.4	565.6	3.0	562.6	0.0
80	570.4	569.9	0.5	568.7	1.7	575.2	4.8	568.8	1.6
90	580.4	580.4	0.0	578.2	2.2	590.6	10.3	577.5	2.9
95	589.8	589.6	0.2	586.6	3.3	605.2	15.4	584.8	5.1
100	600.4	608.3	7.9	603.1	2.7	637.1	36.7	598.4	2.0
AAD (total), K			**1.3**		**1.3**		**6.5**		**2.4**
No. of data used			11		8		3		3
T_o			526		525		530		512
A			0.01634		0.0125		0.03771		0.00627
B			1.67171		1.80881		1.21825		2.50825
RS			0.9994		0.999		1		1
AAD (data used), K			0.25		0.23		0		0
VABP, K	554.7	555.5		555		557		554	

For each case give parameters T_o, A, and B in Eq. (3.35) as well as value of RS and AAD based on all data points and based on data used for the regression. Also calculate VABP from Eq. (3.37) and compare with actual VABP calculated from Eq. (3.6).

Solution—Summary of calculation results for all four cases are given in Table 3.14. For Case A all experimental data given on the distillation curve (second column in Table 3.14) from 5 to 95% points are used for the regression analysis by Eq. (3.36). Volume percentages given in the first column should be converted to cumulative volume fraction, x, (percent values divided by 100) and data are converted to X and Y defined in Eq. (3.36). The first data point used in the regression process is at $x = 0.05$ with $T = 531.5$ K; therefore, the initial guess (T_o) should be less than 531. With a few changes in T_o values, the maximum RS value of 0.9994 is obtained with minimum AAD of 0.25 K (for the 11 data points used in the regression process). The AAD for the entire data set, including the IBP and FBP, is 1.3 K. As mentioned earlier the experimentally reported IBP and especially the value of FBP are not accurate. Therefore, larger errors for prediction of IBP and FBP are expected from Eq. (3.35). Since values of FBP at $x = 1$ are not finite, the value of T at $x = 0.99$ may be used as an approximate predicted value of FBP from the model. These values are given in Table 3.14 as predicted values for each case at 100 vol% vaporized. Estimated VABP from Eq. (3.37) for Case A is 555.5 versus value of 554.7 from actual experimental data and definition of VABP by Eq. (3.6).

For Case B, data from 5 to 70 vol% distilled are used for the regression process and as a result the predicted values up to 70% are more accurate than values above 70% point. However, the overall error (total AAD) is the same as for Case A at 1.3 K. For Case C only three data points at 10, 30, and 50% are used and as a result much larger errors especially for points above 50% are observed. In Case D, data at 30, 50,

and 70% points are used and the predicted values are more accurate than values obtained in Case C. However, for this last case the highest error for the IBP is obtained because the first data point used to obtain the constants is at 30%, which is far from 0% point. Summary of results for predicted distillation curves versus experimental data are also shown in Fig. 3.21. As can be seen from the results presented in both Table 3.14 and Fig. 3.21, a good prediction of the entire distillation curve is possible through use of only three data points at 30, 50, and 70%.

◆

3.3 PREDICTION OF PROPERTIES OF PETROLEUM FRACTIONS

As discussed in Chapter 1, petroleum fractions are mixtures of many hydrocarbon compounds from different families. The most accurate method to determine a property of a mixture is through experimental measurement of that property. However, as this is not possible for every petroleum mixture, methods of estimation of various properties are needed by process or operation engineers. The most accurate method of estimating a property of a mixture is through knowledge of the exact composition of all components existing in the mixture. Then properties of pure components such as those given in Tables 2.1 and 2.2 can be used together with the composition of the mixture and appropriate mixing rules to determine properties of the mixture. If experimental data on properties of pure compounds are not available, such properties should be estimated through the methods presented in Chapter 2. Application of this approach to defined mixtures with very few constituents is practical; however, for petroleum mixtures with many constituents this approach is not feasible as the determination of the exact composition of all components in the mixture is not possible. For this reason appropriate models should be used to represent petroleum mixtures by some limited number of compounds that can best represent the mixture. These limited compounds are different from the real compounds in the mixture and each is called a "pseudocomponent" or a "pseudocompound". Determination of these pseudocompounds and use of an appropriate model to describe a mixture by a certain number of pseudocompounds is an engineering art in prediction of properties of petroleum mixtures and are discussed in this section.

3.3.1 Matrix of Pseudocomponents Table

As discussed in Chapter 2, properties of hydrocarbons vary by both carbon number and molecular type. Hydrocarbon properties for compounds of the same carbon number vary from paraffins to naphthenes and aromatics. Very few fractions may contain olefins as well. Even within paraffins family properties of n-paraffins differ from those of isoparaffins. Boiling points of hydrocarbons vary strongly with carbon number as was shown in Table 2.1; therefore, identification of hydrocarbons by carbon number is useful in property predictions. As discussed in Section 3.1.5.2, a combination of GS-MS in series best separate hydrocarbons by carbon number and molecular type. If a mixture is separated by a distillation column or simulated distillation, each hydrocarbon cut with a single carbon number contains hydrocarbons from different

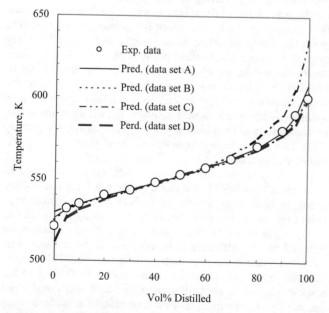

FIG. 3.21—Prediction of distillation curves for the gas oil sample of Example 3.7.

TABLE 3.15—*Presentation of a petroleum fraction (diesel fuel) by a matrix of 30 pseudocomponents.*

Carbon number	n-Paraffins	Isoparaffins	Olefins	Naphthenes	Aromatics
C_{11}	1	2	3	4	5
C_{12}	6	7	8	9	10
C_{13}	11	12	13	14	15
C_{14}	16	17	18	19	20
C_{15}	21	22	23	24	25
C_{16}	26	27	28	29	30

groups, which can be identified by a PIONA analyzer. As an example in Table 1.3 (Chapter 1), carbon number ranges for different petroleum products are specified. For a diesel fuel sample, carbon number varies from C_{11} to C_{16} with a boiling range of 400–550°C. If each single carbon number hydrocarbon cut is further separated into five pseudocomponents from different groups, the whole mixture may be represented by a group of 30 pseudocomponents as shown in Table 3.15. Although each pseudocomponent is not a pure hydrocarbon but their properties are very close to pure compounds from the same family with the same carbon number. If the amounts of all these 30 components are known then properties of the mixture may be estimated quite accurately. This requires extensive analysis of the mixture and a large computation time for estimation of various properties. The number of pseudocomponents may even increase further if the fraction has wider boiling point range such as heptane plus fractions as will be discussed in Chapter 4. However, many petroleum fractions are olefin free and groups of n-paraffins and isoparaffins may be combined into a single group of paraffins. Therefore, the number of different families reduces to three (paraffins, naphthenes, and aromatics). In this case the number of components in Table 3.15 reduces to 6 × 3 or 18. If a fraction is narrow in boiling range then the number of rows in Table 3.15 decreases indicating lower carbon number range. In Chapter 4, boiling points of various single carbon number groups are given and through a TBP curve it would be possible to determine the range of carbon number in a petroleum fraction. In Table 3.15, if every two carbon number groups and all paraffins are combined together, then the whole mixture may be represented by 3 × 3 or 9 components for an olefin-free fraction. Similarly if all carbon numbers are grouped into a single carbon number group, the mixture can be represented by only three pseudocomponents from paraffins (P), naphthenes (N), and aromatics (A) groups all having the same carbon number. This approach is called *pseudocomponent technique.*

Finally the ultimate simplicity is to ignore the difference in properties of various hydrocarbon types and to present the whole mixture by just a single pseudocomponent, which is the mixture itself. The simplicity in this case is that there is no need for the composition of the mixture. Obviously the accuracy of estimated properties decreases as the number of pseudocomponents decreases. However, for narrow boiling range fractions such as a light naphtha approximating the mixture with a single pseudocomponent is more realistic and more accurate than a wide boiling range fraction such as an atmospheric residuum or the C_{7+} fraction in a crude oil sample. As discussed in Chapter 2, the differences between properties of various hydrocarbon families increase with boiling point (or carbon number). Therefore, assumption of a single pseudocomponent for a heavy fraction ($M > 300$) is less accurate than for the case of light fractions. For fractions that are rich in one hydrocarbon type such as coal liquids that may have up to 90% aromatics, it would be appropriate to divide the aromatics into further subgroups of monoaromatics (MA) and polyaromatics (PA). Therefore, creation of a matrix of pseudocomponents, such as Table 3.15, largely depends on the nature and characteristics of the petroleum mixture as well as availability of experimental data.

3.3.2 Narrow Versus Wide Boiling Range Fractions

In general, regardless of molecular type, petroleum fractions may be divided into two major categories: narrow and wide boiling range fractions. A *narrow boiling range fraction* was defined in Section 3.2.1 as a fraction whose ASTM 10–90% distillation curve slope (SL) is less than 0.8°C/%, although this definition is arbitrary and may vary from one source to another. Fractions with higher 10–90% slopes may be considered as wide boiling range. However, for simplicity the methods presented in this section for narrow fractions may also be applied to wider fractions. For narrow fractions, only one carbon number is considered and the whole fraction may be characterized by a single value of boiling point or molecular weight. For such fractions, if molecular type is known (PNA composition), then the number of pseudocomponents in Table 3.15 reduces to three and if the composition is not known the whole mixture may be considered as a single pseudocomponent. For this single pseudocomponent, properties of a pure component whose characteristics, such as boiling point and specific gravity, are the same as that of the fraction can be considered as the mixture properties. For mixtures the best characterizing boiling point is the mean average boiling point (MeABP); however, as mentioned in Section 3.2.1, for narrow fractions the boiling point at 50 vol% distilled may be considered as the characteristic boiling point instead of MeABP. The specific gravity of a fraction is considered as the second characteristic parameter for a fraction represented by a single pseudocomponent. Therefore, the whole mixture may be characterized by its boiling point (T_b) and specific gravity (SG). In lieu of these properties other characterization parameters discussed in Chapter 2 may be used.

Treatment of wide boiling range fractions is more complicated than narrow fractions as a single value for the boiling point, or molecular weight, or carbon number cannot represent the whole mixture. For these fractions the number of constituents in the vertical columns of Table 3.15 cannot be reduced to one, although it is still possible to combine various molecular types for each carbon number. This means that the minimum number of constituents in Table 3.14 for a wide fraction is six rather than one that was considered for narrow fractions. The best example of a wide boiling range fraction is C_{7+} fraction in a crude oil or a reservoir fluid. Characterization of such fractions through the use of a distribution model that reduces the mixture into a number of pseudocomponents with known characterization parameters will be discussed in detail in Chapter 4. However, a simpler approach based on the use of TBP curve is outlined in Ref. [32]. In this approach the mixture property is calculated from the following relation:

$$(3.39) \qquad \theta = \int\limits_0^1 \theta(x)\, dx$$

in which θ is the physical property of mixture and $\theta(x)$ is the value of property at point x on the distillation curve. This approach may be applied to any physical property. The integration should be carried out by a numerical method. The fraction is first divided into a number of pseudocomponents along the entire range of distillation curve with known boiling points and specific gravity. Then for each component physical properties are calculated from methods of Chapter 2 and finally the mixture properties are calculated through a simple mixing rule. The procedure is outlined in the following example.

Example 3.8—For a low boiling naphtha, TBP curve is provided along with the density at 20°C as tabulated below [32]. Estimate specific gravity and molecular weight of this fraction using the wide boiling range approach. Compare the calculated results with the experimental values reported by Lenior and Hipkin and others [1, 11, 32] as SG = 0.74 and $M = 120$.

vol%	0 (IBP)	5	10	20	30	50	70	90	95
TBP, K	283.2	324.8	348.7	369.3	380.9	410.4	436.5	467.6	478.7
d_{20}, g/cm^3	...	0.654	0.689	0.719	0.739	0.765	0.775	0.775	0.785

Solution—For this fraction the 10–90% slope based on TPB curve is about 1.49°C/%. This value is slightly above the slope based on the ASTM D 86 curve but still indicates how wide the fraction is. For this sample based on the ASTM distillation data [1], the 10–90% slope is 1.35°C/%, which is above the value of 0.8 specified for narrow fractions. To use the method by Riazi–Daubert [32] for this relatively wide fraction, first distribution functions for both boiling point and specific gravity should be determined. We use Eqs. (3.35)–(3.38) to determine the distribution functions for both properties. The molecular weight, M, is estimated for all points on the curve through appropriate relations in Chapter 2 developed for pure hydrocarbons. The value of M for the mixture then may be estimated from a simple integration over the entire range of x as given by Eq. (3.39): $M_{av} = \int_0^1 M(x)\,dx$, where $M(x)$ is the value of M at point x determined from $T_b(x)$ and SG(x). M_{av} is the average molecular weight of the mixture. For this fraction values of densities given along the distillation curve are at 20°C and should be converted to specific gravity at 15.5°C (60°F) through use of Eq. (2.112) in Chapter 2: SG = $0.9915 d_{20} + 0.01044$. Parameters of Eq. (3.35) for both temperature and specific gravity have been determined and are given as following.

Parameters in Eq. (3.35)	$T_\circ$, K	SG$_\circ$	A	B	RS
TBP curve	240		1.41285	3.9927	0.996
SG curve		0.5	0.07161	7.1957	0.911

The values of $T_\circ$ and SG$_o$ determined from regression of data through Eq. (3.35) do not match well with the experimental initial values. This is due to the maximizing value of RS with data used in the regression analysis. Actually one can imagine that the actual initial values are lower than experimentally measured values due to the difficulty in such measurements. However, these initial values do not affect subsequent calculations. Predicted values at all other points from 5 up to 95% are consistent with the experimental values. From calculated values of SG$_\circ$, A, and B for the SG curve, one can determine the mixture SG for the whole fraction through use of Eqs. (3.37) and (3.38). For SG, $B = 7.1957$ and from Eq. (3.38), $\Gamma(1 + 1/B) = 0.9355$. From Eq. (3.37) we get $\mathrm{SG}^*_{av} = \left(\frac{0.07161}{7.1957}\right)^{1/7.1957} \Gamma\left(1 + \frac{1}{7.1957}\right) = 0.5269 \times 0.9355 = 0.493$. Therefore, for the mixture: $\mathrm{SG}_{av} = 0.5(1 + 0.493) = 0.746$. Comparing with experimental value of 0.74, the percent relative deviation (%D) with experimental value is 0.8%. In Chapter 4 another method based on a distribution function is introduced that gives slightly better prediction for the density of wide boiling range fractions and crude oils.

To calculate a mixture property such as molecular weight, the mixture is divided to some narrow pseudocomponents, N_P. If the mixture is not very wide such as in this example, even $N_P = 5$ is sufficient, but for wider fractions the mixture may be divided to even larger number of pseudocomponents (10, 20, etc.). If $N_P = 5$, then values of T and SG at $x = 0, 0.2, 0.4, 0.6, 0.8$, and 0.99 are evaluated through Eq. (3.35) and parameters determined above. Value of $x = 0.99$ is used instead of $x = 1$ for the end point as Eq. (3.35) is not defined at $x = 1$. At every point, molecular weight, M, is determined from methods of Chapter 2. In this example, Eq. (2.50) is quite accurate and may be used to calculate M since all components in the mixture have $M < 300$ ($\sim N_C < 20$) and are within the range of application of this method. Equation (2.50) is $M = 1.6604 \times 10^{-4} T_b^{2.1962}\, \mathrm{SG}^{-1.0164}$. Calculations are summarized in the following table.

x	T_b, K	SG	M
0	240.0	0.500	56.7
0.2	367.1	0.718	99.8
0.4	396.4	0.744	114.0
0.6	421.0	0.764	126.7
0.8	448.4	0.785	141.6
0.99	511.2	0.828	178.7

The trapezoidal rule for integration is quite accurate to estimate the molecular weight of the mixture. $M_{av} = (1/5) \times [(56.7 + 178.7)/2 + (99.8 + 114 + 126.7 + 141.6 + 178.7)] = 119.96 \cong 120$. This is exactly the same as the experimental value of molecular weight for this fraction [1, 11].

If the whole mixture is considered as a single pseudocomponent, Eq. (2.50) should be applied directly to the mixture using the MeABP and SG of the mixture. For this fraction the Watson K is given as 12.1 [1]. From Eq. (3.13) using experimental value of SG, average boiling point is calculated as $T_b = (12.1 \times 0.74)^3/1.8 = 398.8$ K. From Eq. (2.50), the mixture molecular weight is 116.1, which is equivalent to %D = -3.25%. For this sample, the difference between 1 and 5 pseudocomponents is not significant, but for wider fraction the improvement of the proposed method is much larger. ♦

Example 3.9—It is assumed for the same fraction of Example 3.8, the only information available is ASTM D 86 data: temperatures of 350.9, 380.9, 399.8, 428.2, and 457.6 K at 10, 30, 50, 70 and 90 vol% distilled, respectively. State how can you apply the proposed approach for wide boiling range fractions to calculate the molecular weight of the fraction.

Solution—Since distillation data are in terms of ASTM D 86, the first step is to convert ASTM to TBP through Eq. (3.14). The second step is to determine the TBP distribution function through Eqs. (3.35) and (3.36). The third step is to generate values of T at $x = 0, 0.2, 0.4, 0.6, 0.8,$ and 0.99 from Eq. (3.35) and parameters determined from TBP distillation curve. Since the specific gravity for this fraction is not known it may be estimated from Eq. (3.17) and constants in Table 3.4 for the ASTM D 86 data as follows: SG = $0.08342 \times (350.9)^{0.10731}(399.8)^{0.26288} = 0.756$. The Watson K is calculated as $K_W = (1.8 \times 399.8)^{1/3}/0.756 = 11.85$. Now we assume that K_W is constant for the entire range of distillation curve and on this basis distribution of SG can be calculated through distribution of true boiling point. At every point that T is determined from Eq. (3.35) the specific gravity can be calculated as SG = $(1.8 \times T)^{1/3}/K_W$, where T is the temperature on the TBP curve. Once TBP temperatures and SG are determined at $x = 0, 0.2, 0.4, 0.6, 0.8,$ and 0.99 points, molecular weight may be estimated from Eq. (2.50). Numerical integration of Eq. (3.39) can be carried out similar to the calculations made in Example 3.8 to estimate the molecular weight. In this approach the result may be less accurate than the result in Example 3.8, as ASTM distillation curve is used as the only available data. ♦

Although the method outlined in this section improves the accuracy of prediction of properties of wide boiling range fraction, generally for simplicity in calculations most petroleum products are characterized by a single value of boiling point, molecular weight, or carbon number regardless of their boiling range. The proposed method is mainly applied to crude oils and C_{7+} fraction of reservoir fluids with an appropriate splitting technique as is shown in the next chapter. However, as shown in the above example, for very wide boiling range petroleum products the method presented in this section may significantly improve the accuracy of the estimated physical properties.

3.3.3 Use of Bulk Parameters (Undefined Mixtures)

An undefined petroleum fraction is a fraction whose composition (i.e., PNA) is not known. For such fractions information on distillation data (boiling point), specific gravity, or other bulk properties such as viscosity, refractive index, CH ratio, or molecular weight are needed. If the fraction is considered narrow boiling range then it is assumed as a single component and correlations suggested in Chapter 2 for pure hydrocarbons may be applied directly to such fractions. All limitations for the methods suggested in Chapter 2 should be considered when they are used for petroleum fractions. As mentioned in Chapter 2, the correlations in terms of T_b and SG are the most accurate methods for the estimation of various properties (molecular weight, critical constants, etc.).

For narrow boiling range fractions, ASTM D 86 temperature at 50 vol% vaporized may be used as the characterizing boiling point for the whole mixture. However, for a wide boiling range fraction if it is treated as a single pesudocomponent the MeABP should be calculated and used as the characterizing parameter for T_b in the correlations of Chapter 2. If for a fraction TBP distillation data are available the average boiling point calculated through Eq. (3.37) with parameters determined from TBP curve would be more appropriate than MeABP determined from ASTM D 86 curve for use as the characterizing boiling point. For cases where only two points on the distillation curve are known the interpolated value at 50% point may be used as the characterizing boiling point of the fraction.

For heavy fractions ($M > 300$) in which atmospheric distillation data (ASTM D 86, SD, or TBP) are not available, if ASTM D 1160 distillation curve is available, it should be converted to ASTM D 86 or TBP through methods outlined in Section 3.2. In lieu of any distillation data, molecular weight or viscosity may be used together with specific gravity to estimate basic parameters from correlations proposed in Chapter 2. If specific gravity is not available, refractive index or carbon-to-hydrogen weight ratio (CH) may be used as the second characterization parameter.

3.3.4 Method of Pseudocomponent (Defined Mixtures)

A *defined mixture* is a mixture whose composition is known. For a petroleum fraction if at least the PNA composition is known it is called a *defined fraction*. Huang [11, 33] used the pseudocompounds approach to estimate enthalpies of narrow and defined petroleum fractions. This technique has been also used to calculate other physical properties by other researchers [34, 35]. According to this method all compounds within each family are grouped together as a single pseudocomponent. An olefin-free fraction is modeled into three pseudocomponents from three homologous groups of n-alkanes (representing paraffins), n-alkylcyclopentanes or n-alkylcyclohexanes (representing naphthenes), and n-alkylbenzenes (representing aromatics) having the same boiling point as that of ASTM D 86 temperature at 50% point. Physical properties of a mixture can be calculated from properties of the model components by the following mixing rule:

$$\theta = x_P\theta_P + x_N\theta_N + x_A\theta_A \tag{3.40}$$

where θ is a physical property for the mixture and θ_P, θ_N, and θ_A are the values of θ for the model pseudocomponents from the three groups. In this equation the composition presented by x_A, x_N, and x_A should be in mole fraction, but because the molecular weights of different hydrocarbon groups having the same boiling point are close to each other, the composition in weight or even volume fractions may also be used with minor difference in the results. If the fraction contains olefinic compounds a fourth term for contribution of this group should be added to Eq. (3.40). Accuracy of Eq. (3.40) can be increased if composition of paraffinic group is known in terms of n-paraffins and isoparaffins. Then another pseudocomponent contributing the isoparaffinic hydrocarbons may be added to the equation. Similarly, the aromatic part may be split into monoaromatics and polyaromatics provided their amount in the fraction is known. However, based on our experience the PNA three-pseudocomponent model is sufficiently accurate for olefin-free petroleum fractions. For coal liquids with a high percentage of aromatic content, splitting aromatics into two subgroups may greatly increase the accuracy of model predictions. In using this method the minimum data needed are at least one characterizing parameter (T_b or M) and the PNA composition.

Properties of pseudocomponents may be obtained from interpolation of values in Tables 2.1 and 2.2 to match boiling point to that of the mixture. As shown in Section 2.3.3, properties of homologous groups can be well correlated to only one characterization parameter such as boiling point, molecular weight, or carbon number, depending on the availability of the parameter for the mixture. Since various properties of pure homologous hydrocarbon groups are given in terms of molecular weight by Eq. (2.42) with constants in Table 2.6, if molecular weight of a fraction is known it can be used directly as the characterizing parameter. But if the boiling point is used as the characterizing parameter, molecular weights of the three model components may be estimated through rearrangement of Eq. (2.42) in terms of boiling point as following:

$$(3.41) \quad M_p = \left\{ \frac{1}{0.02013}[6.98291 - \ln(1070 - T_b)] \right\}^{3/2}$$

$$(3.42) \quad M_N = \left\{ \frac{1}{0.02239}[6.95649 - \ln(1028 - T_b)] \right\}^{3/2}$$

$$(3.43) \quad M_A = \left\{ \frac{1}{0.02247}[6.91062 - \ln(1015 - T_b)] \right\}^{3/2}$$

where M_P, M_N, and M_A are molecular weights of paraffinic, naphthenic, and aromatic groups, respectively. T_b is the characteristic boiling point of the fraction. Predicted values of M_P, M_N, and M_A versus T_b were presented in Fig. 2.15 in Chapter 2. As shown in this figure the difference between these molecular weights increase as boiling point increases. Therefore, the pseudocomponent approach is more effective for heavy fractions. If ASTM D 86 distillation curve is known the temperature at 50% point should be used for T_b, but if TBP distillation data are available an average TBP would be more suitable to be used for T_b. Once M_P, M_N, and M_A are determined, they should be used in Eq. (2.42) to determine properties from corresponding group to calculate other properties. The method is demonstrated in the following example.

Example 3.10—A petroleum fraction has ASTM D 86 50% temperature of 327.6 K, specific gravity of 0.658, molecular weight of 78, and PNA composition of 82, 15.5, and 2.5 in vol% [36]. Estimate molecular weight of this fraction using bulk properties of T_b and SG and compare with the value estimated from the pseudocomponent method. Also estimate the mixture specific gravity of the mixture through the pseudocomponent technique and compare the result with the experimental value.

Solution—For this fraction the characterizing parameters are T_b = 327.6 K and SG = 0.672. To estimate M from these bulk properties, Eq. (2.50) can be applied since the boiling point of the fraction is within the range of 40–360°C (~C_5–C_{22}). The results of calculation is M = 85.0, with relative

deviation of 9%. From the pseudocomponent approach, M_P, M_N, and M_A are calculated from Eqs. (3.41)–(3.43) as 79.8, 76.9, and 68.9, respectively. The mixture molecular weight is calculated through Eq. (3.40) as $M = 0.82 \times 79.8 + 0.155 \times 76.9 + 0.025 \times 68.9 = 79$, with relative deviation of 1.3%. If values of M_P, M_N, and M_A are substituted in Eq. (2.42) for the specific gravity, we get $SG_P = 0.651$, $SG_N = 0.749$, and $SG_A = 0.895$. From Eq. (3.40) the mixture specific gravity is SG = 0.673, with AD of 2.3%. It should be noted that when Eq. (3.40) is applied to molecular weight, it would be more appropriate to use composition in terms of mole fraction rather than volume fraction. The composition can be converted to weight fraction through specific gravity of the three components and then to mole fraction through molecular weight of the components by equations given in Section 1.7.15. The mole fractions are $x_{mP} = 0.785$, $x_{mN} = 0.177$, and $x_{mA} = 0.038$ and Eq. (3.39) yields $M = 78.8$ for the mixture with deviation of 1%. The difference between the use of volume fraction and mole fraction in Eq. (3.40) is minor and within the range of experimental uncertainty. Therefore, use of any form of composition in terms of volume, weight, or mole fraction in the pseudocomponent method is reasonable without significant effect in the results. For this reason, in most cases the PNA composition of petroleum fractions are simply expressed as fraction or percentage and they may considered as weight, mole, or volume. ♦

In the above example the method of pseudocomponent predicts molecular weight of the fraction with much better accuracy than the use of Eq. (2.50) with bulk properties (%AD of 1.3% versus 9%). This is the case for fractions that are highly rich in one of the hydrocarbon types. For this fraction paraffinic content is nearly 80%, but for petroleum fractions with normal distribution of paraffins, naphthenes, and aromatics both methods give nearly similar results and the advantage of use of three pseudocomponents from different groups over the use of single pseudocomponent with mixture bulk properties is minimal. For example, for a petroleum fraction the available experimental data are [36] $M = 170$, $T_b = 487$ K, SG = 0.802, $x_p = 0.42$, $x_N = 0.41$, and $x_A = 0.17$. Equation (2.50) gives $M = 166$, while Eq. (3.40) gives $M = 163$ and SG = 0.792. Equation (3.40) is particularly useful when only one bulk property (i.e., T_b) with the composition of a fraction is available. For highly aromatic (coal liquids) or highly paraffinic mixtures the method of pseudocomponent is recommended over the use of bulk properties.

3.3.5 Estimation of Molecular Weight, Critical Properties, and Acentric Factor

Most physical properties of petroleum fluids are calculated through corresponding state correlations that require pseudocritical properties (T_{pc}, P_{pc}, and V_{pc}) and acentric factor as a third parameter. In addition molecular weight (M) is needed to convert calculated mole-based property to mass-based property. As mentioned in Section 1.3, the accuracy of these properties significantly affects the accuracy of estimated properties. Generally for petroleum fractions these basic characterization parameters are calculated through either the use of bulk properties and correlations of Chapter 2

TABLE 3.16—*Comparison of various methods of predicting pseudocritical properties and acentric factor through enthalpy calculation of eight petroleum fractions [37].*

| Item | Method of estimating input parameters[*a] | | AAD, kJ/kg | |
	T_c, P_c	ω	Liquid (437 data points)	Vapor (273 data points)
1	Pseudocomp.	Pseudocomp.	5.3	7.9
2	RD (80)	LK	5.9	7.7
3	KL	LK	5.8	7.4
4	Winn	LK	9.9	12.8

[a]*Pseudocomp.: The pseudocomponent method by Eqs. (3.40)–(3.43) and (2.42) for T_c, P_c and ω; RD: Riazi–Daubert [38] by Eqs. (2.63) and (2.64); LK: Lee–Kesler [39] by Eq. (2.103); KL: Kesler–Lee [40] by Eqs. (2.69) and (2.70); Winn method [41] by Eqs. (2.94) and (2.95).

or by the pseudocomponent approach as discussed in Sections 3.3.2–3.3.4.

For petroleum fractions, pseudocritical properties are not directly measurable and therefore it is not possible to make a direct evaluation of different methods with experimental data. However, these methods can be evaluated indirectly through prediction of other measurable properties (i.e., enthalpy) through corresponding state correlations. These correlations are discussed in Chapters 6–8. Based on more than 700 data points for enthalpies of eight petroleum fractions over a wide range of temperature and pressure [1], different methods of estimation of pseudocritical temperature, pressure (T_{pc}, P_{pc}), and acentric factor (ω) have been evaluated and compared [37]. These petroleum fractions ranging from naphtha to gas oil all have molecular weights of less than 250. Details of these enthalpy calculations are given in Chapter 7. Summary of evaluation of different methods is given in Table 3.16. As shown in Table 3.16, the methods of pseudocomponent, Lee–Kesler, and Riazi–Daubert have nearly similar accuracy for estimating the critical properties of these light petroleum fractions. However, for heavier fractions as it is shown in Example 3.11, the methods of pseudocomponent provide more accurate results.

Example 3.11—Experimental data on molecular weight and composition of five heavy petroleum fractions are given in Table 3.17. In addition, normal boiling point, specific gravity, density, and refractive index at 20°C are also given [36]. Calculate the molecular weight of these fractions from the following five methods: (1) API method [2, 42] using Eq. (2.51), (2) Twu method [42] using Eqs. (2.89)–(2.92), (3) Goossens method [43] using Eq. (2.55), (4) Lee–Kesler method [40] using Eq. (2.54), and (5) the pseudocomponent method using Eqs. (3.40)–(3.43). Calculate the %AAD for each method.

Solution—Methods 1, 2, and 4 require bulk properties of T_b and SG, while the method of pseudocomponent requires T_b and the PNA composition as it is shown in Example 3.10. Method 2 requires T_b and density at 20°C (d_{20}). Results of calculations are given in Table 3.18.

The Twu method gives the highest error (AAD of 14.3%) followed by the Goossens with average deviation of 11.4%. The Twu and Goossens methods both underestimate the molecular weight of these heavy fractions. The Lee–Kesler method is more accurate for lighter fractions, while the API method is more accurate for heavier fractions. The pseudocomponent method gives generally a consistent error for all fractions and the lowest AAD%. Errors generated by the API, Lee–Kesler, and the pseudocomponent methods are within the experimental uncertainty in the measurement of molecular weight of petroleum fractions. ♦

In summary, for light fractions ($M < 300$) methods recommended by the API for T_c and P_c (Eqs. 2.65 and 2.66) [2] or the simple method of Riazi–Daubert (Eqs. 2.63 and 2.64) [38] are suitable, while for heavier fractions the Lee–Kesler method (Eqs. 2.69 and 2.70) [40] may be used. The pseudocomponent method may also be used for both T_c and P_c when the composition is available. For all fractions methods of calculation of acentric factor from the pseudocomponent or the method of Lee–Kesler [39] presented by Eq. (2.105) may be used. Molecular weight can be estimated from the API method [2] by Eq. (2.51) from the bulk properties; however, if the PNA composition is available the method of pseudocomponent is preferable especially for heavier fractions.

3.3.6 Estimation of Density, Specific Gravity, Refractive Index, and Kinematic Viscosity

Density (d), specific gravity (SG), and refractive index (n) are all bulk properties directly measurable for a petroleum mixture with relatively high accuracy. Kinematic viscosity at 37.8 or 98.9°C ($\nu_{38(100)}$, $\nu_{99(210)}$) are usually reported for heavy fractions for which distillation data are not available. But, for light fractions if kinematic viscosity is not available it should be estimated through measurable properties. Methods of estimation of viscosity are discussed in Chapter 8; however, in this chapter kinematic viscosity at a reference of temperature of 37.8 or 98.9°C (100°F or 210°F) is needed for estimation of viscosity gravity constant (VGC), a parameter required for prediction of composition of petroleum fractions. Generally,

TABLE 3.17—*Molecular weight and composition of five heavy petroleum fractions of Example 3.11 [36].*

No.	M	T_b,°C	SG	d_{20}, g/ml	n_{20}	P%	N%	A%
1	233	298.7	0.9119	0.9082	1.5016	34.1	45.9	20.0
2	267	344.7	0.9605	0.9568	1.5366	30.9	37.0	32.1
3	325	380.7	0.8883	0.8845	1.4919	58.4	28.9	12.7
4	403	425.7	0.9046	0.9001	1.5002	59.0	28.0	13.0
5	523	502.8	0.8760	0.8750	1.4865	78.4	13.3	8.3

TABLE 3.18—*Comparison of various methods of predicting molecular weight of petroleum fractions of Table 3.17 (Example 3.12).*

		(1) API Eq. (2.51)		(2) Twu, Eqs. (2.89)–(2.92)		(3) Goossens, Eq. (2.55)		(4) Lee–Kesler, Eq. (2.54)		(5) Pseudocomp., Eqs. (3.40)–(3.43)	
No.	M, exp.	M, calc	AD%	M, calc	AD%	M, calc	AD%	M, calc	AD%	M, calc	AD%
1	233	223.1	4.2	201.3	13.6	204.6	12.2	231.7	0.5	229.1	1.7
2	267	255.9	4.2	224.0	16.1	235.0	12.0	266.7	0.1	273.2	2.3
3	325	320.6	1.4	253.6	16.8	271.3	11.0	304.3	0.2	321.9	1.0
4	403	377.6	6.3	332.2	17.6	345.8	14.2	374.7	7.0	382.4	5.1
5	523	515.0	1.5	485.1	7.2	483.8	7.5	491.7	6.0	516.4	1.3
Total, AAD%			3.5		14.3		11.4		2.8		2.3

density, which is required in various predictive methods measured at 20°C, is shown by d or d_{20} in g/mL. These properties can be directly estimated through bulk properties of mixtures using the correlations provided in Chapter 2 with good accuracy so that there is no need to use the pseudocomponent approach for their estimation.

Specific gravity (SG) of petroleum fractions may be estimated from methods presented in Sections 2.4.3 and 2.6. If API gravity is known, the specific gravity should be directly calculated from definition of API gravity using Eq. (2.5). If density at one temperature is available, then Eq. (2.110) should be used to estimate the increase in density with decrease in temperature and therefore density at 15.5°C (60°F) may be calculated from the available density. A simpler relation between SG and d_{20} based on the rule of thumb is $SG = 1.005 d_{20}$. If density at 20°C (d_{20}) is available, the following relation developed in Section 2.6.1 can be used:

$$(3.44) \qquad SG = 0.01044 + 0.9915 d_{20}$$

where d_{20} is in g/mL and SG is the specific gravity at 15.5°C. This equation is quite accurate for estimating density of petroleum fractions. If no density data are available, then SG may be estimated from normal boiling point and refractive index or from molecular weight and refractive index for heavy fractions in which boiling point may not be available. Equation (2.59) in Section 2.4.3 gives SG from T_b and I for fractions with molecular weights of less than 300, while for heavier fractions Eq. (2.60) can be used to estimate specific gravity from M and I. Parameter I is defined in terms of refractive index at 20°C, n_{20}, by Eq. (2.36). If viscosity data are available Eq. (2.61) should be used to estimate specific gravity, and finally, if only one type of distillation curves such as ASTM D 86, TBP, or EFV data are available Eq. (3.17) may be used to obtain the specific gravity.

Density (d) of liquid petroleum fractions at any temperature and atmospheric pressure may be estimated from the methods discussed in Section 2.6. Details of estimation of density of petroleum fractions are discussed in Chapter 6; however, for the characterization methods discussed in this chapter, at least density at 20°C is needed. If specific gravity is available then the rule of thumb with $d = 0.995 SG$ is the simplest way of estimating density at 20°C. For temperatures other than 20°C, Eq. (2.110) can be used. Equation (2.113) may also be used to estimate d_{20} from T_b and SG for light petroleum fractions ($M < 300$) provided that estimated density is less than the value of SG used in the equation. This is an accurate way of estimating density at 20°C especially for light fractions. However, the simplest and most accurate method of estimating d_{20} from SG for all types of petroleum fractions is the reverse form of Eq. (3.44), which is equivalent

to Eq. (2.111). If the specific gravity is not available, then it is necessary to estimate the SG at first step and then to estimate the density at 20°C. The liquid density decreases with increase in temperature according to the following relationship [24].

$$d = d_{15.5} - k(T - 288.7)$$

where $d_{15.5}$ is density at 15.5°C, which may be replaced by 0.999SG. T is absolute temperature in kelvin and k is a constant for a specific compound. This equation is accurate within a narrow range of temperature and it may be applied to any other reference temperature instead of 15.5°C. Value of k varies with hydrocarbon type; however, for gasolines it is close to 0.00085 [24].

Refractive index at 20°C, n_{20}, is an important characterization parameter for petroleum fractions. It is needed for prediction of the composition as well as estimation of other properties of petroleum fractions. If it is not available, it may be determined from correlations presented in Section 2.6 by calculation of parameter I. Once I is estimated, n_{20} can be calculated from Eq. (2.114). For petroleum fractions with molecular weights of less than 300, Eq. (2.115) can be used to estimate I from T_b and SG [38]. A more accurate relation, which is also included in the API-TDB [2], is given by Eq. (2.116). For heavier fraction in which T_b may not be available, Eq. (2.117) in terms of M and SG may be used [44]. Most recently Riazi and Roomi [45] made an extensive analysis of predictive methods and application of refractive index in prediction of other physical properties of hydrocarbon systems. An evaluation of these methods for some petroleum fractions is demonstrated in the following example.

Example 3.12—Experimental data on M, T_b, SG, d_{20}, and n_{20} for five heavy petroleum fractions are given in Table 3.17. Estimate SG, d_{20}, and n_{20} from available methods and calculate %AAD for each method with necessary discussion of results.

Solution—The first two fractions in Table 3.17 may be considered light ($M < 300$) and the last two fractions are considered as heavy ($M > 300$). The third fraction can be in either category. Data available on d_{20}, T_b, n_{20}, and M may be used to estimate specific gravity. As discussed in this section, SG can be calculated from d_{20} or from T_b and I or from M and I. In this example specific gravity may be calculated from four methods: (1) rule of thumb using d_{20} as the input parameter; (2) from d_{20} by Eq. (3.44); (3) from T_b and n_{20} by Eq. (2.59); (4) from M and n_{20} by Eq. (2.60). Summary of results is given in Table 3.19. Methods 1 and 2, which use density as the input parameter, give the best results. Method 4 is basically developed for heavy fractions with $M > 300$ and therefore for the last three fractions density is predicted with better accuracy.

TABLE 3.19—*Comparison of various methods of predicting specific gravity of petroleum fractions of Table 3.17 (Example 3.12).*

No.	M, exp	SG, exp	(1) Rule of thumb, SG = 1.005 d_{20}		(2) Use of d_{20}, Eq. (3.44)		(3) Use of T_b & n_{20}, Eq. (2.59) for M < 300		(4) Use of M & n_{20}, Eq. (2.60) for M >300	
			SG, calc	AD%	SG, calc	AD%	SG, calc	AD%	SG, calc	AD%
1	233	0.9119	0.9127	0.09	0.9109	0.11	0.8838	3.09	0.8821	3.27
2	267	0.9605	0.9616	0.11	0.9591	0.15	0.9178	4.44	0.9164	4.59
3	325	0.8883	0.8889	0.07	0.8874	0.10	0.8865	0.20	0.8727	1.76
4	403	0.9046	0.9046	0.00	0.9029	0.19	0.9067	0.23	0.8867	1.98
5	523	0.8760	0.8794	0.39	0.8780	0.23	0.9062	3.45	0.8701	0.67
Total, AAD%				0.13		0.15		2.28		2.45

Method 3, which is recommended for light fractions, gives better results for the specific gravity of heavy fractions. It should be noted that the boundary of 300 for light and heavy fractions is approximate and methods proposed for light fractions can be used well above this boundary limit as shown in Method 3.

Estimation of density is similar to estimation of specific gravity. When both T_b and SG are available Eq. (2.113) is the most accurate method for estimation of density of petroleum fractions. This method gives AAD of 0.09% for the five fractions of Table 3.17 with higher errors for the last two fractions. This equation may be used safely up to molecular weight of 500 but for heavier fractions Eq. (3.44) or the rule of thumb should be used. Predicted value of density at 20°C from Eq. (2.113) is not reliable if it is greater than the value of specific gravity used in the equation. The method of rule of thumb with $d = 0.995$ SG gives an AAD of 0.13% and Eq. (3.44) gives an AAD of 0.15%.

Refractive index is estimated from three different methods and results are given in Table 3.20. In the first method, T_b and SG are used as the input parameters with Eq. (2.115) to estimate I and n is calculated from Eq. (2.114). In the second method Eq. (2.116) is used with the same input data. Equations (2.115) and (2.116) are both developed with data on refractive index of pure hydrocarbons with M < 300. However, Eq. (2.116) in this range of application is more accurate than Eq. (2.115). But for heavier fractions as shown in Table 3.20, Eq. (3.115) gives better result. This is due to the simple nature of Eq. (2.115) which allows its application to heavier fractions. Equation (2.116) does not give very accurate refractive index for fraction with molecular weights of 500 or above. Equation (2.117) in terms of M and SG is developed basically for heavy fractions and for this reason it does not give accurate results for fractions with molecular weights of less than 300. This method is particularly useful when boiling point is not available but molecular weight is available or estimable. However, if boiling point is available, even for heavy fractions Eq. (2.115) gives more accurate results than does Eq. (2.117) as shown in Table 3.20. ♦

Kinematic viscosity of petroleum fractions can be estimated from methods presented in Section 2.7 of the previous chapter. At reference temperatures of 37.8 and 98.9°C (100 and 210°F), $\nu_{38(100)}$ and $\nu_{99(210)}$ can be determined from Eqs. (2.128) and (2.129) or through Fig. 2.12 using API gravity and K_W as the input parameters. In use of these equations attention should be paid to the limitations and to check if API and K_W are within the ranges specified for the method. To calculate kinematic viscosity at any other temperature, Eq. (2.130) or Fig. 2.13 may be used. The procedure is best demonstrated through the following example.

Example 3.13—A petroleum fraction is produced through distillation of a Venezuelan crude oil and has the specific gravity of 0.8309 and the following ASTM D 86 distillation data:

vol% distilled	10	30	50	70	90
ASTM D 86 temperature,°F	423	428	433	442	455

Estimate kinematic viscosity of this fraction at 100 and 140°F (37.8 and 60°C). Compare the calculated values with the experimental values of 1.66 and 1.23 cSt [46].

Solution—Kinematic viscosities at 100 and 210°F, $\nu_{38(100)}$ and $\nu_{99(210)}$, are calculated from Eqs. (2.128) and (2.129), respectively. The API gravity is calculated from Eq. (2.4): API = 38.8. To calculate K_W from Eq. (2.13), MeABP is required. For this fraction since it is a narrow boiling range the MeABP is nearly the same as the mid boiling point or ASTM 50% temperature. However, since complete ASTM D 86 curve is available we use Eqs. (3.6)–(3.12) to estimate this average boiling point. Calculated parameters are VABP = 435.6°F and SL = 0.4°F/vol%. From Eqs. (3.8) and (3.12) we get MeABP = 434°F (223.3°C). As expected this temperature is very close to ASTM 50% temperature of 433°F. From Eq. (2.13), $K_W = 11.59$. Since $0 < API < 80$ and $10 < K_W < 11$, we can use Eqs. (2.128) and (2.129) for calculation of kinematic viscosity and we get $\nu_{38(100)} = 1.8$, $\nu_{99(210)} = 0.82$ cSt. To calculate viscosity at 140°F, $\nu_{60(140)}$, we use Eqs. (2.130)–(2.132). From Eq. (2.131)

TABLE 3.20—*Comparison of various methods of predicting refractive index of petroleum fractions of Table 3.17 (Example 3.12).*

No.	M, exp	n_{20} exp	(1) Use of T_b & SG, Eq. (2.115) for M < 300		(2) Use of T_b & SG, Eq. (2.116) for M < 300		(4) Use of M & SG, Eq. (2.117) for M > 300	
			n_{20} exp.	AD%	n_{20} calc	AD%	n_{20} calc	AD%
1	233	1.5016	1.5122	0.70	1.5101	0.57	1.5179	1.08
2	267	1.5366	1.5411	0.29	1.5385	0.13	1.5595	1.49
3	325	1.4919	1.4960	0.28	1.4895	0.16	1.4970	0.34
4	403	1.5002	1.5050	0.32	1.4952	0.34	1.5063	0.41
5	523	1.4865	1.4864	0.01	1.4746	0.80	1.4846	0.13
Total, AAD%				0.32		0.40		0.69

since $\nu_{38(100)} > 1.5$ and $\nu_{99(210)} < 1.5$ cSt we have $c_{38(100)} = 0$ and $c_{99(210)} = 0.0392$. From Eq. (2.132), $A_1 = 10.4611$, $B_1 = -4.3573$, $D_1 = -0.4002$, $D_2 = -0.7397$, and from Eq. (2.130) at $T = 140°F$ (60°C) we calculate the kinematic viscosity. It should be noted that in calculation of $\nu_{60(140)}$ from Eq. (2.130) trial and error is required for calculation of parameter c. At first it is assumed that $c = 0$ and after calculation of $\nu_{60(140)}$ if it is less than 1.5 cSt, parameter c should be calculated from Eq. (2.131) and substituted in Eq. (2.130). Results of calculations are as follows: $\nu_{38(100)} = 1.8$ and $\nu_{60(140)} = 1.27$ cSt. Comparing with the experimental values, the percent relative deviations for kinematic viscosities at 100 and 140°F are 8.4 and 3.3%, respectively. The result is very good, but usually higher errors are observed for estimation of kinematic viscosity of petroleum fractions from this method. ◆

3.4 GENERAL PROCEDURE FOR PROPERTIES OF MIXTURES

Petroleum fluids are mixtures of hydrocarbon compounds, which in the reservoirs or during processing could be in the form of liquid, gas, or vapor. Some heavy products such as asphalts and waxes are in solid forms. But in petroleum processing most products are in the form of liquid under atmospheric conditions. The same liquid products during processing might be in a vapor form before they are stored as a product. Certain properties such as critical constants, acentric factor, and molecular weight are specifications of a compound regardless of being vapor of liquid. However, physical properties such as density, transport, or thermal properties depend on the state of the system and in many cases separate methods are used to estimate properties of liquid and gases as will be discussed in the following chapters. In this section a general approach toward calculation of such properties for liquids and gases with known compositions is presented. Since density and refractive index are important physical properties in characterization or petroleum fractions they are used in this section to demonstrate our approach for mixture properties. The same approach will be applied to other properties throughout the book.

3.4.1 Liquid Mixtures

In liquid systems the distance between molecules is much smaller than in the case of gases and for this reason the interaction between molecules is stronger in liquids. Therefore, the knowledge of types of molecules in the liquid mixtures is more desirable than in gas mixtures, especially when the mixture constituents differ significantly in size and type. For example, consider two liquid mixtures, one a mixture of a paraffinic hydrocarbon such as *n*-eicosane (*n*-C$_{20}$) with an aromatic compound such as benzene (C$_6$) and the second one a mixture of benzene and toluene, which are both aromatic with close molecular weight and size. The role of composition in the *n*-C$_{20}$–benzene mixture is much more important than the role of composition in the benzene–toluene mixture. Similarly the role of type of composition (weight, mole, or volume fraction) is more effective in mixtures of dissimilar constituents than mixtures of similar compounds. It is for this reason that for narrow-range petroleum fractions, use of the

PNA composition in terms of weight, volume, or mole does not seriously affect the predicted mixture properties. Use of bulk properties such as T_b and SG to calculate mixture properties as described for petroleum fractions cannot be used for a synthetic and ternary mixture of C$_5$–C$_{10}$–C$_{25}$. Another example of a mixture that bulk properties directly cannot be used to calculate its properties is a crude oil or a reservoir fluid. For such mixtures exact knowledge of composition is required and based on an appropriate mixing rule a certain physical property for the mixture may be estimated. The most simple and practical mixing rule that is applicable to most physical properties is as follows:

$$(3.45) \qquad \theta_m = \sum_{i=1}^{N} x_i \theta_i$$

where x_i is the fraction of component i in the mixture, θ_i is a property for pure component i, and θ_m is property of the mixture with N component. This mixing rule is known as *Kay mixing rule* after W. B. Kay of Ohio State, who studied mixture properties, especially the pseudocritical properties in the 1930s and following several decades. Other forms of mixing rules for critical constants will be discussed in Chapter 5 and more accurate methods of calculation of mixture properties are presented in Chapter 6.

Equation (3.45) can be applied to any property such as critical properties, molecular weight, density, refractive index, heat capacity, etc. There are various modified version of Eq. (3.45) when it is applied to different properties. Type of composition used for x_i depends on the type of property. For example, to calculate molecular weight of the mixture ($\theta = M$) the most appropriate type of composition is mole fraction. Similarly mole fraction is used for many other properties such as critical properties, acentric factor, and molar properties (i.e., molar heat capacity). However, when Eq. (3.45) is applied to density, specific gravity, or refractive index parameter $[I = (n^2 - 1)/(n^2 + 2)]$, volume fraction should be used for x_i. For these properties the following mixing rule may also be applied instead of Eq. (3.45) if weight fraction is used:

$$(3.46) \qquad 1/\theta_m = \sum_{i=1}^{N} x_{wi}/\theta_i$$

where x_{wi} is the weight fraction and the equation can be applied to d, SG, or parameter I. In calculation of these properties for a mixture, using Eq. (3.45) with volume fraction and Eq. (3.46) with weight fraction gives similar results. Application of these equations in calculation of mixture properties will be demonstrated in the next chapter to calculate properties of crude oils and reservoir fluids.

For liquid mixtures the mixing rule should be applied to the final desired property rather than to the input parameters. For example, a property such as viscosity is calculated through a generalized correlation that requires critical properties as the input parameters. Equation (3.45) may be applied first to calculate mixture pseudocritical properties and then mixture viscosity is calculated from the generalized correlation. An alternative approach is to calculate viscosity of individual components in the mixture for the generalized correlation and then the mixing rule is directly applied to viscosity. As it is shown in the following chapters the second approach gives more accurate results for properties of liquid mixtures, while for gaseous mixtures there is no significant difference between these two methods.

3.4.2 Gas Mixtures

As discussed earlier the gases at atmospheric pressure condition have much larger free space between molecules than do liquids. As a result the interaction between various like and unlike molecules in a gaseous state is less than the molecular interactions in similar liquid mixtures. Therefore, the role of composition on properties of gas mixtures is not as strong as in the case of liquids. Of course the effect of composition on properties of gas mixtures increases as pressure increases and free space between molecules decreases. The role of composition on properties of dense gases cannot be ignored. Under low-pressure conditions where most gases behave like ideal gases all gas mixtures regardless of their composition have the same molar density at the same temperature and pressure. As it will be discussed in Chapter 5, at the standard conditions (SC) of 1.01325 bar and 298 K (14.7 psia and 60°F), most gases behave like ideal gas and RT/P represents the molar volume of a pure gas or a gas mixture. However, the absolute density varies from one gas to another as following:

$$(3.47) \qquad \rho_{mix} = \frac{M_{mix}P}{83.14T}$$

where ρ_{mix} is the absolute density of gas mixture in g/cm^3, M_{mix} is the molecular weight of the mixture in g/mol, P is pressure in bar, and T is the temperature in kelvin. Equation (3.1) can be used to calculate molecular weight of a gas mixture, M_{mix}. However, the mole fraction of component i in a gas mixture is usually shown as y_i to distinguish from composition of liquid mixtures designated by x_i. From definition of mole and volume fractions in Section 1.7.15 and use of Eq. (3.47) it can be shown that for ideal gas mixtures the mole and volume fractions are identical. Generally volume and mole fractions are used interchangeably for all types of gas mixtures. Composition of gas mixtures is rarely expressed in terms of weight fraction and this type of composition has very limited application for gas systems. Whenever composition in a gas mixture is expressed only in percentage it should be considered as mol% or vol%. Gas mixtures that are mainly composed of very few components, such as natural gases, it is possible to consider them as a single pseudocomponent and to predict properties form specific gravity as the sole parameter available. This method of predicting properties of natural gases is presented in Chapter 4 where characterization of reservoir fluids is discussed. The following example shows derivation of the relation between gas phase specific gravity and molecular weight of gas mixture.

Example 3.14—Specific gravity of gases is defined as the ratio of density of gas to density of dry air both measured at the standard temperature and pressure (STP). Composition of dry air in mol% is 78% nitrogen, 21% oxygen, and 1% argon. Derive Eq. (2.6) for the specific gravity of a gas mixture.

Solution—Equation (2.6) gives the gas specific gravity as

$$(2.6) \qquad SG_g = \frac{M_g}{28.97}$$

where M_g is the gas molecular weight. Density of both a gas mixture and air at STP can be calculated from Eq. (3.47).

$$(3.48) \qquad SG_g = \frac{\rho_{gas}}{\rho_{air}} = \frac{\frac{M_{gas}P_{sc}}{83.14T_{sc}}}{\frac{M_{air}P_{sc}}{83.14T_{sc}}} = \frac{M_{gas}}{M_{air}}$$

where sc indicates the standard condition. Molecular weight of air can be calculated from Eq. (3.48) with molecular weight of its constituents obtained from Table 2.1 as $M_{N2} = 28.01$, $M_{O2} = 32.00$, and $M_{Ar} = 39.94$. With composition given as $y_{N2} = 0.78$, $y_{O2} = 0.21$, and $y_{Ar} = 0.01$, from Eq. (3.1) we get $M_{air} = 28.97$ g/mol. Equation (2.6) can be derived from substituting this value for M_{air} in Eq. (3.49). In practical calculations molecular weight of air is rounded to 29. If for a gas mixture, specific gravity is known its molecular weight can be calculated as

$$(3.49) \qquad M_g = 29SG_g$$

where SG_g is the gas specific gravity. It should be noted that values of specific gravity given for certain gases in Table 2.1 are relative to density of water for a liquefied gas and are different in definition with gas specific gravity defined from Eq. (2.6). ♦

3.5 PREDICTION OF THE COMPOSITION OF PETROLEUM FRACTIONS

As discussed earlier the quality and properties of a petroleum fraction or a petroleum product depend mainly on the mixture composition. As experimental measurement of the composition is time-consuming and costly the predictive methods play an important role in determining the quality of a petroleum product. In addition the pseudocomponent method to predict properties of a petroleum fraction requires the knowledge of PNA composition. Exact prediction of all components available in a petroleum mixture is nearly impossible. In fact there are very few methods available in the literature that are used to predict the composition. These methods are mainly capable of predicting the amounts (in percentages) of paraffins, naphthenes, and aromatic as the main hydrocarbon groups in all types of petroleum fractions. These methods assume that the mixture is free of olefinic hydrocarbons, which is true for most fractions and petroleum products as olefins are unstable compounds. In addition to the PNA composition, elemental composition provides some vital information on the quality of a petroleum fraction or crude oil. Quality of a fuel is directly related to the hydrogen and sulfur contents. A fuel with higher hydrogen or lower carbon content is more valuable and has higher heating value. High sulfur content fuels and crude oils require more processing cost and are less valuable and desirable. Methods of predicting amounts of C, H, and S% are presented in the following section.

3.5.1 Prediction of PNA Composition

Parameters that are capable of identifying hydrocarbon types are called characterization parameters. The best example of such a parameter is the Watson characterization factor, which along with other parameters is introduced and discussed in this section. However, the first known method to predict the PNA composition is the n-d-M method proposed by Van Nes and Van Westen [30] in the 1950s. The n-d-M method is also included in the ASTM manual under ASTM D 3238 test method. The main limitation of this method is that it cannot be applied to light fractions. Later in the 1980s Riazi and Daubert [36, 47] proposed a series of correlations based on careful analysis of various characterization parameters. The unique feature of these correlations is that they are applicable to both light and heavy fractions and identify various types of aromatics in the mixture. In addition various methods are proposed based on different bulk properties of the mixture that might be available. The Riazi–Daubert methods have been adopted by the API Committee on characterization of petroleum fractions and are included in the fourth and subsequent editions of the API-TDB [2] since the early 1980s. The other method that is reported in some literature sources is the Bergamn's method developed in the 1970s [48]. This method is based on the Watson K and specific gravity of the fraction as two main characterization parameters. One common deficiency for all of these methods is that they do not identify *n*-paraffins and *iso*paraffins from each other. In fact compositional types of PIONA, PONA, and PINA cannot be determined from any of the methods available in the literature. These methods provide minimum information on the composition that is predictive of the PNA content. This is mainly due to the complexity of petroleum mixtures and difficulty of predicting the composition from measurable bulk properties. The method of Riazi–Daubert, however, is capable of predicting the monoaromatic (MA) and polyaromatic (PA) content of petroleum fractions.

In general low boiling point fractions have higher paraffinic and lower aromatic contents while as boiling point of the fraction increases the amount of aromatic content also increases. In the direction of increase in boiling point, in addition to

aromatic content, amounts of sulfur, nitrogen, and other heteroatoms also increase as shown in Fig. 3.22.

3.5.1.1 Characterization Parameters for Molecular Type Analysis

A characterization parameter that is useful for molecular type prediction purposes should vary significantly from one hydrocarbon type to another. In addition, its range of variation within a single hydrocarbon family should be minimal. With such specifications an ideal parameter for characterizing molecular type should have a constant value within a single family but different values in different families. Some of these characterization parameters (i.e., SG, *I*, VGC, CH, and K_W), which are useful for molecular type analysis, have been introduced and defined in Section 2.1. As shown in Table 2.4, specific gravity is a parameter that varies with chemical structure particularly from one hydrocarbon family to another. Since it also varies within a single family, it is not a perfect characterizing parameter for molecular type analysis but it is more suitable than boiling point that varies within a single family but its variation from one family to another is not significant. One of the earliest parameters to characterize hydrocarbon molecular type was defined by Hill and Coats in 1928 [49], who derived an empirical relation between viscosity and specific gravity in terms of viscosity gravity constant (VGC), which is defined by Eq. (2.15) in Section 2.1.17. Definition of VGC by Eqs. (2.15) or (2.16) limits its application to viscous oils or fractions with kinematic viscosity at 38°C (100°F) above 38 SUS (~3.8 cSt.). For quick hand estimation of VGC from viscosity and specific gravity, ASTM [4] has provided a nomograph, shown in Fig. 3.23, that gives VGC values close to those calculated from Eq. (2.15). Paraffinic oils have low VGC, while napthenic oils have high VGC values. Watson *K* defined by Eq. (2.13) in terms of MeABP and SG was originally introduced to identify hydrocarbon type [9, 50, 51], but as is shown later, this is not a very suitable parameter to indicate composition of petroleum fractions.

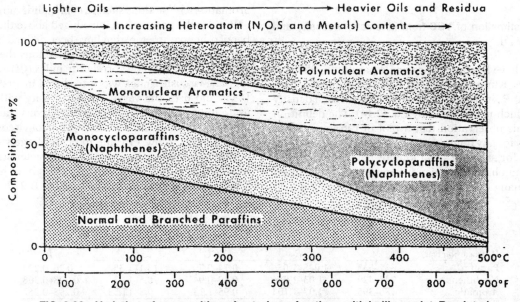

FIG. 3.22—Variation of composition of petroleum fractions with boiling point. Reprinted from Ref. [7], p. 469, by courtesy of Marcel Dekker, Inc.

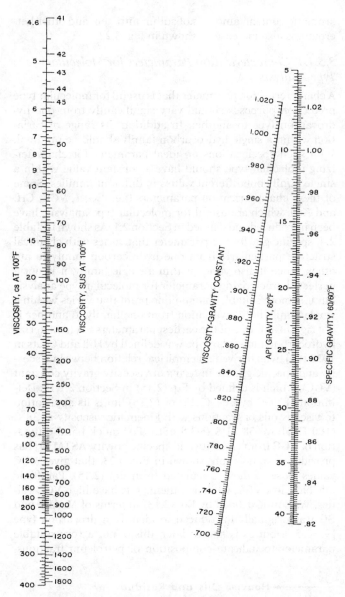

FIG. 3.23—Estimation of VGC from kinematic viscosity and specific gravity [4].

Aromatic oils have low K_W values while paraffinic oils have high K_W values. Kurtz and Ward in 1935 [52] defined refractivity intercept, R_i, in terms of refractive index (n) and density (d) at 20°C, which is presented by Eq. (2.14). The definition is based on this observation that a plot of refractive index against density for any homologous hydrocarbon group is linear. R_i is high for aromatics and low for naphthenic stocks. The most recent characterization parameter was introduced in 1977 by Huang [53] in terms of refractive index and it

is defined by Eq. (2.36). Paraffinic oils have low I values while aromatics have high I values. Carbon-to-hydrogen weight ratio defined in Section 2.1.18 is also a useful parameter that indicates degree of hydrocarbon saturation and its value increases from paraffinic to naphthenic and aromatic oils. Methods of prediction of CH was discussed in Section 2.6.3. Application of the hydrogen-to-carbon ratio in characterization of different types of petroleum products is demonstrated by Fryback [54]. An extensive analysis and comparison of various characterization parameters useful for prediction of the composition of petroleum fractions is presented by Riazi and Daubert [36,47]. Comparison of parameters R_i, VGC, K_W, and I is presented in Table 3.21 and Fig. 3.24. From this analysis it is clear that parameters R_i and VHC best separate hydrocarbon types, while parameters K_W and I show large variations for aromatic and naphthenic compounds making them less suitable for prediction of composition of petroleum fractions.

Another very useful parameter that not only separates paraffins and aromatics but also identifies various hydrocarbon types is defined through molecular weight and refractive index as [36]:

$$(3.50) \qquad m = M(n - 1.475)$$

where n is the refractive index at 20°C. Parameter m was defined based on the observation that refractive index varies linearly with $1/M$ with slope of m for each hydrocarbon group [55]. Values of parameter m for different hydrocarbon groups calculated from Eq. (3.50) are given in Table 3.22.

As shown in Table 3.22, paraffins have low m values while aromatics have high m values. In addition, paraffinic and naphthenic oils have negative m values while aromatic oils have positive m values. Parameter m nicely identifies various aromatic types and its value increases as the number of rings increases in an aromatic compound. A pure hydrocarbon whose m value is calculated as -9 has to be paraffinic, it cannot be naphthenic or aromatic. This parameter is particularly useful in characterizing various aromatic types in aromatic-rich fractions such as coal liquids or heavy residues.

Besides the parameters introduced above there are a number of other parameters that have been defined for the purpose of characterizing hydrocarbon type. Among these parameters viscosity index (VI) and correlation index (CI) are worth defining. The viscosity index was introduced in 1929 by Dean and Davis and uses the variation of viscosity with temperature as an indication of composition of viscous fractions. It is an empirical number indicating variation of viscosity of oil with temperature. A low VI value indicates large variation of viscosity with temperature that is a characteristic of aromatic oils. Similarly, paraffinic hydrocarbons have high VI values. The method is described under ASTM D 2270-64 [4] and in

TABLE 3.21—*Values of characterization factors.*

Hydrocarbon type	M	R_i	VGC	K	I
			Value Range		
Paraffin	337–535	1.048–1.05	0.74–0.75	13.1–1.35	0.26–0.273
Naphthene	248–429	1.03–1.046	0.89–0.94	10.5–13.2	0.278–0.308
Aromatic	180–395	1.07–1.105	0.95–1.13	9.5–12.53	0.298–0.362

Taken with Permission from Ref. [47].

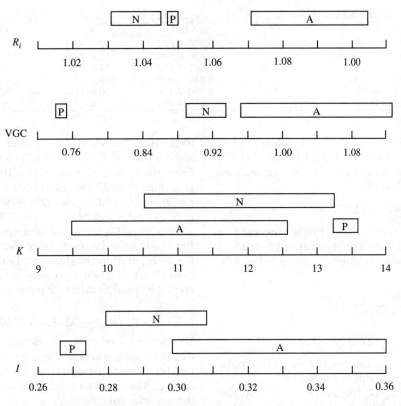

FIG. 3.24—Comparison of different characterization factors for prediction of composition of petroleum fractions (see Table 3.20).

the API-TDB [2]. The VI is defined as [2]

$$VI = \frac{L - U}{D} 100 \tag{3.51}$$

where

L = kinematic viscosity of reference oil at 40°C with 0 VI oil, cSt.

U = kinematic viscosity of oil at 40°C whose VI is to be calculated, cSt.

$D = L - H$, in cSt.

H = kinematic viscosity of reference oil at 40°C with 100 VI oil, cSt.

The reference oils with 0 and 100 VI and the oil whose VI is to be calculated have the same kinematic viscosity at 100°C. In English units of system, VI is defined in terms of viscosity at 37.8 and 98.9°C, which correspond to 100 and 210°F, respectively [2]. However, in the SI units, viscosity at reference temperatures of 40 and 100°C have been used to define the VI [5]. For oils with a kinematic viscosity of 100°C of less than 70 mm²/s (cSt.), the values of L and D are given

in tables in the standard methods (ASTM D 2270, ISO 2909) as well as in the API-TDB [2]. However, for viscous oils with a kinematic viscosity at 100°C of greater than 70 mm²/s, the values of L and D are given by the following relationships [5]:

$$L = 0.8353Y^2 + 14.67Y - 216$$
$$D = 0.6669Y^2 + 2.82Y - 119 \tag{3.52}$$

where Y is the kinematic viscosity of oil whose VI is to be calculated at 100°C in cSt. In English units in which reference temperatures of 37.8 and 98.9°C (100 and 210°F) are used, values of numerical coefficients in Eq. (3.52) are slightly different and are given in the API-TDB [2].

$$L = 1.01523Y^2 + 12.154Y - 155.61$$
$$D = 0.8236Y^2 - 0.5015Y - 53.03 \tag{3.53}$$

Viscosity index defined by Dean and Davis in the form of Eq. (3.51) does not work very well for oils with VI values of greater than 100. For such oils ASTM D 2270 describes the calculation method and it is summarized below for viscosity at reference temperatures of 40 and 100°C [4, 5]:

$$VI = \frac{10^N - 1}{0.00715} + 100 \tag{3.54}$$

where N is given by the following relation:

$$N = \frac{\log H - \log U}{\log Y} \tag{3.55}$$

in which U and Y are kinematic viscosity of oil in cSt whose VI is to be calculated at 40 and 100°C, respectively. Values of

TABLE 3.22—Values of parameter m for different types of hydrocarbons [36].

Hydrocarbon type	m
Paraffins	−8.79
Cyclopentanes	−5.41
Cyclohexanes	−4.43
Benzenes	2.64
Naphthalenes	19.5
Condensed Tricyclics	43.6

TABLE 3.23—*Comparison of VI, VGC, K_W, and CI for several groups of oils [29].*

Type of oil	VI	VGC	K_W	CI
High VI distillate	100	0.800–0.805	12.2–12.5	<15
Medium VI distillate	80	0.830–0.840	11.8–12.0	15–50
Low VI distillate	0	0.865–0.890	11.0–11.5	>50
Solvent Extracts	...	0.880–0.950	10.0–11.0	...
Recycle Stock	...	0.900–0.950	10.0–11.0	...
Cracked Residues	...	0.950–1.000	9.8–11.0	...

H versus kinematic viscosity of oil at 100°C (Y) when Y is less than 70 cSt are tabulated in the ASTM D 2270 method [4]. For oils with values of Y greater than 70 cSt H must be calculated from the relation given below [5].

$$(3.56) \qquad H = 0.1684Y^2 + 11.85Y - 97$$

If kinematic viscosities are available in English units at 37.8 and 98.9°C (100 and 210°F), then Eqs. (3.54) and (3.56) should be replaced with the following relations as given in the API-TDB [2]:

$$(3.57) \qquad VI = \frac{10^N - 1}{0.0075} + 100$$

$$(3.58) \qquad H = 0.19176Y^2 + 12.6559Y - 102.58$$

in which Eq. (3.58) must be used for fractions with kinematic viscosity at 99°C (210°F) greater than 75 cSt [2].

Finally the correlation index (CI) defined by the U.S. Bureau of Mines is expressed by the following equation [7]:

$$(3.59) \qquad CI = \frac{48640}{T_b} + 473.7SG - 456.8$$

in which T_b is the volume average boiling point (VABP) in kelvin. Values of CI between 0 and 15 indicate a predominantly paraffinic oil. A value of CI greater than 50 indicates a predominance of aromatic compounds [7]. It has a tendency to increase with increasing boiling point in a given crude oil. A comparison between values of VI, VGC, K_W, and CI for several types of petroleum fractions and products is presented in Table 3.23. A complete comparison of various characterization parameters indicating composition of petroleum fractions for the three hydrocarbon groups is presented in Table 3.24. All parameters except R_i, K_W, and VI increase in the direction from paraffinic to naphthenic and aromatic oils.

Example 3.15—Calculate viscosity index of an oil having kinematic viscosities of 1000 and 100 cSt at 37.8 and 98.9°C (100 and 210°F), respectively.

Solution—For this oil $v_{99(210)} > 70$ cSt, thus we can use Eqs. (3.51)–(3.57) for calculation of VI. Since the VI is not known we assume it is greater than 100 and we use Eqs. (3.53)–(3.57)

to calculate the VI. However, since kinematic viscosities are given at 38 and 99°C, Eqs. (3.57) and (3.58) should be used. From the information given $U = 1000$ and $Y = 100$. Since Y is greater than 75 cSt, Eq. (3.58) must be used to calculate parameter H, which gives $H = 3080.6$. From Eqs. (3.55) and (3.57) we get $N = 0.2443$ and VI = 200.7. Since calculated VI is greater than 100, the initial assumption is correct. Otherwise, Eq. (3.51) must be used. Since the value of VI is quite high the oil is paraffinic as shown in Table 3.24. If $v_{99(210)}$ was less than 70 cSt then tables provided by ASTM [4] or API-TDB [2] should be used to calculate parameters L and D. ♦

3.5.1.2 API (Riazi–Daubert) Methods

To develop a method for predicting the composition of olefin-free petroleum fractions three equations are required to obtain fractions of paraffins (x_P), naphthenes (x_N), and aromatics (x_A). The first and most obvious equation is known from the material balance:

$$(3.60) \qquad x_P + x_N + x_A = 1$$

Two other equations can be established by applying Eq. (3.40) for two parameters that can characterize hydrocarbon types and are easily measurable. Analysis of various characterization factors shown in Table 3.21 and Fig. 3.24 indicates that R_i and VGC are the most suitable parameters to identify hydrocarbon type. For example, if for a pure hydrocarbon $R_i = 1.04$, it has to be a naphthenic hydrocarbon, it cannot be paraffinic or aromatic since only for the naphthenic group R_i varies from 1.03 to 1.046. Refractivity intercept has been related to the percent of naphthenic carbon atoms (%C_N) as $R_i = 1.05 - 0.0002$ %C_N [7]. Riazi and Daubert [47] used both R_i and VGC to develop a predictive method for the composition of viscous petroleum fractions. Properties of pure hydrocarbons from the API RP-42 [56] have been used to calculate R_i and VGC for a number of heavy hydrocarbons with molecular weights greater than 200 as shown in Table 2.3 and Table 3.21. Based on the values of R_i and VGC for all hydrocarbons, average values of these parameters were determined for the three groups of paraffins, naphthenes, and aromatics. These average values for R_i are as follows: 1.0482 (P), 1.0138 (N), and 1.081 (A). Similar average values for VGC are 0.744 (P),

TABLE 3.24—*Comparison of various characterization parameters for molecular type analysis.*

Parameter	Defined by Eq. (s)	Paraffins	Naphthenes	Aromatics
R_i	(2.14)	medium	low	high
VGC	(2.15) or (2.16)	low	medium	high
m	(3.50)	low	medium	high
SG	(2.2)	low	medium	high
I	(2.3) (2.36)	low	medium	high
CI	(3.59)	low	medium	high
K_W	(2.13)	high	medium	low
VI	(3.51)–(3.58)	high	medium	low

0.915 (N), and 1.04 (A). Applying Eq. (3.40) for R_i and VGC gives the following two relations.

(3.61) $$R_i = 1.0482x_P + 1.038x_N + 1.081x_A$$

(3.62) $$VGC = 0.744x_P + 0.915x_N + 1.04x_A$$

A regression of 33 defined hydrocarbon mixtures changes the numerical constants in the above equations by less than 2% as follows [29, 47]:

(3.63) $$R_i = 1.0486x_P + 1.022x_N + 1.11x_A$$

(3.64) $$VGC = 0.7426x_P + 0.9x_N + 1.112x_A$$

Simultaneous solution of Eqs. (3.60), (3.63), and (3.64) gives the following equations for estimation of the PNA composition of fractions with $M > 200$.

(3.65) $$x_P = -9.0 + 12.53R_i - 4.228\,VGC$$

(3.66) $$x_N = 18.66 - 19.9R_i + 2.973\,VGC$$

(3.67) $$x_A = -8.66 + 7.37R_i + 1.255\,VGC$$

These equations can be applied to fractions with molecular weights in the range of 200–600. As mentioned earlier, x_P, x_N, and x_A calculated from the above relations may represent volume, mole, or weight fractions. Equations (3.65)–(3.67) cannot be applied to light fractions having kinematic viscosity at 38°C of less than 38 SUS ($\sim$3.6 cSt). This is because VGC cannot be determined as defined by Eqs. (2.15) and (2.16). For such fractions Riazi and Daubert [47] defined a parameter similar to VGC called viscosity gravity function, VGF, by the following relations:

(3.68) $$VGF = -1.816 + 3.484SG - 0.1156\ln \nu_{38(100)}$$

(3.69) $$VGF = -1.948 + 3.535SG - 0.1613\ln \nu_{99(210)}$$

where $\nu_{38(100)}$ and $\nu_{99(210)}$ are kinematic viscosity in mm²/s (cSt) at 38 and 99°C (100 and 210°F), respectively. For a petroleum fraction, both Eqs. (3.68) and (3.69) give nearly the same value for VGF; however, if kinematic viscosity at 38°C is available Eq. (3.68) is preferable for calculation of VGF. These equations have been derived based on the observation that at a fixed temperature, plot of SG versus $\ln \nu$ is a linear line for each homologous hydrocarbon group, but each group has a specific slope. Further information on derivation of these equations is provided by Riazi and Daubert [47]. Parameter VGF is basically defined for fractions with molecular weights of less than 200. Based on the composition of 45 defined mixtures (synthetic) and with an approach similar to the one used to develop Eqs. (3.65)–(3.67), three relationships in terms of R_i and VGF have been obtained to estimate the PNA composition (x_P, x_N, x_A) of light ($M < 200$) fractions [47]. These equations were later modified with additional data for both light and heavy fractions and are given below [36].

For fractions with $M \leq 200$

(3.70) $$x_P = -13.359 + 14.4591R_i - 1.41344\,VGF$$

(3.71) $$x_N = 23.9825 - 23.33304R_i + 0.81517\,VGF$$

(3.72) $$x_A = 1 - (x_P + x_N)$$

For fractions with $M > 200$

(3.73) $$x_P = 2.5737 + 1.0133R_i - 3.573\,VGC$$

(3.74) $$x_N = 2.464 - 3.6701R_i + 1.96312\,VGC$$

In these set of equations x_A must be calculated from Eq. (3.72). For cases that calculated x_A is negative it should be set equal to zero and values of x_P and x_N must be normalized in a way that $x_P + x_N = 1$. The same procedure should be applied to x_P or x_N if one of them calculated from the above equations is negative. For 85 samples Eqs. (3.70) and (3.72) give average deviation of 0.04 and 0.06 for x_P and x_N, respectively. For 72 heavy fractions, Eqs. (3.72)–(3.74) predict x_P, x_N, and x_A with average deviations of 0.03, 0.04, and 0.02, respectively [36]. These deviations are within the range of experimental uncertainty for the PNA composition. Equations (3.70)–(3.74) are recommended to be used if experimental data on viscosity are available. For cases that n_{20} and d_{20} are not available, they can be accurately estimated from the methods presented in Chapter 2.

For fractions that kinematic viscosity is not available, Riazi and Daubert [36] developed a series of correlations in terms of other characterization parameters which are readily available or predictable. These parameters are SG, m, and CH and the predictive equations for PNA composition are as follows:

For fractions with $M \leq 200$

(3.75) $$x_P = 2.57 - 2.877SG + 0.02876CH$$

(3.76) $$x_N = 0.52641 - 0.7494x_P - 0.021811m$$

or

(3.77) $$x_P = 3.7387 - 4.0829SG + 0.014772m$$

(3.78) $$x_N = -1.5027 + 2.10152SG - 0.02388m$$

For fractions with $M > 200$

(3.79) $$x_P = 1.9842 - 0.27722R_i - 0.15643CH$$

(3.80) $$x_N = 0.5977 - 0.761745R_i + 0.068048CH$$

or

(3.81) $$x_P = 1.9382 + 0.074855m - 0.19966CH$$

(3.82) $$x_N = -0.4226 - 0.00777m + 0.107625CH$$

In all of these cases x_A must be calculated from Eq. (3.72). Equations (3.75) and (3.76) have been evaluated with PNA composition of 85 fractions in the molecular weight range of 78–214 and give average deviations of 0.05, 0.08, and 0.07 for x_P, x_N, and x_A, respectively. For the same data set Eqs. (3.77) and (3.78) give AAD of 0.05, 0.086, and 0.055 for x_P, x_N, and x_A, respectively. For 72 fractions with molecular weight range of 230–570, Eqs. (3.79)–(3.82) give nearly the same AAD of 0.06, 0.06, and 0.02 for x_P, x_N, and x_A, respectively. In cases that input parameters for the above methods are not available Eqs. (3.77) and (3.78) in terms of SG and m are more suitable than other equations since refractive index and molecular weight can be estimated more accurately than CH. Although Eqs. (3.77) and (3.78) have been derived from a data set on fractions with molecular weights up to 200, they can be safely used up to molecular weight of 300 without serious errors. Most recently, Eqs. (3.77) and (3.78) have been modified to expand the range of application of these equations for heavier fractions, but in general their accuracy is not significantly different from the equations presented here [45]. For example, for fractions with $70 < M < 250$, Riazi and Roomi [45] modified Eqs.

(3.77) and (3.78) as: $x_P = 3.2574 - 3.48148\,SG + 0.011666\,m$ and $x_N = -1.9571 + 2.63853\,SG - 0.03992m$. Most of the data used in development of Eqs. (3.70)–(3.82) were in terms of volume fractions for x_P, x_N, and x_A. Therefore, generally estimated values represent volume fractions; however, they can be used as mole fractions without serious errors.

In all the above equations total aromatic content is calculated from Eq. (3.72). As discussed earlier for cases that aromatic content is high it should be split into two parts for a more accurate representation of hydrocarbon types in a petroleum mixture. Aromatics are divided into monoaromatics (x_{MA}) and polyaromatics (x_{PA}) and the following relations have been derived for fractions with molecular weights of less than 250 [36]:

$$(3.83) \quad x_{MA} = -62.8245 + 59.90816R_i - 0.0248335m$$

$$(3.84) \quad x_{PA} = 11.88175 - 11.2213R_i + 0.023745m$$

$$(3.85) \quad x_A = x_{MA} + x_{PA}$$

The equations may be applied to fractions with total aromatic content in the range of 0.05–0.96 and molecular weight range of 80–250. Based on a data set for aromatic contents of 75 coal liquid sample, Eqs. (3.83)–(3.85) give AAD of about 0.055, 0.065, and 0.063 for x_{MA}, x_{PA}, and x_A, respectively. Maximum AD is about 0.24 for x_{PA}. Equations (3.83) and (3.84) have not been evaluated against petroleum fractions. For heavier fractions no detailed composition on aromatics of fractions were available; however, if such data become available expressions similar to Eqs. (3.83) and (3.84) may be developed for heavier fractions.

Example 3.16—A gasoline sample produced from an Australian crude oil has the boiling range of C_5-65°C, specific gravity of 0.646, and PNA composition of 91, 9, and 0 vol% (Ref. [46], p. 302). Calculate the PNA composition from a suitable method and compare the results with the experimental values.

Solution—For this fraction the only information available are boiling point and specific gravity. From Table 2.1 the boiling point of n-C_5 is 36°C. Therefore, for the fraction the characteristic boiling point is $T_b = (36 + 65)/2 = 50.5$°C. The other characteristic of the fraction is SG = 0.646. This is a light fraction (low T_b) so we use Eq. (2.51) to calculate molecular weight as 84.3. Since viscosity is not known, the most suitable method to estimate composition is through Eqs. (3.77) and (3.78). They require parameter m, which in turn requires refractive index, n. From Eq. (2.115), $I = 0.2216$ and from Eq. (2.114), $n = 1.3616$. With use of M and n and Eq. (3.50), $m = -9.562$. From Eq. (3.77) and (3.78), $x_P = 0.96$ and $x_N = 0.083$. From Eq. (3.72), $x_A = -4.3$. Since x_A is negative thus $x_A = 0$ and x_P, x_N should be normalized as $x_P = 0.96/(0.96 + 0.083) = 0.92$ and $x_N = 1 - 0.92 = 0.08$. Therefore, the predicted PNA composition is 92, 8, 0% versus the experimental values of 91, 9, and 0%.

The aromatic content for this fraction is zero and there is no need to estimate x_{MA} and x_{PA} from Eqs. (3.83) and (3.84); however, to see the performance of these equations for this sample we calculate x_A from Eq. (3.85). From Eq. (2.113), $d = 0.6414$ and from Eq. (2.14), $R_i = 1.0409$. Using these values of R_i and m in Eqs. (3.83) and (3.84), we

get $x_{MA} = -0.228$ and $x_{PA} = -0.025$. Since both numbers are negative the actual estimated values are $x_{MA} = 0$ and $x_{PA} = 0$. From Eq. (3.85), $x_A = 0$, which is consistent with the previous result from Eqs. (3.77), (3.78), and (3.72). ◆

3.5.1.3 API Method

Since 1982 API has adopted the methods developed by Riazi and Daubert [36, 47] for prediction of the composition of petroleum fractions. Equations (3.65)–(3.67) and similar equations developed for light fractions in terms of R_i and VGF by Riazi and Daubert in 1980 [47] were included in the fourth edition of the API-TDB-82. However, after development of Eqs. (3.70)–(3.74) in terms of viscosity and Eqs. (3.83)–(3.84) for prediction of the amount of different types of aromatics in 1986 [36], they were included in the fifth and subsequent editions of the API-TDB [2]. The API methods for prediction of the composition of petroleum fractions require kinematic viscosity at 38 or 99°C and if not available, it should be estimated from Eq. (2.128) or (2.129) in Chapter 2.

3.5.1.4 n-d-M Method

This method requires three physical properties of refractive index (n_{20}), density (d_{20}), and molecular weight (M). For this reason the method is called n-d-M method and is the oldest method for prediction of the composition of petroleum fractions. The method is described in the book by Van Nes and Van Westen in 1951 [30] and it is included in the ASTM manual [4] under ASTM D 3238 test method. The method does not directly give the PNA composition, but it calculates the distribution of carbon in paraffins (%C_P), naphthenes (%C_N), and aromatics (%C_A). However, since carbon is the dominant element in a petroleum mixture it is assumed that the %C_P, %C_N, and %C_A distribution is proportional to %P, %N, and %A distribution. In this assumption the ratio of carbon to hydrogen is considered constant in various hydrocarbon families. Errors caused due to this assumption are within the range of uncertainty in experimental data reported on the PNA composition. Another input data required for this method is sulfur content of the fraction in wt% (%S) and should be known if it exceeds 0.206 wt%. The method should not be applied to fractions with sulfur content of greater than 2%. This method is applicable to fractions with boiling points above gasoline. In addition this method should be applied to fractions with ring percent, %C_R (%C_N + %C_A) up to 75% provided that %C_A (as determined from the n-d-M method) is not higher than 1.5 times %C_N [7]. The n-d-M method also provides equations for calculation of total number of rings (R_T), number of aromatic rings (R_A), and number of naphthenic rings (R_N) in an average molecule in the fractions. The method is expressed in two sets of equations, one for n_{20}, d_{20} (20°C) and another set for n_{70} and d_{70} (70°C) as input data. In this section correlation in terms of n_{20} and d_{20} are presented. The other set of correlations for measurement of n and d at 70°C is given in the literature [7, 24].

$$(3.86) \quad \begin{aligned} \%C_A &= av + 3660/M \\ \%C_N &= \%C_R - \%C_A \\ \%C_P &= 100 - \%C_R \\ R_A &= 0.44 + bvM \\ R_N &= R_T - R_A \end{aligned}$$

where

$$v = 2.51(n - 1.475) - (d - 0.851)$$

$$a = \begin{cases} 430 & \text{if } v > 0 \\ 670 & \text{if } v < 0 \end{cases}$$

$$b = \begin{cases} 0.055 & \text{if } v > 0 \\ 0.080 & \text{if } v < 0 \end{cases}$$

$$w = (d - 0.851) - 1.11(n - 1.475)$$

$$\%C_R = \begin{cases} 820w - 3\%S + 10000/M & \text{if } w > 0 \\ 1440w - 3\%S + 10600/M & \text{if } w < 0 \end{cases}$$

$$R_T = \begin{cases} 1.33 + 0.146M (w - 0.005 \times \%S) & \text{if } w > 0 \\ 1.33 + 0.180M (w - 0.005 \times \%S) & \text{if } w < 0 \end{cases}$$

Once carbon distribution is calculated from Eq. (3.86), the PNA composition can be determined as follows:

$$
\begin{aligned}
x_P &= \%C_P/100 \\
(3.87) \qquad x_N &= \%C_N/100 \\
x_A &= \%C_A/100
\end{aligned}
$$

As mentioned above the n-d-M method cannot be applied to light fractions with molecular weights of less than 200. However, when it was evaluated against PNA composition of 70 fractions for the molecular weight range of 230–570, AAD of 0.064, 0.086, and 0.059 were obtained for x_P, x_N, and x_A, respectively. Accuracy of the n-d-M method for prediction of composition of fractions with $M > 200$ is similar to the accuracy of Eqs. (3.79)–(3.82). But accuracy of Eqs. (3.73) and (3.74) in terms of viscosity (API method) is more than the n-d-M method [30, 36].

In addition to the above methods there are some other procedures reported in the literature for estimation of the PNA composition of petroleum fractions. Among these methods the Bergman's method is included in some references [48]. This method calculates the PNA composition in weight fraction using the boiling point and specific gravity of the fraction as input data. The weight fraction of aromatic content is linearly related to K_W. The x_P and x_N are calculated through simultaneous solution of Eqs. (3.72) and (3.46) when they are applied to specific gravity. Specific gravity of paraffinic, naphthenic, and aromatic pseudocomponents (SG$_P$, SG$_N$, and SG$_A$) are calculated from boiling point of the fraction. Equation (2.42) may be used to calculate SG for different groups from T_b of the fraction. Except in reference [48] this method is not reported elsewhere. There are some other specific methods reported in various sources for each hydrocarbon group. For example, ASTM D 2759 gives a graphical method to estimate naphthene content of saturated hydrocarbons (paraffins and naphthenes only) from refractivity intercept and density at 20°C. In some sources aromatic content of fractions are related to aniline point, hydrogen content, or to hydrogen-to-carbon (HC) atomic ratio [57]. An example of these methods is shown in the next section.

3.5.2 Prediction of Elemental Composition

As discussed earlier, knowledge of elemental composition especially of carbon (%C), hydrogen (%H), and sulfur content (%S) directly gives information on the quality of a fuel. Knowledge of hydrogen content of a petroleum fraction helps to determine the amount of hydrogen needed if it has to go through a reforming process. Petroleum mixtures with higher hydrogen content or lower carbon content have higher heating value and contain more saturated hydrocarbons. Predictive methods for such elements are rare and limited so there is no possibility of comparison of various methods but the presented procedures are evaluated directly against experimental data.

3.5.2.1 Prediction of Carbon and Hydrogen Contents

The amount of hydrogen content of a petroleum mixture is directly related to its carbon-to-hydrogen weight ratio, CH. Higher carbon-to-hydrogen weight ratio is equivalent to lower hydrogen content. In addition aromatics have lower hydrogen content than paraffinic compounds and in some references hydrogen content of a fraction is related to the aromatic content [57] although such relations are approximate and have low degrees of accuracy. The reason for such low accuracy is that the hydrogen content of various types of aromatics varies with molecular type. Within the aromatic family, different compounds may have different numbers of rings, carbon atoms, and hydrogen content. In general more accurate prediction can be obtained from the CH weight ratio method. Several methods of estimation of hydrogen and carbon contents are presented here.

3.5.2.1.1 Riazi Method—This method is based on calculation of CH ratio from the method of Riazi and Daubert given in Section 2.6.3 and estimation of %S from Riazi method in Section 3.5.2.2. The main elements in a petroleum fraction are carbon, hydrogen, and sulfur. Other elements such as nitrogen, oxygen, or metals are in such small quantities that on a wt% basis their presence may be neglected without serious error on the composition of C, H, and S. This is not to say that the knowledge of the amounts of these elements is not important but their weight percentages are negligible in comparison with weight percentages of C, H, and S. Based on this assumption and from the material balance on these three main elements we have

$$(3.88) \qquad \%C + \%H + \%S = 100$$

$$(3.89) \qquad \frac{\%C}{\%H} = CH$$

From simultaneous solution of these two equations, assuming %S is known, the following relations can be obtained for %H and %C:

$$(3.90) \qquad \%H = \frac{100 - \%S}{1 + CH}$$

$$(3.91) \qquad \%C = \left(\frac{CH}{1 + CH}\right) \times (100 - \%S)$$

where %S is the wt% of sulfur in the mixture, which should be determined from the method presented in Section 3.5.2.2 if the experimental value is not available. Value of CH may be determined from the methods presented in Section 2.6.3. In the following methods in which calculation of only %H is presented, %C can be calculated from Eq. (3.88) if the sulfur content is available.

3.5.2.1.2 Goossens' Method—Most recently a simple relation was proposed by Goossens to estimate the hydrogen content of a petroleum fraction based on the assumption of

molar additivity of structural contributions of carbon types [58]. The correlation is derived from data on 61 oil fractions with a squared correlation coefficient of 0.999 and average deviation of 3% and has the following form:

$$(3.92) \qquad \%\mathrm{H} = 30.346 + \frac{82.952 - 65.341n}{d} - \frac{306}{M}$$

where M is the molecular weight and n and d are refractive index and density at 20°C, respectively. This method should be applied to fractions with molecular weight range of 84–459, boiling point range of 60–480°C, refractive index range of 1.38–1.51, and hydrogen content of 12.2–15.6 wt%. In cases that M is not available it should be estimated from the Goossens correlation given by Eq. (2.55).

3.5.2.1.3 ASTM Method—ASTM describes a method to estimate the hydrogen content of aviation fuels under ASTM D 3343 test method based on the aromatic content and distillation data [4]:

$$\%\mathrm{H} = (5.2407 + 0.01448T_b - 7.018x_A)/\mathrm{SG} - 0.901x_A$$
$$(3.93) \qquad + 0.01298x_A T_b - 0.01345T_b + 5.6879$$

where x_A is the fraction of aromatics in the mixture and T_b is an average value of boiling points at 10, 50, and 90 vol% vaporized in kelvin [$T_b = (T_{10} + T_{50} + T_{90})/3$]. This correlation was developed based on 247 aviation fuels and 84 pure hydrocarbons. This method is quite accurate if all the input data are available from experimental measurement.

3.5.2.1.4 Jenkins–Walsh Method—They developed a simple relation in terms of specific gravity and aniline point in the following form [59]:

$$(3.94) \qquad \%\mathrm{H} = 11.17 - 12.89\mathrm{SG} + 0.0389\mathrm{AP}$$

where AP is the aniline point in kelvin and it may be determined from the Winn nomograph (Fig. 2.14) presented in

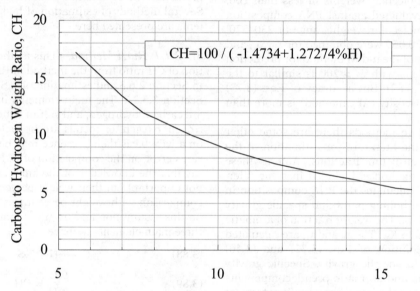

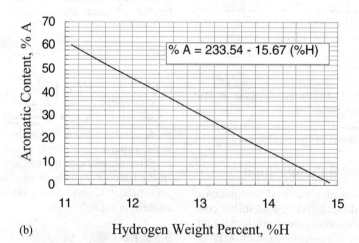

FIG. 3.25—Relationships between fuel hydrogen content, (a) CH weight ratio and (b) aromatic content.

Section 2.8. The correlation is specifically developed for jet fuels with aniline points in the range of 56–77°C.

There are a number of other methods reported in the literature. The Winn nomograph may be used to estimate the CH ratio and then %H can be estimated from Eq. (3.90). Fein–Wilson–Sherman also related %H to aniline point through API gravity [60]. The oldest and simplest method was proposed by Bureau of Standards in terms of specific gravity as given in reference [61]:

$$(3.95) \qquad \%\text{H} = 26 - 15\text{SG}$$

The other simple correlation is derived from data on jet fuels and is in terms of aromatic content (x_A) in the following form [57]:

$$(3.96) \qquad \%\text{H} = 14.9 - 6.38 x_A$$

Finally Fig. 3.25 is based on data from Ref. [57]. Analytical correlation is also presented in Fig. 3.25(a), which represents data with an average deviation of 0.5%. Equation (3.96) is presented in Fig. 3.25(b). When CH ratio is available, %H can be determined from Fig. 3.25(a) and then %A can be determined from Fig. 3.25(b).

3.5.2.2 Prediction of Sulfur and Nitrogen Contents

Sulfur is the most important heteroatom that may be present in a crude oil or petroleum products as dissolved free sulfur and hydrogen sulfide (H_2S). It may also be present as organic compounds such as thiophenes, mercaptanes, alkyl sulfates, sulfides (R—S—R'), disulfides (R—S—S—R'), or sulfoxides (R—SO—R'), where R and R' refer to any aliphatic or aromatic group. Its presence is undesirable for the reasons of corrosion, catalysts poisoning, bad odor, poor burning, and air pollution. In addition presence of sulfur in lubricating oils lowers resistance to oxidation and increases solid deposition on engine parts [62]. New standards and specifications imposed by governments and environmental authorities in industrial countries require very low sulfur content in all petroleum products. For example, reformulated gasolines (RFG) require sulfur content of less than 300 ppm (<0.03 wt%) [63]. Recently a federal court has upheld an Environmental Protection Agency (EPA) rule to cut pollution from tractor-trailers and other large trucks and buses. The rule is expected to reduce tailpipe emissions from tractors, buses, and other trucks up to 90%. The EPA also calls on refineries to reduce the sulfur content in diesel oils to 15 ppm by 2007 from the current level of 500 ppm. The American Lung Association claims that low-sulfur fuel will reduce the amount of soot from larger trucks by 90%. This is expected to prevent 8300 premature deaths, 5500 cases of chronic bronchitis, and another 17600 cases of acute bronchitis in children as provided by the EPA [64]. Products with high-sulfur contents have low quality and heating values. Generally, sulfur is associated with heavy and aromatic compounds [7]. Therefore, high aromatic content or high boiling point fractions (i.e., residues and coal liquids) have naturally higher sulfur contents. Distribution of sulfur in straight-run products of several crude oils and the world average crude with 2 and 5% sulfur contents is shown in Fig. 3.26. Data used to generate this figure are taken from Ref. [61]. As the boiling point of products increases the sulfur content in the products also increases. However, the distribution of sulfur in products may vary from one crude source to another.

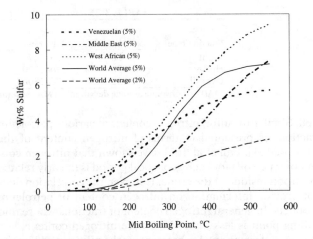

FIG. 3.26—Distribution of sulfur in straight-run products for several crude oils. Numbers in the parentheses indicate sulfur content of crudes.

As boiling point, specific gravity, or aromatic content of a fraction increases the sulfur content also increases [7]. Parameters R_i, m, and SG have been successfully used to predict the PNA composition especially aromatic content of petroleum fractions as shown in Section 3.5.2.1. On this basis the same parameters have been used for the estimation of sulfur content of petroleum fractions in the following form for two ranges of molecular weight.

For fractions with $M < 200$

$$(3.97) \quad \%\text{S} = 177.448 - 170.946 R_i + 0.2258 m + 4.054\text{SG}$$

and for fractions with $M \geq 200$

$$(3.98) \quad \%\text{S} = -58.02 + 38.463 R_i - 0.023 m + 22.4\text{SG}$$

For light fractions in which Eq. (3.96) may give very small negative values, %S would be considered as zero. Squared correlation coefficients (R^2) for these equations are above 0.99. A summary of evaluation of these equations is presented in Table 3.25 as given in Ref. [62].

In using these equations parameters n_{20}, d_{20}, M, and SG are required. For samples in which any of these parameters are not known they can be estimated from the methods discussed earlier in this chapter. In Chapter 4, it is shown how this method can be used to estimate sulfur content of whole crudes. The author is not familiar with any other analytical method for estimation of sulfur content of petroleum fractions reported in the literature so a comparison with other methods is not presented. Generally amount of sulfur in various products is tabulated for various crudes based on the sulfur content of each crude [61].

Another heteroatom whose presence has adverse effect on the stability of the finished product and processing catalysts is nitrogen. High nitrogen content fractions require high hydrogen consumption in hydro processes. Nitrogen content of crudes varies from 0.01 to 0.9 wt%. Most of the compounds having nitrogen have boiling points above 400°C (~750°F) and are generally within the aromatic group. Crudes with higher asphaltene contents have higher nitrogen content as

TABLE 3.25—*Prediction of sulfur content of petroleum fractions [62].*

Fraction type	No. of point	Mol% range	SG range	Sulfur wt% range	Error[a]	
					AAD%	MAD%
Light	76	76–247	0.57–0.86	0.01–1.6	0.09	0.7
Heavy	56	230–1500	0.80–1.05	0.07–6.2	0.24	1.6
Overall	132	76–1500	0.57–1.05	0.01–6.2	0.15	1.6

[a]AAD = Absolute average deviation, %; MAD% = maximum average deviation, %.

well. Similar to sulfur, nitrogen content of various petroleum fractions is presented in terms of nitrogen content of the crude oil [61]. Ball et al. [65] have shown that nitrogen content of crude oils for each geological period is linearly related to carbon residue of the crude. However, the correlation does not provide information on nitrogen content of petroleum products. In general nitrogen content of fractions whose mid boiling point is less than 450°C have nitrogen contents less than that of crude and for heavier cuts the nitrogen wt% in the fraction is greater than that of crude [61]. However, the value of 450°C at which nitrogen content of the fraction is nearly the same as that of crude is approximate and it may vary slightly with the type of the crude. Data reported in Ref. [61] for distribution of nitrogen content of straight run distillates have been correlated in the following form:

$$\frac{\%N_2 \text{ in fraction}}{\%N_2 \text{ in crude}} = -0.4639 + 8.47267T - 28.9448T^2$$
$$+ 27.8155T^3 \quad (3.99)$$

where $T = T_b/1000$ in which T_b is the mid boiling point of the cut in kelvin. This equation is valid for cuts with mid boiling points greater than 220°C and is not applicable to finished petroleum products. Amount of nitrogen in atmospheric distillates is quite small on percent basis. The wt% ratio in Eq. (3.99) can be replaced by ppm weight ratio for small quantities of nitrogen. Estimation of composition of elements is demonstrated in Examples 3.17 and 3.18.

Example 3.17—A petroleum fraction with a boiling range of 250–300°C is produced from a Venezuelan crude oil (Ref. [46], p. 360). Experimentally measured properties are as follows: ASTM distillation 262.2, 268.3, and 278.9°C at 10, 50, and 90 vol% recovered, respectively; specific gravity 0.8597; carbon-to-hydrogen weight ratio 6.69; aniline point 62°C; aromatic content 34.9%; and sulfur wt% 0.8. Estimate sulfur content of the fraction from the method presented in Section 3.5.2.2. Also calculate %C and %H from the following methods: experimental data, Riazi, Goossens, ASTM, Jenkins–Walsh, Bureau of Mines and Eq. (3.96).

Solution—To estimate the sulfur content, parameters M, n_{20}, and d_{20} are required as the input data. The fraction is a narrow fraction and the boiling point at 50% distilled can be considered as the characteristic average boiling point, $T_b = 268.3°C = 541.5$ K. This is a light fraction with $M < 300$; therefore, M, d_{20}, and n_{20} are calculated from Eqs. (2.50) and (2.112)–(2.114) as 195.4, 0.8557, and 1.481, respectively. From

Eqs. (2.15) and (3.50) we get $R_i = 1.0534$, $m = 1.2195$. Since $M < 250$, Eq. (3.97) is used to estimate the sulfur content as %S = 1.1% versus the experimental value of 0.8%. Therefore, the error is calculated as follows: 1.1%–0.8% = 0.3%.

To calculate %C and %H from experimental data, values of CH = 6.69 with %S = 0.8 are used in Eqs. (3.90) and (3.91). This would result in %C = 86.3 and %H = 12.9. According to the general method presented in this book (author's proposed method), CH is calculated from Eq. (2.120) as CH = 6.75 and with estimated value of sulfur content as %S = 1.1, %C and %H are calculated from Eqs. (3.90) and (3.91) as %C = 86.1 and %H = 12.8. In use of Goossens method through Eq. (3.92), estimated values of n, d, and M are required where M should be estimated from Eq. (2.55) as $M = 190$. For this method %C may be calculated from Eq. (3.88) if %S is known. A summary or results for calculation of %H with AD for various methods is given in Table 3.26. The Goossens method gives the highest error because all input data required are predicted values. The ASTM method gives the same value as experimental value because the experimental values on all the input parameters required in Eq. (3.93) are available in this particular example. However, in many cases aromatic content or complete distillation curve as required by the ASTM method are not available. The general method of author presented in this section based on calculation of CH and %S gives good results although 50% ASTM distillation temperature and specific gravity have been used as the only available data. ♦

Example 3.18—A petroleum cut has the boiling range of 370–565°C and is produced from a crude oil from Danish North Sea fields (Ref. [46], p. 353). The nitrogen content of crude is 1235 ppm. Calculate nitrogen content of the fraction and compare with the experimental value of 1625 ppm.

Solution—$T_b = (370 + 565)/2 = 467.5°C = 740.6$ K. $T = T_b/1000 = 0.7406$. Substituting T in Eq. (3.99) gives %N$_2$ in cut = 1.23 × 1235 = 1525. The percent relative deviation with the experimental value is −6%. This is relatively a good prediction, but normally larger errors are obtained especially for lighter cuts. ♦

3.6 PREDICTION OF OTHER PROPERTIES

In this section, predictive methods for some important properties that are useful to determine the quality of certain petroleum products are presented. Some of these properties such as flash point or pour point are useful for safety

TABLE 3.26—*Estimation of hydrogen content of petroleum fraction in Example 3.17.*

Method	Riazi	Goossens	ASTM D 3343	Jenkins–Walsh	Bureau of Mines	Eq. (3.97)
%H, calc.	12.8	12.6	12.9	13.1	13.1	12.7
AD,%	0.1	0.3	0	0.2	0.2	0.2

consideration or storage and transportation of products. One of the most important properties of petroleum products related to volatility after the boiling point is vapor pressure. For petroleum fractions, vapor pressure is measured by the method of Reid. Methods of prediction of true vapor pressure of petroleum fractions are discussed in Chapter 7. However, Reid vapor pressure and other properties related to volatility are discussed in this section. The specific characteristics of petroleum products that are considered in this part are flash, pour, cloud, freezing, aniline, and smoke points as well as carbon residue and octane number. Not all these properties apply to every petroleum fraction or product. For example, octane number applies to gasoline and engine type fuels, while carbon residue is a characteristic of heavy fractions, residues, and crude oils. Freezing, cloud, and pour points are related to the presence of heavy hydrocarbons and are characteristics of heavy products. They are also important properties under very cold conditions. Predictive methods for some of these properties are rare and scatter. Some of these methods are developed based on a limited data and should be used with care and caution.

3.6.1 Properties Related to Volatility

Properties that are related to volatility of petroleum fraction are boiling point range, density, Reid vapor pressure, and flash point. Prediction of boiling point and density of petroleum fractions have been discussed earlier in this chapter. In this

FIG. 3.27—Apparatus to measure RVP of petroleum products by ASTM D 323 test method (courtesy of KISR).

part, methods of prediction of vapor pressure, fuel vapor liquid (*V/L*) ratio, fuel volatility index, and flash points are presented.

3.6.1.1 Reid Vapor Pressure

Reid vapor pressure is the absolute pressure exerted by a mixture at 37.8°C (311 K or 100°F) at a vapor-to-liquid volume ratio of 4 [4]. The RVP is one of the important properties of gasolines and jet fuels and it is used as a criterion for blending of products. RVP is also a useful parameter for estimation of losses from storage tanks during filling or draining. For example, according to Nelson method losses can be approximately calculated as follows: losses in vol% = (14.5 RVP −1)/6, where RVP is in bar [24, 66]. The apparatus and procedures for standard measurement of RVP are specified in ASTM D 323 or IP 402 test methods (see Fig. 3.27). In general, true vapor pressure is higher than RVP because of light gases dissolved in liquid fuel. Prediction of true vapor pressure of pure hydrocarbons and mixtures is discussed in detail in Chapter 7 (Section 7.3). The RVP and boiling range of gasoline governs ease of starting, engine warm-up, mileage economy, and tendency toward vapor lock [63]. Vapor lock tendency is directly related to RVP and at ambient temperature of 21°C (70°F) the maximum allowable RVP is 75.8 kPa (11 psia), while this limit at 32°C (90°F) reduces to 55.2 kPa (8 psia) [63]. RVP can also be used to estimate true vapor pressure of petroleum fractions at various temperatures as shown in Section 7.3. True vapor pressure is important in the calculations related to losses and rate of evaporation of liquid petroleum products. Because RVP does not represent true vapor pressure, the current tendency is to substitute RVP with more modern and meaningful techniques [24]. The more sophisticated instruments for measurement of TVP at various temperatures are discussed in ASTM D 4953 test method. This method can be used to measure RVP of gasolines with oxygenates and measured values are closer to actual vapor pressures [4, 24].

As will be discussed in Chapters 6 and 7, accurate calculation of true vapor pressure requires rigorous vapor liquid equilibrium (VLE) calculations through equations of state. The API-TDB [2] method for calculation of RVP requires a tedious procedure with a series of flash calculations through Soave cubic equation of state. Simple relations for estimation of RVP have been proposed by Jenkins and White and are given in Ref. [61]. These relations are in terms of temperatures along ASTM D 86 distillation curve. An example of these relations in terms of temperatures at 5, 10, 30, and 50 vol% distilled is given below:

$$RVP = 3.3922 - 0.02537(T_5) - 0.070739(T_{10}) + 0.00917(T_{30})$$
$$- 0.0393(T_{50}) + 6.8257 \times 10^{-4}(T_{10})^2$$
(3.100)

where all temperatures are in °C and RVP is in bar. The difficulty with this equation is that it requires distillation data up to 50% point and frequently large errors with negative RVP values for heavier fuels have been observed. Another method for prediction of RVP was proposed by Bird and Kimball [61]. In this method the gasoline is divided into a number (i.e., 28) of cuts characterized by their average boiling points. A blending RVP of each cut is then calculated by the following equation:

$$B_i = \frac{7.641}{\exp(0.03402 T_{bi} + 0.6048)}$$

$$(3.101) \qquad P_a = \sum_{i=1}^{i=28} B_i x_{vi}$$

$$f = 1.0 + 0.003744 \, (\text{VABP} - 93.3)$$

$$\text{RVP} = f P_a$$

where B_i = RVP is the blending number for cut i and T_{bi} = normal boiling point of cut i in °C. x_{vi} is the volume fraction of cut i, VABP is the volume average boiling point in °C, and RVP is the Reid vapor pressure in bars. The constants were obtained from the original constants given in the English units. The average error for this method for 51 samples was 0.12 bar or 1.8 psi [67, 68].

Recently some data on RVP of gasoline samples have been reported by Hatzioznnidis et al. [69]. They measured vapor pressure according to ASTM D 5191 method and related to RVP. They also related their measured vapor pressure data to TVP thus one can obtain RVP from TVP, but their relations have not been evaluated against a wide range of petroleum fractions. Other relations for calculation of TVP from RVP for petroleum fractions and crude oils are given in Section 7.3.3. TVP at 100°F (311 K) can be estimated from Eq. (3.33) as

$$(3.102) \qquad \log_{10}(\text{TVP})_{100} = 3.204 \times \left(1 - 4 \times \frac{T_b - 41}{1393 - T_b}\right)$$

where T_b is the normal boiling point in K and TVP_{100} is the true vapor pressure at 100°F (311 K). Once TVP is calculated it may be used instead of RVP in the case of lack of sufficient data. When this equation is used to estimate RVP of more than 50 petroleum products an average error of 0.13 bar (∼1.9 psi) and a maximum error of 5.9 psi were obtained [67, 68].

RVP data on 52 different petroleum products (light and heavy naphthas, gasolines, and kerosenes) from the *Oil and Gas Journal* data bank [46] have been used to develop a simple relation for prediction of RVP in terms of boiling point and specific gravity in the following form [67]:

$$\text{RVP} = P_c \exp(Y)$$

$$Y = -X \left(\frac{T_b \text{SG}}{T_r}\right) (1 - T_r)^5$$

$$X = -276.7445 + 0.06444 T_b + 10.0245 \text{SG} - 0.129 T_b \text{SG}$$

$$\qquad + \frac{9968.8675}{T_b \text{SG}} + 44.6778 \ln T_b + 63.6683 \ln \text{SG}$$

$$T_r = 311/T_c$$

$$(3.103)$$

where T_b is the mid boiling point and T_c is the pseudocritical temperature of the fraction in kelvin. P_c is the pseudocritical pressure and RVP is the Reid vapor pressure in bars. The basis for development of this equation was to use Miller equation for TVP and its application at 311 K (100°F). The Miller equation (Eq. 7.13) is presented in Section 7.3.1. The constants of vapor pressure correlation were related to boiling point

and specific gravity of the fraction. Critical temperature and pressure may be estimated from T_b and SG using methods presented in Chapter 2. This equation is based on data with RVP in the range of 0.0007–1.207 bar (0.01–17.5 psia), normal boiling point range of 305–494 K, and specific gravity range of 0.65–1.08. The average absolute deviation for 52 samples is 0.061 bar (0.88 psi). The above equation may be used for calculation of RVP to determine quality characteristics of a fuel. The calculated RVP value should not be used for calculation of TVP when very accurate values are needed. (Appropriate methods for direct estimation of TVP of petroleum fractions are discussed in Section 7.3.3.) Vapor pressure of a petroleum mixture depends on the type of its constituents and with use of only two bulk properties to predict RVP is a difficult task. This equation is recommended for a quick and convenient estimation of RVP, but occasionally large errors may be obtained in use of this equation. For more accurate estimation of RVP the sophisticated method suggested in the API-TDB [2] may be used. In this method RVP is calculated through a series of vapor-liquid-equilibrium calculations.

RVP is one of the main characteristics that is usually used to blend a fuel with desired specifications. The desired RVP of a gasoline is obtained by blending naphtha with n-butane ($M = 58$, RVP = 3.58 bar or 52 psia) or another pure hydrocarbon with higher RVPs than the original fuel. For conditions where RVP should be lowered (hot weather), heavier hydrocarbons with lower RVP are used for blending purposes. RVP of several pure hydrocarbons are given as follows: i-C_4: 4.896 bar (71 psia); n-C_4: 3.585 (52); i-C_5: 1.338 (19.4); n-C_5: 1.0135 (14.7); i-C_6: 0.441 (6.4); n-C_6: 0.34 (5.0); benzene: 0.207 (3.0); and toluene: 0.03 (0.5), where all the numbers inside the parentheses are in psia as given in Ref. [63]. However in the same reference in various chapters different values of RVP for a same compound have been used. For example, values of 4.14 bar (60 psi) for n-C_4, 1.1 bar (16 psi) for n-C_5, and 0.48 bar (7 psi) for i-C_6 are also reported by Gary and Handwerk [63]. They also suggested two methods for calculation of RVP of a blend when several components with different RVPs are blended. The first method is based on the simple Kay's mixing rule using mole fraction (x_{mi}) of each component [63]:

$$(3.104) \qquad \text{RVP(blend)} = \sum_i x_{mi} (\text{RVP})_i$$

where $(\text{RVP})_i$ is the RVP of component i in bar or psia. The second approach is to use blending index for RVP as [63]:

$$(\text{RVPBI})_i = (\text{RVP})_i^{1.25}$$

$$(3.105) \qquad \text{RVPBI (blend)} = \sum_i x_{vi} (\text{RVPBI})_i$$

$$\text{RVP (blend)} = \left[\text{RVPBI (blend)}\right]^{0.8}$$

where $(\text{RVPBI})_i$ is the blending index for $(\text{RVP})_i$ and x_{vi} is the volume fraction of component i. Both units of bar or psia may be used in the above equation. This relation was originally developed by Chevron and is also recommended in other industrial manuals under Chevron blending number [61]. Equations (3.104) and (3.105) may also be applied to TVP; however, methods of calculation of TVP of mixtures are discussed in Section 7.3 through thermodynamic relations.

Example 3.19—Estimate RVP of a gasoline sample has molecular weight of 86 and API gravity of 86.

Solution—API = 86 and M = 86. From Eq. (2.4), SG = 0.65 and from Eq. (2.56), T_b = 338 K. Since only T_b and SG are known, Eq. (3.103) is used to calculate the RVP. From Eqs. (2.55) and (2.56) we get T_c = 501.2 K and P_c = 28.82 bar. From Eq. (3.103), T_r = 0.6205, X = 1.3364, and Y = −3.7235. Thus we calculate RVP = 0.696 bar or 10.1 psia. The experimental value is 11.1 psia [63]. ♦

3.6.1.2 V/L Ratio and Volatility Index

Once RVP is known it can be used to determine two other volatility characteristics, namely vapor liquid ratio (V/L) and fuel volatility index (FVI), which are specific characteristics of spark-ignition engine fuels such as gasolines. V/L ratio is a volatility criterion that is mainly used in the United States and Japan, while FVI is used in France and Europe [24]. The V/L ratio at a given temperature represents the volume of vapor formed per unit volume of liquid initially at 0°C. The procedure of measuring V/L ratio is standardized as ASTM D 2533. The volatility of a fuel is expressed as the temperature levels at which V/L ratio is equal to certain values. Usually V/L values of 12, 20, and 36 are of interest. The corresponding temperatures may be calculated from the following relations [24]:

$$T_{(V/L)12} = 88.5 - 0.19E70 - 42.5\,RVP$$
(3.106) $$T_{(V/L)20} = 90.6 - 0.25E70 - 39.2\,RVP$$
$$T_{(V/L)36} = 94.7 - 0.36E70 - 32.3\,RVP$$

where $T_{(V/L)x}$ is the temperature in °C at which $V/L = x$. Parameter E70 is the percentage of volume distilled at 70°C. E70 and RVP are expressed in percent distilled and bar, respectively. Through Lagrange interpolation formula it is possible to derive a general relation to determine temperature for any V/L ratio. E70 can be calculated through a distribution function for distillation curve such as Eq. (3.35) in which by rearrangement of this equation we get

$$(3.107) \quad E70 = 100 - 100\exp\left[-\frac{B}{A}\left(\frac{343.15 - T_\circ}{T_\circ}\right)^B\right]$$

where $T_\circ$ is the initial boiling point in kelvin and together with parameters A and B can be determined from the method discussed in Section 3.2.3. Another simple relation to calculate $T_{(V/L)20}$ is given in terms of RVP and distillation temperatures at 10 and 50% [61]:

$$(3.108) \quad T_{(V/L)20} = 52.5 + 0.2T_{10} + 0.17T_{50} - 33\,RVP$$

where T_{10} and T_{50} are temperatures at 10 and 50 vol% distilled on the ASTM D 86 distillation curve. All temperatures are in °C and RVP is in bar. For cases that T_{10} is not available it may be estimated through reversed form of Eq. (3.17) with T_{50} and SG. Several petroleum refining companies in the United States such as Exxon and Mobil use the critical vapor locking index (CVLI), which is also related to the volatility index [61].

$$(3.109) \quad CVLI = 4.27 + 0.24E70 + 0.069\,RVP$$

The fuel volatility index is expressed by the following relation [24]:

$$(3.110) \quad FVI = 1000\,RVP + 7E70$$

where RVP is in bar. FVI is a characteristic of a fuel for its performance during hot operation of the engine. In France, spec-

ifications require that its value be limited to 900 in summer, 1000 in fall/spring, and 1150 in the winter season. Automobile manufacturers in France require their own specifications that the value of FVI not be exceeded by 850 in summer [24].

3.6.1.3 Flash Point

Flash point of petroleum fractions is the lowest temperature at which vapors arising from the oil will ignite, i.e. flash, when exposed to a flame under specified conditions. Therefore, the flash point of a fuel indicates the maximum temperature that it can be stored without serious fire hazard. Flash point is related to volatility of a fuel and presence of light and volatile components, the higher vapor pressure corresponds to lower flash points. Generally for crude oils with RVP greater than 0.2 bar the flash point is less than 20°C [24]. Flash point is an important characteristics of light petroleum fractions and products under high temperature environment and is directly related to the safe storage and handling of such petroleum products. There are several methods of determining flash points of petroleum fractions. The Closed Tag method (ASTM D 56) is used for petroleum stocks with flash points below 80°C (175°F). The Pensky–Martens method (ASTM D 93) is used for all petroleum products except waxes, solvents, and asphalts. Equipment to measure flash point according to ASTM D 93 test method is shown in Fig. 3.28. The Cleveland Open Cup method (ASTM D 92) is used for petroleum fractions with flash points above 80°C (175°F) excluding fuel oil. This method usually gives flash points 3–6°C higher than the above two methods [61]. There are a number of correlations to estimate flash point of hydrocarbons and petroleum fractions.

Butler et al. [70] noticed that there is a linear relationship between flash point and normal boiling point of hydrocarbons. They also found that at the flash point temperatures, the product of molecular weight (M) and vapor pressure (P^{vap}) for pure hydrocarbons is almost constant and equal to 1.096 bar (15.19 psia).

$$(3.111) \quad MP^{vap} = 1.096$$

Another simple relation for estimation of flash point of hydrocarbon mixtures from vapor pressure was proposed by Walsh and Mortimer [71].

$$(3.112) \quad T_F = 231.2 - 40\log P^{vap}$$

where P^{vap} is the vapor pressure at 37.8°C (100°F) in bar and T_F is the flash point in kelvin. For simplicity RVP may be used for P^{vap}. Methods of calculation of vapor pressure are discussed in Chapter 7. Various oil companies have developed special relations for estimation of flash points of petroleum fractions. Lenoir [72] extended Eq. (3.100) to defined mixtures through use of equilibrium ratios.

The most widely used relation for estimation of flash point is the API method [2], which was developed by Riazi and Daubert [73]. They used vapor pressure relation from Clasius–Clapeyron (Chapter 6) together with the molecular weight relation form Eq. (2.50) in Eq. (3.111) to develop the following relation between flash point and boiling point:

$$(3.113) \quad 1/T_F = a + b/T_b + c\ln T_b + d\ln SG$$

where T_b is the normal boiling point of pure hydrocarbons. It was observed that the coefficient d is very small and T_F is

FIG. 3.28—Equipment for measurement of flash point of petroleum fractions by ASTM D 93 test method (courtesy of Chemical Engineering Department at Kuwait University).

nearly independent of specific gravity. Based on data from pure hydrocarbons and some petroleum fractions, the constants in Eq. (3.113) were determined as

$$(3.114) \quad \frac{1}{T_F} = -0.024209 + \frac{2.84947}{T_{10}} + 3.4254 \times 10^{-3} \ln T_{10}$$

where for pure hydrocarbons T_{10} is normal boiling point, while for petroleum fractions it is distillation temperature at 10 vol% vaporized (ASTM D 86 at 10%) and it is in kelvin. T_F is the flash point in kelvin determined from the ASTM D 93 test method (Pensky–Martens closed cup tester). This equation is presented in Fig. 3.29 for a quick and convenient estimate of flash point. For 18 pure hydrocarbons and 39 fractions, Eq. (3.114) predicts flash points with an average absolute deviation (AD) of 6.8°C (12°F) while Eq. (3.111) predicts the flash points with AD of 18.3°C.

Equation (3.114) should be applied to fractions with normal boiling points from 65 to 590°C (150–1100°F). Equation

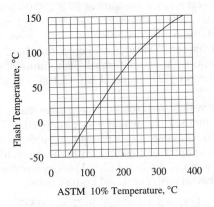

FIG. 3.29—Prediction of flash point of petroleum fractions from Eq. (3.114).

(3.114) is adopted by the API as the standard method to estimate flash point of petroleum fractions [2]. It was shown that Eq. (3.114) can be simplified into the following linear form [73]:

$$(3.115) \quad T_F = 15.48 + 0.70704\, T_{10}$$

where both T_{10} and T_F are in kelvin. This equation is applicable to fractions with normal boiling points (i.e., ASTM D 86 temperature at 50%) less than 260°C (500°F). For such light fractions, Eq. (3.115) is slightly more accurate than Eq. (3.114). For heavier fractions Eq. (3.114) should be used. There are some relations in the literature that correlate flash points to either the initial boiling point (T_{10}) or the distillation temperature at 50% point (T_{50}). Such correlations are not accurate over a wide range of fractions, especially when they are applied to fractions not used in obtaining their coefficients. Generally reported initial boiling points for petroleum fractions are not reliable and if mid boiling point temperature is used as the characteristics boiling point it does not truly represent the boiling point of light components that are initially being vaporized. For this reason the correlations in terms of distillation temperature at 10% point (T_{10}) are more accurate than the other correlations for estimation of flash points of petroleum fractions. Flash points of petroleum fractions may also be estimated from the pseudocomponent method using the PNA composition and values of flash points of pure hydrocarbons from Table 2.2. However, volumetric averaging of component flash point through Eq. (3.40) generally overpredicts the flash point of the blend and the blending index approach described below should be used to estimate flash point of defined mixtures.

If the flash point of a petroleum fraction or a petroleum product does not meet the required specification, it can be adjusted by blending the fraction with other compounds having different flash points. For example in hot regions where the temperature is high, heavy hydrocarbons may be added to a fraction to increase its flash point. The flash point of the blend should be determined from the flash point indexes of the components as given below [74]:

$$(3.116) \qquad \log_{10} BI_F = -6.1188 + \frac{2414}{T_F - 42.6}$$

where log is the logarithm of base 10, BI_F is the flash point blending index, and T_F is the flash point in kelvin. Once BI_F is determined for all components of a blend, the blend flash point index (BI_B) is determined from the following relation:

$$(3.117) \qquad BI_B = \sum x_{vi} BI_i$$

where x_{vi} is the volume fraction and BI_i is the flash point blending index of component i. As it will be shown later, the blending formula by Eq. (3.117) will be used for several other properties. Once BI_{FB} is calculated it should be used in Eq. (3.116) to calculate the flash point of the blend, T_{FB}. Another relation for the blending index is given by Hu–Burns [75]:

$$(3.118) \qquad BI_F = T_F^{1/x}$$

where T_F is the flash point in kelvin and the best value of x is −0.06. However, they suggest that the exponent x be customized for each refinery to give the best results [61]. The following example shows application of these methods.

Example 3.20—A kerosene product with boiling range of 175–260°C from Mexican crude oil has the API gravity of 43.6 (Ref. [46], p. 304). (a) Estimate its flash point and compare with the experimental value of 59°C. (b) For safety reasons it is required to have a minimum flash point of 65°C to be able to store it in a hot summer. How much *n*-tetradecane should be added to this kerosene for a safe storage?

Solution—(a) To estimate flash point we use either Eq. (3.114) or its simplified form Eq. (3.15), which require ASTM 10% temperature, T_{10}. This temperature may be estimated from Eq. (3.17) with use of specific gravity, SG = 0.8081, and ASTM 50% temperature, T_{50}. Since complete ASTM curve is not available it is assumed that the mid boiling point is the same as T_{50}; therefore, $T_{50} = 217.5$°C and from Eq. (3.17) with coefficients in Table 3.4, $T_{10} = 449.9$ K. Since T_{50} is less than 260°C, Eq. (3.115) can be used for simplicity. The result is $T_F = 60.4$°C, which is in good agreement with the experimental value of 59°C considering the fact that an estimated value of ASTM 10% temperature was used.

(b) To increase the flash point from 59 to 65°C, *n*-C$_{14}$ with flash point of 100°C (Table 2.2) is used. If the volume fraction of *n*-C$_{14}$ needed is shown by x_{add}, then using Eq. (3.117) we have $BI_{FB} = (1 - x_{add}) \times BI_{FK} + x_{add} \times BI_{Fadd}$ where BI_{FB}, BI_{FK}, and BI_{Fadd} are the blending indexes for flash points of final blend, kerosene sample, and the additive (*n*-C$_{14}$), respectively. The blending indexes can be estimated from Eq. (3.116) as 111.9, 165.3, and 15.3, respectively, which result in $x_{add} = 0.356$. This means that 35.6% in volume of *n*-C$_{14}$ is required to increase the flash point to 65°C. If the blending indexes are calculated from Eq. (3.118), the amount of *n*-C$_{14}$ required is 30.1%. ♦

3.6.2 Pour Point

The pour point of a petroleum fraction is the lowest temperature at which the oil will pour or flow when it is cooled without stirring under standard cooling conditions. Pour point represents the lowest temperature at which an oil can be stored and still capable of flowing under gravity. Pour point is one of low temperature characteristics of heavy fractions. When temperature is less than pour point of a petroleum product it cannot be stored or transferred through a pipeline. Test procedures for measuring pour points of petroleum fractions are given under ASTM D 97 (ISO 3016 or IP 15) and ASTM D 5985 methods. For commercial formulation of engine oils the pour point can be lowered to the limit of −25 and −40°C. This is achieved by using pour point depressant additives that inhibit the growth of wax crystals in the oil [5]. Presence of wax and heavy compounds increase the pour point of petroleum fractions. Heavier and more viscous oils have higher pour points and on this basis Riazi and Daubert [73] used a modified version of generalized correlation developed in Chapter 2 (Eq. 2.39) to estimate the pour point of petroleum fractions from viscosity, molecular weight, and specific gravity in the following form:

$$T_P = 130.47 [SG^{2.970566}] \times [M^{(0.61235 - 0.47357 SG)}]$$
$$(3.119) \qquad \times [\nu_{38(100)}^{(0.310331 - 0.32834 SG)}]$$

where T_P is the pour point (ASTM D 97) in kelvin, M is the molecular weight, and $\nu_{38(100)}$ is the kinematic viscosity at 37.8°C (100°F) in cSt. This equation was developed with data on pour points of more than 300 petroleum fractions with molecular weights ranging from 140 to 800 and API gravities from 13 to 50 with the AAD of 3.9°C [73]. This method is also accepted by the API and it is included in the API-TDB since 1988 [2] as the standard method to estimate pour point of petroleum fractions. As suggested by Hu and Burns [75, 76], Eqs. (3.117) and (3.118) used for blending index of flash point can also be used for pour point blending index (T_{PB}) with $x = 0.08$:

$$(3.120) \qquad BI_P = T_P^{1/0.08}$$

where T_P is the pour point of fraction or blend in kelvin. The AAD of 2.8°C is reported for use of Eqs. (3.117) and (3.120) to estimate pour points of 47 blends [76].

3.6.3 Cloud Point

The cloud point is the lowest temperature at which wax crystals begin to form by a gradual cooling under standard conditions. At this temperature the oil becomes cloudy and the first particles of wax crystals are observed. The standard procedure to measure the cloud point is described under ASTM D 2500, IP 219, and ISO 3015 test methods. Cloud point is another cold characteristic of petroleum oils under low-temperature conditions and increases as molecular weight of oil increases. Cloud points are measured for oils that contain paraffins in the form of wax and therefore for light fractions, such as naphtha or gasoline, no cloud point data are reported. Cloud points usually occur at 4–5°C (7 to 9°F) above the pour point although the temperature differential could be in the range of 0–10°C (0–18°F) as shown in Table 3.27. The

TABLE 3.27—*Cloud and pour points and their differences for some petroleum products.*

Fraction	API gravity	T_P, °C	T_{CL}, °C	$T_P - T_{CL}$, °C
Indonesian Dist.	33.0	−43.3	−53.9	10.6
Australian GO	24.7	−26.0	−30.0	4.0
Australian HGO	22.0	−8.0	−9.0	1.0
Abu Dhabi LGO	37.6	−19.0	−27.0	8.0
Abu Dhabi HGO	30.3	7.0	2.0	5.0
Abu Dhabi Disst.	21.4	28.0	26.0	2.0
Abu Dhabi Diesel	37.4	−12.0	−12.0	0.0
Kuwaiti Kerosene	44.5	−45.0	−45.0	0.0
Iranian Kerosene	44.3	−46.7	−46.7	0.0
Iranian Kerosene	42.5	−40.6	−48.3	7.8
Iranian GO	33.0	−11.7	−14.4	2.8
North Sea GO	35.0	6.0	6.0	0.0
Nigerian GO	27.7	−32.0	−33.0	1.0
Middle East Kerosene	47.2	−63.3	−65.0	1.7
Middle East Kerosene	45.3	−54.4	−56.7	2.2
Middle East Kerosene	39.7	−31.1	−34.4	3.3
Middle East Disst.	38.9	−17.8	−20.6	2.8

Source: Ref. [46].
T_P: pour point; T_{CL}: cloud point.

difference between cloud and pour point depends on the nature of oil and there is no simplified correlation to predict this difference. Cloud point is one of the important characteristics of crude oils under low-temperature conditions. As temperature decreases below the cloud point, formation of wax crystals is accelerated. Therefore, low cloud point products are desirable under low-temperature conditions. Wax crystals can plug the fuel system lines and filters, which could lead to stalling aircraft and diesel engines under cold conditions. Since cloud point is higher than pour point, it can be considered that the knowledge of cloud point is more important than the pour point in establishing distillate fuel oil specifications for cold weather usage [61]. Table 3.27 shows the difference between cloud and pour points for some petroleum products. Cloud and pour points are also useful for predicting the temperature at which the observed viscosity of an oil deviates from the true (newtonian) viscosity in the low temperature range [7]. The amount of *n*-paraffins in petroleum oil has direct effect on the cloud point of a fraction [8]. Presence of gases dissolved in oil reduces the cloud point which is desirable. The exact calculation of cloud point requires solid–liquid equilibrium calculations, which is discussed in Chapter 9. The blending index for cloud point is calculated from the same relation as for pour point through Eq. (3.118) with $x = 0.05$:

(3.121) $$\mathrm{BI_{CL}} = T_{CL}^{1/0.05}$$

where T_{CL} is the cloud point of fraction or blend in kelvin. Accuracy of this method of calculating cloud point of blends is the same as for the pour point (AAD of 2.8°C). Once the cloud point index for each component of blend, $\mathrm{BI_{CLi}}$, is determined through Eq. (3.121), the cloud point index of the blend, $\mathrm{BI_{CLB}}$, is calculated through Eq. (3.117). Then Eq. (3.121) is used in its reverse form to calculate cloud point of the blend from its cloud point index [76].

3.6.4 Freezing Point

Freezing point is defined in Section 2.1.9 and freezing points of pure hydrocarbons are given in Table 2.2. For a petroleum fraction, freezing point test involves cooling the sample until a slurry of crystals form throughout the sample or it is the temperature at which all wax crystals disappear on rewarming

the oil [61]. Freezing point is one of the important characteristics of aviation fuels where it is determined by the procedures described in ASTM D 2386 (U.S.), IP 16 (England), and NF M 07-048 (France) test methods. Maximum freezing point of jet fuels is an international specification which is required to be at −47°C (−53°F) as specified in the "Aviation Fuel Quantity Requirements for Jointly Operated Systems" [24]. This maximum freezing point indicates the lowest temperature that the fuel can be used without risk of separation of solidified hydrocarbons (wax). Such separation can result in the blockage in fuel tank, pipelines, nozzles, and filters [61]. Walsh–Mortimer suggest a thermodynamic model based on the solubility of *n*-paraffin hydrocarbons in a petroleum mixture to determine the freezing point [71]. Accurate determination of freezing point requires accurate knowledge of the composition of a fuel which is normally not known. However, the method of determination of carbon number distribution along with solid–liquid equilibrium can be used to determine freezing points of petroleum fractions and crude oils as will be discussed in Chapter 9. A simpler but less accurate method to determine freezing points of petroleum fractions is through the pseudocomponent approach as shown in the following example.

Example 3.21—A kerosene sample produced from a crude oil from North Sea Ekofisk field has the boiling range of 150–204.4°C (302–400°F) and API gravity of 48.7. Estimate the freezing point of this kerosene and compare with the experimental value of −65°C (−85°F).

Solution—The mid boiling point is $T_b = 177.2$°C and the specific gravity is SG = 0.785. We use the method of pseudocomponent using predicted composition. From Eq. (2.50), $M = 143$ and since $M > 143$, we use Eqs. (3.77), (3.78), and (3.72) to predict x_P, x_N, and x_A, respectively. From Eqs. (2.114) and (2.115), $n = 1.439$ and from Eq. (3.50), $m = -5.1515$. Using SG and m, we calculate the PNA composition as $x_P = 0.457$, $x_N = 0.27$, and $x_A = 0.273$. From Eqs. (3.41)–(3-43), $M_P = 144.3$, $M_N = 132.9$, and $M_A = 129.3$. Using Eq. (2.42) for prediction of the freezing point for different families we get $T_{FP} = 242.3$, $T_{FN} = 187.8$, and $T_{FA} = 178.6$ K. Using Eq. (3.40) we get $T_F = 210.2$ K or −63.1°C versus the measured

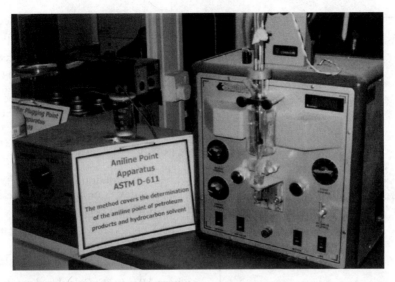

FIG. 3.30—Apparatus to measure aniline point of petroleum fuels by ASTM D 611 test method (courtesy of KISR).

value of −65°C. The result is quite satisfactory considering minimum data on T_b and SG are used as the only available parameters. ♦

3.6.5 Aniline Point

Aniline point of a petroleum fraction is defined as the minimum temperature at which equal volumes of aniline and the oil are completely miscible. Method of determining aniline point of petroleum products is described under ASTM D 611 test method and the apparatus is shown in Fig. 3.30. Aniline point indicates the degree of aromaticity of the fraction. The higher the aniline point the lower aromatic content. For this reason aromatic content of kerosene and jet fuel samples may be calculated from aniline point [59]:

$$(3.122) \quad \%A = 692.4 + 12.15(SG)(AP) - 794(SG) - 10.4(AP)$$

where %A is the percent aromatic content, SG is the specific gravity, and AP is the aniline point in°C. There are a number of methods to estimate aniline point of petroleum fractions. We discuss four methods in this section.

3.6.5.1 Winn Method

Aniline point can be estimated from Winn nomograph (Fig. 2.14) using T_b and SG or M and SG as the input parameters.

3.6.5.2 Walsh–Mortimer

The aniline point can be calculated from the following relation [61, 71]:

$$(3.123) \quad AP = -204.9 - 1.498C_{50} + \frac{100.5C_{50}^{1/3}}{SG}$$

where AP is the aniline point in °C and C_{50} is the carbon number of n-paraffin whose boiling point is the same as the mid boiling point of the fraction. C_{50} may be calculated from the following relation:

$$(3.124) \quad C_{50} = \frac{M_P - 14}{2}$$

in which M_P is the molecular weight of n-paraffin whose boiling point is the same as mid boiling point of the fraction which can be determined from Eq. (3.41).

3.6.5.3 Linden Method

This relation is a mathematical representation of an earlier graphical method and is given as [73]

$$(3.125) \quad AP = -183.3 + 0.27(API)T_b^{1/3} + 0.317T_b$$

where AP is in °C, T_b is the mid boiling point in kelvin and API is API gravity. The blending index for aniline point may be calculated from the following relation developed by Chevron Research [61]:

$$(3.126) \quad BI_{AP} = 1.124[\exp(0.00657AP)]$$

where AP is in °C and BI_{AP} is the blending index for the aniline point. Once the blending indexes of components of a blend are determined, Eq. (3.117) should be used to calculate blending index for aniline point of the blend.

3.6.5.4 Albahri et al. Method

Most recently Albahri et al. [68] developed predictive methods for determination of quality of petroleum fuels. Based on the idea that aniline point is mainly related to the aromatic content of a fuel, the following relation was proposed:

$$(3.127) \quad AP = -9805.269(R_i) + 711.85761(SG) + 9778.7069$$

where AP is in °C and R_i is defined by Eq. (2.14). Equations (3.123), (3.125), and (3.127) were evaluated against data on aniline points of 300 fuels with aniline point range: 45–107°C boiling range: 115–545°C and API gravity range of 14–56. The average absolute deviation (AAD) for Eq. (3.127) was 2.5°C while for Eqs. (3.123) and (3.125) the errors were 4.6 and 6.5°C, respectively [68]. Error distribution for Eq. (3.127) is shown in Fig. 3.31.

3.6.6 Cetane Number and Diesel Index

For diesel engines, the fuel must have a characteristic that favors auto-ignition. The ignition delay period can be evaluated

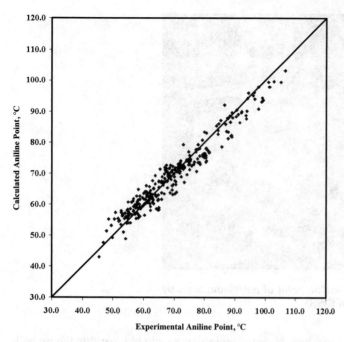

FIG. 3.31—Error distribution for prediction of aniline point from Eq. (3.127). Taken with permission from Ref. [68].

by the fuel characterization factor called *cetane number* (CN). The behavior of a diesel fuel is measured by comparing its performance with two pure hydrocarbons: *n*-cetane or *n*-hexadecane (n-$C_{16}H_{34}$) which is given the number 100 and α-methylnaphthalene which is given the cetane number of 0. A diesel fuel has a cetane number of 60 if it behaves like a binary mixture of 60 vol% cetane and 40 vol% α-methylnaphthalene. In practice heptamethylnonane (HMN) a branched isomer of *n*-cetane with cetane number of 15 is used instead of α-methylnaphthalene [24, 61]. Therefore, in practice the cetane number is defined as:

$$(3.128) \quad CN = vol\%(n\text{-cetane}) + 0.15(vol\% \, HMN)$$

The cetane number of a diesel fuel can be measured by the ASTM D 613 test method. The shorter the ignition delay period the higher CN value. Higher cetane number fuels reduce combustion noise and permit improved control of combustion resulting in increased engine efficiency and power output. Higher cetane number fuels tend to result in easier starting and faster warm-up in cold weather. Cetane number requirement of fuels vary with their uses. For high speed city buses in which kerosene is used as fuel the required CN is 50. For premium diesel fuel for use in high speed buses and light marine engines the required number is 47 while for marine distillate diesel for low speed buses and heavy marine engines the required cetane number is 38 [61]. In France the minimum required CN of fuels by automotive manufacturers is 50. The product distributed in France and Europe have CN in the range of 48–55. In most Scandinavian countries, the United States and Canada the cetane number of diesel fuels are most often less than 50. Higher cetane number fuels in addition to better starting condition can cause reduction in air pollution [24].

Since determination of cetane number is difficult and costly, ASTM D 976 (IP 218) proposed a method of calcu-

lation. Calculated number is called calculated cetane index (CCI) and can be determined from the following relation:

$$CCI = 454.74 - 1641.416SG + 774.74SG^2$$
$$(3.129) \qquad - 0.554T_{50} + 97.083(\log_{10} T_{50})^2$$

where T_{50} is the ASTM D 86 temperature at 50% point in °C. Another characteristic of diesel fuels is called diesel index (DI) defined as:

$$(3.130) \qquad DI = \frac{(API)(1.8AP + 32)}{100}$$

which is a function of API gravity and aniline point in °C. Cetane index is empirically correlated to DI and AP in the following form [24]:

$$(3.131) \qquad CI = 0.72DI + 10$$

$$(3.132) \qquad CI = AP - 15.5$$

where AP is in °C. Calculated cetane index (CI) is also related to *n*-paraffin content (%NP) of diesel fuels in the following from [87].

$$(3.133) \qquad \%NP = 1.45CI - 57.5$$

The relation for calculation of cetane number blending index is more complicated than those for pour and cloud point. Blending indexes for cetane number are tabulated in various sources [61, 75]. Cetane number of diesel fuels can be improved by adding additives such as 2-ethyl-hexyl nitrate or other types of alkyl nitrates. Cetane number is usually improved by 3–5 points once 300–1000 ppm by weight of such additives is added [24]. Equation (3.129) suggested for calculating cetane number does not consider presence of additives and for this reason calculated cetane index for some fuels differ with measured cetane index. Generally, CCI is less than measured CN and for this reason in France automobile manufacturers have established minimum CN for both the calculated CI (49) and the measured CN (50) for the quality requirement of the fuels [24].

3.6.7 Octane Number

Octane number is an important characteristic of spark engine fuels such as gasoline and jet fuel or fractions that are used to produce these fuels (i.e., naphthas) and it represents antiknock characteristic of a fuel. Isooctane (2,2,4-trimethylpentane) has octane number of 100 and *n*-heptane has octane number of 0 on both scales of RON and MON. Octane number of their mixtures is determined by the vol% of isooctane used. As discussed in Section 2.1.13, isoparaffins and aromatics have high octane numbers while *n*-paraffins and olefins have low octane numbers. Therefore, octane number of a gasoline depends on its molecular type composition especially the amount of isoparaffins. There are two types of octane number: *research octane number* (RON) is measured under city conditions while *motor octane number* (MON) is measured under road conditions. The arithmetic average value of RON and MON is known as *posted octane number* (PON). RON is generally greater than MON by 6–12 points, although at low octane numbers MON might be greater than RON by a few points. The difference between RON and MON is known as *sensitivity* of fuel. RON of fuels is determined by ASTM D 908 and MON is measured by ASTM D 357 test methods. Generally there are three kinds of gasolines: regular, intermediate, and premium with PON of 87, 90, and 93, respectively. In France the minimum required RON for superplus gasoline is 98 [24]. Required RON of gasolines vary with parameters such as air temperature, altitude, humidity, engine speed, and coolant temperature. Generally for every 300 m altitude RON required decreases by 3 points and for every 11°C rise in temperature RON required increases by 1.5 points [63]. Improving the octane number of fuel would result in reducing power loss of the engine, improving fuel economy, and a reduction in environmental pollutants and engine damage. For these reasons, octane number is one of the important properties related to the quality of gasolines. There are a number of additives that can improve octane number of gasoline or jet fuels. These additives are tetra-ethyl lead (TEL), alcohols, and ethers such as ethanol, methyl-tertiary-butyl ether (MTBE), ethyl-tertiary-butyl ether (ETBE), or tertiary-amyl methyl ether (TAME). Use of lead in fuels is prohibited in nearly all industrialized countries due to its hazardous nature in the environment, but is still being used in many third world and underdeveloped countries. For a fuel with octane number (ON) of 100, increase in the ON depends on the concentration of TEL added. The following correlations are developed based on the data provided by Speight [7]:

$$\text{TEL} = -871.05 + 2507.81\left(\frac{\text{ON}}{100}\right) - 2415.94\left(\frac{\text{ON}}{100}\right)^2$$
$$(3.134) \quad + 779.12\left(\frac{\text{ON}}{100}\right)^3$$

$$\text{ON} = 100.35 + 11.06(\text{TEL}) - 3.406(\text{TEL})^2$$
$$(3.135) \quad + 0.577(\text{TEL})^3 - 0.038(\text{TEL})^4$$

where ON is the octane number and TEL is milliliter TEL added to one U.S. gallon of fuel. These relations nearly reproduce the exact data given by Speight and valid for ON above 100. In these equations when clear octane number (without TEL) is 100, TEL concentration is zero. By subtracting the calculated ON from 100, the increase in the octane number due to the addition of TEL can be estimated, which may be used to calculate the increase in ON of fuels with clear ON different from 100. Equation (3.134) is useful to calculate amount of TEL required for a certain ON while Eq. (3.135) gives ON of fuel after a certain amount of TEL is added. For example, if 0.3 mL of TEL is added to each U.S. gallon of a gasoline with RON of 95, Eq. (3.135) gives ON of 104.4, which indicates an increase of 4.4 in the ON. This increase is based on the reference ON of 100 which can be used for ON different from 100. Therefore, the ON of gasoline in this example will be 95 + 4.4 or 99.4. Different relations for octane number of various fuels (naphthas, gasolines, and reformates) in terms of TEL concentration are given elsewhere [88].

Octane numbers of some oxygenates (alcohols and ethers) are given in Table 3.28 [24]. Once these oxygenates are added to a fuel with volume fraction of x_{ox} the octane number of product blend is [24]

$$(3.136) \quad \text{ON} = x_{\text{ox}}(\text{ON})_{\text{ox}} + (1 - x_{\text{ox}})(\text{ON})_{\text{clear}}$$

where ON_{clear} is the clear octane number (RON or MON) of a fuel and ON is the corresponding octane number of blend

TABLE 3.28—*Octane numbers of some alcohols and ethers (oxygenates).*

Compound	RON	MON
Methanol	125–135	100–105
MTBE	113–117	95–101
Ethanol	120–130	98–103
ETBE	118–122	100–102
TBA	105–110	95–100
TAME	110–114	96–100

MTBE: methyl-tertiary-butyl ether; ETBE: ethyl-tertiary-butyl ether; TBA: tertiary-butyl alcohol; TAME: tertiary-amyl-methyl ether. Source: Ref. [24].

after addition of an additive. ON_{ox} is the corresponding octane number of oxygenate, which can be taken as the average values for the ranges of RON and MON as given in Table 3.28. For example for MTBE, the range of RON_{ox} is 113–117; therefore, for this oxygenate the value of RON_{ox} for use in Eq. (3.136) is 115. Similarly the value MON_{ox} for this oxygenate is are 98. Equation (3.136) represents a simple linear relation for octane number blending without considering the interaction between the components. This relation is valid for addition of additives in small quantities (low values of x_{ox}, i.e., < 0.15). However, when large quantities of two components are added (i.e., two types of gasolines on 25:75 volume basis), linear mixing rule as given by Eq. (3.136) is not valid and the interaction between components should be taken into account [61]. Du Pont has introduced interaction parameters between two or three components for blending indexes of octane number which are presented in graphical forms [89]. Several other blending approaches are provided in the literature [61]. The simplest form of their tabulated blending indexes have been converted into the following analytical relations:

$$\text{BI}_{\text{RON}} =$$
$$\begin{cases} 36.01 + 38.33X - 99.8X^2 + 341.3X^3 - 507.2X^4 + 268.64X^5 & 11 \le \text{RON} < 76 \\ -299.5 + 1272X - 1552.9X^2 + 651X^3 & 76 \le \text{RON} \le 103 \\ 2206.3 - 4313.64X + 2178.57X^2 & 103 \le \text{RON} \le 106 \\ X = \text{RON}/100 \end{cases}$$
$$(3.137)$$

where BI_{RON} is the blending index for RON and should be used together with Eq. (3.117) to calculate RON of a blend. Equation (3.137) reproduce the tabulated values of RON blending indexes with AAD of 0.06%.

Estimation of octane number of a fuel from its bulk properties is a challenging task, since ON very much depends on the chemical structure of components of the mixture. Figure 3.32 shows variation of RON with boiling point of pure hydrocarbons from different families as produced from data given in Table 2.2. If PIONA composition of a fuel is known, RON of a fuel may be estimated from the pseudocomponent techniques in the following form:

$$\text{RON} = x_{\text{NP}}(\text{RON})_{\text{NP}} + x_{\text{IP}}(\text{RON})_{\text{IP}} + x_{\text{O}}(\text{RON})_{\text{O}}$$
$$(3.138) \quad + x_{\text{N}}(\text{RON})_{\text{N}} + x_{\text{A}}(\text{RON})_{\text{A}}$$

where x is the volume fraction of different hydrocarbon families i.e., n-paraffins (NP), isoparaffins (IP), olefins (O), naphthenes (N), and aromatics (A). RON_{NP}, RON_{IP}, RON_{o}, RON_{N}, and RON_{A} are the values of RON of pseudocomponents from n-paraffin, isoparaffins, olefins, naphthenes, and aromatics families whose boiling points are the same as the

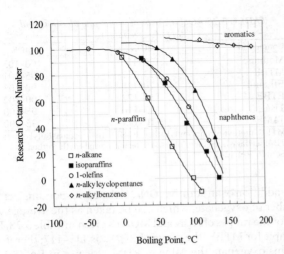

FIG. 3.32—Research octane number of different families of hydrocarbons. Taken with permission from Ref. [68].

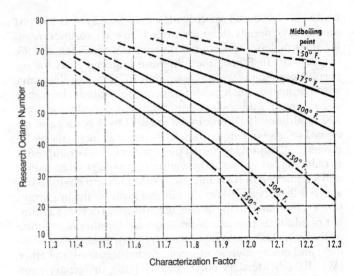

FIG. 3.33—Research octane number of naphthas (°F = 1.8 × °C +32). Taken with permission from Ref. [79].

mid boiling point or the ASTM D 86 temperature at 50% point of the fraction and can be determined from Fig. 3.32 or Table 2.2. Generally petroleum products are free of olefins and the main groups present in a petroleum products are n-paraffins, isoparaffins, naphthenes, and aromatics. The role of isoparaffins on octane number is significant as they have ON values greater than n-paraffins. In addition different types of isoparaffins have different octane numbers at the same boiling point. As the number of branches in an iso-paraffin compound increases the octane number also increases. For this reason it would be more appropriate if RON_{IP} in Eq. (3.138) is an average value of octane numbers of various types of isoparaffins. For convenience and computer calculations, values of RON for these various homologous hydrocarbon groups have been correlated to normal boiling point, T_b in the following form:

$$(3.139) \qquad RON = a + bT + cT^2 + dT^3 + eT^4$$

where RON is the clear research octane number and $T = (T_b - 273.15)/100$ in which T_b is the boiling point in kelvin. Based on the data taken from the API-TDB [2], the coefficients $a - e$ were determined and are given in Table 3.29 [68, 78]. It should be noted that for isoparaffins the coefficients are given for four different groups of 2-methylpentanes, 3-methylpentanes, 2,2-dimethylpentanes, and 2,3-dimethylpentanes. Octane numbers of various isoparaffins vary significantly and for this reason an average value of RON for these four different iso-paraffinic groups is considered as the value of RON_{IP} for use in Eq. (3.138).

Normally when detailed PIONA composition is not available, PNA composition is predicted from the methods presented in Section 3.5.1. For such cases Eq. (3.138) may be simplified by considering $x_O = 0$ and $x_{NP} = x_{IP} = x_P/2$. Because RON of n-paraffins and isoparaffins differs significantly (Fig. 3.32), the assumption of equal amounts of n-paraffins and isoparaffins can lead to substantial errors in calculation of RON for fuels whose normal and iso paraffins contents differ significantly. For such cases this method estimates RON of a fuel with a higher error but requires minimum information on distillation and specific gravity.

Nelson [79] gives graphical relation for estimation of RON of naphthas in terms of K_W characterization factor or paraffin content (wt%) and mid boiling point as given in Figs. 3.33 and 3.34, respectively.

As mentioned earlier if amount of paraffins in wt% is not available, vol% may be used instead of wt% if necessary. Once RON is determined, MON can be calculated from the following relation proposed by Jenkins [80]:

$$MON = 22.5 + 0.83\,RON - 20.0\,SG - 0.12\,(\%O)$$
$$(3.140) \qquad + 0.5\,(TML) + 0.2\,(TEL)$$

where SG is the specific gravity, TML and TEL are the concentrations of tetra methyl lead and tetra ethyl lead in mL/UK gallon, and %O is the vol% of olefins in the gasoline. For olefin- and lead-free fuels (%O = TML = TEL = 0) and Eq. (3.140) reduces to a simple form in terms of RON and SG. From this

TABLE 3.29—*Coefficients for Eq. (3.139) for estimation of RON [68, 78].*

Hydrocarbon family	a	b	c	d	e
n-Paraffins	92.809	−70.97	−53	20	10
*iso*paraffins					
2-Methyl-pentanes	95.927	−157.53	561	−600	200
3-Methyl-pentanes	92.069	57.63	−65	0	0
2,2-Dimethyl-pentanes	109.38	−38.83	−26	0	0
2,3-Dimethyl-pentanes	97.652	−20.8	58	−200	100
Naphthenes	−77.536	471.59	−418	100	0
Aromatics	145.668	−54.336	16.276	0	0

Taken with permission from Ref. [68].

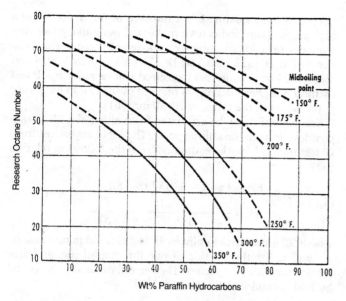

FIG. 3.34—Research octane number versus paraffin content ($°F = 1.8 \times °C + 32$). Taken with permission from Ref. [79].

equation sensitivity of gasoline can be determined. Gasolines with lower sensitivity are desirable.

Example 3.22—A naphtha sample from an Australian crude oil has the following characteristics: boiling point range $15.5 - 70°C$, specific gravity 0.6501, *n*-paraffins 49.33%, isoparaffins 41.45%, naphthenes 9.14%, aromatics 0.08%, clear RON 69.6, and MON 66.2 [Ref. [46], p. 359].

a. Estimate RON from the pseudocomponent method using experimental composition.
b. Estimate RON from the pseudocomponent method using predicted PNA composition.
c. Estimate RON from Fig. 3.33.
d. Estimate RON from Fig. 3.34.
e. Estimate MON from actual reported RON.
f. Estimate MON from predicted RON from Part a.
g. For each case calculate the error (deviation between estimated and reported values).

Solution—For this fraction: $T_b = (15.5 + 70)/2 = 42.8°C$, $SG = 0.6501$, $x_P = 0.4933$, $x_{IP} = 0.4145$, $x_N = 0.0914$, $x_A = 0.008$.

a. RON can be estimated from Eq. (3.138) through pseudocomponent method using RON values for pure hydrocarbons calculated from Eq. (3.139) and Table 3.29 with $T_b = 315.9$ K. Results of calculation are $(RON)_{NP} = 54.63$, $(RON)_{IP} = (90.94 + 104.83 + 88 + 87.05)/4 = 92.7$, $(RON)_N = 55.57$, and $(RON)_A = 125.39$. In calculation of $(RON)_{IP}$, an average value for RON of four families in Table 3.29 is calculated. From Eq. (3.139), clear RON can be calculated as:

$$RON = 0.4933 \times 54.63 + 0.4145 \times 92.7 + 0.0914 \times 55.57$$
$$+ 0.0008 \times 125.39 = 70.55.$$

In comparison with the reported value of 69.6 the error is $70.55 - 69.6 = 0.95$.

b. To predict PNA, we calculate M from Eq. (2.50) as $M = 79.54$. Since $M < 200$ and viscosity is not available we use Eqs. (3.77) and (3.78) and (3.72) to predict the composition. From Eqs. (2.126) and (2.127), $n_{20} = 1.3642$ and from Eq. (3.50), $m = -8.8159$. The predicted composition for %P, %N, and %A is 95.4%, 7.4%, and −2.8%, respectively. Since predicted %A is negative, it is set equal to zero and the normalized composition is $x_P = 0.928$, $x_N = 0.072$, and $x_A = 0.0$. To use Eq. (3.139) we split the paraffin content equally between *n*-paraffins and isoparaffins as $x_{NP} = x_{IP} = 0.928/2 = 0.464$. In this case RON = 72.36. The error on calculation of RON is 2.76.

c. To use Fig. 3.33 we need $T_b = 42.8°C = 109°F$ and K_W, which from Eq. (2.13) is calculated as $K_W = 12.75$. Since the K_W is outside the range of values in Fig. 3.33, accurate reading is not possible, but from value of the boiling point it is obvious that the RON from extrapolation of the curves is above 70.

d. To use Fig. 3.34 we need total paraffins which is % = $49.33 + 41.45 = 90.78$ and $T_b = 109°F$. In this case T_b is outside the range of values on the curves, but with extrapolation a value of about 66 can be read. The error is about −3.6.

e. To calculate MON we use Eq. (3.140) with RON = 69.6, SG = 0.6501, and %O = TML = TEL = 0. The estimated value is $(MON)_{est.} = 67.3$, which is in good agreement with the reported value of 66.2 [46] with error of +1.1.

f. If estimated RON value of 70.55 (from Part a) is used in Eq. (3.140), the predicted MON is 68 with deviation of +1.8.

g. Errors are calculated and given in each part. Equation (3.138) gives the lowest error with deviation of less than 1 when experimental PIONA composition is used. For samples in which the difference between amounts of *n*-paraffins and isoparaffins is small, Eq. (3.138) gives even better results. In the cases that the composition is not available the procedure used in Part b using predicted composition with minimum data on boiling point and specific gravity gives an acceptable value for RON.

♦

3.6.8 Carbon Residue

When a petroleum fraction is vaporized in the absence of air at atmospheric pressure, the nonvolatile compounds have a carbonaceous residue known as *carbon residue*, which is designated by CR. Therefore, heavier fractions with more aromatic contents have higher carbon residues while volatile and light fractions such as naphthas and gasolines have no carbon residues. CR is particularly an important characteristic of crude oils and petroleum residues. Higher CR values indicate low-quality fuel and less hydrogen content. There are two older different test methods to measure carbon residues, Ramsbottom (ASTM D 524) and the Conradson (ASTM D 189). The relationship between these methods are also given by the ASTM D 189 method. Oils that have ash forming compounds have erroneously high carbon residues by both methods. For such oils ash should be removed before the measurement. There is a more recent test method (ASTM D 4530) that requires smaller sample amounts and is often referred as *microcarbon residue* (MCR) and as a result it is less precise in practical technique [7]. In most cases carbon residues are reported in wt% by Conradson method, which is designated by %CCR.

Carbon residue can be correlated to a number of other properties. It increases with an increase in carbon-to-hydrogen ratio (CH), sulfur content, nitrogen content, asphaltenes content, or viscosity of the oil. The most precise relation is between CR and hydrogen content in which as hydrogen content increases the carbon residues decreases [7]. The hydrogen content is expressed in terms of H/C atomic ratio and the following relation may be used to estimate CCR from H/C [81].

$$\text{(3.141)} \qquad \text{\%CCR} = 148.7 - 86.96\,\text{H/C}$$

if H/C $\geq$ 1.71 (%CCR < 0), set %CCR = 0.0 and if H/C < 0.5 (%CCR > 100), set %CCR = 100. H/C ratio can be estimated from CH ratio methods given in Section 2.6.3.

The carbon residue is nearly a direct function of high boiling asphaltic materials and Nelson has reported a linear relation between carbon residue and asphalt yield [82]. One of the main characteristic of residuum is its asphaltene content. Asphaltenes are insoluble in low molecular weight n-alkanes including n-pentane. Knowledge of n-pentane insolubles in residual oils is quite important in determining yields and products qualities for deasphalting, thermal visbreaking, and hydrodesulfurization processing. The relation between the normal pentane insolubles and carbon residue is as follows [61]:

$$\text{(3.142)} \quad \text{\%NC}_5 = 0.74195\,(\text{\%CCR}) + 0.01272\,(\text{\%CCR})^2$$

where %NC$_5$ is the wt% of n-pentane insolubles and %CCR is the wt% of Conradson carbon residue. Once %NC$_5$ is known, the asphaltene content (asphaltene wt%) of a residue can be determined from the following empirical relation:

$$\text{(3.143)} \qquad \text{\%Asphaltene} = a(\text{\%NC}_5)$$

where a is 0.385 for atmospheric residue and 0.455 for vacuum residues [61, 66]. These equations are approximate and do not provide accurate predictions.

Example 3.23—A vacuum residue of an Australian crude oil has carbon-to-hydrogen weight ratio of 7.83. Estimate its carbon residue and asphaltene contents and compare the results with the experimental values of 15.1 and 4.6%, respectively [46].

Solution—With CH=7.83, from Eq. (2.122), HC atomic ratio is calculated as HC = 1.52. From Eq. (3.141), %CCR = 16.4% and from Eq. (3.142), %NC$_5$ = 15.6%. From Eq. (3.143) with a = 0.455 (for vacuum residue) we calculate %Asphaltene = 7.1. The results show that while Eq. (3.141) provides a good prediction for %CCR, prediction of %Asphaltene from Eq. (3.143) is approximate. ◆

3.6.9 Smoke Point

Smoke point is a characteristic of aviation turbine fuels and kerosenes and indicates the tendency of a fuel to burn with a smoky flame. Higher amount of aromatics in a fuel causes a smoky characteristic for the flame and energy loss due to thermal radiation. The smoke point (SP) is a maximum flame height at which a fuel can be burned in a standard wick-fed

lamp without smoking. It is expressed in millimeters and a high smoke point indicates a fuel with low smoke-producing tendency [61]. Measurement of smoke point is described under ASTM D 1322 (U.S.) or IP 57 (UK) and ISO 3014 test methods. For a same fuel measured smoke point by IP test method is higher than ASTM method by 0.5–1 mm for smoke point values in the range of 20–30 mm [61].

Smoke point may be estimated from either the PNA composition or from the aniline point. The SP of kerosenes from IP test method may be estimated from the following relation [90]:

$$\text{(3.144)} \qquad \begin{aligned} SP &= 1.65X - 0.0112X^2 - 8.7 \\ X &= \frac{100}{0.61x_\text{P} + 3.392x_\text{N} + 13.518x_\text{A}} \end{aligned}$$

where SP is the smoke point by IP test method in mm and x_P, x_N, and x_A are the fraction of paraffin, naphthene, and aromatic content of kerosenes. The second method is proposed by Jenkins and Walsh as follows [83]:

$$\text{(3.145)} \quad SP = -255.26 + 2.04AP - 240.8\ln(SG) + 7727(SG/AP)$$

where AP is the aniline point in °C and SG is the specific gravity at 15.5°C. Both Eqs. (3.144) and (3.145) estimate SP according to the IP test method. To estimate SP from the ASTM D 1322 test method, 0.7 mm should be subtracted from the calculated IP smoke point. Equations (3.144) and (3.145) are based on data with specific gravity in the range of 0.76–0.82, and smoke points in the range of 17–39 mm. Based on some preliminary evaluations, Eq. (3.133) is expected to perform better than Eq. (3.144), because smoke point is very much related to the aromatic content of the fuel which is expressed in terms of aniline point in the Jenkins–Walsh method. In addition the specific gravity, which is an indication of molecular type, is also used in the equation. Equation (3.144) may be used for cases that the aniline point is not available but experimental PNA composition is available. Albahri et al [68] also proposed the following relation for prediction of smoke point using API gravity and boiling point:

$$\text{(3.146)} \quad SP = 0.839(\text{API}) + 0.0182634(T_\text{b}) - 22.97$$

where SP is in mm (ASTM method) and T_b is the average boiling point in kelvin. This equation when tested for 136 petroleum fractions gave an average error of about 2 mm [68].

Example 3.24—A Nigerian kerosene has an API gravity of 41.2, aniline point of 55.6°C, and the PNA composition of 36.4, 49.3, and 14.3%. Estimate the smoke point of this fuel from Equations (3.144)–(3.146) and compare with the experimental value of 20 mm (Ref. [46], p. 342).

Solution—To estimate SP from Eq. (3.144), we have x_P = 0.364, x_N = 0.493, and x_A = 0.143 which give X = 26.13. Calculated SP is 26.8 mm according to the IP method or 26.1 mm according to the ASTM method. To use Eq. (3.145) we have from API gravity, SG = 0.819, AP = 55.6°C, the calculated SP is SP = 20 mm. The ASTM smoke point is then 19.3 mm which is in very good agreement with the experimental value of 20 with deviation of −0.7 mm. Predicted value from Eq. (3.146) is 17.6 mm. ◆

3.7 QUALITY OF PETROLEUM PRODUCTS

Methods presented in this chapter can be used to evaluate the quality of petroleum products from available parameters. The quality of a petroleum product depends on certain specifications or properties of the fuel to satisfy required criteria set by the market demand. These characteristics are specified for best use of a fuel (i.e., highest engine performance) or for cleaner environment while the fuel is in use. These specifications vary from one product to another and from one country to another. For example, for gasoline the quality is determined by a series of properties such as sulfur and aromatic contents, octane number, vapor pressure, hydrogen content, and boiling range. Engine warm-up time is affected by the percent distilled at 70°C and the ASTM 90% temperature. For the ambient temperature of 26.7°C (80°F) a gasoline must have ASTM 90% temperature of 188°C and 3% distilled at 70°C to give acceptable warm-up time [63]. Standard organizations such as ASTM give such specifications for various products. For reformulated gasoline sulfur content of less than 300 ppm (0.03 wt%) is required [63]. Amount of particulate emissions is directly related to the aromatic and sulfur content of a fuel. Figure 3.35 shows the influence of sulfur reduction in gasoline from 500 to 50 ppm in the reduction of pollutant emissions [24].

Vapor pressure of gasoline of jet fuel determines their ignition characteristics. While freezing point is important for jet fuels it is not a major characteristic for gas oils. For lubricating oil properties such as viscosity and viscosity index are important in addition to sulfur and PNA composition. Aniline point is a useful characteristic to indicate power of solubility of solvents as well as aromatic contents of certain fuels. For heavy petroleum products knowledge of properties such as carbon residue, pour point, and cloud point are of interest. Some important specifications of jet fuels are given in Table 3.30.

One of the techniques used in refining technology to produce a petroleum product with a certain characteristic is the blending method. Once a certain value for a property (i.e., viscosity, octane number, pour point, etc.) of a petroleum product is required, the mixture may be blended with a certain component, additive or another petroleum fraction to

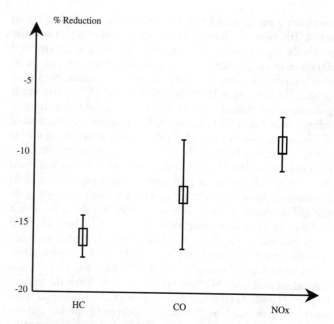

FIG. 3.35—Influence of sulfur content in gasoline (from 500 to 50 ppm) in reduction of pollutant gases. Taken with permission from Ref. [24].

produce the required final product. In Example 3.20, it was shown that to have a product with certain flash point, one can determine the volume of various components in the blend. The same approach can be extended to any other property. For example, to increase vapor pressure of gasoline *n*-butane may be added during winter season to improve engine starting characteristics of the fuel [63]. The amount of required butane to reach a certain vapor pressure value can be determined through calculation of vapor pressure blending index for the components and the product as discussed in Section 3.6.1.1.

3.8 MINIMUM LABORATORY DATA

As discussed earlier measurement of all properties of various petroleum fractions and products in the laboratory is an impossible task due to the required cost and time. However,

TABLE 3.30—Some general characteristics of three fuels [24].

Characteristics	Specifications[a]		
	Gasoline[b]	Jet fuel	Diesel fuel
Max. total sulfur, wt%	0.05	0.2	0.05
Max aromatics content, vol%		20	
Max olefins content, vol%		5	
Distillation at 10 vol%,°C		204	
Max final boiling point, °C	215	300	370[c]
Range of % evaporation at 70°C (E70)	15–47		
Min research octane number (RON)	95		
Min flash point, °C		38	55
Specific gravity range	0.725–0.78	0.775–0.84	0.82–0.86
Min smoke point, °C		25	
Max freezing point, °C		−47	
Range of Reid vapor pressure, bar	0.35–0.90[d]		
Min cetane number			49

Taken with permission from Ref. [24].
[a]European standards in the mid 1990s.
[b]European unleaded Super 98 premium gasoline.
[c]At this temperature minimum of 95 vol% should be evaporated.
[d]Varies with season according to the class of gasoline.

there are a number of basic parameters that must be known for a fraction to determine various properties from the methods presented in this chapter. As more experimental data are available for fraction a better characterization of the fraction is possible. For example, to estimate sulfur content of a fraction from Eqs. (3.97) and (3.98), the input parameters of specific gravity, molecular weight, density and refractive index at 20°C are needed. If experimental values of all these parameters are available a good estimate of sulfur content can be obtained. However, since normally all these data are not available, M, n, and d_{20} should be estimated from SG and T_b. Therefore, a minimum of two parameters that are boiling point and specific gravity are needed to estimate the sulfur content. However, for heavy fractions in which distillation data are not reported, M should be estimated from kinematic viscosity at 38 and 99°C (v_{38} and v_{99}) and specific gravity through Eq. (2.52). Once M is estimated, n can be estimated from M and SG through Eq. (2.127) and d is calculated from SG through Eq. (2.123). With the knowledge of M and SG all other parameters can be estimated from methods presented in Chapter 2. Therefore, at least three parameters of v_{38}, v_{99}, and SG must be known to determine sulfur content or other characteristics. In a case that only one viscosity data is known, i.e., v_{38}, kinematic viscosity at 99°C, v_{99}, can be estimated from Eq. (2.61). In this way estimated value of M is less accurate than the case that three values of v_{38}, v_{99} and SG are known from experimental measurements. We see that again for heavy fractions with knowledge of only two parameters (i.e., v_{38} and SG or v_{99} and SG) all basic properties of the fraction can be estimated. Therefore, to obtain the basic characterization parameters of a petroleum fraction a minimum of two parameters are needed.

If the only information is the distillation curve, then specific gravity can be estimated from T_{10} and T_{50} through Eq. (3.17) and Table 3.4. Having T_{50} and SG, all other parameters can be estimated as discussed above. When only a portion of distillation curve (i.e., T_{20}, T_{40}, and T_{60}) is available, through Eq. (3.35) the complete curve can be predicted and from this equation T_{10} and T_{50} can be determined. Therefore, a portion of distillation curve can also be used to generate all parameters related to properties and quality of petroleum fractions. We showed that with the knowledge of PNA composition a better characterization of a fraction is possible through pseudocomponent technique. Therefore, if the composition along

boiling point is available, nearly all other parameters can be determined through mid boiling point and PNA composition with better accuracy than using only T_b and SG. For heavy fractions in which T_b may not be available, the pseudocomponent technique can be applied through use of M and PNA composition where M may be estimated from viscosity data if it is not available. As there are many scenarios to estimate basic properties of petroleum fractions, use of available data to predict the most accurate characterization parameters is an engineering art which has a direct impact on subsequent prediction of physical properties and eventually on design calculations. The basic laboratory data that are useful in characterization methods based on their significance and simplicity are given below:

1. distillation data, boiling point
2. specific gravity
3. composition (i.e, PNA content)
4. molecular weight
5. refractive index
6. elemental analysis (i.e., CHS composition)
7. kinematic viscosity at 37.8 and 98.9°C (100 and 210°F)

One can best characterize a petroleum fraction if all the above parameters are known from laboratory measurements. However, among these seven items at least two items must be known for characterization purposes. In any case when experimental value for a characterization parameter is available it should be used instead of predicted value. Among these seven items that can be measured in laboratory, refractive index and specific gravity are the most convenient properties to measure. Molecular weight especially for heavy fractions is also very useful to predict other properties. As discussed in Chapter 2, for light fractions ($M < 300$; $N_C < 22$, $T_b < 350°C$) the best two pairs of parameters in the order of their characterizing power are (T_b, SG), (T_b, n), (M, SG), (M, n), (v, SG), (T_b, CH), (M, CH), (v, CH). The most suitable pair is (T_b, SG) and the least one is (v, CH). As it is explained in the next section, for heavy fractions three parameter correlations are more accurate. Therefore, for heavy fractions in which boiling point cannot be measured a minimum of three parameters such as viscosity at two different temperatures and specific gravity (i.e., v_{38}, v_{99}, SG) are needed. For heavy fractions the pseudocomponent method is much more accurate than use of bulk properties for the estimation of various properties. Therefore,

TABLE 3.31—*Standard test methods for measurement of some properties of liquid petroleum products.*

Property	ASTM D	IP	ISO	Property	ASTM D	IP	ISO
Aniline Point	611	2/98	2977	Flash Point	93	34/97	2719
Carbon Residue (Ramsbottom)	524	14/94	4262	Freezing Point	2386	16/98	3013
Carbon Residue (Conradson)	189	13/94	6615	Hydrocarbon Types	1319	156/95	3837
Centane Number	4737	380/98	4264	Heating Value	240	12	
Cloud Point	2500	219/94	3016	Kinematic Viscosity	445	71/97	3104
Color	1500	196/97	2049	Octane Number (Motor)	2700	236	5163
Density/Sp. Gr.	4052	365/97	2185	Refractive Index	1218		
Distillation at Atm. Pressure	86	123/99*	3405	Pour Point	97	15/95	3015
Distillation at Reduced Pressures	1160		6616*	Sulfur Content	1266	107/86	2192
Distillation by Gas Chromatography	2887	406/99*		Thermal Conductivity	2717–95		
Distillation of Crude Oils	2892		8708*	Vapor Pressure (Reid)	323	69/94	3007
				Viscosity (Viscous Oils)	2983	370/85*	

ASTM has test methods for certain properties for which other test methods do not suggest equivalent procedures. Some of these methods include heat of combustion: D 4809; smoke point: D 1322; surface tension: D 3825; vapor–liquid ratio: D 2533; viscosity temperature chart: D 341; autoignition: D 2155 (ISO 3988). Further test methods for some specific properties are given in the text where the property is discussed. ASTM methods are taken from Ref. [4]. IP methods are taken from Ref. [85]. Methods specified by * are similar but not identical to other standard methods. Most IP methods are also used as British Standard under BS2000 methods [85]. The number after IP indicates the year of last approval. ISO methods are taken from Refs. [24] and [85].

the knowledge of PNA composition for prediction of properties of heavy fractions is more useful than for light fractions. For wide boiling range fractions knowledge of complete distillation curve is quite useful to consider nature of different compounds and their effects on the properties of the mixture. As it is shown in Chapter 4, for wide and heavy fractions such as C_{7+} fractions, distribution of carbon number in the fraction is the most useful information besides specific gravity. Further analysis of minimum laboratory data for characterization of petroleum fractions is provided in our previous work [84].

Predictive methods of characterization must be used when experimental data are not available. If possible, one should make maximum use of available experimental data. A summary of standard test methods for some specifications of liquid petroleum products is given in Table 3.31. For some properties equivalent test methods according to the international standards organization (ISO) are also specified in this table [24, 85].

3.9 ANALYSIS OF LABORATORY DATA AND DEVELOPMENT OF PREDICTIVE METHODS

In Chapter 2 and this chapter the predictive methods in terms of readily available parameters are presented for estimation of various properties related to basic characteristics and quality of petroleum fractions. Generally these methods fall within two categories of empirical and semiempirical correlations. In an empirical correlation the structure of the correlation is determined through fitting the data and the type of input parameters in each correlation are determined through analysis of experimental data. While in a semi-empirical correlation, the structure and functionality of the relation is determined from a theoretical analysis of parameters involved and through analysis of existing theoretical relations. Once the main functionality and nature of a correlation between various physical properties is determined, the correlation coefficients can be determined from experimental data. The best example of such a predictive method is development of Eq. (2.38) in Chapter 2, which was developed based on the understanding of the intermolecular forces in hydrocarbon systems. This generalized correlation has been successfully used to develop predictive methods for a variety of physical properties. An example of an empirical correlation is Eq. (2.54) developed for estimation of molecular weight of petroleum fractions. Many other correlations presented in this chapter for estimation of properties such as aniline and smoke points or methods presented for calculation of octane numbers for a blend are also purely empirical in nature. In development of an empirical relation, knowledge of the nature of properties involved in the correlation is necessary. For example, aniline point is a characteristic that depends on the molecular type of hydrocarbons in the fraction. Therefore, it is appropriate to relate aniline point to the parameters that characterize hydrocarbon types (i.e., R_i) rather than boiling point that characterizes carbon number in a hydrocarbon series. Mathematical functions can be expressed in the form of polynomial series; therefore, it is practically possible to develop correlations in the forms of polynomial of various degrees. With powerful computational tools available at present it is possible to find an empirical correlation for any set of

laboratory data for any physical property in terms of some other parameters. As the complexity and the number of parameters increases the accuracy of the correlation also increases with respect to the data used in the development of the correlation. However, the main problem with empirical correlations is their limited power of extrapolation and the large number of numerical constants involved in the correlation. For example, in Chapter 2, several correlations are provided to estimate molecular weight of hydrocarbons in terms of boiling point and specific gravity. Equation (2.50) derived from Eq. (2.38) has only three numerical constants, which are developed from molecular weight of pure hydrocarbons. Tsonopoulos et al. [86] made an extensive analysis of various methods of estimation of molecular weights of coal liquids. Equation (2.50) was compared with several empirical correlations specifically developed for coal liquids having as many as 16 numerical constants. They concluded that Eq. (2.50) is the most accurate method for the estimation of molecular weight of coal liquids. No data on coal liquid were used in development of constants in Eq. (2.50). However, since it was developed with some physical basis and properties of pure hydrocarbons were used to obtain the numerical constants the equation has a wide range of applications from pure hydrocarbons to petroleum fractions and coal liquids, which are mainly aromatics. This indicates the significance of development of correlations based on the physical understanding of the nature of the system and its properties. The main advantage of such correlations is their generality and simplicity.

The main characteristics of an ideal predictive method for a certain property are accuracy, simplicity, generality, and availability of input parameters. The best approach toward the development of such correlations would be to combine physical and theoretical fundamentals with some modifications. An example of such type of correlations is Eq. (2.39), which is an extension of Eq. (2.38) derived from physical basis. A pure empirical correlation might be quite accurate to represent the data used in its development but when it is applied to other systems the accuracy is quite low. In addition characterizing the systems according to their degree of complexity is helpful to develop more accurate correlations. For example, heavy fractions contain heavy and nonpolar compounds, which differ with low-molecular-weight hydrocarbons present in light petroleum fractions. Therefore, in order to increase the degree of accuracy of a predictive method for a certain property it is quite appropriate to develop one correlation for light and one correlation for heavy fractions. For heavy fractions because of the nature of complex compounds in the mixture three input parameters are required. As variation in properties of pure hydrocarbons from one family to another increases with increase in carbon numbers (i.e., see Figs. 2.15 and 2.23), the role of composition on estimation of such properties for heavy fractions is more than its effect on the properties of light fractions. Therefore, including molecular type in the development of predictive methods for such properties of heavy fractions is quite reasonable and useful and would enhance the accuracy of the method.

Once the structure of a correlation is determined from theoretical developments between various properties, experimental data should be used to determine the numerical constants in the correlation. If the data on properties of pure hydrocarbons from different families are used to determine the constants, the resulting correlation would be more general and applicable to various types of petroleum fractions within the same boiling point range. The accuracy of a correlation for a specific group of fractions could be increased if the coefficients in the correlation are obtained from the same group of fractions. In obtaining the constants, many equations may be convertible into linear forms and spreadsheets such as Lotus or Excel programs may be used to obtain the constants by means of least squared method. Non-linear regression of correlations through these spreadsheets is also possible. In analyzing the suitability of a correlation, the best criteria would be the R^2 or correlation parameter defined by Eq. (2.136) in which values of above 0.99 indicate an equation is capable of correlating data. Use of a larger data bank and most recent published data in obtaining the numerical constants would enhance accuracy and applicability of the correlation. A fair way of evaluations and comparison of various correlations to estimate a certain property from the same input parameters would be through a data set not used in obtaining the coefficients in the correlations. In such evaluations AAD or %AAD can be used as the criterion to compare different methods. Average absolute deviation may be used when the range of variation in the property is very large and small values are estimated. For example, in prediction of PNA composition, amount of aromatics varies from 1% in petroleum fractions to more than 90% in coal liquids. AAD of 2 (in terms of percentage) in estimating aromatic content is quite reasonable. This error corresponds to 200% in terms of %AD for fractions with 1% aromatic content. Experience has shown that correlations that have fewer numerical constants and are based on theoretical and physical grounds with constants obtained from a wide range of data set are more general and have higher power of extrapolation.

3.10 CONCLUSIONS AND RECOMMENDATIONS

In this chapter various characterization methods for different petroleum fractions and mixtures have been presented. This is perhaps one of the most important chapters in the book. As the method selected for characterization of a fraction would affect prediction of various properties discussed in the remaining part of the book. As it is discussed in Chapter 4, characterization and estimation of properties of crude oils depend on the characterization of petroleum fractions discussed in this chapter. Through methods presented in this chapter one can estimate basic input data needed for estimation of thermodynamic and physical properties. These input parameters include critical properties, molecular weight, and acentric factor. In addition methods of estimation of properties related to the quality of a petroleum product such as distillation curves, PNA composition, elemental composition, viscosity index, carbon residue, flash, pour, cloud, smoke, and freezing points as well as octane and cetane numbers are presented. Such methods can be used to determine the quality of a fuel or a petroleum product based on the minimum laboratory data available for a fraction. Methods of conversion of various types of distillation curves help to determine necessary information for process design on complete true boiling point distillation curve when it is not available. In addition a method is provided to determine complete distillation curve

based on a minimum of three data points along the distillation curve. For prediction of each characteristic of a petroleum fraction, several methods are provided that have been in use in the petroleum industry. Limitations, advantages, and disadvantages of each method are discussed.

Basically two approaches are proposed in characterization of petroleum fractions. One technique is based on the use of bulk properties (i.e., T_b and SG) considering the whole mixture as a single pseudocomponent. The second approach called pseudocomponent technique considers the fraction as a mixture of three pseudocomponents from the three families. This technique is particularly useful for heavy fractions. A third approach is also provided for wide boiling range fractions. However, since behavior of such fractions is similar to that of crude oils the technique is mainly presented in the next chapter. Fractions are generally divided into light and heavy fractions. For heavy fractions a minimum of three characterization parameters best describe the mixture. Recommendations on the use of various input parameters and advantages of different methods were discussed in Sections 3.8 and 3.9. For light fractions ($M < 300$, $N_C < 22$) and products of atmospheric distillation unit, Eq. (2.38) for M, I, and d is quite accurate. For such light fractions, T_c, P_c, and ω can be estimated from Eqs. (2.65), (2.66), and (2.105), respectively. For prediction of the PNA composition for fractions with $M < 200$, Eqs. (3.77) and (3.78) in terms of m and SG are suitable. For fractions with $M > 200$, Eqs. (3.71)–(3.74) in terms of R_i and VGC are the most accurate relations; however, in cases that viscosity data are not available, Eqs. (3.79) and (3.80) in terms of R_i an CH are recommended. Special recommendations on use of various correlations for estimation of different properties of petroleum fractions from their bulk properties have been given in Section 2.10 and Table 2.16 in Chapter 2.

3.11 PROBLEMS

3.1. List four different types of analytical tools used for compositional analysis of petroleum fractions.

3.2. What are the advantages/disadvantages and the differences between GC, MS, GC–MS, GPC and HPLC instruments?

3.3. A jet naphtha has the following ASTM D 86 distillation curve [1]:

vol% distilled	10	30	50	70	90
ASTM D 86 temperature, °C	151.1	156.1	160.6	165.0	171.7

    a. Calculate VABP, WABP, MABP, CABP, and MeABP for this fraction. Comment on your calculated MeABP.

    b. Estimate the specific gravity of this fraction and compare with reported value of 0.8046.

    c. Calculate the K_W for this fraction and compare with reported value of 11.48.

3.4. A kerosene sample has the following ASTM D 86, TBP, and SG distribution along distillation curve [1]. Convert ASTM D 86 distillation curve to TBP by Riazi–Daubert and Daubert's (API) methods. Draw actual TBP and predicted TBP curves on a single graph in °C. Calculate the average specific gravity of fraction form SG distribution and compare with reported value of 0.8086.

Vol%	IBP	5	10	20	30	40	50	60	70	80	90	95	FBP
ASTM, °F	330	342	350	366	380	390	404	417	433	450	469	482	500
TBP, °F	258	295	312	337	358	...	402	...	442	...	486	499	...
SG	...	0.772	0.778	0.787	0.795	...	0.817	...	0.824	...	0.829	0.830	...

3.5. For the kerosene sample of Problem 3.4, assume ASTM temperatures at 30, 50, and 70% points are the only known information. Based on these data points on the distillation curve predict the entire distillation curve and compare with the actual values in both tabulated and graphical forms.

3.6. For the kerosene sample of Problem 3.4 calculate the Watson K factor and generate the SG distribution by considering constant K_W along the curve. Calculate %AAD for the deviations between predicted SG and actual SG for the 8 data points. Show a graphical evaluation of predicted distribution.

3.7. For the kerosene sample of Problem 3.4 generate SG distribution using Eq. (3.35) and draw predicted SG distribution with actual values. Use Eq. (3.37) to calculate specific gravity of the mixture and compare with the actual value of 0.8086.

3.8. For the kerosene sample of Problem 3.4 calculate density at 75°F and compare with the reported value of 0.8019 g/cm^3 [1].

3.9. For the kerosene sample of Problem 3.4, assume the only data available are ASTM D 86 temperatures at 30, 50, and 70% points. Based on these data calculate the specific gravity at 10, 30, 50, 70, and 90 % and compare with the values given in Problem 3.4.

3.10. A petroleum fraction has the following ASTM D 1160 distillation curve at 1 mm Hg:

vol% distilled	10	30	50	70	90
ASTM D 1160 temperature, °C	104	143	174	202	244

Predict TBP, ASTM D 86, and EFV distillation curves all at 760 mm Hg.

3.11. A gas oil sample has the following TBP and density distribution [32]. Experimental values of M, n_{20}, and SG are 214, 1.4694, and 0.8475, respectively.
  a. Use the method outlined for wide boiling range fractions and estimate M and n_{20}.
  b. Use Eqs. (2.50) and (2.115) to estimate M and n.
  c. Compare %AD from methods a and b.

3.12. A jet naphtha has ASTM 50% temperature of 321°F and specific gravity of 0.8046 [1]. The PNA composition of this fraction is 19, 70, and 11%, respectively. Estimate M, n_{20}, d_{20}, T_c, P_c, and V_c from T_b and SG using the following methods:
  a. Riazi–Daubert (1980) methods [38].
  b. API-TDB methods [2].
  c. Twu methods for M, T_c, P_c, and V_c.
  d. Kesler–Lee methods for M, T_c, and P_c.
  e. Sim–Daubert (computerized Winn nomograph) method for M, T_c, and P_c.
  f. Pseudocomponent method.

3.13. For the fraction of Problem 3.12 calculate acentric factor, ω from the following methods:
  a. Lee–Kesler method with T_c and P_c from Parts (a), (b), (c), and (d) of Problem 3.12.
  b. Edmister method with T_c and P_c from Parts and (a) and (e) of Problem 3.12.
  c. Korsten method with T_c and P_c from Part (a) of Problem 3.12.
  d. Pseudocomponent method

3.14. A pure hydrocarbon has a boiling point of 110.6°C and a specific gravity of 0.8718. What is the type of this hydrocarbon (P, N, MA, or PA)? How can you check your answer? Can you guess the compound? In your analysis it is assumed that you do not have access to the table of properties of pure hydrocarbons.

3.15. Experimental data on sulfur content of some petroleum products along with other basic parameters are given in Table 3.32. It is assumed that the only data available for a petroleum product is its mid boiling point, T_b, (column 1) and refractive index at 20°C, n_{20} (column 2). Use appropriate methods to complete columns 4,6,8,9,10, and 12 in this table.

3.16. ASTM 50% temperature (T_b), specific gravity (SG), and the PNA composition of 12 petroleum fractions are given in Table 3.33. Complete columns of this table by calculating M, n, m, and the PNA composition for each fraction by using appropriate methods.

3.17. A gasoline product from North Sea crude oil has boiling range of C_5-85°C and specific gravity of 0.6771. Predict the PNA composition and compare with experimentally determined composition of 64, 25, and 11 in wt% [46].

3.18. A residue from a North Sea crude has the following experimentally determined characteristics: $v_{99(210)} = 14.77$ cSt., SG = 0.9217. For this fraction estimate the following properties and compare with the experimental values where they are available [46]:
  a. Kinematic viscosity at 38°C (100°F)
  b. Molecular weight and average boiling point

Vol%	IBP	5	10	20	30	40	50	60	70	80	90	95
TBP, °F	420	451	470	495	514	...	544	...	580	...	604	621
d_{20}	...	0.828	0.835	0.843	0.847	...	0.851	...	0.855	...	0.856	0.859

  c. Density and refractive index at 20°C
  d. Pour point (experimental value is 39°C)
  e. Sulfur content (experimental value is 0.63 wt%)
  f. Conradson carbon residue (experimental value is 4.6 wt%)

3.19. A crude oil from Northwest Australian field has total nitrogen content of 310 ppm. One of the products of atmospheric distillation column for this crude has true boiling point range of 190–230°C and the API gravity of 45.5. For this product determine the following properties and

TABLE 3.32—*Basic parameters and sulfur content of some undefined petroleum products (Problem 3.15).*[a]

Fraction	(1) T_b, K exp.	(2) n_{20} exp.	(3) SG, exp.	(4) SG, calc.	(5) M, exp.	(6) M, calc.	(7) d_{20}, exp.	(8) d_{20}, calc.	(9) m, calc.	(10) RI calc.	(11) Sulfur %, exp.	(12) Sulfur %, calc.
Kuwaiti kerosene	468	1.441	0.791	...			...				0.01	
Kuwaiti diesel oil	583	1.480	0.860	...			...				1.3	
US jet naphtha	434	1.444	0.805		144		0.801				1.0	
US high boiling naphtha	435	1.426	0.762		142.4		0.759				0.0	
US kerosene	480	1.444	0.808		162.3		0.804				0.0	
US fuel oil	559	1.478	0.862		227.5		0.858				1.3	

[a]Experimental data for the first two Kuwaiti fractions are taken from Riazi and Roomi [62] and for the US fractions are taken from Lenoir and Hipkin data set [1].

compare with the experimental values as given below [46].

a. What is the type of this product?

b. Molecular weight (experimental value is 160.1)

c. Kinematic viscosities at 20 and 40°C (experimental values are 1.892 and 1.28)

d. Flash point (experimental value is 69°C)

e. PNA composition (the experimental PNA composition from GC–MS analysis in vol% are 46.7, 40.5, and 12.8, respectively)

f. Smoke point (experimental value is 26 mm)

g. Aniline point (experimental value is 64.9°C)

h. Pour point (experimental value is −45°C)

i. Freezing point (experimental value is −44°C)

j. Hydrogen content (experimental value is 14.11 wt%)

3.20. An atmospheric residue produced from the same crude of Problem 3.19 has API gravity of 25 and UOP K factor of 12.0. Predict the following properties and compare with the experimental values [46].

a. Molecular weight (experimental value is 399.8)

b. Total nitrogen content (experimental value is 0.21 wt%)

c. Kinematic viscosity at 100°C (experimental value is 8.082)

d. Kinematic viscosity at 70°C (experimental value is 17.89)

e. Sulfur content (experimental value is 0.17 wt%)

f. Conradson carbon residue (experimental value is 2.2 wt%)

g. Carbon content (experimental value is 86.7 wt%)

h. Hydrogen content (experimental value is 13 wt%)

i. Aniline point (experimental value is 95.2°C)

j. Pour point (experimental value is 39°C)

3.21. A gas oil produced from a crude from Soroosh field (Iran) has boiling range of 520–650°F and the API gravity

of 33 [46]. Reported Watson K factor is 11.72. Calculate the following properties and compare with experimental values.

a. Average boiling point from K_W and compare with mid boiling point

b. Cetane index (reported value is 50.5)

c. Aniline point (reported value is 152.9°F)

3.22. A heavy naphtha sample from Australian crude oil has the boiling range of 140–190°C, specific gravity of 0.7736 and molecular weight of 131.4. The experimentally determined composition for n-paraffins, isoparaffins, naphthenes, and aromatics are 29.97, 20.31, 38.72, and 13%, respectively. For this sample the experimental values of RON and MON are 26 and 28, respectively as reported in Ref. [46], p. 359. Estimate the RON from the pseudocomponent method (Eqs. 3.138 and 3.139) and the Nelson methods (Fig. 3.33 and 3.34). Also calculate the MON from Jenkins method (Eq. 3.140). For each case calculate the error and comment on your results.

3.23. A light naphtha from Abu Dhabi field (UAE) has boiling range of C_5-80°C, the API gravity of 83.1 and K_W of 12.73 [46]. Estimate the following octane numbers and compare with the experimental values.

a. Clear RON from two different methods (experimental value is 65)

b. RON + 1.5 mL of TEL /U.S. Gallon (experimental value is 74.5)

c. Clear MON (experimental value is 61)

d. How much MTBE should be added to this naphtha to increase the RON from 65 to 75.

3.24. A petroleum fraction produced from a Venezuelan crude has ASTM D 86 distillation curve as:

vol% distilled	10	30	50	70	90
ASTM D 86 temperature, °F	504	509	515	523	534

TABLE 3.33—*Estimation of composition of petroleum fractions (Problem 3.16).*[a]

No.	Fraction	T_b, K exp.	SG exp.	n_{20}	M	m	Exp. composition X_P	Exp. composition X_N	Exp. composition X_A	Estimated composition X_P	Estimated composition X_N	Estimated composition X_A
1	China Heavy Naphtha	444.1	0.791				48.9	30.9	20.2			
2	Malaysia Light Naphtha	326.6	0.666				83.0	17.0	0.0			
3	Indonesia Heavy Naphtha	405.2	0.738				62.0	30.0	8.0			
4	Venezuela Kerosene	463.6	0.806				39.8	41.1	19.0			
5	Heavy Iranian Gasoline	323.5	0.647				93.5	5.7	0.8			
6	Qatar Gasoline	309.4	0.649				95.0	3.9	1.1			
7	Sharjah Gasoline	337.7	0.693				78.4	14.4	7.2			
8	American Gasoline	317.2	0.653				92.0	7.3	0.7			
9	Libya Kerosene	465.5	0.794				51.2	34.7	14.1			
10	U.K. North Sea Kerosene	464.9	0.798				42.5	36.4	21.1			
11	U.K. North Sea Gas Oil	574.7	0.855				34.3	39.8	25.9			
12	Mexico Naphtha	324.7	0.677				81.9	13.9	4.2			

[a]Experimental data on T_b, S, and the composition are taken from the *Oil & Gas Journal Data Book* [46].

The specific gravity (at 60/60°F) of fraction is 0.8597 [46]. Predict the following properties for the product and compare with reported values.

a. Kinematic viscosities at 100, 140, and 210°F (experimental values are 3.26, 2.04, and 1.12 cSt.)

b. Molecular weight from viscosity and compare with molecular weight estimated from the boiling point.

c. Boiling point and specific gravity from experimental values of kinematic viscosities at 100 and 210°F (Part a) and compare with actual values.

d. Aniline point (experimental value is 143.5°F)

e. Cetane number (experimental value is 43.2)

f. Freezing point (experimental value is 21°F)

g. Flash point (experimental value is 230°F)

h. Carbon-to-hydrogen weight ratio (experimental value is 6.69)

i. Aromatic content from experimental viscosity at 100°F (experimental value is 34.9)

j. Aromatic content from experimental aniline point (experimental value is 34.9)

k. Refractive index at 75°F (experimental value is 1.4759)

3.25. A vacuum residue has kinematic viscosity of 4.5 mm^2/s at 100°C and specific gravity of 0.854. Estimate viscosity index (VI) of this fraction and compare with reported value of 119 [24].

3.26. A kerosene sample has boiling range of 180–225°C and specific gravity of 0.793. This product has aromatic content of 20.5%. Predict smoke point and freezing point of this product and compare with the experimental values of 19 mm and −50°C [24]. How much 2-methylnonane ($C_{10}H_{22}$) with freezing point of −74°C should be added to this kerosene to reduce the freezing point to −60°C.

REFERENCES

[1] Lenoir, J. M. and Hipkin, H. G., "Measured Enthalpies of Eight Hydrocarbon Fractions," *Journal of Chemical and Engineering Data*, Vol. 18, No. 2, 1973, pp. 195–202.

[2] Daubert, T. E. and Danner, R. P., Eds., *API Technical Data Book—Petroleum Refining*, 6th ed., American Petroleum Institute (API), Washington, DC, 1997.

[3] Curvers, J. and van den Engel, P., "Gas Chromatographic Method for Simulated Distillation up to a Boiling Point of 750°C Using Temperature-Programmed Injection and High Temperature Fused Silica Wide-Bore Columns," *J. High Resolution Chromatography*, Vol. 12, 1989, pp. 16–22.

[4] ASTM, *Annual Book of Standards*, ASTM International, West Conshohocken, PA, 2002.

[5] Denis, J., Briant, J. and Hipeaux, J. C., *Lubricant Properties Analysis and Testing*, Translated to English by G. Dobson, Editions Technip, Paris, 1997.

[6] Cryette, W. R., Model 5009 Wide Range Cryoscope—Instruction Manual Precision Systems Inc., Natick, MA, 1994.

[7] Speight, J. G., *The Chemistry and Technology of Petroleum*, 3rd ed., Marcel Dekker, New York, 1998.

[8] Firoozabadi, A., *Thermodynamics of Hydrocarbon Reservoirs*, Mc-Graw Hill, New York, 1999.

[9] Smith, R. L. and Watson, K. M., "Boiling Points and Critical Properties of Hydrocarbon Mixtures," *Industrial and Engineering Chemistry*, Vol. 29, 1937, pp. 1408–1414.

[10] Zhou, P., "Correlation of the Average Boiling Points of Petroleum Fractions with Pseudocritical Constants," *International Chemistry Engineering*, Vol. 24, 1984, pp. 731–742.

[11] Daubert, T. E., "Property Predictions," *Hydrocarbon Processing*, March, 1980, pp. 107–112.

[12] Van Winkle, M., "Physical Properties of Petroleum Fraction," *Petroleum Refiner*, Vol. 34, No. 6, 1955, pp. 136–138.

[13] Nelson, W. L. and Hansburg, M., "Relation of ASTM and True Boiling Point Distillations," *Oil & Gas Journal*, Vol. 38, No. 12, 1939, pp. 45–47.

[14] Edmister, W. C. and Pllock, D. H., "Phase Relations for Petroleum Fractions," *Chemical Engineering Progress*, Vol. 44, 1948, pp. 905–926.

[15] Edmister, W. C. and Okamoto, K. K., "Applied Hydrocarbon Thermodynamics, Part 12: Equilibrium Flash Vaporization Correlations for Petroleum Fractions," *Petroleum Refiner*, Vol. 38, No. 8, 1959, pp. 117–132.

[16] Edmister, W. C. and Okamoto, K. K., *Applied Hydrocarbon Thermodynamics*, Gulf Publishing, Houston, TX, 1961, pp. 116–133.

[17] Edmister, W. C. and Okamoto, K. K., "Applied Hydrocarbon Thermodynamics, Part 13: Equilibrium Flash Vaporization Correlations for Heavy Oils Under Subatmospheric Pressures," *Petroleum Refiner*, Vol. 38, No. 9, 1959, pp. 271–288.

[18] Okamoto, K. K. and van Winkle, M., "Equilibrium Flash Vaporization of Petroleum Fractions," *Industrial and Engineering Chemistry*, Vol. 45, 1953, pp. 429–439.

[19] House, H. G., Braun, W. G., Thompson, W. T. and Fenske, M. R., "Documentation of the Basis for Selection of the Contents of Chapter 3, ASTM, TBP, and EFV Relationships for Petroleum Fractions in Technical Data Book-Petroleum Refining," Documentation Report No. 3-66, Pennsylvania State University, 1966.

[20] Arnold, V. E., "Microcomputer Program Converts TBP, ASTM, EFV Distillation Curves," *Oil & Gas Journal*, Vol. 83, No. 6, 1985, pp. 55–62.

[21] Ford, D. C., Miller, W. H., Thren, R. C. and Wertzler, R., *Correlation of ASTM Method D 2887-73 Boiling Range Distribution Data with ASTM Method D 86-67 Distillation Data*, STP 577, ASTM International, West Conshohocken, PA, 1973, p. 29.

[22] Riazi, M. R. and Daubert, T. E., "Analytical Correlations Interconvert Distillation Curve Types," *Oil & Gas Journal*, Vol. 84, 1986, August 25, pp. 50–57.

[23] Daubert, T. E., "Petroleum Fraction Distillation Interconversion," *Hydrocarbon Processing*, Vol. 73, No. 9, 1994, pp. 75–78.

[24] Wauquier, J.-P., "Petroleum Refining, Vol. 1: Crude Oil, Petroleum Products, Process Flowsheets," Editions Technip, Paris, 1995.

[25] HYSYS, "Reference Volume 1., Version 1.1," HYSYS Reference Manual for Computer Software, HYSYS Conceptual Design, Hyprotech Ltd., Calgary, Alberta, Canada, 1996.

[26] Daubert, T. E., Riazi, M. R. and Danner, R. P., "Documentation of the Basis for Selection of the Contents of Chapters 2 and 3 in the API Technical Data Book-Petroleum Refining," Documentation Report No. 2,3-86, Pennsylvania State University, 1986.

[27] Maxwell, J. B. and Bonnel, L. S., *Vapor Pressure Charts for Petroleum Engineers*, Esso Research and Engineering Company, Linden, NJ, 1955.

[28] Myers, H. S. and Fenske, M. R., "Measurement and Correlation of Vapor Pressure Data for High-Boiling Hydrocarbons," *Industrial and Engineering Chemistry*, Vol. 47, 1955, pp. 1652–1658.

[29] Riazi, M. R., *Prediction of Thermophysical Properties of Petroleum Fractions*, Doctoral Dissertation, Department of

Chemical Engineering, Pennsylvania State University, University Park, PA, 1979.

[30] Van Nes, K. and Van Western, H. A., *Aspects of the Constitution of Mineral Oils*, Elsevier, New York, 1951.

[31] Riazi, M. R., "Distribution Model for Properties of Hydrocarbon-Plus Fractions," *Industrial and Engineering Chemistry Research*, Vol. 28, 1989, pp. 1731–1735.

[32] Riazi, M. R. and Daubert, T. E., "Improved Characterization of Wide Boiling Range Undefined Petroleum Fractions," *Industrial and Engineering Chemistry Research*, Vol. 26, 1987, pp. 629–632.

[33] Huang, P. K. and Daubert, T. E., "Prediction of the Enthalpy of Petroleum Fractions: Pseudocompound Method," *Industrial and Engineering Chemistry, Process Design and Development*, Vol. 13, No. 4, 1974, pp. 359–362.

[34] Katinas, T. G., *Prediction of the Viscosities of Petroleum Fractions*, M. Sc. Thesis, Department of Chemical Engineering, Pennsylvania State University, University Park, PA, 1977.

[35] Miqueu, C., Satherley, J., Mendiboure, B., Lachaise, J. and Graciaa, A., "The Effect of P/N/A Distribution on the Parachors of Petroleum Fractions," *Fluid Phase Equilibria*, Vol. 180, No.1/2, 2001, pp. 327–344.

[36] Riazi, M. R. and Daubert, T. E., "Prediction of Molecular Type Analysis of Petroleum Fractions and Coal Liquids," *Industrial and Engineering Chemistry, Process Design and Development*, Vol. 25, No. 4, 1986, pp. 1009–1015.

[37] Riazi, M. R. and Al-Sahhaf, T., "Physical Properties of *n*-Alkanes and *n*-Alkyl Hydrocarbons: Application to Petroleum Mixtures," *Industrial and Engineering Chemistry Research*, Vol. 34, 1995, pp. 4145–4148.

[38] Riazi, M. R. and Daubert, T. E., "Simplify Property Predictions," *Hydrocarbon Processing*, Vol. 59, No. 3, 1980, pp. 115–116.

[39] Lee, B. I. and Kesler, M. G., "A Generalized Thermodynamic Correlation Based on Three- Parameter Corresponding States," *American Institute of Chemical Engineers Journal*, Vol. 21, 1975, pp. 510–527.

[40] Kesler, M. G. and Lee, B. I., "Improve Prediction of Enthalpy of Fractions," *Hydrocarbon Processing*, Vol. 55, No. 3, March 1976, pp. 153–158.

[41] Winn, F. W., "Physical Properties by Nomogram," *Petroleum Refiners*, Vol. 36, No. 21, 1957, pp. 157–159.

[42] Twu, C. H., "An Internally Consistent Correlation for Predicting the Critical Properties and Molecular Weights of Petroleum and Coal-Tar Liquids," *Fluid Phase Equilbria*, Vol. 16, 1984, pp. 137–150.

[43] Goossens, A. G., "Prediction of Molecular Weight of Petroleum Fractions," *Industrial and Engineering Chemistry Research*, Vol. 35, 1996, pp. 985–988.

[44] Riazi, M. R. and Daubert, T. E., "Characterization Parameters for Petroleum Fractions," *Industrial and Engineering Chemistry Research*, Vol. 26, 1987, pp. 755–759. (Corrections, p. 1268.)

[45] Riazi, M. R. and Roomi, Y., "Use of the Refractive Index in the Estimation of Thermophysical Properties of Hydrocarbons and Their Mixtures," *Industrial and Engineering Chemistry Research*, Vol. 40, No. 8, 2001, pp. 1975–1984.

[46] *Oil and Gas Journal Data Book*, 2000 Edition, PennWell, Tulsa, OK, 2000.

[47] Riazi, M. R. and Daubert, T. E., "Prediction of the Composition of Petroleum Fractions," *Industrial and Engineering Chemistry, Process Design and Development*, Vol. 19, No. 2, 1980, pp. 289–294.

[48] Ahmed, T., *Hydrocarbon Phase Behavior*, Gulf Publishing, Houston, TX, 1989.

[49] Hill, J. B. and Coats, H. B., "The Viscosity Gravity Constant of Petroleum Lubricating Oils," *Industrial and Engineering Chemistry*, Vol. 20, 1928, pp. 641–644.

[50] Watson, K. M. and Nelson, E. F., "Improved Methods for Approximating Critical and Thermal Properties of Petroleum Fractions," *Industrial and Engineering Chemistry*, Vol. 25, 1933, pp. 880–887.

[51] Watson, K. M., Nelson, E. F., and Murphy, G. B., "Characterization of Petroleum Fractions," *Industrial and Engineering Chemistry*, Vol. 27, 1935, pp. 1460–1464.

[52] Kurtz, Jr., S. S. and Ward, A. L., "The Refractivity Intercept and the Specific Refraction Equation of Newton, I: Development of the Refractivity Intercept and Composition with Specific Refraction Equations," *Journal of Franklin Institute*, Vol. 222, 1936, pp. 563–592.

[53] Huang, P. K., *Characterization and Thermodynamic Correlations for Undefined Hydrocarbon Mixtures*, Ph.D. Dissertation, Pennsylvania State University, University Park, PA, 1977.

[54] Fryback, M. G., "Synthetic Fuels: Promises and Problems," *Chemical Engineering Progress*, Vol. 77, No. 5, 1981, pp. 39–43.

[55] Hersh, R. E., Fenske, M. R., Booser, E. R., and Koch, E. F., "Ring Analysis of Hydrocarbon Mixtures," *Journal of Institute of Petroleum*, Vol. 36, No. 322, 1950, pp. 624–668.

[56] *API Research Project 42: Properties of Hydrocarbons of High Molecular Weight*, American Petroleum Institute, New York, 1966.

[57] Goodger, E. and Vere, R., *Aviation Fuels Technology*, MacMillan, London, 1985.

[58] Goossens, A. G., "Prediction of the Hydrogen Content of Petroleum Fractions," *Industrial and Engineering Chemistry Research*, Vol. 36, 1997, pp. 2500–2504.

[59] Jenkins, G. I. and Walsh, R. P., "Quick Measure of Jet Fuel Properties," *Hydrocarbon Processing*, Vol. 47, No. 5, 1968, pp. 161–164.

[60] Fein, R. S., Wilson, H. I., and Sherman, J., "Net Heat of Combustion of Petroleum Fractions," *Industrial and Engineering Chemistry*, Vol. 45, No. 3, 1953, pp. 610–614.

[61] Baird, C. T., *Crude Oil Yields and Product Properties*, Ch. De la Haute–Belotte 6, Cud Thomas Baird IV, 1222 Vezenaz, Geneva, Switzerland, June, 1981.

[62] Riazi, M. R., Nasimi, N., and Roomi, Y., "Estimating Sulfur Content of Petroleum Products and Crude Oils," *Industrial and Engineering Chemistry Research*, Vol. 38, No. 11, 1999, pp. 4507–4512.

[63] Gary, J. H. and Handwerk, G. E., *Petroleum Refining, Technology and Economic*, 3rd ed., Marcel Dekker, New York, 1994.

[64] Hebert, J., "Court Upholds Tougher Diesel Rules," Associated Press Writer, Yahoo News Website, May 3, 2002.

[65] Ball, S. J., Whisman, M. L., and Wenger, W. J., "Nitrogen Content of Crude Petroleums," *Industrial and Engineering Chemistry*, Vol. 43, No. 11, 1951, pp. 2577–2581.

[66] Nelson, W. L., *Petroleum Refinery Engineering*, 4th ed., McGraw-Hill, New York, 1969.

[67] Riazi, M. R., Albahri, T. A., and Alqattan, A., "Prediction of Reid Vapor Pressure of Petroleum Fuels," *Petroleum Science and Technology*, Vol. 23, No. 1, 2005, pp. 1–12.

[68] Albahri, T., Riazi, M. R., and Alqattan, A., "Analysis of Quality of the Petroleum Fuels," *Journal of Energy and Fuels*, Vol. 17, No. 3, 2003, pp. 689–693.

[69] Hatzioznnidis, I., Voutsas, E. C., Lois, E., and Tassios, D. P., "Measurement and Prediction of Reid Vapor Pressure of Gasoline in the Presence of Additives," *Journal of Chemical and Engineering Data*, Vol. 43, No. 3, 1998, pp. 386–392.

[70] Butler, R. M., Cooke, G. M., Lukk, G. G., and Jameson, B. G., "Prediction of Flash Points of Middle Distillates," *Industrial and Engineering Chemistry*, Vol. 48, No. 4, 1956, pp. 808– 812.

[71] Walsh, R. P. and Mortimer, J. V., "New Way to Test Product Quality," *Hydrocarbon Processing*, Vol. 50, No. 9, 1971, pp. 153–158.

[72] Lenoir, J. M., "Predict Flash Points Accurately," *Hydrocarbon Processing*, Vol. 54, No. 1, 1975, pp. 153–158.

[73] Riazi, M. R. and Daubert, T. E., "Predicting Flash and Pour Points," *Hydrocarbon Processing*, Vol. 66, No. 9, 1987, pp. 81–83.

[74] Wickey, R. O. and Chittenden, D. H., "Flash Points of Blends Correlated," *Hydrocarbon Processing*, Vol. 42, No. 6, 1963, pp. 157–158.

[75] Hu, J. and Burns, A. M., "New Method Predicts Cloud, Pour and Flash Points," *Hydrocarbon Processing*, Vol. 49, No. 11, 1970, pp. 213–216.

[76] Hu, J. and Burns, A. M., "Index Predicts Cloud, Pour and Flash Points of Distillate Fuel Blends," *Oil & Gas Journal*, Vol. 68, No. 45, 1970, pp. 66–69.

[77] Linden, H. R., "The Relationship of Physical Properties and Ultimate Analysis of Liquid Hydrocarbon," *Oil & Gas Journal*, Vol. 48, No. 9, 1949, pp. 60–65.

[78] Albahri, T. A., Riazi, M. R., and Qattan, A., "Octane Number and Aniline Point of Petroleum Fractions," Preprints, Division of Fuel Chemistry, 224 ACS National Meeting, Boston, August 18–22, 2002.

[79] Nelson, W. L., "Octane Numbers of Naphthas," *Oil and Gas Journal*, Vol. 67, No. 25, 1969, p. 122.

[80] Jenkins, G. I., "Calculation of the Motor Octane Number from the Research Octane Number," *Journal of Institute of Petroleum*, Vol. 54, No. 529, 1968, pp. 14–18.

[81] Altgelt, K. H. and Boduszynski, M. M., *Composition and Analysis of Heavy Petroleum Fractions*, Marcel Dekker, New York, 1994.

[82] Nelson, W. L., "Estimating Yields of Asphalt in Crude Oil–Again," *Oil and Gas Journal*, Vol. 66, No. 16, 1968, p. 93.

[83] Jenkins, G. I. and Walsh, R. P., "Quick Measure of Jet Fuel Properties," *Hydrocarbon Processing*, Vol. 47, No. 5, 1968, pp. 161–164.

[84] Riazi, M. R. and Roomi, Y. A., "Minimum Laboratory Data for Properties of Petroleum Fractions and Crude Oils," *Preprints of Division of Petroleum Chemistry Symposia*, American Chemical Society (ACS), Vol. 45, No. 4, 2000, pp. 570–573.

[85] *Standard Methods for Analysis and Testing of Petroleum and Related Products and British Standard 2000 Parts*, IP 1–324 Vol. 1, IP 325–448 Vol. 2, The Institute of Petroleum, London, 1999.

[86] Tsonopoulos, C., Heidman, J. L., and Hwang, S.-C., *Thermodynamic and Transport Properties of Coal Liquids*, An Exxon Monograph, Wiley, New York, 1986.

[87] Knepper, J. I. and Hutton, R. P., "Blend for Lower Pour Point," *Hydrocarbon Processing*, Vol. 54, No. 9, 1975, pp. 129–136.

[88] Auckland, M. H. T. and Charnock, D. J., "The Development of Linear Blending indexes for Petroleum Products," *Journal of Institute of Petroleum*, Vol. 55, No. 545, 1969, pp. 322–329.

[89] Morris, W. E., "The Interaction Approach To Gasoline Blending," *Presented at the NPRA 73rd Annual Meeting*, San Antonio, Texas, March 23–27, 1975.

[90] *Modern Petroleum Technology*, 3rd ed., The Institute of Petroleum, London, 1962.

Characterization of Reservoir Fluids and Crude Oils

NOMENCLATURE

API API Gravity defined in Eq. (2.4)

A, B Coefficients in Eq. (4.56) and other equations

$a, b, \ldots i$ Correlation constants in various equations

CH Carbon-to-hydrogen weight ratio

d_{20} Liquid density at 20°C and 1 atm, g/cm^3

d_c Critical density defined by Eq. (2.9), g/cm^3

E Error function defined in Eqs. (4.41) or (4.42)

$F(P)$ Probability density function in terms of property P

h Difference in molecular weight of successive SCN groups

K_T Equilibrium ratio for a component whose boiling point is T

K_W Watson (UOP) K factor defined by Eq. (2.13)

I Refractive index parameter defined in Eq. (2.36)

J Integration parameter defined in Eq. (4.79), dimensionless

M Molecular weight, g/mol [kg/kmol]

M_b Variable defined by Eq. (4.36)

N_C Carbon number (number of carbon atoms in a single carbon number hydrocarbon group)

N_P Number of pseudocomponents

N_+ Carbon number of the residue or plus fraction in a mixture

n Refers to carbon number in a SCN group

n_{20} Sodium D line refractive index of liquid at 20°C and 1 atm, dimensionless

P A property such as T_b, M, SG, or I used in a probability density function

P^* Dimensionless parameter defined by Eq. (4.56) as $[=(P - P_o)/P_o]$.

P_{pc} Pseudocritical pressure, bar

P^{vap} Vapor (saturation) pressure, bar

R^2 R squared (R^2), defined in Eq. (2.136)

SG Specific gravity of liquid substance at 15.5°C (60°F) defined by Eq. (2.2), dimensionless

SG_g Specific gravity of gas substance at 15.5°C (60°F) defined by Eq. (2.6), dimensionless

T_b Boiling point, K

T_{br} Reduced boiling point ($=T_b/T_c$ in which both T_b and T_c are in K), dimensionless

T_{pc} Pseudocritical temperature, K

V Molar volume, cm^3/gmol

V_c Critical volume (molar), cm^3/mol (or critical specific volume, cm^3/g)

x_c Cumulative mole, weight, or volume fraction

x_{cm} Cumulative mole fraction

x_{mi} Discrete mole fraction of component i

x^* Defined in Eq. (4.56) $[=1 - x_c]$

X Parameter defined in Eq. (4.57) $[= \ln \ln(1/x^*)]$

Y Parameter defined in Eq. (4.57) $[= \ln P^*]$

y_i A Gaussian quadrature point in Section 4.6.1.1

Z_c Critical compressibility factor defined by Eq. (2.8), dimensionless

z_j Predicted mole fraction of pseudocomponent i in a C$_{7+}$ fraction

w_i A weighting factor in Gaussian quadrature splitting scheme

Greek Letters

α Parameter for gamma distribution model, Eq. (4.31)

β Composite parameter for gamma distribution model, Eq. (4.31)

δ Solubility parameter $[= cal/cm^3]^{1/2}$

ε Error parameter

$\Gamma(x)$ Gamma function defined by Eq. (4.43)

$\Gamma(a, q)$ Incomplete gamma function defined by Eq. (4.89)

γ Activity coefficient

η Coefficient in gamma distribution model, Eq. (4.31)

φ Fugacity coefficient

μ_i Chemical potential of component i in a mixture

θ A property of hydrocarbon such as M, T_c, P_c, V_c, I, d, $T_b, \ldots$

ρ Density at a given temperature and pressure, g/cm^3

σ Surface tension, dyn/cm [=mN/m]

ω Acentric factor

Superscript

c Adjusted pseudocritical properties for the effects of nonhydrocarbon compounds in natural gas system as given by Eqs. (4.5) and (4.6).

cal Calculated value

exp Experimental value

i A component in a mixture.

j A pseudocomponent in a C$_{7+}$ fraction

L Liquid phase

V Vapor phase

$-$ Lower value of a property for a SCN group

$+$ Upper value of a property for a SCN group

Subscripts

A Aromatic

av Average value for a property

c Cumulative fraction

M Molecular weight

m	Mole fraction
N	Naphthenic
n	Refers to SCN group with *n* carbon number
P	Paraffinic
pc	Pseudo-Critical
T	A distribution coefficient in Eq. (4.56) for boiling point
v	Volume fraction
w	Weight fraction
o	Value of a property at $x_c = 0$ in Eq. (4.56)
∞	Value of a property at $M \to \infty$
20	Value of a property at 20°C

Acronyms

%AAD	Average absolute deviation percentage defined by Eq. (2.135)
API-TDB	American Petroleum Institute—Technical Data Book
%D	Absolute deviation percentage defined by Eq. (2.134)
EOS	Equation of state
GC	Gas chromatography
KISR	Kuwait Institute for Scientific Research
%MAD	Maximum absolute deviation percentage
OGJ	Oil & Gas Journal
PDF	Probability density function
PNA	Paraffins naphthenes aromatics
RMS	Root mean squares defined by Eq. (4.59)
RVP	Reid vapor pressure
RS	R squared (R^2), defined in Eq. (2.136)
SCN	Single carbon number
TBP	True boiling point
VLE	Vapor–liquid equilibrium

As DISCUSSED IN CHAPTER 1, reservoir fluids are in the forms of natural gases, gas condensates, volatile oils, and black oils. As shown in Table 1.1, these fluids contain hydrocarbons from C_1 to compounds with carbon number greater than 50. Composition of a reservoir fluid is generally expressed in mol% of nonhydrocarbon compounds (i.e., H_2S, CO_2, N_2), C_1, C_2, C_3, nC_4, iC_4, nC_5, iC_5, C_6, and C_{7+}. The boiling range of reservoir fluids can be greater than 550°C ($\sim > 1000°F$). Crude oil is produced by separating light gases from a reservoir fluid and bringing its condition to surface atmospheric pressure and temperature. Therefore, crude oils are generally free from methane and contain little ethane. The main difference between various reservoir fluid and produced crude oil is in their composition, as shown in Table 1.1. Amount of methane reduces from natural gas to gas condensate, volatile oil, black oil, and crude oil while amount of heavier compounds (i.e., C_{7+}) increase in the same direction. Characterization of reservoir fluids and crude oils mainly involves characterization of hydrocarbon-plus fractions generally expressed in terms of C_{7+} fractions. These fractions are completely different from petroleum fractions discussed in Chapter 3. A C_{7+} fraction of a crude oil has a very wide boiling range in comparison with a petroleum product and contains more complex and heavy compounds. Usually the only information available for a C_{7+} fraction is the mole fraction, molecular weight, and specific gravity. The characterization procedure involves how to present this

mixture in terms of arbitrary number of subfractions (pseudocomponents) with known mole fraction, boiling point, specific gravity, and molecular weight. This approach is called pseudoization. The main objective of this chapter is to present methods of characterization of hydrocarbon-plus fractions, which involves prediction of distribution of hydrocarbons in the mixture and to represent the fluid in terms of several narrow range subfractions. However, for natural gases and gas condensate fluids that are rich in low-molecular-weight hydrocarbons simple relations have been proposed in the literature. In this chapter types of data available for reservoir fluids and crude oils are discussed followed by characterization of natural gases. Then physical properties of single carbon number (SCN) groups are presented. Three distribution models for properties of hydrocarbon plus fractions are introduced and their application in characterization of reservoir fluids is examined. Finally, the proposed methods are used to calculate some properties of crude oils. Accuracy of characterization of reservoir fluids largely depends on the distribution model used to express component distribution as well as characterization methods of petroleum fractions discussed in Chapter 2 to estimate properties of the narrow boiling range pseudocomponents.

4.1 SPECIFICATIONS OF RESERVOIR FLUIDS AND CRUDE ASSAYS

Characterization of a petroleum fluid requires input parameters that are determined from laboratory measurements. In this section types of data available for a reservoir fluid or a crude oil are presented. Availability of proper data leads to appropriate characterization of a reservoir fluid or a crude oil.

4.1.1 Laboratory Data for Reservoir Fluids

Data on composition of various reservoir fluids and a crude oil were shown in Table 1.1. Further data on composition of four reservoir fluids from North Sea and South West Texas fields are given in Table 4.1. Data are produced from analysis of the fluid by gas chromatography columns capable of separating hydrocarbons up to C_{40} or C_{45}. Composition of the mixture is usually expressed in terms of mol% for pure hydrocarbons up to C_5 and for heavier hydrocarbons by single carbon number (SCN) groups up to C_{30} or C_{40}. However, detailed composition is available for lower carbon numbers while all heavy hydrocarbons are lumped into a single group called *hydrocarbon-plus* fraction. For example in Table 4.1, data are given up to C_9 for each SCN group while heavier compounds are grouped into a C_{10+} fraction. It is customary in the petroleum industry to lump the hydrocarbons heavier than heptane into a C_{7+} fraction. For this reason the mol% of C_{7+} for the four mixtures is also presented in Table 4.1. For hydrocarbon-plus fractions it is important to report a minimum of two characteristics. These two specifications are generally molecular weight and specific gravity (or API gravity) shown by M_{7+} and SG_{7+}, respectively. In some cases a reservoir fluid is presented in terms of true boiling point (TBP) of each SCN group except for the plus fraction in which boiling point is not available. The plus fractions contain heavy compounds and for this reason their

TABLE 4.1—*Composition of several reservoir fluids.*

Component	North Sea gas condensate			North Sea oil			Texas gas condensate			Texas oil		
	mol%	SG	M	mol%	SG	M	mol%	SG	M	mol%	SG	M
N_2	0.85			0.69			0			0		
CO_2	0.65			3.14			0			52.00		
C_1	83.58			52.81			91.35			3.81		
C_2	5.95			8.87			4.03			2.37		
C_3	2.91			6.28			1.53			0.76		
IC_4	0.45			1.06			0.39			0.96		
nC_4	1.11			2.48			0.43			0.69		
IC_5	0.36			0.87			0.15			0.51		
nC_5	0.48			1.17			0.19			2.06		
C_6	0.60			1.45			0.39			2.63	0.749	99
C_7	0.80	0.7243	95	2.39	0.741	91.7	0.361	0.745	100	2.34	0.758	110
C_8	0.76	0.7476	103	2.67	0.767	104.7	0.285	0.753	114	2.35	0.779	121
C_9	0.47	0.7764	116	1.83	0.787	119.2	0.222	0.773	128	29.52	0.852	221
C_{10+}	1.03	0.8120	167	14.29	0.869	259.0	0.672	0.814	179	36.84	0.841	198.9
C_{7+}	3.06	0.7745	124	21.18	0.850	208.6	1.54	0.787	141.1			

Source: North Sea gas condensate and oil samples are taken from Ref. [1]. South West Texas gas condensate and oil samples are taken from Ref. [2]. Data for C_{7+} have been obtained from data on C_7, C_8, C_9, and C_{10+} components.

boiling point cannot be measured; only molecular weight and specific gravity are available for the plus fractions. Characteristics and properties of SCN groups are given later in this chapter (Section 4.3).

FIG. 4.1—Apparatus to conduct TBP analysis of crude oils and reservoir fluids (courtesy of KISR [5]).

Generation of such data for molecular weight and density distribution from gas chromatography (GC) analysis for crude oils is shown by Osjord et al. [3]. Detailed composition of SCN groups for C_{6+} or C_{7+} fractions can also be obtained by TBP distillation. Experimental data obtained from distillation are the most accurate way of analyzing a reservoir fluid or crude oil, especially when it is combined with measuring specific gravity of each cut. However, GC analysis requires smaller sample quantity, less time, and less cost than does TBP analysis. The ASTM D 2892 procedure is a standard method for TBP analysis of crude oils [4]. The apparatus used in ASTM D 2892, is shown in Fig. 4.1 [5]. A GC for determining SCN distribution in crude oils is shown in Fig. 4.2. The output from this GC for a Kuwaiti crude oil sample is shown in Fig. 4.3. In this figure various SCN from C_5 up to C_{40} are identified and the retention times for each carbon group are given on each pick. A comparison of molecular weight and specific gravity distribution of SCN groups obtained from TBP distillation and GC analysis for the same crude oil is also shown by Osjord et al. [3]. Pedersen et al. [6] have also presented compositional data for many gas condensate samples from the North Sea. An extended composition of a light waxy crude oil is given in Table 4.2 [7]. Distribution of SCN groups for the Kuwait crude determined from Fig. 4.3 is also given in Table 4.2. Other properties of SCN groups are given in Section 4.3. One of the important characteristics of crude oils is the cloud point (CPT). This temperature indicates when the precipitation of wax components in a crude begins. Calculation of CPT requires liquid–solid equilibrium calculations, which are discussed in Chapter 9 (Section 9.3.3).

4.1.2 Crude Oil Assays

Composition of a crude may be expressed similar to a reservoir fluid as shown in Table 1.1. A crude is produced through reducing the pressure of a reservoir fluid to atmospheric pressure and separating light gases. Therefore, a crude oil is usually free of methane gas and has a higher amount of C_{7+} than the original reservoir fluid. However, in many cases information on characteristics of crude oils are given through crude assay. A complete data on crude assay contain information on specification of the whole crude oil as well

FIG. 4.2—A GC for measuring SCN distribution in crude oils and reservoir fluids (courtesy of KISR [5]).

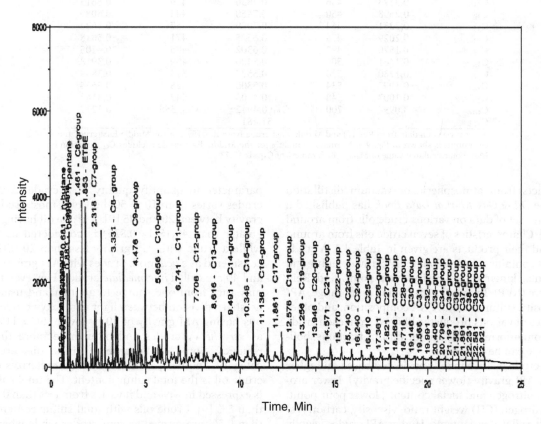

FIG. 4.3—A sample of output from the GC of Fig. 4.2 for a Kuwait crude oil.

TABLE 4.2—*Extended compositional data for a light waxy crude oil and a Middle East crude.*

	Waxy crude oil		Middle East crude oil		
Component	mol%	M, g/mol	wt%	M	Normalized mol%
C_2	0.0041	30	0.0076	30	0.0917
C_3	0.0375	44	0.1208	44	0.9940
iC_4	0.0752	58	0.0921	58	0.5749
NC_4	0.1245	58	0.4341	58	2.7099
iC_5	0.3270	72	0.4318	72	2.1714
NC_5	0.2831	72	0.7384	72	3.7132
C_6	0.3637	86	1.6943	82	7.4812
C_7	3.2913	100	2.2346	95	8.5166
C_8	8.2920	114	2.7519	107	9.3120
C_9	10.6557	128	2.8330	121	8.4772
C_{10}	11.3986	142	2.8282	136	7.5294
C_{11}	10.1595	156	2.3846	149	5.7946
C_{12}	8.7254	170	2.0684	163	4.5945
C_{13}	8.5434	184	2.1589	176	4.4413
C_{14}	6.7661	198	1.9437	191	3.6846
C_{15}	5.4968	212	1.9370	207	3.3881
C_{16}	3.5481	226	1.5888	221	2.6030
C_{17}	3.2366	240	1.5580	237	2.3802
C_{18}	2.1652	254	1.5006	249	2.1820
C_{19}	1.8098	268	1.5355	261	2.1301
C_{20}	1.4525	282	1.5441	275	2.0330
C_{21}	1.2406	296	1.1415	289	1.4301
C_{22}	1.1081	310	1.4003	303	1.6733
C_{23}	0.9890	324	0.9338	317	1.0666
C_{24}	0.7886	338	1.0742	331	1.1750
C_{25}	0.7625	352	1.0481	345	1.1000
C_{26}	0.6506	366	0.9840	359	0.9924
C_{27}	0.5625	380	0.8499	373	0.8250
C_{28}	0.5203	394	0.9468	387	0.8858
C_{29}	0.4891	408	0.8315	400	0.7527
C_{30}	0.3918	422	0.8141	415	0.7103
C_{31}	0.3173	436	0.7836	429	0.6613
C_{32}	0.2598	450	0.7450	443	0.6089
C_{33}	0.2251	464	0.7099	457	0.5624
C_{34}	0.2029	478	0.6528	471	0.5018
C_{35}	0.1570	492	0.6302	485	0.4705
C_{36}	0.1461	506	0.5400	499	0.3918
C_{37}	0.1230	520	0.5524	513	0.3899
C_{38}	0.1093	534	0.5300	528	0.3634
C_{39}	0.1007	548	0.4703	542	0.3142
C_{40+}	3.0994	700	C_{40}: 0.4942	C_{40}: 556	0.3217
C_{41+}			51.481		

Source: waxy oil data from Ref. [7] and Middle East crude from Ref. [5]. For the Middle East crude the GC output is shown in Fig. 4.3. Normalized mole% for the Middle East crude excludes C_{41+} fraction. Based on calculated value of $M_{41+} = 865$, mole% of C_{41+} is 17.73%.

as its products from atmospheric or vacuum distillation columns. The *Oil & Gas Journal Data Book* has published a comprehensive set of data on various crude oils from around the world [8]. Characteristics of seven crude oils from around the world and their products are given in Table 4.3. A crude assay dataset contains information on API gravity, sulfur and metal contents, kinematic viscosity, pour point, and Reid vapor pressure (RVP). In addition to boiling point range, API gravity, viscosity, sulfur content, PNA composition, and other characteristics of various products obtained from each crude are given. From information given for various fractions, boiling point curve can be obtained.

Quality of crude oils are mainly evaluated based on higher value for the API gravity (lower specific gravity), lower aromatic, sulfur, nitrogen and metal contents, lower pour point, carbon-to-hydrogen (CH) weight ratio, viscosity, carbon residue, and salt and water contents. Higher API crudes usually contain higher amount of paraffins, lower CH weight ratio, less sulfur and metals, and have lower carbon residues and viscosity. For this reason API gravity is used as the primary parameter to quantify quality of a crude. API gravity of crudes varies from 10 to 50; however, most crudes have API gravity between 20 and 45 [9]. A crude oil having API gravity greater than 40 (SG < 0.825) is considered as *light* crude, while a crude with API gravity less than 20 (SG > 0.934) is considered as a *heavy* oil. Crudes with API gravity between 20 and 40 are called *intermediate* crudes. However, this division may vary from one source to another and usually there is no sharp division between light and heavy crude oils. Crude oils having API gravity of 10 or lower (SG > 1) are referred as *very heavy* crudes and often have more than 50 wt% residues. Some of Venezuelan crude oils are from this category. Another parameter that characterizes quality of a crude oil is the total sulfur content. The total sulfur content is expressed in wt% and it varies from less than 0.1% to more than 5% [9]. Crude oils with total sulfur content of greater than 0.5% are termed as *sour* crudes while when the sulfur content is less than 0.5% they are referred as "sweet" crude [9]. After sulfur content, lower nitrogen and metal contents signify quality of a crude oil.

TABLE 4.3—Crude assays data for selected oils [8].

Crude 1 — Brent Blend, U.K. North Sea

Whole crude
Density at 15°C, kg/L: 0.8334
Gravity, °API: 38.3
Sulfur, wt%: 0.40
Kinematic viscosity at 20°C, cSt: 6.07
Kinematic viscosity at 30°C, cSt: 14.673
Pour point, °C: −42
Acidity, mg KOH/g: 0.10
Micro carbon res., wt%: 2.13
Asphaltenes, wt%: 0.45
V/Ni, ppm: 6/1
H$_2$S, wt%: <0.0001
Salt (as NaCl), wt%: 0.015
Water content, wt%: 0.38

Light ends (C$_1$–C$_5$)
Yield, wt%: 5.87
Range, °C: IBP–95
Yield, vol%: 12.3
Yield, wt%: 10.3
Density at 15°C, kg/L: 0.6924
Sulfur, wt%: 0.0006
Mercaptan S, ppm: 51
Paraffins, wt%: 82.8
Naphthenes, wt%: 11.9
Aromatics, wt%: 5.3
n-Paraffins, wt%: 30.9

Range, °C: 95–175
Yield, vol%: 16.7
Yield, wt%: 15.4
Density at 15°C, kg/L: 0.7693
Sulfur, wt%: 0.0012
Mercaptan S, ppm: 14
Paraffins, wt%: 44.9
Naphthenes, wt%: 36.7
Aromatics, wt%: 18.4
n-Paraffins, wt%: 20.1

Range, °C: IBP–150
Yield, vol%: 23.8
Yield, wt%: 20.8
Density at 15°C, kg/L: 0.7279
Sulfur, wt%: 0.0007
Mercaptan S, ppm: 21
Paraffins, wt%: 65.3
Naphthenes, wt%: 23.8
Aromatics, wt%: 10.9
n-Paraffins, wt%: 25.9

Range, °C: 150–230
Yield, vol%: 13.9
Yield, wt%: 13.3
Density at 15°C, kg/L: 0.8001
Sulfur, wt%: 0.0048
Mercaptan S, ppm: 42
Kinematic viscosity at 40°, cSt: 1.135
Kinematic viscosity at 60°, cSt: 0.9408
Acidity, mg KOH/g: 0.036
Smoke point, mm: 22
Freeze point, °C: −61.5
Aniline point, °C: 55.0
Cetane index: 36.6
Hydrogen, wt%: 13.45
Color stability: Stable
Naphthalenes, vol%: 2.30

Range, °C: 230–350
Yield, vol%: 22.2
Yield, wt%: 22.5
Density at 15°C, kg/L: 0.8461
Sulfur, wt%: 0.24
Kinematic viscosity at 50°C, cSt: 2.831
Kinematic viscosity at 100°C, cSt: 1.339
Cloud point, °C: −9
Pour point, °C: −9
Wax content, wt%: 6.2
Total N, mg/kg: 107
Acidity, mg KOH/g: 0.028
Aniline point, °C: 70.6
Cetane index: 51
Color stability: Stable

Range, °C: 350–375
Yield, vol%: 3.8
Yield, wt%: 4.0
Density at 15°C, kg/L: 0.8795
Kinematic viscosity at 50°C, cSt: 7.807
Kinematic viscosity at 100°C, cSt: 2.714
Cloud point, °C: 19
Pour point, °C: 15
Wax content, wt%: 17.3
Total N, mg/kg: 532
Aniline point, °C: 80.5
Cetane index: 49.5

Range, °C: 375–550
Yield, vol%: 20.7
Yield, wt%: 22.5
Density at 15°C, kg/L: 0.9059
Sulfur, wt%: 0.61
Kinematic viscosity at 60°C, cSt: 26.23
Kinematic viscosity at 80°C, cSt: 12.69
Kinematic viscosity at 100°C, cSt: 7.711
Wax content, wt%: 20.7
Total N, mg/kg: 1447
Basic N, mg/kg: 354
Acidity, mg KOH/g: 0.052
Aniline point, °C: 90.5
Refractive index at 60°C: 1.4884

Range, °C: 350+
Yield, vol%: 36.7
Yield, wt%: 40.9
Density at 15°C, kg/L: 0.9285
Sulfur, wt%: 0.86
Kinematic viscosity at 50°C, cSt: 126.8
Kinematic viscosity at 60°C, cSt: 77.38
Kinematic viscosity at 80°C, cSt: 31.87
Kinematic viscosity at 100°C, cSt: 16.80
Kinematic viscosity at 120°C, cSt: 9.5*
Kinematic viscosity at 150°C, cSt: 5.5*
Pour point, °C: 36
Wax content, wt%: 14.6
Total N, mg/kg: 2667
Acidity, mg KOH/g: 0.18
Micro carbon res., wt%: 5.0
Asphaltenes, wt%: 0.95
V/Ni, ppm: 12/1
Xylene equivalent, vol%: 5–10+

Range, °C: 550+
Yield, vol%: 12.2
Yield, wt%: 14.4
Density at 15°C, kg/L: 0.9876
Sulfur, wt%: 1.21
Kinematic viscosity at 80°C, cSt: 975.4
Kinematic viscosity at 100°C, cSt: 303.1
Kinematic viscosity at 120°C, cSt: 120*
Kinematic viscosity at 150°C, cSt: 40*
Penetration at 25°C: >400
Softening point, °C: 31.8
Total N, mg/kg: 5242
Micro carbon res., wt%: 2.05
Asphaltenes, wt%: 2.05
Xylene equivalent, vol%: 5–10

Dewaxed oil
Density at 15°C, kg/L: 0.9244
Kinematic viscosity at 40°C, cSt: 104.2
Kinematic viscosity at 60°C, cSt: 37.17
Kinematic viscosity at 100°C, cSt: 9.539
Viscosity index: 54
Pour point, °C: −18

*Extrapolated
+Sample contained sediment

Crude 2 — Pennington, Nigeria

CALM with export from floating storage in offshore producing area

Crude
Gravity, °API: 36.6
Sulfur, wt%: 0.07
Viscosity cSt at 104°F: 3.06
Pour point, °F: 43
C$_4$ & lighter, vol%: 1.5
RVP at 100°F: psi: 5.1
Con. carbon residue, wt%: 0.7
V/Ni, ppm: 2/1

Light naphtha
Range, °F: C$_5$–200
Yield, vol%: 7.4
Sulfur, wt%: 0.001
Paraffins, vol%: 70.5
Naphthenes, vol%: 25.5
Aromatics, vol%: 3.5
RON, clear: 74.9

Heavy naphtha
Range, °F: 200–300
Yield, vol%: 14.7
Sulfur, wt%: 0.001
P/N/A, vol%: 35.0/52.5/12.0
RON, clear: 55.5
Aniline point, °F: 118

Kerosene
Range, °F: 340–470
Yield, vol%: 18.0
Gravity, °API: 40.6
Sulfur, wt%: 0.02
Viscosity cSt at 104°F: 1.45
Smoke point, mm: 19.5
Freeze point, °F: −69
Aniline point, °F: 135
Cloud point, °F < −76

Gas oil
Range, °F: 470–650
Yield, vol%: 31.7
Gravity, °API: 33.7
Sulfur, wt%: 0.05
Viscosity cSt at 104°F: 3.72
Pour point, °F: −10
Aniline point, °F: 158
Cloud point, °F: 6.0

Vacuum gas oil
Range, °F: 650–1000
Yield vol%: 23.4
Gravity, °API: 23.5
Sulfur, wt%: 0.15
Viscosity cSt at 122°F: 28.9
Pour point, °F: 102
Aniline point, °F: 184
V/Ni, ppm: 2/<1

Residue
Range, °F: 650+
Yield, vol%: 26.7
Gravity, °API: 21.8
Sulfur, wt%: 0.17
Viscosity cSt at 122°F: 48.5
Pour point, °F: 91
Aniline point, °F: 179
Con. Carbon residue, wt%: 2.3

TABLE 4.3—(Continued)

Crude 3
Arabian Light, Saudi Arabia

Ras Tanura and Yanbu

Whole crude
SG, 60/60: 0.8581 (33.4 °API gravity)
Sulfur, wt%: 1.77
Viscosity, cSt at 60°F: 12.19
Neut. no. mg/g: 0.00
Nitrogen, wt%: 0.09
Pour point, °F: −65

Range, °F: 68–347
Yield, vol%: 22.73
RON, clear: 48.7
Sulfur, wt%: 0.0370

Range, °F: 347–563
Yield, vol%: 22.72
SG, 60/60: 0.8131
Sulfur, wt%: 0.440
Aniline point, °F: 146.0
Viscosity, cSt at 100°F: 1.83
Freeze point, °F: −21
Pour point, °F: −24
Smoke point, mm: 21.3

Range, °F: 563–650
Yield, vol%: 8.32
SG, 60/60: 0.8599
Sulfur wt%: 1.56
Aniline point, °F: 165.4
Viscosity, cSt at 100°F: 5.57
Freeze point, °F: 39
Pour point, °F: 36

Range, °F: 650–1049
Yield, vol%: 29.74
SG, 60/60: 0.9231
Sulfur, wt%: 2.46
Nitrogen, wt%: 0.080
V/Ni, ppm: 0.07/0.05

Range, °F: 650–1500
Yield, vol%: 44.71
SG, 60/60: 0.9561
Sulfur, wt%: 3.07
Viscosity, cSt at 140°F: 102.8
Pour point, °F: 48
V/Ni, ppm: 27.1/6.7
Iron, ppm: 5.5

Range, °F: 761–1500
Yield, vol%: 35.50
SG, 60/60: 0.9714
Abs. Viscosity, poise at 60°C: 2.79
Viscosity, cSt at 212°F: 45.93

Range, °F: 878–1500
Yield, vol%: 26.80
SG, 60/60: 0.9889
Abs. Viscosity, poise at 60°C: 11.37
Viscosity, cSt at 212°F: 117.9

Range, °F: 1049–1500
Yield, vol%: 14.97
SG, 60/60: 1.0217
Abs. viscosity, poise at 60°C: 330
Viscosity, cSt at 212°F: 1,222
Mina Al Ahmadi

Crude 4
Kuwait Export, Kuwait

Crude
Gravity, °API: 31.4
Viscosity, SUV, sec at 100°F: 58.1
Pour point, °F: +5
Sulfur, wt%: 2.52
Ni/V (calc.), ppm: 8/30
RVP, psi: 6.7
H₂S, wt%: < 0.001
C1–C4 (GC analysis), vol%: 3.03

Naphtha
Range, °F: Over point–140
Yield, vol%: 3.80
Sulfur, wt%: 0.04
RON (micro), clear: 71.8
RON + 3 mL TEL: 89.3

Naphtha
Range, °F: 140–170
Yield, vol%: 2.26
Gravity, °API: 78.6
Sulfur, wt%: 0.02
Aniline point, °F: 147.7
RON, clear/+3 mL TEL: 58.2/80.40
Paraffins, vol%: 88.5
Monocycloparaffins, vol%: 9.9
Aromatics, total, vol%: 1.6

Heavy Naphtha
Range, °F: 170–310
Yield, vol%: 11.91
Gravity, °API: 62.1
Sulfur, wt%: 0.03
Aniline point, °F: 136.2
Paraffins vol%: 70.1
Monocycloparaffins, vol%: 20.4
Dicycloparaffins, vol%: 0.4
Aromatics, total, vol%: 9.1

Kerosine
Range, °F: 310–520
Yield, vol%: 18.43
Gravity, °API: 45.9
Pour point, °F: −45
Sulfur, wt%: 0.39
Aniline point, °F: 142.9
Cetane index: 53.8
Smoke point, mm: 26

Heating oil
Range, °F: 520–680
Yield, vol%: 14.89
Gravity, °API: 32.9
Viscosity, SUV, sec. at 100°F: 43.3
Pour point, °F: +20
Total N, ppm: 100
Sulfur, wt%: 1.98
Aniline point, °F: 160.2
Cetane index: 54.3

Gas oil
Range, °F: 680–1000
Yield, vol%: 24.09
Gravity, °API: 21.4
Pour point, °F: +100
Sulfur, wt%: 3.12
Total N, ppm: 900
Ni/V, ppm: <0.1/<0.1
Viscosity, SUV, sec at 130°F: 175.2
Rams. carbon residue, wt%: 0.47

Residuum
Range (VEFV), °F: 1000+
Yield, vol%: 21.78
Gravity, °API: 5.8
Sulfur, wt%: 5.11
V/Ni, ppm: 116/34
Viscosity, SUV, sec at 250°F: 3394
Rams. carbon residue, wt%: 21.9

Crude 5
Fereidoon, Iran

Kharg Island

Crude
Gravity, °API: 31.3
Sulfur, wt%: 2.50
RVP, psi: 5.7
H₂S (evolved), wt%: 0.015
Pour point, °F: −35
Neut. no., mg KOH/g: 0.70
Rams. carbon, wt%: 5.4
Kinematic viscosity, cSt at
68°F: 15.33
V/Ni, ppm: 36/11
C1–C5, vol%: 6.35

Naphtha
Range, °F: C5–200
Yield, vol%: 9.9
Gravity, °API: 82.6
Sulfur, wt%: 0.017
Charac. Factor: 12.72
Paraffins, vol%: 76.8
Naphthenes, vol%: 21.8
Aromatics, vol%: 1.4
RON/MON, clear: 64.3/62.6

Naphtha
Range, °F: 200–320
Yield, vol%: 10.5
Gravity, °API: 58.1
Sulfur, wt%: 0.056
Charac. Factor: 12.02
P/N/A, vol%: 71.6/17.2/11.3
RON/MON, clear: 30.2/32.0

Kerosine
Range, °F: 320–520
Yield, vol%: 17.7
Gravity, °API: 44.3
Sulfur, wt%: 0.435
Charac. factor: 11.84
Aniline point, °F: 145.2

Diesel index, °F: 64.3
Cetane index: 45.4
Freeze point, °F: −41
Cloud point, °F: −52
Pour point, °F: −52
Smoke point, mm: 22
Viscosity, SUS at 68°F: 32.4

Gas Oil
Range, °F: 520–650
Yield, vol%: 10.8
Gravity, °API: 33.4
Sulfur, wt%: 1.56
Charac. factor: 11.74
Aniline point, °F: 161.2
Diesel index: 53.8
Cetane index: 49.4
Cloud point, °F: 20
Pour point, °F: 19
Viscosity, SUS at 68°F: 48.0

Residuum
Range, °F: 650+
Yield, vol%: 48.8
Gravity, °API: 13.6
Sulfur, wt%: 4.13
Pour point, °F: 34
Rams. Carbon, wt%: 9.86
Viscosity, SUS at 122°F: 3616.0
V/Ni, ppm: 66/20

Residuum
Range, °F: 1050+
Yield, vol%: 20.4
Gravity, °API: 3.8
Sulfur, wt%: 5.6
Pour point, °F: 158
V/Ni, ppm: 143/44

TABLE 4.3—(Continued)

Crude 6	Crude 7
Hondo Monterey, California, USA	**Boscan, Venezuela**

OST, Production Area	
Whole crude	**Bajo Grande**
Gravity, °API: 19.4	**Lake Maracaibo**
SG, 60/60: 0.9377	
Sulfur, wt%: 4.70	**Crude**
Viscosity, cSt at 60°F: 1034	Gravity, °API: 10.10
Neut. no., mg/g: 0.43	Sulfur, wt%: 5.5
Nitrogen, wt%: 0.65	Viscosity, SUS at 100°F: 90000
Pour point, °F: -10	Pour point, °F: +50
Light ends streams (C$_1$-C$_5$)	**Light naphtha**
Yield, vol%: 0.62	Range, °F: 82-200
	Yield, vol%: 0.55
Range, °F: 68-347	
Yield, vol%: 15.12	**Heavy Naphtha**
RON, clear: 59.4	Range, °F: 200-300
Sulfur, wt%: 0.66	Yield, vol%: 1.32
	Sulfur, wt%: 0.68
Range, °F: 347-563	Paraffins, %: 53.28
Yield, vol%: 14.99	
SG, 60/60: 0.8494	**Naphtha**
Sulfur, wt%: 3.14	Range, °F: 300-350
Aniline point, °F: 120.0	Yield, vol%: 0.88
Viscosity, cSt at 100°F: 2.26	Sulfur, wt%: 1.28
Freeze point, °F: -30	Paraffins, %: 34.9
Pour point, °F: -38	
Smoke point, mm: 18.2	**Kerosine**
	Range, °F: 350-400
Range, °F: 563-650	Yield, vol%: 1.28
Yield, vol%: 6.16	Sulfur, wt%: 1.88
SG, 60/60: 0.8911	Aromatics, %: 24.04
Sulfur, wt%: 3.89	Pour point, °F: -141.0
Aniline point, °F: 136.8	Smoke point, mm: 16.0
Viscosity, cSt at 100°F: 7.83	Aniline point, °F: 112.69
Freeze point, °F: 37*	
Pour point, °F: 29	**Gas Oil**
	Range, °F: 400-500
Range, °F: 650-1500	Yield, vol%: 3.96
Yield, vol%: 63.70	Gravity, °API: 34.0
SG, 60/60: 1.0082	Sulfur, wt%: 3.14
Sulfur, wt%: 6.22	Kinematic viscosity at 122°F: 1.81
Viscosity, cSt at 140°F: 17121	Pour point, °F: -77.53
Pour point, °F: 87	Smoke point, °F: 16.11
V/Ni, ppm: 440/210	Aniline point, °F: 113.83
Iron, ppm: 48.0	Cetane index: 37.24
Range, °F: 761-1500	**Gas Oil**
Yield, vol%: 55.92	Range, °F: 500-550
SG, 60/60: 1.0198	Yield, vol%: 2.66
Penetration at 25°C, mm: 238.2*	Gravity, °API: 28.62
Abs. Viscosity, poise at 60°C: 743	Sulfur, wt%: 4.02
Viscosity, cSt at 212°F: 2028	
	Gas Oil
Range, °F: 878-1500	Range, °F: 550-650
Yield, vol%: 48.29	Yield, vol%: 6.46
SG, 60/60: 1.0317	Gravity, °API: 25.54
Penetration at 25°C, mm: 35.6	Sulfur, wt%: 4.47
Abs. Viscosity, poise at 60°C: 8029	Pour point, °F: -10.9
Viscosity, cSt at 212°F: 10145	Cetane no.: 38.0
	Aniline point, °F: 117.3
Range, °F: 1049-1500	Cetane index: 42.27
Yield, vol%: 38.68	
SG, 60/60: 1.0489	**Residue**
Penetration at 25°C, mm: 3.2*	Range, °F: 650+
Abs. Viscosity, poise at 60°C: 650524	Yield, vol%: 82.9
Viscosity, cSt at 212°F: 190077	Sulfur, wt%: 5.86
	Gravity, °API: 6.10
* Extrapolated values.	Sulfur, wt%: 5.86
	Ni/V, ppm: 175/1407
	Kinematic Viscosity at 122°F: 22,502

4.2 GENERALIZED CORRELATIONS FOR PSEUDOCRITICAL PROPERTIES OF NATURAL GASES AND GAS CONDENSATE SYSTEMS

Natural gas is a mixture of light hydrocarbon gases rich in methane. Methane content of natural gases is usually above 75% with C_{7+} fraction less than 1%. If mole fraction of H_2S in a natural gas is less than 4×10^{-6} (4 ppm on gas volume basis) it is called "sweet" gas (Section 1.7.15). A sample composition of a natural gas is given in Table 1.2. Dry gases contain no C_{7+} and have more than 90 mol% methane. The main difference between natural gas and other reservoir fluids is that the amount of C_{7+} or even C_{6+} in the mixture is quite low and the main components are light paraffinic hydrocarbons. Properties of pure light hydrocarbons are given in Tables 2.1 and 2.2. The C_{6+} or C_{7+} fraction of a mixture should be treated as an undefined fraction and its properties may be determined from the correlations given in Chapter 2. If the detailed composition of a natural gas is known the best method of characterization is through Eq. (3.44) with composition in terms of mole fraction (x_i) for calculation of pseudocritical properties, acentric factor, and molecular weight of the mixture. Although the Key's mixing rule is not the most accurate mixing rule for pseudocritical properties of mixtures, but for natural gas systems that mainly contain methane it can be used with reasonable accuracy. More advanced mixing rules are discussed in Chapter 5. Once the basic characterization parameters for the mixture are determined various physical properties can be estimated from appropriate methods.

The second approach is to consider the mixture as a single pseudocomponent with known specific gravity. This method is particularly useful when the exact composition of the mixture is not known. There are a number of empirical correlations in the literature to estimate basic properties of natural gases from their specific gravity. Some of these methods are summarized below.

In cases that the composition of a natural gas is unknown Brown presented a simple graphical method to estimate pseudocritical temperature and pressure from gas specific gravity (SG_g) as shown by Ahmed [10]. Standing [11] converted the graphical methods into the following correlations for estimation of T_{pc} and P_{pc} of natural gases free of CO_2 and H_2S:

$$(4.1) \qquad T_{pc} = 93.3 + 180.6SG_g - 6.94SG_g^2$$

$$(4.2) \qquad P_{pc} = 46.66 + 1.03SG_g - 2.58SG_g^2$$

where T_{pc} and P_{pc} are the pseudocritical temperature and pressure in kelvin and bar, respectively. SG_g is defined in Eq. (2.6). This method is particularly useful when the exact composition of the mixture is not available. This method provides acceptable results since nearly 90% of the mixture is methane and the mixture is close to a pure component. Therefore, assumption of a single pseudocomponent is quite reasonable without significant difference with detailed compositional analysis. Application of these equations to wet gases is less accurate.

Another type of reservoir fluids that are in gaseous phase under reservoir conditions are gas condensate systems. Composition of a gas condensate sample is given in Table 1.2. Its C_{7+} content is more than that of natural gases and it is about few percent, while its methane content is less than that of

natural gas. However, for gas condensate systems simple correlations in terms of specific gravity have been proposed in the following forms similar to the above correlations and are usually used by reservoir engineers [10]:

$$(4.3) \qquad T_{pc} = 103.9 + 183.3SG_g - 39.7SG_g^2$$

$$(4.4) \qquad P_{pc} = 48.66 + 3.56SG_g - 0.77SG_g^2$$

These equations give higher critical temperature and lower critical pressure than do the equations for natural gases since gas condensate samples contain heavier compounds. Because of the greater variation in carbon number, the equations for gas condensate systems are much less accurate than those for natural gas systems. For this reason properties of gas condensate systems may be estimated more accurately from the distribution models presented in Section 4.5.

Equations (4.1)–(4.4) proposed for pseudocritical properties of natural gas and gas condensate systems are based on the assumption that mixtures contain only hydrocarbon compounds. However, these reservoir fluids generally contain components such as carbon dioxide (CO_2), hydrogen sulfide (H_2S), or nitrogen (N_2). Presence of such compounds affects the properties of the gas mixture. For such cases, corrections are added to the calculated pseudocritical properties from Eqs. (4.1)–(4.4). Corrections proposed by Wichert and Aziz [12] and Carr et al. [13] are recommended for the effects of nonhydrocarbons on properties of natural gases [8]. The method of Carr et al. for adjustment of calculated T_{pc} and P_{pc} is given as follows:

$$(4.5) \quad T_{pc}^c = T_{pc} - 44.44y_{CO_2} + 72.22y_{H_2S} - 138.89y_{N_2}$$

$$(4.6) \quad P_{pc}^c = P_{pc} + 30.3369y_{CO_2} + 41.368y_{H_2S} - 11.721y_{N_2}$$

where T_{pc}^c and P_{pc}^c are the adjusted (corrected) pseudocritical temperature and pressure in kelvin and bar, respectively. y_{CO_2}, y_{H_2S} and y_{N_2} are the mole fractions of CO_2, H_2S, and N_2, respectively. T_{pc} and P_{pc} are unadjusted pseudocritical temperature and pressure in kelvin and bar, respectively. These unadjusted properties may be calculated from Eqs. (4.1) and (4.2) for a natural gas. The following example shows calculation of pseudocritical properties for a natural gas sample.

Example 4.1—A natural gas has the following composition in mol%: H_2S 1.2%, N_2 0.2%; CO_2 1%, C_1 90%, C_2 4.8%, C_3 1.7%, iC_4 0.4, nC_4 0.5%, iC_5 0.1, nC_5 0.1%.

a. Calculate T_c, P_c, ω, and M using properties of pure compounds.
b. Calculate the gas specific gravity.
c. Calculate T_c and P_c using Eqs. (4.1) and (4.2) and SG calculated from Part (b).
d. Adjust T_c and P_c for the effects of nonhydrocarbon compounds present in the gas.

Solution—Values of M, T_c, P_c, and ω for pure components present in the gas mixture can be obtained from Table 2.1. These values as well as calculated values of M, T_c, P_c, and ω for the mixture based on Eq. (3.44) are given in Table 4.4. The calculated values of M, T_c, P_c, and ω as shown in Table 4.4 are: $M = 18.17$, $T_{pc} = -68.24°C$, $P_{pc} = 46.74$ bar, and $\omega = 0.0234$. This method should be used for gases with $SG_g > 0.75$ [10].

a. Equation (2.6) can be used to calculate gas specific gravity: $SG_g = 18.17/28.96 = 0.6274$.

4. CHARACTERIZATION OF RESERVOIR FLUIDS AND CRUDE OILS

TABLE 4.4—*Calculation of pseudoproperties of the natural gas system of Example 4.1.*

No.	Component	x_i	M_i	T_{ci}, °C	P_{ci}, bar	ω_i	$x_i \times M_i$	$x_i \times T_{ci}$	$x_i \times P_{ci}$	$x_i \times \omega_i$
1	H_2S	0.012	34.1	100.38	89.63	0.0942	0.41	1.20	1.08	0.0011
2	N_2	0.002	28.0	−146.95	34.00	0.0377	0.06	−0.29	0.07	0.0001
3	CO_2	0.01	44.0	31.06	73.83	0.2236	0.44	0.31	0.74	0.0022
4	C_1	0.90	16.0	−82.59	45.99	0.0115	14.40	−74.33	41.39	0.0104
5	C_2	0.048	30.1	32.17	48.72	0.0995	1.44	1.54	2.34	0.0048
6	C_3	0.017	44.1	96.68	42.48	0.1523	0.75	1.64	0.72	0.0026
7	iC_4	0.004	58.1	134.99	36.48	0.1808	0.23	0.54	0.15	0.0007
8	nC_4	0.005	58.1	151.97	37.96	0.2002	0.29	0.76	0.19	0.0010
9	iC_5	0.001	72.2	187.28	33.81	0.2275	0.07	0.19	0.03	0.0002
10	nC_5	0.001	72.2	196.55	33.70	0.2515	0.07	0.20	0.03	0.0003
Sum	Mixture	1.00					18.17	−68.24	46.74	0.0234

b. The system is a natural gas so Eqs. (4.1) and (4.2) are used to estimate pseudocritical properties using SG_g as the only available input data.

$$T_{pc} = 93.3 + 180.6 \times 0.6274 - 6.9 \times 0.6274^2$$
$$= 203.9 \, K = -69.26°C$$
$$P_{pc} = 46.66 + 1.03 \times 0.6274 - 2.58 \times 0.6274^2 = 46.46 \, bar.$$

c. To calculate the effects of nonhydrocarbons present in the system Eqs. (4.5) and (4.6) are used to calculate adjusted pseudocritical properties. These equations require mole fractions of H_2S, CO_2, and N_2, which are given in Table 4.4 as: $y_{H_2S} = 0.012$, $y_{CO_2} = 0.01$, and $y_{N_2} = 0.002$. Unadjusted T_{pc} and P_{pc} are given in Part c as $T_{pc} = -69.26°C$ and $P_{pc} = -46.46$ bar.

$$T_{pc}^c = -69.26 - 44.44 \times 0.01 + 72.22 \times 0.012$$
$$- 138.89 \times 0.002 = -69.12°C.$$
$$P_{pc}^c = 46.46 + 30.3369 \times 0.01 + 41.368 \times 0.012$$
$$- 11.721 \times 0.002 = 47.24 \, bar.$$

Although use of Eqs. (4.1) and (4.2) together with Eqs. (4.5) and (4.6) gives reliable results for natural gas systems, use of Eqs. (4.3) and (4.4) for gas condensate systems or gases that contain C_{7+} fractions is not reliable. For such systems properties of C_{7+} fractions should be estimated according to the methods discussed later in this chapter. ◆

4.3 CHARACTERIZATION AND PROPERTIES OF SINGLE CARBON NUMBER GROUPS

As shown in Tables 4.1 and 4.2, compositional data on reservoir fluids and crude oils are generally expressed in terms of mol (or wt) percent of pure components (up to C_5) and SCN groups for hexanes and heavier compounds (C_6, C_7, C_8, ..., C_{N+}), where N is the carbon number of plus fraction. In Table 4.1, N is 10 and for the crude oil of Table 4.2, N is 40. Properties of a crude oil or a reservoir fluid can be accurately estimated through knowledge of accurate properties of individual components in the mixture. Properties of pure components up to C_5 can be taken from Tables 2.1 and 2.2. By analyzing the physical properties of some 26 condensates and crude oils, Katz and Firoozabadi [14] have reported boiling point, specific gravity, and molecular weight of SCN groups from C_6 up to C_{45}. Later Whitson [15] indicated that there is inconsistency for the properties of SCN groups reported by Katz and Firoozabadi for $N_C > 22$. Whitson modified properties of SCN groups and reported values of T_b, SG, M, T_c, P_c, and

ω for SCN groups from C_6 to C_{45}. Whitson used Riazi–Daubert correlations (Eqs. (2.38), (2.50), (2.63), and (2.64)) to generate critical properties and molecular weight from T_b and SG. He also used Edmister method (Eq. (2.108)) to generate values of acentric factor. As discussed in Chapter 2 (see Section 2.10), these are not the best methods for calculation of properties of hydrocarbons heavier than C_{22} ($M > 300$). However, physical properties reported by Katz and Firoozabadi have been used in reservoir engineering calculations and based on their tabulated data, analytical correlations for calculation of M, T_c, P_c, V_c, T_b, and SG of SCN groups from C_6 to C_{45} in terms of N_C have been developed [10].

Riazi and Alsahhaf [16] reported a new set of data on properties of SCN from C_6 to C_{50}. They used boiling point and specific gravity data for SCN groups proposed by Katz and Firoozabadi from C_6 to C_{22} to estimate PNA composition of each group using the methods discussed in Section 3.5.1.2. Then Eq. (2.42) was used to generate physical properties of paraffinic, naphthenic, and aromatic groups. Properties of SCN from C_6 to C_{22} have been calculated through Eq. (3.39) using the pseudocomponent approach. These data have been used to obtain coefficients of Eq. (2.42) for properties of SCN groups. The pseudocomponent method produced boiling points for SCN groups from C_6 to C_{22} nearly identical to those reported by Katz and Firoozabadi [14]. Development of Eq. (2.42) was discussed in Chapter 2 and it is given as

$$(4.7) \qquad \theta = \theta_\infty - \exp(a - bM^c)$$

where θ is value of a physical property and θ_∞ is value of θ as $M \to \infty$. Coefficients θ_∞, a, b, and c are specific for each property. Equation (4.7) can also be expressed in terms of carbon number, N_C. Values of T_b and SG from C_6 to C_{22} have been correlated in terms of N_C as

$$(4.8) \qquad T_b = 1090 - \exp(6.9955 - 0.11193 N_C^{2/3})$$

$$(4.9) \qquad SG = 1.07 - \exp(3.65097 - 3.8864 N_C^{0.1})$$

where T_b is in kelvin. Equation (4.8) reproduces Katz–Firoozabadi data from C_6 to C_{22} with an AAD of 0.2% (~1 K) and Eq. (4.9) reproduces the original data with AAD of 0.1%. These equations were used to generate T_b and SG for SCN groups heavier than C_{22}. Physical properties from C_6 to C_{22} were used to obtain the coefficients of Eq. (4.7) for SCN groups. In doing so the internal consistency between T_c and P_c were observed so that as $T_b = T_c$, P_c becomes 1 atm (1.013 bar). This occurs for the SCN group of C_{99} (~$M = 1382$). Coefficients of Eq. (4.7) for $N_C \geq C_{10}$ are given in Table 4.5, which may be used well beyond C_{50}. Data on solubility parameter of SCN groups reported by Won were used to obtain the coefficients for this property [6]. Values of physical properties of SCN groups for C_6 to C_{50} are tabulated in Table 4.6 [16]. Values of P_c calculated from Eq. (4.7) are lower than those reported in other sources [10, 17, 18]. As discussed in Chapter 2, the Lee–Kesler correlations are suitable for prediction of critical properties of heavy hydrocarbons. For this reason T_c and P_c are calculated from Eqs. (2.69)–(2.70) while d_c is calculated through Eq. (2.98) and values. For heavy hydrocarbons Eq. (2.105) is used to estimate ω. Reported values of ω for heavy SCN are lower than those estimated through Eq. (4.7). As discussed in Chapter 2 and recommended by Pan et al. [7], for hydrocarbons with $M > 800$ it is suggested that $\omega = 2.0$. Other properties are calculated through Eq. (4.7) or taken from Ref. [16].

TABLE 4.5—*Coefficients of Eq. (4.7) for physical properties of SCN groups ($\geq C_{10}$) in reservoir fluids and crude oils.*

θ	Constants in Eq. (4.7)				%AAD
	θ_∞	A	b	c	
T_b	1080	6.97996	0.01964	2/3	0.4
SG	1.07	3.56073	2.93886	0.1	0.07
d_{20}	1.05	3.80258	3.12287	0.1	0.1
I	0.34	2.30884	2.96508	0.1	0.1
$T_{br} = T_b/T_c$	1.2	−0.34742	0.02327	0.55	0.15
$-P_c$	0	6.34492	0.7239	0.3	1.0
$-d_c$	−0.22	−3.2201	0.0009	1.0	0.05
$-\omega$	0.3	−6.252	−3.64457	0.1	1.4
σ	30.3	17.45018	9.70188	0.1	1.0
δ	8.6	2.29195	0.54907	0.3	0.1

Taken with permission from Ref. [16].
Units: T_b, T_c in K; P_c in bar; d_{20} and d_c in g/cm^3; σ in dyne/cm; δ in (cal/cm^3)$^{1/2}$.

TABLE 4.6—*Physical properties of SCN groups.*

Carbon number	M	T_b	SG	n_{20}	d_{20}	T_c	P_c	d_c	Z_c	ω	σ	δ
6	84	337	0.690	1.395	0.686	510.3	34.4	0.241	0.275	0.255	18.6	7.25
7	95	365	0.727	1.407	0.723	542.6	31.6	0.245	0.272	0.303	21.2	7.41
8	107	390	0.749	1.417	0.743	570.2	29.3	0.246	0.269	0.346	23.0	7.53
9	121	416	0.768	1.426	0.762	599.0	26.9	0.247	0.265	0.394	24.4	7.63
10	136	440	0.782	1.435	0.777	623.7	25.0	0.251	0.261	0.444	25.4	7.71
11	149	461	0.793	1.442	0.790	645.1	23.5	0.254	0.257	0.486	26.0	7.78
12	163	482	0.804	1.448	0.802	665.5	21.9	0.256	0.253	0.530	26.6	7.83
13	176	500	0.815	1.453	0.812	683.7	20.6	0.257	0.249	0.570	27.0	7.88
14	191	520	0.826	1.458	0.822	700.9	19.6	0.262	0.245	0.614	27.5	7.92
15	207	539	0.836	1.464	0.831	716.5	18.5	0.267	0.241	0.661	27.8	7.96
16	221	556	0.843	1.468	0.839	732.1	17.6	0.269	0.237	0.701	28.1	7.99
17	237	573	0.851	1.472	0.847	745.6	16.7	0.274	0.233	0.746	28.3	8.02
18	249	586	0.856	1.475	0.852	758.8	15.9	0.274	0.229	0.779	28.5	8.05
19	261	598	0.861	1.478	0.857	771.1	15.2	0.275	0.226	0.812	28.6	8.07
20	275	611	0.866	1.481	0.862	782.7	14.7	0.278	0.222	0.849	28.8	8.09
21	289	624	0.871	1.484	0.867	793.8	14.0	0.281	0.219	0.880	28.9	8.11
22	303	637	0.876	1.486	0.872	804.9	13.5	0.283	0.215	0.914	29.0	8.13
23	317	648	0.881	1.489	0.877	814.2	13.0	0.287	0.212	0.944	29.1	8.15
24	331	660	0.885	1.491	0.880	824.1	12.5	0.289	0.209	0.977	29.2	8.17
25	345	671	0.888	1.493	0.884	833.3	12.0	0.291	0.206	1.007	29.3	8.18
26	359	681	0.892	1.495	0.888	841.7	11.7	0.295	0.203	1.034	29.3	8.20
27	373	691	0.896	1.497	0.891	850.2	11.3	0.298	0.200	1.061	29.4	8.21
28	387	701	0.899	1.499	0.895	858.2	10.9	0.301	0.197	1.091	29.4	8.22
29	400	710	0.902	1.501	0.898	865.5	10.6	0.303	0.194	1.116	29.5	8.24
30	415	720	0.905	1.503	0.901	873.5	10.2	0.306	0.191	1.146	29.5	8.25
31	429	728	0.909	1.504	0.904	880.1	10.0	0.310	0.189	1.169	29.6	8.26
32	443	737	0.912	1.506	0.906	887.4	9.7	0.312	0.187	1.195	29.6	8.27
33	457	745	0.915	1.507	0.909	894.0	9.5	0.316	0.184	1.218	29.7	8.28
34	471	753	0.917	1.509	0.912	900.2	9.2	0.319	0.182	1.244	29.7	8.29
35	485	760	0.920	1.510	0.914	906.1	9.0	0.323	0.180	1.263	29.7	8.30
36	499	768	0.922	1.511	0.916	912.2	8.8	0.325	0.177	1.289	29.8	8.31
37	513	775	0.925	1.512	0.918	917.7	8.6	0.328	0.175	1.311	29.8	8.32
38	528	782	0.927	1.514	0.920	923.1	8.3	0.332	0.173	1.333	29.8	8.33
39	542	789	0.929	1.515	0.922	928.6	8.2	0.335	0.171	1.355	29.8	8.34
40	556	795	0.931	1.516	0.924	933.4	8.0	0.338	0.169	1.374	29.9	8.35
41	570	802	0.933	1.517	0.926	938.8	7.8	0.341	0.167	1.396	29.9	8.35
42	584	808	0.934	1.518	0.928	943.6	7.7	0.344	0.165	1.415	29.9	8.36
43	599	814	0.936	1.519	0.930	948.4	7.5	0.348	0.164	1.434	29.9	8.36
44	614	820	0.938	1.520	0.932	952.5	7.4	0.353	0.163	1.448	29.9	8.37
45	629	826	0.940	1.521	0.933	956.9	7.2	0.356	0.160	1.470	29.9	8.38
46	641	831	0.941	1.522	0.935	961.6	7.1	0.358	0.159	1.489	30.0	8.38
47	656	836	0.943	1.523	0.936	965.7	7.0	0.362	0.158	1.504	30.0	8.39
48	670	841	0.944	1.524	0.938	969.4	6.9	0.366	0.156	1.522	30.0	8.39
49	684	846	0.946	1.524	0.939	973.5	6.8	0.369	0.155	1.537	30.0	8.40
50	698	851	0.947	1.525	0.940	977.2	6.6	0.372	0.153	1.555	30.0	8.40

T_c and P_c are calculated from Lee–Kesler correlations (Eqs. 2.69–2.70); d_c is calculated from Eq. (2.98) and Z_c from Eq. (2.8). For $N_C > 20$, ω is calculated from Lee–Kesler, Eq. (2.105). All other properties are taken from Ref. [16] or calculated through Eq. (4.7) for $N_C > 10$. Units: T_b, T_c in K; P_c in bar; d_{20} and d_c in g/cm^3; σ in dyne/cm; δ in (cal/cm^3)$^{1/2}$.
Taken with permission from Ref. [16].

Values of critical properties and acentric factor for SCN groups greater than C_{30} estimated by different methods vary significantly especially for higher SCN groups. Reliability of these values is subject to further research and no concrete recommendation is given in the literature.

Values of M given in Table 4.6 are more consistent with values of M recommended by Pedersen et al. [6] than those suggested by Whitson [15] as discussed by Riazi et al. [16]. Molecular weights of SCN groups recommended by Whitson [15] are based on the Katz–Firoozabadi values. However, more recently Whitson [17] recommends values of M for SCN groups, which are very close to those suggested by Riazi [16]. Although he refers to Katz–Firrozabadi molecular weights, his reported values are much higher than those given in Ref. [14]. In addition the Watson characterization factor (K_W) reported in Ref. [15] from C_6 to C_{45} should be considered with caution as they are nearly constant at 12 while toward heavier hydrocarbon factions as the amount of aromatics increases, the K_W values should decrease. Values of properties of SCN groups given in Table 4.6 or those given in Ref. [15] are approximate as these properties may vary for one fluid mixture to another. As shown in Table 4.1, values of M for the SCN of C_7 in four different reservoir fluids vary from 92 to 100. In Table 4.2, molecular weight of each SCN fraction is determined by adding 14 to molecular weight of preceding SCN group [7]. Thus M for C_6 is specified as 86, which is determined by adding 14 to molecular weight nC_5 that is 72. The Pedersen molecular weight of SCN groups is given in terms of N_C by the following relation [6]:

$$(4.10) \qquad M = 14N_C - 4$$

In very few references the value of 2 is used instead of 4 in Eq. (4.10). To obtain properties of heavier SCN groups ($N_C > 50$), Eq. (4.7) should be used with coefficients given in Table 4.5. However, to use this equation, it is necessary to calculate the molecular weight from boiling point through reversed form of Eq. (4.7). T_b is calculated by Eq. (4.8) using the carbon number. The calculation method is demonstrated in Example 4.2.

Example 4.2—Calculate T_b, SG, d_{20}, n_{20}, T_c, P_c, V_c, σ, and δ for C_{60}, C_{70}, and C_{80} SCN groups.

Solution—The only data needed for calculation of physical properties of SCN groups is the carbon number. For $N_C = 60$, from Eq. (4.8): $T_b = 1090 - \exp(6.9955 - 0.11193 \times 60^{2/3}) = 894$ K. Equation (4.7) in a reversed form can be used to estimate M from T_b with coefficients given in Table 4.5: $M = \left[\frac{1}{0.01964}(6.97996 - \ln(1080 - T_b))\right]^{3/2}$. For $T_b = 894$ K we get $M = 844$. This value of M should be used to calculate other properties. For example SG is calculated as: $SG = 1.07 - \exp(3.56073 - 2.93886 \times 844^{0.1}) = 0.96$. Similarly other properties can be estimated and the results are given in Table 4.7. Values of T_c, P_c, and ω are calculated from Lee–Kesler

correlations and d_c is calculated from Eq. (2.98). Actual values of ω are probably greater than those given in this table. Values of Z_c calculated from its definition by Eq. (2.8) for C_{60}, C_{70}, and C_{80} are 0.141, 0.132, and 0.125, respectively. ♦

4.4 CHARACTERIZATION APPROACHES FOR C_{7+} FRACTIONS

In description of composition of a reservoir fluid, C_6 is a very narrow boiling range fraction and characterization methods discussed in Chapter 2 (i.e., Eq. 2.38 [17]) can be used to estimate various properties of this group. Contrary to C_6 group, the C_{7+} fraction has a very wide boiling range especially for crude oils. Therefore, methods of Chapter 2 or 3 cannot be directly applied to a C_{7+} fraction. However, for a natural gas that its C_{7+} fraction has a narrow boiling range and the amount of C_{7+} is quite small, equations such as Eq. (2.40) in terms of M and SG may be used to estimate various properties. In some references there are specific correlations for properties of C_{7+} fractions. For example, Pedersen et al. [6] suggested use of the following relation for calculation of critical volume of C_{7+} fractions in terms of M_{7+} and SG_{7+}:

$$V_{c7+} = 0.3456 + 2.4224 \times 10^{-4}M_{7+} - 0.443SG_{7+}$$
$$(4.11) \qquad + 1.131 \times 10^{-3}M_{7+}SG_{7+}$$

where V_{c7+} is the critical volume of C_{7+} fraction in cm³/gmol. Standing represented the graphical correlation of Katz for the pseudocritical temperature and pressure of C_{7+} fractions into the following analytical correlations [18]:

$$T_{c7+} = 338 + 202 \times \log(M_{7+} - 71.2)$$
$$(4.12) \qquad + (1361 \times \log M_{7+} - 2111)\log SG_{7+}$$

$$P_{c7+} = 81.91 - 29.7 \times \log(M_{7+} - 61.1) + (SG_{7+} - 0.8)$$
$$(4.13) \qquad \times [159.9 - 58.7 \times \log(M_{7+} - 53.7)]$$

where T_{c7+} and P_{c7+} are in K and bar, respectively. The original development of these correlations goes back to the early 1940s and there is no information on the reliability of these equations. Use of such relations to a C_{7+} fraction as a single pseudocomponent leads to serious errors especially for mixtures with considerable amount of C_{7+}. Properties of C_{7+} fractions have significant effects on estimated properties of the reservoir fluid even when they are present in small quantities [20, 21].

Chorn and Mansoori [21] have documented various methods of characterization of C_{7+} fractions. Generally there are two techniques to characterize a hydrocarbon plus fraction: (i) The pseudocomponent and (ii) the continuous mixture approaches. In the pseudocomponent approach the C_{7+} is split into a number of subfractions with known mole fraction, T_b, SG, and M [22–26]. In this method the TBP curve can also be used to split the mixture into a number of pseudocomponents. Moreover, each subfraction may further be split into

TABLE 4.7—*Calculated values for physical properties of C_{60}, C_{70}, and C_{80} of Example 4.2.*

N_C	M	T_b	SG	n_{20}	d_{20}	T_c	P_c	d_c	ω	σ	δ
60	844	894.0	0.960	1.532	0.952	1010.5	5.7	0.410	1.699	30.1	8.4
70	989	927.0	0.969	1.538	0.961	1035.6	5.1	0.447	1.817	30.1	8.5
80	1134	953.0	0.977	1.542	0.969	1055.7	4.7	0.487	1.909	30.2	8.5

Units: T_b, T_c in K; P_c in bar; d_{20} and d_c in g/cm³; σ in dyne/cm; δ in (cal/cm³)$^{1/2}$.

three pseudocomponents from paraffinic, naphthenic, and aromatic groups. Although a higher number of pseudocomponents leads to more accurate results, the increase in the number of components complicates the calculations as the number of input data required increases significantly. For example, the application of a two-parameter equation of state (such as Peng–Robinson EOS) requires four input parameters for each component: T_c, P_c, ω, and a binary interaction coefficient (k_{ij}), which is a correction factor for a mixture of dissimilar components. The number of variables needed for a 20-component mixture in two-parameter EOS calculations is 290! [27].

The second approach is the continuous mixture characterization method. In this method instead of mole fractions, a distribution function is introduced to describe the composition of many component mixtures [24, 25, 28–32]. Since composition of a reservoir fluid up to C_5 is given in terms of discrete mole fractions, application of this approach to reservoir fluids is also referred as semicontinuous approach in which the distribution function is applied to C_{6+} part of the mixture. Distribution of components in mixtures that consist of many species is presented by a distribution function $F(P)$ whose independent property P is defined in terms of a measurable property such as molecular weight (M), boiling point (T_b), or carbon number (N_C) and varies from a value for the lightest component (P_o) to the value for the heaviest component (P_∞) present in the mixture. Generally the value of P_∞ for M or T_b for a plus fraction is assumed as infinity (∞). Classical thermodynamics for vapor–liquid equilibrium (VLE) calculations of multicomponent systems require equality of temperature, pressure, as well as equality of chemical potential of each component in both phases:

$$(4.14) \qquad \mu_i^L = \mu_i^V \quad i = 1, 2, \ldots, N$$

where μ_i^L and μ_i^V are chemical potential of component i in liquid and vapor phase, respectively. Equation (4.14) should be valid for all N components in the mixture. For VLE calculations of continuous mixtures Eq. (4.12) becomes

$$(4.15) \qquad \mu^L(P) = \mu^V(P) \quad P_o < P < \infty$$

where P is an independent variable such as molecular weight or boiling point. Similarly in calculation of all other thermodynamic properties for the mixture, distribution function is used instead of mole fraction for application of a mixing rule. It should be noted that even when composition of a mixture is expressed in terms of a distribution function, the mixture may be presented in terms of a number of pseudocomponents. Further characteristics of distribution functions and their application to petroleum mixtures are discussed in the next section.

4.5 DISTRIBUTION FUNCTIONS FOR PROPERTIES OF HYDROCARBON-PLUS FRACTIONS

As mentioned before, accurate characterization of a reservoir fluid or a crude oil requires a complete analysis of the mixture with known mole fraction and carbon number such as those shown in Table 4.2. For mixtures that the composition of heavy hydrocarbons is presented by a single hydrocarbon-

plus fraction, it is important to know distribution of carbon numbers to describe the mixture properly. A mathematical function that describes intensity of amount of a carbon number, or value of molecular weight, or boiling point for compounds with $N_C \geq 6$ is referred as *probability density function* (PDF). The PDF can be obtained from a distribution function that describes how various components or their properties are distributed in a mixture. In this section, general characteristics of density functions are discussed and then three different distribution models used to describe properties of hydrocarbon-plus fractions are presented.

4.5.1 General Characteristics

Distribution functions can be applied to determine distribution of compounds from hexane or heavier in a reservoir fluid. However, since the mole fraction of C_6 fraction in reservoir fluid is usually known and heavier hydrocarbons are grouped in a C_{7+} group, distribution functions are generally used to describe properties of C_{7+} fractions. Mole fraction versus molecular weight for SCN groups heavier than C_6 in the West Texas gas condensate sample in Table 4.1 and the waxy and Kuwaiti crude oils of Table 4.2 are shown in Fig. 4.4. Such graphs are known as molar distribution for the hydrocarbon plus (in this case C_{6+}) fraction of reservoir fluids. As can be seen from this figure the molar distribution of gas condensates is usually exponential while for the black oil or crude oil samples it is left-skewed distribution.

For the same three samples shown in Fig. 4.4, the probability density functions (PDF) in terms of molecular weight are shown in Fig. 4.5. Functionality of molecular weight versus cumulative mole fraction, $M(x)$, for the three samples is shown in Fig. 4.6.

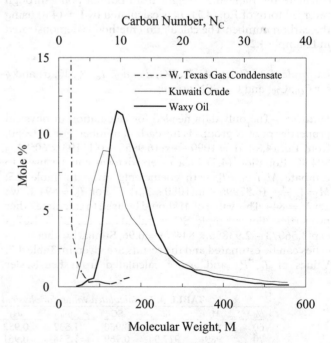

FIG. 4.4—Molar distribution for a gas condensate and a crude oil sample.

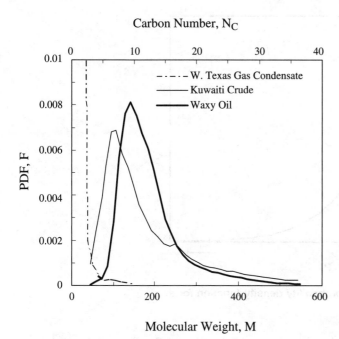

FIG. 4.5—Probability density functions for the gas condensate and crude oil samples of Fig. 4.4.

The continuous distribution for a property P can be expressed in terms of a function such that

$$(4.16) \qquad F(P)dP = dx_c$$

where P is a property such as M, T_b, N_C, SG, or I (defined by Eq. 2.36) and F is the probability density function. If the original distribution of P is in terms of cumulative mole fraction (x_{cm}), then x_c in Eq. (4.16) is the cumulative mole fraction. As mentioned before, parameter P for a continuous mixture varies from the initial value of P_o to infinity. Therefore, for the whole continuous mixture (i.e., C_{7+}), integration of Eq. (4.16)

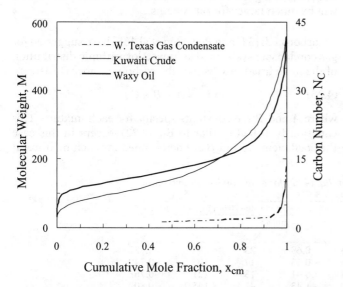

FIG. 4.6—Variation of molecular weight with cumulative mole fraction for the gas condensate and crude oil samples of Fig. 4.4.

based on normalized mole fraction becomes

$$(4.17) \qquad \int_{P_o}^{\infty} F(P)dP = \int_{0}^{1} dx_c = 1$$

If the upper limit of the integral in Eq. (4.17) is at property P, then the upper limit of the right-hand side should be cumulative x_c as shown in the following relation:

$$(4.18) \qquad x_c = \int_{P_o}^{P} F(P)dP$$

Integration of Eq. (4.16) between limits of P_1 and P_2 gives the mole fraction of all components in the mixture whose property P is in the range of $P_1 \le P \le P_2$:

$$(4.19) \qquad \int_{P_1}^{P_2} F(P)dP = x_{c2} - x_{c1} = x_{p_1 \to p_2}$$

where x_{c1} and x_{c2} are the values of x_c at P_1 and P_2, respectively. $x_{p_1 \to p_2}$ is sum of the mole fractions for all components having $P_1 \le P \le P_2$. Equation (4.19) can also be obtained by applying Eq. (4.18) at x_{c2} and x_{c1} and subtracting from each other. Obviously if the PDF is defined in terms of cumulative weight or volume fractions x represents weight or volume fraction, respectively. The average value of parameter P for the whole continuous mixture, P_{av}, is

$$(4.20) \qquad P_{av} = \int_{0}^{1} P(x_c)dx_c = \int_{P_o}^{\infty} PF(P)dP$$

where $P(x)$ is the distribution function for property P in terms of cumulative mole, weight, or volume fraction, x_c. For all the components whose parameters varies from P_1 to P_2 the average value of property P, $P_{av(P_1 \to P_2)}$, is determined as

$$(4.21) \qquad P_{av(P_1 \to P_2)} = \frac{\int_{P_1}^{P_2} PF(P)dP}{\int_{P_1}^{P_2} F(P)dP}$$

This is shown in Fig. 4.7 where the total area under the curve from P_o to ∞ is equal to unity (Eq. 4.17) and the area under curve from P_1 to P_2 represents the fraction of components whose property P is greater than P_1 but less than P_2. Further properties of distribution functions are discussed when different models are introduced in the following sections.

4.5.2 Exponential Model

The exponential model is the simplest form of expressing distribution of SCN groups in a reservoir fluid. Several forms of exponential models proposed by Lohrenz (1964), Katz (1983), and Pedersen (1984) have been reviewed and evaluated by Ahmed [26]. The Katz model [33] suggested for condensate systems gives an easy method of breaking a C_{7+} fraction into various SCN groups as [19, 26, 33]:

$$(4.22) \qquad x_n = 1.38205 \exp(-0.25903 C_N)$$

where x_n is the normalized mole fraction of SCN in a C_{7+} fraction and C_N is the corresponding carbon number of the SCN group. For normalized mole fractions of C_{7+} fraction, the mole fraction of C_{7+} (x_{7+}) is set equal to unity. In splitting a

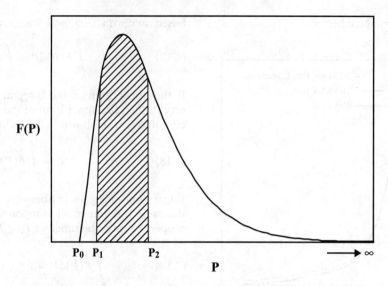

FIG. 4.7—General schematic of a probability density function for a property *P*.

hydrocarbon plus fraction into SCN groups by Eq. (4.22), the last hydrocarbon group is shown by C_{N+} fraction (i.e., 40+ in the waxy oil sample of Table 4.2). Molecular weight and specific gravity of the last fraction can be determined from the following equations

$$(4.23) \qquad \sum_{n=7}^{N} x_n M_n = M_{7+}$$

$$(4.24) \qquad \sum_{n=7}^{N} \frac{x_n M_n}{SG_n} = \frac{M_{7+}}{SG_{7+}}$$

where M_{7+} and SG_{7+} are known information for the C_{7+} fraction. M_n and SG_n are molecular weight and specific gravity of SCN group that may be taken from Table 4.6 or estimated from Eq. (4.7) and coefficients given in Table 4.5. Equation (4.24) is in fact equivalent to Eq. (3.45) when it is applied to SG. The following example shows application of this method to generate SCN groups.

Example 4.3—Use the M_{7+} and SG_{7+} for the North Sea gas condensate mixture of Table 4.1 to generate the composition of SCN groups up to C_{10+} and compare with actual data.

Solution—The C_{7+} has $M_{7+} = 124$, $SG_{7+} = 0.7745$, and $x_{7+} = 0.0306$. At first the mole fractions of SCN groups based on normalized composition are calculated from Eq. (4.22) and

then based on mole fraction of C_{7+} they are converted into mole fractions in the original mixture. For this problem, $N_+ = 10_+$; therefore, mole fractions of C_7, C_8, and C_9 should be estimated. The results are $x_7 = 0.225$, $x_8 = 0.174$, and $x_9 = 0.134$. Mole fraction of C_{10+} is calculated from material balance as $x_{10+} = 1.0 - (0.225 + 0.174 + 0.134) = 0.467$. The mol% of these components in the original mixture can be obtained through multiplying normalized mole fractions by mol% of C_{7+}, 3.06; that is mol% of $C_7 = 0.225 \times 3.06 = 0.69$. Summary of results is given in Table 4.8. The M_{10+} and SG_{10+} are calculated through Eqs. (4.23) and (4.24) using M_n and SG_n from Table 4.5 as $M_{10+} = [124 - (0.225 \times 95 + 0.174 \times 107 + 0.134 \times 121)]/0.467 = 145.2$. Similarly from Eq. (4.24), $M_{10+}/SG_{10+} = 181.5$. Therefore, $SG_{10+} = 145.2/181.5 = 0.80$. Comparison with the actual values from Table 4.1 is presented also in Table 4.8. As is shown in this table this method does not provide a good estimate of SCN distribution. The errors will be much larger for oil systems. ◆

Yarborough [34] and Pedersen et al. [35] have suggested for gas condensate systems to assume a logarithmic distribution of the mole fraction x_n versus the carbon number C_N as

$$(4.25) \qquad \ln x_n = A + B \times C_N$$

where A and B are constants specific for each mixture. This equation is in fact similar to Eq. (4.22), except in this case the coefficients A and B are determined for each mixture. If

TABLE 4.8—*Prediction of SCN groups from Eq. (4.22) for a gas condensate system of Example 4.3.*

| | Actual data from Table 4.1 | | | | Predicted values from Eqs. (4.22)–(4.24)[a] | | | |
| | mol%[b] | | | | mol%[b] | | | |
C_N		Nor.[b]	M	SG		Nor.[b]	M	SG
7	0.80	26	95	0.7243	0.69	23	95	0.727
8	0.76	25	103	0.7476	0.53	17	107	0.749
9	0.47	15	116	0.7764	0.41	13	121	0.768
10+	1.03	34	167	0.8120	1.43	47	145.2	0.800

[a]Values of M and SG for SCN groups are taken from Table 4.6.
[b]Values of mol% in the first column represent composition in the whole original fluid while in the second column under Nor. represent normalized composition ($\sum = 100$) for the C_{7+} fraction.

TABLE 4.9—*Prediction of SCN groups from Eq. (4.27) for a gas condensate system of Example 4.4.*

| | Actual data from Table 4.1 | | | | Predicted values from Eqs. (4.27)–(4.30)[a] | | | |
| | mol%[b] | | | | mol%[b] | | | |
C_N		Nor.[b]	M	SG		Nor.[b]	M	SG
7	0.80	26	95	0.7243	0.95	31	95	0.727
8	0.76	25	103	0.7476	0.69	22	107	0.749
9	0.47	15	116	0.7764	0.45	15	121	0.768
10+	1.03	34	167	0.8120	0.97	32	166	0.819

[a]Values of M and SG for SCN groups up to C_{50} are taken from Table 4.6. C_{10+} fraction represents SCN groups from 10 to 50.
[b]Values of mol% in the first column represent composition in the whole original fluid while in the second column under Nor. represent normalized composition for the C_{7+} fraction.

Eq. (4.10) is combined with Eq. (4.25), an equation for molar distribution of hydrocarbon-plus fractions can be obtained as

$$(4.26) \qquad \ln x_n = A_1 + B_1 \times M_n$$

which may also be written in the following exponential form.

$$(4.27) \qquad x_n = A \exp(B \times M_n)$$

where $B = B_1$ and $A = \exp(A_1)$. It should be noted that A and B in Eq. (2.27) are different from A and B in Eq. (2.25). Parameters A_1 and B_1 can be obtained by regression of data between $\ln(x_n)$ and M_n such as those given in Table 4.2 for the waxy oil. Most common case is that the detailed compositional analysis of the mixture is not available and only M_{7+} is known. For such cases coefficients A_1 and B_1 (or A and B) can be determined by applying Eqs. (4.23) and (4.24) assuming that the mixture contains hydrocarbons to a certain group (i.e., 45, 50, 60, 80, or even higher) without a plus fraction. This is demonstrated in Example 4.4.

Example 4.4—Repeat Example 4.3 using the Pedersen exponential distribution model, Eq. (4.27).

Solution—For this case the only information needed are $M_{7+} = 124$ and $x_{7+} = 0.0306$. We calculate normalized mole fractions from Eq. (4.27) after obtaining coefficients A and B. Since the mixture is a gas condensate we assume maximum SCN group in the mixture is C_{50}. Molecular weights of C_N from 7 to 50 are given in Table 4.6. If Eq. (4.26) is applied to all SCN groups from 7 to 50, since it is assumed there is no N_{50+}, for the whole C_{7+} we have

$$(4.28) \qquad \sum_{n=7}^{50} x_n = \sum_{n=7}^{50} A \exp(BM_n) = 1$$

where x_n is normalized mole fraction of SCN group n in the C_{7+} fraction. From this equation A is found as

$$(4.29) \qquad A = \left(\sum_{n=7}^{50} \exp(BM_n) \right)^{-1}$$

Parameter B can be obtained from M_{7+} as

$$(4.30) \quad M_{7+} = \left(\sum_{n=7}^{50} \exp(BM_n) \right)^{-1} \left(\sum_{n=7}^{50} [\exp(BM_n)]M_n \right)$$

where the only unknown parameter is B. For better accuracy the last SCN can be assumed greater than 50 with M_n calculated as discussed in Example 4.2. For this example we

use values of M_n from Table 4.6 which yields $B = -0.0276$. Parameter A is calculated from Eq. (4.29) as: $A = -4.2943$. Using parameters A and B, normalized mole fractions for SCN from 7 to 50 are calculated from Eq. (4.27). Mole fraction of C_{10+} can be estimated from sum of mole fractions of C_{10} to C_{50}. M_{10+} and SG_{10+} are calculated as in Example 4.3 and the summary of results are given in Table 4.9. In this method M_{10+} and SG_{10+} are calculated as 166 and 0.819, which are close to the actual values of 167 and 0.812. ◆

As shown in Example 4.4, the exponential model works well for prediction of SCN distribution of some gas condensate systems, but generally shows weak performance for crude oils and heavy reservoir fluids. As an example for the North Sea oil described in Table 4.1, based on the procedure described in Example 4.4, M_{10+} is calculated as 248 versus actual value of 259. In this method we have used up to C_{50} and the coefficients of Eq. (4.27) are $A = 0.2215$, $B = -0.0079$. If we use SCN groups up to C_{40}, we get $A = 0.1989$ and $B = -0.0073$ with $M_{10+} = 246.5$, but if we include SCN higher than C_{50} slight improvement will be observed. The exponential distribution model as expressed in terms of Eq. (4.27) is in fact a discrete function, which gives mole fraction of SCN groups. The continuous form of the exponential model will be shown later in this section.

4.5.3 Gamma Distribution Model

The gamma distribution model has been used to express molar distribution of wider range of reservoir fluids including black oils. Characteristics, specifications, and application of this distribution model to molecular weight and boiling point have been discussed by Whitson in details [15, 17, 22, 23, 36, 37]. The PDF in terms of molecular weight for this distribution model as suggested by Whitson has the following form:

$$(4.31) \qquad F(M) = \frac{(M - \eta)^{\alpha - 1} \exp\left(-\frac{M - \eta}{\beta}\right)}{\beta^\alpha \Gamma(\alpha)}$$

where α, β, and η are three parameters that should be determined for each mixture and $\Gamma(\alpha)$ is the gamma function to be defined later. Parameter η represents the lowest value of M in the mixture. Substitution of $F(M)$ into Eq. (4.20) gives the average molecular weight of the mixture (i.e., M_{7+}) as:

$$(4.32) \qquad M_{7+} = \eta + \alpha\beta$$

which can be used to estimate parameter β in the following form.

$$(4.33) \qquad \beta = \frac{M_{7+} - \eta}{\alpha}$$

Whitson et al. [37] suggest an approximate relation between η and α as follows:

$$(4.34) \qquad \eta = 110\left(1 - \frac{1}{1 + 4.043\alpha^{-0.723}}\right)$$

By substituting Eq. (4.31) into Eq. (4.18), cumulative mole fraction, x_{cm} versus molecular weight, M can be obtained which in terms of an infinite series is given as:

$$(4.35) \qquad x_{\mathrm{cm}} = [\exp(-M_{\mathrm{b}})] \sum_{j=0}^{\infty} \left[\frac{M_{\mathrm{b}}^{\alpha+1}}{\Gamma(\alpha + 1 + j)}\right]$$

where parameter M_{b} is a variable defined in terms of M as

$$(4.36) \qquad M_{\mathrm{b}} = \frac{M - \eta}{\beta}$$

Since M varies from η to ∞, parameter M_{b} varies from 0 to ∞. The summation in Eq. (4.35) should be discontinued when $\sum_{j=0}^{J+1} - \sum_{j=0}^{J} \leq 10^{-8}$. For a subfraction i with molecular weight bounds M_{i-1} and M_i, the discrete mole fraction, $x_{m,i}$ is calculated from the difference in cumulative mole fractions calculated from Eq. (4.35) as follows:

$$(4.37) \qquad x_{m,i} = x_{\mathrm{cm},i} - x_{\mathrm{cm},i-1}$$

where $M_{\mathrm{b}i}$ is calculated from Eq. (4.36) at M_i. The average molecular weight of this subfraction, $M_{\mathrm{av},i}$ is then calculated from the following formula:

$$(4.38) \qquad M_{\mathrm{av},i} = \eta + \alpha\beta \times \left(\frac{x_{\mathrm{cm},i}^1 - x_{\mathrm{cm},i-1}^1}{x_{\mathrm{cm},i} - x_{\mathrm{cm},i-1}}\right)$$

where $x_{\mathrm{cm},i}^1$ is evaluated from Eq. (4.35) by starting the summation at $j = 1$ instead of $j = 0$, which is used to evaluate $x_{\mathrm{cm},i}$.

Equation (4.37) can be used to estimate mole fractions of SCN groups in a C_{7+} fraction if lower and upper molecular weight boundaries (M_n^-, M_n^+) for the group are used instead of M_{i-1} and M_i. The lower molecular weight boundary for a SCN group n, M_n^- is the same as the upper molecular weight boundary for the preceding SCN group, that is

$$(4.39) \qquad M_n^- = M_{n-1}^+$$

For a SCN group n, the upper molecular weight boundary M_n^+ may be calculated from the midpoint molecular weights of SCN groups n and $n + 1$ as following:

$$(4.40) \qquad M_n^+ = \frac{M_n + M_{n+1}}{2}$$

where M_n and M_{n+1} are molecular weight of SCN groups n and $n + 1$ as given in Table 4.6. For example, in this table values of molecular weight for M_6, M_7, M_8, and M_9 are given as 82, 95, 107, and 121, respectively. For C_6 the upper molecular weight boundary is $M_6^+ = (82 + 95)/2 = 88.5$, which can be approximated as 88. Similarly, $M_7^+ = (95 + 107)/2 = 101$ and $M_8^+ = 114$. The lower molecular weight boundaries are calculated from Eq. (4.39) as $M_7^- = M_6^+ = 88$ and similarly, $M_8^- = 101$. Therefore for the SCN group of C_8, the lower molecular weight is 101 and the upper boundary is 114. For

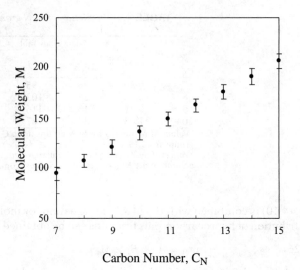

FIG. 4.8—The lower and upper molecular weight boundaries for SCN groups.

SCN groups from C_7 to C_{15} the molecular weight boundaries are shown in Fig. 4.8.

In this distribution model, parameter α can be determined by minimizing only one of the error functions $E_1(\alpha)$ or $E_2(\alpha)$ which for a C_{7+} fraction are defined as follows:

$$(4.41) \qquad E_1(\alpha) = \sum_{i=7}^{N-1}\left(M_{\mathrm{av},i}^{\mathrm{cal}} - M_{\mathrm{av},i}^{\mathrm{exp}}\right)^2$$

$$(4.42) \qquad E_2(\alpha) = \sum_{i=7}^{N-1}\left(x_{m,i}^{\mathrm{cal}} - x_{m,i}^{\mathrm{exp}}\right)^2$$

where $M_{\mathrm{av},i}^{\mathrm{cal}}$ is calculated through Eq. (4.38) and $M_{\mathrm{av},i}^{\mathrm{exp}}$ is experimental value of average molecular weight for the subfraction i. $x_{m,i}^{\mathrm{cal}}$ is the calculated mole fraction of subfraction (or SCN group) from Eq. (4.37). N is the last hydrocarbon group in the C_{7+} fraction and is normally expressed in terms of a plus fraction. Parameter α determines the shape of PDF in Eq. (4.31). For C_{7+} fraction of several reservoir fluids the PDF expressed by Eq. (4.31) is shown in Fig. 4.9. Values of parameters α, β, and η for each sample are given in the figure. As is shown in this figure, when $\alpha \leq 1$, Eq. (4.35) or (4.31) reduces to an exponential distribution model, which is suitable for gas condensate systems. For values of $\alpha > 1$, the system shows left-skewed distribution and demonstrates a maximum in concentration. This peak shifts toward heavier components as the value of α increases. As values of η increase, the whole curve shifts to the right. Parameter η represents the molecular weight of the lightest component in the C_{7+} fraction and it varies from 86 to 95 [23]. However, this parameter is mainly an adjustable mathematical constant rather than a physical property and it may be determined from Eq. (4.34). Whitson [17] suggests that for mixtures that detailed compositional analysis is not available, recommended values for η and α are 90 and 1, respectively, while parameter β should always be calculated from Eq. (4.33). A detailed step-by-step calculation method to determine parameters α, η, and β is given by Whitson [17].

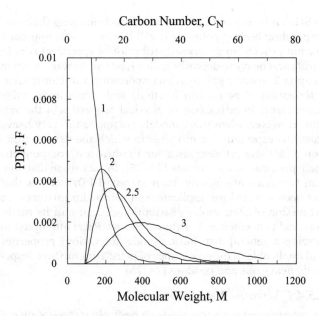

FIG. 4.9—Molar distribution by gamma density function (Eq. 4.31).

In evaluation of the summation in Eq. (4.35), the gamma function is defined as:

$$(4.43) \qquad \Gamma(x) = \int_0^\infty t^{x-1}e^{-t}dt$$

where t is the integration variable. As suggested by Whitson [16], the gamma function can be estimated by the following equation provided in reference [37]:

$$(4.44) \qquad \Gamma(x+1) = 1 + \sum_{i=1}^{8} A_i x^i$$

where for $0 \le x \le 1$, $A_1 = -0.577191652$, $A_2 = 0.988205891$, $A_3 = -0.897056937$, $A_4 = 0.918206857$, $A_5 = -0.756704078$, $A_6 = 0.482199394$, $A_7 = -0.193527818$, and $A_8 = 0.035868343$. And for $x > 1$, the recurrence formula may be used:

$$(4.45) \qquad \Gamma(x+1) = x\Gamma(x)$$

where from Eq. (4.44), $\Gamma(1) = 1$ and thus from the above equation $\Gamma(2) = 1$.

Equation (4.31) with $\alpha = 1$ reduces to an exponential distribution form. From Eq. (4.33) with $\alpha = 1$, $\beta = M_{7+} - \eta$ and substituting these coefficients into Eq. (4.31) the following density function can be obtained:

$$(4.46) \qquad F(M) = \left(\frac{1}{M_{7+} - \eta}\right)\exp\left(-\frac{M-\eta}{M_{7+}-\eta}\right)$$

For a SCN group n, with molecular weight boundaries of M_n^- and M_n^+, substitution of Eq. (4.46) into Eq. (4.19) will result:

$$(4.47) \quad \begin{aligned} x_{m,n} = &-\exp\left(\frac{\eta}{M_{7+}-\eta}\right) \\ &\times \left[\exp\left(-\frac{M_n^+}{M_{7+}-\eta}\right) - \exp\left(-\frac{M_n^-}{M_{7+}-\eta}\right)\right]\end{aligned}$$

where $x_{m,n}$ is the mole fraction of SCN group n. Substituting Eq. (4.46) in Eq. (4.21) for molecular weight gives the

following relation for the average molecular weight of the SCN group n:

$$(4.48) \quad \begin{aligned} M_{av,n} = &-\left(\frac{M_{7+}-\eta}{x_{m,n}}\right)\exp\left(\frac{\eta}{M_{7+}-\eta}\right) \\ &\times \left[\left(\frac{M_n^+}{M_{7+}-\eta}+1\right)\exp\left(-\frac{M_n^+}{M_{7+}-\eta}\right)\right. \\ &\left. -\left(\frac{M_n^-}{M_{7+}-\eta}+1\right)\exp\left(-\frac{M_n^-}{M_{7+}-\eta}\right)\right]\end{aligned}$$

where $M_{av,n}$ is the average molecular weight of SCN group n. Equations (4.47) and (4.48) can also be applied to any group with known lower and upper molecular weight boundaries in a C_{7+} fraction that follows an exponential distribution.

Example 4.5—Show that distribution model expressed by Eq. (4.46) leads to Eq. (4.27) for exponential distribution of SCN groups.

Solution—Equation (4.46) can be written in the following exponential form:

$$(4.49) \qquad F(M) = a\exp(bM)$$

where parameters a and b are given as

$$(4.50) \qquad \begin{aligned} a &= \frac{1}{M_{7+}-\eta}\exp\left(\frac{\eta}{M_{7+}-\eta}\right) \\ b &= -\frac{1}{M_{7+}-\eta}\end{aligned}$$

Substituting Eq. (4.49) into Eq. (4.18) gives the following relation for the cumulative mole fraction, x_{cm} at molecular weight M:

$$(4.51) \quad x_{cm} = \int_\eta^M a\exp(bM)\,dM = \left(\frac{a}{b}\right)[\exp(bM)-\exp(b\eta)]$$

For a SCN group n with lower and upper molecular weights of M_n^- and M_n^+ and use of Eq. (4.19) we get mole fraction of the group, x_n:

$$(4.52) \qquad x_n = \left(\frac{a}{b}\right)[\exp(bM_n^+)-\exp(bM_n^-)]$$

From Eqs. (4.39) and (4.40) we have

$$M_n^+ = \frac{M_n + M_{n+1}}{2} \qquad M_n^- = \frac{M_{n-1}+M_n}{2}$$

Now if we assume the difference between M_n and M_{n-1} is a constant number such as h we have $M_{n+1} = M_n + h$ and $M_{n-1} = M_n - h$, thus $M_n^+ = M_n + h/2$ and $M_n^- = M_n - h/2$. A typical value for h is usually 14. Substituting for M_n^+ and M_n^- in Eq. (4.52) gives

$$\begin{aligned} x_n &= \left(\frac{a}{b}\right)\{\exp[b(M_n+h/2)]-\exp[b(M_n-h/2)]\} \\ (4.53) \quad &= \left(\frac{a}{b}\right)\left[\exp\left(\frac{bh}{2}\right)-\exp\left(-\frac{bh}{2}\right)\right]\exp(bM_n)\end{aligned}$$

This equation can be written as

$$(4.54) \qquad x_n = A\exp(BM_n)$$

TABLE 4.10—*Prediction of molecular weight of SCN groups from Eq. (4.48) for Example 4.6.*

SCN, n	M_n	M_n^-	M_n^+	$x_{av,n}$	M_n, calc.	$M_{n,calc} - M_n$
7	95	88	101	0.0314	94.5	-0.5
8	107	101	114	0.0304	107.5	0.5
9	121	114	128.5	0.0328	121.2	0.2
10	136	128.5	142.5	0.0306	135.5	-0.5
11	149	142.5	156	0.0285	149.2	0.2
12	163	156	169.5	0.0276	162.7	-0.3
13	176	169.5	184	0.0286	176.7	0.7
14	191	184	199	0.0286	191.5	0.5
15	207	199	214	0.0275	206.5	-0.5
16	221	214	229	0.0265	221.5	0.5
17	237	229	243	0.0239	236.0	-1.0
18	249	243	255	0.0199	249.0	0.0
19	261	255	268	0.0209	261.5	0.5
20	275	268	282	0.0217	275.0	0.0

Equation (4.54) is identical to Eq. (4.27) with parameters A and B defined in terms of parameters a and b in the exponential distribution model (Eq. (4.49)) as following:

$$(4.55) \quad A = \left(\frac{a}{b}\right)\left[\exp\left(\frac{bh}{2}\right) - \exp\left(-\frac{bh}{2}\right)\right]$$

$$B = b$$

where a and b are defined in terms of distribution parameters by Eq. (4.50). ◆

Example 4.6—Use the exponential model to estimate average molecular weights of SCN groups from C_7 to C_{20} and compare with values in Table 4.6.

Solution—Average molecular weight of a mixture that follows the exponential distribution model is given by Eq. (4.48). In using this equation, $x_{m,n}$ is needed which should be calculated from Eq. (4.47). Two parameters of η and M_{7+} are needed. Arbitrary values for these parameters may be chosen. Parameter η has no effect on the calculation as long as it is less than M_n^- and M_{7+} does not affect the results as long as is well above M_n. Change in the chosen value for M_{7+} does change value of x_n, but not calculated M_n. For our calculations since we need to estimate M_{20} we choose $M_{7+} = 500$ and $\eta = 90$. Values of M_n^- and M_n^+ for each SCN group are calculated from Eqs. (4.39) and (4.40). Summary or results for calculation of M_n and comparison with values from Table 4.6 is given in Table 4.10. The maximum difference between calculated M_n and values from Table 4.6 is 1, while for most cases both values are identical. ◆

4.5.4 Generalized Distribution Model

The exponential model is the simplest form of expressing distribution of SCN groups in a reservoir fluid but it is mainly applicable to gas condensate systems or at most to volatile oils. For this reason the gamma distribution model has been used to express molar distribution of heavier oils. Although this model also has been applied to express distribution of boiling point but it is not suitable for specific gravity distribution. For this reason the idea of constant Watson K for the whole C_{7+} subfractions has been used [17]. In this approach, based on calculated K_W for C_{7+} from M_{7+} and SG_{7+}, values

to SG can be estimated for each subfraction using their corresponding boiling point. As it will be shown, this approach dos not provide an accurate distribution of specific gravity in a wide and heavy hydrocarbon-plus fraction. As was shown in Chapter 2, specific gravity is an important parameter in characterization of petroleum fractions and errors in its value cause errors in estimation of physical properties of the system. However, when these models are applied to very heavy fractions especially for mixtures in which the density function $F(M)$ sharply decreases for the heaviest components, their performance decreases [24, 25]. In fact these distribution functions are among many standard PDF models that has been selected for application to petroleum mixtures for expression of their molar distributions because of its mathematical convenience. For these reasons, Riazi attempted to develop a general distribution model for various properties and applicable to different types of petroleum mixtures especially heavy oils and residues [24, 25].

4.5.4.1 Versatile Correlation

An extensive analysis was made on basic characterization parameters for C_{7+} fractions of wide range of gas condensate systems and crude oils, light and heavy as well as narrow and wide petroleum fractions. Based on such analysis the following versatile equation was found to be the most suitable fit for various properties of more than 100 mixtures [24]:

$$(4.56) \quad P^* = \left[\frac{A}{B}\ln\left(\frac{1}{x^*}\right)\right]^{\frac{1}{B}}$$

where

$$P^* = \frac{P - P_o}{P_o} \qquad x^* = 1 - x_c$$

P is a property such as absolute boiling point (T_b), molecular weight (M), specific gravity (SG) or refractive index parameter (I) defined by Eq. (2.36). x_c is cumulative weight, mole, or volume fraction. P_o is a parameter specific for each property (T_o, M_o, and SG_o) and each sample. Usually cumulative mole fraction, x_{cm} is used for molecular weight and cumulative weight fraction, x_{cw} is used to express distribution of boiling point. Either cumulative volume fraction, x_{cv} or cumulative weight fraction x_{cw} can be used for presenting distribution of specific gravity, density, or refractive index parameter, I. In Eq. (4.56), P^* is a dimensionless parameter. Equation (4.56) is not defined at $x_c = 1$ ($x^* = 0$). In fact according to this model, it is theoretically assumed that the last component in the mixture is extremely heavy with $P \to \infty$ as $x_c \to 1$. A and B are two other parameters which are specific for each property and may vary from one sample to another. Equation (4.56) has three parameters (P_o, A, B); however, for more than 100 mixtures investigated it was observed that parameter B for each property is the same for most samples [24] reducing the equation into a two-parameter correlation. Parameter P_o corresponds to the value of P at $x_c = 0$, where $x^* = 1$ and $P^* = 0$. Physically P_o represents value of property P for the lightest component in the mixture; however, it is mainly a mathematical constant in Eq. (5.56) that should be determined for each mixture and each property. In fact Eq. (4.56) has been already used in Section (2.2.3) by Eq. (3.34) for prediction of complete distillation curves of petroleum fraction. The main idea behind Eq. (4.56) is to assume every petroleum mixture contains all compounds including extremely heavy compounds up to $M \to \infty$. However, what differs from one mixture to another is the amount of individual components. For low and medium molecular weight range fractions that do not contain high molecular weight compounds, the model expressed by Eq. (4.56) assumes that extremely heavy compounds do exist in the mixture but their amount is infinitely small, which in mathematical calculations do not affect mixture properties.

When sufficient data on property P versus cumulative mole, weight, or volume fraction, x_c, are available constants in Eq. (4.56) can be easily determined by converting the equation into the following linear form:

$$(4.57) \qquad Y = C_1 + C_2 X$$

where $Y = \ln P^*$ and $X = \ln[\ln(1/x^*)]$. By combining Eqs. (4.56) and (4.57) we have

$$(4.58) \qquad B = \frac{1}{C_2}$$

$$A = B \exp(C_1 B)$$

It is recommended that for samples with amount of residues (last hydrocarbon group) greater than 30%, the residue data should not be included in the regression analysis to obtain the coefficients in Eq. (4.57). If a fixed value of B is used for a certain property, then only parameter C_1 should be used to obtain coefficient A from Eq. (4.58).

To estimate P_o in Eq. (4.56), a trial-and-error procedure can be used. By choosing a value for P_o, which must be lower than the first data point in the dataset, parameters A and B can be determined from liner regression of data. Parameter P_o can be determined by minimizing the error function $E(P_o)$ equivalent to the root mean squares (RMS) defined as

$$(4.59) \qquad E(P_o) = \left[\frac{1}{N} \sum_{i=1}^{N} \left(P_i^{calc} - P_i^{exp} \right)^2 \right]^{1/2}$$

where N is the total number of data point used in the regression process and P_i^{calc} is the calculated value of property P for the subfraction i from Eq. (4.56) using estimated parameters P_o, A, and B. As an alternative objective function, best value of P_o can be obtained by maximizing the value of R^2 defined by Eq. (2.136). With spreadsheets such as Microsoft Excel, parameter P_o can be directly estimated from the Solver tool without trial-and-error procedure. However, an initial guess

for the value P_o is always needed. For a C_{7+} fraction value of property P for C_7 or C_6 hydrocarbon group from Table 4.6 may be used as the initial guess. Although linear regression can be performed with spreadsheets such as Excel or Lotus, coefficients C_1 and C_2 in Eq. (4.57) can be determined by hand calculators using the following relation derived from the least squares linear regression method:

$$(4.60) \qquad C_2 = \frac{\sum X_i \sum Y_i - N \sum (X_i Y_i)}{(\sum X_i)^2 - N \sum (X_i^2)}$$

$$C_1 = \frac{\sum Y_i - C_2 \sum X_i}{N}$$

where each sum applies to all data points used in the regression and N is the total number of points used. The least squares linear regression method is a standard method for obtaining the equation of a straight line, such as Eq. (4.57), from a set of data on X_i and Y_i.

Example 4.7—The normalized composition of a C_{7+} fraction derived from a North Sea gas condensate sample (GC) in terms of weight fractions of SCN groups up to C_{17} is given in Table 4.11. M and SG of C_{18+} fraction are 264 and 0.857, respectively. For the whole C_{7+} fraction the M_{7+} and SG_{7+} are 118.9 and 0.7597, respectively. Obtain parameters P_o, A, and B in Eq. (4.56) for M, T_b, and SG and compare calculated values of these properties with data shown in Table 4.6.

Solution—For SCN groups from C_7 to C_{17} values of M, T_b, and SG can be taken from Table 4.6 and are given in Table 4.11. An alternative to this table would be values recommended by Whitson [16] for SCN groups less than C_{25}. Discrete mole fractions, x_{mi} can be calculated from discrete weight fractions, x_{wi} and M_i by a reversed form of Eq. (1.15) as follows:

$$(4.61) \qquad x_{mi} = \frac{x_{wi}/M_i}{\sum_{i=1}^{N} x_{wi}/M_i}$$

where N is the total number of components (including the last plus fraction) and for this example it is 12. Discrete volume fractions x_{vi} can be calculated from x_{wi} and SG_i through Eq. (1.16). Values of x_{mi} and x_{vi} are given in Table 4.11. To obtain parameters in Eq. (4.56), cumulative mole (x_{cm}), weight (x_{cw}), or volume (x_{cv}) fractions are needed. A sample calculation for the estimation of molecular weight versus x_{cm} is shown here. A similar approach can be taken to estimate cumulative weight or volume fractions.

TABLE 4.11—*Sample data on characteristics of a C_{7+} fraction for a gas condensate system in Example 4.7.*

Fraction No.	Carbon No.	x_w	M	T_b, K	SG	x_m	x_v	x_{cm}	x_{cw}	x_{cv}
1	7	0.261	95	365	0.727	0.321	0.273	0.161	0.130	0.137
2	8	0.254	107	390	0.749	0.278	0.259	0.460	0.388	0.403
3	9	0.183	121	416	0.768	0.176	0.181	0.687	0.607	0.622
4	10	0.140	136	440	0.782	0.121	0.137	0.836	0.768	0.781
5	11	0.010	149	461	0.793	0.008	0.009	0.900	0.843	0.854
6	12	0.046	163	482	0.804	0.033	0.043	0.920	0.871	0.880
7	13	0.042	176	500	0.815	0.028	0.040	0.951	0.915	0.922
8	14	0.024	191	520	0.826	0.015	0.022	0.972	0.948	0.953
9	15	0.015	207	539	0.836	0.009	0.014	0.984	0.967	0.971
10	16	0.009	221	556	0.843	0.005	0.008	0.990	0.979	0.982
11	17	0.007	237	573	0.851	0.003	0.006	0.994	0.987	0.988
12	18+	0.010	264	—	0.857	0.004	0.009	0.998	0.995	0.996

x_w, x_m, and x_v are weight, mole, and volume fractions, respectively. Values of M, T_b, and SG are taken from Table 4.6. x_{cm}, x_{cw}, and x_{cv} are cumulative mole, weight, and volume fractions calculated from Eq. (4.62).

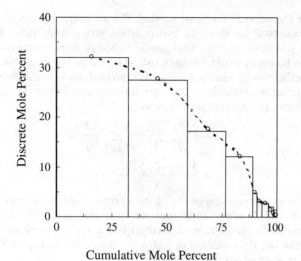

FIG. 4.10—Relation between discrete and cumulative mole fractions for the system of Example 4.7.

For the mixture shown in Table 4.11 there are 12 components each having molecular weight of M_i and mole fraction of x_i ($i = 1, \ldots, 12$). Values of cumulative mole fraction, x_{cmi} corresponding to each value of M_i can be estimated as:

$$(4.62) \qquad x_{cmi} = x_{cmi-1} + \frac{x_{mi-1} + x_{mi}}{2} \quad i = 1, 2, \ldots, N$$

where both x_{cm0} and x_{m0} ($i = 0$) are equal to zero. According to this equation, for the last fraction ($i = N$), $x_{cmN} = 1 - x_{mN}/2$. Equation (4.62) can be applied to weight and volume fractions as well by replacing the subscripts m with w or v, respectively. Values of x_{cmi}, x_{cwi}, and x_{cvi} are calculated from Eq. (4.62) and are given in the last three columns of Table 4.11. Since amount of the last fraction (residue) is very small, x_{cN} is very close to unity. However, in most cases especially for heavy oils the amount of residues may exceed 50% and value of x_c for the last data point is far from unity. The relation between x_{cm} and x_m is shown in Fig. 4.10.

To obtain molar distribution for this system, parameters M_o, A_M, and B_M for Eq. (4.56) should be calculated from the linear relation of Eq. (4.57). Based on the values of M_i and x_{cmi} in Table 4.11, values of Y_i and X_i are calculated from M^* and x^* as defined by Eq. (4.57). In calculation M^* a value of M_o is needed. The first initial guess for M_o should be less than

M_1 (molecular weight of the first component in the mixture). The best value for M_o is the lower molecular weight boundary for C_7 group that is M_7^- in Table 4.10, which is 88. Similarly the best initial guess for T_{bo} and SG_o are 351 K and 0.709, respectively. These numbers can be simplified to 90, 350, and 0.7 for the initial guesses of M_o, T_{bo}, and SG_o, respectively. Similarly for a C_{6+} fraction, the initial guess for its M_o can be taken as the lower molecular weight boundary for C_6 (M_6^-). For this example, based on the value of $M_o = 90$, parameters Y_i and X_i are calculated and are given in Table 4.12. A linear regression gives values of C_1 and C_2 and from Eq. (4.58) parameters A and B are calculated which are given in Table 4.12. For these values of M_o, A, and B, values of M_i are calculated from Eq. (4.56) and the error function $E(M_o)$ and AAD% are calculated as 2.7 and 1.32, respectively. Value of M_o should be changed so that $E(M_o)$ calculated from Eq. (4.59) is minimized. As shown in Table 4.12, the best value for this sample is $M_o = 91$ with $A = 0.2854$ and $B = 0.9429$. These coefficients gives RMS or $E(M_o)$ of 2.139 and AAD of 0.99%, which are at minimum. At $M_o = 91.1$ the value of $E(M_o)$ is calculated as 2.167. The same values for coefficients M_o, A_M, and B_M can be obtained by using Solver tool in Microsoft Excel spreadsheets. Experience has shown that for gas condensate systems and light fractions value of B_M is very close to one like in this case. For such cases B_M can be set equal to unity which is equivalent to $C_2 = 1$. In this example at $M_o = 89.856$, we get $C_1 = -1.1694$ and $C_2 = 1$ which from Eq. (4.58) yields $A_M = 0.3105$ and $B_M = 1$. Use of these coefficients in Eq. (4.56) gives $E(M_o)$ of 2.83 and AAD of 1.39%, which is slightly higher than the error for the optimum value of M_o at 91. Therefore, the final values of coefficients of Eq. (4.56) for M in terms of cumulative mole fraction are determined as: $M_o = 91$, $A_M = 0.2854$, $B_M = 0.9429$. The molar distribution can be estimated from Eq. (4.56) as

$$M^* = \left(\frac{0.2854}{0.9429} \ln \frac{1}{x^*}\right)^{\frac{1}{0.9429}} = 0.28155 \left(\ln \frac{1}{1 - x_{cm}}\right)^{1.06056}$$

From definition of M^* in Eq. (4.56) we can calculate M as

$$(4.63) \qquad M = M_o \times (1 + M^*)$$

and for this example we get:

$$M = 89.86 \left[1 + 0.28155 \left(\ln \frac{1}{1 - x_{cm}}\right)^{1.06056}\right]$$

TABLE 4.12—*Determination of coefficients of Eq. (4.56) for molecular weight from data of Table 4.11.*

M_i	x_i^*	X_i	M_i^*	Y_i	M_i^{calc}	ΔM_i^2	%AD	M_i^*	Y_i	M_i^{calc}	ΔM_i^2	%AD
			$M_o = 90$, $C_1 = -1.1809$, $C_2 = 1.0069$, $A = 0.3074$, $B = 0.9932$, $R^2 = 0.998$, RMS = 2.70, AAD = 1.32%					$M_o = 91$, $C_1 = -1.2674$, $C_2 = 1.0606$, $A = 0.2854$, $B = 0.9429$, $R^2 = 0.999$, RMS = 2.139, AAD = 0.99%				
95	0.839	−1.743	0.056	−2.89	94.8	0.0	0.2	0.044	−3.125	95.0	0.0	0.0
107	0.54	−0.484	0.189	−1.667	107.0	0.0	0.0	0.176	−1.738	106.3	0.4	0.6
121	0.313	0.15	0.344	−1.066	122.1	1.3	0.9	0.330	−1.110	121.1	0.0	0.0
136	0.164	0.591	0.511	−0.671	140.1	16.9	3.0	0.495	−0.704	139.0	8.8	2.2
149	0.1	0.833	0.656	−0.422	153.9	24.5	3.3	0.637	−0.450	153.0	16.1	2.7
163	0.08	0.927	0.811	−0.209	160.3	7.5	1.7	0.791	−0.234	159.5	12.4	2.2
176	0.049	1.101	0.956	−0.045	173.7	5.3	1.3	0.934	−0.068	173.3	7.1	1.5
191	0.028	1.273	1.122	0.115	189.6	2.0	0.7	1.099	0.094	189.9	1.2	0.6
207	0.016	1.413	1.3	0.262	204.6	5.8	1.2	1.275	0.243	205.6	1.8	0.7
221	0.01	1.53	1.456	0.375	219.0	4.0	0.9	1.429	0.357	220.9	0.0	0.1
237	0.006	1.634	1.633	0.491	233.2	14.3	1.6	1.604	0.473	236.0	1.0	0.4
264	0.002	1.814	1.933	0.659	261.6	5.6	0.9	1.901	0.642	266.4	5.9	0.9

$\Delta M_i^2 = (M_i^{calc} - M_i)^2$, %AD = Percent absolute relative deviation.

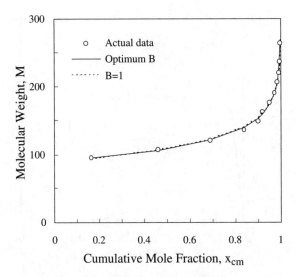

FIG. 4.11—Prediction of molar distribution from Eq. (4.56) for the GC system of Example 4.7.

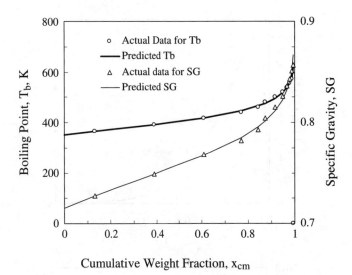

FIG. 4.12—Prediction of boiling point and specific gravity distributions from Eq. (4.56) for the GC system of Example 4.7.

If the fixed value of $B_M = 1$ is used with $M_o = 89.86$ and $A_M = 0.3105$ then the molar distribution is given by a simpler relation

$$M = 89.86\left(1 + 0.3105 \ln \frac{1}{1 - x_{cm}}\right)$$

Prediction of molar distribution based on these two relations ($B_M = 0.9429$ and $B_M = 1$) are shown in Fig. 4.11. The two curves are almost identical except toward the end of the curve where $x_{cm} \rightarrow 1$ and the difference is not visible in the figure.

Using a similar approach, coefficients in Eq. (4.56) for T_b and SG are determined. For SG both cumulative weight and volume fractions can be used. The value of T_b for the residue (C_{18+}) is not known, for this reason only 11 data points are used for the regression analysis. Summary of results for coefficients of Eq. (4.56) for M, T_b, and SG in terms of various x_c is given in Table 4.13. Based on these coefficients T_b and SG distributions predicted from Eq. (4.56) are shown in Fig. 4.12. The linear relation between parameters X and Y defined in Eq. (4.57) for SG is demonstrated in Fig. 4.13. Prediction of PDF for T_b and SG are shown in Figs. 4.14 and 4.15, respectively. Both Eqs. (4.66) and (4.70) have been used to illustrate density function for both T_b and SG. As shown in Table 4.13, the best values of M_o, T_{bo}, and SG_o are 91, 350 K, and 0.705, which are very close to the values of lower boundary properties for the C_{7+} group. ($M_7^- = 88$, $T_{b7}^- = 350$ K, $SG_7^- = 0.709$). For GC and light oils values of B for M are very close to one, for T_b are close to 1.5, and for SG are close

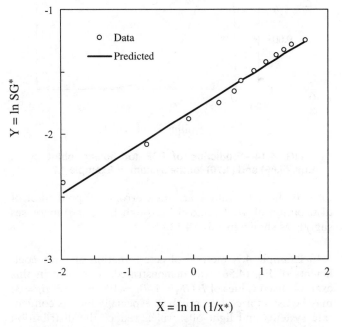

FIG. 4.13—Linearity of parameters *Y* and *X* defined in Eq. (4.57) for specific gravity of the system in Example 4.7.

TABLE 4.13—*Coefficients of Eq. (4.56) for M, T_b, and SG for data of Table 4.11.*

Property	Type of x_c	P_o	A	B	RMS	%AAD	R^2
M	Mole	91	0.2854	0.9429	2.139	0.99	0.999
T_b	Weight	350 (K)	0.1679	1.2586	3.794	0.62	0.998
SG	Volume	0.705	0.0232	1.8110	0.004	0.32	0.997
SG	Weight	0.705	0.0235	1.8248	0.004	0.33	0.997
Coefficients of P_o and A with fixed value of B for each property							
M	Mole	89.86	0.3105	1	2.83	1.39	0.998
T_b	Weight	340 (K)	0.1875	1.5	5.834	1.15	0.993
SG	Volume	0.665	0.0132	3	0.005	0.54	0.984
SG	Weight	0.6661	0.0132	3	0.005	0.53	0.985

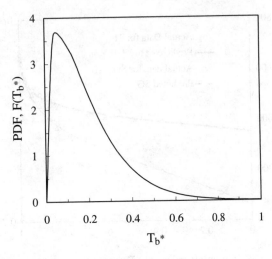

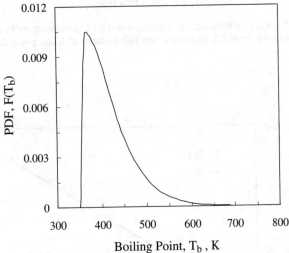

FIG. 4.14—Prediction of PDF for boiling point by Eqs. (4.66) and (4.70) for the system of Example 4.7.

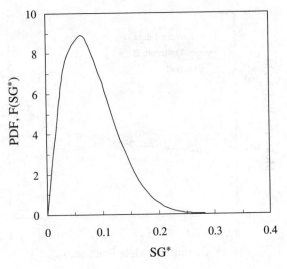

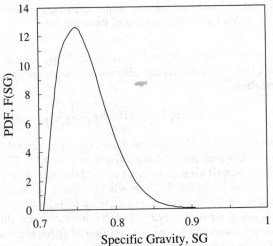

FIG. 4.15—Prediction of PDF for specific gravity by Eqs. (4.66) and (4.70) for the system of Example 4.7.

to 3. If these fix values are used errors for prediction of distribution of M, T_b, and SG through Eq. (4.56) increases slightly as shown in Table 4.13. ♦

In Example 4.7, method of determination of three coefficients of Eq. (4.56) was demonstrated. As shown in this example fixed values of B ($B_M = 1$, $B_T = 1.5$, $B_{SG} = 3$, $B_I = 3$) may be used for certain mixtures especially for gas condensate systems and light oils, which reduce the distribution model into a two-parameter correlation. It has been observed that even for most oil samples the fixed values of $B_T = 1.5$ and $B_{SG} = B_I = 3$ are also valid. Further evaluation of Eq. (4.56) as a three-parameter or a two-parameter correlation and a comparison with the gamma distribution model are shown in Section 4.5.4.5.

4.5.4.2 Probability Density Function for the Proposed Generalized Distribution Model

The distribution model expressed by Eq. (4.56) can be transformed into a probability density function by use of Eq. (4.16). Equation (4.56) can be rearranged as

$$(4.64) \qquad 1 - x_c = \exp\left(-\frac{A}{B}P^{*B}\right)$$

From Eq. (4.16) and in terms P^*, the PDF is given as

$$(4.65) \qquad F(P^*) = \frac{dx_c}{dP^*}$$

where $F(P^*)$ is the PDF in terms of dimensionless parameter P^* which can be determined by differentiation of Eq. (4.64) with respect to P^* according to the above equation:

$$(4.66) \qquad F(P^*) = \frac{B^2}{A}P^{*B-1}\exp\left(-\frac{B}{A}P^{*B}\right)$$

Equation (4.66) is in fact the probability density function for the generalized distribution model of Eq. (4.56) in terms of parameter P^*. In a hydrocarbon plus fraction, parameter P^* varies from 0 to ∞. Application of Eq. (4.17) in terms of P^* gives:

$$(4.67) \qquad \int_0^\infty F(P^*)dP^* = 1$$

and x_c at P^* can be determined from Eq. (4.18) in terms of P^*:

$$(4.68) \qquad x_c = \int_0^{P^*} F(P^*)dP^*$$

It is much easier to work in terms of P^* rather than P, since for any mixture P^* starts at 0. However, based on the definition of P^* in Eq. (4.56), the PDF expressed by Eq. (4.66) can be written in terms of original property P. Since $dx = F(P)dP = F(P^*)dP^*$ and $dP = P_\circ dP^*$, therefore we have

$$(4.69) \qquad F(P) = \frac{1}{P_\circ} F(P^*)$$

Substituting $F(P^*)$ from Eq. (4.66) into the above equation and use of definition of P^* we get

$$F(P) = \left(\frac{1}{P_\circ}\right) \times \left(\frac{B^2}{A}\right) \times \left(\frac{P - P_\circ}{P_\circ}\right)^{B-1} \exp\left[-\frac{B}{A}\left(\frac{P - P_\circ}{P_\circ}\right)^B\right]$$

$$(4.70)$$

with this form of PDF, Eq. (4.18) should be used to calculate cumulative, x_c at P. Obviously it is more convenient to work in terms of P^* through Eq. (4.68) and at the end P^* can be converted to P. This approach is used for calculation of average properties in the next section.

A simple comparison of Eq. (4.70) or (4.66) with the gamma distribution function, Eq. (4.31), indicates that parameter $P_\circ$ is equivalent to parameter η and parameter B is equivalent to parameter α. Parameter A can be related to α and β; however, the biggest difference between these two models is that inside the exponential term in Eq. (4.66), P^* is raised to the exponent B, while in the gamma distribution model, Eq. (4.31), such exponent is always unity. At $B = 1$, the exponential term in Eq. (4.66) becomes similar to that of Eq. (4.31). In fact at $B = 1$, Eq. (4.66) reduces to the exponential distribution model as was the case for the gamma distribution model when $\alpha = 1$. For this reason for gas condensate systems, the molar distribution can be presented by an exponential model as the behavior of two models is the same. However, for molar distribution of heavy oils or for properties other than molecular weight in which parameter B is greater than 1, the difference between two models become more apparent. As it is shown in Section 4.5.4.5, the gamma distribution model fails to present properly the molar distribution of very heavy oils and residues. For the same reason Eq. (4.66) is applicable for presentation of other properties such as specific gravity or refractive index as it is shown in Section 4.5.4.4. A comparison between the gamma distribution model (Eq. 4.31) and generalized model (4.70) when $M_\circ = \eta$ and $B = \alpha$ is shown in Fig. 4.16. As shown in this figure the difference between the proposed model and the gamma model increases as value of parameter B or α (keeping them equal) increases. Effect of parameter B on the form and shape of distribution model by Eq. (4.70) is shown in Fig. 4.17. For both Figs. 4.16 and 4.17, it is assumed that the mixture is a C_{7+} fraction with $M_\circ = 90$ and $M_{7+} = 150$.

4.5.4.3 Calculation of Average Properties of Hydrocarbon-Plus Fractions

Once the PDF for a property is known, the average property for the whole mixture can be determined through application of Eq. (4.20). If the PDF in terms of P^* is used, then Eq. (4.20) becomes

$$(4.71) \qquad P_{av}^* = \int_0^\infty P^* F(P^*) dP^*$$

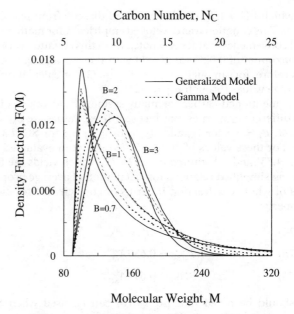

FIG. 4.16—Comparison of Eqs. (4.31) and Eq. (4.66) for $M_\circ = \eta = 90$, $B = \alpha$, and $M_{7+} = 150$.

where P_{av}^* is the average value of P^* for the mixture. Substituting Eq. (4.66) into Eq. (4.71) gives the following relation for P_{av}^*:

$$(4.72) \qquad P_{av}^* = \left(\frac{A}{B}\right)^{\frac{1}{B}} \Gamma\left(1 + \frac{1}{B}\right)$$

where $\Gamma(1 + 1/B)$ is the gamma function defined by Eq. (4.43) and may be evaluated by Eq. (4.44) with $x = 1/B$. A simpler version of Eq. (4.44) was given in Chapter 3 by Eq. (3.37) as

$$\Gamma\left(1 + \frac{1}{B}\right) = 0.992814 - 0.504242 B^{-1} + 0.696215 B^{-2}$$

$$(4.73) \qquad\qquad - 0.272936 B^{-3} + 0.088362 B^{-4}$$

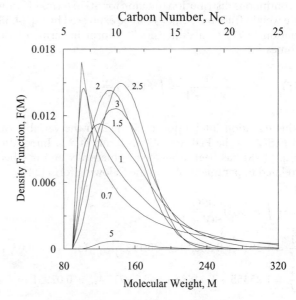

FIG. 4.17—Effect of parameter B on the shape of Eq. (4.70) for $M_\circ = 90$ and $M_{7+} = 150$.

in which $\Gamma(1 + 1/B)$ can be evaluated directly from parameter B. This equation was developed empirically for mathematical convenience. Values estimated from this equation vary by a maximum of 0.02% (at $B = 1$) with those from Eq. (4.43). Therefore, for simplicity we use Eq. (4.73) for calculation of average values through Eq. (4.72).

As mentioned earlier for many systems fixed values of B for different properties may be used. These values are $B_M = 1$ for M, $B_T = 1.5$ for T_b, and $B_{SG} = B_I = B_d = 3$ for SG, I_{20} or d_{20}. For these values of B, $\Gamma(1 + 1/B)$ has been evaluated by Eq. (4.73) and substituted in Eq. (4.72), which yields the following simplified relations for calculation of average properties of whole C_{7+} fraction in terms of coefficient A for each property:

$$(4.74) \qquad M_{av}^* = A_M$$

$$(4.75) \qquad T_{b,av}^* = 0.689 A_M^{2/3}$$

$$(4.76) \qquad SG_{av}^* = 0.619 A_{SG}^{1/3}$$

It should be noted that Eq. (4.76) can be used when SG is expressed in terms of cumulative volume fraction. Equation (4.76) is based on Eq. (4.72), which has been derived from Eq. (4.71). As it was discussed in Chapter 3 (Section 3.4), for SG, d (absolute density), and I (defined by Eq. 2.36) two types of mixing rules may be used to calculate mixture properties. Linear Kay mixing rule in the form of Eq. (3.45) can be used if composition of the mixture is expressed in volume fractions, but when composition if given in terms of weight fractions, Eq. (4.46) must be used. Both equations give similar accuracy; however, for mixtures defined in terms of very few compounds that have SG values with great differences, Eq. (3.46) is superior to Eq. (3.45). Equation (4.46) can be applied to SG in a continuous form as follows:

$$(4.77) \qquad \frac{1}{SG_{av}} = \int_0^1 \frac{dx_{cw}}{SG(x_{cw})}$$

where SG_{av} is the average specific gravity of C_{7+} and $SG(x_{cw})$ is the continuous distribution function for SG in terms of cumulative weight fraction. $SG(x_{cw})$ can be expressed by Eq. (4.56). Equation (4.77) in a dimensionless form in terms of SG^* becomes

$$(4.78) \qquad \frac{1}{SG_{av}^* + 1} = \int_0^\infty F(SG^*) \frac{dSG^*}{SG^* + 1}$$

In this equation integration is carried on the variable SG^* and $F(SG^*)$ is the PDF for SG^* in terms of x_{cw}. Integration in Eq. (4.78) has been evaluated numerically and has been correlated to parameter A_{SG} in the following form [24]:

$$J = \int_0^\infty F(SG^*) \frac{dSG^*}{SG^* + 1}$$

$$\frac{1}{J} = 1.3818 + 0.3503 A_{SG} - 0.1932 A_{SG}^2 \quad \text{for } A_{SG} > 0.05$$

$$\frac{1}{J} = 1.25355 + 1.44886 A_{SG} - 5.9777 A_{SG}^2 + 0.02951 \ln A_{SG}$$

$$(4.79) \qquad \qquad \qquad \qquad \qquad \text{for } A_{SG} \leq 0.05$$

where J is just an integration parameter defined in Eq. (4.79). A_{SG} is the coefficient in Eq. (4.56) when SG is expressed in terms of cumulative weight fraction, x_{cw}. For most samples evaluated, parameter A_{SG} is between 0.05 and 0.4; however, for no system a value greater than 0.4 was observed. By combining Eqs. (4.78) and (4.79) with definition of SG^* by Eq. (4.56), SG_{av} can be calculated from the following relation:

$$(4.80) \qquad SG_{av} = \left(\frac{1}{J}\right) SG_o$$

this equation should be used when SG is expressed in terms of x_{cw} by Eq. (4.56). For analytical integration of Eq. (4.78) see Problem 4.4.

In general, once P_{av}^* is determined from Eq. (4.72), P_{av} can be determined from the definition of P^* by the following relation:

$$(4.81) \qquad P_{av} = P_o(1 + P_{av}^*)$$

Average properties determined by Eqs. (4.74)–(4.76) can be converted to M_{av}, T_{bav}, and SG_{av} by Eq. (4.81). Equation (4.76) derived for SG_{av}^* can also be used for refractive index parameter I or absolute density (d) when they are expressed in terms of x_{cv}. Similarly Eqs. (4.78)–(4.80) can be applied to I_{20} or d_{20} when they are expressed in terms of x_{cw}. The following example shows application of these equations.

Example 4.8—For the gas condensate system of Example 4.7 calculate mixture molecular weight, boiling point, and specific gravity using the coefficients given in Table 4.13. The experimental values are $M_{7+} = 118.9$ and $SG_{7+} = 0.7569$ [24]. Also calculate the boiling point of the residue (component no. 12 in Table 4.11).

Solution—For molecular weight the coefficients of PDF in terms of x_{cm} for Eq. (4.66) as given in Table 4.13 are: $M_o = 91$, $A_M = 0.2854$, and $B_M = 0.9429$. From Eq. (4.73), $\Gamma(1 + 1/B) = 1.02733$ and from Eq. (4.72), $M_{av}^* = 0.2892$. Finally M_{av} is calculated from Eq. (4.81) as 117.3. For this system B_M is very close to unity and we can use the coefficients in Table 4.13 for $M_o = 89.86$, $A_M = 0.3105$, and $B_M = 1$. From Eqs. (4.74) and (4.81) we get $M_{av} = 89.86 \times (1 + 0.3105) = 117.8$. Comparing with the experimental value of 118.9, the relative deviation is -1%.

For specific gravity, the coefficients in terms of x_{cv} are: $SG_o = 0.705$, $A_{SG} = 0.0232$, and $B_{SG} = 1.811$. From Eq. (4.76), $SG_{av}^* = 0.0801$ and from Eq. (4.81), $SG_{av} = 0.7615$. Comparing with experimental value of 0.7597, the relative deviation is 0.24%. If the coefficients in terms of x_{cw} are used, $SG_o = 0.6661$, $A_{SG} = 0.0132$, and from Eq. (4.79) we get $1/J = 1.1439$ using appropriate range for A_{SG}. From Eq. (4.80), $SG_{av} = 0.7619$ which is nearly the same as using cumulative volume fraction.

For T_b the coefficients in terms x_{cw} with fixed value of B_T are $T_o = 340$ K, $A_T = 0.1875$, and $B_T = 1.5$. From Eq. (4.75) and (4.81) we get: $T_{av} = 416.7$ K. To calculate T_b for the residue we use the following relation:

$$(4.82) \qquad T_{bN} = \frac{T_{av} - \sum_{i=1}^{N-1} x_{wi} T_{bi}}{x_{wN}}$$

where T_{bN} is the boiling point of the residue. For this example from Table 4.11, $N = 12$ and $x_{wN} = 0.01$. Using values of x_{wi} and T_{bi} for $i = 1$ to $N - 1$ from Table 4.11, we get $T_{bN} = 787.9$ K. ◆

Example 4.9—Show how Eq. (4.78) can be derived from Eq. (4.77).

Solution—From Eq. (4.16): $dx_w = F(SG)dSG$ and from definition of P^* in Eq. (4.56) we have $SG = SG_oS^* + SG_o$, which after differentiation we get $dSG = SG_odSG^*$. In addition, from Eq. (4.69), $F(SG^*) = SG_oF(SG)$ and from Eq. (4.56), when $x_{cw} = 0$, we have $SG^* = 0$ and at $x_{cw} = 1$, we have $SG^* = \infty$. By combining these basic relations and substituting them into Eq. (4.77) we get

$$\frac{1}{SG_oSG_{av} + SG_o} = \int_0^\infty \frac{F(SG^*)SG_odSG^*}{SG_oSG^* + SG_o}$$

which after simplification reduces to Eq. (4.78). ◆

4.5.4.4 Calculation of Average Properties of Subfractions

In cases that the whole mixture is divided into several pseudocomponents (i.e., SCN groups), it is necessary to calculate average properties of a subfraction i whose property P varies from P_{i-1} to P_i. Mole, weight, or volume fraction of the groups shown by z_i can be calculated through Eq. (4.19), which in terms of P^* becomes

$$(4.83) \qquad z_i = \int_{P_{i-1}^*}^{P_i^*} F(P^*)dP^*$$

Substituting $F(P^*)$ from Eq. (4.66) into the above equation gives

$$(4.84) \qquad z_i = \exp\left(-\frac{B}{A}P_{i-1}^{*B}\right) - \exp\left(-\frac{B}{A}P_i^{*B}\right)$$

Average properties of this subfraction shown by $P_{i,av}^*$ can be calculated from Eq. (4.21), which can be written as

$$(4.85) \qquad P_{i,av}^* = \frac{1}{z_i}\int_{P_{i-1}^*}^{P_i^*} P^*F(P^*)dP^*$$

by substituting $F(P^*)$ from Eq. (4.66) and carrying the integration we get

$$(4.86) \quad P_{i,av}^* = \frac{1}{z_i}\left(\frac{A}{B}\right)^{1/B}\left[\Gamma\left(1 + \frac{1}{B}, q_{i-1}\right) - \Gamma\left(1 + \frac{1}{B}, q_i\right)\right]$$

where

$$(4.87) \qquad q_i = \frac{B}{A}P_i^{*B}$$

z_i should be calculated from Eq. (4.84). $P_{i,av}$ is calculated from $P_{i,av}^*$ through Eq. (4.81) as

$$(4.88) \qquad P_{i,av} = P_o(1 + P_{i,av}^*)$$

In Eq. (4.86), $\Gamma(1 + 1/B, q_i)$ is the incomplete gamma function defined as [38]

$$(4.89) \qquad \Gamma(a, q) = \int_q^\infty t^{a-1}e^{-t}dt$$

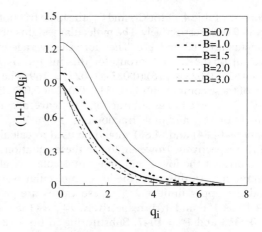

FIG. 4.18—Incomplete gamma function $\Gamma(1 + 1/B, q_i)$ for different values of B. Taken with permission from Ref. [40].

where for the case of Eq. (4.86), $a = 1 + 1/B$. Values of $\Gamma(1 + 1/B, q_i)$ can be determined from various numerical handbooks (e.g., Press et al. [38]) or through mathematical computer software such as *MATHEMATICA*. Values of $\Gamma(1 + 1/B, q_i)$ for $B = 1$, 1.5, 2, 2.5, and 4 versus q_i are shown in Fig. 4.18 [39]. As $q_i \to \infty$, $\Gamma(1 + 1/B, q_i) \to 0$ for any value of B. At $B = 1$, Eq. (4.89) gives the following relation for $\Gamma(1 + 1/B, q_i)$:

$$(4.90) \quad \Gamma(2, q) = \int_q^\infty te^{-t}dt = -(1 + t)e^{-t}\Big|_q^\infty = (1 + q)e^{-q}$$

Further properties of incomplete gamma functions are given in Ref. [39]. Substitution of Eq. (4.90) into Eq. (4.86) we get the following relation to estimate $P_{i,av}^*$ for the case of $B = 1$:

$$P_{i,av}^* =$$
$$\left(\frac{A}{z_i}\right)\left[\left(1 + \frac{P_{i-1}^*}{A}\right)\exp\left(-\frac{P_{i-1}^*}{A}\right) - \left(1 + \frac{P_i^*}{A}\right)\exp\left(-\frac{P_i^*}{A}\right)\right]$$
$$(4.91)$$

where z_i is obtained from Eq. (4.84) which for the case of $B = 1$ becomes:

$$(4.92) \qquad z_i = \exp\left(-\frac{P_{i-1}^*}{A}\right) - \exp\left(-\frac{P_i^*}{A}\right)$$

In these relations, P_i^* and P_{i-1}^* are the upper and lower boundaries of the subfraction i. One can see that if we set $P_i^* = M_n^{+*}$ and $P_{i-1}^* = M_n^{-*}$, then Eq. (4.91) is equivalent to Eq. (4.48) for estimated molecular weight of SCN groups through the exponential model.

Example 4.10—For the C_{7+} fraction of Example 4.7, composition and molecular weight of SCN groups are given in Table 4.11. Coefficients of Eq. (4.56) for the molar distribution of this system are given in Table 4.13 as $M_o = 89.86$, $A = 0.3105$, and $B = 1$. Calculate average molecular weights of C_{12}–C_{13} group and its mole fraction. Compare calculated values from those given in Table 4.11.

***Solution*—**In Table 4.11 for C_{12} and C_{13} the mole fractions are 0.033 and 0.028, respectively. The molecular weights of these components are 163 and 176. Therefore the average molecular weight of the C_{12}–C_{13} group for this mixture is $M_{av} = (0.033 \times 163 + 0.028 \times 176)/(0.033 + 0.028) = 169$. The mole fraction of these components is $0.033 + 0.028$ or 0.061. Group of C_{12}–C_{13} is referred to as subfraction i with average molecular weight of $M_{i,av}$ and mole fraction of z_i.

Equations (4.84) and (4.86) should be used to calculate z_i and $M_{i,av}$, respectively. However, to use these equations, P_{i-1} and P_i represent the lower and upper molecular weights of the subfraction. In this case, the lower molecular weight is M_{12}^- and the upper limit is M_{13}^+. These values are given in Table 4.10 as 156 and 184, respectively. $P_{i-1}^* = (156 - 89.9)/89.9 = 0.7353$ and $P_i^* = 1047$. Substituting in Eq. (4.84) we get: $z_i = 0.059$.

For this system, $A = 0.3501$ and $B = 1$; therefore, from Eq. (4.87), $q_{i-1} = 2.37048$ and $q_i = 3.37401$, which gives [39] $\Gamma(1 + 1/B, q_{i-1}) = 0.3149$ and $\Gamma(1 + 1/B, q_i) = 0.1498$. Substituting these values in Eq. (4.86) gives $M_n^* = 0.8662$ which yields $M_n = 167.7$. Therefore the predicted values for z_i and $M_{i,av}$ for group of C_{12}–C_{13} are 0.059 and 167.7, respectively, versus actual values of 0.061 and 169. ♦

4.5.4.5 Model Evaluations

The distribution model expressed by Eq. (4.56) can be used for M, T_b, SG, d, and refractive index parameter I. The exponential model expressed by Eq. (4.27) or Eq. (4.31) with $\alpha = 1$ can only be used for molecular weight of light oils and gas condensate systems. The gamma distribution model can be applied to both M and T_b, but for SG, the method of constant Watson K is recommended by Whitson [20]. In this method K_W for the whole C_{7+} is calculated from its M_{7+} and SG_{7+} (Eq. 2.133) and it is assumed to be constant for all components. For each component, SG is calculated from Eq. (2.133) using the K_W of the mixture and M for the component.

As mentioned earlier the main advantage of generalized model is its capability to predict distribution of properties of heavy oils. This is demonstrated in Fig. 4.19, for molecular weight distribution of a heavy residue [25]. Experimental data on the molar distribution are taken from Rodgers et al. [41]. The experimentally determined mixture weight averaged molecular weight is 630 [41]. For this sample, parameters M_o, A_M, and B_M for Eq. (4.56) in terms of cumulative weight fraction are calculated as 144, 71.64, and 2.5, respectively. For this heavy oil sample both M_o and parameter B are higher than their typical values for oil mixtures. Predicted mixture molecular weight from Eqs. (4.72) and (4.81) is 632, which is in good agreement with the experimental data. In Fig. 4.19 prediction of molar distribution from the exponential and gamma models are also illustrated. It is obvious that the exponential model cannot be applied to heavy oils. The gamma distribution model tends to predict higher values for M toward heavier components.

Evaluation of these models for boiling point of a North Sea black oil with M_{7+} and SG_{7+} of 177.5 and 0.8067 is shown in Fig. 4.20. This is sample No. 8 in Ref. [25] in which the experimental data on boiling points of 14 subfractions are available. By applying Eq. (4.56), it was found that $T_o = 346$ K, $A_T = 0.5299$, and $B_T = 1.3$, which yields an average error of

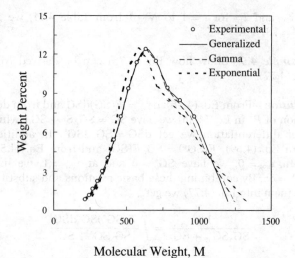

FIG. 4.19—Comparison of various distribution models for molecular weight of a heavy petroleum mixture. Taken with permission from Ref. [25].

$1°C$. Applying the gamma distribution model by Eq. (4.31) gives $\eta_T = 349.9$ K, $\alpha_T = 1.6$, and $\beta_T = 112.4$ K. Use of these coefficients in Eq. (4.31) for prediction of T_b distribution gives average error of $1.6°C$. The exponential model (Eq. (4.56) with $B = 1$) gives an average error of $4°C$. For this mixture with intermediate molecular weight, the generalized and gamma distribution models both are predicting boiling point with a good accuracy. However, the exponential model is the least accurate model for the boiling point distribution since B_T in Eq. (4.56) is greater than unity.

Distribution of specific gravity for the C_{7+} fraction of a black oil system from Ekofisk field of North Sea fields is shown in Fig. 4.21. The generalized model, exponential model, and the

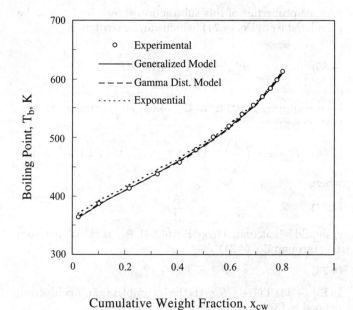

FIG. 4.20—Comparison of various distribution models for prediction of boiling point of C_{7+} of a North Sea Black oil. Taken with permission from Ref. [25].

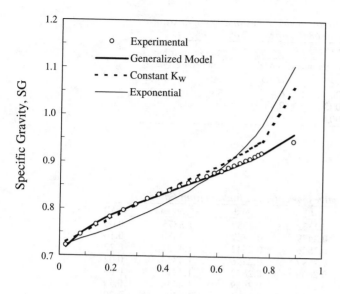

FIG. 4.21—Comparison of three models for prediction of specific gravity distribution of C_{7+} of an oil system.

constant K_W method for estimation of SG distribution are compared in this figure. For the mixture M_{7+} and SG_{7+} are 232.9 and 0.8534, respectively. For the generalized model the coefficients are $SG_o = 0.666$, $A_{SG} = 0.1453$, and $B_{SG} = 2.5528$ which yields an average error of 0.31% for prediction of SG distribution. In the constant K_W approach [15, 23], K_W is calculated from Eq. (2.133) using M_{7+} and SG_{7+} as input data: $K_{W7+} = 11.923$. It is assumed that all components have the same K_W as that of the mixture. Then for each component SG_i is calculated from its M_i and K_{W7+} for the mixture using the same equation. For this system predicted SG distribution gives an average error of 1.7%. The exponential model is the same as Eq. (4.56) assuming $B = 1$, which yields average error of 3.3%. From this figure it is clear that the generalized model of Eq. (4.56) with the fixed value of B ($\cong 3$) generates the best SG distribution. For very light gas condensate systems the Watson K approach and generalized model predict nearly similar SG distribution.

Further evaluation of Eq. (4.56) and gamma distribution model for 45 black oil and 23 gas condensate systems is reported in Ref. [25]. Equation (4.56) can be used as either a two- or a three-parameter model. Summary of evaluations is given in Tables 4.14 and 4.15. As mentioned earlier Eq. (4.56) is more or less equivalent to the gamma distribution model for molar distribution of gas condensate systems and light oils. Molecular weight range for samples evaluated in Table 4.14 is from 120 to 290 and for this reason both models give similar errors for prediction of M distribution (1.2%). However, two-parameter form of Eq. (4.56) is equivalent to the exponential model ($B = 1$) and gives higher error of 2.2%. For these systems, the exponential model does not give high errors since the systems are not quite heavy. For heavy oils exponential model is not applicable for prediction of molar distribution. For T_b distribution both three-parameter form of Eq. (4.56) and the gamma model are equivalent with error of about 0.6% ($\cong 6$ K), while the latter gives slightly higher error. The two-parameter generalized model ($B = 1.5$ in Eq. 4.56) gives an average error of 0.7% for prediction of T_b distribution. For SG distribution through Eq. (4.56), the best value of B is 3 and there is no need for three-parameter model. However, the method is much more accurate than the constant K_W method which gives an error more than twice of the error from the generalized model. Summary of results for prediction of M_{7+} and SG_{7+} for the same systems of Table 4.14 is shown in Table 4.15. The gamma distribution model predicts M_{7+} more accurate than Eq. (4.66) mainly because most of the systems studied are light oil or gas condensate. It is not possible to evaluate predicted T_b of the whole mixture since the experimental data were not available. Refractive index can be accurately predicted by Eq. (4.56) with $B = 3$ as shown in Tables 4.14 and 4.15. Results shown for evaluation of refractive in Table 4.14 are based on about 160 data points for 13 oil samples. Average error for calculation of refractive indices of 13 oils is 0.2% as shown in Table 4.15. Further evaluation of application of the generalized model is demonstrated in Example 4.11.

The generalized distribution model expressed by Eqs. (4.56) and (4.66) can be applied to other physical properties and to petroleum fractions other than C_{7+} fractions. In general they are applicable to any wide boiling range and hydrocarbon-plus fraction. The following example demonstrates how this

TABLE 4.14—*Evaluation of various distribution models for estimation of properties of C_{7+} fractions for 68 mixtures.*[a]

	Generalized model, Eq. (4.56)				Gamma distribution model, Eq. (4.31)	
	Two-parameter model		Three-parameter model			
Property	AAD[b]	%AAD[c]	AAD	%AAD	AAD	%AAD
M	4.09[d]	2.2[d]	2.28	1.2	2.31	1.2
T_b, °C	7.26/1.8	0.7	5.78/1.8	0.56	6.53/1.8	0.63
SG	0.005	0.6	0.005	0.6	0.01[e]	1.24[e]
n_{20}	0.0025	0.18[f]	0.0025	0.18	g	g

[a] Most of samples are from North Sea reservoirs with M ranging from 120 to 290. Systems include 43 black oil and 23 gas condensate systems with total of 941 data points. Full list of systems and reference for data are given in Ref. [24].
[b] AAD = absolute average deviation = $(1/N)\Sigma|$estimated property − experimental property$|$.
[c] %AAD = Percent absolute average deviation = $(1/N)\Sigma[|$(estimated property − experimental property)/experimental property$| \times 100]$.
[d] Same as exponential model for molar distribution.
[e] Method of constant Watson K.
[f] Refractive index was evaluated for 13 oils [42] and total of 161 data points.
[g] The gamma model is not applicable to refractive index.

TABLE 4.15—*Evaluation of various distribution models for estimation of mixture average properties of C_{7+} fractions for 68 mixtures of Table 4.14.[a]*

| | Generalized model, Eq. (4.56) | | | | Gamma distribution model, Eq. (4.31) | |
| | Two-parameter model | | Three-parameter model | | | |
Property	AAD[b]	%AAD[b]	AAD	%AAD	AAD	%AAD
M	12.58	6.8	11.9	6.4	5.4	2.9
SG	0.003	0.35	0.003	0.35	c	c
n_{20}	0.003[d]	0.2	0.003	0.2	c	c

[a] M_{7+} range: 120–290. SG_{7+} range: 0.76–0.905.
[b] Defined in Table 4.14.
[c] The gamma model cannot be applied to SG or n_{20}.
[d] For 13 oil samples.

TABLE 4.16—*Prediction of distribution of refractive index of a C_{6+} fraction from Eq. (4.56).*

N_C	Vol%	n_{20}	x_{cv}	I	I, pred.	n_{20}, pred.	%AD
6	2.50	1.3866	0.013	0.2352	0.2357	1.3875	0.06
7	5.47	1.4102	0.056	0.2479	0.2467	1.4080	0.16
8	4.53	1.4191	0.109	0.2526	0.2542	1.4222	0.22
9	5.06	1.4327	0.161	0.2597	0.2596	1.4324	0.02
10	2.55	1.4407	0.201	0.2639	0.2632	1.4393	0.09
11	3.62	1.4389	0.234	0.2630	0.2659	1.4445	0.39
12	3.70	1.4472	0.274	0.2673	0.2688	1.4502	0.20
13	4.19	1.4556	0.316	0.2716	0.2718	1.4560	0.03
14	3.73	1.4615	0.358	0.2747	0.2747	1.4615	0.00
15	3.96	1.4694	0.399	0.2787	0.2774	1.4668	0.18
16	3.03	1.4737	0.437	0.2809	0.2798	1.4715	0.15
17	3.40	1.4745	0.471	0.2813	0.2819	1.4758	0.09
18	3.13	1.4755	0.506	0.2818	0.2842	1.4802	0.31
19	2.94	1.4808	0.538	0.2845	0.2862	1.4842	0.23
20+	41.70	1.5224	0.777	0.3052	0.3031	1.5182	0.28
Mixture	93.51						0.14

[†] Experimental data on n_{20} are taken from Berge [42].

method can be used to predict distribution of refractive index of a C_{6+} fraction.

Example 4.11—For a C_{6+} of an oil sample experimental data on refractive index at 20°C are given versus vol% of SCN groups from C_6 to C_{20+} in Table 4.16. Refractive index of the whole fraction is 1.483. Use Eq. (4.56) to predict refractive index distribution and obtain the AAD% for the model prediction. Also graphically compare the model prediction with the experimental data and calculate the mixture refractive index.

Solution—Similar to Example 4.7, vol% should be first converted to normalized volume fractions and then to cumulative volume fraction (x_{cv}). For refractive index the characterization parameter is I_{20} instead of n_{20}. Therefore, in Eq. (4.56) we use parameter I (defined by Eq. 2.36) for property P. Values of I versus x_{cv} are also given in Table 4.16. Upon regression of data through Eq. (4.58), we get: $I_o = 0.218$, $A_I = 0.1189$, and $B_I = 3.0$. For these coefficients the RMS is 0.001 and %AAD is 0.14%. Value of I_o for this sample is close to the lower I value of C_6 group and parameter B is same as that of specific gravity. A graphical evaluation of predicted distribution is shown in Fig. 4.22. Since $B = 3$, Eq. (4.76) should be used to calculate I_{av}^* and then from Eq. (4.81) I_{av} is calculated as $I_{av} = 0.2844$. From Eq. (2.114), the mixture refractive index is calculated as $n_{av} = 1.481$, which differs from experimental value of 1.483 by –0.15%.

Further evaluation of Eq. (4.56) for prediction of distribution of refractive index shows that refractive index can be predicted with $B = 3$ with an accuracy of 0.2% as shown in Table 4.14. As discussed in Chapter 2, parameter I is a size parameter similar to density or specific gravity and therefore the average for a mixture should be calculated through Eq. (4.76) or (4.80). ♦

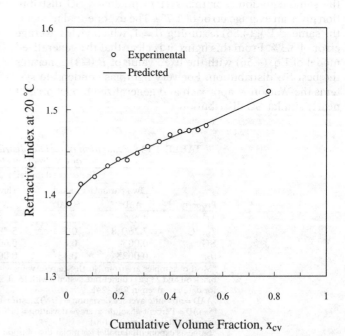

FIG. 4.22—Prediction of distribution of refractive index of C_{6+} of a North Sea oil from Eq. (4.56).

4.5.4.6 Prediction of Property Distributions Using Bulk Properties

As discussed above, Eq. (4.56) can be used as a two-parameter relation with fixed values of B for each property ($B_M = 1$, $B_T = 1.5$, and $B_{SG} = B_I = 3$). In this case Eq. (4.56) is referred as a two-parameter distribution model. In such cases only parameters P_o and A must be known for each property to express its distribution in a hydrocarbon plus fraction. The two-parameter model is sufficient to express property distribution of light oils and gas condensate systems. For very heavy oils two-parameter model can be used as the initial guess to begin calculations for determination of the three parameters in Eq. (4.56). In some cases detailed composition of a C_{7+} fraction in a reservoir fluid is not available and the only information known are M_{7+} and SG_{7+}, while in some other cases in addition to these properties, a third parameter such as refractive index of the mixture or the true boiling point (TBP) curve are also known. For these two scenarios we show how parameters P_o and A can be determined for M, T_b, SG, and I_{20}.

Method A: M_{7+}, SG_{7+}, and n_{7+} are known—Three bulk properties are the minimum data that are required to predict complete distribution of various properties [24, 43]. In addition to M_{7+} and SG_{7+}, refractive index, n_{7+}, can be easily measured and they are known for some 48 C_{7+} fractions [24]. As shown by Eqs. (4.74)–(4.76), if P_{av}^* is known, parameter A can be determined for each property. For example, if M_o is known, M_{av}^* can be determined from definition of M^* as:

$$(4.93) \qquad M_{av}^* = \frac{M_{av} - M_o}{M_o}$$

where M_{av} is the mixture molecular weight of the C_{7+} fraction, which is known from experimental measurement. Similarly, SG_{av}^* and I_{av}^* can be determined from S_{7+} and n_{7+} (or I_{7+}). Parameters A_M, A_{SG}, and A_I are then calculated from Eqs. (4.72) and (4.81). For fixed values of B, Eqs. (4.74)–(4.76) and (4.79) and (4.80) may be used. One should realize that Eq. (4.74) was developed based on cumulative mole fraction, while Eqs. (4.79) and (4.80) are based on cumulative weight fraction. Once distribution of M and SG are known, distribution of T_b can be determined using equations given in Chapter 2, such as Eqs. (2.56) or (2.57), for estimation of T_b from M and SG. Based on data for 48 C_{7+} samples, the following relation has been developed to estimate I_o from M_o and SG_o [23]:

$$
\begin{aligned}
I_o = {}& 0.7454 \exp(-0.01151 M_o - 2.37842 SG_o \\
& (4.94) \qquad + 0.01225 M_o SG_o) M_o^{0.2949} SG_o^{1.53147}
\end{aligned}
$$

This equation can reproduce values of I_o with an average deviation of 0.3%. Furthermore, methods of estimation of parameter I from either T_b and SG or M and SG are given in Section 2.6.2 by Eqs. (2.115)–(2.117). Equation (2.117) may be applied to the molecular weight range of 70–700. However, a more accurate relation for prediction of parameter I from M and SG is Eq. (2.40) with coefficients from Table 2.5 as follows:

$$
\begin{aligned}
I = {}& 0.12399 \exp(3.4622 \times 10^{-4} M + 0.90389 SG \\
& (4.95) \qquad -6.0955 \times 10^{-4} MSG) M^{0.02264} SG^{0.22423}
\end{aligned}
$$

This equation can be used for narrow boiling range fractions with M between 70 and 350. In this molecular weight range this equation is slightly more accurate than Eq. (2.117). Once distribution of I is determined from these equations, if the initial values of M_o and SG_o are correct then I_{av}^* calculated from the distribution coefficients and Eq. (4.72) should be close to the experimental value obtained from n_{7+}. For cases that experimental data on n_{7+} is not available it can be estimated from M_{7+} and SG_{7+} using Eq. (4.95) or (2.117). Equation (2.117) estimates values of n_{7+} for 48 systems [23] with an average error of 0.4%. Steps to predict M, T_b, SG, and I (or n) distributions can be summarized as follows [23]:

1. Read values of M_{7+}, SG_{7+}, and I_{7+} for a given C_{7+} sample. If I_{7+} is not available Eq. (2.117) may be used to estimate this parameter.
2. Guess an initial value for M_o (assume $M_o = 72$) and calculate M_{av}^* from Eq. (4.93).
3. Calculate A_M from Eq. (4.72) or Eq. (4.74) when $B = 1$.
4. Choose 20 (or more) arbitrary cuts for the mixture with equal mole fractions (x_{mi}) of 0.05 (or less). Then calculate M_i for each cut from Eq. (4.56).
5. Convert mole fractions (x_{mi}) to weight fractions (x_{wi}) through Eq. (1.15) using M_i from step 4.
6. Guess an initial value for S_o (assume $S_o = 0.59$ as a starting value).
7. Calculate $1/J$ from Eq. (4.80) using SG_o and SG_{7+}. Then calculate A_{SG} from Eq. (4.79) using Newton's method.
8. Using Eq. (4.56) with A_{SG} and S_o from steps 6 and 7 and $B = 3$, SG distribution in terms of x_{cw} is determined and for each cut SG_i is calculated.
9. Convert x_{wi} to x_{vi} using Eq. (1.16) and SG_i values from step 8.
10. For each cut calculate T_{bi} from M_i and SG_i through Eq. (2.56) or (2.57).
11. For each cut calculate I_i from M_i and SG_i through Eq. (2.95).
12. From distribution of I versus x_{cv} find parameters I_o, A_I and B_I through Eqs. (4.56)–(4.57). Then calculate I_{av} from Eq. (4.72) and (4.81).
13. Calculate $\varepsilon_1 = |(I_{av,calc.} - I_{7+})/I_{7+}|$.
14. If $\varepsilon_1 < 0.005$, continue from step 15, otherwise go back to step 6 with $SG_{o,new} = SG_{o,old} + 0.005$ and repeat steps 7–13.
15. Calculate I_o from Eq. (4.94).
16. Calculate $\varepsilon_2 = |(I_{o,step15} - I_{o,step12})/I_{o,step15}|$.
17. Go back to step 2 with a new guess for M_o (higher than the previous guess). Repeat steps 2–16 until either $\varepsilon_2 < 0.005$ or ε_2 becomes minimum.
18. For heavy oils large value of ε_2 may be obtained, because value of B_M is greater than 1. For such cases values of $B_M = 1.5$, 2.0, and 2.5 should be tried successively and calculations from step 2 to 17 should be repeated to minimize ε_2.
19. Using data for T_b versus x_{cw}, determine parameters T_o, A_T, and B_T from Eqs. (4.56) and (4.57).
20. Print M_o, A_M, B_M, SG_o, A_{SG}, T_o, A_T, B_T, I_o, A_I, and B_I.
21. Generate distributions for M, T_b, SG, and n_{20} from Eq. (4.56).

TABLE 4.17—*Sample calculations for prediction of distribution of properties of the C_{7+} fraction in Example 4.12.*

No.	x_m (1)	x_{cm} (2)	M (3)	x_w (4)	x_{cw} (5)	SG (6)	x_v (7)	x_{cv} (8)	T_b, K (9)	I (10)
1	0.05	0.025	90.7	0.038	0.019	0.719	0.002	0.001	353.9	0.242
2	0.05	0.075	92.3	0.039	0.058	0.727	0.007	0.006	357.9	0.245
3	0.05	0.125	93.9	0.040	0.097	0.732	0.012	0.015	361.6	0.246
I	...	...	...	...	...	...	...	...	...	...

The following example shows application of this method to find property distribution of a C_{7+} when only minimum data of M_{7+} and SG_{7+} are available.

Example 4.12—For the C_{7+} of Example 4.7, M, T_b, and SG distributions are given in Table 4.11. Distribution model coefficients are given in Table 4.13. For this gas condensate system, assume the only data available are $M_{7+} = 118.9$ and $SG_{7+} = 0.7597$. Using the method described above generate M, T_b, SG, and I distributions.

Solution—Since n_{7+} is not available we calculate I_{7+} from Eq. (4.95) using M_{7+} and SG_{7+} as $I_{7+} = 0.2546$ (equivalent to $n_{7+} = 1.4229$). Step-by-step calculations are followed and results of first few points as sample calculations are given in Table 4.17 where calculations are continued up to $i = 20$.

1. $M_{7+} = 118.9$, $SG_{7+} = 0.7597$, $I_{7+} = 0.2546$.
2. For the initial guess of M_o the minimum value of 72 can be used for computer programs. However, the actual value of M_o is very close to M_7^-, which is 88. For this gas condensate system we assume $M_o = 90$. If the calculated error is high then start from 72. By Eq. (4.93), $M_{av}^* = 0.3211$.
3. Assuming $B_M = 1$, from Eq. (4.72) $A_M = M_{av}^* = 0.3211$.
4. The C_{7+} fraction is divided into 20 cuts with equal mole fractions: $x_{mi} = 0.05$ (column 1 in Table 4.17). Now x_{cm} is calculated from x_{wi} as given in column 2. M_i for each cut is estimated through Eq. (4.56) with $M_o = 90$, $A_M = 0.3211$, and $B_M = 1$, and value of x_{cm}. Calculated values of M_i are given in column 3.
5. Weight fractions (x_{wi}) are calculated using x_{mi} and M_i through Eq. (1.15) and are given in column 4.
6. The lowest value of SG_o suitable for computer calculations is 0.59; however, it is usually close to value of the lower limit of SG for C_7 (SG_7^-), which is 0.709. Here it is assumed $SG_o = 0.7$.
7. With $SG_{7+} = 0.7579$ and $SG_o = 0.7$, from Eq. (4.80) we get $1/J = 1.0853$. Using Eq. (4.79) A_{SG} is calculated from $1/J$ as $A_{SG} = 0.0029$ (the second equation is used since $A_{SG} < 0.05$).
8. Cumulative x_{cw} is calculated from x_{wi} and are given in column 5. Using SG_o, A_{SG}, and $B_{SG} = 3$, SG distribution is calculated through use of x_{cw} and Eq. (4.56). Values of SG_i are given in column 6.
9. Volume fractions (x_{vi}) are calculated from x_{wi} and SG_i using Eq. (1.16) and are given in column 7. Cumulative volume fraction is given in column 8.
10. For each cut, T_{bi} is calculated from Eq. (2.56) using M_i and SG_i and is given in column 9.
11. For each cut, I_i is calculated from Eq. (4.95) using M_i and SG_i and is given in column 10.

12. From columns 9 and 10, distribution coefficients of Eq. (4.56) for I are calculated as $I_o = 0.236$, $A_I = 5.3 \times 10^{-5}$, and $B = 4.94$ ($R^2 = 0.99$ and %AAD = 0.16%).
13. From Eqs. (4.72) and (4.81), $I_{av} = 0.2574$ which gives $\varepsilon_1 = 0.01$.
14. ε_1 in step 13 is greater than 0.005; however, a change in SG_o causes a slight change in the error parameter so this value of ε_1 is acceptable.
15. I_o is calculated from Eq. (4.94) using M_o and SG_o as 0.2364.
16. ε_2 is calculated from I_o in steps 15 and 12 as 0.0018, which is less than 0.005.
17. Go to step 18 since $\varepsilon_2 < 0.005$.
18. Since values of ε_1 and ε_2 are acceptable the assumed value of $B_M = 1$ is OK.
19. From columns 5 and 9, distribution coefficients for T_b are calculated as $T_o = 350$ K, $A_T = 0.161$, and $B_T = 1.3$ ($R^2 = 0.998$ and %AAD = 0.3).
20. Final predicted distribution coefficients for M, T_b, SG, and I are given in Table 4.18.
21. Predicted distributions for M, T_b, SG, and I are shown in Figs. 4.20–4.23, respectively. ♦

Method B: M_{7+}, SG_{7+}, and TBP are known—In some cases true boiling point (TBP) distillation curve for a crude or C_{7+} fraction is known through simulated distillation or other methods described in Section 4.1.1. Generally TBP is available in terms of boiling point versus volume or weight fraction. If in addition to TBP, two bulk properties such as M_{7+} and SG_{7+} or M_{7+} and n_{7+} are known, then a better prediction of complete distribution of various properties is possible by applying the generalized distribution model. For these cases an initial guess on SG_o gives complete distribution of SG through Eq. (4.56) along the T_b distribution, which is available from data. Having T_b and SG for each cut, Eq. (2.51) can be used to predict M for each subfraction. Using Eq. (2.115) or (2.116), I_{20} can be estimated for cuts with M values up to 350. For heavier cuts Eq. (2.117) may be used. The procedure can be summarized as follows [24]:

1. Read values of M_{7+}, SG_{7+}, and the TBP distribution (i.e., SD curve) for a given crude oil sample.
2. From TBP determine distribution coefficients in Eq. (4.56) for T_b in terms of x_{cw} or x_{cv}. If simulated distillation is available, x_{cw} should be used.

TABLE 4.18—*Estimated coefficients of Eq. (4.56) for the C_{7+} of Example 4.12.*

Property	P_o	A	B	Type of x_c
M	90	0.3211	1.0	x_{cm}
T_b	350	0.1610	1.3	x_{cw}
SG	0.7	0.0029	3.0	x_{cw}
I	0.236	5.4×10^{-5}	4.94	x_{cv}

TABLE 4.19—*Sample calculations for prediction of distribution of properties of the C_{7+} fraction in Example 4.13.*

No.	x_{wi} (1)	x_{cw} (2)	T_{bi}, K (3)	SG_i (4)	x_{vi} (5)	x_{cv} (6)	M_i (7)	x_{mi} (8)	I_i (9)
1	0.05	0.025	353.8	0.720	0.053	0.026	91.4	0.065	0.244
2	0.05	0.075	359.3	0.730	0.052	0.079	93.9	0.063	0.247
3	0.05	0.125	364.3	0.735	0.052	0.131	96.3	0.062	0.249
i	...	...	...	...	...	...	...	...	...

3. Choose 20 (or more) arbitrary cuts for the mixture with equal weight (or volume) fractions of 0.05 (or less). Then determine T_{bi} for each cut from Eq. (4.56) and coefficients from step 2.

4. Guess an initial value for SG_o (lowest value is 0.59).

5. Calculate $1/J$ from Eq. (4.80) using SG_o and SG_{7+}. Then calculate A_{SG} from Eq. (4.79) using Newton's method or other appropriate procedures. If original TBP is in terms of x_{cv}, then Eq. (4.76) should be used to determine A_{SG} in terms of x_{cv}.

6. If original TBP is in terms of x_{cv}, find SG distribution from Eq. (4.56) in terms of x_{cv}. Then use SG to convert x_v to x_w through Eq. (1.16).

7. Using values of SG and T_b for each cut determine values of M from Eq. (2.56).

8. Use values of M from step 7 to convert x_w into x_m through Eq. (1.15).

9. From data calculated in step 8, find molar distribution by estimating coefficients M_o, A_M, and B_M in Eq. (2.56).

10. Calculate value of I for each cut from T_b and SG through Eq. (2.115) or (2.116).

11. Find coefficients I_o, A_I, and B_I (set $B_I = 3$) from data obtained in step 10. Set $I_1 = I_o$.

12. From SG_o assumed in step 4 and M_o determined in step 9, estimate I_o through Eq. (4.94). Set $I_2 = I_o$.

13. Calculate $\varepsilon_1 = |(I_2 - I_1)/I_1|$.

14. If $\varepsilon_1 \geq 0.005$ go back to step 4 by guessing a new value for SG_o. If $\varepsilon_1 < 0.005$ or it is minimum go to step 15.

15. Print M_o, A_M, B_M, SG_o, A_{SG}, T_o, A_T, B_T, I_o, A_I, and B_I.

16. Generate distributions for M, T_b, SG, and n_{20} from Eq. (4.56).

The following example shows application of Method B to find property distribution of a C_{7+} when data on TBP distillation, M_{7+} and SG_{7+} are available.

Example 4.13—For the C_{7+} of Example 4.7, assume that T_b distillation curve is available, as given in columns 3 and 5 in Table 4.11. In addition assume that for this sample $M_{7+} = 118.9$ and $SG_{7+} = 0.7597$ are also available. Using the method described above (Method B) generate M, T_b, SG, and I distributions. Graphically compare prediction of various distributions by Methods A and B with actual data given in Table 4.11.

Solution—Similar to Example 4.12, step-by-step procedure described under Method B should be followed. Since data on distillation are given in terms of weight fractions (column 3 in Table 4.11) we choose weight fraction as the reference for the composition. From data on T_{bi} versus x_{wi} distribution coefficients in Eq. (4.56) can be determined. This was already done in Example 4.7 and the coefficients are given in Table 4.13 as:

$T_o = 350$ K, $A_T = 0.1679$, and $B_T = 1.2586$. An initial guess value of $SG_o = 0.7$ is used to calculate A_{SG} and SG distribution coefficients. Now we divide the whole fraction into 20 cuts with equal weight fractions as $x_{wi} = 0.05$. Similar to calculations shown in Table 4.17, x_{cw} is calculated and then for each cut values of T_{bi} and SG_i are calculated. From these two parameters M_i and I_i are calculated by Eqs. (2.56) and (2.115), respectively. From x_{wi} and M_i mole fractions (x_{mi}) are calculated. Sample calculation for the first few points is given in Table 4.19 where calculation continues up to $i = 20$. The coefficients of Eq. (4.56) determined from data in Table 4.19 are given in Table 4.20. In this method parameter $\varepsilon_1 = 0.0018$ (step 13), which is less than 0.005 and there is no need to re-guess SG_o. In this set of calculations since initial guess for SG_o is the same as the actual value only one round of calculations was needed. Coefficients given in Table 4.20 have been used to generate distribution for various properties and they are compared with predicted values from Method A as well as actual values given in Table 4.11. Results are shown in Figs. 4.23 and 4.24 for prediction of M and T_b distributions. Methods A and B predict similar distribution curves mainly

TABLE 4.20—*Estimated coefficients of Eq. (4.56) for the C_{7+} of Example 4.13.*

Property	P_o	A	B	Type of x_c
M	90	0.3324	1.096	x_{cm}
T_b	350	0.1679	1.2586	x_{cw}
SG	0.7	0.0029	3.0	x_{cw}
I	0.236	7.4×10^{-4}	3.6035	x_{cv}

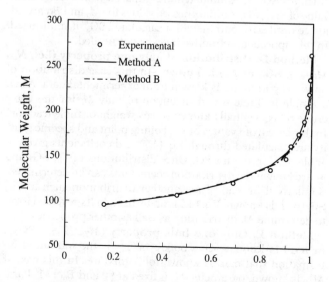

FIG. 4.23—Prediction of molar distribution in Examples 4.12 and 4.13.

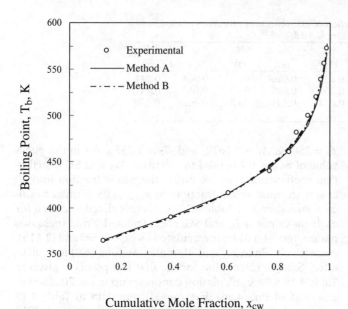

because the system is gas condensate and value of B_M is one. For very heavy oils Method B predicts better prediction. As shown in Fig. 4.24 for T_b, Method B gives better prediction mainly because information on at least one type of distribution was available. ◆

Method C: M_{7+} and SG_{7+} are known—An alternative to method A when only M_{7+} and SG_{7+} are known is to predict M distribution by assuming $B_M = 1$ and a value for M_o as steps 1–5 in method A. For every value of M, SG is estimated from Eq. (4.7) using coefficients given in Table 4.5 for SG. Then parameters SG_o, A_{SG}, and B_{SG} are calculated. From these coefficients SG_{av} is estimated and compared with experimental value of SG_{7+}. The initial guessed values for M_o and B_M are adjusted until error parameter for calculated SG_{av} is minimized. In this approach, refractive index is not needed.

Method D: Distribution of only one property (i.e., N_{ci}, M_i, T_{bi}, SG_i, or N_i) is known—In this case distribution of only one parameter is known from experimental data. As an example in Table 4.2, distribution of only M_i for the waxy crude oil is originally known versus weight or mole fraction. In this case from values of M_i, boiling point and specific gravity are calculated through Eq. (4.7) and coefficients given in Table 4.5 for T_b and SG. Once distributions of T_b, SG, and M are known the distribution coefficients can be determined. Similarly if instead of M_i, another distribution such as T_{bi}, SG_i, or I_i is known, Eq. (4.7) can be used in its reversed form to determine M_i distribution as well as other properties.

Method E: Only one bulk property (M_{7+}, T_{b7+}, SG_{7+}, or n_{7+}) is known—One bulk property is the minimum information that can be known for a mixture. In this case if M_{7+} is known, parameter M_o is fixed at 90 and $B_M = 1$. Parameter A_M is calculated from Eq. (4.74). Once distribution of M is found, SG distribution can be estimated through use of Eq. (4.7) and coefficients in Table 4.5 for SG. Similarly if only SG_{7+} is known, assume $SG_o = 0.7$ and $B_{SG} = 3$. Coefficient

A_{SG} is calculated from Eq. (4.76) and then distribution of SG versus x_{cv} can be obtained through Eq. (4.56). Once SG distribution is known, the reversed form of Eq. (4.7) should be used to estimate M distribution. In a similar approach if n_{7+} is known distributions of M, T_b, and SG can be determined by assuming $I_o = 0.22$ and $B_I = 3$ and use of Eq. (4.7). Obviously this method gives the least accurate distribution since minimum information is used to obtain the distributions. However, this method surprisingly well predicts boiling point distribution from specific gravity (as the only information available) for some crude oils as shown by Riazi et al. [40].

4.6 PSEUDOIZATION AND LUMPING APPROACHES

Generally analytical data for reservoir fluids and crude oils are available from C_1 to C_5 as pure components, group C_6, and all remaining and heavier components are grouped as a C_{7+} fraction as shown in Tables 1.2 and 4.1. As discussed earlier for wide C_{7+} and other petroleum fractions assumption of a single pseudocomponent leads to significant errors in the characterization scheme. In such cases, distribution functions for various characterization parameters are determined through Methods A or B discussed in Section 4.5.4.6. Once the molar distribution is known through an equation such as Eq. (4.56), the mixture (i.e., C_{7+}) can be split into a number of pseudo-components with known x_i, M_i, T_{bi}, and SG_i. This technique is called *pseudoization* or *splitting* and is widely used to characterize hydrocarbon plus fractions, reservoir fluids, and wide boiling range petroleum fractions [15, 17, 18, 23, 24, 26, 36]. In some other cases detailed analytical data on the composition of a reservoir fluid are available for SCN groups such as those shown in Table 4.2. Properties of these SCN groups are determined from methods discussed in Section 4.3. However, when the numbers of SCN components are large (i.e., see Table 4.2) computational methods specially those related to phase equilibrium would be lengthy and cumbersome. In such cases it is necessary to lump some of these components into single groups in order to reduce the number of components in such a way that calculations can be performed smoothly and efficiently. This technique is called *lumping* or *grouping* [24, 26]. In both approaches the mixture is expressed by a number of pseudocomponents with known mole fractions and characterization parameters which effectively describe characteristics of the mixture. These two schemes are discussed in this section in conjunction with the generalized distribution model expressed by Eqs. (4.56) and (4.66).

4.6.1 Splitting Scheme

Generally a C_{7+} fraction is split into 3, 5, or 7 pseudocomponents. For light oils and gas condensate systems C_{7+} is split into 3 components and for black oils it is split into 5 or 7 components. For very heavy oils the C_{7+} may be split to even 10 components. But splitting into 3 for gas condensate and 5 for oils is very common. When the number of pseudocomponents reaches ∞, behavior of defined mixture will be the same as continuous mixture expressed by a distribution model such as Eq. (4.56). Two methods are presented here to generate the pseudocomponents. The first approach is based on the application of Gaussian quadrature technique as discussed by Stroud and Secrest [44]. The second method is based on carbon number range approach in which for each pseudocomponent the lower and higher carbon numbers are specified.

4.6.1.1 The Gaussian Quadrature Approach

The Gaussian quadrature approach is used to provide a discrete representation of continuous functions using different numbers of quadrature points and has been applied to define pseudocomponents in a petroleum mixture [23, 24, 28]. The number of pseudocomponents is the same as the number of quadrature points. Integration of a continuous function such as $F(P)$ can be approximated by a numerical integration as in the following form [44]:

$$(4.96) \quad \int_0^\infty f(y)\exp(-y)dy = \sum_{i=1}^{N_P} w_i f(y_i) = 1$$

where N_P is the number of quadrature points, w_i are weighting factors, y_i are the quadrature points, and $f(y)$ is a continuous function. Sets of values of y_i and w_i are given in various mathematical handbooks [38]. Equation (4.96) can be applied to a probability density function such as Eq. (4.66) used to express molar distribution of a hydrocarbon plus fraction. The left side of Eq. (4.96) should be set equal to Eq. (4.67). In this application we should find $f(y)$ in a way that

$$(4.97) \quad \int_0^\infty F(P^*)dP^* = \int_0^\infty f(y)\exp(-y)dy = 1$$

where $F(P^*)$ is given by Eq. (4.66). Assuming

$$(4.98) \quad y = \frac{B}{A}P^{*B}$$

and integrating both sides

$$(4.99) \quad dy = \frac{B^2}{A}P^{*B-1}dP^*$$

Using Eq. (4.66) we have

$$F(P^*)dP^* = \frac{B^2}{A}P^{*B-1}\exp\left(-\frac{B}{A}P^{*B}\right)dP^*$$
$$(4.100) \qquad = 1 \times \exp(-y)dy$$

By comparing Eqs. (4.97) and (4.100) one can see that

$$(4.101) \quad f(y) = 1$$

and from Eq. (4.96) we get

$$(4.102) \quad z_i = w_i$$

where z_i is the mole fraction of pseudocomponent i. Equation (4.102) indicates that mole fraction of component i is the same as the value of quadrature point w_i. Substituting definition of P^* as $(P - P_o)/P_o$ in Eq. (4.98) gives the following relation for property P_i:

$$(4.103) \quad P_i = P_o\left[1 + \left(\frac{A}{B}\right)^{1/B}y_i^{1/B}\right]$$

Coefficients P_o, A and B for a specific property are known from the methods discussed in Section 4.5.4.6. Table 4.21

TABLE 4.21—*Gaussian quadrature points and weights for 3 and 5 points [38].*

i	Root y_i	Weight w_i
$N_P = 3$		
1	0.41577	7.11093×10^{-1}
2	2.29428	2.78518×10^{-1}
3	6.28995	1.03893×10^{-2}
$N_P = 5$		
1	0.26356	5.21756×10^{-1}
2	1.41340	3.98667×10^{-1}
3	3.59643	7.59424×10^{-2}
4	7.08581	3.61176×10^{-3}
5	12.64080	2.33700×10^{-5}

gives a set of values for roots y_i and weights w_i as given in Ref. [37].

Similarly it can be shown that for the gamma distribution model, Eq. (4.31), $f(y)$ in Eq. (4.96) becomes

$$(4.104) \quad f(y) = \frac{y^{\alpha-1}}{\Gamma(\alpha)}$$

and mole fraction of each pseudocomponent, z_i, is calculated as

$$(4.105) \quad z_i = w_i f(y_i) = w_i \frac{y_i^{\alpha-1}}{\Gamma(\alpha)}$$

Molecular weight M_i for each pseudocomponent is calculated from

$$(4.106) \quad M_i = y_i\beta + \eta$$

where α, β, and η are parameters defined in Eq. (4.31). It should be noted that values of z_i in Eq. (4.102) or (4.105) is based on normalized composition for the C_{7+} fraction (i.e., $z_{7+} = 1$) at which the sum of z_i for all the defined pseudocomponents is equal to unity. For both cases in Eqs. (4.102) and (4.106) we have

$$(4.107) \quad \sum_{i=1}^{N_P} z_i = 1$$

To find mole fraction of pseudocomponent i in the original reservoir fluid these mole fractions should be multiplied by the mole fraction of C_{7+}. Application of this method is demonstrated in Example 4.14.

Example 4.14—For the gas condensate system described in Example 4.13 assume the information available on the C_{7+} are $M_{7+} = 118.9$ and $SG_{7+} = 0.7597$. Based on these data, find three pseudocomponents by applying the Gaussian quadrature method to PDF expressed by Eq. (4.66). Find the mixture M_{7+} based on the defined pseudocomponents and compare with the experimental value. Also determine three pseudocomponents by application of Gaussian quadrature method to the gamma distribution model.

Solution—For Eq. (4.56), the coefficients found for M in Example 4.13 may be used. As given in Table 4.20 we have $M_o = 90$, $A_M = 0.3324$, and $B_M = 1.096$. Values of quadrature points and weights for three components are given in Table 4.21. For each root, y_i, corresponding value of M_i is determined from Eq. (4.103). Mole fractions are equal to the

TABLE 4.22—*Generation of pseudocomponents from Gaussain quadrature method for the C_{7+} sample in Example 4.14.*

i	y_i	w_i	z_i	M_i	$z_i M_i$
1	0.416	0.711	0.711	103.6	73.7
2	2.294	0.279	0.279	154.6	43.1
3	6.290	0.010	0.010	252.2	2.6
Mixture	...	1.000	1.000	...	119.4

weighting factors according to Eq. (4.102). Summary of calculations and mole fractions and molecular weights of the components are given in Table 4.22. As shown in this table average molecular weight of C_{7+} calculated from the 3 pseudocomponents is 119.4, which varies with experimental value of 118.9 by 0.4%. To apply the Gaussian method to the gamma distribution model first we must determine coefficients α, β, and η. Since detailed compositional data are not available as discussed in Section 4.5.3, we assume $\eta = 90$ and $\alpha = 1$. Then β is calculated from Eq. (4.33) as $\beta = (118.9 - 90)/1.0 = 28.9$. Substituting $\alpha = 1$ in Eq. (4.105), considering that $\Gamma(\alpha) = 1$ we get $z_i = w_i$. Values of M_i are calculated from Eq. (4.106) with y_i taken from Table 4.21. Three components have molecular weights of 102, 156.3, and 271.6, respectively. M_{7+} calculated from these values and mole fractions given in Table 4.22 is exactly 118.9 the same as the experimental data. The reason is that this value was used to obtain parameter β. For this sample z_i calculated from the gamma PDF is the same as those obtained from Eq. (4.102) since $\alpha = 1$ and Eq. (2.105) reduces to Eq. (2.102). But for values of α different from unity, the two models generate different mole fractions and different M_i values. ♦

4.6.1.2 Carbon Number Range Approach

In this approach we divide the whole C_{7+} fraction into a number of groups with known carbon number boundaries. As an example if five pseudocomponents are chosen to describe the mixture, then five carbon number ranges must be specified. It was found that for gas condensate systems and light oils the carbon number ranges of C_7–C_{10}, C_{11}–C_{15}, C_{16}–C_{25}, C_{25}–C_{36}, and C_{36+} well describe the mixture [23]. It should be noted that the heaviest component in the first group is C_{10}^+, which is the same as the lightest component in the second group is C_{11}^-. Values of the lower and upper limit molecular weights for each SCN group can be calculated from Eqs. (4.39) and (4.40) and SCN up to C_{20} were calculated and are given in Table 4.10. For example for the C_7–C_{10}, the molecular weight range is from M_0 (initial molecular weight of a C_{7+} fraction) to M_{10}^+ or $M_0 - 142.5$ and for the C_{11}–C_{15}, the molecular weight range is M_{11}^-–M_{15}^+ or 142.5–214. Similarly molecular weight range of other groups can be determined as: 214–352 for C_{16}–C_{25}, 352–492 for C_{26}–C_{35}, and 492–∞ for C_{36+}–∞. For the last group; i.e., C_{36+} the molecular weight range is from M_{36}^- to ∞. Once the lower and upper values of M are known mole fraction and molecular weight for each group can be determined from appropriate equations developed for each distribution model. Mole fraction and molecular weight of each group for the gamma distribution model are determined from Eqs. (4.37) and (4.38), respectively. For the generating distribution model these equations are Eq. (4.84) and (4.86) and for the exponential model mole fractions are calculated from

Eq. (4.92) and molecular weights from Eq. (4.91). Step-by-step calculations for both of these methods with an example are given in the next section.

4.6.2 Lumping Scheme

The lumping scheme is applied when composition of a reservoir fluid or crude oil is given in terms of SCN groups such as those given in Table 4.2. Whitson [15, 17] suggests that the C_{7+} fraction can be grouped into N_P pseudocomponents given by

$$(4.108) \qquad N_P = 1 + 3.3 \log_{10}(N_+ - 7)$$

where N_P is the number of pseudocomponents and N_+ is the carbon number of heaviest fraction in the original fluid description. Obviously N_P is the nearest integer number calculated from the above equation. The groups are separated by molecular weight M_j given by

$$(4.109) \qquad M_j = M_{7+}(M_{N_+}/M_{7+})^{1/N_P}$$

where $j = 1, \ldots, N_P$. SCN groups in the original fluid description that have molecular weights between boundaries M_{j-1} and M_j are included in the group j. This method can be applied only to those C_{7+} fractions that are originally separated by SCN groups and $N_P \geq 20$ [17].

The lumping scheme is very similar to the pseudoization method, except the distribution coefficients are determined for data on distribution of carbon number. For this reason the lumping scheme generates better and more accurate pseudocomponents than does the splitting method when distribution coefficients are determined from only two bulk properties such as M_{7+} and SG_{7+}. Method of lumping is very similar to the calculations made in Example 4.10 in which SCN groups of C_{12} and C_{13} for the C_{7+} sample in Table 4.11 were lumped together and the mole fraction and molecular weight of the group were estimated. Here the two methods that can be used for lumping and splitting schemes are summarized to show the calculations [24]. In these methods the generalized distribution model is used; however, other models (i.e., gamma or exponential) can be used in a similar way.

Method I: Gaussian Quadrature Approach

1. Read properties of SCN groups and properties of plus fractions (e.g., M_{30+} and SG_{30+}). Normalize the mole fractions ($\sum x_{mi} = 1$).
2. If M and SG for each SCN group are not available, obtain these properties from Table 4.6.
3. Determine distribution parameters for molecular weight (M_0, A_M, and B_M) in terms of cumulative mole fraction and for specific gravity (SG_0, A_{SG}, and B_{SG}) in terms of cumulative weight fraction.
4. Choose the number of pseudocomponents (i.e., 5) and calculate their mole fractions (z_i) and molecular weight (M_i) from Eqs. (4.102) and (4.103) or from Eqs. (4.105) and (4.106) for the case of gamma distribution model.
5. Using M_i and z_i in step 4, determine discrete weight fractions, z_{wi}, through Eq. (1.15).
6. Calculate cumulative weight fraction, z_{cw}, from z_{wi} and estimate SG_i for each pseudocomponent through Eq. (4.56) with coefficients determined for SG in step 3. For example, SG_1 can be determined from Eq. (4.56) at $z_{cw1} = z_{w1}/2$.

TABLE 4.23—*Lumping of SCN groups by two methods for the C_{7+} sample in Example 4.15 [24].*

Component	Method I: Gaussian quadrature approach				Method II: Carbon number range approach			
i	Mole fraction	Weight fraction	M_i	SG_i	Mole fraction	Weight fraction	M_i	SG_i
1	0.5218	0.3493	102.1	0.7436	0.532	0.372	106.7	0.7457
2	0.3987	0.4726	180.8	0.8023	0.302	0.328	165.5	0.7957
3	0.0759	0.1645	330.4	0.8591	0.144	0.240	254.4	0.8389
4	0.0036	0.0134	569.5	0.9174	0.019	0.049	392.7	0.8847
5	2.3×10^{-5}	1.4×10^{-4}	950.1	0.9809	0.003	0.011	553.5	0.9214
Mixture	1.0000	1.0000	152.5	0.7905	1.000	1.0000	152.5	0.7908

Taken with permission from Ref. [24].

7. Obtain M_{av} and SG_{av} for the mixture from $M_{av} = \sum_{j=1}^{N_P} z_j M_j$ and $1/SG_{av} = \sum_{j=1}^{N_P} z_{wi}/SG_j$.

Method II: Carbon Number Range Approach

1. Same as Method I.
2. Same as Method I.
3. Same as Method I.
4. Choose number of pseudocomponents (i.e., 5) and corresponding carbon number ranges, e.g., group 1: C_7–C_{10}, group 2: C_{11}–C_{15}, group 3: C_{16}–C_{25}, group 4: C_{25}–C_{36} and C_{36+}.
5. Obtain molecular weight boundaries from Eqs. (4.39) and (4.40). For example for the groups suggested in step 4 the molecular weight ranges are: $(M_o-142.5)$, (142.5–214), (214–352), (352–492) and (492–∞). The number of pseudocomponents (N_P) and molecular weight boundaries may also be determined by Eqs. (4.108) and (4.109).
6. Using the molecular weight boundaries determined in step 5, calculate mole fractions (z_i) and molecular weight (M_i) of these pseudocomponents from Eqs. (4.84) and (4.86) or from Eqs. (4.92) and (4.91) when B_M in Eq. (4.56) is equal to unity.
7. Same as step 5 in Method I.
8. Same as step 6 in Method I.
9. Same as step 7 in Method I.

In this method if the calculated mole fraction for a pseudocomponent in step 6 is too high or too low, we may reduce or increase the corresponding carbon number range chosen for that pseudocomponent in step 4. Application of these methods is shown in the following example.

Example 4.15—Fluid description of a C_{7+} from North Sea fields (sample 42 in Ref. [24]) is given in terms of mole fractions of SCN groups from C_7 to C_{20+} as

N_C	7	8	9	10	11	12	13	14	15	16	17	18	19	20+
x_i	0.178	0.210	0.160	0.111	0.076	0.032	0.035	0.029	0.022	0.020	0.020	0.016	0.013	0.078

where N_C represents carbon number group and x_i is its corresponding normalized mole fraction. For this mixture the $M_{7+} = 151.6$ and $SG_{7+} = 0.7917$. Lump these components into an appropriate number of pseudocomponents and give their molecular weight and specific gravity using the above two methods.

Solution—For this sample $N_+ = 20$ and we may use Eq. (4.108) to determine the number of pseudocomponents. $N_P = 1 + 3.31 \log(20-7) = 4.7$. The nearest integer number is 5, therefore $N_P = 5$, which is the same number as suggested in step 4 of the above methods. For carbon numbers from C_7 to C_{19}, values of M and SG are taken from Table 4.6 and mole fractions are converted into weight fraction (x_{wi}). Distribution coefficients for M in terms of x_{cm} and SG in terms of x_{cw} are then determined from Eqs. (4.56) and (4.57). The results for M are $M_o = 84$, $A_M = 0.7157$, and $B_M = 1$ and for SG the coefficients are $SG_o = 0.655$, $A_{SG} = 0.038$, and $B_{SG} = 3$. For the 5 pseudocomponents, Methods I and II have been applied step by step and for each group j values of z_{mi}, z_{wi}, M_i, and SG_i are given in Table 4.23. Specific gravity and molecular weight of C_{7+} calculated from pseudocomponents generated by Method I are 0.7905 and 152.5, which are very close to experimental values of 0.7917 and 151.6. Method II gives similar results as shown in Table 4.23. Specific gravity differs from the experimental value by 0.1%. Obviously components 1, 2,... generated in Method I are not the same components generated by Method II, but combination of all 5 components by two methods represent the same mixture. That is why M_i and SG_i for the 5 pseudocomponents generated by Methods I and II are not the same. ♦

4.7 CONTINUOUS MIXTURE CHARACTERIZATION APPROACH

A more complicated but more accurate treatment of a C_{7+} fraction is to consider it as a continuous mixture. In this approach the mixture is not expressed in terms of a finite number of pseudocomponent but its properties are given by a continuous function such as Eq. (4.56). This method is in fact equivalent to the pseudocomponent approach but with infinite number of components ($N_P = \infty$). Mansoori and Chorn [27] discussed a general approach toward characterization of continuous mixtures. In this approach instead of specifying a component by i, it is expressed by one of its characteristic parameters such as T_b or M. Formulation of continuous mixtures for phase equilibrium calculations is best expressed by Eq. (4.15), while for the pseudocomponent approach for a defined discrete mixture it is formulated through Eq. (4.14).

To show application of a PDF in characterization of a crude oil by the continuous mixture approach, we use Eq. (4.15) to formulate vapor–liquid equilibrium (VLE) and to obtain species distribution of vapor and liquid products once such distribution is known for the feed during a flash distillation process. Theory of VLE is discussed in Chapter 6 and its application is shown in Chapter 9. In Eq. (4.15), if we take boiling point as the characterization parameter for P the equilibrium relation in terms of fugacity is (see Eq. 6.173)

$$(4.110) \qquad f^V(T) = f^L(T) \qquad T_o \leq T \leq \infty$$

where $f^V(T)$ is the fugacity of a specie in the vapor phase whose boiling point is T. When $T = T_o$ the above equation is applied to the lightest component in the mixture and when $T = \infty$ it is applied to the last and heaviest component whose boiling point may be considered as infinity. For simplicity it is assumed that the vapor phase is ideal gas and the liquid phase is an ideal solution. Under such conditions Eq. (4.110) for a component with boiling point T in a mixture can be written as

$$\text{(4.111)} \qquad \mathrm{d}y_T\, p = \mathrm{d}x_T\, p_{T^s}$$

where $\mathrm{d}y_T$ is the mole fraction of a component having boiling point T in the vapor phase and $\mathrm{d}x_T$ is the mole fraction of the same component in the liquid phase. p_{T^s} is the saturation pressure (or vapor pressure) of components with boiling point T at temperature T^s and p is the total pressure at which vapor and liquid are in equilibrium. T^s is in fact the temperature at which separation occurs and p_{T^s} is a function of T^s and type of component that is characterized by boiling point (see Problem 4.16). This relation is known as the Raoult's law and its derivation will be discussed in Chapter 6. In Eq. (4.111), $\mathrm{d}y_T p$ is the fugacity of components with boiling point T in an ideal gas vapor phase while $\mathrm{d}x_T p_{T^s}$ is the fugacity of components with boiling point T in an ideal liquid solution. To apply Eq. (4.111) for a continuous mixture, we can use Eq. (4.16) to express $\mathrm{d}x_T$ and $\mathrm{d}y_T$ by a PDF in each phase:

$$\text{(4.112)} \qquad \mathrm{d}x_T = F_T^L\, \mathrm{d}T$$

$$\text{(4.113)} \qquad \mathrm{d}y_T = F_T^V\, \mathrm{d}T$$

where F_T^L and F_T^V are the PDF in terms of boiling point T for the liquid and vapor phases, respectively. Equations (4.70) or (4.31) may be used to express F_T^V or F_T^L. Substituting Eqs. (4.112) and (4.113) into Eq. (4.111) we get

$$\text{(4.114)} \qquad F_T^V\, p = F_T^L\, p_T^s$$

Equation (4.114) is the Raoult's law in terms of a PDF applicable to a continuous mixture. If the liquid phase is nonideal, the right-hand side of above equation should be multiplied by activity coefficient $\gamma(T, T^s)$ for those components with boiling point T at temperature T^s. And if the vapor phase is nonideal

gas the left-hand side of Eq. (4.114) should be multiplied by fugacity coefficient $\varphi(T, T^s, p)$ for components with boiling point T at temperature T^s and pressure p. These thermodynamic properties are defined in Chapter 6 and can be obtained from an equation of state for hydrocarbon systems. A more general form of Eq. (4.114) for high-pressure VLE calculations is in terms equilibrium ratio can be written as

$$\text{(4.115)} \qquad F_T^V = F_T^L \times K_T$$

where K_T is the equilibrium ratio for a component with boiling point T at temperature T^s and pressure p. As it will be shown in Chapter 6, K_T depends on vapor pressure p^s.

Now we apply the above equations for design and operation of a separation unit for flash distillation of reservoir fluids and crude oils. As shown in Fig. 4.25 we assume 1 mol of feed enters the unit that is operating at temperature T^s and pressure p. The products are ϕ moles of vapor and $1 - \phi$ moles of liquid in which ϕ is the fraction of the feed vaporized in a single-stage flash distillation unit. Material balance on the distillation unit for a component whose boiling point is T can be written as

$$\text{(4.116)} \qquad \mathrm{d}z_T \times 1 = \mathrm{d}x_T \times (1 - \phi) + \mathrm{d}y_T \times \phi$$

where $\mathrm{d}z_T$ is the mole fraction of all components having boiling point T and can be expressed in terms of a PDF similar to Eq. (4.112). Substituting Eqs. (4.112) and (4.113) for $\mathrm{d}x_T$ and $\mathrm{d}y_T$ and similarly for $\mathrm{d}z_T$ into the above equation gives

$$\text{(4.117)} \qquad F_T^F = (1 - \phi)F_T^V + \phi F_T^L \phi$$

where F_T^F is the density function for the feed in terms of boiling point T. For all three probability density functions, F^F, F^V, and F^L we have

$$\text{(4.118)} \qquad \int_{T_o}^{\infty} F_T^F\, \mathrm{d}T = \int_{T_o}^{\infty} F_T^V\, \mathrm{d}T = \int_{T_o}^{\infty} F_T^L\, \mathrm{d}T = 1$$

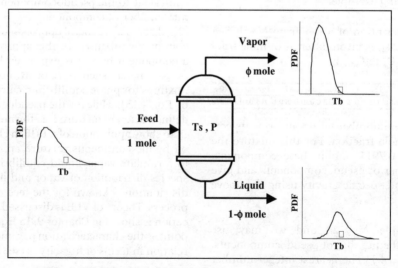

FIG. 4.25—Schematic of a single-stage flash distillation unit.

From Eqs. (4.114), (4.117), and (4.118) one can derive the following relation for calculation of parameter ϕ:

$$(4.119) \qquad \int_{T_o}^{\infty} \frac{p - p_T^s}{(1 - \phi)p + \phi p_T^s} F_T^F dT = 0$$

where the integration should be carried numerically and ϕ may be determined by trial-and-error procedure. As will be shown in Chapter 7, combination of Trouton's rule for the heat of vaporization and the Clasius–Clapeyron equation leads to the following relation for the vapor pressure:

$$(4.120) \qquad p_{T^s} = p_a \exp\left[10.58\left(1 - \frac{T}{T^s}\right)\right]$$

where T is the boiling point of each cut in the distribution model, T^s is the saturation temperature, and p_a is the atmospheric pressure. Both T and T^s must be in K. By combining Eqs. (4.114) and (4.117) we get

$$(4.121) \qquad F_T^L = \frac{p}{(1 - \phi)p + \phi p_T^s} F_T^F$$

$$(4.122) \qquad F_T^V = \frac{p_T^s}{(1 - \phi)p + \phi p_T^s} F_T^F$$

After finding ϕ from Eq. (4.119), it can be substituted in the above equations to find density functions for the vapor and liquid products.

For evaluation and application of these equations, data on boiling point distribution of a Russian crude oil as given by Ratzch et al. [31] were used. In this case TBP distributions for feed, vapor, and liquid streams during flash distillation of the crude are available. Molecular weight, specific gravity, and refractive index of the mixture are 200, 0.8334 and 1.4626, respectively. Applying Method A discussed in Section 4.5.4.6, we obtain distribution coefficients for boiling point of feed as: $T_o = 241.7$ K, $A_T = 1.96$, and $B_T = 1.5$ and F_T^F was determined from Eq. (4.70). Fraction of feed vaporized, ϕ, was determined from Eq. (4.119) as 0.7766. Boiling point distributions for the liquid (F_T^L) and vapor (F_T^V) products were determined from Eqs. (4.121) and (4.122), respectively. Results of calculations for F_T^F, F_T^V, and F_T^L for this crude are shown in Fig. 4.26 and compared with the experimental values provided in Ref. [31]. Since heavier components appear in the liquid product, therefore, the curve for F_T^L is in the right side of both F_T^F and F_T^V corresponding to higher values of boiling points.

Part of errors for predicted distributions of F_T^L and F_T^V is due to assumption of an ideal solution for VLE calculations as well as an approximate relation for the estimation of vapor pressures. For more accurate calculations Eq. (4.115) can be used which would result in the following relations:

$$(4.123) \qquad \int_{T_o}^{\infty} \frac{1 - K_T}{(1 - \phi) + \phi K_T} F_T^F dT = 0$$

$$(4.124) \qquad F_T^L = \frac{1}{(1 - \phi) + \phi K_T} F_T^F$$

$$(4.125) \qquad F_T^V = \frac{K_T}{(1 - \phi) + \phi K_T} F_T^F$$

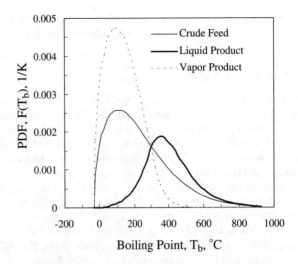

FIG. 4.26—Predicted probability density functions of feed, liquid, and vapor at 300°C for flash vaporization of a Russian crude oil. Actual data are taken from Ref. [31].

where Eqs. (4.123)–(4.125) are equivalent to Eqs. (4.119), (4.121), and (4.122) for ideal systems, respectively. Calculation of equilibrium ratios from equations of state will be discussed in Chapters 5 and 10. Probability density functions in these equations may be expressed in terms of other characterization parameters such as molecular weight or carbon number. However, as discussed in Chapter 2, boiling point is the most powerful characterization parameter and it is preferable to be used once it is available. Similarly the same approach can be used to obtain distribution of any other property (see Problem 4.16).

In treatment of a reservoir fluid, the mixture may be presented in terms of composition of pure hydrocarbon compounds from C_1 to C_5 and nonhydrocarbon compounds such as H_2S and CO_2 as well as grouped C_{6+} or a SCN group of C_6 and C_{7+}. For these mixtures the continuous mixture approach discussed in this section can be applied to the hydrocarbon-plus portion, while the discrete approach can be applied to the lower portion of the mixture containing compounds with known composition. This approach is known as semi-continuous approach and calculation of different properties of reservoir fluids by this approach has been discussed by various researchers [27, 28, 43].

4.8 CALCULATION OF PROPERTIES OF CRUDE OILS AND RESERVOIR FLUIDS

As discussed in Chapter 2, properties of a hydrocarbon compound depend on its carbon number and molecular type. Accurate calculation of properties of a petroleum mixture rely on accurate knowledge of the composition of the mixture by individual constituents, their properties, and an appropriate mixing rule to estimate the mixture properties. In this part based on the methods outlined in this chapter a crude oil or a reservoir fluid is presented by a number of pseudo-components and a general approach is outlined to estimate properties of such mixtures. Application of this approach is shown through estimation of sulfur content of crude oils.

4.8.1 General Approach

For a reservoir fluid accurate properties can be calculated through detailed compositional analysis of pure compounds from C_1 to C_5 and SCN groups from C_6 and heavier groups up to at least C_{50}. The remaining part can be grouped as C_{50+}. For very heavy oils, SCN group separation may be extended up to C_{80} and the residue grouped as C_{80+}. To estimate various properties of these SCN groups at least two characterizing parameters such as T_b and SG or M and SG should be known. This is shown in Table 4.24, where known data are indicated by + sign. Methods outlined in Sections 4.5 and 4.6 can lead to generate such information for a reservoir fluid. For SCN groups of C_6 and heavier, methods in Chapter 2 can be used to estimate various properties (T_b, T_c, P_c, ...) using M and SG as available input parameters. For pure compounds up to C_5, all basic properties are given in Tables 2.1 and 2.2 and no estimation method is required. For more accurate prediction of properties of a reservoir fluid, each SCN groups from C_6 up to C_{50+} may be divided into further three pseudocomponents as paraffinic, naphthenic, and aromatic. Methods of Section 3.5 can be used to determine PNA composition of each SCN group. In this way number of components in Table 4.24 increases to 152. For heavy oils the number of components would be even higher. For each homologous group, different properties may be estimated from molecular weight of individual SCN group through the relations given in Section 2.3.3. Obviously calculation of mixture properties when it is expressed in terms of large number of components is not an easy task. For this reason the number of components in Table 4.24 may be reduced by grouping to SCN components or splitting the C_7 fraction into just 3 or 5 pseudocomponents. Furthermore, iC_4 and nC_4 may be grouped as C_4 and iC_5 and nC_5 could be grouped as C_5. In this way the mixture can be presented by 10–15 components with known specifications. The following example shows how a crude oil can be presented

TABLE 4.24—*Matrix table of components for estimation of properties of reservoir fluids.*

No.	Compound	Mole fraction	M	SG
1	H_2S	+		
2	CO_2	+		
3	N_2	+		
4	H_2O	+		
5	C_1	+		
6	C_2	+		
7	C_3	+		
8	iC_4	+		
9	nC_4	+		
10	iC_5	+		
11	nC_5	+		
12	C_6	+	+	+
13	C_7	+	+	+
14	C_8	+	+	+
15–55[a]	C_9–C_{49}	+	+	+
56	C_{50}	+	+	+
57	C_{50+}	+	+	+

[a]Compounds from 15 to 55 represent SCN groups from C_9 to C_{49}.
For compounds 1–11, properties are given in Tables 2.1 and 2.2.

TABLE 4.25—*Pseudoization of the C_{7+} for the Kuwaiti crude in Example 4.16.*

Pseudocomponent	1	2	3	4	5
Weight fraction	0.097	0.162	0.281	0.197	0.264
Mole fraction	0.230	0.255	0.280	0.129	0.106
Molecular weight	112.0	169.1	267.1	405.8	660.9
Specific gravity	0.753	0.810	0.864	0.904	0.943

by an adequate number of pseudocomponents with known parameters.

Example 4.16—Compositional data on a Kuwaiti crude oil is given as follows:

Component	C_2	C_3	iC_4	nC_4	iC_5	nC_5	C_6	C_{7+}
Wt%	0.03	0.39	0.62	1.08	0.77	1.31	1.93	93.87

The characteristics of the C_{7+} fraction are $M_{7+} = 266.6$ and $SG_{7+} = 0.891$ [44]. Divide the C_{7+} fraction into 5 pseudocomponents and present the crude in terms of mole and weight fractions of representative constituents with known M, SG, and T_b. Estimate M and SG for the whole crude.

Solution—For the C_6 group from Table 4.6 we have $M_6 = 82$, $SG_6 = 0.69$, and $T_{b6} = 337$ K. For pure components from C_2 to C_5, M and SG can be taken from Table 2.1. Using M and x_w, mole fraction x_m can be estimated through Eq. (4.61). Using Method A in Section 4.5.4.6 distribution coefficients in Eq. (4.56) for the C_{7+} fraction are found as $M_o = 90$, $A_M = 1.957$, and $B_M = 1.0$. From Method II outlined in Section 4.6.1.2 and specifying 5 carbon number ranges the C_{7+} can be split into 5 pseudocomponents with known mole fraction (normalized), M and SG as given in Table 4.25. In this table the weight fractions are calculated through Eq. (1.15) using mole fraction and molecular weight. Values of weight fractions in Table 4.25 should be multiplied by wt% of C_{7+} in the whole crude to estimate wt% of each pseudocomponent in the crude. Values of mol% in the original fluid are calculated from wt% and molecular weight of all components present in the mixture as shown in Table 4.26. For the 5 pseudocomponents generated by splitting the C_{7+}, boiling points are calculated from M and SG using Eq. (2.56). From T_b and SG of pseudocomponents given in Table 4.26, one may estimate basic characterization parameters to estimate various

TABLE 4.26—*Characterization of the Kuwait crude oil in Example 4.16.*

Component	Wt%	Mol%	M	SG	T_b, °C
C_2	0.03	0.22	30.1	0.356	
C_3	0.39	1.99	44.1	0.507	
iC_4	0.62	2.40	58.1	0.563	
nC_4	1.08	4.18	58.1	0.584	
iC_5	0.77	2.40	72.2	0.625	
nC_5	1.31	4.08	72.2	0.631	
C_6	1.93	5.29	82	0.690	64
$C_{7+}(1)$	9.1	18.28	112.0	0.753	123
$C_{7+}(2)$	15.2	20.22	169.1	0.810	216
$C_{7+}(3)$	26.4	22.23	267.1	0.864	333
$C_{7+}(4)$	18.5	10.26	405.8	0.904	438
$C_{7+}(5)$	24.8	8.44	660.9	0.943	527
Total	100	100	225.2	0.8469	

properties. For example, for the whole crude the molecular weight is calculated from x_{mi} and M_i using Eq. (3.1) and specific gravity of the whole crude is calculated from x_{wi} and SG_i using Eq. 3.44. Calculated M and SG for the crude are 225 and 0.85, respectively. These values are lower than those for the C_{7+} as the crude contains components lighter than C_7. In the next example estimation of sulfur content of this crude is demonstrated. ♦

4.8.2 Estimation of Sulfur Content of a Crude Oil

Estimation of sulfur content of crude oils is based on the general approach for estimation of various properties of crude oils and reservoir fluids described in Section 4.8.1. Once a crude is presented by a number of pure components and some narrow boiling range pseudocomponents with known mole fraction, T_b, and SG, any physical property may be estimated through methods discussed in Chapters 2 and 3. Properties

TABLE 4.27—*Estimation of sulfur content of crude oil in Example 4.18.*

	Experimental data				Calculated parameters					
No.	wt%	T_b, K	d_{20}, g/cm^3	S% exp.	SG	M	R_I	m	S%, pred.	wt% × S% pre
1	1.7	20	0.566	0.006	0.570	76.9	1.037	−11.970	0.000	0.00
2	0.26	25	0.583	0.006	0.587	77.4	1.038	−11.310	0.000	0.00
3	0.29	30	0.597	0.006	0.602	78.4	1.038	−10.835	0.000	0.00
4	0.31	35	0.610	0.006	0.615	79.5	1.039	−10.395	0.000	0.00
5	0.33	40	0.623	0.006	0.627	80.7	1.040	−10.000	0.000	0.00
6	0.37	45	0.634	0.007	0.639	82.1	1.041	−9.644	0.000	0.00
7	0.38	50	0.645	0.007	0.649	83.5	1.041	−9.313	0.000	0.00
8	0.42	55	0.655	0.007	0.659	85.0	1.042	−9.009	0.002	0.00
9	0.44	60	0.664	0.007	0.669	86.7	1.042	−8.731	0.008	0.00
10	0.47	65	0.673	0.008	0.677	88.4	1.043	−8.480	0.013	0.01
11	0.49	70	0.681	0.008	0.685	90.2	1.043	−8.244	0.017	0.01
12	0.52	75	0.688	0.008	0.693	92.1	1.044	−8.032	0.021	0.01
13	0.55	80	0.695	0.008	0.700	94.0	1.044	−7.829	0.025	0.01
14	0.58	85	0.702	0.008	0.706	96.1	1.045	−7.645	0.028	0.02
15	0.6	90	0.708	0.009	0.713	98.2	1.045	−7.473	0.032	0.02
16	0.63	95	0.714	0.009	0.718	100.3	1.045	−7.314	0.036	0.02
17	0.66	100	0.719	0.009	0.724	102.6	1.045	−7.170	0.040	0.03
18	0.4	105	0.724	0.01	0.729	104.8	1.046	−7.027	0.045	0.02
19	0.52	110	0.729	0.011	0.734	107.2	1.046	−6.893	0.050	0.03
20	0.59	115	0.734	0.016	0.738	109.6	1.046	−6.769	0.055	0.03
21	0.66	120	0.738	0.019	0.743	112.0	1.047	−6.649	0.061	0.04
22	0.71	125	0.743	0.022	0.747	114.5	1.047	−6.532	0.068	0.05
23	0.73	130	0.747	0.026	0.751	117.1	1.047	−6.420	0.076	0.06
24	0.76	135	0.750	0.031	0.755	119.7	1.047	−6.305	0.085	0.06
25	0.76	140	0.754	0.036	0.758	122.3	1.047	−6.196	0.094	0.07
26	0.77	145	0.758	0.041	0.762	125.0	1.047	−6.086	0.104	0.08
27	0.76	150	0.761	0.047	0.766	127.7	1.048	−5.973	0.115	0.09
28	0.75	155	0.765	0.054	0.769	130.5	1.048	−5.859	0.128	0.10
29	0.75	160	0.768	0.061	0.772	133.2	1.048	−5.745	0.141	0.11
30	0.74	165	0.771	0.068	0.775	136.1	1.048	−5.629	0.155	0.11
31	0.73	170	0.774	0.077	0.779	138.9	1.048	−5.505	0.171	0.12
32	0.72	175	0.777	0.086	0.782	141.8	1.048	−5.380	0.188	0.14
33	0.71	180	0.781	0.095	0.785	144.7	1.049	−5.246	0.206	0.15
34	0.71	185	0.784	0.106	0.788	147.7	1.049	−5.102	0.227	0.16
35	0.71	190	0.787	0.117	0.791	150.7	1.049	−4.949	0.249	0.18
36	0.71	195	0.790	0.129	0.794	153.7	1.049	−4.796	0.271	0.19
37	0.71	200	0.793	0.142	0.797	156.7	1.049	−4.633	0.296	0.21
38	1.45	210	0.799	0.17	0.803	162.8	1.049	−4.290	0.350	0.51
39	1.47	220	0.805	0.201	0.809	169.1	1.050	−3.907	0.411	0.60
40	1.51	230	0.810	0.31	0.814	175.4	1.050	−3.484	0.481	0.73
41	1.56	240	0.816	0.46	0.820	181.9	1.050	−3.032	0.556	0.87
42	1.58	250	0.822	0.64	0.826	188.4	1.051	−2.550	0.639	1.01
43	1.6	260	0.828	0.83	0.832	195.1	1.051	−2.027	0.730	1.17
44	1.59	270	0.833	1.03	0.837	201.8	1.051	−1.485	1.191	1.89
45	1.56	280	0.839	1.21	0.842	208.7	1.051	−0.901	1.312	2.05
46	1.49	290	0.844	1.37	0.848	215.7	1.052	−0.313	1.424	2.12
47	1.42	300	0.849	1.5	0.853	222.9	1.052	0.316	1.536	2.18
48	1.32	310	0.854	1.6	0.858	230.2	1.052	0.948	1.640	2.16
49	1.23	320	0.859	1.67	0.862	237.6	1.052	1.593	1.739	2.14
50	1.18	330	0.863	1.75	0.867	245.2	1.053	2.249	1.832	2.16
51	1.18	340	0.868	1.86	0.871	252.9	1.053	2.930	1.922	2.27
52	1.29	350	0.872	2.05	0.875	260.8	1.053	3.606	2.005	2.59
53	1.56	360	0.875	2.4	0.879	269.1	1.053	4.210	2.072	3.23
54	26	449	0.915	2.81	0.918	343.4	1.055	13.025	2.847	74.02
55	28.1	678	1.026	5.2	1.028	561.1	1.064	57.206	4.631	130.13
Sum	100.0									233.94

of pure components can be directly obtained from Tables 2.1 and 2.2. For each physical property an appropriate mixing rule should be applied. For example, SG of the mixture should be calculated from Eq. (3.44) as was shown in Example 4.16. For the sulfur content of a crude oil the appropriate mixing rule is [45]:

$$(4.126) \quad \text{sulfur wt\% of crude} = \sum_i x_{wi}(\text{sulfur wt\%})_i$$

in which x_{wi} is the weight fraction of pseudocomponent i in the crude. The method is well demonstrated in the following examples for calculation of sulfur content of crude oils.

Example 4.17—For the crude oil of Example 4.16 estimate the total sulfur content in wt%. The whole crude has API gravity of 31 and sulfur content of 2.4 wt% [45].

Solution—The crude is presented in terms of 12 components (6 pure compounds and 6 pseudocompounds) in Table 4.26. Sulfur content of the crude should be estimated through Eq. (4.126). For pure hydrocarbons from C_2 to nC_5 the sulfur content is zero; however, sulfur content of pseudocomponents from C_6 to $C_{7+}(5)$ should be calculated from Eqs. (3.96) and (3.97). Parameters n_{20} and d_{20} needed in these equations have been estimated from methods discussed in Sections 2.6.1 and 2.6.2. Estimated sulfur content of C_6 and the 5 C_{7+} pseudocomponents are 0.2, 0.1, 0.7, 1.9, 2.9, and 3.8, respectively. Substituting these values into Eq. (4.126) would estimate sulfur content of the whole crude as 2.1% versus experimental value of 2.4% with −0.3 wt% error. ♦

Example 4.18—For the crude oil of Example 4.17 a complete TBP, SG and sulfur wt% curves versus weight fraction are available as given in the first five columns of Table 4.27. Estimate the sulfur content curve and graphically compare with the experimental values. Also estimate the sulfur content of whole crude from predicted sulfur content curve and compare with the experimental value of 2.4 wt%.

Solution—A complete characterization dataset on a crude oil include two suitable characterization parameters such as T_b and SG versus cumulative weight or volume fraction with low residue. When such data are available, properties of the crude may be estimated quite accurately. In Table 4.27, T_b, d_{20}, and sulfur wt% of 55 cuts are given with known wt%. The boiling point of last cut (residue) was not originally known from experimental data. Based on the fractions with T_b greater than 100°C (cuts 21–54), weight fractions were normalized and cumulative weight fractions were calculated. Temperature of 100°C is near the boiling point of nC_7. For the C_{7+} portion of the crude distribution, coefficients in Eq. (4.56) were determined as $T_o = 360$ K, $A_T = 1.6578$, and $B_T = 1.485$. Using these values, boiling point of the residue (cut 55) was determined from Eq. (4.56) as 678.3°C. Specific gravity of cuts were determined from d_{20} and parameters M, R_i, and m were determined using the methods discussed in Chapters 2 and 3. Sulfur content of each cut were determined from Eq. (3.96) for cuts with $M < 200$ and from Eq. (3.97) for cuts 44–55 with $M > 200$. For cuts 1–7 calculated values of S% from Eq. (3.96) were slightly less than zero and they are set as zero as discussed in Section 3.5.2.2. Finally sulfur content of the whole

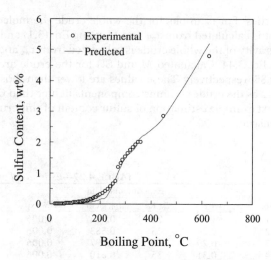

FIG. 4.27—Distribution of sulfur content in the crude oil of Example 4.18. Taken with permission from Ref. [45].

crude is calculated from Eq. (4.126) as shown in the last column of Table 4.27. The estimated sulfur content of the crude is 2.34 wt%, which is near the experimental value of 2.4%. A graphical comparison between predicted and experimental sulfur distribution along distillation curve is presented in Fig. 4.27. ♦

Calculations made in Examples 4.17 and 4.18 show that as more characterization data for a crude are available better property prediction is possible. In many cases characterization data on a crude contain only the TBP curve without SG distribution. In such cases M and SG distributions can be determined from Eq. (4.7) and coefficients given in Table 4.5. Equation (4.7) can be used in its reversed form using T_b as input instead of M. Once M is determined it can be used to estimate SG, n_{20}, and d_{20} from Eq. (4.7) with corresponding coefficients in Table 4.5. This approach has been used to estimate sulfur content of 7 crudes with API gravity in the range of 31–40. An average deviation of about 0.3 wt% was observed [45].

4.9 CONCLUSIONS AND RECOMMENDATIONS

In this chapter methods of characterization of reservoir fluids, crude oils, natural gases and wide boiling range fractions have been presented. Crude assay data for seven different crudes from around the world are given in Section 4.1.2. Characterization of reservoir fluids mainly depends on the characterization of their C_{7+} fractions. For natural gases and gas condensate samples with little C_{7+} content, correlations developed directly for C_{7+}, such as Eqs. (4.11)–(4.13), or the correlations suggested in Chapters 2 and 3 for narrow-boiling range fractions may be used. However, this approach is not applicable to reservoir fluids with considerable amount of C_{7+} such as volatile or black oil samples. The best way of characterizing a reservoir fluid or a crude oil is to apply a distribution model to its C_{6+} or C_{7+} portion and generate a distribution of SCN groups or a number of pseudocomponents that represent the C_{7+} fraction. Various characterization parameters and basic properties of SCN groups from C_6 to C_{50} are given in Table 4.6 and in the form of Eq. (4.7) for computer applications.

Characterization of C_{7+} fraction is presented through application of a distribution model and its parameters may be determined from bulk properties with minimum required data on M_{7+} and SG_{7+}. Three types of distribution models have been presented in this chapter: exponential, gamma, and a generalized model. The exponential model can be used only to molecular weight and is suitable for light reservoir fluids such as gas condensate systems and wet natural gases. The gamma distribution model can be applied to both molecular weight and boiling point of gas condensate systems. However, the model does not accurately predict molar distribution of very heavy oils and residues. This model also cannot be applied to other properties such as specific gravity or refractive index. The third model is the most versatile distribution model that can be applied to all major characterization parameters of M, T_b, SG, and refractive index parameter I. Furthermore, the generalized distribution model predicts molar distribution of heavy oils and residues with reasonable accuracy. Application of the generalized distribution model (Eq. 4.56) to phase behavior prediction of complex petroleum fluids has been reported in the literature [46]. Both the gamma and the generalized distribution models can be reduced to exponential in the form of a two-parameter model.

Once a distribution model is known for a C_{7+} fraction, the mixture can be considered as a continuous mixture or it could be split into a number of pseudocomponents. Examples for both cases are presented in this chapter. The method of continuous distribution approach has been applied to flash distillation of a crude oil and the method of pseudocomponent approach has been applied to predict sulfur content of an oil. Several characterization schemes have been outlined for different cases when different types of data are available. Methods of splitting and grouping have been presented to represent a crude by a number of representative pseudocomponents. A good characterization of a crude oil or a reservoir fluid is possible when TBP distillation curve is available in addition to M_{7+} and SG_{7+}. The most complete and best characterization data on a crude oil or a C_{7+} fraction would be TBP and SG distribution in terms of cumulative weight or volume fraction such as those shown in Table 4.27. Knowledge of carbon number distribution up to C_{40} and specification of residue as C_{40+} fraction is quite useful and would result in accurate property prediction provided the amount of the residue (hydrocarbon plus) is not more than a few percent. For heavy oils separation up to C_{60+} or C_{80+} may be needed. When the boiling point of the residue in a crude or a C_{7+} fraction is not known, a method is proposed to predict this boiling point from the generalized distribution model. When data on characterization of a crude are available in terms of distribution of carbon number such as those shown in Table 4.2, the method of grouping should be used to characterize the mixture in terms of a number of subfractions with known mole fraction, M, T_b and SG. Further information on options available for crude oil characterization from minimum data is given by Riazi et al. [40]. Properties of subfractions or pseudocomponents can be estimated from T_b and SG using methods presented in Chapters 2 and 3. For light portion of a crude or a reservoir fluid whose composition

is presented in terms weight, volume, or mole fraction of pure compounds, the basic characterization parameters and properties may be taken from Tables 2.1 and 2.2. Once a crude is expressed in term of a number of components with known properties, a mixture property can be determined through application of an appropriate mixing rule for the property as it will be shown in the next chapter.

4.10 PROBLEMS

4.1. Consider the dry natural gas, wet natural gas, and gas condensate systems in Table 1.2. For each reservoir fluid estimate the following properties:
   a. SG_g for and the API gravity.
   b. Estimate T_{pc} and P_{pc} from methods of Section 4.2.
   c. Estimate T_{pc}, P_{pc}, and V_{pc} from Eq. (3.44) using pure components properties from Table 2.1 and C_{7+} properties from Eqs. (4.12) and (4.13).
   d. Compare the calculated values for T_{pc} and P_{pc} in parts b and c and comment on the results.

4.2. Calculate T_b, SG, d_{20}, n_{20}, T_c, P_c, V_c, σ, and δ for C_{55}, C_{65}, and C_{75} SCN groups.

4.3. Predict SCN distribution for the West Texas oil sample in Table 4.1, using Eq. (4.27) and M_{7+} and x_{7+} (mole fraction of C_{7+}) as the available data.

4.4. Derive an analytical expression for Eq. (4.78), and show that when SG is presented in terms of x_{cw} we have

$$\frac{1}{SG_{av}^* + 1} \equiv J_0 = \sum_{k=0}^{\infty} (-1)^{k+1} \left(\frac{B}{A}\right)^{\left(\frac{1-k}{B}\right)} \Gamma\left(1 + \frac{k-1}{B}\right)$$

4.5. Basic characterization data, including M, T_b, and SG, versus weight fraction for seven subfractions of a C_{7+} fluid are given in Table 4.28. Available experimental bulk properties are $M_{7+} = 142.79$, and $SG_{7+} = 0.7717$ [47]. Make the following calculations:
   a. Calculate x_m and x_v.
   b. Estimate distribution parameter I from T_b and SG using methods of Chapter 2.
   c. Using experimental data on M, T_b, SG and I distributions calculate distribution coefficients P_o, A and B in Eq. (4.56) for these properties. Present M in terms of x_{cm} and T_b, SG and I in terms of x_{cw}.
   d. Calculate PDF from Eq. (4.66) and show graphical presentation of $F(M)$, $F(T)$, $F(SG)$, and $F(I)$.
   e. Find refractive index distribution
   f. Calculate mixture M, T_b, SG, and n_{20} based on the coefficients obtained in part c.
   g. For parts b and f calculate errors for M, T_b, and SG in terms of AAD.

TABLE 4.28—*Characterization parameters for the C_{7+} fraction of the oil system in Problem 4.5 [47].*

x_{wi}	M_i	T_{bi}, K	SG
0.1269	98	366.5	0.7181
0.0884	110	394.3	0.7403
0.0673	121	422.1	0.7542
0.1216	131	449.8	0.7628
0.1335	144	477.6	0.7749
0.2466	165	505.4	0.7859
0.2157	216	519.3	0.8140

4.6. Repeat Problem 4.5 for the gamma and exponential distribution models to
  a. find the coefficients of Eq. (4.31): η, α, and β in Eq. (4.31) for M and T_b.
  b. estimate SG distribution based on exponential model and constant K_W approach.
  c. calculate mixture M, T_b, and SG and compare with experimental data.
  d. make a graphical comparison between predicted distributions for M, T_b, and SG from Eq. (4.56) as obtained in Problem 4.5, gamma and exponential models with each other and experimental data.

4.7. For the C_{7+} of Problem 4.5 find distributions of M, T_b, and SG assuming:
  a. Only information available are $M_{7+} = 142.79$ and $SG_{7+} = 0.7717$.
  b. Only information available is $M_{7+} = 142.79$.
  c. Only information available is $SG_{7+} = 0.7717$.
  d. Graphically compare predicted distributions from parts a, b and c with data given in Table 4.28.
  e. Estimate M_{7+} and SG_{7+} form distribution parameters obtained in parts a, b and c and compare with the experimental data.

4.8. Using the Guassian Quadrature approach, split the C_{7+} fraction of Problem 4.7 into three pseudocomponents. Determine, x_m, M, T_b, and SG for each component. Calculate the mixture M and SG from the three pseudocomponents. Repeat using carbon number range approach with 15 pseudocomponents and appropriate boundary values of M_i.

4.9. For the C_{7+} fraction of Problem 4.5 estimate total sulfur content in wt%.

4.10. For the waxy oil in Table 4.2 present the oil in six groups as C_2–C_3, C_4–C_6, C_7–C_{10}, C_{11}–C_{20}, C_{21}–C_{30}, and C_{31+}. Determine M and SG for each group and calculate M and the API gravity of the oil. Compare estimated M from the six groups with M calculated for the crude based on the detailed data given in Table 4.2.

4.11. Use the crude assay data for crude number 7 in Table 4.3 to
  a. determine T_b and SG distributions.
  b. estimate T_b for the residue based on the distribution found in Part a.
  c. estimate M for the residue from T_b in Part b and SG.
  d. estimate M for the residue from viscosity and SG and compare with value from c.
  e. Determine distribution of sulfur for the crude and graphically evaluate variation of S% versus cumulative wt%.
  f. Estimate sulfur content of the crude based on the predicted S% distribution.

4.12. For the crude sample in Problem 4.8 find distribution of melting point and estimate average melting point of the whole crude.

4.13. Estimate molecular weights of SCN groups from 7 to 20 using Eqs. (4.91) and (4.92) and compare your results with those calculated in Example 4.6 as given in Table 4.10.

4.14. Construct the boiling point and specific gravity curves for the California crude based on data given in Table 4.3

(crude number 6). In constructing this figure the mid-volume points may be used for the specific gravity. Determine the distribution coefficients in Eq. (4.56) for T_b and SG in terms of x_{cv} and compare with the experimental values. Also estimate crude sulfur content.

4.15. Show how Eqs. (4.104), (4.105), and (4.106) have been derived.

4.16. As it will be shown in Chapter 7, Lee and Kesler have proposed the following relation for estimation of vapor pressure (P_{vap}) of pure compounds, which may be applied to narrow boiling range fractions (Eq. 7.18).

$$\ln P_r^{vap} = 5.92714 - 6.09648/T_{br} - 1.28862 \ln T_{br}$$
$$+ 0.169347 T_{br}^6 + \omega(15.2518 - 15.6875/T_{br}$$
$$- 13.4721 \ln T_{br} + 0.43577 T_{br}^6)$$

where $P_r^{vap} = P^{vap}/P_c$ and $T_{br} = T_b/T_c$ in which both T_b and T_c must be in K. Use the continuous mixture approach (Section 4.7) to predict distribution of vapor pressure at 311 and 600 K for the waxy crude oil in Table 4.2 and graphically show the vapor pressure distribution versus cumulative mol% and carbon number.

4.17. Minimum information that can be available for a crude oil is its API or specific gravity. A Saudi light crude has API gravity of 33.4 (SG = 0.8581), and experimental data on boiling point and specific gravity of its various cuts are given in the following table as given in the *Oil and Gas Journal Data Book* (2000) (p. 318 in Ref. [8]).

Vol%	SG	T_b, K	SG (calc)	T_b, K (calc)
23.1	...	370.8	?	?
23.1	0.8131	508.3	?	?
8.5	0.8599	592.5	?	?
30.2	0.9231	727.5	?	?
15.1	1.0217	...	?	?

  a. Using the minimum available data (API gravity), estimate values of T_b and SG in the above table and compare with given experimental data graphically.
  b. Similar data exist for a Saharan crude oil from Algeria (page 320 in Ref. [8]) with API gravity of 43.7. Construct T_b and SG distribution diagram in terms of cumulative volume fraction.

4.18. Similar to the continuous mixture approach introduced in Section 4.7, calculate vapor and liquid product distributions for flash distillation of the same crude at 1 atm and 400°C. Present the results in a fashion similar to Fig. 4.26 and calculate the vapor to feed ratio (ϕ).

4.19. Repeat Problem 4.18 but instead of Eq. (4.120) for the vapor pressure, use the Lee–Kesler correlation given in Problem 4.16. Compare the results with those obtained in Problem 4.17 and discuss the results.

REFERENCES

[1] Pedersen, K. S., Thomassen, P., and Fredenslund, Aa., "Characterization of Gas Condensate Mixtures," *C7+ Fraction Characterization*, L. G. Chorn and G. A. Mansoori, Eds., Taylor & Francis, New York, 1989, pp. 137–152.

[2] Hoffmann, A. E., Crump, J. S., and Hocott, C. R. P., "Equilibrium Constants for a Gas-Condensate System," *Petroleum Transactions AIME*, Vol. 198, 1953, pp. 1–10.

[3] Osjord, E. H., Ronnisgsen, H. P., and Tau, L., "Distribution of Weight, Density, and Molecular Weight in Crude Oil Derived from Computerized Capillary GC Analysis," *Journal of High Resolution Chromatography*, Vol. 8, 1985, pp. 683–690.

[4] ASTM, *Annual Book of Standards*, Section Five, Petroleum Products, Lubricants, and Fossil Fuels (in 5 Vol.), ASTM International, West Conshohocken, PA, 2002.

[5] KISR, Private Communication, Petroleum Research Studies Center, Kuwait Institute for Scientific Research (KISR), Ahmadi, Kuwait, May 2002.

[6] Pedersen, K. S., Fredenslund, Aa., and Thomassen, P., *Properties of Oils and Natural Gases*, Gulf Publishing, Houston, TX, 1989.

[7] Pan, H., Firoozabadi, A., and Fotland, P., "Pressure and Composition Effect on Wax Precipitation: Experimental Data and Model Results," *Society of Petroleum Engineers Production and Facilities*, Vol. 12, No. 4, 1997, pp. 250–258.

[8] *Oil and Gas Journal Data Book*, 2000 edition, PennWell, Tulsa, OK, 2000, pp. 295–365.

[9] Gary, J. H. and Handwerk, G. E., *Petroleum Refining, Technology and Economic*, 3rd ed., Marcel Dekker, New York, 1994.

[10] Ahmed, T., *Reservoir Engineering Handbook*, Gulf Publishing, Houston, TX, 2000.

[11] Standing, M. B., *Volumetric and Phase Behavior of Oil Field Hydrocarbon Systems*, Society of Petroleum Engineers, Dallas, TX, 1977.

[12] Wichert, E. and Aziz, K., "Calculation of Z's for Sour Gases," *Hydrocarbon Processing*, Vol. 51, No. 5, 1972, pp. 119–122.

[13] Carr, N., Kobayashi, R., and Burrows, D.,"Viscosity of Hydrocarbon Gases Under Pressure," *Transactions of AIME*, Vol. 201, 1954, pp. 23–25.

[14] Katz, D. L. and Firoozabadi, A. "Predicting Phase Behavior of Condensate Crude/Oil Systems Using Methane Interaction Coefficients," *Journal of Petroleum Technology*, November 1978, pp. 1649–1655.

[15] Whitson, C. H., "Characterizing Hydrocarbon Plus Fractions," *Society of Petroleum Engineers Journal*, August 1983, pp. 683–694.

[16] Riazi, M. R. and Al-Sahhaf, T. A., "Physical Properties of Heavy Petroleum Fractions and Crude Oils," *Fluid Phase Equilibria*, Vol. 117, 1996, pp. 217–224.

[17] Whitson, C. H. and Brule, M. R., *Phase Behavior*, Monograph Volume 20, SPE, Richardson, TX, 2000.

[18] Danesh, A., *PVT and Phase Behavior of Petroleum Reservoir Fluids*, Elsevier, Amsterdam, 1998.

[19] Riazi, M. R. and Daubert, T. E., "Characterization Parameters for Petroleum Fractions," *Industrial and Engineering Chemistry Research*, Vol. 26, 1987, pp. 755–759. Corrections, p. 1268.

[20] Whitson, C. H., "Effect of C_{7+} Properties on Equation-of-State Predictions," *Society of Petroleum Engineers Journal*, December 1984, pp. 685–696.

[21] Chorn, L. G. and Mansoori, G. A., *C_{7+} Fraction Characterization*, Taylor & Francis, New York, 1989, 235 p.

[22] Benmekki, H. and Mansoori, G. A., "Pseudization techniques and heavy fraction characterization with equation of state models, *Advances in Thermodynamics, Vol. 1: C_{7+} Fraction Characterization*, L. G. Chorn and G. A. Mansoori, Eds., Taylor & Francis, New York, 1989, pp. 57–78.

[23] Soreide, I., *Improved Phase Behavior Predictions of Petroleum Reservoir Fluids from a Cubic Equation of State*, Doctoral Dissertation, Norwegian Institute of Technology (NTH), Trondheim, Norway, 1989.

[24] Riazi, M. R. "A Distribution Model for C_{7+} Fractions Characterization of Petroleum Fluids," *Industrial and Engineering Chemistry Research*, Vol. 36, 1997, pp. 4299–4307.

[25] Riazi, M. R., "Distribution Model for Properties of Hydrocarbon-Plus Fractions," *Industrial and Engineering Chemistry Research*, Vol. 28, 1989, pp. 1731–1735.

[26] Ahmed, T., *Hydrocarbon Phase Behavior*, Gulf Publishing, Houston, TX, 1989.

[27] Mansoori, G. A. and Chorn, L. G., "Multicomponent Fractions Characterization: Principles and Theories," *C_{7+} Characterization*, L. G. Chorn and G. A. Mansoori, Eds.,Taylor & Francis, New York, 1989, pp. 1–10.

[28] Cotterman, R. L. and Prausnitz, J. M., "Flash Calculation for Continuous or Semicontinuous Mixtures Using An Equation of State," *Industrial Engineering Chemistry, Process Design and Development*, Vol. 24, 1985, pp. 434–443.

[29] Kehlen, H., Ratsch, M. T., and Berhmann, J., "Continuous Thermodynamics of Multicomponent Systems," *AIChE Journal*, Vol. 31, 1985, pp. 1136–1148.

[30] Ratsch, M. T. and Kehlen, H., "Continuous Thermodynamics of Complex Mixtures," *Fluid Phase Equilibria*, Vol. 14, 1983, pp. 225–234.

[31] Ratzch, M. T., Kehlen, H., and Schumann, J., "Flash Calculations for a Crude Oil by Continuous Thermodynamics," *Chemical Engineering Communications*, Vol. 71, 1988, pp. 113–125.

[32] Wang, S. H. and Whiting, W. B., "A Comparison of Distribution Functions for Calculation of Phase Equilibria of Continuous Mixtures," *Chemical Engineering Communications*, Vol. 71, 1988, pp. 127–143.

[33] Katz, D. L., "Overview of Phase Behavior of Oil and Gas Production," *Journal of Petroleum Technology*, June 1983, pp. 1205–1214.

[34] Yarborough, L., "Application of a Generalized Equation of State to Petroleum Reservoir Fluids," Paper presented at the *176th National Meeting of the American Chemical Society*, Miami Beach, FL, 1978.

[35] Pedersen, K. S., Thomassen, P., and Fredenslund, Aa., "Thermodynamics of Petroleum Mixtures Containing Heavy Hydrocarbons, 1: Phase Envelope Calculations by Use of the Soave-Redlich-Kwong Equation of State," *Industrial Engineering Chemisty, Process Design and Development*, Vol. 23, 1984, pp. 163–170.

[36] Whitson, C. H., Anderson, T. F., and Soreide, I., "C_{7+} Characterization of Related Equilibrium Fluids Using the Gamma Distribution," *C_{7+} Fraction Characterization*, L. G. Chorn and G. A. Mansoori, Eds., Taylor & Francis, New York, 1989, pp. 35–56.

[37] Whitson, C. H., Anderson, T. F., and Soreide, I., "Application of the Gamma Distribution Model to Molecular Weight and Boiling Point Data for Petroleum Fractions," *Chemical Engineering Communications*, Vol. 96, 1990, pp. 259–278.

[38] Abramowitz, M. and Stegun, I. A., *Handbook of Mathematical Functions*, Dover Publication, New York, 1970.

[39] Press, W. H., Flannery, B. P., Teukolsky, S. A., and Vetterling, W. T., *Numerical Recipes, The Art of Scientific Computing*, Cambridge University Press, Cambridge, London, 1986, pp. 160–161.

[40] Riazi, M. R., Al-Adwani, H. A., and Bishara, A., "The Impact of Characterization Methods on Properties of Reservoir Fluids and Crude Oils: Options and Restrictions," *Journal of Petroleum Science and Engineering*, Vol. 42., No. 2–4, 2004, pp. 195–207.

[41] Rodgers, P. A., Creagh, A. L., Prauge, M. M., and Prausnitz, J. M., "Molecular Weight Distribution for Heavy Fossil Fuels from Gel-Permeation Chromatograph,," *Industrial and Engineering Chemistry Research*, Vol. 26, 1987, pp. 2312–2321.

[42] Berge, O., *Damp/Vaeske-Likevekter i Raoljer: Karakterisering av Hydrokarbonfraksjon*, M.Sc. Thesis, Department of Chemical

Engineering, Norwegian Institute of Technology, Trondheim, Norway, 1981.

[43] Manafi, H., Mansoori, G. A., and Ghotbi, S., "Phase Bahavior Prediction of Petroleum Fluids with Minimum Characterization Data," *Journal of Petroleum Science and Engineering*, Vol. 22, 1999, pp. 67–93.

[44] Stroud, A. H. and Secrest, D., *Gaussian Quadrature Formulas*, Prentice-Hall, Englewood Cliffs, NJ, 1966.

[45] Riazi, M. R., Nasimi, N., and Roomi, Y., "Estimating Sulfur Content of Petroleum Products and Crude Oils," *Industrial and Engineering Chemistry Research*, Vol. 38, No. 11, 1999, pp. 4507–4512.

[46] Fazlali, A., Modarress, H., and Mansoori, G. A., "Phase Behavior Prediction of Complex Petroleum Fluids," *Fluid Phase Equilibra*, Vol. 179, 2001, pp. 297–317.

[47] Jacoby, R. H., Koeller, R. C., and Berry Jr., V. J., "Effect of Composition and Temperature on Phase Behavior and Depletion Performance of Rich Gas-Condensate Systems," *Transactions of AIME*, Vol. 216, 1959, pp. 406–411.

PVT Relations and Equations of State

5

NOMENCLATURE

API	API Gravity defined in Eq. (2.4)
$A, B, C, \ldots$	Coefficients in various equations
a_c	Parameter defined in Eq. (5.41) and given in Table 5.1
$a, b, c, \ldots$	Constants in various equations
B	Second virial coefficient
C	Third virial coefficient
c	Volume translation for use in Eq. (5.50), cm³/mol
d_{20}	Liquid density of liquid at 20°C and 1 atm, g/cm³
d_c	Critical density defined by Eq. (2.9), g/cm³
e	Correlation parameter, exponential function
exp	Exponential function
F	Degrees of freedom in Eq. (5.4)
f_ω	A function defined in terms of ω for parameter a in PR and SRK equations as given in Table 5.1 and Eq. (5.53)
h	Parameter defined in Eq. (5.99), dimensionless
k_B	Boltzman constant ($= R/N_A = 1.2 \times 10^{-20}$ J/K)
k_{ij}	Binary interaction parameter (BIP), dimensionless
I	Refractive index parameter defined in Eq. (2.36)
M	Molecular weight, g/mol [kg/kmol]
m	Mass of system, g
N_A	Avogadro number = number of molecules in one mole (6.022×10^{23} mol^{-1})
N	Number of components in a mixture
n	Number of moles
n_{20}	Sodium D line refractive index of liquid at 20°C and 1 atm, dimensionless
P	Pressure, bar
P^{sat}	Saturation pressure, bar
P_c	Critical pressure, bar
P_r	Reduced pressure defined by Eq. (5.100) ($= P/P_c$), dimensionless
R	Gas constant = 8.314 J/mol·K (values are given in Section 1.7.24)
R_m	Molar refraction defined by Eq. (5.133), cm³/mol
r	Reduced molar refraction defined by Eq. (5.129), dimensionless
r	Intermolecular distance in Eqs. (5.10)–(5.12), Å (10^{-10} m)
r	A parameter specific for each substance in Eq. (5.98), dimensionless
u_1, u_2	Parameters in Eqs. (5.40) and (5.42)
SG	Specific gravity of liquid substance at 15.5°C (60°F) defined by Eq. (2.2), dimensionless
T	Absolute temperature, K

T_c	Critical temperature, K
T_{cric}	Cricondentherm temperature, K
T_r	Reduced temperature defined by Eq. (5.100) ($= T/T_c$), dimensionless
V	Molar volume, cm³/gmol
V^L	Saturated liquid molar volume, cm³/gmol
V^{sat}	Saturation molar volume, cm³/gmol
V^V	Saturated vapor molar volume, cm³/gmol
V_c	Critical volume (molar), cm³/mol (or critical specific volume, cm³/g)
V_r	Reduced volume ($= V/V_c$)
x_i	Mole fraction of i in a mixture (usually used for liquids)
y_i	Mole fraction of i in a mixture (usually used for gases)
Z	Compressibility factor defined by Eq. (5.15)
Z_c	Critical compressibility factor defined by Eq. (2.8), dimensionless
Z_{RA}	Rackett parameter, dimensionless
z_j	Mole fraction of i in a mixture
$Z_1, Z_2,$ and Z_3	Roots of a cubic equation of state

Greek Letters

α	Parameter defined by Eq. (5.41), dimensionless
α	Polarizability factor defined by Eq. (5.134), cm³
α, γ	Parameters in BWR EOS defined by Eq. (5.89)
β	A correction factor for b parameter in an EOS defined by Eq. (5.55), dimensionless
Δ	Difference between two values of a parameter
δ_{ij}	Parameter defined in Eq. (5.70), dimensionless
ε	Energy parameter in a potential energy function
Γ	Potential energy function defined by Eq. (5.10)
ϕ_i	Volume fraction of i in a liquid mixture defined by Eq. (5.125)
Π	Number of phases defined in Eq. (5.4)
μ	Dipole moment in Eq. (5.134)
θ	A property in Eq. (5.1), such as volume, enthalpy, etc.
θ	Degrees in Eq. (5.47)
ρ	Density at a given temperature and pressure, g/cm³ (molar density unit: cm³/mol)

ρ^o Value of density at low pressure (atmospheric pressure), g/cm^3

σ Size parameter in a potential energy function, Å (10^{-10} m)

ω Acentric factor defined by Eq. (2.10)

ξ Packing fraction defined by Eq. (5.91), dimensionless

Superscript

bp   Value of a property for a defined mixture at its bubble point

c   Value of a property at the critical point

cal   Calculated value

exp   Experimental value

g   Value of a property for gas phase

HS   Value of a property for hard sphere molecules

ig   Value of a property for an ideal gas

L   Saturated liquid

l   Value of a property for liquid phase

V   Saturated vapor

sat   Value of a property at saturation pressure

(0)   A dimensionless term in a generalized correlation for a property of simple fluids

(1)   A dimensionless term in a generalized correlation for a property of acentric fluids

Subscripts

c   Value of a property at the critical point

i A component in a mixture

j A component in a mixture

i, j Effect of binary interaction on a property

m   Value of a property for a mixture

P Value of a property at pressure P

p   Pseudoproperty for a mixture

P, N, A   Value of parameter c in Eq. (5.52) for paraffins, naphthenes, and aromatics

t   Value of a property for the whole (total) system

Acronyms

API-TDB   American Petroleum Institute—Technical Data Book

BIP   Binary interaction parameter

BWRS   Starling modification of Benedict–Webb–Rubin EOS (see Eq. 5.89)

COSTALD   **co**rresponding **sta**te **l**iquid **d**ensity (given by Eq. 5.130)

CS   Carnahan–Starling EOS (see Eq. 5.93)

EOS   Equations of state

GC   Generalized correlation

HC   Hydrocarbon

HS   Hard sphere

HSP   Hard sphere potential given by Eq. (5.13)

KISR   Kuwait Institute for Scientific Research

IAPWS   International Association for the Properties of Water and Steam

LJ   Lennard–Jones potential given by Eq. (5.11)

LJ EOS   Lennard–Jones EOS given by Eq. (5.96)

LK GC   Lee–Kesler generalized correlation for Z (Eqs. 5.107–5.113)

LK EOS   Lee–Kesler EOS given by Eq. (5.109)

MRK   Modified Redlich–Kwong EOS given by Eqs. (5.38) and (5.137)–(5.140)

NIST   National Institute of Standards and Technology

OGJ   Oil and Gas Journal

PHCT   Perturbed Hard Chain Theory (see Eq. 5.97)

PR   Peng–Robinson EOS (see Eq. 5.39)

RHS   Right-hand side of an equation

RK   Redlich–Kwong EOS (see Eq. 5.38)

RS R squared (R^2), defined in Eq. (2.136)

SRK   Soave–Redlich–Kwong EOS given by Eq. (5.38) and parameters in Table 5.1

SAFT   Statistical associating fluid theory (see Eq. 5.98)

SW   Square–Well potential given by Eq. (5.12).

vdW   van der Waals (see Eq. 5.21)

VLE   Vapor–liquid equilibrium

%AAD   Average absolute deviation percentage defined by Eq. (2.135)

%AD   Absolute deviation percentage defined by Eq. (2.134)

%MAD   Maximum absolute deviation percentage

As discussed in Chapter 1, the main application of characterization methods presented in Chapters 2–4 is to provide basic data for estimation of various thermophysical properties of petroleum fractions and crude oils. These properties are calculated through thermodynamic relations. Although some of these correlations are empirically developed, most of them are based on sound thermodynamic and physical principles. The most important thermodynamic relation is pressure–volume–temperature (PVT) relation. Mathematical PVT relations are known as *equations of state*. Once the PVT relation for a fluid is known various physical and thermodynamic properties can be obtained through appropriate relations that will be discussed in Chapter 6. In this chapter we review principles and theory of property estimation methods and equations of states that are needed to calculate various thermophysical properties.

5.1 BASIC DEFINITIONS AND THE PHASE RULE

The state of a system is fixed when it is in a thermodynamic or phase equilibrium. A system is in equilibrium when it has no tendency to change. For example, pure liquid water at 1 atm and 20°C is at stable equilibrium condition and its state is perfectly known and fixed. For a mixture of vapor and liquid water at 1 atm and 20°C the system is not stable and has a tendency to reach an equilibrium state at another temperature or pressure. For a system with two phases at equilibrium only temperature or pressure (but not both) is sufficient to determine its state. The state of a system can be determined by its properties. A property that is independent of size or mass of the system is called *intensive* property. For example, temperature, pressure, density, or molar volume are intensive properties, while total volume of a system is an extensive property. All molar properties are intensive properties and are related to total property as

$$\theta = \frac{\theta^t}{n} \tag{5.1}$$

where n is the number of moles, θ^t is a total property such as volume, V^t, and θ is a molar property such as molar volume, V. The number of moles is related to the mass of the system, m, through molecular weight by Eq. (1.6) as

$$n = \frac{m}{M} \tag{5.2}$$

If total property is divided by mass of the system (m), instead of n, then θ is called *specific* property. Both molar and specific properties are intensive properties and they are related to each other through molecular weight.

$$\text{Molar Property} = \text{Specific Property} \times M \tag{5.3}$$

Generally thermodynamic relations are developed among molar properties or intensive properties. However, once a molar property is calculated, the total property can be calculated from Eq. (5.1).

The phase rule gives the minimum number of independent variables that must be specified in order to determine thermodynamic state of a system and various thermodynamic properties. This number is called *degrees of freedom* and is shown by F. The phase rule was stated and formulated by the American physicist J. Willard Gibbs in 1875 in the following form [1]:

$$F = 2 + N - \Pi \tag{5.4}$$

where Π is the number of phases and N is the number of components in the system. For example for a pure component ($N = 1$) and a single phase ($\Pi = 1$) system the degrees of freedom is calculated as 2. This means when two intensive properties are fixed, the state of the systems is fixed and its properties can be determined from the two known parameters. Equation (5.4) is valid for nonreactive systems. If there are some reactions among the components of the systems, degrees of freedom is reduced by the number of reactions within the system. If we consider a pure gas such as methane, at least two intensive properties are needed to determine its thermodynamic properties. The most easily measurable properties are temperature (T) and pressure (P). Now consider a mixture of two gases such as methane and ethane with mole fractions x_1 and x_2 ($x_2 = 1 - x_1$). According to the phase rule three properties must be known to fix the state of the system. In addition to T and P, the third variable could be mole fraction of one of the components (x_1 or x_2). Similarly, for a mixture with single phase and N components the number of properties that must be known is $N + 1$ (i.e., $T, P, x_1, x_2, \ldots, x_{N-1}$). When the number of phases is increased the degrees of freedom is decreased. For example, for a mixture of certain amount of ice and liquid water ($\Pi = 2$, $N = 1$) from Eq. (5.4) we have $F = 1$. This means when only a single variable such as temperature is known the state of the system is fixed and its properties can be determined. Minimum value of F is zero. A system of pure component with three phases in equilibrium with each other, such as liquid water, solid ice, and vapor, has zero degrees of freedom. This means the temperature and pressure of the system are fixed and only under unique conditions of

T and P three phases of a pure component can coexist all the time. This temperature and pressure are known as *triple point temperature* and *triple point pressure* and are characteristics of any pure compound and their values are given for many compounds [2, 3]. For example, for water the triple point temperature and pressure are $0.01\,^\circ\text{C}$ and 0.6117 kPa ($\sim$0.006 bar), respectively [3]. The most recent tabulation and formulation of properties of water recommended by International Association for the Properties of Water and Steam (IAPWS) are given by Wagner and Pruss [4].

A thermodynamic property that is defined to formulate the first law of thermodynamics is called *internal energy* shown by U and has the unit of energy per mass or energy per mole (i.e., J/mol). Internal energy represents both kinetic and potential energies that are associated with the molecules and for any pure substance it depends on two properties such as T and V. When T increases the kinetic energy increases and when V increases the potential energy of molecules also increases and as a result U increases. Another useful thermodynamic property that includes PV energy in addition to the internal energy is *enthalpy* and is defined as

$$H = U + PV \tag{5.5}$$

where H is the molar enthalpy and has the same unit as U. Further definition of thermodynamic properties and basic relations are presented in Chapter 6.

5.2 PVT RELATIONS

For a pure component system after temperature and pressure, a property that can be easily determined is the volume or molar volume. According to the phase rule for single phase and pure component systems V can be determined from T and P:

$$V = f_1(T, P) \tag{5.6}$$

where V is the molar volume and f_1 represents functional relation between V, T, and P for a given system. This equation can be rearranged to find P as

$$P = f_2(T, V) \tag{5.7}$$

where the forms of functions f_1 and f_2 in the above two relations are different. Equation (5.6) for a mixture of N components with known composition is written as

$$P = f_3(T, V, x_1, x_2, \ldots, x_{N-1}) \tag{5.8}$$

where x_i is the mole fraction of component i. Any mathematical relation between P, V, and T is called an *equation of state* (EOS). As will be seen in the next chapter, once the PVT relation is known for a system all thermodynamic properties can be calculated. This indicates the importance of such relations. In general the PVT relations or any other thermodynamic relation may be expressed in three forms of (1) mathematical equations, (2) graphs, and (3) tables. The graphical approach is tedious and requires sufficient data on each substance to construct the graph. Mathematical or analytical forms are the most important and convenient relations as they can be

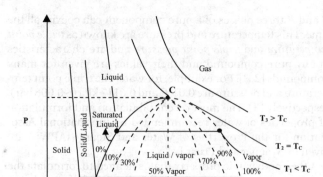

FIG. 5.1—Typical PV diagram for a pure substance.

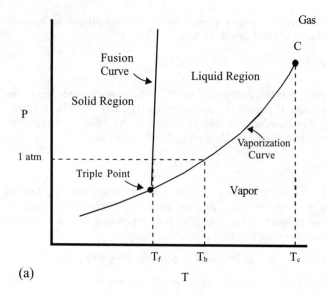

(a)

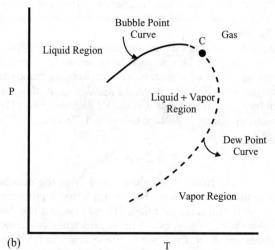

(b)

FIG. 5.2—Typical PT diagrams for a pure substances and mixtures.

used in computer programs for accurate estimation of various properties. Graphical and tabulated relations require interpolation with hand calculations, while graphical relations were quite popular in the 1950s and 1960s. With the growth of computers in recent decades mathematical equations are now the most popular relations.

A typical PVT relation in the form of PV and PT diagrams for a pure substance is shown in Figs. 5.1 and 5.2a, respectively. The solid, liquid, and vapor phases are clearly specified in the PT diagram. The two-phase region of vapor and liquid is best shown in the PV diagram. In Fig. 5.1, three isotherms of $T_1 < T_2 < T_3$ are shown where isotherm T_2 passes through the critical point, that is $T_c = T_2$. In the PV diagram lines of saturated liquid (solid line) and saturated vapor (dotted line) meet each other at the *critical point*. At this point properties of vapor phase and liquid phase become identical and two phases are indistinguishable. Since the critical isotherm exhibits a horizontal inflection at the critical point we may impose the following mathematical conditions at this point:

$$(5.9) \quad \left(\frac{\partial P}{\partial V}\right)\bigg|_{T_c, P_c} = \left(\frac{\partial^2 P}{\partial V^2}\right)\bigg|_{T_c, P_c} = 0$$

The first and second partial derivatives of P with respect to V (at constant T) may be applied to any EOS in the form of Eq. (5.7) and at the critical point they should be equal to zero. Simultaneous solution of resulting two equations will give relations for calculation of EOS parameters in terms of critical constants as will be seen later in Section 5.5.1.

The two-phase region in the PV diagram of Fig. 5.1 is under the envelope. As is seen from this figure the slope of an isotherm in the liquid region is much greater than its slope in the vapor phase. This is due to the greater change of volume of a gas with pressure in comparison with liquids that show less dependency of volume change with pressure under constant temperature condition. The dotted lines inside the envelope indicate percentage of vapor in a mixture of liquid and vapor, which is called *quality* of vapor. On the saturated vapor curve (right side) this percentage is 100% and on the saturated liquid curve (left side) this percentage is zero. Vapor region is part of a greater region called *gas phase*. *Vapor* is usually referred to a gas that can be liquefied under pressure. A vapor at a temperature above T_c cannot be liquefied no matter how

high the pressure is and it is usually referred as a gas. When T and P of a substance are greater than its T_c and P_c the substance is neither liquid nor vapor and it is called *supercritical fluid* or simply *fluid*. However, the word fluid is generally used for either a liquid or a vapor because of many similarities that exist between these two phases to distinguish them from solids.

As is seen in Fig. 5.1, lines of saturated liquid and vapor are identical in the *PT* diagram. This line is also called *vapor pressure* (or vaporization) curve where it begins from the triple point and ends at the critical point. The saturation line between solid and liquid phase is called *fusion* curve while between solid and vapor is called *sublimation* curve. In Fig. 5.2 typical *PT* diagrams for pure substances (a) and mixtures (b) are shown.

In Fig. 5.2a the freezing point temperature is almost the same as triple point temperature but they have different corresponding pressures. The normal boiling point and critical point both are on the vaporization line. A comparison between PV and PT diagrams for pure substances (Figs. 5.1 and 5.2a) shows that the two-phase region, which is an area in the PV diagram, becomes a line in the PT diagram. Similarly triple point, which is a point on the PT diagram, becomes a line on the PV diagram. For a mixture, as shown in Fig. 5.2*b*, the two-phase region is under the envelope and bubble point and dew points curves meet each other at the critical point. The main application of PT diagram is to determine the phase of a system under certain conditions of temperature and pressure as will be discussed in Chapter 9 (Section 9.2.3). Figure 5.1 shows that as temperature of a pure substance increases, at constant pressure, the following phase changes occur:

Subcooled solid (1) → Saturated solid at sublimation temperature (2) → Saturated liquid at sublimation temperature (3) → Subcooled liquid (4) → Saturated liquid at vaporization temperature (5) → Saturated vapor at vaporization temperature (6)→ Superheated vapor (7)

The process from (2) to (3) is called fusion or melting and the heat required is called *heat of fusion*. The process from (5) to (6) is called vaporization or boiling and the heat required is called *heat of vaporization*. Fusion and vaporization are two-phase change processes at which both temperature and pressure remain constant while volume, internal energy, and enthalpy would increase. A gas whose temperature is greater than T_c cannot be liquefied no matter how high the pressure is. The term *vapor* usually refers to a gas whose temperature is less than T_c and it can be converted to liquid as pressure exceeds the vapor pressure or saturation pressure at temperature T.

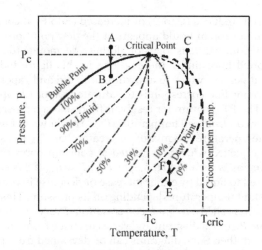

FIG. 5.3—Typical PT diagram for a reservoir fluid mixture.

An extended version of Fig. 5.2*b* is shown in Fig. 5.3 for a typical PT diagram of a reservoir fluid mixture. Lines of constant quality in the two-phase region converge at the critical point. The saturated vapor line is called *dewpoint* curve (dotted line) and the line of saturated liquid is usually called *bubblepoint* curve (solid line) as indicated in Figs. 5.2*b* and 5.3. In Fig. 5.3 when pressure of liquid is reduced at constant temperature (A to B), vaporization begins at the bubble point pressure. The bubblepoint curve is locus of all these bubble points. Similarly for temperatures above T_c when gas

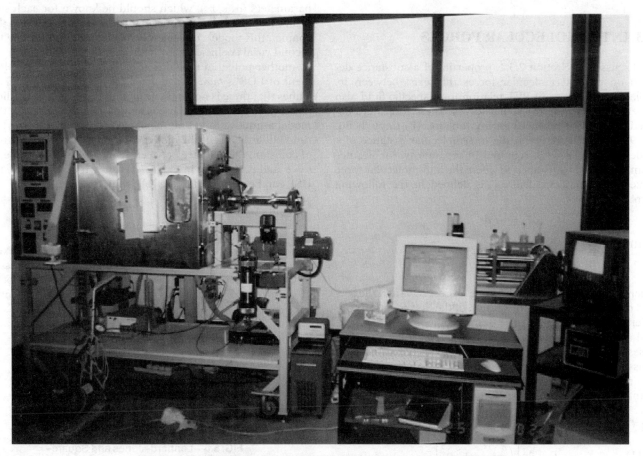

FIG. 5.4—A DB Robinson computerized PVT cell (courtesy of KISR) [5].

pressure is reduced (C to D) or increased (E to F) at constant T, the first drop of liquid appears at the dew point pressure. The dewpoint curve is locus of all these dewpoints (dotted). The dotted lines under the envelope in this figure indicate constant percent vapor in a mixture of liquid and vapor. The 100% vapor line corresponds to saturated vapor (dewpoint) curve. The PT diagram for reservoir fluids has a temperature called *cricondentherm* temperature (T_{cric}) as shown in Fig. 5.3. When temperature of a mixture is greater than T_{cric} a gas cannot be liquefied when pressurized at constant temperature. However, as is seen in Fig. 5.3, at $T_c < T < T_{cric}$ a gas can be converted to liquid by either increase or decrease in pressure at constant temperature depending on its pressure. This phenomenon is called *retrograde condensation*. Every mixture has a unique *PT* or *PV* diagram and varies in shape from one mixture to another. Such diagrams can be developed from phase equilibrium calculations that require composition of the mixture and is discussed in Chapter 9.

Accurate measurement of fluid phase behavior and related physical properties can be obtained from a PVT apparatus. The central part of this equipment is a transparent cylindrical cell of about 2.0–2.5 cm diameter and 20 cm length sealed by a piston that can be moved to adjust desired volume. A typical modern and mercury-free PVT system made by D B Robinson, courtesy of KISR [5], is shown in Fig. 5.4. Variation of P and V can be determined at various isotherms for different systems of pure compounds and fluid mixtures. The PVT cell is particularly useful in the study of phase behavior of reservoir fluids and construction of *PT* diagrams as will be discussed in Chapter 9.

5.3 INTERMOLECULAR FORCES

As discussed in Section 2.3.1, properties of a substance depend on the intermolecular forces that exist between its molecules. The type of PVT relation for a specific fluid also depends on the intermolecular forces. These forces are defined in terms of potential energy function (Γ) through Eq. (2.19). Potential energy at the intermolecular distance of r is defined as the work required to separate two molecules from distance r to distance ∞ where the intermolecular force is zero and mathematically Γ is defined in the following forms:

$$d\Gamma = -F\,dr$$

(5.10)
$$\Gamma(r) = \int_r^{\infty} F(r)\,dr$$

where the first equation is the same as Eq. (2.19) and the second one is derived from integration of the first equation considering the fact that $\Gamma(\infty) = 0$. Γ is composed of repulsive and attractive terms where the latter is negative. For ideal gases where the distance between the molecules is large, it is assumed that $\Gamma = 0$ as shown in Fig. 5.5 [6]. For nonpolar compounds such as hydrocarbon systems for which the dominant force is London dispersion force, the potential energy may be expressed by Lennard–Jones (LJ) model given by Eq. (2.21) as

(5.11)
$$\Gamma = 4\varepsilon\left[\left(\frac{\sigma}{r}\right)^{12} - \left(\frac{\sigma}{r}\right)^{6}\right]$$

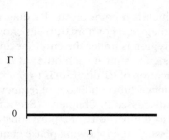

FIG. 5.5—Potential energy for ideal gases.

where ε and σ are energy and size parameters, which are characteristics of each substance. The significance of this function is that (a) at $r = \sigma$, $\Gamma = 0$ (i.e., at $r = \sigma$ repulsion and attraction forces are just balanced) and (b) $F = -d\Gamma/dr = 0$ at $\Gamma = -\varepsilon$. In fact $\Gamma = -\varepsilon$ is the minimum potential energy, which defines equilibrium separation where force of attraction is zero. The potential model is illustrated in Fig. 5.6.

Since the LJ potential is not mathematically convenient to use, the following potential model called *Square–Well potential* (SWP) is proposed to represent the LJ model for nonpolar systems:

(5.12)
$$\Gamma(r) = \begin{cases} \infty & r \leq \sigma \\ -\varepsilon & \sigma \leq r \leq r^*\sigma \\ 0 & r \geq r^*\sigma \end{cases}$$

where in the region $1 < r/\sigma < r^*$ we have Square–Well (SW). This model is also shown in Fig. 5.6. The SW model has three parameters (σ, ε, r^*), which should be known for each substance from molecular properties. As will be seen later in this chapter, this model conveniently can be used to estimate the second virial coefficients for hydrocarbon systems.

Another potential model that has been useful in development of EOS is *hard-sphere potential* (HSP). This model assumes that there is no interaction until the molecules collide. At the time of collision there is an infinite interaction. In this model attractive forces are neglected and molecules are like rigid billiard balls. If the molecular diameter is σ, at the time of collision, the distance between centers of two molecules is $r = \sigma$ and it is shown in Fig. 5.7. As shown in this figure, the HSP can be expressed in the following form:

(5.13)
$$\Gamma = \begin{cases} \infty & \text{at } r \leq \sigma \\ 0 & \text{at } r > \sigma \end{cases}$$

It is assumed that as $T \to \infty$ all gases behave like hard sphere molecules. Application of this model will be discussed in Chapter 6 for the development of EOS based on velocity of sound. In all models according to definition of potential

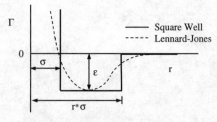

FIG. 5.6—Lennard–Jones and Square–Well potential models.

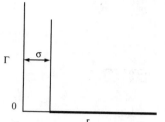

FIG. 5.7—Hard sphere potential model.

energy we have, as $r \to \infty$, $\Gamma \to 0$. For example, in the Sutherland model it is assumed that the repulsion force is ∞ but the attraction force is proportional with $1/r^n$, that is for $r > \sigma$, $\Gamma = -D/r^6$, where D is the model parameter [6]. Potential energy models presented in this section do not describe molecular forces for heavy hydrocarbons and polar compounds. For such molecules, additional parameters must be included in the model. For example, dipole moment is a parameter that characterizes degree of polarity of molecules and its knowledge for very heavy molecules is quite useful for better property prediction of such compounds. Further discussion and other potential energy functions and intermolecular forces are discussed in various sources [6, 7].

5.4 EQUATIONS OF STATE

An EOS is a mathematical equation that relates pressure, volume, and temperature. The simplest form of these equations is the ideal gas law that is only applicable to gases. In 1873, van der Waals proposed the first cubic EOS that was based on the theory of continuity of liquids and gases. Since then many modifications of cubic equations have been developed and have found great industrial application especially in the petroleum industry because of their mathematical simplicity. More sophisticated equations are also proposed in recent decades that are useful for certain systems [8]. Some of these equations particularly useful for petroleum fluids are reviewed and discussed in this chapter.

5.4.1 Ideal Gas Law

As discussed in the previous section the intermolecular forces depend on the distance between the molecules. With an increase in molar volume or a decrease in pressure the intermolecular distance increases and the intermolecular forces decrease. Under very low-pressure conditions, the intermolecular forces are so small that they can be neglected ($\Gamma = 0$). In addition since the empty space between the molecules is so large the volume of molecules may be neglected in comparison with the gas volume. Under these conditions any gas is considered as an *ideal gas*. Properties of ideal gases can be accurately estimated based on the kinetic theory of gases [9, 10]. The universal form of the EOS for ideal gases is

$$(5.14) \qquad PV^{ig} = RT$$

where T is absolute temperature, P is the gas absolute pressure, V^{ig} is the molar volume of an ideal gas, and R is the

universal gas constant in which its values in different units are given in Section 1.7.24. The conditions that Eq. (5.14) can be used depend on the substance and its critical properties. But approximately this equation may be applied to any gas under atmospheric or subatmospheric pressures with an acceptable degree of accuracy. An EOS can be nondimensionalized through a parameter called *compressibility factor*, Z, defined as

$$(5.15) \qquad Z \equiv \frac{V}{V^{ig}} = \frac{PV}{RT}$$

where for an ideal gas $Z = 1$ and for a real gas it can be greater or less than unity as will be discussed later in this chapter. Z in fact represents the ratio of volume of real gas to that of ideal gas under the same conditions of T and P. As the deviation of a gas from ideality increases, so does deviation of its Z factor from unity. The application of Z is in calculation of physical properties once it is known for a fluid. For example, if Z is known at T and P, volume of gas can be calculated from Eq. (5.15). Application of Eq. (5.15) at the critical point gives *critical compressibility factor*, Z_c, which was initially defined by Eq. (2.8).

In ideal gases, molecules have mass but no volume and they are independent from each other with no interaction. An ideal gas is mathematically defined by Eq. (5.14) with the following relation, which indicates that the internal energy is only a function of temperature.

$$(5.16) \qquad U^{ig} = f_4(T)$$

Substitution of Eqs. (5.14) and (5.16) into Eq. (5.5) gives

$$(5.17) \qquad H^{ig} = f_5(T)$$

where H^{ig} is the ideal gas enthalpy and it is only a function of temperature. Equations (5.14), (5.16), and (5.17) simply define ideal gases.

5.4.2 Real Gases—Liquids

Gases that do not follow ideal gas conditions are called *real gases*. At a temperature below critical temperature as pressure increases a gas can be converted to a liquid. In real gases, volume of molecules as well as the force between molecules are not zero. A comparison among an ideal gas, a real gas, and a liquid is demonstrated in Fig. 5.8. As pressure increases behavior of real gases approaches those of their liquids. The space between the molecules in liquids is less than real gases and in real gases is less than ideal gases. Therefore, the intermolecular forces in liquids are much stronger than those in real gases. Similarly the molecular forces in real gases are higher than those in ideal gases, which are nearly zero. It is for this reason that prediction of properties of liquids is more difficult than properties of gases.

Most gases are actually real and do not obey the ideal gas law as expressed by Eqs. (5.14) and (5.16). Under limiting conditions of $P \to 0$ ($T > 0$) or at T and $V \to \infty$ (finite P) we can obtain a set of constraints for any real gas EOS. When $T \to \infty$ translational energy becomes very large and other energies are negligible. Any valid EOS for a real gas should obey the following constraints:

$$(5.18) \qquad \lim_{P \to 0} (PV) = RT$$

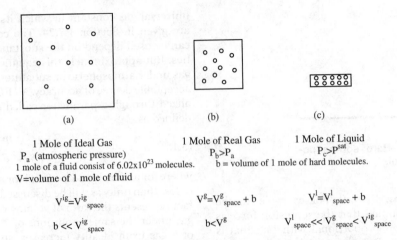

1 Mole of Ideal Gas
P_a (atmospheric pressure)
1 mole of a fluid consist of 6.02×10^{23} molecules.
V = volume of 1 mole of fluid

1 Mole of Real Gas
$P_b > P_a$
b = volume of 1 mole of hard molecules.

1 Mole of Liquid
$P_c > P^{sat}$

$$V^{ig} = V^{ig}_{space}$$

$$V^g = V^g_{space} + b$$

$$V^l = V^l_{space} + b$$

$$b \ll V^{ig}_{space}$$

$$b < V^g$$

$$V^l_{space} \ll V^g_{space} < V^{ig}_{space}$$

FIG. 5.8—Difference between an ideal gas, a real gas, and a liquid.

(5.19)
$$\lim_{T \to \infty} \left(\frac{\partial V}{\partial T} \right)_P = \frac{R}{P}$$

(5.20)
$$\lim_{T \to \infty} \left(\frac{\partial^2 V}{\partial T^2} \right)_P = 0$$

In general for any gas as $P \to 0$ (or $V \to \infty$) it becomes an ideal gas; however, as $T \to \infty$ it is usually assumed that gas behavior approaches those of hard sphere gases. Constraints set by the above equations as well as Eq. (5.9) may be used to examine validity of an EOS for real fluids.

5.5 CUBIC EQUATIONS OF STATE

The ideal gas law expressed by Eq. (5.14) is neither applicable to real gases (high pressure) nor to liquids where the volume of molecules cannot be ignored in comparison with the volume of gas (see Fig. 5.8). Cubic EOS are designed to overcome these two shortcomings of ideal gas law with mathematical convenience. Several commonly used equations, their solution, and characteristics are discussed in this section.

5.5.1 Four Common Cubic Equations (vdW, RK, SRK, and PR)

The behavior of high-pressure gases approaches the behavior of liquids until the critical point where both gas and liquid behavior become identical. van der Waal (vdW) proposed the idea of continuity of gases and liquids and suggested that a single equation may represent the PVT behavior of both gases and liquids. He modified Eq. (5.14) by replacing P and V with appropriate modifications to consider real gas effects in the following form [1]:

(5.21)
$$\left(P + \frac{a}{V^2} \right) (V - b) = RT$$

where a and b are two constants specific for each substance but independent of T and P. The above equation is usually written as

(5.22)
$$P = \frac{RT}{V - b} - \frac{a}{V^2}$$

To find V from T and P, the above equation may be rearranged as

(5.23)
$$V^3 - \left(b + \frac{RT}{P} \right) V^2 + \left(\frac{a}{P} \right) V - \frac{ab}{P} = 0$$

where it is a cubic equation in terms of V. For this reason the vdW EOS, Eq. (5.22), is known as a cubic EOS. As a matter of fact any EOS that can be converted into a cubic form is called a *cubic* EOS. In Eq. (5.22), parameters a and b have physical meanings. Parameter b also called co-volume or repulsive parameter represents volume of 1 mol of hard cores of molecules and has the same unit as the molar volume (V). Parameter a is also referred to as attraction parameter and has the same unit as that of PV^2 (i.e., bar·cm^6/mol^2). In Eq. (5.22), the term $RT/(V - b)$ represents the repulsive term of a molecule, while a/V^2 represents attractive term and accounts for nonideal behavior of gas. $V - b$ is in fact the space between molecules (Figs. 5.8b and 5.8c). When parameters a and b are zero Eq. (5.22) reduces to ideal gas law. Mathematically it can be shown from Eq. (5.22) that as $P \to \infty$, $V \to b$ and the free volume between molecules disappears.

Since Eq. (5.21) has only two parameters it is also known as a two-parameter EOS. Parameters a and b in the vdW EOS can be best determined from experimental data on PVT. However, mathematically these constants can be determined by imposing Eq. (5.9) as shown in the following example.

Example 5.1—Obtain vdW parameters in terms of T_c and P_c using Eq. (5.9) and (5.21). Also determine Z_c for fluids that obey vdW EOS.

Solution—$\partial P/\partial V$ and $\partial^2 P/\partial V^2$ are calculated from Eq. (5.22) by keeping T constant and set equal to zero at $T = T_c$, $P = P_c$, and $V = V_c$ as

(5.24)
$$\left. \frac{\partial P}{\partial V} \right|_{T_c} = -\frac{RT_c}{(V_c - b)^2} + \frac{2a}{V_c^3} = 0$$

(5.25)
$$\left. \frac{\partial^2 P}{\partial V^2} \right|_{T_c} = \frac{2RT_c}{(V_c - b)^3} - \frac{6a}{V_c^4} = 0$$

By taking the second terms to the right-hand side in each equation and dividing Eq. (5.24) by Eq. (5.25) we get

$$(5.26) \qquad b = \frac{V_c}{3}$$

By substituting Eq. (5.26) into Eq. (5.24) we get

$$(5.27) \qquad a = \frac{9}{8} T_c V_c$$

Since T_c and P_c are usually available, it is common to express parameters a and b in terms of T_c and P_c rather than T_c and V_c. For this reason V_c can be found from Eq. (5.21) in terms of T_c and P_c and replaced in the above equations. Similar results can be obtained by a more straightforward approach. At the critical point we have $V = V_c$ or $V - V_c = 0$, which can be written as follows:

$$(5.28) \qquad (V - V_c)^3 = 0$$

Application of Eq. (5.23) at T_c and P_c gives

$$(5.29) \qquad V^3 - \left(b + \frac{RT_c}{P_c}\right) V^2 + \left(\frac{a}{P_c}\right) V - \frac{ab}{P_c} = 0$$

Expansion of Eq. (5.28) gives

$$(5.30) \qquad (V - V_c)^3 = V^3 - 3V_c V^2 + 3V_c^2 V - V_c^3 = 0$$

Equations (5.29) and (5.30) are equivalent and the corresponding coefficients for V^3, V^2, V^1, and V^0 must be equal in two equations. This gives the following set of equations for the coefficients:

$$(5.31) \qquad -\left(b + \frac{RT_c}{P_c}\right) = -3V_c \qquad \text{coefficients of } V^2$$

$$(5.32) \qquad \frac{a}{P_c} = 3V_c^2 \qquad \text{coefficients of } V$$

$$(5.33) \qquad -\frac{ab}{P_c} = -V_c^3 \qquad \text{coefficients of } V^0$$

By dividing Eq. (5.33) by (5.32), Eq. (5.26) can be obtained. By substituting $V_c = 3b$ (Eq. 5.26) to the right-hand side (RHS) of Eq. (5.31) the following relation for b is found:

$$(5.34) \qquad b = \frac{RT_c}{8P_c}$$

Combining Eqs. (5.26) and (5.34) gives

$$(5.35) \qquad V_c = \frac{3RT_c}{8P_c}$$

Substituting Eq. (5.35) into Eq. (5.27) gives

$$a = \frac{9}{8} RT_c \left(\frac{3RT_c}{8P_c}\right) = \frac{27R^2 T_c^2}{64P_c}$$

Therefore, the final relation for parameter a in terms of T_c and P_c is as follows:

$$(5.36) \qquad a = \frac{27R^2 T_c^2}{64P_c}$$

In calculation of parameters a and b unit of R should be consistent with the units chosen for T_c and P_c. Another useful result from this analysis is estimation of critical compressibility factor through Eq. (5.35). Rearranging this equation and using definition of Z_c from Eq. (2.8) gives

$$(5.37) \qquad Z_c = \frac{P_c V_c}{RT_c} = \frac{3}{8} = 0.375$$

Equation (5.37) indicates that value of Z_c is the same for all compounds. Values of Z_c given in Table 2.1 varies from 0.28 to 0.21 for most hydrocarbons. Therefore, vdW EOS significantly overpredicts values of Z_c (or V_c) and its performance in the critical region is quite weak. Similar approaches can be used to determine EOS parameters and Z_c for any other EOS.

♦

Since the introduction of the vdW EOS as the first cubic equation 130 years ago, dozens of cubic EOSs have been proposed, many of them developed in recent decades. The mathematical simplicity of a cubic EOS in calculation of thermodynamic properties has made it the most attractive type of EOS. When van der Waals introduced Eq. (5.21) he indicated that parameter a is temperature-dependent. It was in 1949 when Redlich and Kwong (RK) made the first modification to vdW EOS as [11]

$$(5.38) \qquad P = \frac{RT}{V - b} - \frac{a}{V(V + b)}$$

where parameter a depends on temperature as $a_c/T^{0.5}$ in which a_c is related to T_c and P_c. Parameters a and b in Eq. (5.38) are different from those in Eq. (5.22) but they can be obtained in a similar fashion as in Example 5.1 (as shown later). The repulsive terms in Eqs. (5.38) and (5.22) are identical. Performance of RK EOS is much better than vdW EOS; however, it is mainly applicable to simple fluids and rare gases such as Kr, CH_4, or O_2, but for heavier and complex compounds it is not a suitable PVT relation.

The RK EOS is a source of many modifications that began in 1972 by Soave [12]. The Soave modification of Redlich–Kwong equation known as SRK EOS is actually a modification of parameter a in terms of temperature. Soave obtained parameter a in Eq. (5.38) for a number of pure compounds using saturated liquid density and vapor pressure data. Then he correlated parameter a to reduced temperature and acentric factor. Acentric factor, ω, defined by Eq. (2.10) is a parameter that characterizes complexity of a molecule. For more complex and heavy compounds value of ω is higher than simple molecules as given in Table 2.1. SRK EOS has been widely used in the petroleum industry especially by reservoir engineers for phase equilibria calculations and by process engineers for design calculations. While RK EOS requires T_c and P_c to estimate its parameters, SRK EOS requires an additional parameter, namely a third parameter, which in this case is ω. As it will be seen later that while SRK EOS is well capable of calculating vapor–liquid equilibrium properties, it seriously underestimates liquid densities.

Another popular EOS for estimation of phase behavior and properties of reservoir fluids and hydrocarbon systems is Peng-Robinson (PR) proposed in the following form [13]:

$$(5.39) \qquad P = \frac{RT}{V - b} - \frac{a}{V(V + b) + b(V - b)}$$

where a and b are the two parameters for PR EOS and are calculated similar to SRK parameters. Parameter a was correlated in terms of temperature and acentric factor and later it was modified for properties of heavy hydrocarbons [14]. The original idea behind development of PR EOS was to improve liquid density predictions. The repulsive term in all four cubic equations introduced here is the same. In all these equations

TABLE 5.1—*Constants in Eq. (5.40) for four common cubic EOS (with permission from Ref. [15]).*

Equation	u_1	u_2	a_c	α	b	Z_c
vdW	0	0	$\frac{27}{64}\frac{R^2 T_c^2}{P_c}$	1	$\frac{RT_c}{8P_c}$	0.375
RK	1	0	$\frac{0.42748R^2 T_c^2}{P_c}$	$T_r^{-1/2}$	$\frac{0.08664RT_c}{P_c}$	0.333
SRK	1	0	$\frac{0.42748R^2 T_c^2}{P_c}$	$\left[1 + f_\omega\left(1 - T_r^{1/2}\right)\right]^2$ $f_\omega = 0.48 + 1.574\omega - 0.176\omega^2$	$\frac{0.08664RT_c}{P_c}$	0.333
PR	2	−1	$\frac{0.45724R^2 T_c^2}{P_c}$	$\left[1 + f_\omega\left(1 - T_r^{1/2}\right)\right]^2$ $f_\omega = 0.37464 + 1.54226\omega - 0.2699\omega^2$	$\frac{0.07780RT_c}{P_c}$	0.307

when $a = b = 0$, the equation reduces to ideal gas law, Eq. (5.14). In addition, all equations satisfy the criteria set by Eqs. (5.18)–(5.20) as well as Eq. (5.9). For example, consider the PR EOS expressed by Eq. (5.39). To show that criteria set by Eq. (5.18) are satisfied, the limits of all terms as $V \to \infty$ (equivalent to $P \to 0$) should be calculated. If both sides of Eq. (5.39) are multiplied by V/RT and taking the limits of all terms as $V \to \infty$ (or $P \to 0$), the first term in the RHS approaches unity while the second term approaches zero and we get $Z \to 1$, which is the EOS for the ideal gases.

Reid et al. [15] have put vdW, RK, SRK, and PR two-parameter cubic EOS into a practical and unified following form:

$$(5.40) \qquad P = \frac{RT}{V - b} - \frac{a}{V^2 + u_1 bV + u_2 b^2}$$

where u_1 and u_2 are two integer values specific for each cubic equation and are given in Table 5.1. Parameter a is in general temperature-dependent and can be expressed as

$$(5.41) \qquad a = a_c \alpha$$

where α is a dimensionless temperature-dependent parameter and usually is expressed in terms of reduced temperature ($T_r = T/T_c$) and acentric factor as given in Table 5.1. For both vdW and RK equations this parameter is unity. Parameters u_1 and u_2 in Eq. (5.40) are the same for both RK and SRK equations, as can be seen in Table 5.1, but vdW and PR equations have different values for these parameters. Equation (5.40) can be converted into a cubic form equation similar to Eq. (5.23) but in term of Z rather than V:

$$(5.42) \qquad \begin{aligned} &Z^3 - (1 + B - u_1 B)Z^2 + (A + u_2 B^2 - u_1 B - u_1 B^2)Z \\ &\quad - AB - u_2 B^2 - u_2 B^3 = 0 \end{aligned}$$

$$\text{where } A = \frac{aP}{R^2 T^2} \quad \text{and} \quad B = \frac{bP}{RT}$$

in which parameters A and B as well as all terms in Eq. (5.42) are dimensionless. Parameters a and b and Z_c have been determined in a way similar to the methods shown in Example 5.1. Z_c for both RK and SRK is the same as 1/3 or 0.333 while for the PR it is lower and equal to 0.307 for all compounds. As it will be shown later performance of all these equations near the critical region is weak and leads to large errors for calculation of Z_c. Prediction of an isotherm by a cubic EOS is shown in Fig. 5.9. As is seen in this figure, pressure prediction for an isotherm by a cubic EOS in the two-phase region is not reliable. However, isotherms outside the two-phase envelope may be predicted by a cubic EOS with a reasonable accuracy. In calculation of Z for saturated liquid and saturated vapor at the same T and P, Eq. (5.42) should be solved at once, which

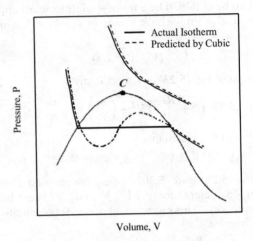

FIG. 5.9—Prediction of isotherms by a cubic EOS.

gives three roots for Z. The lowest value of Z corresponds to saturated liquid, the highest root gives Z for the saturated vapor, and the middle root has no physical meaning.

5.5.2 Solution of Cubic Equations of State

Equation (5.42) can be solved through solution of the following general cubic equation [16, 17]:

$$(5.43) \qquad Z^3 + a_1 Z^2 + a_2 Z + a_3 = 0$$

Let's define parameters Q, L, D, S_1, and S_2 as

$$(5.44) \qquad \begin{aligned} Q &= \frac{3a_2 - a_1^2}{9} \\ L &= \frac{9a_1 a_2 - 27a_3 - 2a_1^3}{54} \\ D &= Q^3 + L^2 \\ S_1 &= (L + \sqrt{D})^{1/3} \\ S_2 &= (L - \sqrt{D})^{1/3} \end{aligned}$$

The type and number of roots of Eq. (5.43) depends on the value of D. In calculation of $X^{1/3}$ if $X < 0$, one may use $X^{1/3} = -(-X)^{1/3}$.

If $D > 0$ Eq. (5.43) has one real root and two complex conjugate roots. The real root is given by

$$(5.45) \qquad Z_1 = S_1 + S_2 - a_1/3$$

If $D = 0$ all roots are real and at least two are equal. The unequal root is given by Eq. (5.45) with $S_1 = S_2 = L^{1/3}$. The two equal roots are

(5.46) $$Z_2 = Z_3 = -L^{1/3} - a_1/3$$

If $D < 0$ all roots are real and unequal. In this case S_1 and S_2 (Eq. 5.44) cannot be calculated and the computation is simplified by use of trigonometry as

(5.47)
$$Z_1 = 2\sqrt{-Q}\, \text{Cos}\left(\frac{1}{3}\theta + 120°\right) - \frac{a_1}{3}$$
$$Z_2 = 2\sqrt{-Q}\, \text{Cos}\left(\frac{1}{3}\theta + 240°\right) - \frac{a_1}{3}$$
$$Z_3 = 2\sqrt{-Q}\, \text{Cos}\left(\frac{1}{3}\theta\right) - \frac{a_1}{3}$$

where $\text{Cos}\,\theta = \dfrac{L}{\sqrt{-Q^3}}$

where θ is in degrees. To check validity of the solution, the three roots must satisfy the following relations

(5.48)
$$Z_1 + Z_2 + Z_3 = -a_1$$
$$Z_1 \times Z_2 + Z_2 \times Z_3 + Z_3 \times Z_1 = a_2$$
$$Z_1 \times Z_2 \times Z_3 = -a_3$$

A comparison of Eq. (5.42) and (5.43) indicates that the following relations exist between coefficients $a_i(s)$ and EOS parameters

(5.49)
$$a_1 = -(1 + B - u_1 B)$$
$$a_2 = A + u_2 B^2 - u_1 B - u_1 B^2$$
$$a_3 = -AB - u_2 B^2 - u_2 B^3$$

For the case that there are three different real roots ($D < 0$), Z^{liq} is equal to the lowest root (Z_1) while Z^{vap} is equal to the highest root (Z_3). The middle root (Z_2) is disregarded as physically meaningless. Equation (5.42) may also be solved by successive substitution methods; however, appropriate forms of the equation and initial values are different for vapor and liquid cases. For example, for gases the best initial value for Z is 1 while for liquids a good initial guess is bP/RT [1]. Solution of cubic equations through Eq. (5.42) is shown in the following example.

Example 5.2—Estimate molar volume of saturated liquid and vapor for *n*-octane at 279.5°C and pressure of 19.9 bar from the RK, SRK, and PR cubic EOS. Values of V^L and V^V extracted from the experimental data are 304 and 1216 cm³/mol, respectively [18]. Also estimate the critical volume.

Solution—To use SRK and PR EOS pure component data for *n*-C$_8$ are taken from Table 2.1 as $T_c = 295.55°C$ (568.7 K), $P_c = 24.9$ bar, $\omega = 0.3996$, and $V_c = 486.35$ cm³/mol. When T is in K, P is in bar, and V is in cm³/mol, value of R from Section 1.7.24 is 83.14 cm³·bar/mol·K. Sample calculation is shown here for SRK EOS. $T_r = (279.5 + 273.15)/568.7 = 0.972$.

From Table 5.1, $u_1 = 1, u_2 = 0$ and a_{SRK} and b_{SRK} are calculated as

$$f_\omega = 0.48 + 1.574 \times 0.3996 - 0.176 \times (0.3996)^2 = 1.08087$$

$$a_{SRK} = \frac{0.42748 \times (83.14)^2 \times (568.7)^2}{24.9}$$
$$\times [1 + 1.08087 \times (1 - 0.9717^{1/2})]^2$$
$$= 3.957 \times 10^7 \text{ cm}^6/\text{mol}^2.$$

$$b_{SRK} = \frac{0.08664 \times 83.14 \times 568.7}{24.9} = 164.52 \text{ cm}^3/\text{mol}.$$

Parameters A and B are calculated from Eq. (5.42):

$$A = \frac{3.957 \times 10^7 \times 19.9}{(83.14)^2 \times (552.65)^2} = 0.373$$

and

$$B = \frac{164.52 \times 19.9}{83.14 \times 552.65} = 0.07126$$

Coefficients $a_1, a_2,$ and a_3 are calculated from Eq. (5.49) as

$$a_1 = -(1 + 0.07126 - 1 \times 0.07126) = -1$$
$$a_2 = 0.373 + 0 \times 0.07126^2 - 1 \times 0.07126 - 1 \times 0.07126^2$$
$$= 0.29664$$
$$a_3 = -0.37305 \times 0.07126 - 0 \times 0.07126^2 - 0 \times 0.07126^3$$
$$= -0.026584$$

From Eq. (5.44), $Q = -0.01223$, $L = 8.84 \times 10^{-4}$, and $D = -1.048 \times 10^{-6}$. Since $D < 0$, the solution is given by Eq. (5.47). $\theta = \text{Cos}^{-1}(8.84 \times 10^{-4}/\sqrt{-(-0.01223)^3}) = 492°$ and the roots are $Z_1 = 0.17314$, $Z_2 = 0.28128$, and $Z_3 = 0.54553$. Acceptable results are the lowest and highest roots while the intermediate root is not useful: $Z^L = Z_1 = 0.17314$ and $Z^V = Z_2 = 0.54553$. Molar volume, V, can be calculated from Eq. (5.15): $V = ZRT/P$ in which $T = 552.65$ K, $P = 19.9$ bar, and $R = 83.14$ cm³·bar/mol·K; therefore, $V^L = 399.9$ cm³/mol and $V^V = 1259.6$ cm³/mol. From Table 5.1, $Z_c = 0.333$ and V_c is calculated from Eq. (2.8) as $V_c = (0.333 \times 83.14 \times 568.7)/24.9 = 632.3$ cm³/mol. Errors for V^L, V^V, and V_c are 31.5, 3.6, and 30%, respectively. It should be noted that Z_c can also be found from the solution of cubic equation with $T = T_c$ and $P = P_c$. However, for this case $D > 0$ and there is only one solution which is obtained by Eq. (5.45) with similar answer. As is seen in this example, liquid and critical volumes are greatly overestimated. Summary of results for all four cubic equations are given in Table 5.2. ♦

5.5.3 Volume Translation

In practice the SRK and PR equations are widely used for VLE calculations in industrial applications [19–21]. However, their ability to predict volumetric data especially for liquid systems

TABLE 5.2—*Prediction of saturated liquid, vapor and critical molar volumes for n-octane in Example 5.2.*

Equation	V^L, cm³/mol	%D	V^V, cm³/mol	%D	V_c, cm³/mol	%D
Data*	304.0	...	1216.0	...	486.3	...
RK	465.9	53.2	1319.4	8.5	632.3	30
SRK	399.9	31.5	1259.6	3.6	632.3	30
PR	356.2	17.2	1196.2	-1.6	583.0	19.9

Source: V^L and V^V from Ref. [18]; V_c from Table 2.1.

is weak. Usually the SRK equation predicts densities more accurately for compounds with low acentric values, while the PR predicts better densities for compounds with acentric factors near 0.33 [21]. For this reason a correction term, known as volume translation, has been proposed for improving volumetric prediction of these equations [8, 15, 19, 22]:

$$(5.50) \qquad V = V^{\text{EOS}} - c$$

where c is the volume translation parameter and has the same unit as the molar volume. Equation (5.50) can be applied to both vapor and liquid volumes. Parameter c mainly improves liquid volume predictions and it has no effect on vapor pressure and VLE calculations. Its effect on vapor volume is negligible since V^{V} is very large in comparison with c, but it greatly improves prediction of liquid phase molar volumes. Values of c have been determined for a number of pure components up to C_{10} for both SRK and PR equations and have been included in references in the petroleum industry [19]. Peneloux et al. [22] originally obtained values of c for some compounds for use with the SRK equation. They also suggested the following correlation for estimation of c for SRK equation:

$$(5.51) \qquad c = 0.40768 \left(0.29441 - Z_{\text{RA}}\right) \frac{RT_{\text{c}}}{P_{\text{c}}}$$

where Z_{RA} is the Rackett parameter, which will be discussed in Section 5.8.1. Similarly Jhaveri and Yougren [23] obtained parameter c for a number of pure substances for use with PR EOS and for hydrocarbon systems have been correlated to molecular weight for different families as follows:

$$(5.52) \qquad \begin{aligned} c_{\text{P}} &= b_{\text{PR}} \left(1 - 2.258 M_{\text{P}}^{-0.1823}\right) \\ c_{\text{N}} &= b_{\text{PR}} \left(1 - 3.004 M_{\text{N}}^{-0.2324}\right) \\ c_{\text{A}} &= b_{\text{PR}} \left(1 - 2.516 M_{\text{A}}^{-0.2008}\right) \end{aligned}$$

where b_{PR} refers to parameter b for the PR equation as given in Table 5.1. Subscripts P, N, and A refer to paraffinic, naphthenic, and aromatic hydrocarbon groups. The ratio of c/b is also called *shift parameter*. The following example shows application of this method.

Example 5.3—For the system of Example 5.2, estimate V^{L} and V^{V} for the PR EOS using the volume translation method.

Solution—For n-C_8, from Table 2.1, $M = 114$ and b_{PR} are calculated from Table 5.1 as 147.73 cm³/mol. Since the hydrocarbon is paraffinic Eq. (5.51) for c_{P} should be used, $c = 7.1$ cm³/mol. From Table 5.2, $V^{\text{L(PR)}} = 356.2$ and $V^{\text{V(PR)}} = 1196.2$ cm³/mol. From Eq. (5.50) the corrected molar volumes are $V^{\text{L}} = 356.2 - 7.1 = 3491$ and $V^{\text{V}} = 1196.2 - 7.1 = 1189.1$ cm³/mol. By the volume translation correction, error for V^{L} decreases from 17.2 to 14.8% while for V^{V} it has lesser effect and it increases error from –1.6% to –2.2%. ♦

As is seen in this example improvement of liquid volume by volume translation method is limited. Moreover, estimation of c by Eq. (5.51) is limited to those compounds whose Z_{RA} is known. With this modification at least four parameters namely T_{c}, P_{c}, ω, and c must be known for a compound to determine its volumetric properties.

5.5.4 Other Types of Cubic Equations of State

In 1972 Saove for the first time correlated parameter a in a cubic EOS to both T_{r} and ω as given in Table 5.1. Since then this approach has been used by many researchers who tried to improve performance of cubic equations. Many modifications have been made on the form of f_ω for either SRK or PR equations. Graboski and Daubert modified the constants in the f_ω relation for the SRK to improve prediction of vapor pressure of hydrocarbons [24]. Robinson and Peng [14] also proposed a modification to their f_ω equation given in Table 5.1 to improve performance of their equation for heavier compounds. They suggested that for the PR EOS and for compounds with $\omega > 0.49$ the following relation should be used to calculate f_ω:

$$(5.53) \quad f_\omega = 0.3796 + 1.485\omega - 0.1644\omega^2 + 0.01667\omega^3$$

Some other modifications give different functions for parameter α in Eq. (5.41). For example, Twu et al. [25] developed the following relation for the PR equation.

$$(5.54) \quad \begin{aligned} \alpha =\ & T_{\text{r}}^{-0.171813} \exp\left[0.125283\left(1 - T_{\text{r}}^{1.77634}\right)\right] \\ & + \omega\left\{T_{\text{r}}^{-0.607352} \exp\left[0.511614\left(1 - T_{\text{r}}^{2.20517}\right)\right]\right. \\ & \left. - T_{\text{r}}^{-0.171813} \exp\left[0.125283\left(1 - T_{\text{r}}^{1.77634}\right)\right]\right\} \end{aligned}$$

Other modifications of cubic equations have been derived by suggesting different integer values for parameters u_1 and u_2 in Eq. (5.40). One can imagine that by changing values of u_1 and u_2 in Eq. (5.40) various forms of cubic equations can be obtained. For example, most recently a modified two-parameter cubic equation has been proposed by Moshfeghian that corresponds to $u_1 = 2$ and $u_2 = -2$ and considers both parameters a and b as temperature-dependent [26]. Poling et al. [8] have summarized more than two dozens types of cubic equations into a generalized equation similar to Eq. (5.42). Some of these modifications have been proposed for special systems. However, for hydrocarbons systems the original forms of SRK and PR are still being used in the petroleum industry. The most successful modification was proposed by Zudkevitch and Joffe [27] to improve volumetric prediction of RK EOS without sacrificing VLE capabilities. They suggested that parameter b in the RK EOS may be modified similar to Eq. (5.41) for parameter a as following:

$$(5.55) \qquad b = b_{\text{RK}}\beta$$

where β is a dimensionless correction factor for parameter b and it is a function of temperature. Later Joffe et al. [28] determined parameter α in Eq. (5.41) and β in Eq. (5.55) by matching saturated liquid density and vapor pressure data over a range of temperature for various pure compounds. In this approach for every case parameters α and β should be determined and a single dataset is not suitable for use in all cases. SRK and ZJRK are perhaps the most widespread cubic equations being used in the petroleum industry, especially for phase behavior studies of reservoir fluids [19]. Other researchers have also tried to correlate parameters α and β in Eqs. (5.41) and (5.55) with temperature. Feyzi et al. [29] correlated $\alpha^{1/2}$ and $\beta^{1/2}$ for PR EOS in terms T_{r} and ω for heavy reservoir fluids and near the critical region. Their correlations particularly improve liquid density prediction in comparison with SRK and PR equations while it has similar VLE

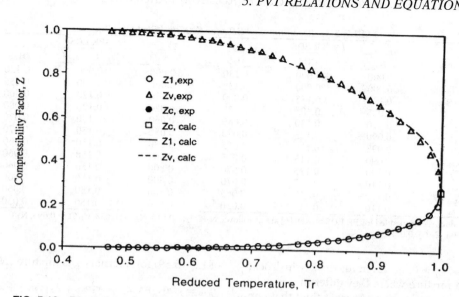

FIG. 5.10—Prediction of saturation curves for ethane using a modified PR EOS [29].

prediction capabilities. Another improvement in their correlation was prediction of saturation curves near the critical region. This is shown for prediction of compressibility factor of saturated liquid and vapor curves as well as the critical point for methane and ethane in Fig. 5.10.

Prediction of isotherms by a cubic EOS is shown on *PV* diagram in Fig. 5.9. As shown in this figure in the two-phase region the prediction of isotherm is not consistent with true behavior of the isotherm. In addition, performance of these cubic equations in calculation of liquid densities and derived thermodynamic properties such as heat capacity is weak. This indicates the need for development of other EOS. Further information on various types of cubic EOS and their characteristics are available in different sources [30–34].

5.5.5 Application to Mixtures

Generally when a PVT relation is available for a pure substance, the mixture property may be calculated in three ways when the mixture composition (mole fraction, x_i) is known. The first approach is to use the same equation developed for pure substances but the input parameters (T_c, P_c, and ω) are estimated for the mixture. Estimation of these pseudocritical properties for petroleum fractions and defined hydrocarbon mixtures were discussed in Chapter 3. The second approach is to estimate desired physical property (i.e., molar volume or density) for all pure compounds using the above equations and then to calculate the mixture property using the mixture composition through an appropriate mixing rule for the property (i.e., Eq. (3.44) for density). This approach in some cases gives good estimate of the property but requires large calculation time especially for mixtures containing many components. The third and most widely used approach is to calculate EOS parameters (parameters a and b) for the mixture using their values for pure components and mixture composition.

The simplest EOS for gases is the ideal gas law given by Eq. (5.14). When this is applied to component i with n_i moles in a mixture we have

$$(5.56) \qquad PV_i^t = n_i RT$$

where V_i^t is the total volume occupied by i at T and P of the mixture. For the whole mixture this equation becomes

$$(5.57) \qquad PV^t = n^t RT$$

where V^t is the total volume of mixture ($V^t = \Sigma V_i^t$) and n^t is the total number of moles ($n^t = \Sigma n_i$). By dividing Eq. (5.56) by Eq. (5.57) we get

$$(5.58) \qquad y_i = \frac{n_i}{n} = \frac{V_i^t}{V^t}$$

where y_i is the mole fraction of i in the gas mixture. The above equation indicates that in an ideal gas mixture the mole fractions and volume fractions are the same (or mol% of i = vol% of i). This is an assumption that is usually used for gas mixtures even when they are not ideal.

For nonideal gas mixtures, various types of mixing rules for determining EOS parameters have been developed and presented in different sources [6, 8]. The mixing rule that is commonly used for hydrocarbon and petroleum mixtures is called *quadratic mixing rule*. For mixtures (vapor or liquid) with composition x_i and total of N components the following equations are used to calculate a and b for various types of cubic EOS:

$$(5.59) \qquad a_{mix} = \sum_{i=1}^{N} \sum_{j=1}^{N} x_i x_j a_{ij}$$

$$(5.60) \qquad b_{mix} = \sum_{i=1}^{N} x_i b_i$$

where a_{ij} is given by the following equation:

$$(5.61) \qquad a_{ij} = (a_i a_j)^{1/2} (1 - k_{ij})$$

For the volume translation c, the mixing rule is the same as for parameter b:

$$(5.62) \qquad c_{mix} = \sum_{i=1}^{N} x_i c_i$$

In Eq. (5.61), k_{ij} is a dimensionless parameter called *binary interaction parameter* (BIP), where $k_{ii} = 0$ and $k_{ij} = k_{ji}$. For most

TABLE 5.3—Recommended BIPs for SRK and PR EOS [19].

Comp.	PR EOS			SRK EOS		
	N2	CO2	H2S	N2	CO2	H2S
N2	0.000	0.000	0.130	0.000	0.000	0.120
CO2	0.000	0.000	0.135	0.000	0.000	0.135
H2S	0.130	0.135	0.000	0.120	0.120	0.000
C1	0.025	0.105	0.070	0.020	0.120	0.080
C2	0.010	0.130	0.085	0.060	0.150	0.070
C3	0.090	0.125	0.080	0.080	0.150	0.070
iC4	0.095	0.120	0.075	0.080	0.150	0.060
nC4	0.090	0.115	0.075	0.800	0.150	0.060
iC5	0.100	0.115	0.070	0.800	0.150	0.060
nC5	0.110	0.115	0.070	0.800	0.150	0.050
C6	0.110	0.115	0.055	0.800	0.150	0.030
C7+	0.110	0.115	0.050	0.800	0.150	0.030

Values recommended for PR EOS by Ref. [6] are as follows: N_2/CO_2: -0.013; N_2/C_1: 0.038; C_1/CO_2: 0.095; N_2/C_2: 0.08; C_1/C_2: 0.021.

hydrocarbon systems, $k_{ij} = 0$; however, for the key hydrocarbon compounds in a mixture where they differ in size value of k_{ij} is nonzero. For example, for a reservoir fluid that contains a considerable amount of methane and C_{7+} the BIP for C_1 and C_7 fractions cannot be ignored. For nonhydrocarbon–hydrocarbon pairs k_{ij} values are nonzero and have a significant impact on VLE calculations [20, 35]. Values of k_{ij} for a particular pair may be determined from matching experimental data with predicted data on a property such as vapor pressure. Values of k_{ij} are specific to the particular EOS being used. Some researchers have determined k_{ij} for SRK or PR equations. Values of BIP for N_2, CO_2, and methane with components in reservoir fluids from C_1 to C_6 and three subfractions of C_{7+} for PR and SRK are tabulated by Whitson [19]. Values that he has recommended for use with SRK and PR equations are given in Table 5.3. There are some general correlations to estimate BIPs for any equation [36, 37]. The most commonly used correlation for estimating BIPs of hydrocarbon–hydrocarbon (HC–HC) systems is given by Chueh and Prausnitz [37]:

$$(5.63) \quad k_{ij} = A \left\{ 1 - \left[\frac{2(V_{ci} V_{cj})^{1/6}}{(V_{ci})^{1/3} + (V_{cj})^{1/3}} \right]^B \right\}$$

where V_{ci} and V_{cj} are critical molar volume of components i and j in cm^3/mol. Originally $A = 1$ and $B = 3$; however, in practical cases B is set equal to 6 and A is adjusted to match saturation pressure and other variable VLE data [20, 38]. For most reservoir fluids, A is within 0.2–0.25; however, as is seen in Chapter 9 for a Kuwaiti oil value of A was found as 0.18. As discussed by Poling et al. [8], Tsonopoulos recommends the original Chueh–Prausnitz relation ($A = 1$ and $B = 3$) for nonpolar compounds. Pedersen et al. [39] proposed another relation for calculation of BIPs for HC–HC systems. Their correlation is based on data obtained from North Sea reservoir fluids and it is related to molecular weights of components i and j as $k_{ij} \cong 0.001 M_i / M_j$ where $M_i > M_j$. Another correlation was proposed by Whitson [40] for estimation of BIPs of methane and C_{7+} fraction components based on the data presented by Katz and Firoozabadi [36] for use with PR EOS. His correlation is as: $k_{1j} = 0.14\,SG_j - 0.0688$, where 1 refers to methane and j refers to the $C_{7+}(j)$ fraction, respectively.

Equations (5.59)–(5.62) can be applied to either liquid or vapor mixtures. However, for the case of vapor mixtures with N components, mole fraction y_i should be used. Expansion of Eq. (5.59) for a ternary gas mixture ($N = 3$) becomes

$$a_{mix} = y_1^2 a_{11} + y_2^2 a_{22} + y_3^2 a_{33} + 2y_1 y_2 a_{12} + 2y_1 x_3 a_{13} + 2y_2 y_3 a_{23}$$

(5.64)

where $a_{11} = a_1$, $a_{22} = a_2$, and $a_{33} = a_3$. Interaction coefficients such as a_{12} can be found from Eq. (5.61): $a_{12} = \sqrt{a_1 a_2}(1 - k_{12})$ where k_{12} may be taken from Table 5.3 or estimated from Eq. (5.63). a_{13} and a_{23} can be calculated in a similar way.

5.6 NONCUBIC EQUATIONS OF STATE

The main reason for wide range application of cubic EOS is their application to both phases of liquids and vapors, mathematical simplicity and convenience, as well as possibility of calculation of their parameters through critical constants and acentric factor. However, these equations are mainly useful for density and phase equilibrium calculations. For other thermodynamic properties such as heat capacity and enthalpy, noncubic equations such as those based on statistical associating fluid theory (SAFT) or perturbed hard chain theory (PHCT). Some of these equations have been particularly developed for special mixtures, polar molecules, hard sphere molecules, and near critical regions. Summary of these equations is given by Poling et al. [8]. In this section three important types of noncubic EOS are presented: (1) virial, (2) Carnahan–Starling, and (3) modified Benedict–Webb–Rubin.

5.6.1 Virial Equation of State

The most widely used noncubic EOS is the virial equation or its modifications. The original virial equation was proposed in 1901 by Kammerlingh–Onnes and it may be written either in the form of polynomial series in inverse volume (pressure explicit) or pressure expanded (volume explicit) as follows:

$$(5.65) \quad Z = 1 + \frac{B}{V} + \frac{C}{V^2} + \frac{D}{V^3} + \cdots$$

$$Z = 1 + \left(\frac{B}{RT} \right) P + \left(\frac{C - B^2}{R^2 T^2} \right) P^2$$

$$(5.66) \quad + \left(\frac{D - 3BC + 2B^3}{R^3 T^3} \right) P^3 + \cdots$$

TABLE 5.4—*Second virial coefficients for several gases [41].*

Compound	Temperature, K			
	200	300	400	500
N_2	−35.2	−4.2	9	16.9
CO_2	...	−122.7	60.5	−29.8
CH_4	−105	−42	−15	−0.5
C_2H_6	−410	−182	−96	−52
C_3H_8	...	−382	−208	−124

Note: Values of B are given in cm^3/mol.

where $B, C, D, \ldots$ are called second, third, and fourth virial coefficients and they are all temperature-dependent. The above two forms of virial equation are the same and the second equation can be derived from the first equation (see Problem 5.7). The second form is more practical to use since usually T and P are available and V should be estimated. The number of terms in a virial EOS can be extended to infinite terms but contribution of higher terms reduces with increase in power of P. Virial equation is perhaps the most accurate PVT relation for gases. However, the difficulty with use of virial equation is availability of its coefficients especially for higher terms. A large number of data are available for the second virial coefficient B, but less data are available for coefficient C and very few data are reported for the fourth coefficient D. Data on values of virial coefficients for several compounds are given in Tables 5.4 and 5.5. The virial coefficient has firm basis in theory and the methods of statistical mechanics allow derivation of its coefficients.

B represents two-body interactions and C represented three-body interactions. Since the chance of three-body interaction is less than two-body interaction, therefore, the importance and contribution of B is much greater than C. From quantum mechanics it can be shown that the second virial coefficient can be calculated from the knowledge of potential function (Γ) for intermolecular forces [6]:

$$(5.67) \qquad B = 2\pi N_A \int_0^\infty \left(1 - e^{-\Gamma(r)/kT}\right) r^2 dr$$

where N_A is the Avogadro's number (6.022×10^{23} mol^{-1}) and k is the Boltzman's constant ($k = R/N_A$). Once the relation for Γ is known, B can be estimated. For example, if the fluid follows hard sphere potential function, one by substituting Eq. (5.13) for Γ into the above equation gives $B = (2/3)\pi N_A \sigma^3$. Vice versa, constants in a potential relation (ε and σ) may be estimated from the knowledge of virial coefficients. For mixtures, B_{mix} can be calculated from Eq. (5.59) with a being replaced by B. For a ternary system, B can be calculated from Eq. (5.64). B_{ij} is calculated from Eq. (5.67) using Γ_{ij} with σ_{ij} and ε_{ij} given as [6]

$$(5.68) \qquad \sigma_{ij} = \frac{1}{2}(\sigma_i + \sigma_j)$$

$$(5.69) \qquad \varepsilon_{ij} = (\varepsilon_i \varepsilon_j)^{1/2}$$

TABLE 5.5—*Sample values of different virial coefficients for several compounds [1].*

Compound	T, °C	B, cm^3/mol	C, cm^6/mol^2
Methane[a]	0	−53.4	2620
Ethane	50	−156.7	9650
Steam (H_2O)	250	−152.5	−5800
Sulfur dioxide (SO_2)	157.5	−159	9000

[a] For methane at 0°C the fourth virial coefficient D is 5000 cm^9/mol^3.

Another form of Eq. (5.59) for calculation of B_{mix} can be written as following:

$$B_{mix} = \sum_{i=1}^N y_i B_{ii} + \frac{1}{2}\sum_{i=1}^N \sum_{j=1}^N y_i y_j \delta_{ij} \quad \text{where} \quad \delta_{ij} = 2B_{ij} - B_{ii} - B_{jj}$$

(5.70)

There are several correlations developed based on the theory of corresponding state principles to estimate the second virial coefficients in terms of temperature. Some of these relations correlate B/V_c to T_r and ω. Prausnitz et al. [6] reviewed some of these relations for estimation of the second virial coefficients. The relation developed by Tsonopoulos [42] is useful to estimate B from T_c, P_c, and ω.

$$\frac{BP_c}{RT_c} = B^{(0)} + \omega B^{(1)}$$

$$B^{(0)} = 0.1445 - \frac{0.330}{T_r} - \frac{0.1385}{T_r^2} - \frac{0.0121}{T_r^3} - \frac{0.000607}{T_r^8}$$

$$B^{(1)} = 0.0637 + \frac{0.331}{T_r^2} - \frac{0.423}{T_r^3} - \frac{0.008}{T_r^8}$$

(5.71)

where $T_r = T/T_c$. There are simpler relations that can be used for normal fluids [1].

$$\frac{BP_c}{RT_c} = B^{(0)} + \omega B^{(1)}$$

$$(5.72) \qquad B^{(0)} = 0.083 - \frac{0.422}{T_r^{1.6}}$$

$$B^{(1)} = 0.139 - \frac{0.172}{T_r^{4.2}}$$

Another relation for prediction of second virial coefficients of simple fluids is given by McGlashan [43]:

$$(5.73) \qquad \frac{BP_c}{RT_c} = 0.597 - 0.462 e^{0.7002/T_r}$$

A graphical comparison of Eqs. (5.71)–(5.73) for prediction of second virial coefficient of ethane is shown in Fig. 5.11. Coefficient B at low and moderate temperatures is negative and increases with increase in temperature; however, as is seen from the above correlations as $T \to \infty$, B approaches a positive number.

To predict B_{mix} for a mixture of known composition, the interaction coefficient B_{ij} is needed. This coefficient can be calculated from B_{ii} and B_{jj} using the following relations [1, 15]:

$$B_{ij} = \frac{RT_{cij}}{P_{cij}} \left(B^{(0)} + \omega_{ij} B^{(1)} \right)$$

$B^{(0)}$ and $B^{(1)}$ are calculated through $T_{rij} = T/T_{cij}$

$$\omega_{ij} = \frac{\omega_i + \omega_j}{2}$$

$$T_{cij} = (T_{ci}T_{cj})^{1/2}(1 - k_{ij})$$

$$(5.74) \qquad P_{ij} = \frac{Z_{ij}RT_{cij}}{V_{cij}}$$

$$Z_{cij} = \frac{Z_{ci} + Z_{cj}}{2}$$

$$V_{cij} = \left(\frac{V_{ci}^{1/3} + V_{ci}^{1/3}}{2}\right)^3$$

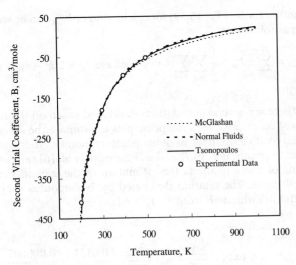

FIG. 5.11—Prediction of second virial coefficient for ethane from different methods. Experimental data from Table 5.4: McGlashan, Eq. (5.73); Normal fluids, Eq. (5.72); Tsopoulos, Eq. (5.71).

where k_{ij} is the interaction coefficient and for hydrocarbons of similar size it is zero. $B^{(0)}$ and $B^{(1)}$ must be calculated from the same relations used to calculate B_{ii} and B_{jj}. Another simpler method that is fairly accurate for light, nonpolar gases is the geometric mean:

$$B_{ij} = \left(B_i B_j\right)^{1/2}$$

$$B_{mix} = \left(\sum_{i=1}^{N} y_i B_i^{1/2}\right)^2$$

The importance of these relations is that at moderate pressures, Eq. (5.66) may be truncated after the second term as follows:

$$(5.75) \qquad Z = 1 + \frac{BP}{RT}$$

This equation is usually referred to as the truncated virial equation and may be used with a reasonable degree of accuracy in certain ranges of reduced temperature and pressure: $V_r > 2.0$ [i.e., $(P_r < 0.5, T_r > 1)$, $(0.5 < P_r < 1, T_r > 1.2)$, $(1 < P_r < 1.7, T_r > 1.5)$]. At low-pressure range $(P_r < 0.3)$, Eq. (5.72) provides good prediction for the second virial coefficients for use in Eq. (5.75) [1].

A more accurate form of virial equation for gases is obtained when Eq. (5.65) or (5.66) are truncated after the third term:

$$(5.76) \qquad Z = 1 + \frac{B}{V} + \frac{C}{V^2}$$

An equivalent form of this equation in terms of P can be obtained by Eq. (5.66) with three terms excluding fourth virial coefficient and higher terms. Because of lack of sufficient data, a generalized correlation to predict the third virial coefficient, C, is less accurate and is based on fewer data. The generalized correlation has the following form [6]:

$$\frac{C}{V_c^2} = \left(0.232 T_r^{-0.25} + 0.468 T_r^{-5}\right) \times \left[1 - e^{\left(1 - 1.89 T_r^2\right)}\right]$$

$$(5.77) \qquad + d e^{-\left(2.49 - 2.30 T_r + 2.70 T_r^2\right)}$$

where V_c is the molar critical volume in cm³/mol and d is a parameter that is determined for several compounds, i.e., $d = 0.6$ for methane, 1 for ethane, 1.8 for neopentane, 2.5 for benzene, and 4.25 for n-octane. In general when $T_r > 1.5$ the second term in the above equation is insignificant. A more practical and generalized correlation for third virial coefficient was proposed by Orbey and Vera [44] for nonpolar compounds in a form similar to Eq. (5.71), which was proposed for the second virial coefficients:

$$\frac{CP_c^2}{(RT_c)^2} = C^{(0)} + \omega C^{(1)}$$

$$(5.78) \quad C^{(0)} = 0.01407 + \frac{0.02432}{T_r^{2.8}} - \frac{0.00313}{T_r^{10.5}}$$

$$C^{(1)} = -0.02676 + \frac{0.0177}{T_r^{2.8}} + \frac{0.040}{T_r^3} - \frac{0.003}{T_r^6} - \frac{0.00228}{T_r^{10.5}}$$

where $C^{(0)}$ and $C^{(1)}$ are dimensionless parameters for simple and correction terms in the generalized correlation. Estimation of the third virial coefficients for mixtures is quite difficult as there are three-way interactions for C and it should be calculated from [6]:

$$(5.79) \qquad C_{mix} = \sum\sum\sum y_i y_j y_k C_{ijk}$$

Methods of estimation of cross coefficients C_{ijk} are not reliable [6]. For simplicity, generally it is assumed that $C_{iij} = C_{iji} = C_{jii}$ but still for a binary system at least two cross coefficients of C_{112} and C_{122} must be estimated. In a binary system, C_{112} expresses interaction of two molecules of component 1 with one molecule of component 2. Orbey and Vera [44] suggest the following relation for calculation of C_{ijk} as

$$(5.80) \qquad C_{ijk} = \left(C_{ij} C_{ik} C_{jk}\right)^{1/3}$$

where C_{ij} is evaluated from Eq. (5.78) using T_{cij}, P_{cij} and ω_{ij} obtained from Eq. (5.74). This approach gives satisfactory estimates for binary systems.

There are certain specific correlations for the virial coefficients of some specific gases. For example, for hydrogen the following correlations for B and C are suggested [6]:

$$B = \sum_1^4 b_i x^{(2i-1)/4}$$

$$(5.81) \quad C = 1310.5 x^{1/2}\left(1 + 2.1486 x^3\right) \times \left[1 - \exp\left(1 - x^{-3}\right)\right]$$

$$\text{where } x = \frac{109.83}{T}, b_1 = 42.464, b_2 = -37.1172,$$

$$b_3 = -2.2982, \text{ and } b_4 = -3.0484$$

where T is in K, B is in cm³/mol, and C is in cm⁶/mol². The range of temperature is 15–423 K and the average deviations for B and C are 0.07 cm³/mol and 17.4 cm⁶/mol², respectively [6].

As determination of higher virial coefficients is difficult, application of truncated virial EOS is mainly limited to gases and for this reason they have little application in reservoir fluid studies where a single equation is needed for both liquid and vapor phases. However, they have wide applications in estimation of properties of gases at low and moderate pressures. In addition, special modifications of virial equation has industrial applications, as discussed in the next section. From mathematical relations it can be shown that any EOS can be converted into a virial form. This is shown by the following example.

Example 5.4—Convert RK EOS into the virial form and obtain coefficients B and C in terms of EOS parameters.

Solution—The RK EOS is given by Eq. (5.38). If both sides of this equation are multiplied by V/RT we get

$$(5.82) \qquad Z = \frac{PV}{RT} = \frac{V}{V-b} - \frac{a}{RT(V+b)}$$

Assume $x = b/V$ and $A = a/RT$, then the above equation can be written as

$$(5.83) \qquad Z = \frac{1}{1-x} - A\frac{1}{V} \times \frac{1}{1+x}$$

Since $b < V$, therefore, $x < 1$ and the terms in the RHS of the above equation can be expanded through Taylor series [16, 17]:

$$(5.84) \qquad f(x) = \sum_{n=0}^{\infty} \frac{f^{(n)}(x_o)}{n!}(x-x_o)^n$$

where $f^{(n)}(x_o)$ is the nth order derivative $d^n f(x)/dx^n$ evaluated at $x = x_o$. The zeroth derivative of f is defined to be f itself and both 0! and 1! are equal to 1. Applying this expansion rule at $x_o = 0$ we get:

$$(5.85) \qquad \begin{aligned} \frac{1}{1-x} &= 1 + x + x^2 + x^3 + x^4 + \cdots \\ \frac{1}{1+x} &= 1 - x + x^2 - x^3 + x^4 - \cdots \end{aligned}$$

It should be noted that the above relations are valid when $|x| < 1$. Substituting the above two relations in Eq. (5.83) we get

$$(5.86) \qquad Z = (1 + x + x^2 + x^3 + \cdots) - A\frac{1}{V} \times (1 - x + x^2 - x^3 + \cdots)$$

If x is replaced by its definition b/V and A by a/RT we have

$$(5.87) \qquad Z = 1 + \frac{b - a/RT}{V} + \frac{b^2 + ab/RT}{V^2} + \frac{+b^3 - ab^2/RT}{V^3} + \cdots$$

A comparison with Eq. (5.65) we get the virial coefficients in terms of RK EOS parameters as follows:

$$(5.88) \qquad B = b - \frac{a}{RT} \qquad C = b^2 + \frac{ab}{RT} \qquad D = b^3 - \frac{ab^2}{RT}$$

Considering the fact that a is a temperature-dependent parameter one can see that the virial coefficients are all temperature-dependent parameters. With use of SRK EOS, similar coefficients are obtained but parameter a also depends on the acentric factor as given in Table 5.1. This gives better estimation of the second and third virial coefficients (see Problem 5.10) ♦

The following example shows application of truncated virial equation for calculation of vapor molar volumes.

Example 5.5—Propane has vapor pressure of 9.974 bar at 300 K. Saturated vapor molar volume is $V^V = 2036.5$ cm^3/mol [Ref. 8, p. 4.24]. Calculate (a) second virial coefficient from Eqs. (5.71)–(5.73), (b) third virial coefficient from Eq. (5.78), (c) V^V from virial EOS truncated after second term using Eqs. (5.65) and (5.66), (d) V^V from virial EOS truncated after third term using Eqs. (5.65) and (5.66), and (e) V^V from ideal gas law.

Solution—(a) and (b): For propane from Table 2.1 we get $T_c = 96.7°C$ (369.83 K), P_c 42.48 bar, and $\omega = 0.1523$. $T_r = 0.811$, $P_r = 0.23$, and $R = 83.14$ cm$^3 \cdot$ bar/mol $\cdot$ K. Second virial coefficient, B, can be estimated from Eqs. (5.71) or (5.72) or (5.73) and the third virial coefficient from Eq. (5.78). Results are given in Table 5.6. (c) Truncated virial equation after second term from Eq. (5.65) is $Z = 1 + B/V$, which is referred to as V expansion form, and from Eq. (5.66) is $Z = 1 + BP/RT$, which is the same as Eq. (5.75) and it is referred to as P expansion form. For the V expansion (Eq. 5.65), V should be calculated through successive substitution method or from mathematical solution of the equation, while in P expansion form (Eq. 5.66) Z can be directly calculated from T and P. Once Z is determined, V is calculated from Eq. (5.15): $V = ZRT/P$. In part (d) virial equation is truncated after the third term. The V expansion form reduces to Eq. (5.76). Summary of calculations for molar volume is given in Table 5.6. The results from V expansion (Eq. 5.65) and P expansion (5.66) do not agree with each other; however, the difference between these two forms of virial equation reduces as the number of terms increases. When the number of terms becomes infinity (complete equation), then the two forms of virial equation give identical results for V. Obviously for truncated virial equation, the V expansion form, Eq. (5.65), gives more accurate result for V as the virial coefficients are originally determined from this equation. As can be seen from Table 5.6, when B is calculated from Eq. (5.71) better

TABLE 5.6—*Prediction of molar volume of propane at 300 K and 9.974 bar from virial equation with different methods for second virial coefficient (Example 5.5).*

| Method of estimation of second virial coefficient (B) | B, cm^3/mol | Virial equation with two terms | | | | Virial equation with three terms[a] | | | | |
| | | P expansion[b] | | V expansion[c] | | P expansion[d] | | V expansion[e] | |
		V, cm^3/mol	%D	V, cm^3/mol	%D	V, cm^3/mol	%D	V, cm^3/mol	%D
Tsonopoulos (Eq. 5.71)	−390.623	2110.1	3.6	2016.2	−1.0	2056.8	1.0	2031.6	−0.2
Normal fluids (Eq. 5.72)	−397.254	2103.5	3.3	2005.3	−1.5	2048.1	0.6	2021.0	−0.7
McGlashan (Eq. 5.73)	−360.705	2140.0	5.1	2077.8	2.0	2095.7	2.9	2063.6	1.3

The experimental value of vapor molar volume is: $V = 2036.5$ cm^3/mol (Ref. [8], p. 4.24).
[a] In all calculations with three terms, the third virial coefficient C is calculated from Eq. (5.78) as $C = 19406.21$ cm^6/mol^2.
[b] Truncated two terms (P expansion) refers to pressure expansion virial equation (Eq. 5.66) truncated after second term (Eq. 5.75): $Z = 1 + BP/RT$.
[c] Truncated two terms (V expansion) refers to volume expansion virial equation (Eq. 5.65) truncated after second term: $Z = 1 + B/V$.
[d] Truncated three terms (P expansion) refers to pressure expansion virial equation (Eq. 5.66) truncated after third term: $Z = 1 + BP/RT + (C - B^2)P^2/(RT)^2$.
[e] Truncated three terms (V expansion) refers to volume expansion virial equation (Eq. 5.65) truncated after third term (Eq. 5.76): $Z = 1 + B/V + C/V^2$.

TABLE 5.7—*Coefficients for the BWRS EOS—Eq. (5.89) [21].*

$B_o/V_c = 0.44369 + 0.115449\omega$	$E_o/(RT_c^5 V_c) = 0.00645 - 0.022143\omega \times \exp(-3.8\omega)$	$d/(RT_c^2 V_c^2) = 0.0732828 + 0.463492\omega$
$A_o/(RT_c V_c) = 1.28438 - 0.920731\omega$		$\alpha/V_c^3 = 0.0705233 - 0.044448\omega$
$C_o/(RT_c^3 V_c) = 0.356306 + 1.7087\omega$	$b/(V_c^2) = 0.528629 + 0.349261\omega$	$c/(RT_c^3 V_c^2) = 0.504087 + 1.32245\omega$
$D_o/(RT_c^4 V_c) = 0.0307452 + 0.179433\omega$	$a/(RT_c V_c^2) = 0.484011 + 0.75413\omega$	$\gamma/V_c^2 = 0.544979 - 0.270896\omega$

predictions are obtained. Equation (5.72) also gives reasonable results but Eq. (5.73) gives a less accurate estimate of B. The best result is obtained from Eq. (5.76) with Eqs. (5.71) and (5.78), which give a deviation of 0.2%. (e) The ideal gas law ($Z = 1$) gives $V^V = 2500.7$ cm^3/mol with a deviation of $+22.8\%$. ♦

5.6.2 Modified Benedict–Webb–Rubin Equation of State

Another important EOS that has industrial application is the Benedict–Webb–Rubin (BWR) EOS [45]. This equation is in fact an empirical expansion of virial equation. A modification of this equation by Starling [46] has found successful applications in petroleum and natural gas industries for properties of light hydrocarbons and it is given as

$$
P = RT \frac{1}{V} + \left(B_o RT - A_o - \frac{C_o}{T^2} + \frac{D_o}{T^3} - \frac{E_o}{T^4} \right) \frac{1}{V^2}
$$

(5.89)
$$
+ \left(bRT - a - \frac{d}{T} \right) \frac{1}{V^3} + \alpha \left(a + \frac{d}{T} \right) \frac{1}{V^6}
$$

$$
+ \frac{c}{T^2 V^3} \left(1 + \frac{\gamma}{V^2} \right) \exp \left(\frac{-\gamma}{V^2} \right)
$$

where the 11 constants $A_o, B_o, \ldots, a, b, \ldots, \alpha$ and γ are given in Table 5.7 in terms of V_c, T_c, and ω as reported in Ref. [21]. This equation is known as BWRS EOS and may be used for calculation of density of light hydrocarbons and reservoir fluids. In the original BWR EOS, constants D_o, E_o, and d were all zero and the other constants were determined for each specific compound separately. Although better volumetric data can be obtained from BWRS than from cubic-type equations, but prediction of phase equilibrium from cubic equations are quite comparable in some cases (depending on the mixing rules used) or better than this equation in some other cases. Another problem with the BWRS equation is large computation time and mathematical inconvenience to predict various physical properties. To find molar volume V from Eq. (5.89), a successive substitutive method is required. However, as it will be discussed in the next section, this type of equations can be used to develop generalized correlations in the graphical or tabulated forms for prediction of various thermophysical properties.

5.6.3 Carnahan–Starling Equation of State and Its Modifications

Equations of state are mainly developed based on the understanding of intermolecular forces and potential energy functions that certain fluids follow. For example, for hard sphere fluids where the potential energy function is given by Eq. (5.13) it is assumed that there are no attractive forces. For such fluids, Carnahan and Starling proposed an EOS that has been used extensively by researchers for development of more accurate EOS [6]. For hard sphere fluids, the smallest possible

volume that be can occupied by N molecules of diameter σ is

(5.90)
$$
V_{oN} = N \left(\frac{V_o}{N_A} \right)
$$

$$
V_o = \left(\frac{1}{\sqrt{2}} \sigma^3 \right) N_A
$$

where N_A is the Avogadro's number and V_o is the volume of 1 mol (N_A molecules) of hard spheres as packed molecules without empty space between the molecules. V_{oN} is the total volume of packed N molecules. If the molar volume of fluid is V, then a dimensionless reduced density, ξ, is defined in the following form:

(5.91)
$$
\xi = \left(\frac{\sqrt{2}}{6} \pi \right) \times \left(\frac{V_o}{V} \right)
$$

Parameter ξ is also known as *packing fraction* and indicates fraction of total volume occupied by hard molecules. Substituting V_o from Eq. (5.90) into Eq. (5.91) gives the following relation for packing fraction:

(5.92)
$$
\xi = \left(\frac{\pi}{6} \right) \times \left(\frac{N_A \sigma^3}{V} \right)
$$

The Carnahan–Starling EOS is then given as [6]

(5.93)
$$
Z^{HS} = \frac{PV}{RT} = \frac{1 + \xi + \xi^2 - \xi^3}{(1 - \xi)^3}
$$

where Z^{HS} is the compressibility factor for hard sphere molecules. For this EOS there is no binary constant and the only parameter needed is molecular diameter σ for each molecule. It is clear that as $V \to \infty$ ($P \to 0$) from Eq. (5.93) $\zeta \to 0$ and $Z^{HS} \to 1$, which is in fact identical to the ideal gas law. Carnahan and Starling extended the HS equation to fluids whose spherical molecules exert attractive forces and suggested two equations based on two different attractive terms [6]:

(5.94)
$$
Z = Z^{HS} - \frac{a}{RTV}
$$

or

(5.95)
$$
Z = Z^{HS} - \frac{a}{RT} (V - b)^{-1} T^{-1/2}
$$

where Z^{HS} is the hard sphere contribution given by Eq. (5.93). Obviously Eq. (5.94) is a two-parameter EOS (a, σ) and Eq. (5.95) is a three-parameter EOS (a, b, σ). Both Eqs. (5.94) and (5.95) reduce to ideal gas law ($Z \to Z^{HS} \to 1$) as $V \to \infty$ (or $P \to 0$), which satisfies Eq. (5.18). For mixtures, the quadratic mixing rule can be used for parameter a while a linear rule can be applied to parameter b. Application of these equations for mixtures has been discussed in recent references [8, 47]. Another modification of CS EOS is through LJ EOS in the following form [48, 49]:

(5.96)
$$
Z = Z^{HS} - \frac{32\varepsilon\xi}{3k_B T}
$$

where ε is the molecular energy parameter and ζ (see Eq. 5.92) is related to σ the size parameter. ε and σ are two parameters in the LJ potential (Eq. 5.11) and k_B is the Boltzman constant. One advanced noncubic EOS, which has received significant attention for property calculations specially derived properties (i.e., heat capacity, sonic velocity, etc.), is that of SAFT originally proposed by Chapman et al. [50] and it is given in the following form [47]:

$$(5.97) \quad Z^{SAFT} = 1 + Z^{HS} + Z^{CHAIN} + Z^{DISP} + Z^{ASSOC}$$

where HS, CHAIN, DISP, and ASSOC refer to contributions from hard sphere, chain formation molecule, dispersion, and association terms. The relations for Z^{HS} and Z^{CHAIN} are simple and are given in the following form [47]:

$$Z^{SAFT} = 1 + r\left[\frac{4\xi - 2\xi^2}{(1-\xi)^3}\right] + (1-r)\left[\frac{5\xi - 2\xi^2}{(1-\xi)(2-\xi)}\right]$$
$$(5.98) \qquad + Z^{DISP} + Z^{ASSOC}$$

where r is a specific parameter characteristic of the substance of interest. ζ in the above relation is segment packing fraction and is equal to ζ from Eq. (5.92) multiplied by r. The relations for Z^{DISP} and Z^{ASSOC} are more complex and are in terms of summations with adjusting parameters for the effects of association. There are other forms of SAFT EOS. A more practical, but much more complex, form of SAFT equation is given by Li and Englezos [51]. They show application of SAFT EOS to calculate phase behavior of systems containing associating fluids such as alcohol and water. SAFT EOS does not require critical constants and is particularly useful for complex molecules such as very heavy hydrocarbons, complex petroleum fluids, water, alcohol, ionic, and polymeric systems. Parameters can be determined by use of vapor pressure and liquid density data. Further characteristics and application of these equations are given by Prausnitz et al. [8, 47]. In the next chapter, the CS EOS will be used to develop an EOS based on the velocity of sound.

5.7 CORRESPONDING STATE CORRELATIONS

One of the simplest forms of an EOS is the two-parameter RK EOS expressed by Eq. (5.38). This equation can be used for fluids that obey a two-parameter potential energy relation. In fact this equation is quite accurate for simple fluids such as methane. Rearrangement of Eq. (5.38) through multiplying both sides of the equation by V/RT and substituting parameters a and b from Table 5.1 gives the following relation in terms of dimensionless variables [1]:

$$(5.99) \quad Z = \frac{1}{1-h} - \frac{4.934}{T_r^{1.5}}\left(\frac{h}{1+h}\right) \quad \text{where } h \equiv \frac{0.08664P_r}{ZT_r}$$

where T_r and P_r are called *reduced temperature* and *reduced pressure* and are defined as:

$$(5.100) \qquad T_r \equiv \frac{T}{T_c} \qquad P_r \equiv \frac{P}{P_c}$$

where T and T_c must be in absolute degrees (K), similarly P and P_c must be in absolute pressure (bar). Both T_r and P_r are dimensionless and can be used to express temperature and

pressure variations from the critical point. By substituting parameter h into the first equation in Eq. (5.99) one can see that

$$(5.101) \qquad Z = f(T_r, P_r)$$

This equation indicates that for all fluids that obey a two-parameter EOS, such as RK, the compressibility factor, Z, is the only function of T_r and P_r. This means that at the critical point where $T_r = P_r = 1$, the critical compressibility factor, Z_c, is constant and same for all fluids (0.333 for RK EOS). As can be seen from Table 2.1, Z_c is constant only for simple fluids such as N_2, CH_4, O_2, or Ar, which have Z_c of 0.29, 0.286, 0.288, and 0.289, respectively. For this reason RK EOS is relatively accurate for such fluids. Equation (5.101) is the fundamental of *corresponding states principle* (CSP) in classical thermodynamics. A correlation such as Eq. (5.101) is also called *generalized correlation*. In this equation only two parameters (T_c and P_c) for a substance are needed to determine its PVT relation. These types of relations are usually called two-parameter *corresponding states correlations* (CSC). The functionality of function f in Eq. (5.101) can be determined from experimental data on PVT and is usually expressed in graphical forms rather than mathematical equations. The most widely used two-parameter CSC in a graphical form is the Standing–Katz generalized chart that is developed for natural gases [52]. This chart is shown in Fig. 5.12 and is widely used in the petroleum industry [19, 21, 53, 54]. Obviously this chart is valid for light hydrocarbons whose acentric factor is very small such as methane and ethane, which are the main components of natural gases.

Hall and Yarborough [55] presented an EOS that was based on data obtained from the Standing and Katz Z-factor chart. The equation was based on the Carnahan-Starling equation (Eq. 5.93), and it is useful only for calculation of Z-factor of light hydrocarbons and natural gases. The equation is in the following form:

$$(5.102) \quad Z = 0.06125P_rT_r^{-1}y^{-1}\exp\left[-1.2\left(1-T_r^{-1}\right)^2\right]$$

where T_r and P_r are reduced temperature and pressure and y is a dimensionless parameter similar to parameter ξ defined in Eq. (5.91). Parameter y should be obtained from solution of the following equation:

$$F(y) = -0.06125P_rT_r^{-1}\exp\left[-1.2\left(1-T_r^{-1}\right)^2\right]$$
$$+ \frac{y+y^2+y^3-y^4}{(1-y)^3} - \left(14.76T_r^{-1}-9.76T_r^{-2}+4.58T_r^{-3}\right)y^2$$
$$+ \left(90.7T_r^{-1} - 242.2T_r^{-2} + 42.4T_r^{-3}\right)y^{\left(2.18+2.82T_r^{-1}\right)} = 0$$
$$(5.103)$$

The above equation can be solved by the Newton–Raphson method. To find y an initial guess is required. An approximate relation to find the initial guess is obtained at $Z = 1$ in Eq. (5.102):

$$(5.104) \quad y^{(k)} = 0.06125P_rT_r^{-1}\exp\left[-1.2\left(1-T_r^{-1}\right)^2\right]$$

Substituting $y^{(k)}$ in Eq. (5.103) gives $F^{(k)}$, which must be used in the following relation to obtain a new value of y:

$$(5.105) \qquad y^{(k+1)} = y^{(k)} - \frac{F^{(k)}}{\frac{dF^{(k)}}{dy}}$$

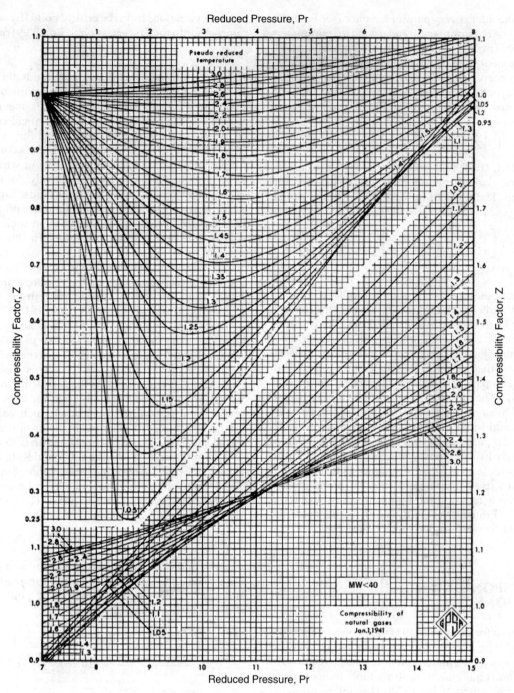

FIG. 5.12—Standing–Katz generalized chart for compressibility factor of natural gases (courtesy of GPSA and GPA [53]).

where $\mathrm{d}F^{(k)}/\mathrm{d}y$ is the derivative of F with respect to y at $y = y^{(k)}$ and it is given by the following relation:

$$
\begin{aligned}
\frac{\mathrm{d}F}{\mathrm{d}y} =\ & \frac{1 + 4y + 4y^2 - 4y^3 + y^4}{(1-y)^4} \\
& - \left(29.52 T_r^{-1} - 19.52 T_r^{-2} + 9.16 T_r^{-3}\right) y \\
& + \left(2.18 + 2.82 T_r^{-1}\right) \times \left(90.7 T_r^{-1} - 242.2 T_r^{-2} + 42.4 T_r^{-3}\right) \\
& \times y^{(2.18 + 2.82 T_r^{-1})}
\end{aligned}
$$

(5.106)

Calculations must be continued until the difference between $y^{(k+1)} - y^{(k)}$ becomes smaller than a tolerance (e.g., 10^{-10}).

As mentioned before, the Standing–Katz chart or its equivalent Hall–Yarborough correlation is applicable only to light hydrocarbons and they are not suitable to heavier fluids such as gas condensates, ω of which is not near zero. For this reason a modified version of two-parameter CSC is needed. As it can be seen from Table 2.1, for more complex compounds, value of Z_c decreases from those for simple fluids and Eq. (5.101) with constant Z_c is no longer valid. A parameter that indicates complexity of molecules is acentric factor that was defined by Eq. (2.10). Acentric factor, ω, is defined in a way that for simple fluids it is zero or very small. For example, N_2, CH_4, O_2, or Ar have acentric factors of 0.025, 0.011, 0.022, and 0.03, respectively. Values of ω increase with complexity of molecules. In fact as shown in Section 2.5.3, Z_c can be correlated to ω and both indicate deviation from simple fluids. Acentric factor was originally introduced by Pitzer [56, 57] to extend application of two-parameter CSC to more complex fluids. Pitzer and his coworkers realized the linear relation between Z_c and ω (i.e., see Eq. (2.103)) and assumed that such linearity exists between ω and Z at temperatures other than T_c. They introduced the concept of three-parameter corresponding states correlations in the following form:

$$(5.107) \qquad Z = Z^{(0)} + \omega Z^{(1)}$$

where both $Z^{(0)}$ and $Z^{(1)}$ are functions of T_r and P_r. For simple fluids ($\omega \cong 0$), this equation reduces to Eq. (5.101). $Z^{(0)}$ is the contribution of simple fluids and $Z^{(1)}$ is the correction term for complex fluids. It can be shown that as $P \to 0$, $Z^{(0)} \to 1$ while $Z^{(1)} \to 0$, therefore, $Z \to 1$. The original three-parameter CSC developed by Pitzer was in the form of two graphs similar to Fig. (5.12): one for $Z^{(0)}$ and the other for $Z^{(1)}$, both in terms of T_r and P_r. Pitzer correlations found wide application and were extended to other thermodynamic properties. They were in use for more than two decades; however, they were found to be inaccurate in the critical region and for liquids at low temperatures [58].

The most advanced and accurate three-parameter corresponding states correlations were developed by Lee and Kesler [58] in 1975. They expressed Z in terms of values of Z for two fluids: simple and a reference fluid assuming linear relation between Z and ω as follows:

$$(5.108) \qquad Z = Z^{(0)} + \frac{\omega}{\omega^{(r)}}(Z^{(r)} - Z^{(0)})$$

where $Z^{(r)}$ and $\omega^{(r)}$ represent compressibility factor and acentric factor of the reference fluid. A comparison between Eqs. (5.107) and (5.108) indicates that $[Z^{(r)} - Z^{(0)}]/\omega^{(r)}$ is equivalent to $Z^{(1)}$. The simple fluid has acentric factor of zero, but the reference fluid should have the highest value of ω to cover a wider range for application of the correlation. However, the choice of reference fluid is also limited by availability of PVT and other thermodynamic data. Lee and Kesler chose n-octane with ω of 0.3978 (this number is slightly different from the most recent value of 0.3996 given in Table 2.1) as the reference fluid. The same EOS was used for both the simple and reference fluid, which is a modified version of BWR EOS as given in the following reduced form:

$$(5.109) \quad Z = 1 + \frac{B}{V_r} + \frac{C}{V_r^2} + \frac{D}{V_r^5} + \frac{c_4}{T_r^3 V_r^2}\left(\beta + \frac{\gamma}{V_r^2}\right)\exp\left(\frac{-\gamma}{V_r^2}\right)$$

where V_r is the reduced volume defined as

$$(5.110) \qquad V_r = \frac{V}{V_c}$$

Coefficients B, C, and D are temperature-dependent as

$$B = b_1 - \frac{b_2}{T_r} - \frac{b_3}{T_r^2} - \frac{b_4}{T_r^3} \quad C = c_1 - \frac{c_2}{T_r} + \frac{c_3}{T_r^3} \quad D = d_1 + \frac{d_2}{T_r}$$
$$(5.111)$$

TABLE 5.8—*Constants for the Lee-Kesler modification of BWR EOS—Eq. (5.109) [58].*

Constant	Simple fluid	Reference fluid
b_1	0.1181193	0.2026579
b_2	0.265728	0.331511
b_3	0.154790	0.027655
b_4	0.030323	0.203488
c_1	0.0236744	0.0313385
c_2	0.0186984	0.0503618
c_3	0.0	0.016901
c_4	0.042724	0.041577
$d_1 \times 10^4$	0.155488	0.48736
$d_2 \times 10^4$	0.623689	0.0740336
β	0.65392	1.226
γ	0.060167	0.03754

In determining the constants in these equations the constraints by Eq. (5.9) and equality of chemical potentials or fugacity (Eq. 6.104) between vapor and liquid at saturated conditions were imposed. These coefficients for both simple and reference fluids are given in Table 5.8.

In using Eq. (5.108), both $Z^{(0)}$ and $Z^{(r)}$ should be calculated from Eq. (5.109). Lee and Kesler also tabulated values of $Z^{(0)}$ and $Z^{(1)}$ versus T_r and P_r for use in Eq. (5.107). The original Lee–Kesler (LK) tables cover reduced pressure from 0.01 to 10. These tables have been widely used in major texts and references [1, 8, 59]. However, the API-TDB [59] gives extended tables for $Z^{(0)}$ and $Z^{(1)}$ for the P_r range up to 14. Lee–Kesler tables and their extension by the API-TDB are perhaps the most accurate method of estimating PVT relation for gases and liquids. Values of $Z^{(0)}$ and $Z^{(1)}$ as given by LK and their extension by API-TDB are given in Tables 5.9–5.11. Table 5.11 give values of $Z^{(0)}$ and $Z^{(1)}$ for $P_r > 10$ as provided in the API-TDB [59]. In Tables 5.9 and 5.10 the dotted lines separate liquid and vapor phases from each other up to the critical point. Values above and to the right are for liquids and below and to the left are gases. The values for liquid phase are highlighted with bold numbers. Graphical representations of these tables are given in the API-TDB [59]. For computer applications, Eqs. (5.108)–(5.111) should be used with coefficients given in Table 5.8. Graphical presentation of $Z^{(0)}$ and $Z^{(1)}$ versus P_r and T_r with specified liquid and vapor regions is shown in Fig. 5.13. The two-phase region as well as saturated curves are also shown in this figure. For gases, as $P_r \to 0$, $Z^{(0)} \to 1$ and $Z^{(1)} \to 0$. It is interesting to note that at the critical point ($T_r = P_r = 1$), $Z^{(0)} = 0.2901$, and $Z^{(1)} = -0.0879$, which after substitution into Eq. (5.107) gives the following relation for Z_c:

$$(5.112) \qquad Z_c = 0.2901 - 0.0879\omega$$

This equation is slightly different from Eq. (2.93) and gives different values of Z_c for different compounds. Therefore, in the critical region the LK correlations perform better than cubic equations, which give a constant value for Z_c of all compounds. Graphical presentations of both $Z^{(0)}$ and $Z^{(1)}$ for calculation of Z from Eq. (5.107) are given in other sources [60].

For the low-pressure region where the truncated virial equation can be used, Eq. (5.75) may be written in a generalized dimensionless form as

$$(5.113) \qquad Z = 1 + \frac{BP}{RT} = 1 + \left(\frac{BP_c}{RT_c}\right)\frac{P_r}{T_r}$$

TABLE 5.9—*Values of $Z^{(0)}$ for use in Eq. (5.107) from the Lee-Kesler modification of BWR EOS (Eq. 5.109) [58].*

$P_r \rightarrow$	0.01	0.05	0.1	0.2	0.4	0.6	0.8	1	1.2	1.5	2	3	5	7	10
0.30	0.0029	0.0145	0.0290	0.0579	0.1158	0.1737	0.2315	0.2892	0.3479	0.4335	0.5775	0.8648	1.4366	2.0048	2.8507
0.35	0.0026	0.0130	0.0261	0.0522	0.1043	0.1564	0.2084	0.2604	0.3123	0.3901	0.5195	0.7775	1.2902	1.7987	2.5539
0.40	0.0024	0.0119	0.0239	0.0477	0.0953	0.1429	0.1904	0.2379	0.2853	0.3563	0.4744	0.7095	1.1758	1.6373	2.3211
0.45	0.0022	0.0110	0.0221	0.0442	0.0882	0.1322	0.1762	0.2200	0.2638	0.3294	0.4384	0.6551	1.0841	1.5077	2.1338
0.50	0.0021	0.0103	0.0207	0.0413	0.0825	0.1236	0.1647	0.2056	0.2465	0.3077	0.4092	0.6110	1.0094	1.4017	1.9801
0.55	0.9804	0.0098	0.0195	0.0390	0.0778	0.1166	0.1553	0.1939	0.2323	0.2899	0.3853	0.5747	0.9475	1.3137	1.8520
0.60	0.9849	0.0093	0.0186	0.0371	0.0741	0.1109	0.1476	0.1842	0.2207	0.2753	0.3657	0.5446	0.8959	1.2398	1.7440
0.65	0.9881	0.9377	0.0178	0.0356	0.0710	0.1063	0.1415	0.1765	0.2113	0.2634	0.3495	0.5197	0.8526	1.1773	1.6519
0.70	0.9904	0.9504	0.8958	0.0344	0.0687	0.1027	0.1366	0.1703	0.2038	0.2538	0.3364	0.4991	0.8161	1.1341	1.5729
0.75	0.9922	0.9598	0.9165	0.0336	0.0670	0.1001	0.1330	0.1656	0.1981	0.2464	0.3260	0.4823	0.7854	1.0787	1.5047
0.80	0.9935	0.9669	0.9319	0.8539	0.0661	0.0985	0.1307	0.1626	0.1942	0.2411	0.3182	0.4690	0.7598	1.0400	1.4456
0.85	0.9946	0.9725	0.9436	0.8810	0.0661	0.0983	0.1301	0.1614	0.1924	0.2382	0.3132	0.4591	0.7388	1.0071	1.3943
0.90	0.9954	0.9768	0.9528	0.9015	0.7800	0.1006	0.1321	0.1630	0.1935	0.2383	0.3114	0.4527	0.7220	0.9793	1.3496
0.93	0.9959	0.9790	0.9573	0.9115	0.8059	0.6635	0.1359	0.1664	0.1963	0.2405	0.3122	0.4507	0.7138	0.9648	1.3257
0.95	0.9961	0.9803	0.9600	0.9174	0.8206	0.6967	0.1410	0.1705	0.1998	0.2432	0.3138	0.4501	0.7092	0.9561	1.3108
0.97	0.9963	0.9815	0.9625	0.9227	0.8338	0.7240	0.5580	0.1779	0.2055	0.2474	0.3164	0.4504	0.7052	0.9480	1.2968
0.98	0.9965	0.9821	0.9637	0.9253	0.8398	0.7360	0.5887	0.1844	0.2097	0.2503	0.3182	0.4508	0.7035	0.9442	1.2901
0.99	0.9966	0.9826	0.9648	0.9277	0.8455	0.7471	0.6138	0.1959	0.2154	0.2538	0.3204	0.4514	0.7018	0.9406	1.2835
1.00	0.9967	0.9832	0.9659	0.9300	0.8509	0.7574	0.6355	0.2901	0.2237	0.2583	0.3229	0.4522	0.7004	0.9372	1.2772
1.01	0.9968	0.9837	0.9669	0.9322	0.8561	0.7671	0.6542	0.4648	0.2370	0.2640	0.3260	0.4533	0.6991	0.9339	1.2710
1.02	0.9969	0.9842	0.9679	0.9343	0.8610	0.7761	0.6710	0.5146	0.2629	0.2715	0.3297	0.4547	0.6980	0.9307	1.2650
1.05	0.9971	0.9855	0.9707	0.9401	0.8743	0.8002	0.7130	0.6026	0.4437	0.3131	0.3452	0.4604	0.6956	0.9222	1.2481
1.10	0.9975	0.9874	0.9747	0.9485	0.8930	0.8323	0.7649	0.6880	0.5984	0.4580	0.3953	0.4770	0.6950	0.9110	1.2232
1.15	0.9978	0.9891	0.9780	0.9554	0.9081	0.8576	0.8032	0.7443	0.6803	0.5798	0.4760	0.5042	0.6987	0.9033	1.2021
1.20	0.9981	0.9904	0.9808	0.9611	0.9205	0.8779	0.8330	0.7858	0.7363	0.6605	0.5605	0.5425	0.7069	0.8990	1.1844
1.30	0.9985	0.9926	0.9852	0.9702	0.9396	0.9083	0.8764	0.8438	0.8111	0.7624	0.6908	0.6344	0.7358	0.8998	1.1580
1.40	0.9988	0.9942	0.9884	0.9768	0.9534	0.9298	0.9062	0.8827	0.8595	0.8256	0.7753	0.7202	0.7761	0.9112	1.1419
1.50	0.9991	0.9954	0.9909	0.9818	0.9636	0.9456	0.9278	0.9103	0.8933	0.8689	0.8328	0.7887	0.8200	0.9297	1.1339
1.60	0.9993	0.9964	0.9928	0.9856	0.9714	0.9575	0.9439	0.9308	0.9180	0.9000	0.8738	0.8410	0.8617	0.9518	1.1320
1.70	0.9994	0.9971	0.9943	0.9886	0.9775	0.9667	0.9563	0.9463	0.9367	0.9234	0.9043	0.8809	0.8984	0.9745	1.1343
1.80	0.9995	0.9977	0.9955	0.9910	0.9823	0.9739	0.9659	0.9583	0.9511	0.9413	0.9275	0.9118	0.9297	0.9961	1.1391
1.90	0.9996	0.9982	0.9964	0.9929	0.9861	0.9796	0.9735	0.9678	0.9624	0.9552	0.9456	0.9359	0.9557	1.0157	1.1452
2.00	0.9997	0.9986	0.9972	0.9944	0.9892	0.9842	0.9796	0.9754	0.9715	0.9664	0.9599	0.9550	0.9772	1.0328	1.1516
2.20	0.9998	0.9992	0.9983	0.9967	0.9937	0.9910	0.9886	0.9865	0.9847	0.9826	0.9806	0.9827	1.0094	1.0600	1.1635
2.40	0.9999	0.9996	0.9991	0.9983	0.9969	0.9957	0.9948	0.9941	0.9936	0.9935	0.9945	1.0011	1.0313	1.0793	1.1728
2.60	1.0000	0.9998	0.9997	0.9994	0.9991	0.9990	0.9990	0.9993	0.9998	1.0010	1.0040	1.0137	1.0463	1.0926	1.1792
2.80	1.0000	1.0000	1.0001	1.0002	1.0007	1.0013	1.0021	1.0031	1.0042	1.0063	1.0106	1.0223	1.0565	1.1016	1.1830
3.00	1.0000	1.0002	1.0004	1.0008	1.0018	1.0030	1.0043	1.0057	1.0074	1.0101	1.0153	1.0284	1.0635	1.1075	1.1848
3.50	1.0001	1.0004	1.0008	1.0017	1.0035	1.0055	1.0075	1.0097	1.0120	1.0156	1.0221	1.0368	1.0723	1.1138	1.1834
4.00	1.0001	1.0005	1.0010	1.0021	1.0043	1.0066	1.0090	1.0115	1.0140	1.0179	1.0249	1.0401	1.0747	1.1136	1.1773

where BP_c/RT_c can be estimated from Eq. (5.71) or (5.72) through T_r and ω. Equation (5.114) may be used at low P_r and $V_r > 2$ or $T_r > 0.686 + 0.439 P_r$ [60] instead of complex Eqs. (5.108)–(5.111). The API-TDB [59] also recommends the following relation, proposed by Pitzer et al. [56], for calculation of Z for gases at $P_r \leq 0.2$.

$$Z = 1 + \frac{P_r}{T_r}\big[(0.1445 + 0.073\omega) - (0.33 - 0.46\omega)T_r^{-1}$$
$$- (0.1385 + 0.5\omega)T_r^{-2} - (0.0121 + 0.097\omega)T_r^{-3}$$
$$(5.114) \quad - 0.0073\omega T_r^{-8}\big]$$

Obviously neither Eq. (5.113) nor (5.114) can be applied to liquids.

The LK corresponding states correlations expressed by Eq. (5.107) and Tables 5.9–5.11 can also be applied to mixtures. Such correlations are sensitive to the input data for the pseudocritical properties. The mixing rules used to calculate mixture critical temperature and pressure may greatly affect calculated properties specially when the mixture contains dissimilar compounds. Lee and Kesler proposed special set of equations for mixtures for use with their correlations.

These equations are equivalent to the following equations as provided by the API-TDB [59].

$$V_{ci} = Z_{ci} RT_{ci}/P_{ci}$$

$$Z_{ci} = 0.2905 - 0.085\omega_i$$

$$V_{mc} = \frac{1}{4}\left[\sum_{i=1}^{N} x_i V_{ci} + 3\left(\sum_{i=1}^{N} x_i V_{ci}^{2/3}\right)\left(\sum_{i=1}^{N} x_i V_{ci}^{2/3}\right)\right]$$

$$T_{mc} = \frac{1}{4V_{mc}}\left[\sum_{i=1}^{N} x_i V_{ci} T_{ci} + 3\left(\sum_{i=1}^{N} x_i V_{ci}^{2/3}\sqrt{T_{ci}}\right)\left(\sum_{i=1}^{N} x_i V_{ci}^{2/3}\sqrt{T_{ci}}\right)\right]$$

$$\omega_m = \sum_{i=1}^{N} x_i \omega_i$$

$$P_{mc} = Z_{mc} RT_{mc}/V_{mc}c = (0.2905 - 0.085\omega_m)RT_{mc}/V_{mc}$$

$$(5.115)$$

where x_i is the mole fraction of component i, N is the number of compounds in the mixture, and T_{mc}, P_{mc}, and V_{mc} are the mixture pseudocritical temperature, pressure, and volume respectively. ω_m is the mixture acentric factor and it is calculated

TABLE 5.10—*Values of $Z^{(1)}$ for use in Eq. (5.107) from the Lee-Kesler modification of BWR EOS—Eq. (5.109) [58].*

$P_r \rightarrow$	0.01	0.05	0.1	0.2	0.4	0.6	0.8	1	1.2	1.5	2	3	5	7	10
0.30	**-0.0008**	**-0.0040**	**-0.0081**	**-0.0161**	**-0.0323**	**-0.0484**	**-0.0645**	-0.0806	-0.0966	-0.1207	-0.1608	-0.2407	-0.3996	-0.5572	-0.7915
0.35	**-0.0009**	**-0.0046**	**-0.0093**	**-0.0185**	**-0.0370**	**-0.0554**	**-0.0738**	-0.0921	-0.1105	-0.1379	-0.1834	-0.2738	-0.4523	-0.6279	-0.8863
0.40	**-0.0010**	**-0.0048**	**-0.0095**	**-0.0190**	**-0.0380**	**-0.0570**	**-0.0758**	-0.0946	-0.1134	-0.1414	-0.1879	-0.2799	-0.4603	-0.6365	-0.8936
0.45	**-0.0009**	**-0.0047**	**-0.0094**	**-0.0187**	**-0.0374**	**-0.0560**	**-0.0745**	-0.0929	-0.1113	-0.1387	-0.1840	-0.2734	-0.4475	-0.6162	-0.8606
0.50	**-0.0009**	**-0.0045**	**-0.0090**	**-0.0181**	**-0.0360**	**-0.0539**	**-0.0716**	-0.0893	-0.1069	-0.1330	-0.1762	-0.2611	-0.4253	-0.5831	-0.8099
0.55	**-0.0314**	**-0.0043**	**-0.0086**	**-0.0172**	**-0.0343**	**-0.0513**	**-0.0682**	-0.0849	-0.1015	-0.1263	-0.1669	-0.2465	-0.3991	-0.5446	-0.7521
0.60	**-0.0205**	**-0.0041**	**-0.0082**	**-0.0164**	**-0.0326**	**-0.0487**	**-0.0646**	-0.0803	-0.0960	-0.1192	-0.1572	-0.2312	-0.3718	-0.5047	-0.6928
0.65	-0.0137	**-0.0772**	**-0.0078**	**-0.0156**	**-0.0309**	**-0.0461**	**-0.0611**	-0.0759	-0.0906	-0.1122	-0.1476	-0.2160	-0.3447	-0.4653	-0.6346
0.70	-0.0093	-0.0507	**-0.1161**	**-0.0148**	**-0.0294**	**-0.0438**	**-0.0579**	-0.0718	-0.0855	-0.1057	-0.1385	-0.2013	-0.3184	-0.4270	-0.5785
0.75	-0.0064	-0.0339	-0.0744	**-0.0143**	**-0.0282**	**-0.0417**	**-0.0550**	-0.0681	-0.0808	-0.0996	-0.1298	-0.1872	-0.2929	-0.3901	-0.5250
0.80	-0.0044	-0.0228	-0.0487	**-0.1160**	**-0.0272**	**-0.0401**	**-0.0526**	-0.0648	-0.0767	-0.0940	-0.1217	-0.1736	-0.2682	-0.3545	-0.4740
0.85	-0.0029	-0.0152	-0.0319	-0.0715	**-0.0268**	**-0.0391**	**-0.0509**	-0.0622	-0.0731	-0.0888	-0.1138	-0.1602	-0.2439	-0.3201	-0.4254
0.90	-0.0019	-0.0099	-0.0205	-0.0442	**-0.1118**	**-0.0396**	**-0.0503**	-0.0604	-0.0701	-0.0840	-0.1059	-0.1463	-0.2195	-0.2862	-0.3788
0.93	-0.0015	-0.0075	-0.0154	-0.0326	-0.0763	**-0.1662**	**-0.0514**	-0.0602	-0.0687	-0.0810	-0.1007	-0.1374	-0.2045	-0.2661	-0.3516
0.95	-0.0012	-0.0062	-0.0126	-0.0262	-0.0589	-0.1110	**-0.0540**	-0.0607	-0.0678	-0.0788	-0.0967	-0.1310	-0.1943	-0.2526	-0.3339
0.97	-0.0010	-0.0050	-0.0101	-0.0208	-0.0450	-0.0770	**-0.1647**	-0.0623	-0.0669	-0.0759	-0.0921	-0.1240	-0.1837	-0.2391	-0.3163
0.98	-0.0009	-0.004	-0.0090	-0.0184	-0.0390	-0.0641	-0.1100	-0.0641	-0.0661	-0.0740	-0.0893	-0.1202	-0.1783	-0.2322	-0.3075
0.99	-0.0008	-0.0039	-0.0079	-0.0161	-0.0335	-0.0531	-0.0796	-0.0680	-0.0646	-0.0715	-0.0861	-0.1162	-0.1728	-0.2254	-0.2989
1.00	-0.0007	-0.0034	-0.0069	-0.0140	-0.0285	-0.0435	-0.0588	-0.0879	-0.0609	-0.0678	-0.0824	-0.1118	-0.1672	-0.2185	-0.290
1.01	-0.0006	-0.0030	-0.0060	-0.0120	-0.0240	-0.0351	-0.0429	-0.0223	-0.0473	-0.0621	-0.0778	-0.1072	-0.1615	-0.2116	-0.2816
1.02	-0.0005	-0.0026	-0.0051	-0.0102	-0.0198	-0.0277	-0.0303	-0.0062	-0.0227	-0.0524	-0.0722	-0.1021	-0.1556	-0.2047	-0.2731
1.05	-0.0003	-0.0015	-0.0029	-0.0054	-0.0092	-0.0097	-0.0032	0.0220	0.1059	0.0451	-0.0432	-0.0838	-0.1370	-0.1835	-0.2476
1.10	0.0000	0.0000	0.0001	0.0007	0.0038	0.0106	0.0236	0.0476	0.0897	0.1630	0.0698	-0.0373	-0.1021	-0.1469	-0.2056
1.15	0.0002	0.0011	0.0023	0.0052	0.0127	0.0237	0.0396	0.0625	0.0943	0.1548	0.1667	0.0332	-0.0611	-0.1084	-0.1642
1.20	0.0004	0.0019	0.0039	0.0084	0.0190	0.0326	0.0499	0.0719	0.0991	0.1477	0.1990	0.1095	-0.0141	-0.0678	-0.1231
1.30	0.0006	0.0030	0.0061	0.0125	0.0267	0.0429	0.0612	0.0819	0.1048	0.1420	0.1991	0.2079	0.0875	0.0176	-0.0423
1.40	0.0007	0.0036	0.0072	0.0147	0.0306	0.0477	0.0661	0.0857	0.1063	0.1383	0.1894	0.2397	0.1737	0.1008	0.0350
1.50	0.0008	0.0039	0.0078	0.0158	0.0323	0.0497	0.0677	0.0864	0.1055	0.1345	0.1806	0.2433	0.2309	0.1717	0.1058
1.60	0.0008	0.0040	0.0080	0.0162	0.0330	0.0501	0.0677	0.0855	0.1035	0.1303	0.1729	0.2381	0.2631	0.2255	0.1673
1.70	0.0008	0.0040	0.0081	0.0163	0.0329	0.0497	0.0667	0.0838	0.1008	0.1259	0.1658	0.2305	0.2788	0.2628	0.2179
1.80	0.0008	0.0040	0.0081	0.0162	0.0325	0.0488	0.0652	0.0814	0.0978	0.1216	0.1593	0.2224	0.2846	0.2871	0.2576
1.90	0.0008	0.0040	0.0079	0.0159	0.0318	0.0477	0.0635	0.0792	0.0947	0.1173	0.1532	0.2144	0.2848	0.3017	0.2876
2.00	0.0008	0.0039	0.0078	0.0155	0.0310	0.0464	0.0617	0.0767	0.0916	0.1133	0.1476	0.2069	0.2819	0.3097	0.3096
2.20	0.0007	0.0037	0.0074	0.0147	0.0293	0.0437	0.0579	0.0719	0.0857	0.1057	0.1374	0.1932	0.2720	0.3135	0.3355
2.40	0.0007	0.0035	0.0070	0.0139	0.0276	0.0411	0.0544	0.0675	0.0803	0.0989	0.1285	0.1812	0.2602	0.3089	0.3459
2.60	0.0007	0.0033	0.0066	0.0131	0.0260	0.0387	0.0512	0.0634	0.0754	0.0929	0.1207	0.1706	0.2484	0.3009	0.3475
2.80	0.0006	0.0031	0.0062	0.0124	0.0245	0.0365	0.0483	0.0598	0.0711	0.0876	0.1138	0.1613	0.2372	0.2915	0.3443
3.00	0.0006	0.0029	0.0059	0.0117	0.0232	0.0345	0.0456	0.0565	0.0672	0.0828	0.1076	0.1529	0.2268	0.2817	0.3385
3.50	0.0005	0.0026	0.0052	0.0103	0.0204	0.0303	0.0401	0.0497	0.0591	0.0728	0.0949	0.1356	0.2042	0.2584	0.3194
4.00	0.0005	0.0023	0.0046	0.0091	0.0182	0.0270	0.0357	0.0443	0.0527	0.0651	0.0849	0.1219	0.1857	0.2378	0.2994

TABLE 5.11—*Values of $Z^{(0)}$ and $Z^{(1)}$ for use in Eq. (5.107) from the Lee–Kesler modification of BWR EOS—Eq. (5.109) [59].*

$P_r \rightarrow$ T_r	$Z^{(0)}$				$Z^{(1)}$			
↓	10	11	12	14	10	11	12	14
0.30	2.851	3.131	3.411	3.967	−0.792	−0.869	−0.946	−1.100
0.35	2.554	2.804	3.053	3.548	−0.886	−0.791	−1.056	−1.223
0.40	2.321	2.547	2.772	3.219	−0.894	−0.978	−1.061	−1.225
0.45	2.134	2.340	2.546	2.954	−0.861	−0.940	−1.019	−1.173
0.50	1.980	2.171	2.360	2.735	−0.810	−0.883	−0.955	−1.097
0.55	1.852	2.029	2.205	2.553	−0.752	−0.819	−0.885	−1.013
0.60	1.744	1.909	2.073	2.398	−0.693	−0.753	−0.812	−0.928
0.65	1.652	1.807	1.961	2.266	−0.635	−0.689	−0.742	−0.845
0.70	1.573	1.720	1.865	2.152	−0.579	−0.627	−0.674	−0.766
0.75	1.505	1.644	1.781	2.053	−0.525	−0.568	−0.610	−0.691
0.80	1.446	1.578	1.708	1.966	−0.474	−0.512	−0.549	−0.621
0.85	1.394	1.520	1.645	1.890	−0.425	−0.459	−0.491	−0.555
0.90	1.350	1.470	1.589	1.823	−0.379	−0.408	−0.437	−0.493
0.95	1.311	1.426	1.540	1.763	−0.334	−0.360	−0.385	−0.434
0.98	1.290	1.402	1.513	1.731	−0.308	−0.331	−0.355	−0.401
0.99	1.284	1.395	1.504	1.721	−0.299	−0.322	−0.345	−0.390
1.00	1.277	1.387	1.496	1.710	−0.290	−0.313	−0.335	−0.379
1.01	1.271	1.380	1.488	1.701	−0.282	−0.304	−0.326	−0.368
1.02	1.265	1.373	1.480	1.691	−0.273	−0.295	−0.316	−0.357
1.03	1.259	1.367	1.473	1.682	−0.265	−0.286	−0.307	−0.347
1.04	1.254	1.360	1.465	1.672	−0.256	−0.277	−0.297	−0.337
1.05	1.248	1.354	1.458	1.664	−0.248	−0.268	−0.288	−0.326
1.06	1.243	1.348	1.451	1.655	−0.239	−0.259	−0.278	−0.316
1.07	1.238	1.342	1.444	1.646	−0.231	−0.250	−0.269	−0.306
1.08	1.233	1.336	1.438	1.638	−0.222	−0.241	−0.260	−0.296
1.09	1.228	1.330	1.431	1.630	−0.214	−0.233	−0.251	−0.286
1.10	1.223	1.325	1.425	1.622	−0.206	−0.224	−0.242	−0.276
1.11	1.219	1.319	1.419	1.614	−0.197	−0.215	−0.233	−0.267
1.12	1.214	1.314	1.413	1.606	−0.189	−0.207	−0.224	−0.257
1.13	1.210	1.309	1.407	1.599	−0.181	−0.198	−0.215	−0.247
1.15	1.202	1.299	1.395	1.585	−0.164	−0.181	−0.197	−0.228
1.20	1.184	1.278	1.370	1.552	−0.123	−0.139	−0.154	−0.183
1.25	1.170	1.259	1.348	1.522	−0.082	−0.098	−0.112	−0.139
1.30	1.158	1.244	1.328	1.496	−0.042	−0.058	−0.072	−0.097
1.40	1.142	1.220	1.298	1.453	0.035	0.019	0.005	−0.019
1.50	1.134	1.205	1.276	1.419	0.106	0.090	0.076	0.052
1.60	1.132	1.197	1.262	1.394	0.167	0.152	0.138	0.116
1.70	1.134	1.193	1.253	1.374	0.218	0.204	0.192	0.171
1.80	1.139	1.192	1.247	1.359	0.258	0.247	0.237	0.218
2.00	1.152	1.196	1.243	1.339	0.310	0.305	0.300	0.290
2.50	1.176	1.210	1.244	1.316	0.348	0.356	0.362	0.371
3.00	1.185	1.213	1.241	1.300	0.338	0.353	0.365	0.385
3.50	1.183	1.208	1.233	1.284	0.319	0.336	0.350	0.376
4.00	1.177	1.200	1.222	1.268	0.299	0.316	0.332	0.360

High Pressure Range: Value of $Z^{(0)}$ and $Z^{(1)}$ for $10 \leq P_r \leq 14$.

similar to the Kay's mixing rule. Application of Kay's mixing rule, expressed by Eq. (3.39), gives the following relations for calculation of pseudocritical temperature and pressure:

$$(5.116) \qquad T_{pc} = \sum_{i=1}^{N} x_i T_{ci} \qquad P_{pc} = \sum_{i=1}^{N} x_i P_{ci}$$

where T_{pc} and P_{pc} are the pseudocritical temperature and pressure, respectively. Generally for simplicity pseudocritical properties are calculated from Eqs. (5.116); however, use of Eqs. (5.115) for the LK correlations gives better property predictions [59].

Example 5.6—Repeat Example 5.2 using LK generalized correlations to estimate V^V and V^L for *n*-octane at 279.5°C and 19.9 bar.

Solution—For *n*-octane, from Example 5.2, $T_c = 295.55°C$ (568.7 K), $P_c = 24.9$ bar, $\omega = 0.3996$. $T_r = 0.972$, and $P_r = 0.8$. From Table 5.9 it can be seen that the point (0.972 and 0.8) is on the saturation line; therefore, there are both liquid and vapor phases at this condition and values of $Z^{(0)}$ and $Z^{(1)}$ are separated by dotted lines. For the liquid phase at $P_r = 0.8$, extrapolation of values of $Z^{(0)}$ at $T_r = 0.90$ and $T_r = 0.95$ to $T_r = 0.972$ gives $Z^{(0)} = 0.141 + [(0.972 − 0.93)/(0.95 − 0.93)] \times (0.141 − 0.1359) = 0.1466$, similarly we get $Z^{(1)} = −0.056$. Substituting $Z^{(0)}$ and $Z^{(1)}$ into Eq. (5.107) gives $Z^L = 0.1466 + 0.3996 \times (−0.056) = 0.1242$. Similarly for the vapor phase, values of $Z^{(0)}$ and $Z^{(1)}$ below the dotted line should be used. For this case linear interpolations between the values for $Z^{(0)}$ and $Z^{(1)}$ at $T_r = 0.97$ and $T_r = 0.98$ for the gas phase give $Z^{(0)} = 0.5642$, $Z^{(1)} = −0.1538$. From Eq. (5.107) we get $Z^V = 0.503$. From Eq. (5.15) corresponding volumes are $V^L = 286.8$ and

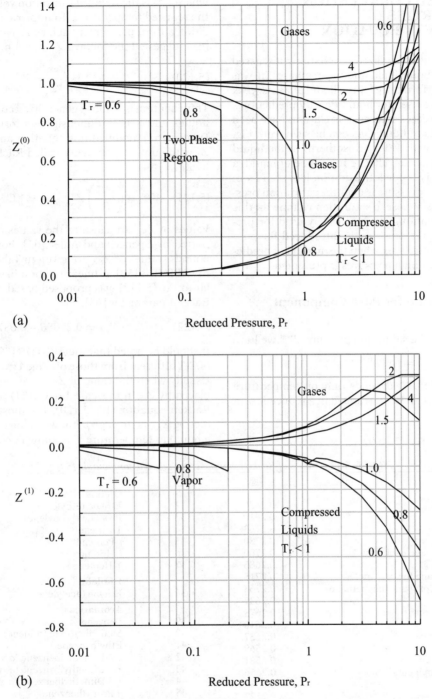

FIG. 5.13—Compressibility factor (a) $Z^{(0)}$ and (b) $Z^{(1)}$ from Tables 5.9 and 5.10.

$V^V = 1161.5$ cm³/mol, which give errors of −5.6 and −4.5% for the liquid and vapor volumes, respectively. ♦

The corresponding states correlation expressed by Eq. (5.107) is derived from principles of classical thermodynamics. However, the same theory can be derived from microscopic thermodynamics. Previously the relation between virial coefficients and intermolecular forces was shown through Eq. (5.67). From Eq. (5.11), Γ can be written in a

dimensionless form as

$$(5.117) \qquad \frac{\Gamma}{\varepsilon} = f\left(\frac{\sigma}{r}\right)$$

which is the basis for the development of microscopic (molecular) theory of corresponding states. Substitution of Eq. (5.117) into Eq. (5.67) would result into a generalized correlation for the second virial coefficient [6].

CHARACTERIZATION AND PROPERTIES OF PETROLEUM FRACTIONS

5.8 GENERALIZED CORRELATION FOR PVT PROPERTIES OF LIQUIDS—RACKETT EQUATION

Although cubic EOS and generalized correlations discussed above can be used for both liquid and vapor phases, it was mentioned that their performance for the liquid phase is weak especially when they are used for liquid density predictions. For this reason in many cases separate correlations have been developed for properties of liquids. As can be seen from Fig. 5.1, the variation of P with V for an isotherm in the liquid phase is very steep and a small change in volume of liquid, a big change in pressure is needed. In addition it is seen from this figure that when the pressure is near the saturation pressure, liquid volume is very close to saturation volume. In this section the Rackett equation, which is widely used for prediction of saturated liquid densities, is introduced for pure substances and defined mixtures. Then the method of prediction of liquid densities at high pressures is presented.

5.8.1 Rackett Equation for Pure Component Saturated Liquids

If Eq. (5.6) is applied at the saturation pressure, P_r^{sat} we have

$$(5.118) \qquad V^{sat} = f_1(T, P^{sat})$$

Since for any substance, P^{sat} depends only on temperature thus the above equation can be rearranged in a reduced form as

$$(5.119) \qquad V_r^{sat} = f_2(T_r)$$

where V_r^{sat} is the reduced saturation volume (V^{sat}/V_c) and T_r is the reduced temperature. To improve this generalized correlation a third parameter such as Z_c can be used and Rackett [61] suggested the following simple form for V_r^{sat} versus T_r:

$$(5.120) \qquad V_r^{sat} = \frac{V^{sat}}{V_c} = Z_c^{(1-T_r)^{2/7}}$$

This equation is in fact a generalized correlation for saturated liquids and it is in dimensionless form. Later Spencer and Danner [62] modified this equation and replaced parameter Z_c with another parameter called Rackett parameter shown by Z_{RA}:

$$(5.121) \quad V^{sat} = \left(\frac{RT_c}{P_c}\right) Z_{RA}^n \quad n = 1.0 + (1.0 - T_r)^{2/7}$$

Values of Z_{RA} are close to the values of Z_c and they are reported by Spencer and Adler [63]. For some selected compounds, values of Z_{RA} are given in Table 5.12 as reported by the API-TDB [59]. A linear relation between Z_{RA} and ω similar to Eq. (5.112) was proposed based on the initial values of Rackett parameter [64].

$$(5.122) \qquad Z_{RA} = 0.29056 - 0.08775\omega$$

It should be noted that the API-TDB [59] recommends values of Z_{RA} different from those obtained from the above equation. Usually when the value of Z_{RA} is not available, it may be replaced by Z_c. In this case Eq. (5.121) reduces to the original Rackett equation (Eq. 5.120). The most accurate way of predicting Z_{RA} is through a known value of density. If density of a liquid at temperature T is known and is shown by d_T, then

TABLE 5.12—*Values of Rackett parameter for selected compounds [59].*

No.		Z_{RA}	No.		Z_{RA}
	Paraffins			**Olefins**	
1	Methane	0.2880	32	Ethene (ethylene)	0.2813
2	Ethane	0.2819	33	Propene (propylene)	0.2783
3	Propane	0.2763	34	1-Butene	0.2735
4	n-Butane	0.2730	35	i-Pentene	0.2692
5	2-Methylpropane (isobutane)	0.2760	36	i-Hexene	0.2654
6	n-Pentane	0.2685	37	1-Heptene	0.2614
7	2-Methylbutane (isopentane)	0.2718		**Di-olefin**	
8	2,2-Dimethylpropane (neopentane)	0.2763	38	Ethyne (acetylene)	0.2707
9	n-Hexane	0.2637			
10	2-Methylpentane	0.2673		**Aromatics**	
11	n-Heptane	0.2610	39	Benzene	0.2696
12	2-Methylhexane	0.2637	40	Methylbenzene (toluene)	0.2645
13	n-Octane	0.2569	41	Ethylbenzene	0.2619
14	2-Methylheptane	0.2581	42	1,2-Dimethylbenzene (o-xylene)	0.2626
15	2,3,4-Trimethylpentane	0.2656	43	1,3-Dimethylbenzene (m-xylene)	0.2594
16	n-Nonane	0.2555	44	1,4-Dimethylbenzene (p-xylene)	0.2590
17	n-Decane	0.2527	45	n-Propylbenzene	0.2599
18	n-Undecane	0.2500	46	Isopropylbenzene (cumene)	0.2616
19	n-Dodecane	0.2471	47	n-Butylbenzene	0.2578
20	n-Tridecane	0.2468	48	Naphthalene	0.2611
21	n-Tetradecane	0.2270	49	Aniline	0.2607
22	n-Pentadecane	0.2420			
23	n-Hexadecane	0.2386		**Nonhydrocarbons**	
24	n-Heptadecane	0.2343	59	Ammonia	0.2466
25	n-Octadecane	0.2292	51	Carbon dioxide	0.2729
26	n-Nonadecane	0.2173[a]	52	Hydrogen	0.3218
27	n-Eicosane	0.2281	53	Hydrogen sulfide	0.2818
			54	Nitrogen	0.2893
	Naphthenes		55	Oxygen	0.2890
28	Cyclopentane	0.2709	56	Water	0.2374
29	Methylcyclopentane	0.2712	57	Methanol	0.2334
30	Cyclohexane	0.2729	58	Ethanol	0.2502
31	Methylcyclohexane	0.2702	59	Diethylamine (DEA)	0.2568

[a]Calculated from Eq. (5.123) using specific gravity.

Eq. (5.121) can be rearranged to get Z_{RA}:

$$(5.123) \qquad Z_{RA} = \left(\frac{MP_c}{RT_c d_T} \right)^{1/n}$$

where n is calculated from Eq. (5.121) at temperature T at which density is known. For hydrocarbon systems and petroleum fractions usually specific gravity (SG) at 15.5°C is known and value of 288.7 K should be used for T. Then d_T (in g/cm^3) is equal to 0.999SG according to the definition of SG by Eq. (2.2). In this way predicted values of density are quite accurate at temperatures near the reference temperature at which density data are used. The following example shows the procedure.

Example 5.7—For *n*-octane of Example 5.2, calculate saturated liquid molar volume at 279.5°C from Rackett equation using predicted Z_{RA}.

Solution—From Example 5.2, $M = 114.2$, SG $= 0.707$, $T_c = 295.55$°C (568.7 K), $P_c = 24.9$ bar, $R = 83.14$ cm$^3 \cdot$ bar/mol · K, and $T_r = 0.972$. Equation (5.123) should be used to predict Z_{RA} from SG. The reference temperature is 288.7 K, which gives $T_r = 0.5076$. This gives $n = 1.8168$ and from Eq. (5.123) we calculate $Z_{RA} = 0.2577$. ($Z_{RA} = 0.2569$ from Table 5.12). From Eq. (5.121), V^{sat} is calculated: $n = 1 + (1 - 0.972)^{2/7} = 1.36$, $V^{sat} = (83.14 \times 568.7/24.9) \times 0.2577^{1.36} = 300$ cm^3/mol. Comparing with actual value of 304 cm^3/mol gives the error of -1.3%. Calculated density is $\rho = 114.2/300 = 0.381$ g/cm^3. ♦

5.8.2 Defined Liquid Mixtures and Petroleum Fractions

Saturation pressure for a mixture is also called *bubble point* pressure and saturation molar volume is shown by V^{bp}. Liquid density at the bubble point is shown by ρ^{bp}, which is related to V^{bp} by the following relation:

$$(5.124) \qquad \rho^{bp} = \frac{M}{V^{bp}}$$

where ρ^{bp} is absolute density in g/cm^3 and M is the molecular weight. V^{bp} can be calculated from the following set of equations recommended by Spencer and Danner [65]:

$$V^{bp} = R \left(\sum_{i=1}^{N} x_i \frac{T_{ci}}{P_{ci}} \right) Z_{RAm}^n$$

$$n = 1 + (1 - T_r)^{2/7}$$

$$Z_{RAm} = \sum_{i=1}^{N} x_i Z_{RAi}$$

$$T_r = T/T_{cm}$$

$$(5.125) \qquad T_{cm} = \sum_{i=1}^{N} \sum_{j=1}^{N} \phi_i \phi_j T_{cij}$$

$$\phi_i = \frac{x_i V_{ci}}{\sum_{i=1}^{N} x_i V_{ci}}$$

$$T_{cij} = \sqrt{T_{ci} T_{cj}} \left(1 - k_{ij} \right)$$

$$k_{ij} = 1.0 - \left[\frac{\sqrt{V_{ci}^{1/3} V_{cj}^{1/3}}}{(V_{ci}^{1/3} + V_{cj}^{1/3})/2} \right]^3$$

This method is also included in the API-TDB [59]. Another approach to estimate density of defined liquid mixtures at its bubble point pressure is through the following mixing rule:

$$(5.126) \qquad \frac{1}{\rho^{bp}} = \sum_{i=1}^{N} \frac{x_{wi}}{\rho_i^{sat}}$$

where x_{wi} is weight fraction of i in the mixture. $\rho_i^{sat} (= M/V_i^{sat})$ is density of pure saturated liquid i and should be calculated from Eq. (5.121) using T_{ci} and Z_{RAi}.

For petroleum fractions in which detailed composition is not available Eq. (5.121) developed for pure liquids may be used. However, Z_{RA} should be calculated from specific gravity using Eq. (5.123) while T_c and P_c can be calculated from methods given in Chapter 2 through T_b and SG.

5.8.3 Effect of Pressure on Liquid Density

As shown in Fig. 5.1, effect of pressure on volume of liquids is quite small specially when change in pressure is small. When temperature is less than normal boiling point of a liquid, its saturation pressure is less than 1.0133 bar and density of liquid at atmospheric pressure can be assumed to be the same as its density at saturation pressure. For temperatures above boiling point where saturation pressure is not greatly more than 1 atm, calculated saturated liquid density may be considered as liquid density at atmospheric pressure. Another simple way of calculating liquid densities at atmospheric pressures is through Eq. (2.115) for the slope of density with temperature. If the only information available is specific gravity, SG, the reference temperature would be 15.5°C (288.7 K) and Eq. (2.115) gives the following relation:

$$\rho_T^o = 0.999SG - 10^{-3} \times (2.34 - 1.898SG) \times (T - 288.7)$$

$$(5.127)$$

where SG is the specific gravity at 15.5°C (60°F/60°F) and T is absolute temperature in K. ρ_T^o is liquid density in g/cm^3 at temperature T and atmospheric pressure. If instead of SG at 15.5°C (288.7 K), density at another temperature is available a similar equation can be derived from Eq. (2.115). Equation (5.127) is not accurate if T is very far from the reference temperature of 288.7 K.

The effect of pressure on liquid density or volume becomes important when the pressure is significantly higher than 1 atm. For instance, volume of methanol at 1000 bar and 100°C is about 12% less than it is at atmospheric pressure. In general, when pressure exceeds 50 bar, the effect of pressure on liquid volume cannot be ignored. Knowledge of the effect of pressure on liquid volume is particularly important in the design of high-pressure pumps in the process industries. The following relation is recommended by the API-TDB [59] to calculate density of liquid petroleum fractions at high pressures:

$$(5.128) \qquad \frac{\rho^o}{\rho} = 1.0 - \frac{P}{B_T}$$

where ρ^o is the liquid density at low pressures (atmospheric pressure) and ρ is density at high pressure P (in bar). B_T is called isothermal *secant bulk modulus* and is defined as $-(1/\rho^o)(\Delta P/\Delta V)_T$. Parameter B_T indicates the slope of change of pressure with unit volume and has the unit of pressure. Steps to calculate B_T are summarized in the following set of equations:

$$B_T = mX + B_I$$

$$m = 1492.1 + 0.0734P + 2.0983 \times 10^{-6}P^2$$

$$X = \left(B_{20} - 10^5\right)/23170$$

(5.129)
$$\log B_{20} = -1.098 \times 10^{-3}T + 5.2351 + 0.7133\rho^\circ$$

$$B_I = 1.0478 \times 10^3 + 4.704\,P - 3.744 \times 10^{-4}P^2$$

$$+\, 2.2331 \times 10^{-8}P^3$$

where B_T is in bar and ρ° is the liquid density at atmospheric pressure in g/cm^3. In the above equation T is absolute temperature in kelvin and P is the pressure in bar. The average error from this method is about 1.7% except near the critical point where error increases to 5% [59]. This method is not recommended for liquids at $T_r > 0.95$. In cases that ρ° is not available it may be estimated from Eq. (5.121) or (5.127). Although this method is recommended for petroleum fractions but it gives reasonable results for pure hydrocarbons ($\geq C_5$) as well.

For light and medium hydrocarbons as well as light petroleum fractions the Tait-COSTALD (**co**rresponding **sta**te **l**iquid **d**ensity) correlation originally proposed by Hankinson and Thomson may be used for the effect of pressure on liquid density [66]:

(5.130)
$$\rho_P = \rho_{P^\circ}\left[1 - C\ln\left(\frac{B+P}{B+P^\circ}\right)\right]^{-1}$$

where ρ_P is density at pressure P and ρ_{P° is liquid density at reference pressure of P° at which density is known. When ρ_{P° is calculated from the Rackett equation, $P^\circ = P^{sat}$ where P^{sat} is the saturation (vapor) pressure, which may be estimated from methods of Chapter 7. Parameter C is a dimensionless constant and B is a parameter that has the same unit as pressure. These constants can be calculated from the following equations:

$$\frac{B}{P_c} = -1 - 9.0702\left(1 - T_r\right)^{1/3} + 62.45326\left(1 - T_r\right)^{2/3}$$

$$-\, 135.1102\left(1 - T_r\right) + e\left(1 - T_r\right)^{4/3}$$

(5.131)
$$e = \exp\left(4.79594 + 0.250047\omega + 1.14188\omega^2\right)$$

$$C = 0.0861488 + 0.0344483\omega$$

where T_r is the reduced temperature and ω is the acentric factor. All the above relations are in dimensionless forms. Obviously Eq. (5.130) gives very accurate result when P is close to P°; however, it should not be used at $T_r > 0.95$. The COSTALD correlation has been recommended for industrial applications [59, 67]. However, in the API-TDB [59] it is recommended that special values of acentric factor obtained from vapor pressure data should be used for ω. These values for some hydrocarbons are given by the API-TDB [59]. The following example demonstrates application of these methods. The most recent modification of the Thomson method for polar and associating fluids was proposed by Garvin in

the following form [68]:

$$V = V^{sat}\left[1 - C\ln\left(\frac{B + P/P_{ce}}{B + P^{sat}/P_{ce}}\right)\right]$$

$$\kappa = -\frac{1}{V^{sat}}\left(\frac{\partial V}{\partial P}\right)_T = \frac{C}{(BP_{ce} + P)}$$

(5.132)
$$B = -1 - 9.070217\tau^{1/3} + 62.45326\tau^{2/3}$$

$$-\, 135.1102\tau + e\tau^{4/3}$$

$$C = 0.0861488 + 0.0344483\omega$$

$$e = \exp\left(4.79594 + 0.250047\omega + 1.14188\omega^2\right)$$

$$\tau = 1 - T/T_c$$

where V^{sat} is the saturation molar volume and P^{sat} is the saturation pressure at T. V is liquid molar volume at T and P and κ is the isothermal bulk compressibility defined in the above equation (also see Eq. 6.24). T_c is the critical temperature and ω is the acentric factor. P_{ce} is equivalent critical pressure, which for all alcohols was near the mean value of 27.0 bar. This value for diols is about 8.4 bar. For other series of compounds P_{ce} would be different. Garvin found that use of P_{ce} significantly improves prediction of V and κ for alcohols. For example, for estimation of κ of methanol at 1000 bar and 100°C, Eq. (5.132) predicts κ value of 7.1×10^{-7} bar^{-1}, which gives an error of 4.7% versus experimental value of 6.8×10^{-7} bar^{-1}, while using P_c the error increases to 36.6%. However, one should note that the numerical coefficients for B, C, and e in Eq. (5.132) may vary for other types of polar liquids such as coal liquids.

Another correlation for calculation of effect of pressure on liquid density was proposed by Chueh and Prausnitz [69] and is based on the estimation of isothermal compressibility:

$$\rho_P = \rho_{P^\circ}[1 + 9\beta(P - P^\circ)]^{1/9}$$

$$\beta = \alpha\left(1 - 0.89\sqrt{\omega}\right)\exp\left(6.9547 - 76.2853T_r + 191.306T_r^2\right.$$

$$\left. -\, 203.5472T_r^3 + 82.7631T_r^4\right)$$

$$\alpha = \frac{V_c}{RT_c} = \frac{Z_c}{P_c}$$

(5.133)

The parameters are defined the same as were defined in Eqs. (5.132) and (5.133). V_c is the molar critical volume and the units of P, V_c, R, and T_c must be consistent in a way that PV_c/RT_c becomes dimensionless. This equation is applicable for T_r ranging from 0.4 to 0.98 and accuracy of Eq. (5.134) is just marginally less accurate than the COSTALD correlation [67].

Example 5.8—Propane has vapor pressure of 9.974 bar at 300 K. Saturated liquid and vapor volumes are $V^L = 90.077$ and $V^V = 2036.5$ cm^3/mol [Ref. 8, p. 4.24]. Calculate saturated liquid molar volume using (a) Rackett equation, (b) Eqs. (5.127)–(5.129), (c) Eqs. (5.127) and (5.130), and (d) Eq. (5.133).

Solution—(a) Obviously the most accurate method to estimate V^L is through Eq. (5.121). From Table 2.1, $M = 44.1$, $SG = 0.507$, $T_c = 96.7°C$ (369.83 K), P_c 42.48 bar, and $\omega = 0.1523$. From Table 5.12, $Z_{RA} = 0.2763$. $T_r = 0.811$ so from Eq. (5.121), $V^{sat} = 89.961$ cm³/mol (-0.1% error). (b) Use of Eqs. (5.127)–(5.129) is not suitable for this case that Rackett equation can be directly applied. However, to show the application of method V^{sat} is calculated to see their performance. From Eq. (5.127) and use of $SG = 0.507$ gives $\rho° = 0.491$ g/cm³. From Eq. (5.129), $m = 1492.832$, $B_{20} = 180250.6$, $X = 3.46356$, $B_I = 1094.68$, and $B_T = 6265.188$ bar. Using Eq. (5.128), $0.491/\rho = 1 - 9.974/6265.188$. This equation gives density at T (300 K) and P (9.974 bar) as $\rho = 0.492$ g/cm³. $V^{sat} = M/\rho = 44.1/0.492 = 89.69$ cm³/mol (error of -0.4%). (c) Use of Eqs. (5.127) and (5.130) is not a suitable method for density of propane, but to show its performance, saturated liquid volume is calculated in a way similar to part (b): From Eq. (5.131), $B = 161.5154$ bar and $C = 0.091395$. For Eq. (5.130) we have $\rho_{P°} = 0.491$ g/cm³, $P° = 1.01325$ bar, $P = 9.974$ bar, and calculated density is $\rho_P = 0.4934$ g/cm³. Calculated V^{sat} is 89.4 cm³/mol, which gives a deviation of -0.8% from experimental value of 90.077 cm³/mol. (d) Using the Chueh–Prausnitz correlation (Eq. 5.133) we have $Z_c = 0.276$, $\alpha = 0.006497$, $\beta = 0.000381$, $\rho_P = 0.49266$ g/cm³, and $V^{sat}_{calc} = 89.5149$ cm³/mol, which gives an error of -0.62% from the actual value. ◆

5.9 REFRACTIVE INDEX BASED EQUATION OF STATE

From the various PVT relations and EOS discussed in this chapter, cubic equations are the most convenient equations that can be used for volumetric and phase equilibrium calculations. The main deficiency of cubic equations is their inability to predict liquid density accurately. Use of volume translation improves accuracy of SRK and PR equations for liquid density but a fourth parameter specific of each equation is required. The shift parameter is not known for heavy compounds and petroleum mixtures. For this reason some specific equations for liquid density calculations are used. As an example Alani–Kennedy EOS is specifically developed for calculation of liquid density of oils and reservoir fluids and is used by some reservoir engineers [19, 21]. The equation is in van der Waals cubic EOS form but it requires four numerical constants for each pure compound, which are given from C_1 to C_{10}. For the C_{7+} fractions the constants should be estimated from M_{7+} and SG_{7+}. The method performs well for light reservoir fluids and gas condensate samples. However, as discussed in Chapter 4, for oils with significant amount of heavy hydrocarbons, which requires splitting of C_{7+} fraction, the method cannot be applied to C_{7+} subfractions. In addition the method is not applicable to undefined petroleum fractions with a limited boiling range.

Generally constants of cubic equations are determined based on data for hydrocarbons up to C_8 or C_9. As an example, the LK generalized correlations is based on the data for the reference fluid of n-C_8. The parameter that indicates complexity of a compound is acentric factor. In SRK and PR EOS parameter a is related to ω in a polynomial form of at least second order (see f_ω in Table 5.1). This indicates that extrapolation of such equations for compounds having acentric factors greater than those used in development of EOS parameters is not accurate. And it is for this reason that most cubic equations such as SRK and PR equations break down when they are applied for calculation of liquid densities for C_{10} and heavier hydrocarbons. For this reason Riazi and Mansoori [70] attempted to improve capability of cubic equations for liquid density prediction, especially for heavy hydrocarbons.

Most modifications on cubic equations is on parameter a and its functionality with temperature and ω. However, a parameter that is inherent to volume is the co-volume parameter b. RK EOS presented by Eq. (5.38) is the simplest and most widely used cubic equation that predicts reasonably well for prediction of density of gases. In fact as shown in Table 5.13 for simple fluids such as oxygen or methane (with small ω) RK EOS works better than both SRK and PR regarding liquid densities.

For liquid systems in which the free space between molecules reduces, the role of parameter b becomes more important than that of parameter a. For low-pressure gases, however, the role of parameter b becomes less important than a because the spacing between molecules increases and as a result the attraction energy prevails. Molar refraction was defined by Eq. (2.34) as

$$(5.134) \qquad R_m = VI = \frac{M}{d_{20}} \left(\frac{n^2 - 1}{n^2 + 2} \right)$$

where R_m is the molar refraction and V is the molar volume both in cm³/mol. R_m is nearly independent of temperature but is normally calculated from density and refractive index at 20°C (d_{20} and n_{20}). R_m represents the actual molar volume of molecules and since b is also proportional to molar volume of molecules (excluding the free space); therefore, one can conclude that parameter b must be proportional to R_m. In fact the polarizability is related to R_m in the following form:

$$(5.135) \qquad \alpha = \frac{3}{4\pi N_A} R_m - \mu(T)$$

where N_A is the Avogadro's number and $\mu(T)$ is the dipole moment, which for light hydrocarbons is zero [7]. Values of R_m calculated from Eq. (5.134) are reported by Riazi et al. [70, 71] for a number of hydrocarbons and are given in Table 5.14. Since the original RK EOS is satisfactory for methane we choose this compound as the reference substance. Parameter

TABLE 5.13—*Evaluation of RK, SRK, and PR EOS for prediction of density of simple fluids.*

Compound	No. of data points	Temperature range, K	Pressure range, bar	%AAD			Data source
				RK	SRK	PR	
Methane	135	90–500	0.7–700	0.88	1.0	4.5	Goodwin [72]
Oxygen	120	80–1000	1–500	1.1	1.4	4.0	TRC [73]

TABLE 5.14—*Data source for development of Eq. (5.139), values of parameter r and predicted Z_c from MRK EOS [70].*

No.	Compound	R_m, at 20°C, cm³/mol	R	No. of data points	Temp. range, K	Pressure range, bar	Ref.	Critical compressibility, Z_c		
								Table 2.1	Pred. MRK	%AD
1	Methane (C₁)	6.987	1.000	135	90–500	0.5–700	Goodwin [72]	0.288	0.333	15.6
2	**Ethane (C₂)**	11.319	1.620	157	90–700	0.1–700	Goodwin et al. [74]	0.284	0.300	5.6
3	**Ethylene**	10.508	1.504	90	100–500	1–400	McCarty and Jacobsen [75]	0.276	0.295	6.9
5	**Propane (C₃)**	15.784	2.259	130	85–700	0.1–700	Goodwin and Haynes [76]	0.280	0.282	0.7
6	**Isobutane**	20.647	2.955	115	110–700	0.1–700	Goodwin and Haynes [76]	0.282	0.280	0.7
7	**n-Butane (C₄)**	20.465	2.929	183	130–700	0.1–700	Haynes and Goodwin [77]	0.274	0.278	1.5
8	**n-Pentane (C₅)**	25.265	3.616	...	...	...	...	0.269	0.271	0.7
9	**n-Hexane (C₆)**	29.911	4.281	100	298–1000	1–500	TRC Tables [73]	0.264	0.266	0.7
10	**Cyclohexane**	27.710	3.966	140	320–1000	1–500	TRC Tables [73]	0.273	0.269	1.5
11	**Benzene**	26.187	3.748	110	310–1000	1–500	TRC Tables [73]	0.271	0.270	0.4
12	**Toluene**	31.092	4.450	110	330–1000	1–500	TRC Tables [73]	0.264	0.265	0.4
13	**n-Heptane (C₇)**	34.551	4.945	100	300–1000	1–500	TRC Tables [73]	0.263	0.262	0.4
14	**n-Octane (C₈)**	39.183	5.608	80	320–1000	1–500	TRC Tables [73]	0.259	0.258	0.4
15	**i-Octane**	39.260	5.619	70	340–1000	1–500	TRC Tables [73]	0.266	0.256	3.8
16	n-Heptane (C₇)[a]	34.551	4.945	35	303–373	50–500	Doolittle [78]	...	...	...
17	n-Nonane (C₉)[a]	43.836	6.274	35	303–373	50–500	Doolittle [78]	0.255	0.254	0.4
18	n-Decane (C₁₀)[a]	48.497	6.941	...	...	...	...	0.249	0.250	0.4
19	n-Undecane (C₁₁)[a]	53.136	7.605	35	303–373	50–500	Doolittle [78]	0.243	0.247	1.6
20	n-Dodecane (C₁₂)[a]	57.803	8.273	...	...	...	...	0.238	0.245	2.9
21	n-Tridecane (C₁₃)[a]	62.478	8.942	30	303–373	50–500	Doolittle [78]	0.236	0.242	2.6
22	n-Tetradecane (C₁₄)[a]	67.054	9.597	...	...	...	...	0.234	0.240	2.5
23	n-Pentadecane (C₁₅)[a]	71.708	10.263	...	...	...	...	0.228	0.238	4.3
24	n-Hexadecane (C₁₆)[a]	76.389	10.933	...	...	...	...	0.225	0.235	4.2
25	n-Heptadecane (C₁₇)[a]	81.000	11.593	30	323–573	50–500	Doolittle [78]	0.217	0.233	7.4
26	n-Eicosane (C₂₀)[a]	95.414	13.656	20	373–573	50–500	Doolittle [78]	0.213	0.227	6.6
27	n-Triacosane (C₃₀)[a]	141.30	20.223	20	373–573	50–500	Doolittle [78]	...	0.213	...
28	n-Tetracontane (C₄₀)[a]	187.69	26.862	20	423–573	50–500	Doolittle [78]	...	...	...
	Overall	...	...	1745	90–1000	0.1–700		...	...	3.0

Density data for compounds 16–28 are all only for liquids [78]. Compounds specified by bold are used in development of Eq. (5.139). Calculated values of Z_c from SRK and PR EOSs for all compounds are 0.333 and 0.307, respectively. These give average errors of 28.2 and 18.2%, respectively.
[a]PVT data for the following compounds were not used in development of Eq. (5.139).

β is defined as

$$(5.136) \qquad \beta = \frac{b_{actual}}{b_{RK}}$$

where b_{actual} is the optimum value of b and b_{RK} is the value of b obtained for RK EOS and is calculated through the relation given in Table 5.1. For the reference fluid, $\beta_{ref} = 1$. We now assume that

$$(5.137) \qquad \frac{\beta}{\beta_{ref}} = \frac{\alpha}{\alpha_{ref}} = f\left(\frac{R_m}{R_{m,ref}}, T_r\right)$$

Parameter r is defined as

$$(5.138) \qquad r = \frac{R_m}{R_{m,ref.}} = \frac{R_m}{6.987}$$

r is a dimensionless parameter and represents reduced molecular size. Values of r calculated from Eq. (5.138) are also given in Table 5.14. By combining Eqs. (5.137) and (5.138) and based on data for densities of hydrocarbons from C₂ to C₈, the following relation was found for calculation of parameter b in the RK EOS:

$$\frac{1}{\beta} = 1 + \{0.02\,[1 - 0.92\exp{(-1000\,|T_r - 1|)}] - 0.035\,(T_r - 1)\}$$
$$\times\,(r - 1)$$

$$(5.139)$$

Once β is determined from the above relation, the co-volume parameter b for the RK can be calculated by substituting b_{RK}

from Table 5.1 into Eq. (5.136) as

$$(5.140) \qquad b = \left(\frac{0.08664RT_c}{P_c}\right)\beta$$

Parameter a for the RK EOS is given in Table 5.1 as

$$(5.141) \qquad a = \frac{0.42748R^2T_c^2}{P_c}$$

Therefore, the modified RK EOS is composed of Eq. (5.38) and Eqs. (5.138)–(5.141) for calculation of the parameters a and b. Equation (5.39) for the PVT relation and Eq. (5.141) for parameter a are the same as the original RK EOS. This modified version of RK EOS is referred as MRK. In fact when $\beta = 1$ the MRK EOS reduces to RK EOS. The exponential term in Eq. (5.139) is the correction for the critical region. At $T_r = 1$ this equation reduces to

$$(5.142) \qquad b_{at\,T_c} = 1 + 0.0016(r - 1)$$

This equation indicates that the MRK EOS does not give a constant Z_c for all compounds but different values for different compounds. For this reason this EOS does not satisfy the constraints set by Eq. (5.9). But calculations show that $(\partial P/\partial V)_{T_c}$ and $(\partial^2 P/\partial V^2)_{T_c}$ are very small. For hydrocarbons from C₁ to C₂₀ the average values for these derivatives are 0.0189 and 0.001, respectively [70]. In summary 1383 data points on densities of liquids and gases for hydrocarbons from C₂ to C₈ with pressure range of 0.1–700 bar and temperature up to 1000 K were used in development of Eq. (5.139). The

TABLE 5.15—*Evaluation of various EOS for prediction of liquid density of heavy hydrocarbons [70].*

Compound	No. of data points	%AAD			
		MRK	RK	SRK	PR
n-Heptane (n-C_7)	35	0.6	12.1	10.5	1.4
n-Nonane (n-C_9)	35	0.6	15.5	13.4	3.4
n-Undecane (n-C_{11})	35	1.7	18.0	15.5	5.4
n-Tridecane (n-C_{13})	30	2.8	20.3	17.7	7.9
n-Heptadecane (n-C_{17})	30	1.2	27.3	24.8	16.0
n-Eicosane (n-C_{20})	20	2.8	29.5	26.7	18.2
n-Triacontane (n-C_{30})	20	0.6	41.4	39.4	32.5
n-Tetracontane (n-C_{40})	20	4.1	50.9	49.4	44.4
Total	225	1.6	24.3	22.1	13.3

MRK: Eqs. (5.38), (5.138), and (5.141). Note none of these data were used in development of Eq. (5.139).

interesting point about this equation is that it can be used up to C_{40} for density estimations. Obviously this equation is not designed for VLE calculations as no VLE data were used to develop Eq. (5.139). Prediction of Z_c from MRK EOS is shown in Table 5.14. Evaluation of MRK with PR and SRK equations for prediction of liquid density of heavy hydrocarbons is given in Table 5.15. Data sources for these compounds are given in Table 5.14. Overall results for prediction of density for both liquid and gaseous hydrocarbon compounds from C_1 to C_{40} is shown in Table 5.15. The overall error for the MRK EOS for more than 1700 data points is about 1.3% in comparison with 4.6 for PR and 7.3 for SRK equations.

To apply this EOS to defined mixtures a set of mixing rules are given in Table 5.17 [70]. For petroleum fractions parameters can be directly calculated for the mixture. For binary and ternary liquid mixtures containing compounds from C_1 to C_{20} an average error of 1.8% was obtained for 200 data points [70]. For the same dataset RK, SRK, and PR equations gave errors of 15, 13, and 6%, respectively. Further characteristics and evaluations of this modified RK EOS are discussed by Riazi and Roomi [71]. Application of this method in calculation of density is shown in the following example.

Example 5.9—Repeat Example 5.2 for prediction of liquid and vapor density of n-octane using MRK EOS.

Solution—The MRK EOS is to use Eq. (5.38) with parameters obtained from Eqs. (5.139)–(5.141). The input data needed to use MRK EOS are T_c, P_c, and r. From Example 5.2, $T_c = 568.7$ K, $P_c = 24.9$ bar, and $T_r = 0.9718$ K. From Table 5.14 for n-C_8, $r = 5.608$. From Eq. (5.139), $\beta = 1.5001 \times 10^{-4}$. From Eq. (5.139), $b = 150.01$ cm^3/mol and from Eq. (5.141), $a = 3.837982 \times 10^7$ cm^6/mol^2. Solving Eq. (5.42) with $u_1 = 1$ and $u_2 = 0$ (Table 5.1) and in a way similar to that performed in Example 5.2 we get $V^L = 295.8$ and $V^V = 1151.7$ cm^3/mol. Deviations of predicted values from experimental data are −2.7% and −5.3% for liquid and vapor molar volume, respectively.

TABLE 5.16—*Comparison of various EOSs for prediction of density of liquid and gaseous hydrocarbons.*

Compound	No. of data points	%AAD			
		MRK	RK	SRK	PR
$C_1C_8^a$	1520	1.3	4.9	5.1	3.3
C_7–C_{40}^b	225	1.6	24.3	22.1	13.3
Total	1745	1.33	7.38	7.28	4.59

[a]These are the compounds that have been marked as bold in Table 5.14 and are used in development of Eq. (5.139).
[b]These are the same compounds as in Table 5.15.

TABLE 5.17—*Mixing rules for MRK EOS parameters (Eqs. (5.38) and (5.137)–(5.140)).*

$$T_{cm} = \frac{\sum_i \sum_j x_i x_j T_{cij}^2 / P_{cij}}{\sum_i \sum_j x_i x_j T_{cij} / P_{cij}} \qquad T_{cij} = \left(T_{ci}T_{cj}\right)^{1/2}\left(1 - k_{ij}\right)$$

$$P_{cm} = \frac{\left(\sum_i \sum_j x_i x_j T_{cij}^2 / P_{cij}\right)}{\left(\sum_i \sum_j x_i x_j T_{cij} / P_{cij}\right)^2} \qquad P_{cij} = \frac{8T_{cij}}{\left[(T_{cii}/P_{cii})^{1/3} + (T_{cjj}/P_{cjj})^{1/3}\right]^3}$$

$$R_m = \sum_i \sum_j x_i x_j r_{ij} \qquad r_{ij} = \frac{\left(r_{ii}^{1/3} + r_{jj}^{1/3}\right)^3}{8}$$

Predicted liquid densities from SRK and PR equations (Example 5.2) deviate from experimental data by +31.5 and 17.2%, respectively. Advantage of MRK over other cubic equations for liquid density is greater for heavier compounds as shown in Table 5.15. ♦

This modified version of RK EOS is developed only for density calculation of hydrocarbon systems and their mixtures. It can be used directly to calculate density of petroleum fractions, once M, d_{20}, n_{20}, T_c, and P_c are calculated from methods discussed in Chapters 2 and 3. Moreover parameter r can be accurately estimated for heavy fractions, while prediction of acentric factor for heavy compounds is not reliable (see Figs. 2.20–2.22). The main characteristic of this equation is its application to heavy hydrocarbons and undefined petroleum fractions. The fact that Eq. (5.139) was developed based on data for hydrocarbons from C_2 to C_8 and it can well be used up to C_{40} shows its extrapolation capability. The linear relation that exists between $1/\beta$ and parameter r makes its extrapolation to heavier hydrocarbons possible. In fact it was found that by changing the functionality of $1/\beta$ with r, better prediction of density is possible but the relation would no longer be linear and its extrapolation to heavier compounds would be less accurate. For example, for C_{17} and C_{18}, if the constant 0.02 in Eq. (5.139) is replaced by 0.018, the %AAD for these compounds reduces from 2 to 0.5%. The following example shows application of this method.

Analysis of various EOS shows that use of refractive index in obtaining constants of an EOS is a promising approach. Further work in this area should involve use of saturation pressure in addition to liquid density data to obtain relations for EOS parameters that would be suitable for both liquid density and VLE calculations.

5.10 SUMMARY AND CONCLUSIONS

In this chapter the fundamental of PVT relations and mathematical EOS are presented. Once the PVT relation for a fluid is known various physical and thermodynamic properties can be determined as discussed in Chapters 6 and 7. Intermolecular forces and their importance in property predictions were discussed in this chapter. For light hydrocarbons two-parameter potential energy relations such as LJ describes the intermolecular forces and as a result two-parameter EOS are sufficient to describe the PVT relation for such fluids. It is shown that EOS parameters can be directly calculated from the potential energy relations. Criteria for correct EOS are given so that validity of any EOS can be analyzed. Three category of EOSs are presented in this chapter: (1) cubic type, (2) noncubic type, and (3) generalized correlations.

Four types of cubic equations vdW, RK, SRK, and PR and their modifications have been reviewed. The main advantage of cubic equations is simplicity, mathematical convenience, and their application for both vapor and liquid phases. The main application of cubic equations is in VLE calculations as will be discussed in Chapters 6 and 9. However, their ability to predict liquid phase density is limited and this is the main weakness of cubic equations. PR and SRK equations are widely used in the petroleum industry. PR equation gives better liquid density predictions, while SRK is used in VLE calculations. Use of volume translation improves capability of liquid density prediction for both PR and SRK equations; however, the method of calculation of this parameter for heavy petroleum fractions is not available and generally these equations break down at about C_{10}. Values of input parameters greatly affect EOS predictions. For heavy hydrocarbons, accurate prediction of acentric factor is difficult and for this reason an alternative EOS based on modified RK equation is presented in Section 5.9. The MRK equation uses refractive index parameter instead of acentric factor and it is recommended for density calculation of heavy hydrocarbons and undefined petroleum fraction. This equation is not suitable for VLE and vapor pressure calculations. In Chapter 6, use of velocity of sound data to obtain EOS parameters is discussed [79].

Among noncubic equations, virial equations provide more accurate PVT relations; however, prediction of fourth and higher virial coefficients is not possible. Any EOS can be converted into a virial form. For gases at moderate pressures, truncated virial equation after third term (Eq. 5.75) is recommended. Equation (5.71) is recommended for estimation of the second virial coefficient and Eq. (5.78) is recommended for prediction of the third virial coefficients. For specific compounds in which virial coefficients are available, these should be used for more accurate prediction of PVT data at certain moderate conditions such as those provided by Gupta and Eubank [80].

Several other noncubic EOS such as BWRS, CS, LJ, SPHC, and SAFT are presented in this chapter. As will be discussed in the next chapter, recent studies show that cubic equations are also weak in predicting derivative properties such as enthalpy, Joule Thomson coefficient, or heat capacity. For this reason, noncubic equations such as simplified perturbed hard chain (SPHC) or statistical associating fluid theory (SAFT) are being investigated for prediction of such derived properties [81]. For heavy hydrocarbons in which two-parameter potential energy functions are not sufficient to describe the intermolecular forces, three- and perhaps four-parameter EOS must be used. The most recent reference on the theory and application of EOSs for pure fluids and fluid mixtures is provided by Sengers et al. [82]. In addition, for a limited number of fluids there are highly accurate EOS that generally take on a modified MBWR form or a Helmholtz energy representation like the IAPWS water standard [4]. Some of these equations are even available free on the webs [83].

The theory of corresponding state provides a good PVT relation between Z-factor and reduced temperature and pressure. The LK correlation presented by Eqs. (5.107)–(5.111) is based on BWR EOS and gives the most accurate PVT relation if accurate input data on T_c, P_c, and ω are known. While the cubic equations are useful for phase behavior calculations, the LK corresponding states correlations are recommended for

calculation of density, enthalpy, entropy, and heat capacity of hydrocarbons and petroleum fractions. Analytical form of LK correlation is provided for computer applications, while the tabulated form is given for hand calculations. Simpler two-parameter empirical correlation for calculation of Z-factor of gases, especially for light hydrocarbons and natural gases, is given in a graphical form in Fig. 5.12 and Hall–Yarborough equation can be used for computer applications..

For calculation of liquid densities use of Rackett equation (Eq. 5.121) is recommended. For petroleum fractions in which Racket parameter is not available it should be determined from specific gravity through Eq. (5.123). For the effect of pressure on liquid density of light pure hydrocarbons, defined hydrocarbon mixtures and light petroleum fractions, the COSTALD correlation (Eq. 5.130) may be used. For petroleum fractions effect of pressure on liquid density can be calculated through Eq. (5.128).

For defined mixtures the simplest approach is to use Kay's mixing rule (Eqs. 3.39 and 5.116) to calculate pseudocritical properties and acentric factor of the mixture. However, when molecules in a mixture are greatly different in size (i.e., C_5 and C_{20}), more accurate results can be obtained by using appropriate mixing rules given in this chapter for different EOS. For defined mixtures liquid density can be best calculated through Eq. (5.126) when pure component densities are known at a given temperature and pressure. For undefined narrow boiling range petroleum fractions T_c, P_c, and ω should be estimated according to the methods described in Chapters 2 and 3. Then the mixture may be treated as a single pseudo-component and pure component EOS can be directly applied to such systems. Some other graphical and empirical methods for the effect of temperature and pressure on density and specific gravity of hydrocarbons and petroleum fractions are given in Chapter 7. Further application of methods presented in this chapter for calculation of density of gases and liquids especially for wide boiling range fractions and reservoir fluids will be presented in Chapter 7. Theory of prediction of thermodynamic properties and their relation with PVT behavior of a fluid are discussed in the next chapter.

5.11 PROBLEMS

5.1. Consider three phases of water, oil, and gas are in equilibrium. Also assume the oil is expressed in terms of 10 components (excluding water) with known specifications. The gas contains the same compounds as the oil. Based on the phase rule determine what is the minimum information that must be known in order to determine oil and gas properties.

5.2. Obtain coefficients a and b for the PR EOS as given in Table 5.1. Also obtain $Z_c = 0.307$ for this EOS.

5.3. Show that the Dieterici EOS exhibits the correct limiting behavior at $P \rightarrow 0$ (finite T) and $T \rightarrow \infty$ (finite P)

$$P = \frac{RT}{V-b}e^{-a/RTV}$$

where a and b are constants.

5.4. The Lorentz EOS is given as

$$\left(P + \frac{a}{V^2}\right)\left(V - \frac{bV}{V+b}\right) = RT$$

where a and b are the EOS constants. Is this a valid EOS?

5.5. A graduate student has come up with a cubic EOS in the following form:

$$\left[P + \frac{aV^2}{(V+b)(V-b)}\right](V-b) = RT$$

Is this equation a correct EOS?

5.6. Derive a relation for the second virial coefficient of a fluid that obeys the SWP relation. Use data on B for methane in Table 5.4 to obtain the potential energy parameters, σ and ε. Compare your calculated values with those obtained from LJ Potential as $\sigma = 4.01$ Å and $\varepsilon/k = 142.87$ K [6, 79].

5.7. Derive Eq. (5.66) from Eq. (5.65) and discuss about your derivation.

5.8. Show that for the second virial coefficient, Eq. (5.70) can be reduced to a form similar to Eq. (5.59). Also show that these two forms are identical for a binary system.

5.9. Derive the virial form of PR EOS and obtain the virial coefficients B, C, and D in terms of PR EOS parameters.

5.10. With results obtained in Example 5.4 and Problem 5.9 for the virial coefficients derived from RK, SRK, and PR equations estimate the following:

a. The second virial coefficient for propane at temperatures 300, 400 and 500 K and compare the results with those given in Table 5.4. Also predict B from Eqs. (5.71)–(5.73).

b. The third virial coefficients for methane and ethane and compare with those given in Table 5.5.

c. Compare predicted third virial coefficients from (b) with those predicted from Eq. (5.78).

5.11. Specific volume of steam at 250°C and 3 bar is 796.44 cm³/g [1]. The virial coefficients (B and C) are given in Table 5.5. Estimate specific volume of this gas from the following methods:

a. RK, SRK, and PR equations.

b. Both virial forms by Eqs. (5.65) and (5.66). Explain why the two results are not the same.

c. Virial equation with coefficients estimated from Eqs. (5.71), (5.72), and (5.78)

5.12. Estimate molar volume of n-decane at 373 K and 151.98 bar from LK generalized correlations. Also estimate the critical compressibility factor. The actual molar volume is 206.5 cm³/mol.

5.13. For several compounds liquid density at one temperature is given in the table below.

Component[a]	N₂	H₂O	C₁	C₂	C₃	n-C₄
T, K	78	293	112	183	231	293
ρ, g/cm³	0.804	0.998	0.425	0.548	0.582	0.579

[a] Source: Reid et al. [15].

For each compound calculate the Rackett parameter from reference density and compare with those given in Table 5.12. Use estimated Rackett parameter to calculate specific gravity of C₃ and n-C₄ at 15.5°C and compare with values of SG given in Table 2.1.

5.14. For a petroleum fraction having API gravity of 31.4 and Watson characterization factor of 12.28 estimate liquid density at 68°F and pressure of 5400 psig using the following methods. The experimental value is 0.8838 g/cm³ (Ref. [59] Ch 6)

a. SRK EOS

b. SRK using volume translation

c. MRK EOS

d. Eq. (5.128)

e. COSTALD correlation (Eq. 5.130)

f. LK generalized correlation

g. Compare errors from different methods

5.15. Estimate liquid density of n-decane at 423 K and 506.6 bar from the following methods:

a. PR EOS

b. PR EOS with volume translation

c. PR EOS with Twu correlation for parameter a (Eq. 5.54)

d. MRK EOS

e. Racket equation with COSTALD correlation

f. Compare the values with the experimental value of 0.691 g/cm³

5.16. Estimate compressibility factor of saturated liquid and vapor (Z^L and Z^V) methane at 160 K (saturation pressure of 15.9 bar) from the following methods:

a. Z^L from Racket equation and Z^V from Standing–Katz chart

b. PR EOS

c. PR EOS with Twu correlation for parameter a (Eq. 5.54)

d. MRK EOS

e. LK generalized correlation

f. Compare estimated values with the values from Fig. 6.12 in Chapter 6.

5.17. Estimate Z^V of saturated methane in Problem 5.16 from virial EOS and evaluate the result.

5.18. A liquid mixture of C₁ and n-C₅ exists in a PVT cell at 311.1 K and 69.5 bar. The volume of liquid is 36.64 cm³. Mole fraction of C₁ is 0.33. Calculate mass of liquid in grams using the following methods:

a. PR EOS with and without volume translation

b. Rackett equation and COSTALD correlation

c. MRK EOS

5.19. A natural gas has the following composition:

Component	CO₂	H₂S	N₂	C₁	C₂	C₃
mol%	8	16	4	65	4	3

Determine the density of the gas at 70 bar and 40°C in g/cm³ using the following methods:

a. Standing–Katz chart

b. Hall–Yarborough EOS

c. LK generalized correlation

5.20. Estimate Z^L and Z^V of saturated liquid and vapor ethane at $T_r = 0.8$ from MRK and virial EOSs. Compare calculated values with values obtained from Fig. 5.10.

REFERENCES

[1] Smith, J. M., Van Ness, H. C., and Abbott, M. M., *Introduction to Chemical Engineering Thermodynamics*, 5th ed., McGraw-Hill, New York, 1996.

[2] Daubert, T. E., Danner, R. P., Sibul, H. M., and Stebbins, C. C., *Physical and Thermodynamic Properties of Pure Compounds: Data Compilation*, DIPPR-AIChE, Taylor & Francis, Bristol, PA, 1994 (extant) (www.aiche.org/dippr). Updated reference: Rowley, R. L., Wilding, W. V., Oscarson, J. L., Zundel, N. A., Marshall, T. L., Daubert, T. E., and Danner, R. P., *DIPPR Data Compilation of Pure Compound Properties*, Design Institute for Physical Properties (DIPPR), Taylor & Francis, New York, 2002 (http://dippr.byu.edu).

[3] AIChE DIPPR® Database, *Design Institute for Physical Property Data* (DIPPR), EPCON International, Houston, TX, 1996 (also see http://dippr.byu.edu).

[4] Wagner, W. and Pruss, A., "The IAPWS Formulation for the Thermodynamic Properties of Ordinary Water Substance for General and Scientific Use," *Journal of Physical and Chemical Reference Data*, Vol. 31, No. 2, 2002, pp. 387–535 (also see http://www.iapws.org/).

[5] KISR, Private Communication, Petroleum Research Studies Center, Kuwait Institute for Scientific Research (KISR), Ahmadi, Kuwait, May 2002.

[6] Prausnitz, J. M., Lichtenthaler, R. N., and de Azevedo, E.G., *Molecular Thermodynamics of Fluid-Phase Equilibria*, Prentice-Hall, New Jersey, 1986.

[7] Hirschfelder, O. J., Curtiss, C. F., and Bird, R. B., *Molecular Theory of Gases and Liquids*, Wiley, New York, 1964.

[8] Poling, B. E., Prausnitz, J. M., and O'Connell, J. P., *Properties of Gases & Liquids*, 5th ed., Mc-Graw Hill, New York, 2000.

[9] Alberty, R. A. and Silbey, R. J., *Physical Chemistry*, 2nd ed., Wiley, New York, 1999.

[10] Levine, I. N., *Physical Chemistry*, 4th ed., McGraw-Hill, New York, 1995.

[11] Redlich, O. and Kwong, J. N. S., "On the Thermodynamics of Solutions," *Chemical Review*, Vol. 44, 1949, pp. 233–244.

[12] Soave, G., "Equilibrium Constants from a Modified Redlich-Kwong Equation of State," *Chemical Engineering Science*, Vol. 27, 1972, pp. 1197–1203.

[13] Peng, D. Y. and Robinson, D. B., "A New Two-Constant Equation of State," *Industrial and Engineering Chemistry Fundamentals*, Vol. 15, No. 1, 1976, pp. 59–64.

[14] Robinson, D. B. and Peng, D. Y., The Characterization of the Heptanes and Heavier Fractions for the GPA Peng-Robinson Programs, Research Report 28, Gas Producers Association (GPA), Tulsa, 1978.

[15] Reid, R. C., Prausnitz, J. M., and Poling, B. E., *Properties of Gases & Liquids*, 4th ed., McGraw-Hill, New York, 1987.

[16] Perry, R. H. and Green, D. W., Eds., *Chemical Engineers' Handbook*, 7th ed., McGraw-Hill, New York, 1997.

[17] Abramowitz, M. and Stegun, I. A., *Handbook of Mathematical Functions*, Dover Publication, New York, 1970.

[18] Guo, T. M., Du, L., Pedersen, K. S., and Fredenslund, Aa., "Application of the Du-Guo and the SRK Equations of State to Predict the Phase Behavior of Chinese Reservoir Fluids," *SPE Reservoir Engineering*, Vol. 6, No. 3, 1991, pp. 379–388.

[19] Whitson, C. H. and Brule, M. R., *Phase Behavior*, Monograph Volume 20, SPE, Richardson, Texas, 2000.

[20] Soreide, I., *Improved Phase Behavior Predictions of Petroleum Reservoir Fluids from a Cubic Euation of State*, Dr. Engineering Dissertation, Norwegian Institute of Technology, Trondheim, Norway, 1989.

[21] Danesh, A., *PVT and Phase Behavior of Petroleum Reservoir Fluids*, Elsevier, Amsterdam, 1998.

[22] Peneloux, A., Rauzy, E., and Freze, R., "A Consistent Correction for Redlich-Kwong-Soave Volumes," *Fluid Phase Equilibria*, Vol. 8, 1982, pp. 7–23.

[23] Jhaveri, B. S. and Youngren, G. K., "Three-Parameter Modification of the Peng-Robinson Equation of State to Improve Volumetric Predictions," *SPE Reservoir Engineering*, Trans., AIME, Vol. 285, 1998, pp. 1033–1040.

[24] Graboski, M. S. and Daubert, T. E., "A Modified Soave Equation of State for Phase Equilibrium Calculations, 1: Hydrocarbon Systems," *Industrial Engineering and Chemistry, Process Design and Development*, Vol. 17, No. 4, 1978, pp. 443–448.

[25] Twu, C. H., Coon, J. E., and Cunningham, J. R., "A New Alpha-Function for a Cubic Equation of State, 1: Peng-Robinson Equation," *Fluid Phase Equilibria*, Vol. 105, No. 1, 1995, pp. 49–59.

[26] Nasrifar, K. and Moshfeghian, M., "A New Cubic Equation of State for Simple Fluids, Pure and Mixtures," *Fluid Phase Equilibria*, Vol. 190, No. 1–2, 2001, pp. 73–88.

[27] Zudkevitch, D. and Joffe, E., "Correlation and Prediction of Vapor–Liquid Equilibria with the Redlich–Kwong Equation of State," *American Institute of Chemical Engineers Journal*, Vol. 16, No. 1, 1970, pp. 112–119.

[28] Joffe, E., Schroeder, G. M., and Zudkevitch, D., "Vapor–Liquid Equilibrium with the Redlich–Kwong Equation of State," *American Institute of Chemical Engineers Journal*, Vol. 16, No. 5, 1970, pp. 496–498.

[29] Feyzi, F., Riazi, M. R., Shaban, H. I., and Ghotbi, S., "Improving Cubic Equations of State for Heavy Reservoir Fluids and Critical Region," *Chemical Engineering Communications*, Vol. 167, 1998, pp. 147–166.

[30] Abbott, M. M., "Cubic Equations of State: An Interpretive Review," Chapter 3 in *Equation of State in Engineering and Research*, K. C. Chao and R. L. Robinson, Jr., Eds., *Advances in Chemistry Series*, American Chemical Society, Washington, DC, Vol. 182, 1979, pp. 47–97.

[31] Martin, J. J., "Cubic Equation of State—Which?," *Industrial and Engineering Chemistry*, Vol. 18, No. 2, 1979, pp. 81–97.

[32] Tsonopoulos, C. and Heidman, J. L., "From Redlich–Kwong to the Present," *Fluid Phase Equilbria*, Vol. 24, 1985, pp. 1–23.

[33] Firoozabadi, A., "Reservoir Fluid Phase Behavior and Volumetric Prediction with Equation of State," *Journal of Petroleum Technology*, Vol. 40, No. 4, 1988, pp. 397–406.

[34] Patel, N. C. and Teja, A. S., "A New Cubic Equation of State for Fluids and Fluid Mixtures," *Chemical Engineering Science*, Vol. 77, No. 3, 1982, pp. 463–473.

[35] Manafi, H., Mansoori, G. A., and Ghotbi, S., "Phase Behavior Prediction of Petroleum Fluids with Minimum Characterization Data," *Journal of Petroleum Science and Engineering*, Vol. 22, 1999, pp. 67–93.

[36] Katz, D. L. and Firoozabadi, A., "Predicting Phase Behavior of Condensate/Crude Oil Systems Using Methane Interaction Coefficients," *Journal of Petroleum Technology*, Trans., AIME, Vol. 265, November 1978, pp. 1649–1655.

[37] Chueh, P. L. and Prausnitz, J. M., "Vapor-Liquid Equilibria at High Pressures, Calculation of Partial Molar Volume in Non-Polar Liquid Mixtures," *American Institute of Chemical Engineers Journal*, Vol. 13, No. 6, 1967, pp. 1099–1113.

[38] Whitson, C. H., Anderson, T. F., and Soreide, I., "C_{7+} Characterization of Related Equilibrium Fluids Using the Gamma Distribution," C_{7+} *Fraction Characterization*, L. G. Chorn and G. A. Mansoori, Eds., Taylor & Francis, New York, 1989, pp. 35–56.

[39] Pedersen, K. S., Thomassen, P., and Fredenslund, Aa., "On the Dangers of "Tuning" Equation of State Parameters," *Chemical Engineering Science*, Vol. 43, No. 2, pp. 269–278.

[40] Whitson, C. H., "Characterizing Hydrocarbon Plus Fractions," *Society of Petroleum Engineers Journal*, Trans., AIME, Vol. 275, 1983, pp. 683–694.

[41] Dymond, J. H. and Smith, E. B., *The Virial Coefficients of Pure Gases and Mixtures*, Clarendon Press, Oxford, UK, 1980.

[42] Tsonopoulos, C., "Empirical Correlation of Second Virial Coefficients," *American Institute of Chemical Engineers Journal*, Vol. 20, 1974, pp. 263–272.

[43] McGlashan, M. L., *Chemical Thermodynamics*, Academic Press, New York, 1980.

[44] Orbey, H. and Vera, J. H., "Correlation for the Third Virial Coefficient Using T_c, P_c and ω as Parameters," *American Institute of Chemical Engineers Journal*, Vol. 29, No. 1, 1983, pp. 107–113.

[45] Bendict, M., Webb, G. B., and Rubin, L. C., "An Empirical Equation for Thermodynamic Properties of Light Hydrocarbons and Their Mixtures, Methane, Ethane, Propane and n-Butane," *Journal of Chemical Physics*, Vol. 8, 1940, pp. 334–345.

[46] Starling, K. E., *Fluid Thermodynamic Properties for Light Petroleum Systems*, Gulf Publishing, Houston, TX, 1973.

[47] Prausnitz, J. M., Lichtenthaler, R. N., and de Azevedo, E. G., *Molecular Thermodynamics of Fluid-Phase Equilibria*, 3rd ed., Prentice-Hall, New Jersey, 1999.

[48] Rao, R. V. G. and Dutta, S. K., "An Equation of State of L-J Fluids and Evaluation of Some Thermodynamic Properties," *Zeitschrift für Physikalische Chemie*, Vol. 264, 1983, pp. 771–783.

[49] Rao, R. V. G. and Gupta, B. D., "Visco-Elastic Properties of Ionic Liquids," *Acoustics Letter*, Vol. 8, No. 9, 1985, pp. 147–149.

[50] Chapman, W. G., Gubbins, K. E., Jackson, G., and Radosz, M., "New Reference Equation of State for Associating Liquids," *Industrial and Engineering Chemistry Research*, Vol. 29, 1990, pp. 1709–1721.

[51] Li, X.-S. and Englezos, P., "Vapor–Liquid Equilibrium of Systems Containing Alcohols Using the Statistical Associating Fluid Theory Equation of State," *Industrial and Engineering Chemistry Research*, Vol. 42, 2003, pp. 4953–4961.

[52] Standing, M. B. and Katz, D. L., "Density of Natural Gases," *AIME Transactions*, Vol. 146, 1942, pp. 140–149.

[53] *Engineering Data Book*, Gas Processors (Suppliers) Association, 10th ed., Tulsa, OK, 1987.

[54] Ahmed, T. *Reservoir Engineering Handbook*, Gulf Publishing, Houston, TX, 2000.

[55] Hall, K. R. and Yarborough, L., "A New Equation of State for Z-Factor Calculations," *Oil and Gas Journal*, June 18, 1973, pp. 82–92.

[56] Pitzer, K. S., Lipmann, D. Z., Curl, R. F., Huggins, C. M., and Petersen, D. E., "The Volumetric and Thermodynamic Properties of Fluids, II: Compressibility Factor, Vapor Pressure, and Entropy of Vaporization," *Journal of American Chemical Society*, Vol. 77, 1955, pp. 3433–3440.

[57] Pitzer, K. S., *Thermodynamics*, 3rd ed., MacGraw-Hill, New York, 1995.

[58] Lee, B. I. and Kesler, M. G., "A Generalized Thermodynamic Correlation Based on Three-Parameter Corresponding States," *American Institute of Chemical Engineers Journal*, Vol. 21, 1975, pp. 510–527.

[59] Daubert, T. E. and Danner, R. P., Eds., *API Technical Data Book—Petroleum Refining*, 6th ed., American Petroleum Institute (API), Washington, DC, 1997.

[60] Elliott, J. R. and Lira, C. T., *Introductory Chemical Engineering Thermodynamics*, Prentice Hall, New Jersey, 1999 (www.phptr.com).

[61] Rackett, H. G., "Equation of State for Saturated Liquids," *Journal of Chemical and Engineering Data*, Vol. 15, No. 4, 1970, pp. 514–517.

[62] Spencer, C. F. and Danner, R. P., "Improved Equation for Prediction of Saturated Liquid Density," *Journal of Chemical and Engineering Data*, Vol. 17, No. 2, 1972, pp. 236–241.

[63] Spencer, C. F. and Adler, S. B., "A Critical Review of Equations for Predicting Saturated Liquid Density," *Journal of Chemical and Engineering Data*, Vol. 23, No. 1, 1978, pp. 82–89.

[64] Yamada, T. G., "Saturated Liquid Molar Volume: the Rackett Equation," *Journal of Chemical and Engineering Data*, Vol. 18, No. 2, 1973, pp. 234–236.

[65] Spencer, C. F. and Danner, R. P., "Prediction of Bubble Point Pressure of Mixtures," *Journal of Chemical and Engineering Data*, Vol. 18, No. 2, 1973, pp. 230–234.

[66] Thomson, G. H., Brobst, K. R., and Hankinson, R. W., "An Improved Correlation for Densities of Compressed Liquids and Liquid Mixtures," *American Institute of Chemical Engineers Journal*, Vol. 28, No. 4, 1982, pp. 671–676.

[67] Tsonopoulos, C., Heidman, J. L., and Hwang, S.-C., *Thermodynamic and Transport Properties of Coal Liquids*, An Exxon Monograph, Wiley, New York, 1986.

[68] Garvin, J., "Estimating Compressed Liquid Volumes for Alcohols and Diols," *Chemical Engineering Progress*, 2004, Vol. 100, No. 3, 2004, pp. 43–45.

[69] Chueh, P. L. and Prausnitz, J. M., "A Generalized Correlation for the Compressibilities of Normal Liquids," *American Institute of Chemical Engineers Journal*, Vol. 15, 1969, p. 471.

[70] Riazi, M. R. and Mansoori, G. A., "Simple Equation of State Accurately Predicts Hydrocarbon Densities," *Oil and Gas Journal*, July 12, 1993, pp. 108–111.

[71] Riazi, M. R. and Roomi, Y., "Use of the Refractive Index in the Estimation of Thermophysical Properties of Hydrocarbons and Their Mixtures," *Industrial and Engineering Chemistry Research*, Vol. 40, No. 8, 2001, pp. 1975–1984.

[72] Goodwin, R. D., *The Thermophysical Properties of Methane from 90 to 500 K at Pressures to 700 Bar*, National Bureau of Standards, NBS Technical Note 653, April 1974.

[73] Hall, K. R., Ed., *TRC Thermodynamic Table—Hydrocarbons*, Thermodynamic Research Center, Texas A&M University System, 1986.

[74] Goodwin, R. D., Roder, H. M., and Starty, G. C., *The Thermophysical Properties of Ethane from 90 to 600 K at Pressures to 700 Bar*, National Bureau of Standards, NBS technical note 684, August 1976.

[75] McCarty, R. D. and Jaconsen, R. T., *An Equation of State for Fluid Ethylene*, National Bureau of Standards, NBS Technical Note 1045, July 1981.

[76] Goodwin, R. D. and Hayens, W. M., *The Thermophysical Properties of Propane from 85 to 700 K at Pressures to 70 MPa*, National Bureau of Standards, NBS Monograph 170, April 1982.

[77] Haynes, W. M. and Goodwin, R. D., *The Thermophysical Properties of Propane from 85 to 700 K at Pressures to 70 MPa*, National Bureau of Standards, NBS Monograph 170, April 1982.

[78] Doolittle, A. K., "Specific Volumes of Normal Alkanes," *Journal of Chemical and Engineering Data*, Vol. 9, 1964, pp. 275–279.

[79] Riazi, M. R. and Mansoori, G. A., "Use of the Velocity of Sound in Predicting the PVT Relations," *Fluid Phase Equilibria*, Vol. 90, 1993, pp. 251–264.

[80] Gupta, D. and Eubank, P. T., "Density and Virial Coefficients of Gaseous Butane from 265 to 450 K Pressures to 3.3 Mpa," *Journal of Chemical and Engineering Data*, Vol. 42, No. 5, Sept.–Oct. 1997, pp. 961–970.

[81] Konttorp, K., Peters, C. J., and O'Connell, J. P., "Derivative Properties from Model Equations of State," Paper presented at the *Ninth International Conference on Properties and Phase Equilibria for Product and Process Design (PPEPPD 2001)*, Kurashiki, Japan, May 20–25, 2001.

[82] Sengers, J. V., Kayser, R. F., Peters, C. J., and White, Jr., H. J., Eds., *Experimental Thermodynamics, Vol. V: Equations of State for Fluids and Fluid Mixtures*, IUPAC, Elsevier, Amsterdam, 2000.

[83] National Institute of Standards and Technology (NIST), Boulder, CO, 2003, http://webbook.nist.gov/chemistry/fluid/.

Thermodynamic Relations for Property Estimations

6

NOMENCLATURE

A Helmholtz free energy defined in Eq. (6.7), J/mol

API   API gravity defined in Eq. (2.4)

$A, B, C, \ldots$ Coefficients in various equations

a, b Cubic EOS parameters given in Table 5.1

a_c Parameter defined in Eq. (5.41) and given in Table 5.1

a_i Activity of component i defined in Eq. (6.111), dimensionless

$a, b, c, \ldots$ Constants in various equations

b A parameter defined in the Standing correlation, Eq. (6.202), K

B Second virial coefficient, cm^3/mol

B', B'' First- and second-order derivatives of second virial coefficient with respect to temperature

C Third virial coefficient, (cm^3/mol)^2

C', C'' First- and second-order derivatives of third virial coefficient with respect to temperature

c Velocity of sound, m/s

c_{PR} Velocity of sound calculated from PR EOS

C_P Heat capacity at constant pressure defined by Eq. (6.17), J/mol·K

C_V Heat capacity at constant volume defined by Eq. (6.18), J/mol·K

d_{20} Liquid density at 20°C and 1 atm, g/cm^3

$F(x, y)$ A mathematical function of independent variables x and y.

f Fugacity of a pure component defined by Eq. (6.45), bar

$\hat{f}_i$ Fugacity of component i in a mixture defined by Eq. (6.109), bar

$f_i^{\circ L}$ Fugacity of pure liquid i at standard pressure (1.01 bar) and temperature T, bar

f_i^{S} Fugacity of pure solid i at P and T (Eq. 6.155), bar

$f^{\circ L}$ Fugacity of pure hypothetical liquid at temperature T ($T > T_c$), bar

$f_r^{\circ L}$ Reduced fugacity of pure hypothetical liquid at temperature T ($= f^{\circ L}/P_c$), dimensionless

f_ω A function defined in terms of ω for parameter a in the PR and SRK equations as given in Table 5.1 and Eq. (5.53)

G Molar Gibbs free energy defined in Eq. (6.6), J/mol

G^R Molar residual Gibbs energy ($= G - G^{ig}$), J/mol

H Molar enthalpy defined in Eq. (6.1), J/mol

CH   Carbon-to-hydrogen weight ratio

K_i Equilibrium ratio in vapor–liquid equilibria ($K_i = y_i/x_i$) defined in Eq. (6.196), dimensionless

K_i^{SL} Equilibrium ratio in solid–liquid equilibria ($K_i^{SL} = x_i^S/x_i^L$) defined in Eq. (6.208), dimensionless

K_W Watson characterization factor defined by Eq. (2.13)

k_B Boltzman constant ($= R/N_A = 1.381 \times 10^{-23}$ J/K)

k_i Henry's law constant defined by Eq. (6.184), bar

$k_{i,M}$ Henry's law constant of component i in a multicomponent solvent, bar

k_{ij} Binary interaction parameter (BIP), dimensionless

M A molar property of system (i.e., S, V, H, S, G, ...)

M^E Excess property ($= M - M^{id}$)

Mt Total property of system ($= n^t M$)

$\bar{M}_i$ Partial molar property for M defined by Eq. (6.78)

M Molecular weight, g/mol [kg/kmol]

M^g Gas molecular weight, g/mol [kg/kmol]

N_A Avogadro number = number of molecules in 1 mol (6.022×10^{23} mol^{-1})

N Number of components in a mixture

N_C Number of carbon atoms in a hydrocarbon compound

n Number of moles (g/molecular wt), mol

n_i Number of moles of component i in a mixture, mol

P Pressure, bar

P^{sat} Saturation pressure, bar

P_a Atmospheric pressure, bar

P_c Critical pressure, bar

P_r Reduced pressure defined by Eq. (5.100) ($= P/P_c$), dimensionless

P^{sub} Sublimation pressure, bar

P_1, P_2, P_3 Derivative parameters defined in Table 6.1

Q_{rev} Heat transferred to the system by a reversible process, J/mol

R Gas constant = 8.314 J/mol·K (values in different units are given in Section 1.7.24)

RC   An objective function defined in Eq. (6.237)

U Molar internal energy, J/mol

u_1, u_2 Parameters in Eqs. (5.40) and (5.42) as given in Table 5.1 for a cubic EOS

V_i^L Liquid molar volume, cm^3/mol

S Molar entropy defined by Eq. (6.2), J/mol·K

S Shrinkage factor defined by Eq. (6.95), dimensionless

SG^g Specific gravity of gas fluid (pure or mixture) [$= M^g/29$], dimensionless

SG Specific gravity of liquid substance at 15.5°C (60°F) defined by Eq. (2.2), dimensionless

T Absolute temperature, K

T_c Critical temperature, K

T_r Reduced temperature defined by Eq. (5.100) ($= T/T_c$), dimensionless

T_B A parameter in the Standing correlation (Eq. 6. 202), K

T_M Freezing (melting) point for a pure component at 1.013 bar, K

T_{tp} Triple point temperature, K

V Molar volume, cm³/gmol

V^L Saturated liquid molar volume, cm³/gmol

V^{sat} Saturation molar volume, cm³/gmol

V^V Saturated vapor molar volume, cm³/gmol

V_c Critical volume (molar), cm³/mol (or critical specific volume, cm³/g)

V_r Reduced volume ($= V/V_c$)

V_{25} Liquid molar volume at 25°C, cm³/mol

x_i Mole fraction of component i in a mixture (usually used for liquids), dimensionless

x_{wi} Weight fraction of component i in a mixture (usually used for liquids), dimensionless

y_i Mole fraction of i in mixture (usually used for gases), dimensionless

Z Compressibility factor defined by Eq. (5.15), dimensionless

Z^L Compressibility factor of liquid phase, dimensionless

Z^V Compressibility factor of vapor phase, dimensionless

Greek Letters

α Parameter defined by Eq. (5.41), dimensionless

α_S, β_S Parameters defined based on velocity of sound for correction of EOS parameters a and b defined by Eq. (6.242), dimensionless

β Coefficient of thermal expansion defined by Eq. (6.24), K⁻¹.

Δ Difference between two values of a parameter

δ_i Solubility parameter for i defined in Eq. (6.147), (J/cm³)^{1/2}

δ_i Parameter used in Eq. (6.126), dimensionless

δ_{ij} Parameter defined in Eq. (5.70)

ε Energy parameter in a potential energy function

ε Error parameter defined by Eq. (106), dimensionless

Φ_i Volume fraction of i in a liquid mixture defined by Eq. (6.146)

ϕ Fugacity coefficient of pure i at T and P defined by Eq. (6.49), dimensionless

ϕ_i Fugacity coefficient of component i at T and P in an ideal solution mixture, dimensionless

$\hat{\phi}_i$ Fugacity coefficient of component i in a mixture at T and P defined by Eq. (6.110)

θ A parameter defined in Eq. (6.203), dimensionless

ρ Density at a given temperature and pressure, g/cm³ (molar density unit: cm³/mol)

σ Diameter of hard sphere molecule, Å (10⁻¹⁰ m)

σ Molecular size parameter, Å (10⁻¹⁰ m)

ω Acentric factor defined by Eq. (2.10)

ξ Packing fraction defined by Eq. (5.86), dimensionless

κ Isothermal compressibility defined by Eq. (6.25), bar⁻¹

η Joule–Thomson coefficient defined by Eq. (6.27), K/bar

γ Heat capacity ratio ($= C_P/C_V$), dimensionless

γ_i Activity coefficient of component i in liquid solution defined by Eq. (6.112), dimensionless

γ_1^S Activity coefficient of a solid solute (component 1) in the liquid solution defined by Eq. (6.161), dimensionless

γ_i^∞ Activity coefficient of component i in liquid solution at infinite dilution ($x_i \to 0$), dimensionless

$\hat{\mu}_i$ Chemical potential of component i defined in Eq. (6.115)

ΔC_{Pi} Difference between heat capacity of liquid and solid for pure component i ($= C_{Pi}^L - C_{Pi}^S$), J/mol·K

ΔH_i^f Heat of fusion (or latent heat of melting) for pure component i at the freezing point and 1.013 bar, J/mol

ΔH^{vap} Heat of vaporization (or latent heat of melting) at 1.013 bar defined by Eq. (6.98), J/mol

ΔH_{mix} Heat of mixing. J/mol

ΔM Property change for M due to mixing defined by Eq. (6.84)

ΔS_i^f Entropy of fusion for pure component i at the freezing point and 1.013 bar, J/mol·K

ΔS^{vap} Entropy of vaporization at 1.013 bar defined by Eq. (6.97), J/mol

ΔT_{b2} Boiling point elevation for solvent 2 (Eq. 6.214), K

ΔT_{M2} Freezing point depression for solvent 2 (Eq. 6.213), K

ΔV_{mix} Volume change due to mixing defined by Eq. (6.86)

Superscript

E Excess property defined for mixtures (with respect to ideal solution)

exp Experimental value

HS Value of a property for hard sphere molecules

ig Value of a property for a component as ideal gas at temperature T and $P \to 0$

id Value of a property for an ideal solution

L Value of a property for liquid phase

R A residual property (with respect to ideal gas property)

V Value of a property for vapor phase

vap Change in value of a property due to vaporization

S Value of a property for solid phase

sat Value of a property at saturation pressure

t Value of a property for the whole (total) system

[]⁽⁰⁾ A dimensionless term in a generalized correlation for a property of simple fluids

[]⁽¹⁾ A dimensionless term in a generalized correlation for a property of acentric fluids

[]⁽ʳ⁾ A dimensionless term in a generalized correlation for a property of reference fluids

α, β Value of a property for phase α or phase β

∞ Value of a property for i in the liquid solution at infinite dilution as $x_i \to 0$

$\circ$ Value of a property at standard state, usually the standard state is chosen at pure component at T and P of the mixture according to the Lewis/Randall rule

$\wedge$ Value of molar property of a component in the mixture

Subscripts

c   Value of a property at the critical point

i   A component in a mixture

j   A component in a mixture

i, j   Effect of binary interaction on a property

m   Value of a property for a mixture

mix   Change in value of a property due to mixing at constant T and P

PR   Value of a property determined from PR EOS

SRK   Value of a property determined from SRK EOS

Acronyms

API-TDB   American Petroleum Institute—Technical Data Book

BIP   Binary interaction parameter

bbl   Barrel, unit of volume of liquid as given in Section 1.7.11

CS   Carnahan–Starling EOS (see Eq. 5.93)

DIPPR   Design Institute for Physical Property Data

EOS   Equation of state

GC   Generalized correlation

GD   Gibbs–Duhem equation (see Eq. 6.81)

HS   Hard sphere

HSP   Hard sphere potential given by Eq. (5.13)

IAPWS   International Association for the Properties of Water and Steam

LJ   Lennard–Jones potential given by Eq. (5.11)

LJ EOS   Lennard–Jones EOS given by Eq. (5.96)

LK EOS   Lee–Kesler EOS given by Eq. (5.104)

LLE   Liquid–liquid equilibria

NIST   National Institute of Standards and Technology

PVT   Pressure–volume–temperature

PR   Peng–Robinson EOS (see Eq. 5.39)

RHS   Right-hand side of an equation

RK   Redlich–Kwong EOS (see Eq. 5.38)

SRK   Soave–Redlich–Kwong EOS given by Eq. (5.38) and parameters in Table 5.1

SAFT   Statistical associating fluid theory (see Eq. 5.98)

SLE   Solid–liquid equilibrium

SLVE   Solid–liquid–vapor equilibrium

VLE   Vapor–liquid equilibrium

VLS   Vapor–liquid–solid equilibrium

VS   Vapor–solid equilibrium

%AAD   Average absolute deviation percentage defined by Eq. (2.135)

%AD   Absolute deviation percentage defined by Eq. (2.134)

IN CHAPTER 5 THE PVT relations and theory of intermolecular forces were discussed. The PVT relations and equations of states are the basis of property calculations as all physical and thermodynamic properties can be related to PVT properties. In this chapter we review principles and theory of property estimation methods and basic thermodynamic relations that will be used to calculate physical and thermodynamic properties.

The PVT relations and equations of state are perhaps the most important thermodynamic relations for pure fluids and their mixtures. Once the PVT relation is known, various physical and thermodynamic properties needed for design and operation of units in the petroleum and related industries can be calculated. Density can be directly calculated from knowledge of molar volume or compressibility factor through Eq. (5.15). Various thermodynamic properties such heat capacity, enthalpy, vapor pressure, phase behavior and vapor liquid equilibrium (VLE), equilibrium ratios, intermolecular parameters, and transport properties all can be calculated through accurate knowledge of PVT relation for the fluid. Some of these relations are developed in this chapter through fundamental thermodynamic relations. Once a property is related to PVT, using an appropriate EOS, it can be estimated at any temperature and pressure for pure fluids and fluid mixtures. Development of such important relations is discussed in this chapter, while their use to estimate thermophysical properties for petroleum mixtures are discussed in the next chapter.

6.1 DEFINITIONS AND FUNDAMENTAL THERMODYNAMIC RELATIONS

In this section, thermodynamic properties such as entropy, Gibbs energy, heat capacity, residual properties, and fugacity are defined. Thermodynamic relations that relate these properties to PVT relation of pure fluids are developed.

6.1.1 Thermodynamic Properties and Fundamental Relations

Previously two thermodynamic properties, namely internal energy (U) and enthalpy (H), were defined in Section 5.1. The enthalpy is defined in terms of U and PV (Eq. 5.5) as

$$(6.1) \qquad H = U + PV$$

Another thermodynamic property that is used to formulate the second law of thermodynamics is called *entropy* and it is defined as

$$(6.2) \qquad dS = \frac{\delta Q_{rev}}{T}$$

where S is the entropy and δQ_{rev} is the amount of heat transferred to the system at temperature T through a reversible process. The symbol δ is used for the differential heat Q to indicate that heat is not a thermodynamic property such as H or S. The unit of entropy is energy per absolute degrees, e.g. J/K, or on a molar basis it has the unit of J/mol · K in the SI unit system. The first law of thermodynamics is derived based on the law of conservation of energy and for a closed system (constant composition and mass) is given as follows [1, 2]:

$$(6.3) \qquad dU = \delta Q - P dV$$

Combining Eqs. (6.2) and (6.3) gives the following relation:

(6.4)
$$dU = TdS - PdV$$

This relation is one of the fundamental thermodynamic relations. Differentiating Eq. (6.1) and combining with Eq. (6.4) gives

(6.5)
$$dH = TdS + VdP$$

Two other thermodynamic properties known as auxiliary functions are *Gibbs free energy* (G) and *Helmholtz free energy* (A) that are defined as

(6.6)
$$G \equiv H - TS$$

(6.7)
$$A \equiv U - TS$$

G and A are mainly defined for convenience and formulation of useful thermodynamic properties and are not measurable properties. Gibbs free energy also known as Gibbs energy is particularly a useful property in phase equilibrium calculations. These two parameters both have units of energy similar to units of U, H, or PV. Differentiating Eqs. (6.6) and (6.7) and combining with Eqs. (6.4) and (6.5) lead to the following relations:

(6.8)
$$dG = VdP - SdT$$

(6.9)
$$dA = -PdV - SdT$$

Equations (6.4), (6.5), (6.8), and (6.9) are the four fundamental thermodynamic relations that will be used for property calculations for a homogenous fluid of constant composition. In these relations either molar or total properties can be used.

Another set of equations can be obtained from mathematical relations. If $F = F(x, y)$ where x and y are two independent variables, the total differential of F is defined as

(6.10)
$$dF = \left(\frac{\partial F}{\partial x}\right)_y dx + \left(\frac{\partial F}{\partial y}\right)_x dy$$

which may also be written as

(6.11)
$$dF = M(x, y)dx + N(x, y)dy$$

where $M(x, y) = (\partial F/\partial x)_y$ and $N(x, y) = (\partial F/\partial y)_x$. Considering the fact that $\partial^2 F/\partial x \partial y = \partial^2 F/\partial y \partial x$, the following relation exists between M and N:

(6.12)
$$\left(\frac{\partial M}{\partial y}\right)_x = \left(\frac{\partial N}{\partial x}\right)_y$$

Applying Eq. (6.12) to Eqs. (6.4), (6.5), (6.8), and (6.9) leads to the following set of equations known as Maxwell's equations [1, 2]:

(6.13)
$$\left(\frac{\partial T}{\partial V}\right)_S = -\left(\frac{\partial P}{\partial S}\right)_V$$

(6.14)
$$\left(\frac{\partial T}{\partial P}\right)_S = \left(\frac{\partial V}{\partial S}\right)_P$$

(6.15)
$$\left(\frac{\partial V}{\partial T}\right)_P = -\left(\frac{\partial S}{\partial P}\right)_T$$

(6.16)
$$\left(\frac{\partial P}{\partial T}\right)_V = \left(\frac{\partial S}{\partial V}\right)_T$$

Maxwell's relations are the basis of property calculations by relating a property to PVT relation. Before showing application of these equations, several measurable properties are defined.

6.1.2 Measurable Properties

In this section some thermodynamic properties that are directly measurable are defined and introduced. Heat capacity at constant pressure (C_P) and heat capacity at constant volume (C_V) are defined as:

(6.17)
$$C_P = \left(\frac{\delta Q}{dT}\right)_P$$

(6.18)
$$C_V = \left(\frac{\delta Q}{dT}\right)_V$$

Molar heat capacity is a thermodynamic property that indicates amount of heat needed for 1 mol of a fluid to increase its temperature by 1 degree and it has unit of J/mol · K (same as J/mol · °C) in the SI unit system. Since temperature units of K or °C represent the temperature difference they are both used in the units of heat capacity. Similarly specific heat is defined as heat required to increase temperature of one unit mass of fluid by 1° and in the SI unit systems has the unit of kJ/kg · K (or J/g · °C). In all thermodynamic relations molar properties are used and when necessary they are converted to specific property using molecular weight and Eq. (5.3). Since heat is a path function and not a thermodynamic property, amount of heat transferred to a system in a constant pressure process differs from the amount of heat transferred to the same system under constant volume process for the same amount of temperature increase. Combining Eq. (6.3) with (6.18) gives the following relation:

(6.19)
$$C_V = \left(\frac{\partial U}{\partial T}\right)_V$$

similarly C_P can be defined in terms of enthalpy through Eqs. (6.2), (6.5), and (6.17):

(6.20)
$$C_P = \left(\frac{\partial H}{\partial T}\right)_P$$

For ideal gases since U and H are functions of only temperature (Eqs. 5.16 and 5.17), from Eqs. (6.20) and (6.19) we have

(6.21)
$$dH^{ig} = C_P^{ig} dT$$

(6.22)
$$dU^{ig} = C_V^{ig} dT$$

where superscript ig indicates ideal gas properties. In some references ideal gas properties are specified by superscript ° or * (i.e., C_P° or C_P^* for ideal gas heat capacity). As will be seen later, usually C_P^{ig} is correlated to absolute temperature T in the form of polynomial of degrees 3 or 5 and the correlation coefficients are given for each compound [1–5]. Combining Eqs. (6.1), (5.14), (6.21), and (6.22) gives the following relation between C_P^{ig} and C_V^{ig} through universal gas constant R:

(6.23)
$$C_P^{ig} - C_V^{ig} = R$$

For ideal gases C_P^{ig} and C_V^{ig} are both functions of only temperature, while for a real gas C_P is a function of both T and P as it is clear from Eqs. (6.20) and (6.28). The ratio of C_P/C_V is called *heat capacity ratio* and usually in thermodynamic texts is shown by γ and it is greater than unity. For monoatomic gases (i.e., helium, argon, etc.) it can be assumed that $\gamma = 5/3$, and for diatomic gases (nitrogen, oxygen, air, etc.) it is assumed that $\gamma = 7/5 = 1.4$.

There are two other measurable properties: *coefficient of thermal expansion*, β, and the bulk *isothermal compressibility*, κ. These are defined as

$$(6.24) \qquad \beta = \frac{1}{V}\left(\frac{\partial V}{\partial T}\right)_P$$

$$(6.25) \qquad \kappa = -\frac{1}{V}\left(\frac{\partial V}{\partial P}\right)_T$$

since $\partial V/\partial P$ is negative, the minus sign in the definition of κ is used to make it a positive number. The units of β and κ in SI system are K^{-1} and Pa^{-1}, respectively. Values of β and κ can be calculated from these equations with use of an equation of state. For example, with use of Lee–Kesler EOS (Eq. 5.104), the value of κ is 0.84×10^{-9} Pa^{-1} for liquid benzene at temperature of 17°C and pressure of 6 bar, while the actual measured value is 0.89×10^{-9} Pa^{-1} [6]. Once β and κ are known for a fluid, the PVT relation can be established for that fluid (see Problem 6.1). Through the above thermodynamic relations and definitions one can show that

$$(6.26) \qquad C_P - C_V = \frac{TV\beta^2}{\kappa}$$

Applying Eqs. (6.24) and (6.25) for ideal gases (Eq. 5.14) gives $\beta^{ig} = 1/T$ and $\kappa^{ig} = 1/P$. Substituting β^{ig} and κ^{ig} into Eq. (6.26) gives Eq. (6.23). From Eq. (6.26) it is clear that $C_P > C_V$; however, for liquids the difference between C_P and C_V is quite small and most thermodynamic texts neglect this difference and assume $C_P \cong C_V$. Most recently Garvin [6] has reviewed values of constant volume specific heats for liquids and concludes that in some cases $C_P - C_V$ for liquids is significant and must not be neglected. For example, for saturated liquid benzene when temperature varies from 300 to 450 K, the calculated heat capacity ratio, C_P/C_V, varies from 1.58 to 1.41 [6]. Although these values are not yet confirmed as they have been calculated from Lee–Kesler equation of state, but one should be careful that assumption of $C_P \cong C_V$ for liquids in general may not be true in all cases. In fact for ideal incompressible liquids, $\beta \to 0$ and $\kappa \to 0$ and according to Eq. (6.26), $(C_P - C_V) \to 0$, which leads to $\gamma = C_P/C_V \to 1$. There is an EOS with high accuracy for benzene [7]. It gives C_p/C_v for saturated liquids having a calculated heat capacity ratio of 1.43–1.38 over a temperature range of 300–450 K.

Another useful property is *Joule–Thomson coefficient* that is defined as

$$(6.27) \qquad \eta = \left(\frac{\partial T}{\partial P}\right)_H$$

This property is useful in throttling processes where a fluid passes through an expansion valve at which enthalpy is nearly constant. Such devices are useful in reducing the fluid pressure, such as gas flow in a pipeline. η expresses the change of temperature with pressure in a throttling process and can be related to C_P and may be calculated from an equation of state (see Problem 6.10).

6.1.3 Residual Properties and Departure Functions

Properties of ideal gases can be determined accurately through kinetic theory. In fact all properties of ideal gases are known or they can be estimated through the ideal gas law.

Values of C_P^{ig} are known for many compounds and they are given in terms of temperature in various industrial handbooks [5]. Once C_P^{ig} is known, C_V^{ig}, U^{ig}, H^{ig}, and S^{ig} can also be determined from thermodynamic relations discussed above. To calculate properties of a real gas an auxiliary function called *residual property* is defined as the difference between property of real gas and its ideal gas property (i.e., $H - H^{ig}$). The difference between property of a real fluid and ideal gas is also called *departure* from ideal gas. All fundamental relations also apply to residual properties. By applying basic thermodynamic and mathematical relations, a residual property can be calculated through a PVT relation of an equation of state. If only two properties such as H and G or H and S are known in addition to values of V at a given T and P, all other properties can be easily determined from basic relations given in this section. For example from H and G, entropy can be calculated from Eq. (6.6). Development of relations for calculation of enthalpy departure is shown here. Other properties may be calculated through a similar approach.

Assume that we are interested to relate residual enthalpy $(H - H^{ig})$ into PVT at a given T and P. For a homogenous fluid of constant composition (or pure substance), H can be considered as a function of T and P:

$$(6.28) \qquad H = H(T, P)$$

Applying Eq. (6.10) gives

$$(6.29) \qquad dH = \left(\frac{\partial H}{\partial T}\right)_P dT + \left(\frac{\partial H}{\partial P}\right)_T dP$$

Dividing both sides of Eq. (6.5) to ∂P at constant T gives

$$(6.30) \qquad \left(\frac{\partial H}{\partial P}\right)_T = V + T\left(\frac{\partial S}{\partial P}\right)_T$$

Substituting for $(\partial S/\partial P)_T$ from Eq. (6.15) into Eq. (6.30) and substitute resulting $(\partial H/\partial P)_T$ into Eq. (6.29) with use of Eq. (6.20) for $(\partial H/\partial T)_P$, Eq. (6.29) becomes

$$(6.31) \qquad dH = C_P dT + \left[V - T\left(\frac{\partial V}{\partial T}\right)_P\right]dP$$

where the right-hand side (RHS) of this equation involves measurable quantities of C_P and PVT, which can be determined from an equation of state. Similarly it can be shown that

$$(6.32) \qquad dS = C_P\frac{dT}{T} - \left(\frac{\partial V}{\partial T}\right)_P dP$$

Equations (6.31) and (6.32) are the basis of calculation of enthalpy and entropy and all other thermodynamic properties of a fluid from its PVT relation and knowledge of C_P. As an example, integration of Eq. (6.31) from (T_1, P_1) to (T_2, P_2) gives change of enthalpy (ΔH) for the process. The same equation can be used to calculate departure functions or residual properties from PVT data or an equation of state at a given T and P. For an ideal gas the second term in the RHS of Eq. (6.32) is zero. Since any gas as $P \to 0$ behaves like an ideal gas, at a fixed temperature of T, integration of Eq. (6.31) from $P \to 0$ to a desired pressure of P gives

$$(6.33) \qquad (H - H^{ig})_T = \int_0^P \left[V - T\left(\frac{\partial V}{\partial T}\right)_P\right]dP \quad \text{(at constant } T\text{)}$$

For practical applications the above equation is converted into dimensionless form in terms of parameters Z defined by Eq. (5.15). Differentiating Z with respect to T at constant P, from Eq. (5.15) we get

$$(6.34) \qquad \left(\frac{\partial Z}{\partial T}\right)_P = \frac{P}{RT}\left(\frac{\partial V}{\partial T}\right)_P + \frac{PV}{R}\left(-\frac{1}{T^2}\right)$$

Dividing both sides of Eq. (6.33) by RT and combining with Eq. (6.34) gives

$$(6.35) \qquad \frac{H - H^{ig}}{RT} = -T\int_0^P \left(\frac{\partial Z}{\partial T}\right)_P \frac{dP}{P} \quad \text{(at constant } T\text{)}$$

It can be easily seen that for an ideal gas where $Z = 1$, Eq. (6.35) gives the expected result of $H - H^{ig} = 0$. Similarly for any equation of state the residual enthalpy can be calculated. Using definitions of T_r and P_r by Eq. (5.100), the above equation may be written as

$$(6.36) \qquad \frac{H - H^{ig}}{RT_c} = -T_r^2 \int_0^{P_r} \left(\frac{\partial Z}{\partial T_r}\right)_{P_r} \frac{dP_r}{P_r} \quad \text{(at constant } T\text{)}$$

where the term in the left-hand side and all parameters in the RHS of the above equation are in dimensionless forms. Once the residual enthalpy is calculated, real gas enthalpy can be determined as follows:

$$(6.37) \qquad H = H^{ig} + RT_c\left(\frac{H - H^{ig}}{RT_c}\right)$$

In general, absolute values of enthalpy are of little interest and normally the difference between enthalpies in two different conditions is useful. Absolute enthalpy has meaning only with respect to a reference state when the value of enthalpy is assigned as zero. For example, tabulated values of enthalpy in steam tables are with respect to the reference state of saturated liquid water at 0°C [1]. As the choice of reference state changes so do the values of absolute enthalpy; however, this change in the reference state does not affect change in enthalpy of systems from one state to another.

A relation similar to Eq. (6.33) can be derived in terms of volume where the gas behavior becomes as an ideal gas as $V \to \infty$:

$$(6.38) \quad (H - H^{ig})_{T,V} = \int_{V \to \infty}^V \left[T\left(\frac{\partial P}{\partial T}\right)_V - P\right]dV + PV - RT$$

Similar relation for the entropy departure is

$$(6.39) \qquad (S - S^{ig})_{T,V} = \int_{V \to \infty}^V \left[\left(\frac{\partial P}{\partial T}\right)_V - \frac{R}{V}\right]dV$$

Once H is known, U can be calculated from Eq. (6.1). Similarly all other thermodynamic properties can be calculated from basic relations and definitions.

Example 6.1—Derive a relation for calculation of C_P from PVT relation of a real fluid at T and P.

Solution—By substituting the Maxwell's relation of Eq. (6.15) into Eq. (6.30) we get

$$(6.40) \qquad \left(\frac{\partial H}{\partial P}\right)_T = V - T\left(\frac{\partial V}{\partial T}\right)_P$$

differentiating this equation with respect to T at constant P gives

$$\left[\frac{\partial}{\partial T}\left(\frac{\partial H}{\partial P}\right)_T\right]_P = \left(\frac{\partial V}{\partial T}\right)_P - \left[\left(\frac{\partial V}{\partial T}\right)_P + T\left(\frac{\partial^2 V}{\partial T^2}\right)_P\right]$$

$$(6.41) \qquad = -T\left(\frac{\partial^2 V}{\partial T^2}\right)_P$$

From mathematical identity we have

$$(6.42) \qquad \left[\frac{\partial}{\partial T}\left(\frac{\partial H}{\partial P}\right)_T\right]_P = \left[\frac{\partial}{\partial P}\left(\frac{\partial H}{\partial T}\right)_P\right]_T$$

Using definition of C_P through Eq. (6.20) and combining the above two equations we get

$$(6.43) \qquad \left(\frac{\partial C_P}{\partial P}\right)_T = -T\left(\frac{\partial^2 V}{\partial T^2}\right)_P$$

Upon integration from $P = 0$ to the desired pressure of P at constant T we get

$$(6.44) \qquad C_P - C_P^{ig} = \int_{P=0}^P \left[-T\left(\frac{\partial^2 V}{\partial T^2}\right)_P\right]_T dP$$

Once C_P^{ig} is known, C_P can be determined at T and P of interest from an EOS, PVT data, or generalized corresponding states correlations. ♦

6.1.4 Fugacity and Fugacity Coefficient for Pure Components

Another important auxiliary function that is defined for calculation of thermodynamic properties, especially Gibbs free energy, is called *fugacity* and it is shown by f. This parameter is particularly useful in calculation of mixture properties and formulation of phase equilibrium problems. Fugacity is a parameter similar to pressure, which indicates deviation from ideal gas behavior. It is defined to calculate properties of real gases and it may be defined in the following form:

$$(6.45) \qquad \lim_{P \to 0}\left(\frac{f}{P}\right) = 1$$

With this definition fugacity of an ideal gas is the same as its pressure. One main application of fugacity is to calculate Gibbs free energy. Application of Eq. (6.8) at constant T to an ideal gas gives

$$(6.46) \qquad dG^{ig} = RT d\ln P$$

For a real fluid a similar relation can be written in terms of fugacity

$$(6.47) \qquad dG = RT d\ln f$$

where for an ideal gas $f^{ig} = P$. Subtracting Eq. (6.46) from (6.47), the residual Gibbs energy, G^R, can be determined through fugacity:

$$(6.48) \qquad \frac{G^R}{RT} = \frac{G - G^{ig}}{RT} = \ln\frac{f}{P} = \ln\phi$$

TABLE 6.1—*Calculation of thermodynamic properties from cubic equations of state [8].*

	RK EOS	SRK EOS	PR EOS
Equation of state	$Z^3 - Z^2 + (A - B - B^2)Z - AB = 0$		$Z^3 - (1-B)Z^2 + (A - 2B - 3B^2)Z - AB + B^2 + B^3 = 0$
Definition of dimensionless parameters	$Z = \dfrac{PV}{RT} \qquad A = \dfrac{aP}{R^2T^2} \qquad B = \dfrac{bP}{RT}$		
Definition of various parameters	$a_1 = \dfrac{da}{dT} \qquad a_2 = \dfrac{d^2a}{dT^2} \qquad P_1 = \left(\dfrac{\partial P}{\partial T}\right)_V \qquad P_2 = \left(\dfrac{\partial P}{\partial V}\right)_T \qquad P_3 = \int\limits_\infty^V \left(\dfrac{\partial^2 P}{\partial T^2}\right)_V dV$		
Residual heat capacity and heat capacity ratio	$C_P - C_P^{ig} = TP_3 - \dfrac{TP_1^2}{P_2} - R \qquad C_V - C_V^{ig} = TP_3 \qquad \gamma = \dfrac{C_P}{C_V}$		
$\dfrac{H - H^{ig}}{RT}$	$(Z-1) + \dfrac{1}{bRT}(a - Ta_1)\ln\dfrac{Z}{Z+B}$		$(Z-1) + \dfrac{1}{2\sqrt{2}(bRT)}(a - Ta_1) \times \ln\dfrac{Z+B(1-\sqrt{2})}{Z+B(1+\sqrt{2})}$
$\ln\left(\dfrac{f}{P}\right)$	$Z - 1 - \ln(Z - B) + \dfrac{A}{B}\ln\dfrac{Z}{Z+B}$		$Z - 1 - \ln(Z - B) + \dfrac{A}{2\sqrt{2}B}\ln\dfrac{Z+B(1-\sqrt{2})}{Z+B(1+\sqrt{2})}$
P_1	$\dfrac{R}{V-b} - \dfrac{a_1}{V(V+b)}$		$\dfrac{R}{V-b} - \dfrac{a_1}{V^2+2bV-b^2}$
P_2	$\dfrac{-RT}{(V-b)^2} + \dfrac{a(2V+b)}{V^2(V+b)^2}$		$\dfrac{-RT}{(V-b)^2} + \dfrac{2a(V+b)}{V^2(2bV+b^2)^2}$
P_3	$-\dfrac{1}{b}a_2\ln\dfrac{V}{V+b}$		$-\dfrac{1}{2\sqrt{2}b}a_2\ln\dfrac{V+b-\sqrt{2}}{V+b}$
a_1	$-\dfrac{1}{2}\dfrac{a}{T}$	$-\dfrac{a_c\alpha^{1/2}}{T_c T_r^{1/2}}f_\omega$	$-\dfrac{a_c\alpha^{1/2}}{T_c T_r^{1/2}}f_\omega$
a_2	$\dfrac{3}{4}\dfrac{a}{T^2}$	$\dfrac{a_c}{2T_c^2 T_r^{3/2}}f_\omega(1 + f_\omega)$	$\dfrac{a_c}{2T_c^2 T_r^{3/2}}f_\omega(1 + f_\omega)$

$a = a_c\alpha$, where a_c, α, f_ω, and b for RK, SRK, and PR EOS are given in Table 5.1.

The ratio of f/P is a dimensionless parameter called *fugacity coefficient* and it is shown by ϕ:

$$(6.49) \qquad \phi = \frac{f}{P}$$

where for an ideal gas, $\phi = 1$. Once ϕ is known, G can be calculated through Eq. (6.48) and from G and H, S may be calculated from Eq. (6.6). *Vice versa* when H and S are known, G and eventually f can be determined.

6.1.5 General Approach for Property Estimation

Similar to the method used in Example 6.1, every thermodynamic property can be related to PVT relation either at a given T and P or at a given T and V. These relations for $(H - H^{ig})$ are given by Eqs. (6.33) and (6.39). For the residual entropy an equivalent relation in terms of pressure is (see Problem 6.3)

$$(6.50) \qquad (S - S^{ig})_{T,P} = \int\limits_0^P \left[\frac{R}{P} - \left(\frac{\partial V}{\partial T}\right)_P\right]dP$$

This equation can be written in terms of Z as

$$(S - S^{ig})_{T,P} = -R\int\limits_0^P \frac{(Z-1)}{P}dP - RT\int\limits_0^P \left(\frac{\partial Z}{\partial T}\right)_P \frac{dP}{P}$$

$$(6.51) \qquad\qquad\qquad\qquad\qquad \text{(at constant } T)$$

Once residual enthalpy and entropy are calculated, residual Gibbs energy is calculated from the following relation based on Eq. (6.6):

$$(6.52) \qquad (G - G^{ig})_{T,P} = (H - H^{ig})_{T,P} - T(S - S^{ig})_{T,P}$$

Substituting Eqs. (6.33) and (6.50) into Eq. (6.52) and combining with Eq. (6.48) gives the following equation which can be used to calculate fugacity coefficient for a pure component:

$$(6.53) \qquad \ln\phi = \int\limits_0^P (Z-1)\frac{dP}{P}$$

For the residual heat capacity $(C_P - C_P^{ig})$, the relation at a fixed T and P is given by Eq. (6.44). In general when fugacity coefficient is calculated through Eq. (6.53), residual Gibbs energy can be calculated from Eq. (6.48). Properties of ideal gases can be calculated accurately as will be discussed later in this chapter. Once H and G are known, S can be calculated from Eq. (6.6). Therefore, either H and S or H and ϕ are needed to calculate various properties. In this chapter, methods of calculation of H, C_P, and ϕ are presented.

When residual properties are related to PVT, any equation of state may be used to calculate properties of real fluids and departure functions. Calculation of $(H - H^{ig})$, $(C_P - C_P^{ig})$, and $\ln\phi$ from RK, SRK, and PR equations of state are given in Table 6.1. RK and SRK give similar results while the only difference is in parameter a, as given in Table 5.1. EOS parameters needed for use in Table 6.1 are given in Table 5.1. Relations presented in Table 6.1 are applicable to both vapor and liquid phases whenever the EOS can be applied. However, when they are used for the liquid phase, values of Z and V must be obtained for the same phase as discussed in Chapter 5. It should be noted that relations given in Table 6.1 for various properties are based on assuming that parameter b in the corresponding EOS is independent of temperature as for RK, SRK, and PR equations. However, when parameter b is considered temperature-dependent, then its derivative with respect to temperature is not zero and derived relations for residual properties are significantly more complicated than those given in Table 6.1. As will be discussed in the next section, cubic equations do not provide accurate values for enthalpy and heat capacity of fluids unless their constants are adjusted for such calculations.

6.2 GENERALIZED CORRELATIONS FOR CALCULATION OF THERMODYNAMIC PROPERTIES

It is generally believed that cubic equations of state are not suitable for calculation of heat capacity and enthalpy and in some cases give negative heat capacities. Cubic equations are widely used for calculation of molar volume (or density) and fugacity coefficients. Usually BWR or its various modified versions are used to calculate enthalpy and heat capacity. The Lee–Kesler (LK) modification of BWR EOS is given by Eq. (5.109). Upon use of this PVT relation, residual properties can be calculated. For example, by substituting Z from Eq. (5.109) into Eqs. (6.53) and (6.36) the relations for the fugacity coefficient and enthalpy departure are obtained and are given by the following equations [9]:

$$(6.54) \quad \ln\left(\frac{f}{P}\right) = Z - 1 - \ln(Z) + \frac{B}{V_r} + \frac{C}{2V_r^2} + \frac{D}{5V_r^5} + E$$

$$(6.55) \quad \frac{H - H^{ig}}{RT_c} = T_r\left(Z - 1 - \frac{b_2 + 2b_3/T_r + 3b_4/T_r^2}{T_r V_r}\right.$$
$$\left. - \frac{c_2 - 3c_3/T_r^2}{2T_r V_r^2} - \frac{d_2}{5T_r V_r^5} + 3E\right)$$

where parameter E in these equations is given by:

$$E = \frac{c_4}{2T_r^3 \gamma}\left[\beta + 1 - \left(\beta + 1 + \frac{\gamma}{V_r^2}\right)\exp\left(-\frac{\gamma}{V_r^2}\right)\right]$$

The coefficients in the above equations for the simple fluid and reference fluid of n-octane are given in Table 5.8. Similar equations for estimation of $(C_P - C_P^{ig})$, $(C_V - C_V^{ig})$, and $(S - S^{ig})$ are given by Lee and Kesler [9]. To make use of these equations for calculation of properties of all fluids a similar approach as used to calculate Z through Eq. (5.108) is recommended. For practical calculations Eq. (6.55) and other equations for fugacity and heat capacity can be converted into the following corresponding states correlations:

$$(6.56) \quad \left[\frac{H - H^{ig}}{RT_c}\right] = \left[\frac{H - H^{ig}}{RT_c}\right]^{(0)} + \omega\left[\frac{H - H^{ig}}{RT_c}\right]^{(1)}$$

$$(6.57) \quad \left[\frac{C_P - C_P^{ig}}{R}\right] = \left[\frac{C_P - C_P^{ig}}{R}\right]^{(0)} + \omega\left[\frac{C_P - C_P^{ig}}{R}\right]^{(1)}$$

$$(6.58) \quad \left[\ln\left(\frac{f}{P}\right)\right] = \left[\ln\left(\frac{f}{P}\right)\right]^{(0)} + \omega\left[\ln\left(\frac{f}{P}\right)\right]^{(1)}$$

where for convenience Eq. (6.58) may also be written as [1]

$$(6.59) \quad \phi = (\phi^{(0)})(\phi^{(1)})^\omega$$

Simple fluid terms such as $[(H - H^{ig})/RT_c]^{(0)}$ can be estimated from Eq. (6.55) using coefficients given in Table 5.8 for simple fluid. A graphical presentation of $[(H - H^{ig})/RT_c]^{(0)}$ and $[(H - H^{ig})/RT_c]^{(1)}$ is demonstrated in Fig. 6.1 [2].

The correction term $[(H - H^{ig})/RT_c]^{(1)}$ is calculated from the following relation:

$$(6.60) \quad \left[\frac{H - H^{ig}}{RT_c}\right]^{(1)} = \left(\frac{1}{\omega_r}\right)\left\{\left[\frac{H - H^{ig}}{RT_c}\right]^{(r)} - \left[\frac{H - H^{ig}}{RT_c}\right]^{(0)}\right\}$$

where $[(H - H^{ig})/RT_c]^{(r)}$ should be calculated from Eq. (6.55) using coefficients in Table 5.8 for the reference fluid (n-octane). ω_r is the acentric factor of reference fluid in which for n-C_8 the value of 0.3978 was originally used. A similar approach can be used to calculate other thermodynamic

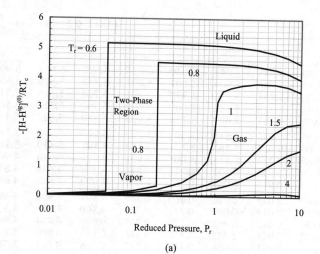

(a)

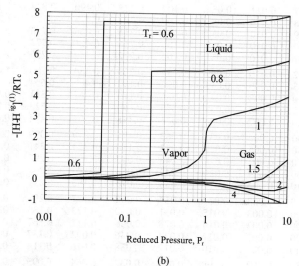

(b)

FIG. 6.1—The Lee–Kesler correlation for (a) $[(H - H^{ig})/RT_c]^{(0)}$ and (b) $[(H - H^{ig})/RT_c]^{(1)}$ in terms of T_r and P_r.

properties. While this method is useful for computer calculations, it is of little use for practical and quick hand calculations. For this reason tabulated values similar to $Z^{(0)}$ and $Z^{(1)}$ are needed. Values of residual enthalpy, heat capacity, and fugacity in dimensionless forms for both $[]^{(0)}$ and $[]^{(1)}$ terms are given by Lee and Kesler [9] and have been included in the API-TDB [5] and other references [1, 2, 10]. These values for enthalpy, heat capacity, and fugacity coefficient are given in Tables 6.2–6.7. In use of values for enthalpy departure it should be noted that for simplicity all values in Tables 6.2 and 6.3 have been multiplied by the negative sign and this is indicated in the titles of these tables. In Tables 6.4 and 6.5, for heat capacity departure there are certain regions of maximum uncertainty that have been specified by the API-TDB [5]. In Table 6.4, when $P_r > 0.9$ and values of $[(C_P - C_P^{ig})/R]^{(0)}$ are greater than 1.6 there is uncertainty as recommended by the API-TDB. In Table 6.5, when $P_r > 0.72$ and values of $[(C_P - C_P^{ig})/R]^{(1)}$ are greater than 2.1 the uncertainty exists as recommended by the API-TDB. In these regions values of heat capacity departure are less accurate. Tables 6.6 and 6.7 give values of $\phi^{(0)}$ and $\phi^{(1)}$ that are calculated from ($\ln\phi^{(0)}$) as given by Smith et al. [1].

TABLE 6.2—*Values of* $-\left[\frac{H-H^0}{RT_c}\right]^{(0)}$ *for use in Eq. (6.56).*

T_r	\multicolumn{15}{c}{P_r}
	0.01	0.05	0.1	0.2	0.4	0.6	0.8	1	1.2	1.5	2	3	5	7	10
0.30	**6.045**	**6.043**	**6.040**	**6.034**	**6.022**	**6.011**	**5.999**	5.987	5.975	5.957	5.927	5.868	5.748	5.628	5.446
0.35	**5.906**	**5.904**	**5.901**	**5.895**	**5.882**	**5.870**	**5.858**	5.845	5.833	5.814	5.783	5.721	5.595	5.469	5.278
0.40	**5.763**	**5.761**	**5.757**	**5.751**	**5.738**	**5.726**	**5.713**	5.700	5.687	5.668	5.636	5.572	5.442	5.311	5.113
0.45	**5.615**	**5.612**	**5.609**	**5.603**	**5.590**	**5.577**	**5.564**	5.551	5.538	5.519	5.486	5.421	5.288	5.154	5.950
0.50	**5.465**	**5.469**	**5.459**	**5.453**	**5.440**	**5.427**	**5.414**	5.401	5.388	5.369	5.336	5.270	5.135	4.999	4.791
0.55	0.032	**5.312**	**5.309**	**5.303**	**5.290**	**5.278**	**5.265**	5.252	5.239	5.220	5.187	5.121	4.986	4.849	4.638
0.60	0.027	**5.162**	**5.159**	**5.153**	**5.141**	**5.129**	**5.116**	5.104	5.091	5.073	5.041	4.976	4.842	4.704	4.492
0.65	0.023	0.118	**5.008**	**5.002**	**4.991**	**4.980**	**4.968**	4.956	4.945	4.927	4.896	4.833	4.702	4.565	4.353
0.70	0.020	0.101	0.213	**4.848**	**4.838**	**4.828**	**4.818**	4.808	4.797	4.781	4.752	4.693	4.566	4.432	4.221
0.75	0.017	0.088	0.183	**4.687**	**4.679**	**4.672**	**4.664**	4.655	4.646	4.632	4.607	4.554	4.434	4.303	4.095
0.80	0.015	0.078	0.160	0.345	**4.507**	**4.504**	**4.499**	4.494	4.488	4.478	4.459	4.413	4.303	4.178	3.974
0.85	0.014	0.069	0.141	0.300	**4.309**	**4.313**	**4.316**	4.316	4.316	4.312	4.302	4.269	4.173	4.056	3.857
0.90	0.012	0.062	0.126	0.264	0.596	**4.074**	**4.094**	4.108	4.118	4.127	4.132	4.119	4.043	3.935	3.744
0.93	0.011	0.058	0.118	0.246	0.545	0.960	**3.920**	3.953	3.976	4.000	4.020	4.024	3.963	3.863	3.678
0.95	0.011	0.056	0.113	0.235	0.516	0.885	**3.763**	3.825	3.865	3.904	3.940	3.958	3.910	3.815	3.634
0.97	0.011	0.054	0.109	0.225	0.490	0.824	1.356	3.658	3.732	3.796	3.853	3.890	3.856	3.767	3.591
0.98	0.010	0.053	0.107	0.221	0.478	0.797	1.273	3.544	3.652	3.736	3.806	3.854	3.829	3.743	3.569
0.99	0.010	0.052	0.105	0.216	0.466	0.773	1.206	3.376	3.558	3.670	3.758	3.818	3.801	3.719	3.548
1.00	0.010	0.052	0.105	0.216	0.466	0.773	1.206	2.593	3.558	3.670	3.758	3.818	3.801	3.719	3.548
1.01	0.010	0.051	0.103	0.212	0.455	0.750	1.151	1.796	3.441	3.598	3.706	3.782	3.774	3.695	3.526
1.02	0.010	0.049	0.099	0.203	0.434	0.708	1.060	1.627	3.039	3.422	3.595	3.705	3.718	3.647	3.484
1.05	0.009	0.046	0.094	0.192	0.407	0.654	0.955	1.359	2.034	3.030	3.398	3.583	3.632	3.575	3.420
1.10	0.008	0.042	0.086	0.175	0.367	0.581	0.827	1.120	1.487	2.203	2.965	3.353	3.484	3.453	3.315
1.15	0.008	0.039	0.079	0.160	0.334	0.523	0.732	0.968	1.239	1.719	2.479	3.091	3.329	3.329	3.211
1.20	0.007	0.036	0.073	0.148	0.305	0.474	0.657	0.857	1.076	1.443	2.079	2.807	3.166	3.202	3.107
1.30	0.006	0.031	0.063	0.127	0.259	0.399	0.545	0.698	0.860	1.116	1.560	2.274	2.825	2.942	2.899
1.40	0.005	0.027	0.055	0.110	0.224	0.341	0.463	0.588	0.716	0.915	1.253	1.857	2.486	2.679	2.692
1.50	0.005	0.024	0.048	0.097	0.196	0.297	0.400	0.505	0.611	0.774	1.046	1.549	2.175	2.421	2.486
1.60	0.004	0.021	0.043	0.086	0.173	0.261	0.350	0.440	0.531	0.667	0.894	1.318	1.904	2.177	2.285
1.70	0.004	0.019	0.038	0.076	0.153	0.231	0.309	0.387	0.446	0.583	0.777	1.139	1.672	1.953	2.091
1.80	0.003	0.017	0.034	0.068	0.137	0.206	0.275	0.344	0.413	0.515	0.683	0.996	1.476	1.751	1.908
1.90	0.003	0.015	0.031	0.062	0.123	0.185	0.246	0.307	0.368	0.458	0.606	0.880	1.309	1.571	1.736
2.00	0.003	0.014	0.028	0.056	0.111	0.167	0.222	0.276	0.330	0.411	0.541	0.782	1.167	1.411	1.577
2.20	0.002	0.012	0.023	0.046	0.092	0.137	0.182	0.226	0.269	0.334	0.437	0.629	0.937	1.143	1.295
2.40	0.002	0.010	0.019	0.038	0.076	0.114	0.150	0.187	0.222	0.275	0.359	0.513	0.761	0.929	1.058
2.60	0.002	0.008	0.016	0.032	0.064	0.095	0.125	0.155	0.185	0.228	0.297	0.422	0.621	0.756	0.858
2.80	0.001	0.007	0.014	0.027	0.054	0.080	0.105	0.130	0.154	0.190	0.246	0.348	0.508	0.614	0.689
3.00	0.001	0.006	0.011	0.023	0.045	0.067	0.088	0.109	0.129	0.159	0.205	0.288	0.415	0.495	0.545
3.50	0.001	0.004	0.007	0.015	0.029	0.043	0.056	0.069	0.081	0.099	0.127	0.174	0.239	0.270	0.264
4.00	0.000	0.002	0.005	0.009	0.017	0.026	0.033	0.041	0.048	0.058	0.072	0.095	0.116	0.110	0.061

Taken with permission from Ref. [9]. The value at the critical point ($T_r = P_r = 1$) is taken from the API-TDB [5]. Bold numbers indicate liquid region.

For low and moderate pressures where truncated virial equation in the form of Eq. (5.113) is valid, the relation for fugacity coefficient can be derived from Eq. (6.53) as

$$(6.61) \qquad \ln(\phi) = \frac{BP}{RT}$$

This relation may also be written as

$$(6.62) \qquad \phi = \exp\left[\frac{P_r}{T_r}\left(\frac{BP_c}{RT_c}\right)\right]$$

where (BP_c/RT_c) can be calculated from Eq. (5.71) or (5.72). Similarly enthalpy departure based on the truncated virial equation is given as [1]

$$(6.63) \qquad \frac{H - H^{ig}}{RT_c} = P_r\left[B^{(0)} - T_r\frac{dB^{(0)}}{dT_r} + \omega\left(B^{(1)} - T_r\frac{dB^{(1)}}{dT_r}\right)\right]$$

where $B^{(0)}$ and $B^{(1)}$ are given by Eq. (5.72) with $dB^{(0)}/dT_r = 0.675/T_r^{2.6}$ and $dB^{(1)}/dT_r = 0.722/T_r^{5.2}$. Obviously $B^{(0)}$ and $B^{(1)}$

may be used from Eq. (5.71), but corresponding derivatives must be used. The above equation may be applied at the same region that Eq. (5.75) or (5.114) were applicable, that is, $V_r > 2.0$ or $T_r > 0.686 + 0.439 P_r$ [2].

For real gases that follow truncated virial equation with three terms (coefficients D and higher assumed zero in Eq. 5.76), the relations for C_P and C_V are givens as

$$\frac{C_P - C_P^{ig}}{R} = -\left[\frac{T^2 B''}{V} - \frac{(B - T B'')^2 - C + T C' - T^2 C''/2}{V^2}\right]$$
$$(6.64)$$

$$(6.65) \qquad \frac{C_V - C_V^{ig}}{R} = -\left[\frac{2T B' + T^2 B''}{V} - \frac{T C' + T^2 C''/2}{V^2}\right]$$

where B' and C' are the first-order derivatives of B and C with respect to temperature, while B'' and C'' are the second-order derivatives of B and C with respect to temperature.

TABLE 6.3—*Values of* $-\left[\frac{H-H^{ig}}{RT_c}\right]^{(1)}$ *for use in Eq. (6.56).*

T_r	0.01	0.05	0.1	0.2	0.4	0.6	0.8	1	1.2	1.5	2	3	5	7	10
								P_r							
0.30	**11.098**	**11.096**	**11.095**	**11.091**	**11.083**	**11.076**	**11.069**	11.062	11.055	11.044	11.027	10.992	10.935	10.872	10.781
0.35	**10.656**	**10.655**	**10.654**	**10.653**	**10.650**	**10.646**	**10.643**	10.640	10.637	10.632	10.624	10.609	10.581	10.554	10.529
0.40	**10.121**	**10.121**	**10.121**	**10.121**	**10.121**	**10.121**	**10.121**	10.121	10.121	10.121	10.122	10.123	10.128	10.135	10.150
0.45	**9.515**	**9.515**	**9.516**	**9.516**	**9.519**	**9.521**	**9.523**	9.525	9.527	9.531	9.537	9.549	9.576	9.611	9.663
0.50	**8.868**	**8.869**	**8.870**	**8.870**	**8.876**	**8.880**	**8.884**	8.888	8.892	8.899	8.909	8.932	8.978	9.030	9.111
0.55	0.080	**8.211**	**8.212**	**8.215**	**8.221**	**8.226**	**8.232**	8.238	8.243	8.252	8.267	8.298	8.360	8.425	8.531
0.60	0.059	**7.568**	**7.570**	**7.573**	**7.579**	**7.585**	**7.591**	7.596	7.603	7.614	7.632	7.669	7.745	7.824	7.950
0.65	0.045	0.247	**6.949**	**6.952**	**6.959**	**6.966**	**6.973**	6.980	6.987	6.997	7.017	7.059	7.147	7.239	7.381
0.70	0.034	0.185	0.415	**6.360**	**6.367**	**6.373**	**6.381**	6.388	6.395	6.407	6.429	6.475	6.574	6.677	6.837
0.75	0.027	0.142	0.306	**5.796**	**5.802**	**5.809**	**5.816**	5.824	5.832	5.845	5.868	5.918	6.027	6.142	6.318
0.80	0.021	0.110	0.234	0.542	**5.266**	**5.271**	**5.278**	5.285	5.293	5.306	5.330	5.385	5.506	5.632	5.824
0.85	0.017	0.087	0.182	0.401	**4.753**	**4.754**	**4.758**	4.763	4.771	4.784	4.810	4.872	5.008	5.149	5.358
0.90	0.014	0.070	0.144	0.308	0.751	**4.254**	**4.248**	4.249	4.255	4.268	4.298	4.371	4.530	4.688	4.916
0.93	0.012	0.061	0.126	0.265	0.612	1.236	**3.942**	3.934	3.937	3.951	3.987	4.073	4.251	4.422	4.662
0.95	0.011	0.056	0.115	0.241	0.542	0.994	**3.737**	3.712	3.713	3.730	3.773	3.873	4.068	4.248	4.497
0.97	0.010	0.052	0.105	0.219	0.483	0.837	1.616	3.470	3.467	3.492	3.551	3.670	3.885	4.077	4.336
0.98	0.010	0.050	0.101	0.209	0.457	0.776	1.324	3.332	3.327	3.363	3.434	3.568	3.795	3.992	4.257
0.99	0.009	0.048	0.097	0.200	0.433	0.722	1.154	3.164	3.164	3.223	3.313	3.464	3.705	3.909	4.178
1.00	0.009	0.046	0.093	0.191	0.410	0.675	1.034	2.348	2.952	3.065	3.186	3.358	3.615	3.825	4.100
1.01	0.009	0.044	0.089	0.183	0.389	0.632	0.940	1.375	2.595	2.880	3.051	3.251	3.525	3.742	4.023
1.02	0.008	0.042	0.085	0.175	0.370	0.594	0.863	1.180	1.723	2.650	2.906	3.142	3.435	3.661	3.947
1.05	0.007	0.037	0.075	0.153	0.318	0.498	0.691	0.877	0.878	1.496	2.381	2.800	3.167	3.418	3.722
1.10	0.006	0.030	0.061	0.123	0.251	0.381	0.507	0.617	0.673	0.617	1.261	2.167	2.720	3.023	3.362
1.15	0.005	0.025	0.050	0.099	0.199	0.296	0.385	0.459	0.503	0.487	0.604	1.497	2.275	2.641	3.019
1.20	0.004	00.020	0.040	0.080	0.158	0.232	0.297	0.349	0.381	0.381	0.361	0.934	1.840	2.273	2.692
1.30	0.003	0.013	0.026	0.052	0.100	0.142	0.177	0.203	0.218	0.218	0.178	0.300	1.066	1.592	2.086
1.40	0.002	0.008	0.016	0.032	0.060	0.083	0.100	0.111	0.115	0.108	0.070	0.044	0.504	1.012	1.547
1.50	0.001	0.005	0.009	0.018	0.032	0.042	0.048	0.049	0.046	0.032	−0.008	−0.078	0.142	0.556	1.080
1.60	0.000	0.002	0.004	0.007	0.012	0.013	0.011	0.005	−0.004	−0.023	−0.065	−0.151	−0.082	0.217	0.689
1.70	0.000	0.000	0.000	0.000	−0.003	−0.009	−0.017	−0.027	−0.040	−0.063	−0.109	−0.202	−0.223	−0.028	0.369
1.80	−0.000	−0.001	−0.003	−0.006	−0.015	−0.025	−0.037	−0.051	−0.067	−0.094	−0.143	−0.241	−0.317	−0.203	0.112
1.90	−0.001	−0.003	−0.005	−0.011	−0.023	−0.037	−0.053	−0.070	−0.088	−0.117	−0.169	−0.271	−0.381	−0.330	−0.092
2.00	−0.001	−0.003	−0.007	−0.015	−0.030	−0.047	−0.065	−0.085	−0.105	−0.136	−0.190	−0.295	−0.428	−0.424	−0.255
2.20	−0.001	−0.005	−0.010	−0.020	−0.040	−0.062	−0.083	−0.106	−0.128	−0.163	−0.221	−0.331	−0.493	−0.551	−0.489
2.40	−0.001	−0.006	−0.012	−0.023	−0.047	−0.071	−0.095	−0.120	−0.144	−0.181	−0.242	−0.356	−0.535	−0.631	−0.645
2.60	−0.001	−0.006	−0.013	−0.026	−0.052	−0.078	−0.104	−0.130	−0.156	−0.194	−0.257	−0.376	−0.567	−0.687	−0.754
2.80	−0.001	−0.007	−0.014	−0.028	−0.055	−0.082	−0.110	−0.137	−0.164	−0.204	−0.269	−0.391	−0.591	−0.729	−0.836
3.00	−0.001	−0.007	−0.014	−0.029	−0.058	−0.086	−0.114	−0.142	−0.170	−0.211	−0.278	−0.403	−0.611	−0.763	−0.899
3.50	−0.002	−0.008	−0.016	−0.031	−0.062	−0.092	−0.122	−0.152	−0.181	−0.224	−0.294	−0.425	−0.650	−0.827	−1.015
4.00	−0.002	−0.008	−0.016	−0.032	−0.064	−0.096	−0.127	−0.158	−0.188	−0.233	−0.306	−0.442	−0.680	−0.874	−1.097

Taken with permission from Ref. [9]. The value at the critical point ($T_r = P_r = 1$) is taken from the API-TDB [5]. Bold numbers indicate liquid region.

6.3 PROPERTIES OF IDEAL GASES

Calculation of thermodynamic properties through the methods outlined above requires properties of ideal gases. Based on definition of ideal gases, U^{ig}, H^{ig}, and C_P^{ig} are functions of only temperature. Kinetic theory shows that the molar translational energy of a monoatomic ideal gas is $\frac{3}{2}RT$, where R is the universal gas constant and T is the absolute temperature [10]. Since for ideal gases the internal energy is independent of pressure thus $U^{ig} = \frac{3}{2}RT$. This leads to $H^{ig} = \frac{5}{2}RT$, $C_P^{ig} = \frac{5}{2}R$, $C_V^{ig} = \frac{3}{2}R$, and $\gamma = \frac{C_P^{ig}}{C_V^{ig}} = \frac{5}{2} = 1.667$. Ideal gas heat capacity of monoatomic gases such as argon, helium, etc. are constant with respect to temperature [10]. Similarly for diatomic gases such as N_2, O_2, air, etc., $H^{ig} = \frac{7}{2}R$, which leads to $C_P^{ig} = \frac{7}{2}R$, $C_V^{ig} = \frac{5}{2}R$, and $\gamma = 7/5 = 1.4$. In fact variation of heat capacities of ideal diatomic gases with temperature is very moderate. For multiatomic molecules such as hydrocarbons, ideal gas properties do

change with temperature appreciably. As the number of atoms in a molecule increases, dependency of ideal gas properties to temperature also increases. Data on properties of ideal gases for a large number of hydrocarbons have been reported by the API-TDB [5]. These data for ideal gas heat capacity have been correlated to temperature in the following form [5]:

$$(6.66) \qquad \frac{C_P^{ig}}{R} = A + BT + CT^2 + DT^3 + ET^4$$

where R is the gas constant (Section 1.7.24), C_P^{ig} is the molar heat capacity in the same unit as R, and T is the absolute temperature in kelvin. Values of the constants for a number of nonhydrocarbon gases as well as some selected hydrocarbons are given in Table 6.8. The temperature range at which these constants can be used is also given for each compound in Table 6.8. For a compound with known chemical structure, ideal gas heat capacity is usually predicted from group

TABLE 6.4—*Values of* $\left[\frac{C_p - C_p^{ig}}{R}\right]^{(0)}$ *for use in Eq. (6.57).*

T_r	P_r														
	0.01	0.05	0.1	0.2	0.4	0.6	0.8	1	1.2	1.5	2	3	5	7	10
0.30	**2.805**	**2.807**	**2.809**	**2.814**	**2.830**	**2.842**	**2.854**	2.866	2.878	2.896	2.927	2.989	3.122	3.257	3.466
0.35	**2.808**	**2.810**	**2.812**	**2.815**	**2.823**	**2.835**	**2.844**	2.853	2.861	2.875	2.897	2.944	3.042	3.145	3.313
0.40	**2.925**	**2.926**	**2.926**	**2.928**	**2.933**	**2.935**	**2.940**	2.945	2.951	2.956	2.965	2.979	3.014	3.085	3.164	3.293
0.45	**2.989**	**2.990**	**2.990**	**2.991**	**2.993**	**2.995**	**2.997**	2.999	3.002	3.006	3.014	3.032	3.079	3.135	3.232
0.50	**3.006**	**3.005**	**3.004**	**3.003**	**3.001**	**3.000**	2.998	2.997	2.996	2.995	2.995	2.999	3.019	3.054	3.122
0.55	0.118	**3.002**	**3.000**	2.997	2.990	2.984	2.978	2.973	2.968	2.961	2.951	2.938	2.934	2.947	2.988
0.60	0.089	**3.009**	**3.006**	2.999	2.986	2.974	2.963	2.952	2.942	2.927	2.907	2.874	2.840	2.831	2.847
0.65	0.069	0.387	**3.047**	3.036	3.014	2.993	2.973	2.955	2.938	2.914	2.878	2.822	2.753	2.720	2.709
0.70	0.054	0.298	0.687	**3.138**	3.099	3.065	3.033	3.003	2.975	2.937	2.881	2.792	2.681	2.621	2.582
0.75	0.044	0.236	0.526	**3.351**	3.284	3.225	3.171	3.122	3.076	3.015	2.928	2.795	2.629	2.537	2.469
0.80	0.036	0.191	0.415	1.032	**3.647**	**3.537**	**3.440**	3.354	3.277	3.176	3.038	2.838	2.601	2.473	2.373
0.85	0.030	0.157	0.336	0.794	**4.404**	**4.158**	**3.957**	3.790	3.647	3.470	3.240	2.931	2.599	2.427	2.292
0.90	0.025	0.131	0.277	0.633	1.858	**5.679**	**5.095**	4.677	4.359	4.000	3.585	3.096	2.626	2.399	2.227
0.93	0.023	0.118	0.249	0.560	1.538	4.208	**6.720**	5.766	5.149	4.533	3.902	3.236	2.657	2.392	2.195
0.95	0.021	0.111	0.232	0.518	1.375	3.341	**9.316**	7.127	6.010	5.050	4.180	3.351	2.684	2.391	2.175
0.97	0.020	0.104	0.217	0.480	1.240	2.778	9.585	10.011	7.451	5.785	4.531	3.486	2.716	2.393	2.159
0.98	0.019	0.101	0.210	0.463	1.181	2.563	7.350	13.270	8.611	6.279	4.743	3.560	2.733	2.395	2.151
0.99	0.019	0.098	0.204	0.447	1.126	2.378	6.038	21.948	10.362	6.897	4.983	3.641	2.752	2.398	2.144
1.00	0.018	0.095	0.197	0.431	1.076	2.218	5.156	∞	13.182	7.686	5.255	3.729	2.773	2.401	2.138
1.01	0.018	0.092	0.191	0.417	1.029	2.076	4.516	22.295	18.967	8.708	5.569	3.821	2.794	2.405	2.131
1.02	0.017	0.089	0.185	0.403	0.986	1.951	4.025	13.183	31.353	10.062	5.923	3.920	2.816	2.408	2.125
1.05	0.016	0.082	0.169	0.365	0.872	1.648	3.047	6.458	20.234	16.457	7.296	4.259	2.891	2.425	2.110
1.10	0.014	0.071	0.147	0.313	0.724	1.297	2.168	3.649	6.510	13.256	9.787	4.927	3.033	2.462	2.093
1.15	0.012	0.063	0.128	0.271	0.612	1.058	1.670	2.553	3.885	6.985	9.094	5.535	3.186	2.508	2.083
1.20	0.011	0.055	0.113	0.237	0.525	0.885	1.345	1.951	2.758	4.430	6.911	5.710	3.326	2.555	2.079
1.30	0.009	0.044	0.089	0.185	0.400	0.651	0.946	1.297	1.711	2.458	3.850	4.793	3.452	2.628	2.077
1.40	0.007	0.036	0.072	0.149	0.315	0.502	0.711	0.946	1.208	1.650	2.462	3.573	3.282	2.626	2.068
1.50	0.006	0.029	0.060	0.122	0.255	0.399	0.557	0.728	0.912	1.211	1.747	2.647	2.917	2.525	2.038
1.60	0.005	0.025	0.050	0.101	0.210	0.326	0.449	0.580	0.719	0.938	1.321	2.016	2.508	2.347	1.978
1.70	0.004	0.021	0.042	0.086	0.176	0.271	0.371	0.475	0.583	0.752	1.043	1.586	2.128	2.130	1.889
1.80	0.004	0.018	0.036	0.073	0.150	0.229	0.311	0.397	0.484	0.619	0.848	1.282	1.805	1.907	1.778
1.90	0.003	0.016	0.031	0.063	0.129	0.196	0.265	0.336	0.409	0.519	0.706	1.060	1.538	1.696	1.656
2.00	0.003	0.014	0.027	0.055	0.112	0.170	0.229	0.289	0.350	0.443	0.598	0.893	1.320	1.505	1.531
2.20	0.002	0.011	0.021	0.043	0.086	0.131	0.175	0.220	0.265	0.334	0.446	0.661	0.998	1.191	1.292
2.40	0.002	0.009	0.017	0.034	0.069	0.104	0.138	0.173	0.208	0.261	0.347	0.510	0.779	0.956	1.086
2.60	0.001	0.007	0.014	0.028	0.056	0.084	0.112	0.140	0.168	0.210	0.278	0.407	0.624	0.780	0.917
2.80	0.001	0.006	0.012	0.023	0.046	0.070	0.093	0.116	0.138	0.172	0.227	0.332	0.512	0.647	0.779
3.00	0.001	0.005	0.010	0.020	0.039	0.058	0.078	0.097	0.116	0.144	0.190	0.277	0.427	0.545	0.668
3.50	0.001	0.003	0.007	0.013	0.027	0.040	0.053	0.066	0.079	0.098	0.128	0.187	0.289	0.374	0.472
4.00	0.000	0.002	0.005	0.010	0.019	0.029	0.038	0.048	0.057	0.071	0.093	0.135	0.209	0.272	0.350

Taken with permission from Ref. [9]. The value at the critical point ($T_r = P_r = 1$) is taken from the API-TDB [5]. Bold numbers indicate liquid region.

contribution methods [4, 11]. Once C_P^{ig} is known, C_V^{ig}, H^{ig}, and S^{ig} can be determined from the following relations:

$$(6.67) \quad \frac{C_V^{ig}}{R} = \frac{C_P^{ig}}{R} - 1 = A - 1 + BT + CT^2 + DT^3 + ET^4$$

$$(6.68) \quad H^{ig} = A_H + AT + \frac{B}{2}T^2 + \frac{C}{3}T^3 + \frac{D}{4}T^4 + \frac{E}{5}T^5$$

$$(6.69) \quad S^{ig} = A_S + A\ln T + BT + \frac{C}{2}T^2 + \frac{D}{3}T^3 + \frac{E}{4}T^4$$

The relation for C_V^{ig} is obtained through Eqs. (6.23) and (6.66), while relations for H^{ig} and S^{ig} have been obtained from Eqs. (6.21) and (6.32), respectively. Constants A_H and A_S are obtained from integration of relations for dH^{ig} and dS^{ig} and can be determined based on the reference state for ideal gas enthalpy and entropy. These parameters are not necessary for calculation of H and S as they are omitted during calculations with respect to the arbitrary reference state chosen for H and S. Usually the choice of reference state is on H and not on H^{ig}. For example, Lenoir and Hipkin [12] reported experimental data on enthalpy of some petroleum fractions with reference

state of saturated liquid at 75°F at which $H = 0$. In steam tables where properties of liquid water and steam are reported [1] the reference state at which $H = S = 0$ is saturated liquid at 0.01°C. Therefore there is no need for the values of integration constants A_H and A_S in Eqs. (6.68) and (6.69) as they cancel in the course of calculations. There are several other forms of Eq. (6.66) for C_P^{ig}, as an example the following simple form is given for ideal gas heat capacity of water [1]:

$$(6.70) \quad \frac{C_P^{ig}}{R} = 3.47 + 1.45 \times 10^{-3}T + 0.121 \times 10^5 T^{-2}$$

another relation for water is given by DIPPR [13]:

$$(6.71) \quad \frac{C_P^{ig}}{R} = 4.0129 + 3.222\left[\frac{2610.5/T}{\sinh(2610.5/T)}\right]^2 + 1.07\left[\frac{1169/T}{\cosh(1169/T)}\right]^2$$

where in both equations T is in kelvin and they are valid up to 2000°C. A graphical comparison of C_P^{ig}/R for water from Eqs. (6.66), (6.70), and (6.71) is shown in Fig. 6.2. Equations (6.66)

TABLE 6.5—*Values of* $\left[\frac{C_p-C_p^{ig}}{R}\right]^{(1)}$ *for use in Eq. (6.57).*

T_r	0.01	0.05	0.1	0.2	0.4	0.6	0.8	1	1.2	1.5	2	3	5	7	10
0.30	**8.462**	**8.445**	**8.424**	**8.381**	**8.281**	**8.192**	**8.102**	8.011	7.921	7.785	7.558	7.103	6.270	5.372	4.020
0.35	**9.775**	**9.762**	**9.746**	**9.713**	**9.646**	**9.568**	**9.499**	9.430	9.360	9.256	9.080	8.728	8.013	7.290	6.285
0.40	**11.494**	**11.484**	**11.471**	**11.438**	**11.394**	**11.343**	**11.291**	11.240	11.188	11.110	10.980	10.709	10.170	9.625	8.803
0.45	**12.651**	**12.643**	**12.633**	**12.613**	**12.573**	**12.532**	**12.492**	12.451	12.409	12.347	12.243	12.029	11.592	11.183	10.533
0.50	**13.111**	**13.106**	**13.099**	**13.084**	**13.055**	**13.025**	**12.995**	12.964	12.933	12.886	12.805	12.639	12.288	11.946	11.419
0.55	0.511	**13.035**	**13.030**	**13.021**	**13.002**	**25.981**	**12.961**	12.939	12.917	12.882	12.823	12.695	12.407	12.103	11.673
0.60	0.345	**12.679**	**12.675**	**12.668**	**12.653**	**12.637**	**12.620**	12.589	12.574	12.550	12.506	12.407	12.165	11.905	11.526
0.65	0.242	1.518	**12.148**	**12.145**	**12.137**	**12.128**	**12.117**	12.105	12.092	12.060	12.026	11.943	11.728	11.494	11.141
0.70	0.174	1.026	2.698	**11.557**	**11.564**	**11.563**	**11.559**	11.553	11.536	11.524	11.495	11.416	11.208	10.985	10.661
0.75	0.129	0.726	1.747	**10.967**	**10.995**	**11.011**	**11.019**	11.024	11.022	11.013	10.986	10.898	10.677	10.448	10.132
0.80	0.097	0.532	1.212	3.511	**10.490**	**10.536**	**10.566**	10.583	10.590	10.587	10.556	10.446	10.176	9.917	9.591
0.85	0.075	0.399	0.879	2.247	**9.999**	**10.153**	**10.245**	10.297	10.321	10.324	10.278	10.111	9.740	9.433	9.075
0.90	0.058	0.306	0.658	1.563	5.486	**9.793**	**10.180**	10.349	10.409	10.401	10.279	9.940	9.389	8.999	8.592
0.93	0.050	0.263	0.560	1.289	3.890	...	**10.285**	10.769	10.875	10.801	10.523	9.965	9.225	8.766	8.322
0.95	0.046	0.239	0.505	1.142	3.215	9.389	**9.993**	11.420	11.607	11.387	10.865	10.055	9.136	8.621	8.152
0.97	0.042	0.217	0.456	1.018	2.712	6.588	...	13.001	...	12.498	11.445	10.215	9.061	8.485	7.986
0.98	0.040	0.207	0.434	0.962	2.506	5.711	*20.918*	*14.884*	*14.882*	*13.420*	11.856	10.323	9.037	8.420	7.905
0.99	0.038	0.198	0.414	0.863	2.324	5.027	...	...	...	...	12.388	10.457	9.011	8.359	7.826
1.00	0.037	0.189	0.394	0.863	2.162	4.477	10.511	∞	*25.650*	*16.895*	13.081	10.617	8.990	8.293	7.747
1.01	0.035	0.181	0.376	0.819	2.016	4.026	8.437	...	...	...	...	10.805	8.973	8.236	7.670
1.02	0.034	0.173	0.359	0.778	1.884	3.648	7.044	*15.109*	*115.101*	*26.192*	*15.095*	11.024	8.960	8.182	7.595
1.05	0.30	0.152	0.313	0.669	1.559	2.812	4.679	7.173	2.277	...	...	11.852	8.939	8.018	7.377
1.10	0.024	0.123	0.252	0.528	1.174	1.968	2.919	3.877	4.002	3.927	...	...	8.933	7.759	7.031
1.15	0.020	0.101	0.205	0.424	0.910	1.460	2.048	2.587	2.844	2.236	7.716	12.812	8.849	7.504	6.702
1.20	0.016	0.083	0.168	0.345	0.722	1.123	1.527	1.881	2.095	1.962	2.965	9.494	8.508	7.206	6.384
1.30	0.012	0.058	0.118	0.235	0.476	0.715	0.938	1.129	1.264	1.327	1.288	3.855	6.758	6.365	5.735
1.40	0.008	0.042	0.083	0.166	0.329	0.484	0.624	0.743	0.833	0.904	0.905	1.652	4.524	5.193	5.035
1.50	0.006	0.030	0.061	0.120	0.235	0.342	0.437	0.517	0.580	0.639	0.666	0.907	2.823	3.944	4.289
1.60	0.005	0.023	0.045	0.089	0.173	0.249	0.317	0.374	0.419	0.466	0.499	0.601	1.755	2.871	3.545
1.70	0.003	0.017	0.034	0.068	0.130	0.187	0.236	0.278	0.312	0.349	0.380	0.439	1.129	2.060	2.867
1.80	0.003	0.013	0.027	0.052	0.100	0.143	0.180	0.212	0.238	0.267	0.296	0.337	0.764	1.483	2.287
1.90	0.002	0.011	0.021	0.041	0.078	0.111	0.140	0.164	0.185	0.209	0.234	0.267	0.545	1.085	1.817
2.00	0.002	0.008	0.017	0.032	0.062	0.088	0.110	0.130	0.146	0.166	0.187	0.217	0.407	0.812	1.446
2.20	0.001	0.005	0.011	0.021	0.042	0.057	0.072	0.085	0.096	0.110	0.126	0.150	0.256	0.492	0.941
2.40	0.001	0.004	0.007	0.014	0.028	0.039	0.049	0.058	0.066	0.076	0.089	0.109	0.180	0.329	0.644
2.60	0.001	0.003	0.005	0.010	0.020	0.028	0.035	0.042	0.048	0.056	0.066	0.084	0.137	0.239	0.466
2.80	0.000	0.002	0.004	0.008	0.014	0.021	0.026	0.031	0.036	0.042	0.051	0.067	0.110	0.187	0.356
3.00	0.000	0.001	0.003	0.006	0.011	0.016	0.020	0.024	0.028	0.033	0.041	0.055	0.092	0.153	0.285
3.50	0.000	0.001	0.002	0.003	0.006	0.009	0.012	0.015	0.017	0.021	0.026	0.038	0.067	0.108	0.190
4.00	0.000	0.001	0.001	0.001	0.002	0.004	0.006	0.008	0.010	0.012	0.015	0.019	0.029	0.054	0.146

Taken with permission from Ref. [9]. The value at the critical point ($T_r = P_r = 1$) is taken from the API-TDB [5]. Bold numbers indicate liquid region.

and (6.71) are almost identical and Eq. (6.70) is not valid at very low temperatures. The most accurate formulation and tabulation of properties of water and steam is made by IAPWS [14].

To calculate ideal gas properties of petroleum fractions, the pseudocomponent method discussed in Section 3.3.4 may be used. Kesler and Lee [15] provide an equation for direct calculation of ideal gas heat capacity of petroleum fractions in terms of Watson K_W, and acentric factor, ω:

$$C_P^{ig} = M\left[A_0 + A_1 T + A_2 T^2 - C\left(B_0 + B_1 T + B_2 T^2\right)\right]$$
$$A_0 = -1.41779 + 0.11828 K_W$$
$$A_1 = -\left(6.99724 - 8.69326 K_W + 0.27715 K_W^2\right) \times 10^{-4}$$
$$A_2 = -2.2582 \times 10^{-6}$$
$$(6.72)\quad B_0 = 1.09223 - 2.48245\omega$$
$$B_1 = -(3.434 - 7.14\omega) \times 10^{-3}$$
$$B_2 = -(7.2661 - 9.2561\omega) \times 10^{-7}$$
$$C = \left[\frac{(12.8 - K_W) \times (10 - K_W)}{10\omega}\right]^2$$

where C_P^{ig} is in J/mol·K, M is the molecular weight (g/mol), T is in kelvin, K_W is defined by Eq. (2.13), and ω may be determined from Eq. (2.10). Tsonopoulos et al. [16] suggested that the correction term C in the above equation should equal to zero when K_W is less than 10 or greater than 12.8. But our evaluations show that the equation in its original form predicts values of C_P^{ig} for hydrocarbons in this range of K_W close to those reported by DIPPR [13]. This equation may also be applied to pure hydrocarbons with carbon number greater than or equal to C_5. Ideal gas heat capacities of several hydrocarbons from paraffinic group predicted from Eqs. (6.66) and (6.72) are shown in Fig. 6.3. As expected heat capacity and enthalpy increase with carbon number or molecular weight. Equation (6.72) generally predicts C_P^{ig} of pure hydrocarbons with errors of 1–2% as evaluated by Kesler and Lee [15] and can be used in the temperature range of 255–922 K (0–1200°F). There are similar other correlations for estimation of ideal gas heat capacity of natural gases and petroleum fractions [17, 18]. The relation reported by Firoozabadi [17] for calculation of heat capacity of natural gases is in the form

TABLE 6.6—*Values of $\phi^{(0)}$ for use in Eq. (6.59).*

T_r	P_r														
	0.01	0.05	0.1	0.2	0.4	0.6	0.8	1	1.2	1.5	2	3	5	7	10
0.30	**0.0002**	**0.0000**	**0.0000**	**0.0000**	**0.0000**	**0.0000**	**0.0000**	0.0000	0.0000	0.0000	0.0000	0.0000	0.0000	0.0000	0.0000
0.35	**0.0094**	**0.0007**	**0.0009**	**0.0002**	**0.0001**	**0.0001**	**0.0001**	0.0000	0.0000	0.0000	0.0000	0.0000	0.0000	0.0000	0.0000
0.40	**0.0272**	**0.0055**	**0.0028**	**0.0014**	**0.0007**	**0.0005**	**0.0004**	0.0003	0.0003	0.0003	0.0002	0.0002	0.0002	0.0002	0.0003
0.45	**0.1921**	**0.0266**	**0.0195**	**0.0069**	**0.0096**	**0.0025**	**0.0020**	0.0016	0.0014	0.0012	0.0010	0.0008	0.0008	0.0009	0.0012
0.50	**0.4529**	**0.0912**	**0.0461**	**0.0295**	**0.0122**	**0.0085**	**0.0067**	0.0055	0.0048	0.0041	0.0034	0.0028	0.0025	0.0027	0.0034
0.55	0.9817	**0.2492**	**0.1227**	**0.0625**	**0.0925**	**0.0225**	**0.0176**	0.0146	0.0127	0.0107	0.0089	0.0072	0.0063	0.0066	0.0080
0.60	0.9840	**0.5989**	**0.2716**	**0.1984**	**0.0718**	**0.0497**	**0.0986**	0.0321	0.0277	0.0234	0.0193	0.0154	0.0132	0.0135	0.0160
0.65	0.9886	0.9419	**0.5212**	**0.2655**	**0.1974**	**0.0948**	**0.0798**	0.0611	0.0527	0.0445	0.0364	0.0289	0.0244	0.0245	0.0282
0.70	0.9908	0.9528	0.9057	**0.4560**	**0.2960**	**0.1626**	**0.1262**	0.1045	0.0902	0.0759	0.0619	0.0488	0.0406	0.0402	0.0453
0.75	0.9931	0.9616	0.9226	**0.7178**	**0.3715**	**0.2559**	**0.1982**	0.1641	0.1413	0.1188	0.0966	0.0757	0.0625	0.0610	0.0673
0.80	0.9931	0.9683	0.9354	0.8730	**0.5445**	**0.9750**	**0.2904**	0.2404	0.2065	0.1738	0.1409	0.1102	0.0899	0.0867	0.0942
0.85	0.9954	0.9727	0.9462	0.8933	**0.7594**	**0.5188**	**0.4018**	0.3319	0.2858	0.2399	0.1945	0.1517	0.1227	0.1175	0.1256
0.90	0.9954	0.9772	0.9550	0.9099	0.8204	**0.6829**	**0.5297**	0.4375	0.3767	0.3162	0.2564	0.1995	0.1607	0.1524	0.1611
0.95	0.9954	0.9817	0.9616	0.9226	0.8472	0.7709	0.6668	0.5521	0.4764	0.3999	0.3251	0.2523	0.2028	0.1910	0.2000
1.00	0.9977	0.9840	0.9661	0.9333	0.8690	0.8035	0.7379	0.6668	0.5781	0.8750	0.3972	0.3097	0.2483	0.2328	0.2415
1.05	0.9977	0.9863	0.9705	0.9441	0.8872	0.8318	0.7762	0.7194	0.6607	0.5728	0.4710	0.3690	0.2958	0.2773	0.2844
1.10	0.9977	0.9886	0.9750	0.9506	0.9016	0.8531	0.8072	0.7586	0.7112	0.6412	0.5408	0.4285	0.3451	0.3228	0.3296
1.15	0.9977	0.9886	0.9795	0.9572	0.9141	0.8730	0.8318	0.7907	0.7499	0.6918	0.6026	0.4875	0.3954	0.3690	0.3750
1.20	0.9977	0.9908	0.9817	0.9616	0.9247	0.8892	0.8531	0.8166	0.7834	0.7328	0.6546	0.5420	0.4446	0.4150	0.4198
1.30	0.9977	0.9931	0.9863	0.9705	0.9419	0.9141	0.8872	0.8590	0.8318	0.7943	0.7345	0.6383	0.5383	0.5058	0.5093
1.40	0.9977	0.9931	0.9886	0.9772	0.9550	0.9333	0.9120	0.8892	0.8690	0.8395	0.7925	0.7145	0.6237	0.5902	0.5943
1.50	1.0000	0.9954	0.9908	0.9817	0.9638	0.9462	0.9290	0.9141	0.8974	0.8730	0.8375	0.7745	0.6966	0.6668	0.6714
1.60	1.0000	0.9954	0.9931	0.9863	0.9727	0.9572	0.9441	0.9311	0.9183	0.8995	0.8710	0.8222	0.7586	0.7328	0.7430
1.70	1.0000	0.9977	0.9954	0.9886	0.9772	0.9661	0.9550	0.9462	0.9354	0.9204	0.8995	0.8610	0.8091	0.7907	0.8054
1.80	1.0000	0.9977	0.9954	0.9908	0.9817	0.9727	0.9661	0.9572	0.9484	0.9376	0.9204	0.8913	0.8531	0.8414	0.8590
1.90	1.0000	0.9977	0.9954	0.9931	0.9863	0.9795	0.9727	0.9661	0.9594	0.9506	0.9376	0.9162	0.8872	0.8831	0.9057
2.00	1.0000	0.9977	0.9977	0.9954	0.9886	0.9840	0.9795	0.9727	0.9683	0.9616	0.9528	0.9354	0.9183	0.9183	0.9462
2.20	1.0000	1.0000	0.9977	0.9977	0.9931	0.9908	0.9886	0.9840	0.9817	0.9795	0.9727	0.9661	0.9616	0.9727	1.0093
2.40	1.0000	1.0000	1.0000	0.9977	0.9977	0.9954	0.9931	0.9931	0.9908	0.9908	0.9886	0.9863	0.9931	1.0116	1.0568
2.60	1.0000	1.0000	1.0000	1.0000	1.0000	0.9977	0.9977	0.9977	0.9977	0.9977	0.9977	1.0023	1.0162	1.0399	1.0889
2.80	1.0000	1.0000	1.0000	1.0000	1.0000	1.0000	1.0023	1.0023	1.0023	1.0046	1.0069	1.0116	1.0328	1.0593	1.1117
3.00	1.0000	1.0000	1.0000	1.0000	1.0023	1.0023	1.0046	1.0046	1.0069	1.0069	1.0116	1.0209	1.0423	1.0740	1.1298
3.50	1.0000	1.0000	1.0000	1.0023	1.0023	1.0046	1.0069	1.0093	1.0116	1.0139	1.0186	1.0304	1.0593	1.0914	1.1508
4.00	1.0000	1.0000	1.0000	1.0023	1.0046	1.0069	1.0093	1.0116	1.0139	1.0162	1.0233	1.0375	1.0666	1.0990	1.1588

Taken with permission from Ref. [9]. The value at the critical point ($T_r = P_r = 1$) is taken from the API-TDB [5]. Bold numbers indicate liquid region.

of $C_P^{ig} = A + BT + C(SG^g) + D(SG^g)^2 + E[T(SG^g)]$, where T is temperature and SG^g is gas specific gravity ($M^g/29$). Although the equation is very useful for calculation of C_P^{ig} of undefined natural gases but using the reported coefficients we could not obtain reliable values for C_P^{ig}. In another correlation, ideal heat capacities of hydrocarbons ($N_C > C_5$) were related to boiling point and specific gravity in the form of Eq. (2.38) at three temperatures of 0°F (~255 K), 600°F (~589 K), and 1200°F (922 K) [18]. For light gases based on the data generated through Eq. (6.66) for compounds from C_1 to C_5 with H_2S, CO_2, and N_2 the following relation has been determined:

$$(6.73) \qquad \frac{C_P^{ig}}{R} = \sum_{i=0}^{2}(A_i + B_i M)T^i$$

for natural and light gases with $16 \leq M \leq 60$

where

$A_0 = 3.3224$ $B_0 = -2.5379 \times 10^{-2}$
$A_1 = -7.3308 \times 10^{-3}$ $B_1 = 7.5939 \times 10^{-4}$
$A_2 = 4.3235 \times 10^{-6}$ $B_2 = -2.6565 \times 10^{-7}$

in which T is the absolute temperature in kelvin. This equation is based on more than 500 data points generated in the temperature range of 50–1500 K and molecular weight

range of 16–60. The average deviation for this equation for these ranges is 5%; however, when it is applied in the temperature range of 200–1000 K and molecular weight range of 16–50, the error reduces to 2.5%. Use of this equation is recommended for undefined light hydrocarbon gas mixtures when gas specific gravity (SG^g) is known ($M = 29 \times SG^g$). For defined hydrocarbon mixtures of known composition the following relation may be used to calculate mixture ideal gas heat capacity:

$$(6.74) \qquad \frac{C_{P,\text{mix}}^{ig}}{R} = \sum_i y_i \frac{C_{Pi}^{ig}}{R}$$

where C_{Pi}^{ig} is the molar ideal heat capacity of component i (with mole fraction y_i) and may be calculated from Eq. (6.66).

Example 6.2—Calculate C_P for saturated liquid benzene at 450 K and 9.69 bar using generalized correlation and SRK EOS and compare with the value of 2.2 kJ/kg·K as given in Ref. [6]. Also calculate heat capacity at constant volume, heat capacity ratio, and residual enthalpy ($H - H^{ig}$) from both SRK EOS and generalized correlations of LK.

Solution—From Table 2.1 for benzene we have $T_c = 562$ K, $P_c = 49$ bar, $\omega = 0.21$, $M = 78.1$, and $K_W = 9.72$ and from

TABLE 6.7—*Values of $\phi^{(1)}$ for use in Eq. (6.59).*

T_r	0.01	0.05	0.1	0.2	0.4	0.6	0.8	1	1.2	1.5	2	3	5	7	10
0.30	**0.0000**	**0.0000**	**0.0000**	**0.0000**	**0.0000**	**0.0000**	**0.0000**	0.0000	0.0000	0.0000	0.0000	0.0000	0.0000	0.0000	0.0000
0.35	**0.0000**	**0.0000**	**0.0000**	**0.0000**	**0.0000**	**0.0000**	**0.0000**	0.0000	0.0000	0.0000	0.0000	0.0000	0.0000	0.0000	0.0000
0.40	**0.0000**	**0.0000**	**0.0000**	**0.0000**	**0.0000**	**0.0000**	**0.0000**	0.0000	0.0000	0.0000	0.0000	0.0000	0.0000	0.0000	0.0000
0.45	**0.0002**	**0.0002**	**0.0002**	**0.0002**	**0.0002**	**0.0002**	**0.0002**	0.0000	0.0000	0.0000	0.0000	0.0000	0.0000	0.0000	0.0000
0.50	**0.0014**	**0.0014**	**0.0014**	**0.0014**	**0.0014**	**0.0014**	**0.0013**	0.0013	0.0013	0.0013	0.0012	0.0011	0.0009	0.0008	0.0006
0.55	0.9705	**0.0069**	**0.0068**	**0.0068**	**0.0066**	**0.0065**	**0.0064**	0.0063	0.0062	0.0061	0.0058	0.0053	0.0045	0.0039	0.0031
0.60	0.9795	**0.0227**	**0.0226**	**0.0223**	**0.0220**	**0.0216**	**0.0213**	0.0210	0.0207	0.0202	0.0194	0.0179	0.0154	0.0133	0.0108
0.65	0.9863	0.9311	**0.0572**	**0.0568**	**0.0559**	**0.0551**	**0.0543**	0.0535	0.0527	0.0516	0.0497	0.0461	0.0401	0.0350	0.0289
0.70	0.9908	0.9528	0.9036	**0.1182**	**0.1163**	**0.1147**	**0.1131**	0.1116	0.1102	0.1079	0.1040	0.0970	0.0851	0.0752	0.0629
0.75	0.9931	0.9683	0.9332	**0.2112**	**0.2078**	**0.2050**	**0.2022**	0.1994	0.1972	0.1932	0.1871	0.1754	0.1552	0.1387	0.1178
0.80	0.9954	0.9772	0.9550	0.9057	**0.3302**	**0.3257**	**0.3212**	0.3168	0.3133	0.3076	0.2978	0.2812	0.2512	0.2265	0.1954
0.85	0.9977	0.9863	0.9705	0.9375	**0.4774**	**0.4708**	**0.4654**	0.4590	0.4539	0.4457	0.4325	0.4093	0.3698	0.3365	0.2951
0.90	0.9977	0.9908	0.9795	0.9594	0.9141	**0.6323**	**0.6250**	0.6165	0.6095	0.5998	0.5834	0.5546	0.5058	0.4645	0.4130
0.95	0.9977	0.9931	0.9885	0.9750	0.9484	0.9183	**0.7888**	0.7797	0.7691	0.7568	0.7379	0.7063	0.6501	0.6026	0.5432
1.00	1.0000	0.9977	0.9931	0.9863	0.9727	0.9594	0.9440	0.9311	0.9204	0.9078	0.8872	0.8531	0.7962	0.7464	0.6823
1.05	1.0000	0.9977	0.9977	0.9954	0.9885	0.9863	0.9840	0.9840	0.9954	0.9078	0.8872	0.8531	0.7962	0.7464	0.6823
1.10	1.0000	1.0000	1.0000	1.0000	1.0023	1.0046	1.0093	1.0163	1.0280	1.0593	1.0162	0.9886	0.9354	0.8872	0.8222
1.15	1.0000	1.0000	1.0023	1.0046	1.0116	1.0186	1.0257	1.0375	1.0520	1.0814	1.1376	1.1015	1.0617	1.0186	0.9572
1.20	1.0000	1.0023	1.0046	1.0069	1.0163	1.0280	1.0399	1.0544	1.0691	1.0990	1.1588	1.2388	1.1722	1.1403	1.0864
1.30	1.0000	1.0023	1.0069	1.0116	1.0257	1.0399	1.0544	1.0716	1.0914	1.1194	1.1776	1.2853	1.3868	1.2411	1.2050
1.40	1.0000	1.0046	1.0069	1.0139	1.0304	1.0471	1.0642	1.0815	1.0990	1.1298	1.1858	1.2942	1.4488	1.5171	1.5524
1.50	1.0000	1.0046	1.0069	1.0163	1.0328	1.0496	1.0666	1.0865	1.0990	1.1298	1.1858	1.2942	1.4488	1.5171	1.5524
1.60	1.0000	1.0046	1.0069	1.0163	1.0328	1.0496	1.0691	1.0865	1.1041	1.1350	1.1858	1.2883	1.4689	1.5996	1.7140
1.70	1.0000	1.0046	1.0093	1.0163	1.0328	1.0496	1.0691	1.0865	1.1041	1.1350	1.2883	1.4689	1.5996	1.7140	
1.80	1.0000	1.0046	1.0069	1.0163	1.0328	1.0496	1.0666	1.0865	1.1041	1.1324	1.1803	1.2794	1.4622	1.6033	1.7458
1.90	1.0000	1.0046	1.0069	1.0163	1.0328	1.0496	1.0666	1.0815	1.1015	1.1298	1.1749	1.2706	1.4488	1.5959	1.7620
2.00	1.0000	1.0046	1.0069	1.0163	1.0304	1.0471	1.0642	1.0815	1.0990	1.1272	1.1695	1.2618	1.4355	1.5849	1.7620
2.20	1.0000	1.0046	1.0069	1.0139	1.0304	1.0447	1.0593	1.0765	1.0965	1.1220	1.1641	1.2503	1.4191	1.5704	1.7539
2.40	1.0000	1.0046	1.0069	1.0139	1.0280	1.0423	1.0568	1.0716	1.0914	1.1143	1.1535	1.2331	1.3900	1.5346	1.7219
2.60	1.0000	1.0023	1.0069	1.0139	1.0257	1.0399	1.0544	1.0666	1.0864	1.1066	1.1429	1.2190	1.3614	1.4997	1.6866
2.80	1.0000	1.0023	1.0069	1.0116	1.0257	1.0375	1.0496	1.0642	1.0814	1.1015	1.1350	1.2023	1.3397	1.4689	1.6482
3.00	1.0000	1.0023	1.0069	1.0116	1.0233	1.0352	1.0471	1.0593	1.0765	1.0940	1.1272	1.1912	1.3183	1.4388	1.6144
3.50	1.0000	1.0023	1.0046	1.0023	1.0209	1.0304	1.0423	1.0520	1.0715	1.0889	1.1194	1.1803	1.3002	1.4158	1.5813
4.00	1.0000	1.0023	1.0046	1.0093	1.0186	1.0280	1.0375	1.0471	1.0617	1.0789	1.1041	1.1561	1.2618	1.3614	1.5101

Taken with permission from Ref. [9]. The value at the critical point ($T_r = P_r = 1$) is taken from the API-TDB [5]. Bold numbers indicate liquid region.

Table 5.12, $Z_{RA} = 0.23$. From Eq. (6.72), $C_P^{ig} = 127.7$ J/mol·K, $T_r = T/T_c = 0.8$, and $P_r = 0.198 \cong 0.2$. From Tables 6.4 and 6.5, $[(C_P - C_P^{ig})/R]^{(0)} = 3.564$, $[(C_P - C_P^{ig})/R]^{(1)} = 10.377$. It is important to note that the system is saturated liquid and extrapolation of values for the liquid region from $T_r = 0.7$ and 0.75–0.8 at $P_r = 0.2$ is required for both (0) and (1) terms. Direct data given in the tables at $T_r = 0.8$ and $P_r = 0.2$ correspond to saturated vapor and special care should be taken when the system is at saturated conditions. From Eq. (6.57), $[(C_P - C_P^{ig})/R] = 5.74317$, $C_P - C_P^{ig} = 5.74317 \times 8.314 = 47.8$ J/mol·K, and $C_P = 47.8 + 127.7 = 175.5$ J/mol·K. The specific heat is calculated through molecular weight using Eq. (5.3) as $C_P = 175.5/78.1 = 2.25$ J/g·°C. This value is basically the same as the reported value. For SRK EOS, $a = 1.907 \times 10^7$ bar(cm³/mol)², $b = 82.69$ cm³/mol, $Z^L = 0.033$, $V^L = 126.1$ cm³/mol, $c = 9.6$ cm³/mol, and $V^L = 116.4$ (corrected), which gives Z^L (corrected) = 0.0305, where c is calculated from Eq. (5.51). From generalized correlations $Z^{(0)} = 0.0328$ and $Z^{(1)} = -0.0138$, which gives $Z = 0.0299$ that is very close to the value calculated from SRK EOS. Using units of kelvin for temperature, bar for pressure, and cm³ for volume, $R = 83.14$ cm³·bar/mol·K and $V = 116.4$ cm³/mol. From relations given in Table 6.1 we calculate $a_1 = -33017$ and $a_2 = 60.9759$. $P_1 = 4.13169$, $P_2 = -38.2557$, and $P_3 = 0.402287$. From Table 6.1 for SRK EOS we have $C_P - C_P^{ig} = 450 \times 0.39565 - 450 \times (3.887)^2/(-32.8406) - 83.14 = 301.936$ cm³·bar/mol·K. Since 1 J = 10 cm³·bar, thus $C_P - C_P^{ig} = 301.936/10 = 30.194$ J/mol·K. $C_P = (C_P - C_P^{ig}) + C_P^{ig} = 30.194 + 127.7 = 157.9$ J/mol·K or $C_P = 157.9/78.1 = 2.02$ J/g·°C. The deviation with generalized correlation is –8.1%. Effect of considering volume translation c on volume in calculation of C_P is minor and in this problem if V^L directly calculated from SRK equation (126.1 cm³/mol) is used, value of calculated C_P would be still the same as 2.02 J/g·°C. For calculation of C_V, SRK equation is used with relations given in Table 6.1. $C_V - C_V^{ig} = TP_3 = 450 \times 0.402287 = 178.04$ cm³·bar/mol·K = 178.04/10 = 17.8 J/mol·K. $C_V^{ig} = C_P^{ig} - R = 127.7 - 8.314 = 119.4$ J/mol·K. Thus, $C_V = 119.4 + 17.8 = 137.2$ J/mol·K = 137.2/78.1 = 1.75 J/g·K. The heat capacity ratio is $\gamma = C_P/C_V = 2.02/1.75 = 1.151$.

To calculate $H - H^{ig}$ from generalized correlations we get from Tables 6.2 and 6.3 as $[(H - H^{ig})/RT_c]^{(0)} = -4.518$ and $[(H - H^{ig})/RT_c]^{(1)} = -5.232$. From Eq. (6.56), $[(H - H^{ig})/RT_c] = -5.6167$. Again it should be noted that the values of $[\,]^{(0)}$ and $[\,]^{(1)}$ terms are taken for saturated liquid by extrapolation of T_r from 0.7 and 0.75 to 0.8. Values in the tables for saturated vapor (at $T_r = 0.8$, $P_r = 0.2$) should be avoided.

TABLE 6.8—*Constants for Eqs. (6.66)–(6.69) for ideal gas heat capacity, enthalpy, and entropy.*

No.	Compound name	Formula	M	A	$B \times 10^3$	$C \times 10^6$	$D \times 10^{10}$	$E \times 10^{14}$	T_{min}, K	T_{max}, K
	Paraffins									
1	Methane	CH_4	16.043	4.34610	−6.14488	26.62607	−219.2998	588.89965	50	1500
2	Ethane	C_2H_6	30.070	4.00447	−1.33847	42.86416	−452.2446	1440.4853	50	1500
3	Propane	C_3H_8	44.096	3.55751	10.07312	39.13602	−475.7220	1578.1656	50	1500
4	n-Butane	C_4H_{10}	58.123	2.91601	28.06907	15.37435	−292.9255	1028.0462	200	1500
5	Isobutane	C_4H_{10}	58.123	2.89796	25.14031	26.04226	−405.3691	1396.6324	50	1500
6	n-Pentane	C_5H_{12}	72.150	4.06063	29.87141	30.46993	−461.3523	1559.8971	200	1500
7	Isopentane	C_5H_{12}	72.150	0.61533	49.99361	−9.72605	−121.1597	563.52870	200	1500
8	Neopentane	C_5H_{12}	72.150	6.60029	24.43268	32.52759	−402.96336	1258.46299	220	1500
9	n-Hexane	C_6H_{14}	86.177	3.89054	41.42970	24.35860	−457.5222	1599.4100	200	1500
10	n-Heptane	C_7H_{12}	100.204	4.52739	47.36877	31.09932	−570.22085	1999.68224	200	1500
11	n-Octane	C_8H_{18}	114.231	4.47277	57.81747	29.07465	−621.09106	2265.33690	200	1500
12	n-Nonane	C_9H_{20}	128.258	3.96754	68.72207	31.85998	−758.47191	2875.17975	200	1000
13	n-Decane	$C_{10}H_{22}$	142.285	14.56771	−9.12133	283.5241	−3854.9259	16158.7933	200	1000
14	n-Undecane	$C_{11}H_{24}$	156.312	15.72269	−8.39015	308.0195	−4205.1509	17634.1470	200	1000
15	n-Dodecane	$C_{12}H_{26}$	170.338	16.87761	−7.65919	332.5144	−4555.3529	19109.3961	200	1000
16	n-Tridecane	$C_{13}H_{28}$	184.365	30.63938	−107.2144	632.4036	−8053.8502	33377.9390	200	1000
17	n-Tetradecane	$C_{14}H_{30}$	198.392	−2.95801	−6.19822	381.5291	−5256.0878	22061.2353	200	1000
18	n-Pentadecane	$C_{15}H_{32}$	212.419	−2.65315	−5.09511	404.3408	−5576.7343	−23366.9827	200	1000
19	n-Hexadecane	$C_{16}H_{34}$	226.446	−36.57941	−4.73820	430.5450	−5956.8328	25013.0768	200	1000
20	n-Heptadecane	$C_{17}H_{36}$	240.473	23.25896	−4.00829	431.5163	−6307.0717	26488.4354	200	1000
21	n-Octadecane	$C_{18}H_{38}$	254.500	−2.20866	−3.27840	479.5420	−6657.3161	27963.8136	200	1000
22	n-Nonadecane	$C_{19}H_{40}$	268.527	25.68345	−2.54834	504.0402	−7007.5535	29439.1464	200	1000
23	n-Eicosane	$C_{20}H_{42}$	282.553	26.82718	−1.81886	528.5571	−7358.0352	30915.5067	200	1500
24	2-Methylpentane	C_6H_{14}	86.177	0.44073	60.77573	−10.93570	−180.70573	833.40865	200	1500
25	3-Methylpentane	C_6H_{14}	86.177	−0.07902	63.31181	−18.82562	−90.58759	510.28364	200	1500
26	2,2-Dimethylbutane	C_6H_{14}	86.177	1.00342	56.10078	−1.05011	−237.84301	956.53142	200	1500
27	2-Methylhexane	C_7H_{16}	100.204	0.57808	70.71556	−15.00679	−187.48705	899.92106	200	1500
28	3-Methylhexane	C_7H_{16}	100.204	−0.37490	75.26096	−22.63052	−131.65268	744.03635	200	1500
29	2,4-Dimethylpentane	C_7H_{16}	100.204	−3.20582	98.77224	−72.48550	293.10189	−500.62465	200	1500
30	2-Methylheptane	C_8H_{18}	114.231	0.92650	78.42561	−11.24742	−281.97592	1265.61161	200	1500
31	2,2,4-Trimethyl-pentane	C_8H_{18}	114.231	−1.85230	96.08105	−47.77416	68.93159	137.05449	200	1500
	Olefins									
32	Ethylene	C_2H_4	28.054	2.11112	8.32103	11.24746	−155.42099	516.14291	200	1500
33	Propylene	C_3H_6	42.081	2.15234	17.76767	8.26700	−166.33525	582.37236	200	1500
34	1-Butene	C_4H_8	56.107	4.25402	10.78298	47.84869	−597.94406	2039.55839	50	1500
35	1-Pentene	C_5H_{10}	70.135	2.04789	37.52066	5.51442	−254.11404	993.58809	300	1200
36	1-Hexene	C_6H_{12}	84.162	0.00610	63.24725	−35.49665	88.16710	−57.32517	200	1500
37	1-Heptene	C_7H_{14}	98.189	3.47887	48.09877	23.25712	−531.31261	1997.9072	200	1500
38	1-Octene	C_8H_{16}	112.216	3.98703	54.62745	29.77494	−652.07494	2453.5780	200	1500
39	1-Nonene	C_9H_{18}	126.243	4.54519	60.98631	36.26160	−769.45635	2891.5695	200	1500
40	1-Decene	$C_{10}H_{20}$	140.270	4.95682	68.28457	40.89381	−873.20972	3296.6252	200	1500
41	1-Undecene	$C_{11}H_{22}$	154.219	5.68918	73.39300	49.96821	−1012.1464	3796.2346	200	1500
42	1-Dodecene	$C_{12}H_{24}$	168.310	5.94633	81.98315	51.37713	−1083.4987	4090.6826	200	1500
43	1-Tridecene	$C_{13}H_{26}$	182.337	−0.32099	129.0630	−25.20577	−505.45836	2487.2522	200	1500
44	1-Tetradecene	$C_{14}H_{28}$	196.364	−0.29904	138.5541	−25.09881	−568.90191	2764.1344	200	1500
45	1-Pentadecene	$C_{15}H_{30}$	210.391	0.09974	145.9734	−20.68363	−670.89438	3163.1228	200	1500
46	1-Hexadecene	$C_{16}H_{32}$	224.418	0.54495	152.863	−15.10253	−781.81088	3585.1609	200	1500
47	1-Heptadecene	$C_{17}H_{34}$	238.445	0.41533	163.244	−16.44658	−838.48469	3862.2068	200	1500
48	1-Octadecene	$C_{18}H_{36}$	252.472	31.69585	−73.95796	647.4299	−8000.0231	29069.9319	200	1500
49	1-Nonadecene	$C_{19}H_{38}$	266.490	0.77613	180.596	−13.04267	−993.43903	4506.8347	200	1500
50	1-Eicosene	$C_{20}H_{40}$	280.517	−0.20146	196.7237	−27.69713	−923.83278	4366.8186	200	1500
	Naphthenes									
51	Cyclopentane	C_5H_{10}	70.134	−7.43795	69.82174	−43.64337	122.59611	−92.59304	300	1500
52	Methylcyclopentane	C_6H_{12}	84.161	−6.81073	80.58175	−50.42977	141.93915	−107.77270	300	1500
53	Ethylcyclopentane	C_7H_{14}	98.188	−7.51027	93.72668	−58.81706	167.05996	−127.54803	300	1500
54	n-Propylcyclopentane	C_8H_{16}	112.216	−7.61363	105.1051	−65.42900	183.75014	−138.61493	300	1500
55	n-Butylcyclopentane	C_9H_{18}	126.243	−7.58208	115.9123	−71.30847	197.13396	−146.30090	300	1500
56	n-Pentylcyclopentane	$C_{10}H_{20}$	140.270	−8.03062	128.8771	−80.20325	225.20999	−169.73008	300	1500
57	n-Hexylcyclopentane	$C_{11}H_{22}$	154.290	−5.33508	122.4990	−48.15565	−101.53331	871.78359	300	1500
58	n-Heptylcyclopentane	$C_{12}H_{24}$	168.310	−8.17951	151.3485	−93.06028	256.93490	−190.08988	300	1500
59	n-Octylcyclopentane	$C_{13}H_{26}$	182.340	−8.20466	162.3974	−99.23258	271.5150	−199.08614	300	1500
60	n-Nonylcyclopentane	$C_{14}H_{28}$	196.360	−8.27104	173.6250	−105.7060	287.84321	−209.81363	300	1500
61	n-Decylcyclopentane	$C_{15}H_{30}$	210.390	−8.70424	186.5104	−114.4444	314.93805	−232.04645	300	1500
62	n-Unoecylcyclopentane	$C_{16}H_{32}$	224.420	−8.71319	197.4859	−120.5348	329.23444	−240.60241	300	1500
63	n-Dodecylcyclopentane	$C_{17}H_{34}$	238.440	−8.81568	208.8487	−127.1364	345.90292	−251.65896	300	1500

(Continued)

TABLE 6.8—*(Continued)*

No.	Compound name	Formula	M	A	$B \times 10^3$	$C \times 10^6$	$D \times 10^{10}$	$E \times 10^{14}$	T_{min}, K	T_{max}, K
64	n-Tridecylcyclopentane	$C_{18}H_{36}$	252.470	−8.82057	219.8119	−133.2056	360.15605	−260.35782	300	1500
65	n-Tetradecylcyclopentane	$C_{19}H_{38}$	266.490	−8.81992	227.8540	−139.2730	374.44617	−268.88152	300	1500
66	n-Pentadecylcyclopentane	$C_{20}H_{40}$	280.520	−9.29147	243.8142	−148.2795	402.88398	−292.64837	300	1500
67	n-Hexadecylcyclopentane	$C_{21}H_{42}$	294.550	−9.34807	254.9133	−154.7201	419.05364	−303.20007	300	1500
68	Cyclohexane	C_6H_{12}	84.161	−7.66115	77.46123	−31.65303	−45.48807	456.29714	300	1500
69	Methylcyclohexane	C_7H_{14}	98.188	−8.75751	100.2054	−62.47659	169.33320	−123.27361	300	1500
70	Ethylcyclohexane	C_8H_{16}	112.215	−5.50074	91.59292	−26.04906	−192.84542	1021.80248	300	1500
71	n-Propylcyclohexane	C_9H_{18}	126.243	−8.87526	124.6789	−76.99183	180.70008	20.22888	300	1500
72	n-Butylcyclohexane	$C_{10}H_{20}$	140.270	−7.38694	127.4674	−67.63120	73.28814	355.51905	300	1500
73	n-Pentylcyclohexane	$C_{11}H_{22}$	154.290	−10.16016	152.5757	−98.38009	265.14011	−106.39559	300	1500
74	n-Hexylcyclohexane	$C_{12}H_{24}$	168.310	−9.58825	161.8750	−104.4133	302.29623	−236.75537	300	1500
75	n-Heptylcyclohexane	$C_{13}H_{26}$	182.340	−12.53870	188.4588	−138.5801	523.83412	−881.68097	300	1500
76	n-Octylcyclohexane	$C_{14}H_{28}$	196.360	−7.88711	178.2886	−112.8765	330.77533	−271.09270	300	1500
77	n-Nonylcycloi-iexane	$C_{15}H_{30}$	210.390	−8.48961	187.0067	−105.2157	192.47573	192.71532	300	1500
78	n-Decylcyclohexane	$C_{16}H_{32}$	224.420	−10.58196	209.8953	−134.2136	385.00443	−297.79779	300	1500
79	n-Undecylcyclohexane	$C_{17}H_{34}$	238.440	−9.25980	214.8824	−131.9175	357.79740	−261.20128	300	1500
80	n-Dodecylcyclohexane	$C_{18}H_{36}$	252.470	−9.94518	228.7293	−141.7915	389.80571	−289.05527	300	1500
81	n-Tridecylcyclohexane	$C_{19}H_{38}$	266.490	−10.06895	240.3258	−148.9432	410.23509	−304.86051	300	1500
82	n-Tetradecylcyclohexane	$C_{20}H_{40}$	280.520	−10.98687	255.5423	−161.2184	455.82197	−347.77976	300	1500
83	n-Hexadecylcyclohexane	$C_{22}H_{44}$	308.570	−8.96825	268.1151	−159.9818	417.54247	−292.26560	300	1500
	Aromatics									
84	Benzene	C_6H_6	78.114	−7.29786	75.33056	−69.66390	336.46848	−660.39655	300	1500
85	Toluene	C_7H_8	92.141	−2.46286	57.69575	−19.66557	−106.61110	654.52596	200	1500
86	Ethylbenzene	C_8H_{10}	106.167	4.72510	9.02760	141.1887	−1989.2347	8167.1805	50	1000
87	m-Xylene	C_8H_{10}	106.167	−4.00149	76.37388	−44.21568	82.57499	90.13866	260	1500
88	o-Xylene	C_8H_{10}	106.167	−1.51679	68.03181	−33.61164	24.37900	206.82729	260	1500
89	p-Xylene	C_8H_{10}	106.167	−4.77265	80.94644	−51.89215	136.1966	−45.64845	260	1500
90	n-Propylbenzene	C_9H_{12}	120.195	4.42447	33.21919	74.42459	−1045.5561	3656.7834	50	1500
91	n-Butylbenzene	$C_{10}H_{14}$	134.222	−6.24190	110.6923	−74.17854	221.3160	−178.64701	300	1500
92	m-Cymene	$C_{10}H_{14}$	134.222	−4.41825	103.1174	−65.46564	182.7512	−138.15307	300	1500
93	o-Cymene	$C_{10}H_{14}$	134.222	−2.40242	96.87475	−58.63517	154.5568	−109.45170	300	1500
94	p-Cymene	$C_{10}H_{14}$	134.222	−4.47668	102.5377	−64.61930	179.2371	−134.70678	300	1500
95	n-Pentylbenzene	$C_{11}H_{16}$	148.240	−6.89760	124.5723	−84.11348	251.1513	−201.56517	300	1500
96	n-Hexylbenzene	$C_{12}H_{18}$	162.260	−7.66975	139.1540	−95.04913	285.7856	−230.69678	300	1500
97	n-Heptylbenzene	$C_{13}H_{20}$	176.290	−8.36450	153.2807	−105.2641	316.7510	−255.85057	300	1500
98	n-Octylbenzene	$C_{14}H_{22}$	190.320	−9.35221	168.8057	−117.4996	357.7099	−292.04881	300	1500
99	Styrene	C_8H_8	104.152	−6.20755	91.11255	−83.45606	411.3630	−842.07179	300	1500
	Dienes and acetylenes									
100	Propadiene	C_3H_4	40.065	1.30128	23.37745	−13.57151	26.91489	26.81000	200	1500
101	1,2-Butadiene	C_4H_6	54.092	3.43878	19.01555	11.36858	−212.98223	751.33700	50	1500
102	Acetylene	C_2H_2	26.038	1.04693	21.20409	−29.08273	203.04028	−533.31364	50	1500
	Diaromatics									
103	Naphthalene	$C_{10}H_8$	128.174	−5.74112	86.70543	−46.55922	−1.47621	531.58512	200	1500
	Nonhydrocarbons									
104	Water	H_2O	18.015	4.05852	−0.71473	2.68748	−11.97480	13.19231	50	1500
105	Carbon dioxide	CO_2	44.01	3.51821	−2.68807	31.88523	−499.2285	2410.9439	50	1000
106	Hydrogen sulfide	H_2S	34.08	4.07259	−1.43459	6.47044	−45.32724	103.38528	50	1500
107	Nitrogen	N_2	28.014	3.58244	−0.84375	2.09697	−10.19404	11.22372	50	1500
108	Oxygen	O_2	32	3.57079	−1.18951	4.79615	−40.80219	110.40157	50	1500
109	Ammonia	NH_3	17.03	0.98882	−0.68636	3.61604	−32.60481	96.53173	50	1500
110	Carbon monoxide	CO	28.01	3.56423	−0.78152	2.20313	−11.29291	13.00233	50	1500
111	Hydrogen	H_2	2.016	3.24631	1.43467	−2.89398	25.8003	−73.9095	160	1220
112	Nitrogen dioxide	NO_2	46.01	3.38418	3.13875	3.98534	−58.69776	197.35202	200	1500
113	Nitrous oxide	NO	30.01	4.18495	−4.19791	9.45630	−72.74068	192.33738	50	1500

T_{min} and T_{max} are approximated to nearest 10. Data have been determined from Method 7A1.2 given in the API-TDB [5].

$(H - H^{ig}) = -5.6167 \times 8.314 \times (1/78.1) \times 562 = -336$ kJ/kg. To use SRK EOS, Eq. (5.40) should be used, which gives $Z^L = 0.0304$ and $B = 0.02142$. From Table 6.1 $(H - H^{ig})/RT = -7.438$, which gives $(H - H^{ig}) = -7.438 \times 8.314 \times (1/78.1) \times 450 = -356$ kJ/kg. The difference with the generalized correlation is about 6%. The generalized correlation gives more accurate result than a cubic EOS for calculation of enthalpy and heat capacity.

♦

6.4 THERMODYNAMIC PROPERTIES OF MIXTURES

Thermodynamics of mixtures also known as solution thermodynamics is particularly important in estimation of properties of petroleum mixtures especially in relation with phase equilibrium calculations. In this section we discuss partial molar quantities, calculation of properties of ideal and real

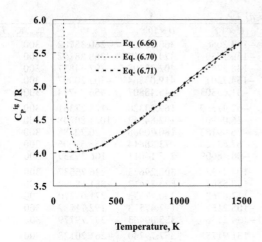

FIG. 6.2—Prediction of ideal gas heat capacity of water from various methods.

solutions, and volume change due to mixing and blending of petroleum mixtures.

6.4.1 Partial Molar Properties

Consider a homogeneous phase mixture of N components at T and P with number of moles of $n_1, n_2, \ldots, n_N$. A total property is shown by M^t where superscript t indicates total (extensive) property and M can be any intensive thermodynamic property (i.e., V, H, S, G). In general from the phase rule discussed in Chapter 5 we have

$$(6.75) \qquad M^t = M^t(T, P, n_1, n_2, n_3, \ldots, n_N)$$

$$(6.76) \qquad n = \sum_{i=1}^{N} n_i$$

$$(6.77) \qquad M = \frac{M^t}{n}$$

where n is the total number of moles and M is the molar property of the mixture. *Partial molar property* of component

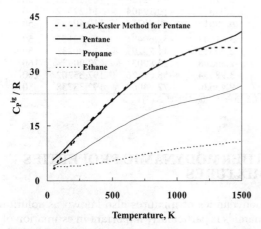

FIG. 6.3—Prediction of ideal gas heat capacity of some hydrocarbons from Eq. (6.66) and Lee–Kesler method (Eq. 6.72).

i in a mixture is shown by $\bar{M}_i$ and is defined as

$$(6.78) \qquad \bar{M}_i = \left(\frac{\partial M^t}{\partial n_i} \right)_{T,P,n_{j \neq i}}$$

$\bar{M}_i$ indicates change in property M^t per infinitesimal addition of component i at constant T, P, and amount of all other species. This definition applies to any thermodynamic property and $\bar{M}_i$ is a function of T, P, and composition. Partial molar volume ($\bar{V}_i$) is useful in calculation of volume change due to mixing for nonideal solutions, partial molar enthalpy ($\bar{H}_i$) is useful in calculation of heat of mixing, and $\bar{G}_i$ is particularly useful in calculation of fugacity and formulation of phase equilibrium problems. The main application of partial molar quantities is to calculate mixture property from the following relation:

$$(6.79) \qquad M^t = \sum_{i=1}^{N} n_i \bar{M}_i$$

or on the molar basis we have

$$(6.80) \qquad M = \sum_{i=1}^{N} x_i \bar{M}_i$$

where x_i is mole fraction of component i. Similar equations apply to specific properties (quantity per unit mass) with replacing mole fraction by mass or weight fraction. In such cases $\bar{M}_i$ is called *partial specific property*.

Partial molar properties can be calculated from the knowledge of relation between M and mole fraction at a given T and P. One relation that is useful for calculation of $\bar{M}_i$ is the Gibbs–Duhem (GD) equation. This equation is also a useful relation for obtaining a property of one component in a mixture from properties of other components. This equation can be derived by total differentiation of M^t in Eq. (6.75) and equating with total differential of M^t from Eq. (6.79), which at constant T and P can be reduced to the following simplified form [1]:

$$(6.81) \qquad \sum_i x_i \, \mathrm{d}\bar{M}_i = 0 \quad \text{(at constant } T, P)$$

This equation is the constant T and P version of the GD equation. As an example for a binary system ($x_2 = 1 - x_1$) we can show that Eqs. (6.80) and (6.81) give the following relations for calculation of $\bar{M}_i$:

$$(6.82) \qquad \begin{aligned} \bar{M}_1 &= M + x_2 \frac{\mathrm{d}M}{\mathrm{d}x_1} \\ \bar{M}_2 &= M - x_1 \frac{\mathrm{d}M}{\mathrm{d}x_1} \end{aligned}$$

Based on these relations it can be shown that when graphical presentation of M versus x_1 is available, partial molar properties can be determined from the interceptions of the tangent line (at x_1) with the Y axis. As shown in Fig. 6.4 the interception of tangent line at $x_1 = 0$ gives $\bar{M}_2$ and at $x_1 = 1$ gives $\bar{M}_1$ according to Eq. (6.82).

Example 6.3—Based on the graphical data available on enthalpy of aqueous solution of sulfuric acid (H_2SO_4) [1], the following relation for molar enthalpy of acid solution at 25°C is obtained:

$$H = 123.7 - 1084.4x_{w1} + 1004.5x_{w1}^2 - 1323.2x_{w1}^3 + 1273.7x_{w1}^4$$

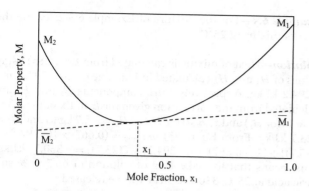

FIG. 6.4—Graphical method for calculation of partial molar properties.

where H is the specific enthalpy of solution in kJ/kg and x_{w1} is the weight fraction of H_2SO_4. Calculate $\bar{H}_1$ and $\bar{H}_2$ for a solution of 66.7 wt% sulfuric acid. Also calculate H for the mixture from Eq. (6.78) and compare with the value from the above empirical correlation.

Solution—Equation (6.82) is used to calculate $\bar{H}_1$ and $\bar{H}_2$. By direct differentiation of H with respect to x_{w1} we have $dH/dx_{w1} = -1084.4 + 2009x_{w1} - 3969.6x_{w1}^2 + 5094.8x_{w1}^3$. At $x_{w1} = 0.667$ we calculate $H = -293.3$ kJ/kg and $dH/dx_{w1} = -1.4075$ kJ/kg. From Eq. (6.82) we have $\bar{H}_1 = -293.3 + (0.333) \times (1.4075) = -292.8$ and $\bar{H}_2 = -294.2$ kJ/kg. Substituting the values in Eq. (6.80) we get H(at $x_{w1} = 0.667$) = $0.667 \times (-293.3) + 0.333 \times (-294.2) = -293.3$ kJ/kg, which is the same value as obtained from the original relation for H. Graphical calculation of partial specific enthalpies $\bar{H}_1$ and $\bar{H}_2$ is shown in Fig. 6.5. The tangent line at $x_1 = 0.667$ is almost horizontal and it gives equal values for $\bar{H}_1$ and $\bar{H}_2$ as -295 kJ/kg. ◆

6.4.2 Properties of Mixtures—Property Change Due to Mixing

Calculation of properties of a mixture from properties of its pure components really depends on the nature of the mixture.

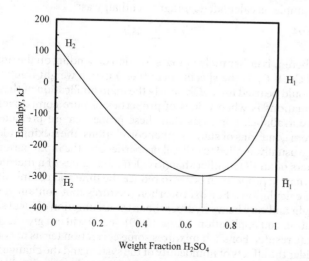

FIG. 6.5—Specific enthalpy of sulfuric acid solution at 25°C (part of Example 6.3).

In general the mixtures are divided into two groups of ideal solutions and real solutions. An ideal solution is a homogenous mixture in which all components (like and unlike) have the same molecular size and intermolecular forces, while real solutions have different molecular size and intermolecular forces. This definition applies to both gas mixtures and liquid mixtures likewise; however, the terms normally are applied to liquid solutions. Obviously all ideal gas mixtures are ideal solutions but not all ideal solutions are ideal gas mixtures. Mixtures composed of similar components especially with similar molecular size and chemical structure are generally ideal solutions. For example, benzene and toluene form an ideal solution since both are aromatic hydrocarbons with nearly similar molecular sizes. A mixture of polar component with a nonpolar component (i.e., alcohol and hydrocarbon) obviously forms a nonideal solution. Mixtures of hydrocarbons of low-molecular-weight hydrocarbons with very heavy hydrocarbons (polar aromatics) cannot be considered ideal solutions. If molar property of an ideal solution is shown by M^{id} and real solution by M the difference is called *excess property* shown by M^E

$$ (6.83) \qquad M^E = M - M^{id} $$

M^E is a property that shows nonideality of the solution and it is zero for ideal solutions. All thermodynamic relations that are developed for M also apply to M^E as well. Another important quantity is property change due to mixing which is defined as

$$ (6.84) \qquad \Delta M_{mix} = M - \sum_i x_i M_i = \sum_i x_i (\bar{M}_i - M_i) $$

During mixing it is assumed that temperature and pressure remain constant. From the first law it is clear that at constant T and P, the heat of mixing is equal to ΔH_{mix}, therefore

$$ (6.85) \qquad \text{Heat of mixing} = \Delta H_{mix} = \sum_i x_i (\bar{H}_i - H_i) $$

Similarly the volume change due to mixing is given by the following relation:

$$ \text{Volume change due to mixing} = \Delta V_{mix} = \sum_i x_i (\bar{V}_i - V_i) $$

$$ (6.86) $$

where H_i and V_i are molar enthalpy and volume of pure components at T and P of the mixture. For ideal solutions both the heat of mixing and the volume change due to the mixing are zero [19]. This means that in an ideal solution, partial molar volume of component i in the mixture is the same as pure component specific volume ($\bar{V}_i = V_i$) and $\bar{V}_i$ nor $\bar{H}_i$ vary with composition. Figure 6.6 shows variation of molar volume of binary mixture with mole fraction for both a real and an ideal solution (dotted line) for two cases. In Fig. 6.6*a* the real solution shows positive deviation, while in Fig. 6.6*b* the solution shows negative deviation from ideal solution. Systems with positive deviation from ideality have an increase in volume due to mixing, while systems with negative deviation have decrease in volume upon mixing.

Equations (6.85) and (6.86) are useful when pure components are mixed to form a solution. If two solutions are mixed then the volume change due to mixing can be calculated from

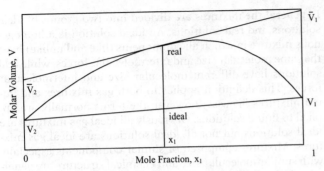

(a) Systems with increase in volume due to mixing

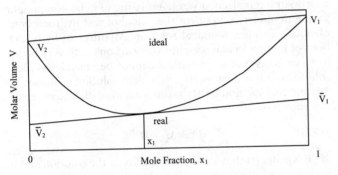

(b) Systems with decrease in volume due to mixing

FIG. 6.6—Variation of molar volume of a binary mixture with composition.

the following relation [17]:

$$\Delta V^t_{\text{mixing}} = \sum_i n_{i,\text{after}} \bar{V}_i(T, P, n_{i,\text{after}}) - \sum_i n_{i,\text{before}} \bar{V}_i(T, P, n_{i,\text{before}})$$
(6.87)

where $n_{i,\text{before}}$ is the moles of i before mixing and $n_{i,\text{after}}$ represents moles of i in the solution after the mixing. Obviously since the mixture composition before and after the mixing are not the same, $\bar{V}_i$ for i in the solution before the mixing and its value for i in the solution after the mixing are not the same. The same equation may be applied to enthalpy by replacing H with V to calculate heat of mixing when two solutions are mixed at constant T and P. Partial molar volume and enthalpy may be calculated from their definition, Eq. (6.78) through an EOS. For example in deriving the relation for $\bar{V}_i$, derivative $[\partial(nV)/\partial n_i]_{T,P,nj\neq i}$ should be determined from the EOS. For the PR EOS the partial molar volume is given as [20]

$$(6.88) \qquad \bar{V}_i = \frac{X_1 + X_2}{X_3 + X_4}$$

where

$$X_1 = (RT + b_i P) \times (V^2 + 2bV - b^2)$$
$$X_2 = \left[2b_i RT - 2\sum_j x_j a_{ij} - 2b_i P(V - b)\right] \times (V - b) + b_i a$$
$$X_3 = P(V^2 + 2bV - b^2) + 2P(V - b)(V + b)$$
$$X_4 = -2RT(V + b) + a$$

where V is the mixture molar volume calculated from PR EOS. For more accurate calculation of $\bar{V}_i$, corrected V through use of volume translation concept (Eq. 5.50) may be used. Similar relation for $\bar{H}_i$ can be obtained (see Problem 6.5).

Example 6.4—For the mixture of Example 6.3 calculate the heat of mixing at 25°C.

Solution—Heat of mixing is calculated from Eq. (6.85) using values of $\bar{H}_1$ and $\bar{H}_2$ calculated in Example 6.3 as −292.8 and −294.2 kJ/kg, respectively. Pure components H_1 and H_2 are calculated from the correlation given for H in Example 6.3 at $x_1 = 1$ (for H_1) and $x_1 = 0$ (for H_2) as $H_1 = -5.7$ kg/kJ and $H_2 = 123.7$ kJ/kg. From Eq. (6.85), $\Delta H_{\text{mix}} = (0.667) \times [(-292.8) - (-5.7)] + (1 - 0.667) \times [(-294.2) - (123.7)] = -330.7$ kJ/kg. This means that to make 1 kg of solution of 66.7 wt% sulfuric acid at 25°C, 330.7 kJ heat will be released. ◆

For the ideal solutions, H, V, G, and S of the mixture may be calculated from pure component properties through the following relations [1, 21]:

$$(6.89) \qquad H^{\text{id}} = \sum_i x_i H_i$$

$$(6.90) \qquad V^{\text{id}} = \sum_i x_i V_i$$

$$(6.91) \qquad G^{\text{id}} = \sum_i x_i G_i + RT \sum_i x_i \ln x_i$$

$$(6.92) \qquad S^{\text{id}} = \sum_i x_i S_i - R \sum_i x_i \ln x_i$$

where H^{id}, V^{id}, G^{ig}, and S^{ig} can be either molar or specific enthalpy, volume, Gibbs energy, and entropy of mixture. In case of specific property, x_i is weight fraction. For example, if V is specific volume ($= 1/\rho$), Eq. (6.90) can be written in the following form for density:

$$(6.93) \qquad \frac{1}{\rho} = \sum_i \frac{x_{wi}}{\rho_i}$$

where x_{wi} is the weight fraction of i and ρ_i is the density of pure i. This equation was previously introduced in Chapter 3 (Eq. 3.45). Although all hydrocarbon mixtures do not really behave like ideal solutions, mixtures that do not contain nonhydrocarbons or very heavy hydrocarbons, may be assumed as ideal solutions. For simplicity, application of Eqs. (6.89) and (6.90) is extended to many thermodynamic properties as it was shown in Chapters 3 and 4. Mixture heat capacity, for example, is calculated similar to enthalpy as:

$$(6.94) \qquad C_P = \sum_i x_i C_{Pi}$$

where x_i is either mole or mass fraction depending on the unit of C_P. If C_P is the specific heat (i.e., J/g·°C), weight fraction should be used for x_i. Obviously the main application of these equations is when values of properties of pure components are available. For cases that these properties are predicted from equations of state or other correlations, the mixing rules are usually applied to critical properties and the input parameters of an EOS rather than to calculated values of a thermodynamic property in order to reduce the time and complexity of calculations. For hydrocarbon mixtures that contain very light and very heavy hydrocarbons the assumption of ideal solution and application of Eqs. (6.89)–(6.93) will not give accurate results. For such mixtures some correction terms to consider the effects of nonideality of the system and the change in molecular behavior in presence of unlike molecules should be added to the RHS of such equations. The following empirical method recommended by the API for calculation of volume of petroleum blends is based on such theory [5].

6.4.3 Volume of Petroleum Blends

One of the applications of partial molar volume is to calculate volume change due to mixing as shown by Eq. (6.86). However, for practical applications a simpler empirical method has been developed for calculation of volume change when petroleum products are blended.

Consider two liquid hydrocarbons or two different petroleum fractions (products) which are being mixed to produce a blend of desired characteristics. If the mixture is an ideal solution, volume of the mixture is simply the sum of volumes of the components before the mixing. This is equivalent to "no volume change due to mixing." Experience shows that when a low-molecular-weight hydrocarbon is added to a heavy molecular weight crude oil there is a shrinkage in volume. This is particularly the case when a crude oil API gravity is improved by addition of light products such as gasoline or lighter hydrocarbons (i.e., butane, propane). Assume the volume of light and heavy hydrocarbons before mixing are V_{light} and V_{heavy}, respectively. The volume of the blend is then calculated from the following relation [5]:

$$
\begin{aligned}
&V_{blend} = V_{heavy} + V_{light}(1 - S) \\
&S = 2.14 \times 10^{-5} C^{-0.0704} G^{1.76} \\
&G = \text{API}_{light} - \text{API}_{heavy} \\
&C = \text{vol\% of light component in the mixture}
\end{aligned}
\tag{6.95}
$$

where S is called shrinkage factor and G is the API gravity difference between light and heavy component. The amount of shrinkage of light component due to mixing is $V_{light}(1 - S)$. The following example shows application of this method.

Example 6.5—Calculate volume of a blend and its API gravity produced by addition of 10000 bbl of light naphtha with API gravity of 90 to 90000 bbl of a crude oil with API gravity of 30.

Solution—Equation (6.95) is used to calculate volume of blend. The vol% of light component is 10% so $C = 10$. $G = 90 - 30 = 60$. $S = 2.14 \times 10^{-5} \times (10^{-0.0704}) \times 60^{1.76} = 0.025$. $V_{Blend} = 90000 + 10000(1 - 0.025) = 99750$ bbl. The amount of shrinkage of naphtha is $10000 \times 0.025 = 250$ bbl. As can be seen from Eq. (6.95) as the difference between densities of two components reduces the amount of shrinkage also decreases and for two oils with the same density there is no shrinkage. The percent shrinkage is 100S or 2.5% in this example. It should be noted that for calculation of density of mixtures a new composition should be calculated as: $x_{vi} = 9750/99750 = 0.0977$ which is equivalent to 9.77% instead of 10% originally assumed. For this example the mixture API gravity is calculated as: $\text{SG}_L = 0.6388$ and $\text{SG}_H = 0.8762$ where L and H refer to light and heavy components. Now using Eq. (3.45): $\text{SG}_{Blend} = (1 - 0.0977) \times 0.8762 + 0.0977 \times 0.6388 = 0.853$ which gives API gravity of blend as 34.4 while direct application of mixing rule to the API gravity with original composition gives $\text{API}_{Blend} = (1 - 0.1) \times 30 + 0.1 \times 90 = 36$. Obviously the more accurate value for the API gravity of blend is 34.4. ◆

6.5 PHASE EQUILIBRIA OF PURE COMPONENTS—CONCEPT OF SATURATION PRESSURE

As discussed in Section 5.2, a pure substance may exist in a solid, liquid or vapor phases (i.e., see Fig. 5.1). For pure substances four types of equilibrium exist: vapor–liquid (VL), vapor–solid (VS), liquid–solid (LS) and vapor–liquid–solid (VLS) phases. As shown in Fig. 5.2 the VLS equilibrium occurs only at the triple point, while VL, VS, and LS equilibrium exist over a range of temperature and pressure. One important type of phase equilibria in the thermodynamics of petroleum fluids is *vapor–liquid equilibria* (VLE). The VLE line also called *vapor pressure* curve for a pure substance begins from triple point and ends at the critical point (Fig. 5.2a). The equilibrium curves between solid and liquid is called *fusion* line and between vapor and solid is called *sublimation* line. Now we formulate VLE; however, the same approach may be used to formulate any type of multiphase equilibria for single component systems.

Consider vapor and liquid phases of a substance coexist in equilibrium at T and P (Fig. 6.7a). The pressure is called saturation pressure or vapor pressure and is shown by P^{sat}. As shown in Fig. 5.2a, vapor pressure increases with temperature and the critical point, normal boiling point and triple point are all located on the vapor pressure curve. As was shown in Fig. 2.1, for hydrocarbons the ratio T_b/T_c known as *reduced boiling point* varies from 0.6 to more than one for very heavy compounds. While the triple point temperature is almost the same as the freezing point temperature, but the triple point pressure is much lower than atmospheric pressure at which

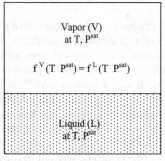

a. Pure Component System

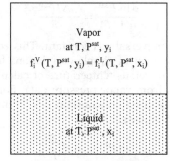

b. Multi Component System

FIG. 6.7—General criteria for vapor–liquid equilibrium.

freezing occurs. When a system is in equilibrium its energy is in minimum level ($dG = 0$), which for a system with only vapor and liquid is $d(G^V - G^L) = 0$, which can be written as [1]:

$$(6.96) \qquad dG^V(T, P^{sat}) = dG^L(T, P^{sat})$$

where P^{sat} indicates that the relation is valid at the saturation temperature and pressure. Similar equation applies to solid–liquid or solid–vapor phases.

During a phase change (i.e., vapor to liquid or *vice versa*), temperature and pressure of the system remain constant and therefore from Eq. (6.5) we have:

$$(6.97) \qquad \Delta S^{vap} = \frac{\Delta H^{vap}}{T}$$

where ΔH^{vap} is *heat of vaporization* and ΔS^{vap} is the *entropy of vaporization*. ΔH^{vap} is defined as:

$$(6.98) \qquad \begin{aligned} \Delta H^{vap} = {} & H(T, P^{sat}, \text{saturated vapor}) \\ & - H(T, P^{sat}, \text{saturated liquid}) \end{aligned}$$

Similarly ΔS^{vap} and ΔV^{vap} are defined. For a phase change from solid to liquid instead of heat of vaporization ΔH^{vap}, heat of fusion or melting ΔH^{fus} is defined by the difference between enthalpy of saturated liquid and solid. Since P^{sat} is only a function of temperature, ΔS^{vap} and ΔV^{vap} are also functions of temperature only for any pure substance. ΔH^{vap} and ΔV^{vap} decrease with increase in temperature and at the critical point they approach zero as vapor and liquid phases become identical. While ΔV^{vap} can be calculated from an equation of state as was discussed in Chapter 5, methods of calculation ΔH^{vap} will be discussed in Chapter 7. By applying Eq. (6.8) to both dG^V and dG^L and use of Eqs. (6.97) and (6.98) the following relation known as *Clapeyron equation* can be derived:

$$(6.99) \qquad \frac{dP^{sat}}{dT} = \frac{\Delta H^{vap}}{T \Delta V^{vap}}$$

This equation is the basis of development of predictive methods for vapor pressure versus temperature. Now three simplifying assumptions are made: (1) over a narrow range of temperature, ΔH^{vap} is constant, (2) volume of liquid is small in comparison with vapor volume ($\Delta V^{vap} = V^V - V^L \cong V^V$), and (3) volume of vapor can be calculated from ideal gas law (Eq. 5.14). These assumptions are not true in general but at a narrow range of temperature and low pressure conditions they can be used for simplicity. Upon application of assumptions 2 and 3, Eq. (6.99) can be written in the following form known as *Clausius–Clapeyron equation*:

$$(6.100) \qquad \frac{d \ln P^{sat}}{d(1/T)} = -\frac{\Delta H^{vap}}{R}$$

where R is the universal gas constant. This equation is the basis of development of simple correlations for estimation of vapor pressure versus temperature or calculation of heat of vaporization from vapor pressure data. For example, by using the first assumption (constant ΔH^{vap}) and integrating the above equation we get

$$(6.101) \qquad \ln P^{sat} = A - \frac{B}{T}$$

where T is absolute temperature and A and B are two positive constants specific for each pure substance. This equation suggests that $\ln P^{vap}$ versus $1/T$ is a straight line with slope of $-B$.

Constant B is in fact same as $\Delta H^{vap}/R$. Because of three major simplifying assumptions made above, Eq. (6.101) is very approximate and it may be used over a narrow temperature range when minimum data are available. Constants A and B can be determined from minimum two data points on the vapor pressure curve. Usually the critical point (T_c, P_c) and normal boiling point (1.01325 bar and T_b) are used to obtain the constants. If ΔH^{vap} is known, then only one data point (T_b) would be sufficient to obtain the vapor pressure correlation. A more accurate vapor pressure correlation is the following three-constant correlation known as Antoine equation:

$$(6.102) \qquad \ln P^{sat} = A - \frac{B}{T + C}$$

A, B, and C, known as Antoine constants, have been determined for a large number of compounds. Antoine proposed this simple modification of the Clasius–Clapeyron equation in 1888. Various modifications of this equation and other correlations for estimation of vapor pressure are discussed in the next chapter.

Example 6.6—For pure water, estimate vapor pressure of water at 151.84°C. What is its heat of vaporization? The actual values as given in the steam tables are 5 bar and 2101.6 kJ/kg, respectively [1]. Assume that only T_b, T_c, and P_c are known.

Solution—From Table 2.1 for water we have $T_c = 647.3$ K, $P_c = 220.55$ bar, and $T_b = 100$°C. Applying Eq. (6.101) at the critical point and normal boiling point gives $\ln P_c = A - B/T_c$ and $\ln(1.01325) = A - B/T_b$. Simultaneous solution of these equations gives the following relations to calculate A and B from T_b, T_c, and P_c.

$$(6.103) \qquad \begin{aligned} B &= \frac{\ln \left(\frac{P_c}{1.01325} \right)}{\frac{1}{T_b} - \frac{1}{T_c}} \\ A &= 0.013163 + \frac{B}{T_b} \end{aligned}$$

where T_c and T_b must be in kelvin and P_c must be in bar. The same units must be used in Eq. (6.101). In cases that a value of vapor pressure at one temperature is known it should be used instead of T_c and P_c so the resulting equation will be more accurate between that point and the boiling point. As the difference between temperatures of two reference points used to obtain constants in Eq. (6.101) reduces, the accuracy of resulting equation for the vapor pressure between two reference temperatures increases. For water from Eq. (6.103), $A = 12.7276$ bar and $B = 4745.66$ bar·K. Substituting A and B in Eq. (6.101) at $T = 151.84 + 273.15 = 425$ K gives $\ln P = 1.5611$ or $P = 4.764$ bar. Comparing predicted value with the actual value of 5 bar gives an error of −4.9%, which is acceptable considering simple relation and minimum data used. Heat of vaporization is calculated as follows: $\Delta H^{vap} = RB = 8.314 \times 4745.66 = 39455.4$ J/mol = 39455.4/18 = 2192 kJ/kg. This value gives an error of +4.3%. Obviously more accurate method of estimation of heat of vaporization is through $\Delta H^{vap} = H^{sat,vap} - H^{sat,liq}$, where $H^{sat,vap}$ and $H^{sat,liq}$ can be calculated through generalized correlations. Empirical methods of calculation of heat of vaporization are given in Chapter 7. ♦

An alternative method for formulation of VLE of pure substances is to combine Eqs. (6.47) and (6.96), which gives the following relation in terms of fugacity:

$$f^V = f^L \tag{6.104}$$

where f^V and f^L are fugacity of a pure substance in vapor and liquid phases at T and P^{sat}. Obviously for solid–liquid equilibrium, superscript V in the above relation is replaced by S indicating fugacity of solid is the same as fugacity of liquid. Since at VLE pressure of both phases is the same, an alternative form of Eq. (6.104) is

$$\phi^V(T, P^{sat}) = \phi^L(T, P^{sat}) \tag{6.105}$$

An equation of state or generalized correlation may be used to calculate both ϕ^V and ϕ^L if T and P^{sat} are known. To calculate vapor pressure (P^{sat}) from the above equation a trial-and-error procedure is required. Value of P^{sat} calculated from Eq. (6.101) may be used as an initial guess. To terminate calculations an error parameter can be defined as

$$\varepsilon = \left| 1 - \frac{\phi^L}{\phi^V} \right| \tag{6.106}$$

when ε is less than a small value (i.e., 10^{-6}) calculations may be stopped. In each round of calculations a new guess for pressure may be calculated as follows: $P^{new} = P^{old}(\phi^L/\phi^V)$. The following example shows the procedure.

Example 6.7—Repeat Example 6.6 using Eq. (6.105) and the SRK EOS to estimate vapor pressure of water at 151.84°C. Also calculate V^L and V^V at this temperature.

Solution—For water $T_c = 647.3$ K, $P_c = 220.55$ bar, $\omega = 0.3449$, and $Z_{RK} = 0.2338$. Using the units of bar, cm^3/mol, and kelvin for P, V, and T with $R = 83.14$ cm$^3 \cdot$ bar/mol$\cdot$K and $T = 423$ K, SRK parameters are calculated using relations given in Tables 5.1 and 6.1 as follows: $a_c = 5.6136 \times 10^{-6}$ bar (cm^3/mol)^2, $\alpha = 1.4163$, $a = 7.9504 \times 10^{-6}$ bar (cm^3/mol)^2, $b = 21.1$ cm^3/mol, $A = 0.030971$, and $B = 0.00291$. Relation for calculation of ϕ for SRK is given in Table 6.1 as follows: $\ln \phi = Z - 1 - \ln(Z - B) + \frac{A}{B} \ln \left(\frac{z}{z+B} \right)$, where Z for both saturated liquid and vapor is calculated from solution of cubic equation (SRK EOS): $Z^3 - Z^2 + (A - B - B^2)Z - AB = 0$. The first initial guess is to use the value of P calculated in Example 6.6 from Eq. (6.101): $P = 4.8$ bar, which results in $\varepsilon = 1.28 \times 10^{-2}$ (from Eq. (6.106)) as shown in Table 6.9. The second guess for P is calculated as $P = 4.86 \times (0.9848/0.97235) = 4.86$, which gives a lower value for ε. Summary of results is shown in Table 6.9. The final answer is $P^{sat} = 4.8637$ bar, which differs by -2.7% from the actual value of 5 bar. Values of specific volumes of liquid and vapor are calculated from Z^L and Z^V: $Z^L = 0.003699$ and $Z^V = 0.971211$. Molar volume is calculated from $V = ZRT/P$, where $R = 83.14$, $T = 425$ K, and $P = 4.86$ bar. $V^L = 26.9$

TABLE 6.9—*Estimation of vapor pressure of water at 151.8°C from SRK EOS (Example 6.7).*

P, bar	Z^L	Z^V	ϕ^L	ϕ^V	ε
4.8	0.00365	0.9716	0.9848	0.97235	1.2×10^{-2}
4.86	0.003696	0.9712	0.9727	0.972	7.4×10^{-4}
4.8637	0.003699	0.971211	0.971981	0.971982	1.1×10^{-6}

and $V^V = 7055.8$ cm^3/mol. The volume translation parameter c is calculated from Eq. (5.51) as $c = 6.03$ cm^3/mol, which through use of Eq. (6.50) gives $V^L = 20.84$ and $V^V = 7049.77$ cm^3/mol. The specific volume is calculated as V(molar)/M, where for water $M = 18$. Thus, $V^L = 1.158$ and $V^V = 391.653$ cm^3/g. Actual values of V^L and V^V are 1.093 and 374.7 cm^3/g, respectively [1]. The errors for calculated V^L and V^V are $+5.9$ and $+4.6\%$, respectively. For a cubic EOS these errors are acceptable, although without correction factor by volume translation the error for V^L is 36.7%. However, for calculation of volume translation a fourth parameter, namely Racket parameter is required. It is important to note that in calculation of fugacity coefficients through a cubic EOS use of volume translation, c, for both vapor and liquid does not affect results of vapor pressure calculation from Eq. (6.105). This has been shown in various sources [20]. ◆

Equation (6.105) is the basis of determination of EOS parameters from vapor pressure data. For example, coefficients given in Table 5.8 for the LK EOS (Eqs. 5.109–5.111) or the f_ω relations for various cubic EOSs given in Table 5.1 were found by matching predicted P^{sat} and saturated liquid density with the experimental data for pure substances for each equation.

The same principle may be applied to any two-phase system in equilibrium, such as VSE or SLE, in order to derive a relation between saturation pressure and temperature. For example, by applying Eq. (6.96) for solid and vapor phases, a relation for vapor pressure curve for sublimation (i.e., see Fig. 5.2a) can be derived. The final resulting equation is similar to Eq. (6.101), where parameter B is equal to $\Delta H^{sub}/R$ in which ΔH^{sub} is the heat of sublimation in J/mol as shown by Eq. (7.27). Then A and B can be determined by having two points on the sublimation curve. One of these points is the triple point (Fig. 5.2a) as discussed in Section 7.3.4. The same approach can be applied to SLE and derive a relation for melting (or freezing) point line (see Fig. 5.2a) of pure components. This is shown in the following example.

Example 6.8—*Effect of pressure on the melting point:* Derive a general relation for melting point of pure components versus pressure in terms of heat of melting (or fusion), ΔH^M, and volume change due to melting ΔV^M, assuming both of these properties are constant with respect to temperature. Use this equation to predict

a. melting point of n-octadecane (n-C$_{18}$) at 300 bar and
b. triple point temperature.

The following data are available from DIPPR data bank [13]: Normal melting point, $T_{M°} = 28.2°$C; heat of melting at normal melting point, $\Delta H^M = 242.4597$ kJ/kg; liquid density at T_{Mo}, $\rho^L = 0.7755$ g/cm^3; solid density at T_M, $\rho^S = 0.8634$ g/cm^3; and triple point pressure, $P_{tp} = 3.39 \times 10^{-5}$ kPa.

Solution—To derive a general relation for saturation pressure versus temperature for melting/freezing point of pure compounds we start by applying Eq. (6.96) between solid and liquid. Then Eq. (6.99) can be written as

$$\frac{dP}{dT_M} = \frac{\Delta H^M}{T_M \Delta V^M}$$

where T_M is the melting point temperature at pressure P. If atmospheric pressure is shown by P^o (1.01325 bar) and the melting point at P^o is shown by T_{Mo} (normal melting point), integration of the above equation from P^o to pressure P gives

$$(6.107) \qquad T_M = T_{Mo} \exp \left[\frac{\Delta V^M \times (P - P_o)}{\Delta H^M} \right]$$

where in deriving this equation it is assumed that both ΔV^M and ΔH^M are constants with respect to temperature (melting point). This is a reasonable assumption since variation of T_M with pressure is small (see Fig. 5.2a). Since this equation is derived for pure substances, T_M is the same as freezing point (T_f) and ΔH^M is the same as heat of fusion (ΔH_f).

(a) To calculate melting point of $n\text{-}C_{18}$ at 300 bar, we have $P = 300$ bar, $P_o = 1.01325$ bar, $T_{Mo} = 301.4$ K, and $\Delta V^M = 1/\rho^V - 1/\rho^S = 0.1313$ cm^3/g. $\Delta H^M = 242.4597$ J/g, $1/J = 10$ bar $\cdot$ cm^3, thus from Eq. (6.107) we have $T_M = 301.4 \times \exp$ [$0.1313 \times (300 - 1.013)/10 \times 242459.7$] $= 301.4 \times 1.0163 =$ $301.4 \times 1.0163 = 306.3$ K or $T_M = 33.2°C$. This indicates that when pressure increases to 300 bar, the melting point of $n\text{-}C_{18}$ increases only by 5°C. In this temperature range assumption of constant ΔV^M and ΔH^M is quite reasonable. (b) To calculate the triple point temperature, Eq. (6.107) must be applied at $P = P_{tp} = 3.39 \times 10^{-5}$ kPa $= 3.39 \times 10^{-7}$ bar. This is a very low number in comparison with $P_o = 1$ bar, thus $T_M = 301.4 \times \exp(-0.1313 \times 1.013/10 \times 242459.7) = 301.4 \times 0.99995 \cong 301.4$ K. Thus, we get triple point temperature same as melting point. This is true for most of pure substances as P_{tp} is very small. It should be noted that Eq. (6.107) is not reliable to calculate pressure at which melting point is known because a small change in temperature causes significant change in pressure. This example explains why melting point of water decreases while for n-octadecane it increases with increase in pressure. As it is shown in Section 7.2 density of ice is less than water, thus ΔV^M for water is negative and from Eq. (6.107), T_M is less than T_{Mo} at high pressures. ◆

6.6 PHASE EQUILIBRIA OF MIXTURES—CALCULATION OF BASIC PROPERTIES

Perhaps one of the biggest applications of equations of state and thermodynamics of mixtures in the petroleum science is formulation of phase equilibrium problems. In petroleum production phase equilibria calculations lead to the determination of the composition and amount of oil and gas produced at the surface facilities in the production sites, PT diagrams to determine type of hydrocarbon phases in the reservoirs, solubility of oil in water and water in oils, compositions of oil and gas where they are in equilibrium, solubility of solids in oils, and solid deposition (wax and asphaltene) or hydrate formation due to change in composition or T and P. In petroleum processing phase equilibria calculations lead to the determination of vapor pressure and equilibrium curves needed for design and operation of distillation, absorption, and stripping columns.

A system is at equilibrium when there is no tendency to change. In fact for a multicomponent system of single phase to be in equilibrium, there must be no change in T, P, and $x_1, x_2, \ldots, x_{N-1}$. When several phases exist together while at

equilibrium similar criteria must apply to every phase. In this case every phase has different composition but all have the same T and P. We know for mechanical equilibrium, total energy (i.e., kinetic and potential) of the system must be minimum. The best example is oscillation of hanging object that it comes to rest when its potential and kinetic energies are minimum at the lowest level. For thermodynamic equilibrium the criterion is minimum Gibbs free energy. As shown by Eq. (6.73) a mixture molar property such as G varies with T and P and composition. A mathematical function is minimum when its total derivative is zero:

$$(6.108) \qquad dG(T, P, x_i) = 0$$

Schematic and criteria for VLE of multicomponent systems are shown in Fig. 6.7b. Phase equilibria calculations lead to determination of the conditions of T, P, and composition at which the above criteria are satisfied. In this section general formulas for phase equilibria calculations of mixtures are presented. These are required to define new parameters such as activity, activity coefficient, and fugacity coefficient of a component in a mixture. Two main references for thermodynamics of mixtures in relation with equilibrium are Denbigh [19] and Prausnitz et al. [21].

6.6.1 Definition of Fugacity, Fugacity Coefficient, Activity, Activity Coefficient, and Chemical Potential

In this section important properties of fugacity, activity, and chemical potential needed for formulation of solution thermodynamics are defined and methods of their calculation are presented. Consider a mixture of N components at T and P and composition y_i. Fugacity of component i in the mixture is shown by $\hat{f}_i$ and defined as

$$(6.109) \qquad \lim \left(\frac{\hat{f}_i}{y_i P} \right)_{P \to 0} \to 1$$

where sign $\wedge$ indicates the fact that component i is in a mixture. When $y_i \to 1$ we have $\hat{f}_i \to f_i$, where f_i is fugacity of pure i as defined in Eq. (6.45). The fugacity coefficient of i in a mixture is defined as

$$(6.110) \qquad \hat{\phi}_i = \frac{\hat{f}_i}{y_i P}$$

where for an ideal gas, $\hat{\phi}_i = 1$ or $\hat{f}_i = y_i P$. In a gas mixture $y_i P$ is the same as partial pressure of component i. Activity of component i, $\hat{a}_i$, is defined as

$$(6.111) \qquad \hat{a}_i = \frac{\hat{f}_i}{f_i^\circ}$$

where f_i° is fugacity of i at a standard state. One common standard state for fugacity is pure component i at the same T and P of mixture, that is to assume $f_i^\circ = f_i$, where f_i is the fugacity of pure i at T and P of mixture. This is usually known as standard state base on Lewis rule. Choice of standard state for fugacity and chemical potential is best discussed by Denbigh [19]. Activity is a parameter that indicates the degree of nonideality in the system. The activity coefficient of component i in a mixture is shown by γ_i and is defined as

$$(6.112) \qquad \gamma_i = \frac{\hat{a}_i}{x_i}$$

where x_i is mole fraction of component i in the mixture. Both $\hat{a}_i$ and γ_i are dimensionless parameters. With the above definitions one may calculate $\hat{f}_i$ from one of the following relations:

$$ (6.113) \qquad \hat{f}_i = \hat{\phi}_i y_i P $$

$$ (6.114) \qquad \hat{f}_i = x_i \gamma_i f_i $$

Although generally $\hat{\phi}_i$ and γ_i are defined for any phase, but usually $\hat{\phi}_i$ is used to calculate fugacity of i in a gas mixture and γ_i is used to calculate fugacity of component i in a liquid or solid solution. However, for liquid mixtures at high pressures, i.e., high pressure VLE calculations, $\hat{f}_i$ is calculated from $\hat{\phi}_i$ through Eq. (6.113). In such calculations as it will be shown later in this section, for the sake of simplicity and convenience, $\hat{\phi}_i$ for both phases are calculated through an equation of state. Both $\hat{\phi}_i$ and γ_i indicate degree of nonideality for a system. In a gas mixture, $\hat{\phi}_i$ indicates deviation from an ideal gas and in a liquid solution, γ_i indicates deviation from an ideal solution. To formulate phase equilibrium of mixtures a new parameter called *chemical potential* must be defined.

$$ (6.115) \qquad \hat{\mu}_i \equiv \left(\frac{\partial G^t}{\partial n_i} \right)_{T,P,n_{j \neq i}} = \bar{G}_i $$

where $\hat{\mu}_i$ is the chemical potential of component i in a mixture and $\bar{G}_i$ is the partial molar Gibbs energy. General definition of partial molar properties was given by Eq. (6.78). For a pure component both partial molar Gibbs energy and molar Gibbs energy are the same: $\bar{G}_i = G_i$. For a pure ideal gas and an ideal gas mixture from thermodynamic relations we have

$$ (6.116) \qquad dG_i = RT d\ln P $$

$$ (6.117) \qquad d\bar{G}_i = RT d\ln(y_i P) $$

where in Eq. (6.117) if $y_i = 1$, it reduces to Eq. (6.116) for pure component systems. For real gases these equations become

$$ (6.118) \qquad dG_i = d\mu_i = RT d\ln f_i $$

$$ (6.119) \qquad d\bar{G}_i = d\hat{\mu}_i = RT d\ln \hat{f}_i $$

Equation (6.118) is the same as Eq. (6.46) derived for pure components. Equation (6.119) reduces to Eq. (6.118) at $y_i = 1$. Subtracting Eq. (6.118) from Eq. (6.119) and using Eq. (6.114) for $\hat{f}_i$ one can derive the following relation for $\hat{\mu}_i$ in a solution:

$$ (6.120) \qquad \hat{\mu}_i - \mu_i^\circ = RT \ln \gamma_i x_i $$

where μ_i° is the pure component chemical potential at T and P of mixture and x_i is the mole fraction of i in liquid solution. For ideal solutions where $\gamma_i = 1$, Eq. (6.120) reduces to $\hat{\mu}_i - \mu_i^\circ = RT \ln x_i$. In fact this is another way to define an ideal solution. A solution that is ideal over the entire range of composition is called *perfect solution* and follows this relation.

6.6.2 Calculation of Fugacity Coefficients from Equations of State

Through thermodynamic relations and definition of G one can derive the following relation for the mixture molar Gibbs free energy [21].

$$ (6.121) \qquad G = \int_0^P \left(V - \frac{RT}{P} \right) dP + RT \sum_i y_i \ln(y_i P) + \sum_i y_i G_i^\circ $$

where G_i° is the molar Gibbs energy of pure i at T of the system and pressure of 1 atm (ideal gas state). By replacing $G = \sum y_i \hat{\mu}_i$, and $V = \sum y_i \bar{V}_i$ in the above equation and removing the summation sign we get

$$ (6.122) \qquad \hat{\mu}_i = \int_0^P \left(\bar{V}_i - \frac{RT}{P} \right) dP + RT \ln(y_i P) + G_i^\circ $$

Integration of Eq. (6.119) from pure ideal gas at T and $P = 1$ atm to real gas at T and P gives

$$ (6.123) \qquad \hat{\mu}_i - \mu_i^\circ = RT \ln \frac{\hat{f}_i}{1} $$

where μ_i° is the chemical potential of pure component i at T and pressure of 1 atm (ideal gas as a standard state). For a pure component at the same T and P we have: $\mu_i^\circ = G_i^\circ$. Combining Eqs. (6.122) and (6.123) gives

$$ (6.124) \qquad RT \ln \left(\frac{\hat{f}_i}{y_i P} \right) = RT \ln \hat{\phi}_i = \int_0^P \left(\bar{V}_i - \frac{RT}{P} \right) dP $$

where $\bar{V}_i$ is the partial molar volume of component i in the mixture. It can be seen that for a pure component ($\bar{V}_i = V_i$ and $y_i = 1$) this equation reduces to Eq. (6.53) previously derived for calculation of fugacity coefficient of pure components. There are other forms of this equation in which integration is carried over volume in the following form [21]:

$$ (6.125) \qquad RT \ln \hat{\phi}_i = \int_{V^t}^\infty \left[\left(\frac{\partial P}{\partial n_i} \right)_{T,V,n_{j \neq i}} - \frac{RT}{V^t} \right] dV^t - \ln Z $$

where V^t is the total volume ($V^t = nV$). In using these equations one should note that n is the sum of n_i and is not constant when derivative with respect to n_i is carried. These equations are the basis of calculation of fugacity of a component in a mixture. Examples of such derivations are available in various texts [1, 4, 11, 20–22]. One can use an EOS to obtain $\bar{V}_i$ and upon substitution in Eq. (6.124) a relation for calculation of $\hat{\phi}_i$ can be obtained. For the general form of cubic equations given by Eqs. (5.40)–(5.42), $\hat{\phi}_i$ is given as [11]

$$ (6.126) \qquad \begin{aligned} \ln \hat{\phi}_i &= \frac{b_i}{b}(Z-1) - \ln(Z-B) + \frac{A}{B\sqrt{u_1^2 - u_2^2}} \left(\frac{b_i}{b} - \delta_i \right) \\ &\times \ln \frac{2Z + B\left(u_1 + \sqrt{u_1^2 - 4u_2}\right)}{2Z + B\left(u_1 - \sqrt{u_1^2 - 4u_2}\right)} \end{aligned} $$

where

$$ \frac{b_i}{b} = \frac{T_{ci}/P_{ci}}{\sum_j y_j T_{cj}/P_{cj}} $$

and

$$ \delta_i = \frac{2a_i^{1/2}}{a} \sum_i x_j a_j^{1/2}(1 - k_{ij}) $$

if all $k_{ij} = 0$ then $\delta_i = 2\left(\frac{a_i}{a} \right)^{1/2}$

All parameters in the above equation for vdW, RK, SRK, and PR equations of state are defined in Tables 5.1 and 6.1. Parameters a and b for the mixture should be calculated from Eqs. (5.59)–(5.61). Equation (6.126) can be used for calculation of fugacity of i in both liquid and vapor phases provided appropriate Z values are used as for the case of pure component systems that was shown in Example 6.7. For calculation of $\hat{\phi}_i$ from PR and SRK equations through the above relation, use of volume translation is not required.

If truncated virial equation (Eq. 5.75) is used, $\hat{\phi}_i$ is calculated from the following relation as derived from Eq. (6.124):

$$(6.127) \qquad \ln \hat{\phi}_i = \left(2 \sum_j y_j B_{ij} - B\right) \frac{P}{RT}$$

where B (for whole mixture) and B_{ij} (interaction coefficients) should be calculated from Eqs. (5.70) and (5.74), respectively. As discussed earlier Eq. (5.70) is useful for gases at moderate pressures. Equation (6.127) is not valid for liquids.

Example 6.9—Suppose that fugacity coefficient of the whole mixture, ϕ_{mix}, is defined similar to that of pure components. Through mixture Gibbs energy, derive a relation between f_{mix} and f_i for mixtures.

Solution—Applying Eq. (6.80) to residual molar Gibbs free energy ($G^R = G - G^{\text{ig}}$) gives $G^R = \sum y_i \bar{G}_i^R$ and since $\mu_i = \bar{G}_i$ from Eq. (6.119) $d\bar{G}_i = RT d \ln \hat{f}_i$ and for ideal gases from Eq. (6.117) we have $dG_i^{\text{ig}} = RT d \ln y_i P$. Subtracting these two relations from each other gives $d\bar{G}_i^R = RT d \ln \hat{\phi}_i$, which after integration gives $\bar{G}_i^R = RT \ln \hat{\phi}_i$. Therefore for the whole mixture we have

$$(6.128) \qquad G^R = RT \sum x_i \ln \hat{\phi}_i$$

where after comparing with Eq. (6.48) for the whole mixture we get

$$(6.129) \qquad \ln \phi_{\text{mix}} = \sum x_i \ln \hat{\phi}_i$$

or in terms of fugacity for the whole mixture, f_{mix}, it can be written as

$$(6.130) \qquad \ln f_{\text{mix}} = \sum x_i \ln \frac{\hat{f}_i}{x_i}$$

This relation can be applied to both liquid and gases. f_{mix} is useful for calculation of properties of only real mixtures but is not useful for phase equilibrium calculation of mixtures except under certain conditions (see Problem 6.19). ◆

6.6.3 Calculation of Fugacity from Lewis Rule

Lewis rule is a simple method of calculation of fugacity of a component in mixtures and it can be used if the assumptions made are valid for the system of interest. The main assumption in deriving the Lewis fugacity rule is that the molar volume of the mixture at constant temperature and pressure is a linear function of the mole fraction (this means $\bar{V}_i = V_i = $ constant). This assumption leads to the following simple rule for $\hat{f}_i$ known as *Lewis/Randall* or simply *Lewis rule* [21, 22]:

$$(6.131) \qquad \hat{f}_i(T, P) = y_i f_i(T, P)$$

where $f_i(T, P)$ is the fugacity of pure i at T and P of mixture. Lewis rule simply says that in a mixture $\hat{\phi}_i$ is only a function of T and P and not a function of composition. Direct conclusion of Lewis rule is

$$(6.132) \qquad \hat{\phi}_i(T, P) = \phi_i(T, P)$$

which can be obtained by dividing both sides of Eq. (6.131) by $y_i P$. The Lewis rule may be applied to both gases and liquids with the following considerations [21]:

—Good approximation for gases at low pressure where the gas phase is nearly ideal.
—Good approximation at any pressure whenever i is present in large excess (say, $y_i > 0.9$). The Lewis rule becomes exact in the limit of $y_i \to 1$.
—Good approximation over all range of pressure and composition whenever physical properties of all components present in the mixture are the same as (i.e., benzene and toluene mixture).
—Good approximation for liquid mixtures whose behavior is like an ideal solution.
—A poor approximation at moderate and high pressures whenever the molecular properties of components in the mixture are significantly different from each other (i.e., a mixture of methane and a heavy hydrocarbon).

Lewis rule is attractive because of its simplicity and is usually used when the limiting conditions are applied in certain situations. Therefore when the Lewis rule is used, fugacity of i in a mixture is calculated directly from its fugacity as pure component. When Lewis rule is applied to liquid solutions, Eq. (6.114) can be combined with Eq. (6.131) to get $\gamma_i = 1$ (for all components).

6.6.4 Calculation of Fugacity of Pure Gases and Liquids

Calculation of fugacity of pure components using equations of state was discussed in Section 6.5. Generally fugacity of pure gases and liquids at moderate and high pressures may be estimated from equations given in Table 6.1 or through generalized correlations of LK as given by Eq. (6.59). For pure gases at moderate and low pressures Eq. (6.62) derived from virial equation can be used.

To calculate fugacity of i in a liquid mixture through Eq. (6.114) one needs fugacity of pure liquid i in addition to the activity coefficient. To calculate fugacity of a pure liquid i at T and P, first its fugacity is calculated at T and corresponding saturation P^{sat}. Under the conditions of T and P^{sat} both vapor and liquid phases of pure i are in equilibrium and thus

$$(6.133) \qquad f_i^L(T, P^{\text{sat}}) = f_i^V(T, P^{\text{sat}}) = \phi_i^{\text{sat}} P^{\text{sat}}$$

where ϕ_i^{sat} is the fugacity coefficient of pure vapor at T and P^{sat}. Effect of pressure on liquid fugacity should be considered to calculate $f_i^L(T, P)$ from $f_i^L(T, P^{\text{sat}})$. This is obtained by combining Eq. (6.8) (at constant T) and Eq. (6.47):

$$(6.134) \qquad dG_i = RT d \ln f_i = V_i dP$$

Integration of this equation from P^{sat} to desired pressure of P for the liquid phase gives

$$(6.135) \qquad \ln \frac{f_i^L(T, P)}{f_i^L(T, P^{\text{sat}})} = \int_{P_i^{\text{at}}}^{P} \frac{V_i^L}{RT} dP$$

Combining Eqs. (6.133) and (6.135) leads to the following relation for fugacity of pure i in liquid phase.

$$(6.136) \qquad f_i^L(T, P) = P_i^{\text{sat}} \phi_i^{\text{sat}} \exp\left(\int_{P_i^{\text{at}}}^{P} \frac{V_i^L}{RT} dP\right)$$

P^{sat} is the saturation pressure or vapor pressure of pure i at T and methods of its calculation are discussed in the next chapter. ϕ_i^{sat} is the vapor phase fugacity coefficient of pure component i at P_i^{sat} and can be calculated from methods discussed in Section 6.2. The exponential term in the above equation is called *Poynting correction* and is calculated from liquid molar volume. Since variation of V_i^L with pressure is small, usually it is assumed constant versus pressure and the Poynting factor is simplified as $\exp[V_i^L(P - P_i^{sat})/RT]$. In such cases V_i^L may be taken as molar volume of saturated liquid at temperature T and it may be calculated from Racket equation (Section 5.8). At very low pressures or when $(P - P_i^{sat})$ is very small, the Poynting factor approaches unity and it could be removed from Eq. (6.136). In addition, when P_i^{sat} is very small (~ 1 atm or less), ϕ_i^{sat} may be considered as unity and f_i^L is simply equal to P_i^{sat}. Obviously this simplification can be used only in special situations when the above assumptions can be justified. For calculation of Poynting factor when V_i^L is in cm^3/mol, P in bar, and T in kelvin, then the value of R is 83.14.

6.6.5 Calculation of Activity Coefficients

Activity coefficient γ_i is needed in calculation of fugacity of i in a liquid mixture through Eq. (6.114). Activity coefficients are related to excess molar Gibbs energy, G^E, through thermodynamic relations as [21]

$$(6.137) \qquad RT \ln \gamma_i = \bar{G}_i = \left[\frac{\partial (nG^E)}{\partial n_i} \right]_{T,P,n_{j \neq i}}$$

where $\bar{G}_i$ is the partial molar excess Gibbs energy as defined by Eq. (6.78) and may be calculated by Eq. (6.82). This equation leads to another equally important relation for the activity coefficient in terms of excess Gibbs energy, G^E:

$$(6.138) \qquad G^E = RT \sum_i x_i \ln \gamma_i$$

where this equation is obtained by substitution of Eq. (6.137) into Eq. (6.79). Therefore, once the relation for G^E is known it can be used to determine γ_i. Similarly, when γ_i is known G^E can be calculated. Various models have been proposed for G^E of binary systems. Any model for G^E must satisfy the condition that when $x_1 = 0$ or 1 ($x_2 = 0$), G^E must be equal to zero; therefore, it must be a factor of $x_1 x_2$. One general model for G^E of binary systems is called Redlich–Kister expansion and is given by the following power series form [1, 21]:

$$(6.139) \qquad \frac{G^E}{RT} = x_1 x_2 [A + B(x_1 - x_2) + C(x_1 - x_2)^2 + \cdots]$$

where $A, B, \ldots$ are empirical temperature-dependent coefficients. If all these coefficients are zero then the solution is ideal. The simplest nonideal solution is when only coefficient A is not zero but all other coefficients are zero. This is known as two-suffix Margules equation and upon application of Eq. (6.137) the following relations can be obtained for γ_1 and γ_2:

$$(6.140) \qquad \begin{aligned} \ln \gamma_1 &= \frac{A}{RT} x_2^2 \\ \ln \gamma_2 &= \frac{A}{RT} x_1^2 \end{aligned}$$

According to the definition of γ_i when $x_i = 1$ (pure i) then $\gamma_i = 1$. Generally for binary systems when a relation for activity coefficient of one component is known the relation for activity coefficient of other components can be determined from the following relation:

$$(6.141) \qquad x_1 \frac{d \ln \gamma_1}{dx_1} = x_2 \frac{d \ln \gamma_2}{dx_2}$$

which is derived from Gibbs–Duhem equation. One can obtain γ_2 from γ_1 by applying the above equation with use of $x_2 = 1 - x_1$ and $dx_2 = -dx_1$. Constant A in Eq. (6.140) can be obtained from data on the activity coefficient at infinite dilution (γ_i^∞), which is defined as $\lim_{x_i \to 0}(\gamma_i)$. This will result in $A = RT \ln \gamma_1^\infty = RT \ln \gamma_2^\infty$. This simple model applies well to simple mixtures such as benzene–cyclohexane; however, for more complex mixtures other activity coefficient models must be used. A more general form of activity coefficients for binary systems that follow Redlich–Kister model for G^E are given as

$$(6.142) \qquad \begin{aligned} RT \ln \gamma_1 &= a_1 x_2^2 + a_2 x_2^3 + a_3 x_2^4 + a_4 x_2^5 + \cdots \\ RT \ln \gamma_2 &= b_1 x_1^2 + b_2 x_1^3 + b_3 x_1^4 + b_4 x_1^5 + \cdots \end{aligned}$$

If in Eq. (6.139) coefficient C and higher order coefficients are zero then resulting activity coefficients correspond to only the first two terms of the above equation. This model is called four-suffix Margules equation. Since data on γ_i^∞ are useful in obtaining the constants for an activity coefficient model, many researchers have measured such data for various systems. Figure 6.8 shows values of γ_i^∞ for n-C$_4$ and n-C$_8$ in various n-alkane solvents from C$_{15}$ to C$_{40}$ at 100°C based on data available from C$_{20}$ to C$_{36}$ [21]. As can be seen from this figure, as the size of solvent molecule increases the deviation of activity coefficients from unity also increases.

Another popular model for activity coefficient of binary systems is the van Laar model proposed by van Laar during 1910–1913. This model is particularly useful for binaries whose molecular sizes vary significantly. Van Laar model is

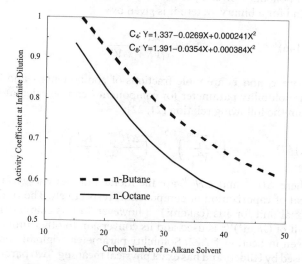

FIG. 6.8—Values of γ_i^∞ for *n*-butane and *n*-octane in *n*-paraffin solvents at 100°C.

based on the Wohl's model for the excess Gibbs energy [21]. The G^E relation for the van Laar model is given by

$$(6.143) \qquad \frac{G^E}{RT} = x_1 x_2 [A + B(x_1 - x_2)]^{-1}$$

Upon application of Eq. (6.137), the activity coefficients are obtained as

$$(6.144) \qquad \begin{aligned} \ln \gamma_1 &= A_{12} \left(1 + \frac{A_{12} x_1}{A_{21} x_2} \right)^{-2} \\ \ln \gamma_2 &= A_{21} \left(1 + \frac{A_{21} x_2}{A_{12} x_1} \right)^{-2} \end{aligned}$$

where coefficients A_{12} and A_{21} are related to A and B in Eq. (6.143) as $A - B = 1/A_{12}$ and $A + B = 1/A_{21}$. Coefficients A_{12} and A_{21} can be determined from the activity coefficients at infinite dilutions ($A_{12} = \ln \gamma_1^\infty$, $A_{21} = \ln \gamma_2^\infty$). Once for a given system VLE data are available, they can be used to calculate activity coefficients through Eqs. (6.179) or (6.181) and then G^E/RT is calculated from Eq. (6.138). From the knowledge of G^E/RT versus $(x_1 - x_2)$ the best model for G^E can be found. Once the relation for G^E has been determined the activity coefficient model will be found.

For *regular solutions* where different components have the same intermolecular forces it is generally assumed that $V^E = S^E = 0$. Obviously systems containing polar compounds generally do not fall into the category of regular solutions. Hydrocarbon mixtures may be considered as regular solutions. The activity coefficient of component i in a binary liquid solution according to the regular solution theory can be calculated from the Scatchard–Hildebrand relation [21, 22]:

$$(6.145) \qquad \begin{aligned} \ln \gamma_1 &= \frac{V_1^L (\delta_1 - \delta_2)^2 \, \Phi_2^2}{RT} \\ \ln \gamma_2 &= \frac{V_2^L (\delta_1 - \delta_2)^2 \, \Phi_1^2}{RT} \end{aligned}$$

where V^L is the liquid molar volume of pure components (1 or 2) at T and P and δ are the solubility parameter of pure components 1 or 2. Φ_1 is the volume fraction of component 1 and for a binary system it is given by

$$(6.146) \qquad \Phi_1 = \frac{x_1 V_1^L}{x_1 V_1^L + x_2 V_2^L}$$

where x_1 and x_2 are mole fractions of components 1 and 2. The solubility parameter for component i can be calculated from the following relation [21, 22]:

$$(6.147) \qquad \delta_i = \left(\frac{\Delta U_i^{vap}}{V_i^L} \right)^{1/2} = \left(\frac{\Delta H_i^{vap} - RT}{V_i^L} \right)^{1/2}$$

where ΔU_i^{vap} and ΔH_i^{vap} are the molar internal energy and heat of vaporization of component i, respectively. The traditional unit for δ is (cal/cm^3)$^{1/2}$; however, in this chapter the unit of (J/cm^3)$^{1/2}$ is used and its conversion to other units is given in Section 1.7.22. Solubility parameter originally proposed by Hildebrand has exact physical meaning. Two parameters that are used to define δ are energy of vaporization and molar volume. In Chapter 5 it was discussed that for nonpolar

compounds two parameters, namely energy parameter and size parameter describe the intermolecular forces. Energy of vaporization is directly related to the energy required to overcome forces between molecules in the liquid phase and molar volume is proportional to the molecular size. Therefore, when two components have similar values of δ their molecular size and forces are very similar. Molecules with similar size and intermolecular forces easily can dissolve in each other. The importance of solubility parameter is that when two components have δ values close to each other they can dissolve in each other appreciably. It is possible to use an EOS to calculate δ from Eq. (6.147) (see Problem 6.20). According to the theory of regular solutions, excess entropy is zero and it can be shown that for such solutions $RT \ln \gamma_i$ is constant at constant composition and does not change with temperature [11]. Values of V_i^L and δ_i at a reference temperature of 298 K is sufficient to calculate γ_i at other temperatures through Eq. (145). Values of solubility parameter for single carbon number components are given in Table 4.6. Values of V_i^L and δ_i at 25°C for a number of pure substances are given in Table 6.10 as provided by DIPPR [13]. In this table values of δ have the unit of (J/cm^3)$^{1/2}$. In Table 6.10 values of freezing point and heat of fusion at the freezing point are also given. These values are needed in calculation of fugacity of solids as will be seen in the next section.

Based on the data given in Table 6.10 the following relations are developed for estimation of liquid molar volume of n-alkanes (P), n-alkylcyclohexanes (N), and n-alkylbenzenes (A) at 25°C, V_{25} [23]: It gives Cp/Cv for saturated liquids having a calculated heat capacity ratio of 1.43 to 1.38 over a temperature range of 300–450 K.

$$(6.148) \qquad \begin{aligned} \ln V_{25} &= -0.51589 + 2.75092 M^{0.15} \\ &\qquad \text{for } n\text{-alkanes (C}_1 - \text{C}_{36}) \\ V_{25} &= 10.969 + 1.1784 M \\ &\qquad \text{for } n\text{-alkylcyclohexanes (C}_6 - \text{C}_{16}) \\ \ln V_{25} &= -96.3437 + 96.54607 M^{0.01} \\ &\qquad \text{for } n\text{-alkylbenzenes (C}_6 - \text{C}_{24}) \end{aligned}$$

where V_{25} is in cm^3/mol. These correlations can reproduce data in Table 6.10 with average deviations of 0.9, 0.4, and 0.2% for n-alkanes, n-alkylcyclohexanes, and n-alkylbenzens, respectively. Similarly the following relations are developed for estimation of solubility parameter at 25°C [23]:

$$(6.149) \qquad \begin{aligned} \delta &= 16.22609 \, [1 + \exp(0.65263 - 0.02318 M)]^{-0.4007} \\ &\qquad \text{for } n\text{-alkanes (C}_1 - \text{C}_{36}) \\ \delta &= 16.7538 + 7.2535 \times 10^{-5} M \\ &\qquad \text{for } n\text{-alkylcyclohexanes (C}_6 - \text{C}_{16}) \\ \delta &= 26.8557 - 0.18667 M + 1.36926 \times 10^{-3} M^2 \\ &\quad - 4.3464 \times 10^{-6} M^3 + 4.89667 \times 10^{-9} M^4 \\ &\qquad \text{for } n\text{-alkylbenzenes (C}_6 - \text{C}_{24}) \end{aligned}$$

where δ is in (J/cm^3)$^{1/2}$. The conversion factor from this unit to the traditional units is given in Section 1.7.22: 1 (cal/cm^3)$^{1/2}$ = 2.0455 (J/cm^3)$^{1/2}$. Values predicted from these equations give average deviation of 0.2% for n-alkanes, 0.5% for n-alkylcyclohexanes, and 1.4% for n-alkylbenzenes. It

TABLE 6.10—*Freezing point, heat of fusion, molar volume, and solubility parameters for some selected compounds [DIPPR].*

No.	Compound	Formula	N_C	M	T_M, K	$\Delta H_f/RT_M$ at T_M	V_{25}, cm³/mol	δ_{25}, (J/cm³)$^{1/2}$
	n-Paraffins							
1	Methane	CH_4	1	16.04	90.69	1.2484	37.969(52)[a]	11.6
2	Ethane	C_2H_6	2	30.07	90.35	3.8059	55.229(68)[a]	12.4
3	Propane	C_3H_8	3	44.09	85.47	4.9589	75.700(84)[a]	13.1
4	n-Butane	C_4H_{10}	4	58.12	134.86	4.1568	96.48(99.5)[a]	13.7
5	n-Pentane	C_5H_{12}	5	72.15	143.42	7.0455	116.05	14.4
6	n-Hexane	C_6H_{14}	6	86.17	177.83	8.8464	131.362	14.9
7	n-Heptane	C_7H_{16}	7	100.20	182.57	9.2557	147.024	15.2
8	n-Octane	C_8H_{18}	8	114.22	216.38	11.5280	163.374	15.4
9	n-Nonane	C_9H_{20}	9	128.25	219.66	8.4704	179.559	15.6
10	n-Decane	$C_{10}H_{22}$	10	142.28	243.51	14.1801	195.827	15.7
11	n-Undecane	$C_{11}H_{24}$	11	156.30	247.57	10.7752	212.243	15.9
12	n-Dodecane	$C_{12}H_{26}$	12	170.33	263.57	16.8109	228.605	16.0
13	n-Tridecane	$C13H_{28}$	13	184.35	267.76	12.8015	244.631	16.0
14	n-Tetradecane	$C_{14}H_{30}$	14	198.38	279.01	19.4282	261.271	16.1
15	n-Pentadecane	$C_{15}H_{32}$	15	212.41	283.07	14.6966	277.783	16.1
16	n-Hexadecane	$C_{16}H_{34}$	16	226.43	291.31	22.0298	294.213	16.2
17	n-Heptadecane	$C_{17}H_{36}$	17	240.46	295.13	16.3674	310.939	16.2
18	n-Octadecane	$C_{18}H_{38}$	18	254.48	301.31	24.6307	328.233	16.2
19	n-Nonadecane	$C_{19}H_{40}$	19	268.51	305.04	18.0620	345.621	16.2
20	n-Eicosane	$C_{20}H_{42}$	20	282.54	309.58	27.1445	363.69	16.2
21	n-Heneicosane	$C_{21}H_{44}$	21	296.56	313.35	18.3077	381.214	16.2
22	n-Docosane	$C_{22}H_{46}$	22	310.59	317.15	18.5643	399.078	16.2
23	n-Triacosane	$C_{23}H_{48}$	23	324.61	320.65	20.2449	416.872	16.3
24	n-Tetracosane	$C_{24}H_{50}$	24	338.64	323.75	20.3929	434.942	16.3
25	n-Hexacosane	$C_{26}H_{54}$	26	366.69	329.25	22.1731	469.975	16.3
26	n-Heptacosane	$C_{27}H_{56}$	27	380.72	332.15	21.8770	488.150	16.2
27	n-Octacosane	$C_{28}H_{58}$	28	394.74	334.35	23.2532	506.321	16.2
28	n-Nonacosane	$C_{29}H_{60}$	29	408.77	336.85	23.6034	523.824	16.2
29	n-Triacontane	$C_{30}H_{62}$	30	422.80	338.65	24.4439	540.500	16.2
30	n-Docontane	$C_{32}H_{66}$	32	450.85	342.35	26.8989	576.606	16.2
31	n-Hexacontane	$C_{36}H_{74}$	36	506.95	349.05	30.6066	648.426	16.2
	Isoparaffins							
32	Isobutane	C_4H_{10}	4	58.12	113.54	4.8092	105.238	12.57
33	Isopentane	C_5H_{12}	5	72.15	113.25	5.4702	117.098	13.86
34	Isooctane (2,2,4-trimethylpentane)	C_8H_{18}	8	114.23	165.78	6.6720	165.452	14.08
	n-Alkylcyclopentanes (naphthenes)							
35	Cyclopentane	C_5H_{10}	5	70.14	179.31	0.4084	94.6075	16.55
36	Methylcyclopentane	C_6H_{12}	6	84.16	146.58	4.7482	128.1920	16.06
37	Ethylcyclopentane	C_7H_{14}	7	98.19	134.71	6.1339	128.7490	16.25
38	n-Propylcyclopentane	C_8H_{16}	8	112.22	155.81	7.7431	145.1930	16.36
39	n-Butylcyclopentane	C_9H_{18}	9	126.24	165.18	8.2355	161.5720	16.39
	n-Alkylcyclohexanes (naphthenes)							
40	Cyclohexane	C_6H_{12}	6	84.16	279.69	1.1782	108.860	16.76
41	Methylcyclohexane	C_7H_{14}	7	98.18	146.58	5.5393	128.192	16.06
42	Ethylcyclohexane	C_8H_{16}	8	112.21	161.839	6.1935	143.036	16.34
43	n-Propylcyclohexane	C_9H_{18}	9	126.23	178.25	6.9970	159.758	16.35
44	n-Butylcyclohexane	$C_{10}H_{20}$	10	140.26	198.42	8.5830	176.266	16.40
45	n-Decylcyclohexane	$C_{16}H_{32}$	16	224.42	271.42	17.1044	275.287	16.65
	n-Alkylbenzenes (aromatics)							
46	Benzene	C_6H_6	6	78.11	278.65	4.2585	89.480	18.70
47	Methylbenzene (Toluene)	C_7H_8	7	92.14	178.15	4.4803	106.650	18.25
48	Ethylbenzene	C_8H_{10}	8	106.17	178.15	6.1983	122.937	17.98
49	Propylbenzene	C_9H_{12}	9	120.20	173.55	6.4235	139.969	17.67
50	n-Butylbenzene	$C_{10}H_{14}$	10	134.22	185.25	7.2849	156.609	17.51
51	n-Pentylbenzene	$C_{11}H_{16}$	11	148.25	198.15	9.2510	173.453	17.47
52	n-Hexylbenzene	$C_{12}H_{18}$	12	162.28	211.95	10.4421	189.894	17.43
53	n-Heptylbenzene	$C_{13}H_{20}$	13	176.30	225.15	11.6458	206.428	17.37
54	n-Octylbenzene	$C_{14}H_{22}$	14	190.33	237.15	13.1869	223.183	17.37
55	n-Nonylbenzene	$C_{15}H_{24}$	15	204.36	248.95	13.9487	239.795	17.39
56	n-Decylbenzene	$C_{16}H_{26}$	16	218.38	258.77	15.1527	256.413	17.28
57	n-Undecylbenzene	$C_{17}H_{28}$	17	232.41	268.00	16.1570	272.961	17.21
58	n-Dodecylbenzene	$C_{18}H_{30}$	18	246.44	275.93	17.5238	289.173	17.03
59	n-Tridecylbenzene	$C_{19}H_{32}$	19	260.47	283.15	18.6487	306.009	16.87
60	n-Tetradecylbenzene	$C_{20}H_{34}$	20	274.49	289.15	19.8420	322.197	16.64
61	n-Pentadecylbenzene	$C_{21}H_{36}$	21	288.52	295.15	20.9874	339.135	16.49

(Continued)

TABLE 6.10—(*Continued*)

No.	Compound	Formula	N_C	M	T_M, K	$\Delta H_f/RT_M$ at T_M	V_{25}, cm³/mol	δ_{25}, (J/cm³)$^{1/2}$
62	*n*-Hexadecylbenzene	$C_{22}H_{38}$	22	302.55	300.15	22.1207	356.160	16.39
63	*n*-Heptadecylbenzene	$C_{23}H_{40}$	23	316.55	305.15	22.9782	373.731	16.30
64	*n*-Octadecylbenzene	$C_{24}H_{42}$	24	330.58	309.00	23.7040	390.634	16.24
	1-*n*-Alkylnaphthalenes (aromatics)							
65	Naphthalene	$C_{10}H_8$	10	128.16	353.43	6.4588	123.000	19.49
66	1-Methylnaphthalene	$C_{11}H_{10}$	11	142.19	242.67	3.4420	139.899	19.89
67	1-Ethylnaphthalene	$C_{12}H_{12}$	12	156.22	259.34	7.5592	155.579	19.85
68	1-*n*-Propylnaphthalene	$C_{13}H_{14}$	13	170.24	264.55	7.9943	172.533	19.09
69	1-*n*-Butylnaphthalene	$C_{14}H_{16}$	14	184.27	253.43	11.9117	189.358	19.10
70	1-*n*-Pentylnaphthalene	$C_{15}H_{18}$	15	198.29	248.79	11.3121	205.950	18.85
71	1-*n*-Hexylnaphthalene	$C_{16}H_{20}$	16	212.32	255.15	...	224.155	18.72
72	1-*n*-Nonylnaphthalene	$C_{19}H_{26}$	19	254.40	284.15	...	272.495	17.41
73	1-*n*-Decylnaphthalene	$C_{20}H_{28}$	20	268.42	288.15	...	289.211	17.20
	Other organic compounds							
74	Benzoic acid	$C_7H_6O_2$	7	122.12	395.52	5.4952	112.442	24.59
75	Diphenylmethane	$C_{13}H_{12}$	13	168.24	298.39	7.3363	167.908	19.52
76	Antheracene	$C_{14}H_{10}$	14	190.32	488.93	7.7150	182.900	17.75
	Nonhydrocarbons							
77	Water	H_2O	...	18.02	273.15	2.6428	18.0691	47.81
78	Methanol	CH_3OH	1	32.04	−97.68	0.2204	40.58	29.59
79	Ethanol	C_2H_5OH	2	46.07	−114.1	0.3729	58.62	26.13
80	Isobutanol	C_4H_9OH	4	74.12	−108.0	0.4634	...	22.92
81	Carbon dioxide	CO_2	1	44.01	216.58	5.0088	37.2744	14.56
82	Hydrogen sulfide	H_2S	...	34.08	187.68	1.5134	35.8600	18.00
83	Nitrogen	N_2	...	28.01	63.15	1.3712	34.6723	9.082
84	Hydrogen	H_2	...	2.02	13.95	1.0097	28.5681	6.648
85	Oxygen	O_2	...	32.00	54.36	0.9824	28.0225	8.182
86	Ammonia	NH_3	...	17.03	195.41	3.4819	24.9800	29.22
87	Carbon monoxide	CO	1	28.01	68.15	1.4842	35.4400	6.402

[a]API-TDB [11] gives different values for V_{25} of light hydrocarbons. These values are given in parentheses and seem more accurate, as also given in Table 6.11. Values in this table are obtained from a program in Ref. [13].

should be noted that the polynomial correlation given for *n*-alkylbenzenes cannot be used for compounds heavier than C_{24}. The other two equations may be extrapolated to heavier compounds. Equations (6.148) and (6.149) may be used together with the pseudocomponent method described in Chapter 3 to estimate V_{25} and δ for petroleum fractions whose molecular weights are in the range of application of these equations. Values of V_i^L and δ given in Table 6.10 are taken from Ref. [13] at temperature of 25°C. It seems that for some light gases (i.e., CH_4), there are some discrepancies with reported values in other references. Values of these properties for some compounds as recommended by Pruasnitz et al. [21] are given in Table 6.11. Obviously at 25°C, for light gases such as CH_4 or N_2 values of liquid properties represent extrapolated values and for this reason they vary from one source to another. It seems that values given in Table 6.10 for light gases correspond to temperatures lower than 25°C. For this reason for compounds such as C_1, C_2, H_2S, CO_2, N_2, and O_2 values of V_i^L and δ at 25°C as given in Table 6.11 are recommended to be used.

For multicomponent solutions, Eqs. (6.145) and (6.146) are replaced by the following relation:

$$\ln \gamma_i = \frac{V_i^L (\delta_i - \delta_{mix})^2}{RT}$$
$$\delta_{mix} = \sum_j \Phi_j \delta_j$$
(6.150)
$$\Phi_j = \frac{x_j V_j^L}{\sum_k x_k V_k^L}$$

where the summation applies to all components in the mixture. Regular solution theory is in fact equivalent to van

Laar theory since by replacing $A_{12} = (V_1^L/RT)(\delta_1 - \delta_2)^2$ and $A_{21} = (V_2^L/RT)(\delta_1 - \delta_2)^2$ into Eq. (6.144), it becomes identical to Eq. (6.145). However, the main advantage of Eq. (6.145) over Eq. (6.144) is that parameters V_i^L and δ_i are calculable from thermodynamic relations. Riazi and Vera [23] have shown that predicted values of solubility are sensitive to the values of V_i^L and δ_i and they have recommended some specific values for δ_i of various light gases in petroleum fractions.

Other commonly used activity coefficient models include Wilson and NRTL (nonrandom two-liquid) models, which are applicable to systems of heavy hydrocarbons, water, and

TABLE 6.11—*Values of liquid molar volume and solubility parameters for some pure compounds at 90 and 298 K.*

Compound	V_i^L, (cm³/mol)	δ_i, (J/cm³)$^{1/2}$
N_2 (at 90 K)	38.1	10.84
N_2 (at 298 K)	32.4	5.28
CO (at 90 K)	37.1	11.66
CO (at 298 K)	32.1	6.40
O_2 (at 90 K)	28.0	14.73
O_2 (at 298 K)	33.0	8.18
CO_2 (at 298 K)	55.0	12.27
CH_4 (at 90 K)	35.3	15.14
CH_4 (at 298 K)	52.0	11.62
C_2H_6 (at 90 K)	45.7	19.43
C_2H_6 (at 298 K)	70.0	13.50

Taken from Ref. [21]. Components N_2, CO, O_2, CO_2, CH_4, and C_2H_6 at 298 K are in gaseous phase ($T_c < 298$ K) and values of V_i^L and δ_i are hypothetical liquid values which are recommended to be used. Values given at 90 K are for real liquids. All other components are in liquid form at 298 K. Values reported for hydrocarbons heavier than C_5 are similar to the values given in Table 6.10. For example, for *n*-C_{16} it provides values of 294 and 16.34 for V_i^L and δ_i, respectively. Similarly for benzene values of 89 and 18.8 were provided in comparison with 89.48 and 18.7 given in Table 6.10.

alcohol mixtures [21]. For hydrocarbon systems, UNIQUAC (**uni**versal **qua**si **c**hemical) model that is based on a group contribution model is often used for calculation of activity coefficient of compounds with known structure. More details on activity coefficient models and their applications are discussed in available references [4, 21]. The major application of activity coefficient models is in liquid–liquid and solid–liquid equilibria as well as low pressure VLE calculations when cubic equations of state do not accurately estimate liquid fugacity coefficients.

6.6.6 Calculation of Fugacity of Solids

In the petroleum industry solid fugacity is used for SLE calculations. Solids are generally heavy organics such as waxes and asphaltenes that are formed under certain conditions. Solid–liquid equilibria follows the same principles as VLE. Generally fugacity of solids are calculated similar to the methods that fugacity of liquids are calculated. In the study of solubility of solids in liquid solvents usually solute (solid) is shown by component 1 and solvent (liquid) is shown by component 2. Mole fraction of solute in the solution is x_1, which is the main parameter that must be estimated in calculation of solubility of solids in liquids. We assume that the solid phase is pure component 1. In such a case fugacity of solid in the solution is shown by $\hat{f}_1^S$, which is given by

$$(6.151) \qquad \hat{f}_1^S \text{ (solid in liquid solution)} = x_1 \gamma_1^S f_1^\circ$$

where f_1° is the fugacity of solute at a standard state but temperature T of solution. γ_1^S is the activity coefficient of solid component in the solution. Obviously for ideal solutions γ_1^S is unity. Model to calculate γ_1^S is similar to liquid activity coefficients, such as two-suffix Margules equation:

$$(6.152) \qquad \ln \gamma_1^S = \frac{A}{RT}(1 - x_1)^2$$

A more accurate activity coefficient model for nonpolar solutes and solvents is given by the Scatchard–Hildebrand relation (Eq. 6.145):

$$(6.153) \qquad \ln \gamma_1^S = \frac{V_1^L (\delta_1 - \delta_2)^2 \Phi_2^2}{RT}$$

where V_1^L is the liquid molar volume of pure component 1 at T and P, δ_2 is solubility of solvent, δ_1 is the solubility parameter of subcooled component 1, and Φ_2 is the volume fraction of solvent and is given by Eq. (6.146). Methods of calculation of δ_1 and Φ_1 have been discussed in Section 6.6.5. δ_1 can be calculated from Eq. (6.147) from the knowledge of heat of vaporization of solute, ΔH_1^{vap}. Values of the solubility parameter for heavy single carbon number components are given in Table 4.6. When the liquid solvent is a mixture δ_2 is replaced by δ_{mix} and γ_1^S is calculated through Eq. (6.150). It should be noted that when Eq. (6.153) is used to calculate fugacity of a solid in a liquid solution value of δ can be obtained from Table 6.10 from liquid solubility data. However, when this equation is applied for calculation of fugacity of a solid component i in a homogeneous solid phase mixture (i.e., wax) then solid solubility, δ^S, should be used for value of δ as recommended by Won [24]. If a value of liquid solubility given in Tables 6.10

and 6.11 is shown by δ^L, then δ^S may be calculated from the following relation [24]:

$$(6.154) \qquad \left(\delta_i^S\right)^2 = \left(\delta_i^L\right)^2 + \frac{\Delta H_i^f}{V_i}$$

in which δ is in $(J/cm^3)^{1/2}$, ΔH_i^f is in J/mol, and V_i is in cm^3/mol.

Calculation of fugacity of solids through Eq. (6.151) requires calculation of f_i°. For convenience the standard state for calculation of f_i° is considered subcooled liquid at temperature T and for this reason we show it by f_i^L. In the following discussion solute component 1 is replaced by component i to generalize the equation for any component. Based on the SLE for pure i at temperature T it can be shown that [21, 25]

$$f_i^S(T, P) = f_i^L(T, P)$$

$$\times \exp\left[\frac{\Delta H_i^f}{RT_{Mi}}\left(1 - \frac{T_{Mi}}{T}\right) - \frac{\Delta C_{Pi}}{R}\left(1 - \frac{T_{Mi}}{T}\right) - \frac{\Delta C_{Pi}}{R}\ln\frac{T_{Mi}}{T}\right]$$

$$(6.155)$$

where $f_i^S(T, P)$ is the fugacity of pure solid at T and P, ΔH_i^f is the molar *heat of fusion* of solute, T_{Mi} is the melting or freezing point temperature, and $\Delta C_{Pi} = C_{Pi}^L - C_{Pi}^S$, which is the difference between heat capacity of liquid and solid solute at average temperature of $(T + T_{Mi})/2$. Derivation of Eq. (6.155) is similar to the derivation of Eq. (6.136) for calculation of fugacity of pure liquids but in this case equilibrium between solid and liquid is used to develop the above relation. Firoozabadi [17] clearly describes calculation of fugacity of solids. Since methods of calculation of f_i^L were discussed in the previous section, f_i^S can be calculated from the above equation. Values of T_M and ΔH_i^f for some selected compounds are given in Table 6.10 along with liquid molar volume and solubility parameter. Estimation of freezing point T_M for pure hydrocarbons was discussed in Section 2.6.4. From Eq. (2.42) and coefficients given in Table 2.6 we have

$$T_M = 397 - \exp(6.5096 - 0.1487M^{0.47})$$
$$n\text{-alkanes (C}_5 - \text{C}_{40})$$

$$(6.156) \qquad T_M = 370 - \exp(6.52504 - 0.04945M^{2/3})$$
$$n\text{-alkylcyclopentanes (C}_7 - \text{C}_{40})$$

$$T_M = 375 - \exp(6.53599 - 0.04912M^{2/3})$$
$$n\text{-alkanes (C}_9 - \text{C}_{42})$$

where T_M is in kelvin. Average deviation for these equations are 1.5, 1.2, and 0.9%, for n-alkanes, n-alkylcyclopentanes, and n-alkylbenzenes, respectively. Similarly based on the data given for ΔH_i^f in Table 6.10 the following relations are developed for estimation of heat of fusion of pure hydrocarbons for the PNA homologous families.

$$\ln\frac{\Delta H_f}{RT_M} = -71.9215 + 70.7847M^{0.01}$$
$$\text{for } n\text{-alkanes (C}_2 - \text{C}_{36})$$

$$\ln\frac{\Delta H_f}{RT_M} = 0.8325 + 0.009M$$
$$\text{for } n\text{-alkylcyclohexanes (C}_7 - \text{C}_{16})$$

$$\frac{\Delta H_f}{RT_M} = 1.1556 + 0.009M + 0.000396M^2 - 6.544 \times 10^{-7}M^3$$
$$\text{for } n\text{-alkylbenzenes(C}_6 - \text{C}_{24})$$

$$(6.157)$$

where T_{Mi} is the melting point in kelvin and R is the gas constant. The ratio $\Delta H_i^f / RT_M$ is dimensionless and represents entropy of fusion. Unit of ΔH_i^f depends on the unit of R. ΔH_i^f may also be calculated from entropy change of fusion, ΔS_i^f, whenever it is available.

$$(6.158) \qquad \Delta H_i^f = T_{Mi} \Delta S_i^f$$

The above equation may also be used to estimate ΔS_i^f from ΔH_i^f calculated through Eq. (6.157). Equation (6.157) can reproduce data with average deviations of 12.5, 5.4, and 3.8% for *n*-alkanes, *n*-alkylcyclohexanes, and *n*-alkylbenzens, respectively. Firoozabadi and co-workers [17, 24–26] have provided the following equations for calculation of ΔH_i^f:

$$(6.159) \quad \begin{aligned} \frac{\Delta H_i^f}{RT_{Mi}} &= 0.07177 M_i \quad \text{for paraffins} \\[2mm] \frac{\Delta H_i^f}{RT_{Mi}} &= 0.02652 M_i \quad \text{for naphthenes and isoparaffins} \\[2mm] \frac{\Delta H_i^f}{RT_{Mi}} &= 5.63664 \quad\quad \text{for aromatics} \end{aligned}$$

where $\Delta H_i^f / RT_M$ is dimensionless. The relation given for calculation of ΔH_i^f of aromatics (Eq. 6.159) suggests that the entropy of fusion is constant for all aromatics. While this may be true for some multiring aromatics, it certainly is not true for *n*-alkylbenzenes. Graphical comparisons of Eqs. (6.157) and (6.159) for calculation of entropy of fusion of *n*-alkanes and *n*-alkylbenzenes and evaluation with data given in Table 6.10 are shown in Figs. 6.9 and 6.10. As is seen from Fig. 6.10, the entropy of *n*-alkylbenzenes does change with carbon number.

Calculation of fugacity of solids also requires ΔC_{pi}. The following relation developed for all types of hydrocarbons (P, N, and A) by Pedersen et al. [26] is recommended by Firoozabadi for calculation of ΔC_{Pi} [17]:

$$(6.160) \quad \Delta C_{Pi} = R(0.1526 M_i - 2.3327 \times 10^{-4} M_i T)$$

where T is the absolute temperature in kelvin and M_i is molecular weight of i. The unit of ΔC_{pi} is the same as the unit of R. Evaluation of this equation with data from DIPPR [13] for

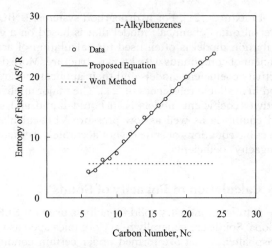

FIG. 6.10—Prediction of entropy of fusion of *n*-alkylbenzens. Proposed equation: Eq. (6.157); Won method: Eq. (6.159); data from DIPPR [13].

n-alkanes at two different temperature of 298 K and freezing point is shown in Fig. 6.11. As is seen from this figure, Eq. (6.160) gives values higher than actual values of ΔC_{pi}. Generally, actual values of ΔC_{pi} are small and as will be seen later they may be neglected in the calculation of f_i^S from Eq. (6.155) with good approximation.

Another type of SLE that is important in the petroleum industry is precipitation of heavy organics, such as asphaltenes and waxes, that occurs under certain conditions. Wax and asphaltene precipitation can plug the well bore formations and it can restrict or plug the tubing and facilities, such as flowlines and production handling facilities, which can lead to major economic problems. For this reason, knowledge of the conditions at which precipitation occurs is important. In formulation of this phase transition, the solid phase is considered as a solution of mixtures of components that fugacity of i is shown by $\hat{f}_i^S$ and can be calculated from the following relation:

$$(6.161) \quad \hat{f}_i^S \,(\text{solid } i \text{ in solid mixture}) = x_i^S \gamma_i^S f_i^S$$

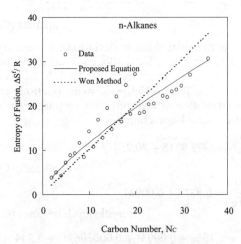

FIG. 6.9—Prediction of entropy of fusion of *n*-alkanes. Proposed equation: Eq. (6.157); Won method: Eq. (6.159); data from DIPPR [13].

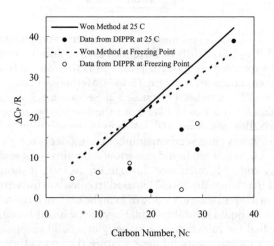

FIG. 6.11—Values of ΔC_{pi} ($= C_{Pi}^L - C_{Pi}^S$) for *n*-alkanes. Won method: Eq. (6.160); data from DIPPR [13].

where f_i^S is the fugacity of pure i at T and P of the system. In wax precipitation usually the solid solution is considered ideal and γ_i^S is assumed as unity [17]. x_i^S is the mole fraction of solid i in the solid phase solution. Here the term solution means homogeneous mixture of solid phase. As it will be seen in the next chapter these relations can also be used to determine the conditions at which hydrates are formed.

Calculation of fugacity of pure solids through Eq. (6.155) is useful for SLE calculations where the temperature is above the triple-point temperature (T_{tp}). When temperature is less than T_{tp} we have solid–vapor equilibrium as shown in Fig. 5.2a. For such cases the relation for calculation of fugacity of pure solids can be derived from fugacity of pure vapor and effect of pressure on vapor phase fugacity similar to derivation of Eq. 6.136, where f_i^L, P_i^{sat}, and V_i^L should be replaced by f_i^S, P_i^{sub}, and V_i^S, respectively. However at $T < T_{tp}$, P_i^{sub} or solid–vapor pressure is very low and ϕ_i^{sat} is unity. Furthermore molar volume of solid, V_i^S is constant with respect to pressure (see Problem 6.15).

6.7 GENERAL METHOD FOR CALCULATION OF PROPERTIES OF REAL MIXTURES

Two parameters have been defined to express nonideality of a system, fugacity coefficient and activity coefficient. Fugacity coefficient indicates deviation from ideal gas behavior and activity coefficient indicates deviation from ideal solution behavior for liquid solutions. Once residual properties (deviation from ideal gas behavior) and excess properties (deviation from ideal solution behavior) are known, properties of real mixtures can be calculated from properties of ideal gases or real solutions. Properties of real gas mixtures can be calculated through residual properties. For example, applying the definition of residual property to G we get

$$(6.162) \qquad G = G^{ig} + G^R$$

where G^R is the residual Gibbs energy (defined as $G - G^{ig}$). G^R is related to $\hat{\phi}_i$ by Eq. (6.128), which when combined with the above equation gives

$$(6.163) \qquad G = G^{ig} + RT \sum y_i \ln \hat{\phi}_i$$

Furthermore from thermodynamic relations one can show that [1]

$$(6.164) \quad \begin{aligned} H &= \sum y_i H_i^{ig} - RT^2 \sum y_i \left(\frac{\partial \ln \hat{\phi}_i}{\partial T} \right)_{P,y_i} \\ V &= \sum y_i V_i^{ig} + RT \sum y_i \left(\frac{\partial \ln \hat{\phi}_i}{\partial P} \right)_{T,y_i} \end{aligned}$$

Calculation of properties of ideal gases have been discussed in Section 6.3, therefore, from the knowledge of fugacity coefficients one can calculate properties of real gases.

Similarly for real liquid solutions a property can be calculated from the knowledge of excess property. Properties of ideal solutions are given by Eqs. (6.89)–(6.92). Property of a real solution can be calculated from knowledge of excess property and ideal solution property using Eq. (6.83):

$$(6.165) \qquad M = M^{ig} + M^E$$

where M^E is the excess property and can be calculated from activity coefficients. For example, G^E can be calculated from

Eq. (6.138). Similarly V^E and H^E can be calculated from γ_i and H and V of the solution may be calculated from the following relations:

$$(6.166) \quad \begin{aligned} H &= H^{id} - RT^2 \sum x_i \left(\frac{\partial \ln \gamma_i}{\partial T} \right)_{P,x_i} \\ V &= V^{id} + RT \sum x_i \left(\frac{\partial \ln \gamma_i}{\partial P} \right)_{T,x_i} \end{aligned}$$

Once G, H, and V are known, all other properties can be calculated from appropriate thermodynamic relations discussed in Section 6.1.

Another common way of determining thermophysical properties is through thermodynamic diagrams. In these diagrams various properties such as H, S, V, T, and P for both liquid and vapor phases of a pure substance are graphically shown. One type of these diagrams is the P–H diagram that is shown in Fig. 6.12 for methane as given by the GPA [28]. Such diagrams are available for many industrially important pure compounds [28]. Most of these thermodynamic charts and computer programs were developed by NIST [29]. Values used to construct such diagrams are calculated through thermodynamic models discussed in this chapter. While these diagrams are easy to use, but it is hard to determine an accurate value from the graph because of difficulty in reading the values. In addition they are not suitable for computer applications. However, these figures are useful for the purpose of evaluation of an estimated property from a thermodynamic model. Other types of these diagrams are also available. The H–S diagram known as Mollier diagram is usually used to graphically correlate properties of refrigerant fluids.

6.8 FORMULATION OF PHASE EQUILIBRIA PROBLEMS FOR MIXTURES

In this section equations needed for various phase equilibrium calculations for mixtures are presented. Two cases of vapor–liquid equilibria (VLE) and liquid–solid equilibria (LSE) are considered due to their wide application in the petroleum industry, as will be seen in Chapter 9.

6.8.1 Criteria for Mixture Phase Equilibria

The criteria for phase equilibrium is set by minimum Gibbs free energy, which requires derivative of G to be zero at the conditions where the system is in thermodynamic equilibrium as shown by Eq. (6.108). Gibbs energy varies with T, P, and x_i. At fixed T and P, one can determine x_i that is when G is minimized or at a fixed T (or P) and x_i, equilibrium pressure (or temperature) can be found by minimizing G. At different pressures functionality of G with x_i at a fixed temperature varies. Baker et al. [29] have discussed variation of Gibbs energy with composition. A typical curve is shown in Fig. 6.13. To avoid a false solution to find equilibrium conditions, there is a second constraint set by the second derivative of G as [17, 20, 30]

$$(6.167) \quad \begin{aligned} (\partial G)_{T,P} &= 0 \\ (\partial^2 G)_{T,P} &> 0 \end{aligned}$$

This discussion is known as stability criteria and it has received significant attention by reservoir engineers in analysis

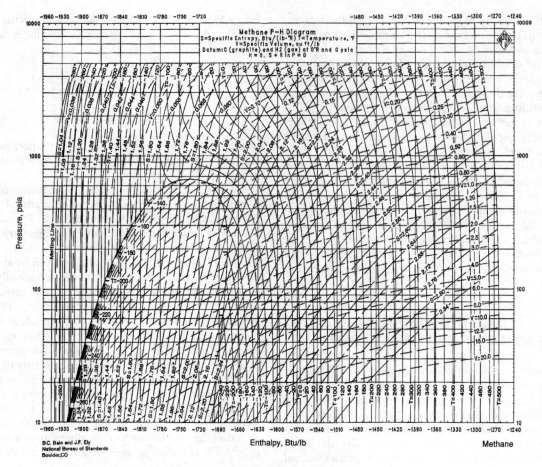

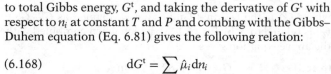

FIG. 6.12—The *P–H* diagram for methane. Unit conversion: °F = °C × 1.8 + 32 psia = 14.504 × bar. Taken with permission from Ref. [27].

of fluid phase equilibrium of petroleum mixtures. Further discussion regarding phase stability is given in a number of recent references [17, 20, 31].

Derivation of the general formula for equilibrium conditions in terms of chemical potential and fugacity for multi-component systems is shown here. Consider a mixture of N

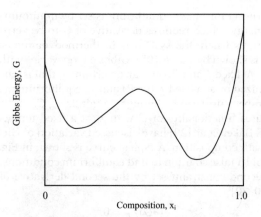

FIG. 6.13—A sample variation of Gibbs energy versus composition for a binary system at constant *T* and *P*.

components with two phases of α and β. Applying Eq. (6.79) to total Gibbs energy, G^t, and taking the derivative of G^t with respect to n_i at constant T and P and combing with the Gibbs–Duhem equation (Eq. 6.81) gives the following relation:

$$(6.168) \qquad dG^t = \sum_i \hat{\mu}_i dn_i$$

where $\hat{\mu}_i$ is the chemical potential defined by Eq. (6.115). Combining Eqs. (6.167) and (6.168) for all phases of the system at equilibrium gives

$$(6.169) \qquad \sum_i \hat{\mu}_i^\alpha dn_i^\alpha + \sum_i \hat{\mu}_i^\beta dn_i^\beta = 0$$

Since $n_i = n_i^\alpha + n_i^\beta$ and n_i is constant (closed system without chemical reaction), therefore, $dn_i^\alpha = -dn_i^\beta$, which by substituting into the above equation leads to the following conclusion:

$$(6.170) \qquad \hat{\mu}_i^\alpha = \hat{\mu}_i^\beta \quad \text{(at constant } T \text{ and } P\text{)}$$

This relation must apply to all components when the system is in equilibrium. If there are more than two phases (i.e., α, β, γ, ...) the same approach leads to the following conclusion:

$$(6.171) \quad \hat{\mu}_i^\alpha = \hat{\mu}_i^\beta = \hat{\mu}_i^\gamma = \cdots \quad \text{for every } i \text{ at constant } T \text{ and } P$$

Using the relation between $\hat{\mu}_i$ and $\hat{f}_i$ given by Eq. (6.119) we get

$$(6.172) \quad \hat{f}_i^\alpha = \hat{f}_i^\beta = \hat{f}_i^\gamma = \cdots \quad \text{for every } i \text{ at constant } T \text{ and } P$$

Equations (6.171) or (6.172) are the basis for formulation of mixture phase equilibrium calculations. Application of Eq. (6.172) to VLE gives

$$(6.173) \quad \hat{f}_i^V(T, P, y_i) = \hat{f}_i^L(T, P, x_i)$$

for all i components at constant T and P

For SLE, Eq. (6.172) becomes

$$(6.174) \quad \hat{f}_i^S(T, P, x_i^S) = \hat{f}_i^L(T, P, x_i^L)$$

for all i components at constant T and P

where x_i^S is the mole fraction of i in the solid phase. Similarly Eq. (6.172) can be used in liquid–liquid equilibria (LLE), solid–liquid–vapor equilibria (SLVE), or vapor–liquid–liquid equilibria (VLLE).

6.8.2 Vapor–Liquid Equilibria—Gas Solubility in Liquids

In this section general relations for VLE and specific relations developed for certain systems such as Raoult's and Henry's laws are presented. For high pressure VLE calculations equilibrium ratio (K_i) is defined and its methods of estimation for hydrocarbon systems are presented.

6.8.2.1 Formulation of Vapor–Liquid Equilibria Relations

Formulation of VLE calculations requires substitution of relations for $\hat{f}_i^V$ and $\hat{f}_i^L$ from Eqs. (6.113) and (6.114). Combining Eqs. (6.114) and (6.173) gives the following relation:

$$(6.175) \quad y_i \hat{\phi}_i^V P = x_i \gamma_i f_i^L$$

where f_i^L is the fugacity of pure liquid i at T and P of the mixture and it may be calculated through Eq. (6.136). The activity coefficient γ_i is also a temperature-dependent parameter in addition to x_i. Another general VLE relation may be obtained when both $\hat{f}_i^V$ and $\hat{f}_i^L$ are expressed in terms of fugacity coefficients $\hat{\phi}_i^V$ and $\hat{\phi}_i^L$ through Eq. (6.113) and are substituted in Eq. (6.173):

$$(6.176) \quad y_i \hat{\phi}_i^V(T, P, y_i) = x_i \hat{\phi}_i^L(T, P, x_i)$$

where pressure P from both sides of the equation is dropped. Equation (6.176) is essentially the same as Eq. (6.175) and the activity coefficient can be related to fugacity coefficient as [17]

$$(6.177) \quad \ln \gamma_i = \ln \hat{\phi}_i(T, P, x_i) - \ln \phi_i(T, P)$$

where $\hat{\phi}_i(T, P, x_i)$ is the fugacity coefficient of i in the liquid mixture and $\phi_i(T, P)$ is fugacity coefficient of pure liquid i at T and P of mixture. In fact one may use an EOS to calculate γ_i by calculating $\hat{\phi}_i$ and ϕ_i for the liquid phase through Eq. (6.126). Application of PR EOS in the above equation, at

$x_i \to 0$, will result in the following relation for calculation of activity coefficient at infinite dilution for component 1 (γ_1^∞) in a binary system of components 1 and 2 at T and P [17]:

$$
\begin{aligned}
(6.178) \quad \ln \gamma_1^\infty =\ & \frac{b_1}{b_2}(Z_2 - 1) - (Z_1 - 1) + \ln\left(\frac{Z_1 - B_1}{Z_2 - B_2}\right) \\
& + \frac{A_1}{2\sqrt{2}B_1}\ln\left(\frac{Z_1 + 2.414B_1}{Z_1 - 0.414B_1}\right) \\
& - \left(\frac{a_{12}P}{R^2T^2}\right)\frac{1}{\sqrt{2}B_2}\ln\left(\frac{Z_2 + 2.414B_2}{Z_2 - 0.414B_2}\right) \\
& + \frac{b_1 A_2}{2\sqrt{2}B_2 b_2}\ln\left(\frac{Z_2 + 2.414B_2}{Z_2 - 0.414B_2}\right)
\end{aligned}
$$

where $a_{12} = a_1^{1/2}a_2^{1/2}(1 - k_{12})$ in which k_{12} is the binary interaction parameter. Parameters a, b, A, and B for PR EOS are given in Table 5.1. Z_1 and Z_2 are the compressibility factor for components 1 and 2 calculated from the PR EOS.

The main difference between Eqs. (6.175) and (6.176) for VLE calculations is in their applications. Equation (6.176) is particularly useful when both $\hat{\phi}_i^V$ and $\hat{\phi}_i^L$ are calculated from equations of state. Cubic EOSs generally work well in the VLE calculation of petroleum systems at high pressures through this equation. $\hat{\phi}_i^V$ and $\hat{\phi}_i^L$ may be calculated through Eq. (6.126) with use of appropriate composition and Z; that is, x_i^L and Z^L must be used in calculation of $\hat{\phi}_i^L$, while y_i^V and Z^V are used in calculation of $\hat{\phi}_i^V$. Binary interaction coefficients (BIPs) given in Table 5.3 must be used when dissimilar (very light and very heavy or nonhydrocarbon and hydrocarbon) molecules exist in a mixture. However, as mentioned earlier there is no need for use of volume translation or shift parameter in calculation of $\hat{\phi}_i^V$ and $\hat{\phi}_i^L$ for use in Eq. (6.176).

At low and moderate pressures use of Eq. (6.175) with activity coefficient models is more accurate than use of Eq. (6.176) with an EOS. Assuming constant V_i^L and substituting Eq. (6.136) into Eq. (6.175) we have

$$(6.179) \quad y_i \hat{\phi}_i^V P = x_i \gamma_i \phi_i^{sat} P_i^{sat} \exp\left[\frac{V_i^L(P - P_i^{sat})}{RT}\right]$$

where the effect of pressure on the liquid molar volume is neglected and saturated liquid molar volume V_i^{sat} may be used for V_i^L. As discussed in Section 6.5, the vapor pressure P_i^{sat} is a function of temperature and the highest temperature at which P_i^{sat} can be calculated is T_c, where $P_i^{sat} = P_c$. Therefore, Eq. (6.179) cannot be applied to a component in a mixture at which $T > T_c$. For ideal liquid solutions or those systems that follow Lewis rule (Section 6.6.3), the activity coefficient for all components is unity ($\gamma_i = 1$). If pressure P and saturation pressure P_i^{sat} are low and the gas phase can be considered as an ideal gas, then $\hat{\phi}_i^V$ and ϕ_i^{sat} are unity and the Poynting factor is also unity; therefore, the above relation reduces to the following simple form:

$$(6.180) \quad y_i P = x_i P_i^{sat}$$

This is the simplest VLE relation and is known as the *Raoult's law*. This rule only applies to ideal solutions such as benzene–toluene mixture at pressures near or below 1 atm. If the gas phase is ideal gas, but the liquid is not ideal solution then

Eq. (6.180) reduces to

$$(6.181) \qquad y_i P = x_i \gamma_i P_i^{\text{sat}}$$

This relation also known as *modified* Raoult's law is valid for nonideal systems but at pressures of 1 atm or less where the gas phase is considered ideal gas. We know that as $x_i \to 1$ (toward a pure component) thus $\gamma_i \to 1$ and therefore Eq. (6.181) reduces to Raoult's law even for a real solution. Nonideal systems with $\gamma_i > 1$ show positive deviation while with $\gamma_i < 1$ show negative deviation from the Raoult's law. One direct application of modified Raoult's law is to calculate composition of a compound in the air when it is vaporized from its pure liquid phase ($x_i = 1$, $\gamma_i = 1$).

$$(6.182) \qquad y_i P = P_i^{\text{sat}}$$

Since for ideal gas mixtures volume and mole fractions are the same therefore we have

$$(6.183) \qquad \text{vol\% of } i \text{ in air} = \frac{P_i^{\text{sat}}}{P_a}$$

(for vaporization of pure liquid i)

where P_a is atmospheric pressure. This is the same as Eq. (2.11) that was used to calculate amount of a gas in the air for flammability test. Behavior of ideal and nonideal systems is shown in Fig. 6.14 through Txy and Pxy diagrams. Calculation of bubble and dew point pressures and generation of such diagrams will be discussed in Chapter 9.

6.8.2.2 Solubility of Gases in Liquids—Henry's Law

Another important VLE relation is the relation for gas solubility in liquids. Many years ago it has been observed that solubility of gases in liquids (x_i) is proportional to partial pressure of component in the gas phase ($y_i P$), which can be formulated as [21]

$$(6.184) \qquad y_i P = k_i x_i$$

This relation is known as *Henry's law* and the proportionality constant k_i is called Henry's constant. $k_{i\text{-solvent}}$ has the unit of pressure per mole (or weight) fraction and for any given solute and solvent system is a function of temperature. Henry's law is a good approximation when pressure is low (not exceeding 5–10 bar) and the solute concentration in the solvent, x_i, is low (not exceeding 0.03) and the temperature is well below the critical temperature of solvent [21]. Henry's law is exact as $x_i \to 0$. In fact through application of Gibbs–Duhem equation in terms of γ_i (Eq. 6.141), it can be shown that for a binary system when Henry's law is valid for one component the Raoult's law is valid for the other component (see Problem 6.32). Equation (6.184) may be applied to gases at higher pressures by multiplying the left side of equation by $\hat{\phi}_i^V$.

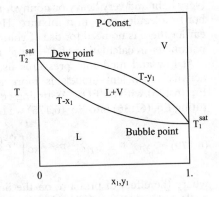

(a)Txy diagram for an ideal binary system

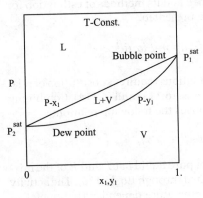

(b)Pxy diagram for an ideal binary system

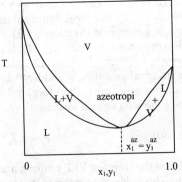

(c)Txy diagram for a real binary system

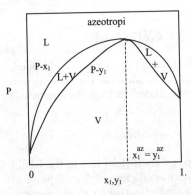

(d)Pxy diagram for a real binary system

FIG. 6.14—Txy and Pxy diagrams for ideal and nonideal systems.

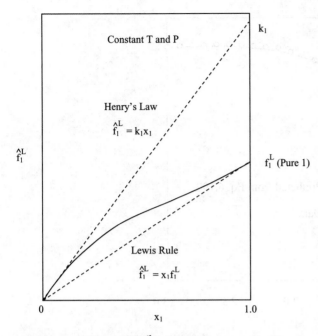

FIG. 6.15—Variation of f_1^L with x_1 in a binary liquid solution and comparison with its values from Henry's law and Lewis rule.

The RHS of Eq. (6.184) is $\hat{f}_i^L$ and in fact the exact definition of Henry's constant is [1, 21]

$$(6.185) \qquad k_i \equiv \lim it_{x_i \to 0}\left(\frac{\hat{f}_i^L}{x_i}\right)$$

Therefore, k_i is in fact the slope of $\hat{f}_i^L$ versus x_i at $x_i = 0$. This is demonstrated in Fig. 6.15 for a binary system. The Henry's law is valid at low values of x_1 ($\sim< 0.03$) while as $x_1 \to 1$, the system follows Raoult's law. Henry's constant generally decreases with increase in temperature and increases with increase in pressure. However, there are cases that that k_i increases with increase in temperature such as Henry's constant for H_2S and NH_3 in water [21]. Generally with good approximation, effect of pressure on Henry's constant is neglected and k_i is considered only as a function of temperature. Henry's law constant for a solute (component i) in a solvent can be estimated from an EOS through liquid phase fugacity coefficient at infinite dilution ($\hat{\phi}_i^{L,\infty} = \lim_{x_i \to 0} \hat{\phi}_i^L$) [21].

$$(6.186) \qquad k_i = \hat{\phi}_i^{L,\infty} P$$

Plöcker et al. [33] calculated k_i using Lee–Kesler EOS through calculation of $\hat{\phi}_i^{L,\infty}$ and the above equation for solute hydrogen (component 1) in various solvents versus in temperature range of 295–475 K. Their calculated values of k_i for H_2 in n-C_{16} are presented in Fig. 6.16 for the temperature range of 0–200°C. These calculated values are in good agreement with the measured values. The equation used for extrapolation of data is also given in the same figure that reproduce original data with an average deviation of 1%. Another useful relation for the Henry's constant is obtained by combining Eqs. (6.177) and (6.186):

$$(6.187) \qquad k_i = \gamma_i^\infty f_i^L$$

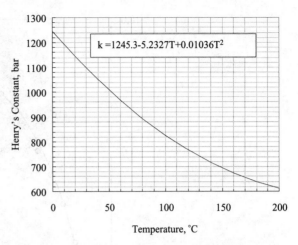

FIG. 6.16—Henry's constant for hydrogen in n-hexadecane (n-$C_{16}H_{34}$).

where γ_i^∞ is the activity coefficient at infinite dilution and f_i^L is the fugacity of pure liquid i at T and P of the system. Where if γ_i^∞ is calculated through Eq. (6.178) and the PR EOS is used to calculate liquid fugacity coefficient ($f_i^L = \phi_i^L P$), Henry's constant can be calculated from the PR EOS.

The general mixing rule for calculation of Henry's constant for a solute in a mixed solvent is given by Prausnitz [21]. For ternary systems, Henry's law constant for component 1 into a mixed solvent (2 and 3) is given by the following relation:

$$\ln k_{1,M} = x_2 \ln k_{1,2} + x_3 \ln k_{1,3} - \alpha_{23} x_2 x_3$$
$$(6.188) \qquad \alpha_{23} \approx \frac{(\delta_2 - \delta_3)^2 (V_2^L + V_3^L)}{2RT}$$

where δ is the solubility parameter, V is molar volume, and x is the mole fraction. This relation may be used to calculate activity coefficient of component 1 in a ternary mixture. Herein we assume that the mixture is a binary system of components 1 and M, where M represents components 2 and 3 together ($x_M = 1 - x_1$). Activity coefficient at infinite dilution is calculated through Eq. (6.187) as $\gamma_{1,M}^\infty = k_{1,M}/f_1^L$. Once $\gamma_{1,M}^\infty$ is known, it can be used to calculate parameters in an activity coefficient model as discussed earlier.

The main application of Henry's law is to calculate solubility of gases in liquids where the solubility is limited (small x_1). For example, solubilities of hydrocarbons in water or light hydrocarbons in heavy oils are very limited and Henry's law may be used to estimate the solubility of a solute in a solvent. The general relation for calculation of solubility is through Eq. (6.147). For various homologous groups, Eq. (6.149) may be used to estimate solubility parameter at 25°C. One major problem in using Eq. (6.179) occurs when it is used to calculate solubility of light gases (C_1, C_2, or C_3) in oils at temperatures greater than T_c of these components. In such cases calculation of P_i^{sat} is not possible since the component is not in a liquid form. For such situations Eq. (6.175) must be used and f_i^L represents fugacity of component i in a hypothetical liquid state. If solute (light gas) is indicated as component 1, the following equation should be used to calculate fugacity

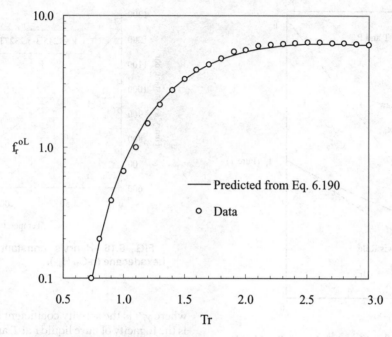

FIG. 6.17—Fugacity of hypothetical liquid at 1.013 bar. Data taken from Ref. [21] for CH_4, C_2H_4, C_2H_6, and N_2.

of pure component 1 as a hypothetical liquid when $T_r > 1$ [21]:

$$(6.189) \qquad f_1^L = f_r^{\circ L} P_c \exp\left[\frac{V_1^L(P - 1.013)}{RT}\right]$$

where $f_r^{\circ L}$ is the reduced hypothetical liquid fugacity at pressure of 1 atm ($f_r^{\circ L} = f^{\circ L}/P_c$) and it should be calculated from Fig. 6.17 as explained in Ref. [21]. Data on $f^{\circ L}$ of C_1, C_2, N_2, CO, and CO_2 have been used to construct this figure. For convenience, values obtained from Fig. 6.17 are represented by the following equation [23]:

$$(6.190) \qquad f_r^{\circ L} = \exp\left(7.902 - \frac{8.19643}{T_r} - 3.08\ln T_r\right)$$

where T_r is the reduced temperature. Data obtained from Fig. 6.17 in the range of $0.95 < T_r < 2.6$ are used to generate the above correlation and it reproduces the graph with %AAD of 1.3. This equation is not valid for $T_r > 3$ and for compounds such as H_2.

If the vapor phase is pure component 1 and is in contact with solvent 2 at pressure P and temperature T, then its solubility in terms of mole fraction, x_1, is found from Eq. (6.168) as

$$(6.191) \qquad x_1 = \frac{\phi_1^V P}{\gamma_1 f_1^L}$$

where ϕ_1^V is the fugacity coefficient of pure gas (component 1) at T and P. γ_1 is the activity coefficient of solute 1 in solvent 2, which is a function of x_1. f_1^L is the fugacity of pure component 1 as liquid at T and P and it may be calculated from Eq. (6.189) for light gases when $T > 0.95T_{c1}$. It is clear that to find x_1 from Eq. (6.191) a trial-and-error procedure is required since γ_1 is a function of x_1. To start the calculations

an initial value of x_1 is normally obtained from Eq. (6.191) by assuming $\gamma_1 = 1$. As an alternative method, since values of x_1 are normally small, initial value of x_1 can be assumed as zero. For hydrocarbon systems γ_1 may be calculated from regular solution theory. The following example shows the method.

Example 6.10—Estimate solubility of methane in *n*-pentane at 100°C when the pressure of methane is 0.01 bar.

Solution—Methane is considered as the solute (component 1) and *n*-pentane is the solvent (component 2). Properties of methane are taken from Table 2.1 as $M = 16$, $T_c = 190.4$ K, and $P_c = 46$ bar. $T = 373.15$ K ($T_r = 1.9598$) and $P = 0.01$ bar. Since the pressure is quite low the gas phase is ideal gas, thus $\phi_1^V = 1.0$. In Eq. (6.191) only γ_1 and f_1^L must be calculated. For C_1–C_5 system, the regular solution theory can be used to calculate γ_1 through Eq. (6.145). From Table 6.11, at 298 K, $V_1^L = 52$ and $V_2^L = 116$ cm³/mol, $\delta_1 = 11.6$, and $\delta_2 = 14.52$ (J/cm³)$^{1/2}$. Assuming $\Phi_1 \cong 0$ ($\Phi_2 \cong 1$), from Eq. (6.145), $\ln \gamma_1 = 0.143$ or $\gamma_1 = 1.154$. Since $T_r > 1$, f_1^L is calculated from Eq. (6.189). From Eq. (6.190), $f_r^{\circ L} = 5.107$ and from Eq. (6.189): $f_1^L = 234.8$ bar. Therefore, the solubility is $x_1 = 0.01/(1.1519 \times 234.8) = 3.7 \times 10^{-5}$. Since x_1 is very small, the initial guess for $\Phi_1 \cong 0$ is acceptable and there is no need for recalculation of γ_1. Therefore, the answer is 3.7×10^{-5}, which is close to value of 4×10^{-5} as given in Ref. [21]. ◆

One type of useful data is correlation of mole fraction solubility of gases in water at 1.013 bar (1 atm). Once this information is available, it can be used to determine solubility at other elevated pressures through Henry's law. Mole fraction solubility is given in the following correlations for a number of gases in water versus temperature as given by Sandler [22]:

$$
\begin{aligned}
\text{methane (275–328)} \quad & \ln x = -416.159289 + 15557.5631/T + 65.2552591 \ln T - 0.0616975729T \\
\text{ethane (275–323)} \quad & \ln x = -11268.4007 + 221617.099/T + 2158.421791 \ln T - 7.18779402T + 4.0501192 \times 10^{-3} T^2 \\
\text{propane (273–347)} \quad & \ln x = -316.46 + 15921.2/T + 44.32431 \ln T \\
n\text{-butane (276–349)} \quad & \ln x = -290.238 + 15055.5/T + 40.1949 \ln T \\
(6.192) \quad i\text{-butane (278–343)} \quad & \ln x = 96.1066 - 2472.33/T - 17.3663 \ln T \\
\text{H}_2\text{S (273–333)} \quad & \ln x = -149.537 + 8226.54/T + 20.2308 \ln T + 0.00129405T \\
\text{CO}_2 \text{ (273–373)} \quad & \ln x = -4957.824 + 105{,}288.4/T + 933.17 \ln T - 2.854886T + 1.480857 \times 10^{-3} T^2 \\
\text{N}_2 \text{ (273–348)} \quad & \ln x = -181.587 + 8632.129/T + 24.79808 \ln T \\
\text{H}_2 \text{ (274–339)} \quad & \ln x = -180.054 + 6993.54/T + 26.3121 \ln T - 0.0150432T
\end{aligned}
$$

For each gas the range of temperature (in kelvin) at which the correlation is applicable is given in parenthesis. T is the absolute temperature in kelvin and x is the mole fraction of dissolved gas in water at 1.013 bar. Henry's constant of light hydrocarbon gases (C_1, C_2, C_3, C_4, and i–C_4) in water may be estimated from the following correlation as suggested by the API-TDB [5]:

$$
(6.193) \qquad \ln k_{\text{gas–water}} = A_1 + A_2 T + \frac{A_3}{T} + A_4 \ln T
$$

where $k_{\text{gas–water}}$ is the Henry's constant of a light hydrocarbon gas in water in the unit of bar per mole fraction and T is the absolute temperature in kelvin. The coefficients A_1–A_4 and the range of T and P are given in Table 6.12.

To calculate solubility of a hydrocarbon liquid mixture in the aqueous phase, the following relation may be used:

$$
\hat{x}_i = x_i \left(\frac{\hat{f}_i^{\text{L}}}{f_i^{\text{L}}} \right)
$$

where $\hat{x}_i$ is the solubility of component i in the water when it is in a liquid mixture. x_i is the solubility of pure i in the water. $\hat{f}_i^{\text{L}}$ is the fugacity of i in the mixture of liquid hydrocarbon phase and f_i^{L} is the fugacity of pure i in the liquid phase. More accurate calculations can be performed through liquid–liquid phase equilibrium calculations.

For calculation of solubility of water in hydrocarbons the following correlation is proposed by the API-TDB [5]:

$$
\log_{10} x_{\text{H}_2\text{O}} = -\left(\frac{4200}{\text{CH weight ratio}} + 1050 \right) \times \left(\frac{1}{T} - 0.0016 \right)
$$

(6.194)

where T is in kelvin and $x_{\text{H}_2\text{O}}$ is the mole fraction of water in liquid hydrocarbon at 1.013 bar. CH weight ratio is the carbon-to-hydrogen weight ratio. This equation is known as Hibbard correlation and should be used for pentanes and heavier hydrocarbons (C_{5+}). The reliability of this method is ±20% [5]. If this equation is applied to undefined hydrocarbon fractions, the CH weight ratio may be estimated from the methods discussed in Section 2.6.3 of Chapter 2. However, API-TDB [5] recommends the following equation for calcula-

tion of solubility of water in some undefined petroleum fractions:

$$
(6.195) \quad
\begin{aligned}
\text{naphtha} \quad & \log_{10} x_{\text{H}_2\text{O}} = 2.94 - \frac{1841.3}{T} \\
\text{kerosene} \quad & \log_{10} x_{\text{H}_2\text{O}} = 2.74 - \frac{2387.3}{T} \\
\text{paraffinic oil} \quad & \log_{10} x_{\text{H}_2\text{O}} = 2.69 - \frac{1708.3}{T} \\
\text{gasoline} \quad & \log_{10} x_{\text{H}_2\text{O}} = 2.63 - \frac{1766.8}{T}
\end{aligned}
$$

In the above equations T is in kelvin and $x_{\text{H}_2\text{O}}$ is the mole fraction of water in the petroleum fraction. Obviously these correlations give approximate values of water solubility as composition of each fraction vary from one source to another.

6.8.2.3 Equilibrium Ratios (K_i Values)

The general formula for VLE calculation is obtained through definition of a new parameter called *equilibrium ratio* shown by K_i:

$$
(6.196) \qquad K_i \equiv \frac{y_i}{x_i}
$$

K_i is a dimensionless parameter and in general varies with T, P, and composition of both liquid and vapor phases. In many references, equilibrium ratios are referred as K_i value and can be calculated from combining Eq. (6.176) with Eq. (6.196) as in the following form:

$$
(6.197) \qquad K_i = \frac{\hat{\phi}_i^{\text{L}}(T, P, x_i)}{\hat{\phi}_i^{\text{V}}(T, P, y_i)}
$$

In high-pressure VLE calculations, K_i values are calculated from Eq. (6.197) through Eq. (6.126) for calculation of fugacity coefficients with use of cubic equations (SRK or PR). In calculation of K_i values from a cubic EOS use of binary interaction parameters (BIPs) introduced in Chapter 5 is required specially when components such as N_2, H_2S, and CO_2 exist in the hydrocarbon mixture. Also in mixtures when the difference in molecular size of components is appreciable

TABLE 6.12—*Constants for Eq. (6.193) for estimation of Henry's constant for light gases in water [5].*

Gas	T range, K	Pressure range, bar	A_1	A_2	A_3	A_4	%AAD
Methane	274–444	1–31	569.29	0.107305	−19537	−92.17	3.6
Ethane	279–444	1–28	109.42	−0.023090	−8006.3	−11.467	7.5
Propane	278–428	1–28	1114.68	0.205942	−39162.2	−181.505	5.3
n-Butane	277–444	1–28	182.41	−0.018160	−11418.06	−22.455	6.2
i-Butane	278–378	1–10	1731.13	0.429534	−52318.06	−293.567	5.3

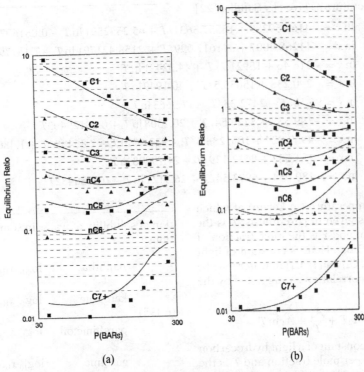

FIG. 6.18—Comparison of predicted equilibrium ratios (K_i values) from PR EOS without (a) and with (b) use of interaction parameters. ■ Experimental data for a crude oil. Taken with permission from Ref. [32].

(i.e., C_1 and some heavy compounds) use of BIPs is required. Effect of BIPs in calculation of K_i values is demonstrated in Fig. 6.18. If both the vapor and liquid phases are assumed as ideal solutions, then by applying Eq. (6.132) the Lewis rule, Eq. (6.197) becomes

$$(6.198) \qquad K_i = \frac{\phi_i^{\mathrm{L}}(T, P)}{\phi_i^{\mathrm{V}}(T, P)}$$

where ϕ_i^{V} and ϕ_i^{L} are pure component fugacity coefficients and K_i is independent of composition and depends only on T and P. The main application of this equation is for light hydrocarbons where their mixtures may be assumed as ideal solution. For systems following Raoult's law (Eq. 6.180) the K_i values can be calculated from the relation:

$$(6.199) \qquad K_i(T, P) = \frac{P_i^{\mathrm{sat}}(T)}{P}$$

Equilibrium ratios may also be calculated from Eq. (6.181) through calculation of activity coefficients for the liquid phase.

Another method for calculation of K_i values of nonpolar systems was developed by Chao and Seader in 1961 [34]. They suggested a modification of Eq. (6.197) by replacing $\hat{\phi}_i^{\mathrm{L}}$ with $(\gamma_i \phi_i^{\mathrm{L}})$, where ϕ_i^{L} is the fugacity coefficient of pure liquid i and γ_i is the activity coefficient of component i.

$$(6.200) \qquad K_i = \frac{y_i}{x_i} = \frac{\gamma_i \phi_i^{\mathrm{L}}(T_{\mathrm{ri}}, P_{\mathrm{ri}}, \omega_i)}{\hat{\phi}_i^{\mathrm{V}}}$$

γ_i must be evaluated from Eq. (6.150)
$\hat{\phi}_i^{\mathrm{V}}$ must be evaluated from the Redlich–Kwong EOS
ϕ_i^{L} empirically developed correlation in terms of $T_{\mathrm{ri}}, P_{\mathrm{ri}}, \omega_i$

γ_i must be evaluated with the Scatchard–Hildebrand regular solution relationship (Eq. 6.150). $\hat{\phi}_i^{\mathrm{V}}$ must be evaluated with the original Redlich–Kwong equation of state. Furthermore, Chao and Seader developed a generalized correlation for calculation of ϕ_i^{L} in terms of reduced temperature, pressure, and acentric factor of pure component i ($T_{\mathrm{ri}}, P_{\mathrm{ri}}$, and ω_i). Later Grayson and Streed [35] reformulated the correlation for ϕ_i^{L} to temperatures about 430°C (~ 800°F). Some process simulators (i.e., PRI/II [36]) use the Greyson–Streed expression for ϕ_i^{L}. This method found wide industrial applications in the 1960s and 1970s; however, it should not be used for systems containing polar compounds or compounds with close boiling points (i.e., i-C_4/n-C_4). It should not be used for temperatures below -17°C (0°F) nor near the critical region where it does not recognize $x_i = y_i$ at the critical point [37]. For systems composed of complex molecules such as very heavy hydrocarbons, water, alcohol, ionic (i.e., salt, surfactant), and polymeric systems, SAFT EOS may be used for phase equilibrium calculations. Relations for convenient calculation of fugacity coefficients and compressibility factor are given by Li and Englezos [38].

Once K_i values for all components are known, various VLE calculations can be made from the following general relationship between x_i and y_i:

$$(6.201) \qquad y_i = K_i x_i$$

Assuming ideal solution for hydrocarbons, K_i values at various temperature and pressure have been calculated for n-paraffins from C_1 to C_{10} and are presented graphically for quick estimation. These charts as given by Gas Processor Association (GPA) [28] are given in Figs. 6.19–6.31 for various

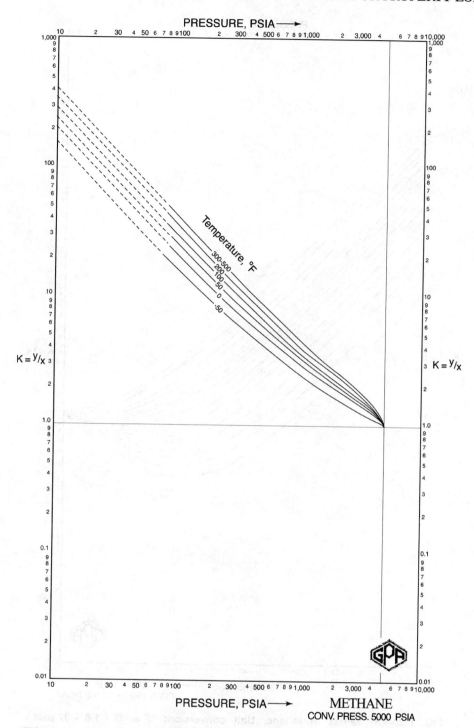

FIG. 6.19—K_i values of methane. Unit conversion: °F = °C × 1.8 + 32 psia = 14.504 × bar. Taken with permission from Ref. [28].

components from methane to decane and hydrogen sulfide. Equilibrium ratios are perhaps the most important parameter for high-pressure VLE calculations as described in Chapter 9.

For hydrocarbon systems and reservoir fluids there are some empirical correlations for calculation of K_i values. The correlation proposed by Hoffman et al. [39] is widely used in the industry. Later Standing [40] used values of K_i reported by Katz and Hachmuth [41] on crude oil and natural gas systems to obtain the following equations based on the

Hoffman original correlation:

$$K_i = \left(\frac{1}{P}\right) \times 10^{(a+cF)}$$

$$(6.202) \quad F = b\left(\frac{1}{T_B} - \frac{1}{T}\right)$$

$$a = 0.0385 + 6.527 \times 10^{-3}P + 3.155 \times 10^{-5}P^2$$

$$c = 0.89 - 2.4656 \times 10^{-3}P - 7.36261 \times 10^{-6}P^2$$

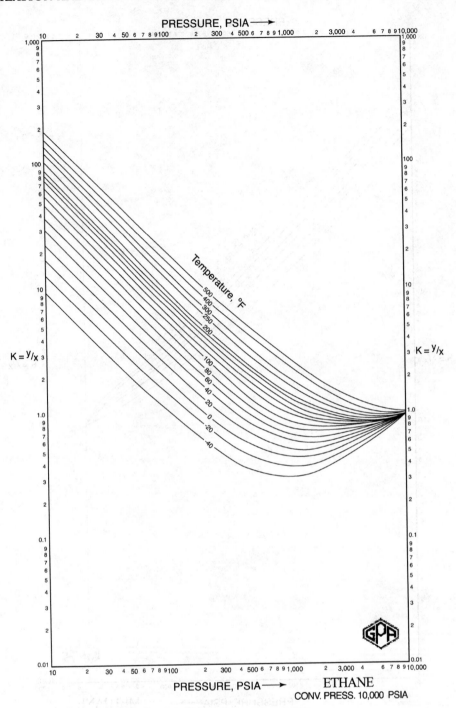

FIG. 6.20—K_i values of ethane. Unit conversion: °F = °C × 1.8 + 32 psia = 14.504 × bar. Taken with permission from Ref. [28].

where P is the pressure in bar and T is the temperature in kelvin. These equations are restricted to pressures below 69 bar (~1000 psia) and temperatures between 278–366 K (40–200°F). Values of b and T_B for these T and P ranges are given in Tables 6.13 for some pure compounds and lumped C_6 group. These equations reproduce original data within 3.5% error. For C_{7+} fractions the following equations are provided by Standing [40]:

$$\theta = 3.85 + 0.0135T + 0.02321P$$

(6.203)
$$b_{7+} = 562.78 + 180\theta - 2.364\theta^2$$

$$T_{B,7+} = 167.22 + 33.25\theta - 0.5394\theta^2$$

where T is in kelvin and P is in bar. It should be noted that all the original equations and constants in Table 6.11 were given in the English units and have been converted to the SI units as presented here. As it can be seen in these equations K_i is related only to T and P and they are independent of composition and are based on the assumption that mixtures behave like ideal solutions. These equations are referred as Standing method and they are recommended for gas condensate systems and are useful in calculations for surface separators. Katz and Hachmuth [41] originally recommended that $K_{7+} = 0.15K_{n-C_7}$, which has been used by Glaso [42] with satisfactory results. As will be seen in Chapter 9, in VLE

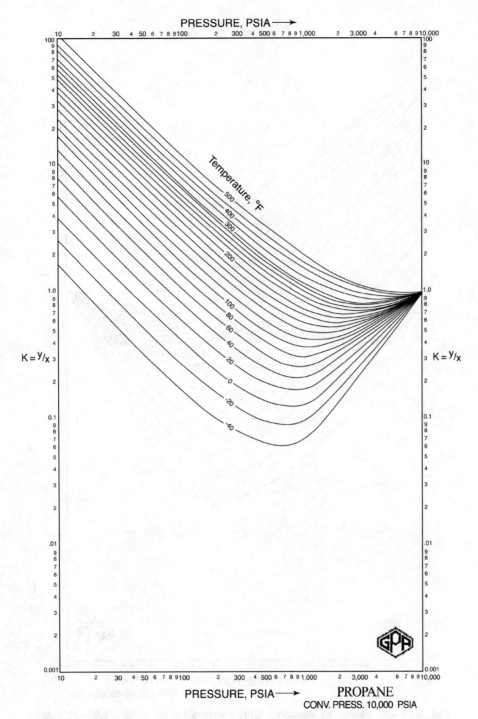

FIG. 6.21—K_i values of propane. Unit conversion: °F = °C × 1.8 + 32 psia = 14.504 × bar. Taken with permission from Ref. [28].

calculations some initial K_i values are needed. Whitson [31] suggests use of Wilson correlation for calculation of initial K_i values:

(6.204) $$K_i = \frac{\exp\left[5.37\left(1 + \omega_i\right)\left(1 - T_{ri}^{-1}\right)\right]}{P_{ri}}$$

where T_{ri} and P_{ri} are the reduced temperature and pressure as defined in Eq. (5.100) and ω_i is the acentric factor. It can be shown that Wilson equation reduces to Hoffman-type

equation when the Edmister equation (Eq. 2.108) is used for the acentric factor (see Problem 6.39).

Example 6.11—Pure propane is in contact with a nonvolatile oil ($M = 550$) at 134°C and pressure of 10 bar. Calculate K_i value using the regular solution theory and Standing correlation.

Solution—Consider the system as a binary system of component 1 (propane) and component 2 (oil). Component 2 is

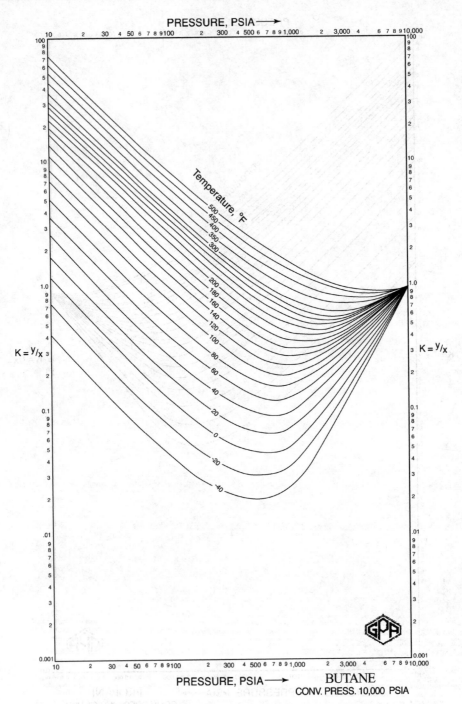

FIG. 6.22—K_i values of i-butane. Unit conversion: °F = °C × 1.8 + 32 psia = 14.504 × bar. Taken with permission from Ref. [28].

in fact solvent for component 1, which can be considered as solute. Also for simplicity consider oil as a single carbon number with molecular weight of 550. This assumption does not cause major error in the calculations as properties of propane are needed for the calculation. K_i is defined by Eq. (6.196) as $K_1 = y_1/x_1$. Since the oil is nonvolatile thus the vapor phase is pure propane and $y_1 = 1$, therefore, $K_1 = 1/x_1$. To calculate x_1, a similar method as used in Example 6.10 is followed. In this example since $P = 10$ bar the gas phase is not ideal and to use Eq. (6.179), ϕ_1^V must be calculated for propane

at 10 bar and $T = 134$°C (407 K). From Table 2.1 for C$_3$ we have $M = 44.1$, SG = 0.5063, $T_c = 369.83$ K, $P_c = 42.48$ bar, and $\omega = 0.1523$. $T_r = 1.125$, $P_r = 0.235$. Since P_r is low, ϕ can be conveniently calculated from virial EOS by Eq. (6.62) together with Eq. (5.72) for calculation of the second virial coefficient. The result is $\phi_1^V = 0.94$. Calculation of γ_1 is similar to Example 6.10 with use of Eq. (6.145). It requires parameters V^L and δ for both C$_3$ and the oil. Value of V^L for C$_3$ as given in Table 6.10 seems to be lower than extrapolated value at 298 K. The molar volume of propane at 298 K can

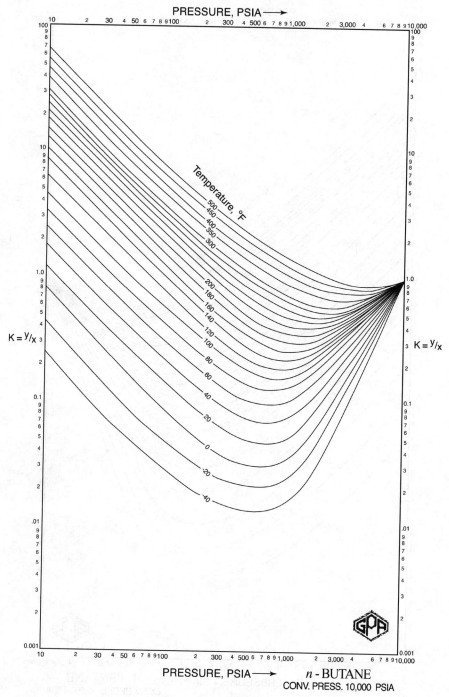

PRESSURE, PSIA ➤

$K = y/x$

$K = y/x$

PRESSURE, PSIA ➤   *n* - BUTANE
CONV. PRESS. 10,000 PSIA

FIG. 6.23—K_i **values of** *n***-butane. Unit conversion:** °F = °C × 1.8 + 32 psia = 14.504 × bar. Taken with permission from Ref. [28].

be calculated from its density. Substituting SG = 0.5063 and T = 298 K in Eq. (5.127) gives density at 25°C as 0.493 g/cm³ and the molar volume is V_1^L = 44.1/0.493 = 89.45 cm³/mol. Similarly at 134°C we get V_1^L = 128.7 cm³/mol. Value of δ for C₃ is given in Table 6.10 as δ = 13.9 (J/cm³)¹ᐟ². From Eq. (4.7) and coefficients in Table 4.5 for oil of M = 550, we get d_{20} = 0.9234 g/cm³ and δ_2 = 8.342 (cal/cm³)¹ᐟ². These values are very approximate as oil is assumed as a single carbon number. Density is corrected to 25°C through Eq. (2.115) as d_{25} = 0.9123 g/cm³. Thus at 298 K for component 2 (solvent)

we have V_2^L = 550/0.9123 = 602.9 cm³/mol. To calculate γ_1 from Eq. (6.145), x_1 is required. The initial value of x_1 is calculated through Eq. (6.191) assuming γ_1 = 1. Since $T_r >$ 1, the value of f_1^L is calculated through Eqs. (6.189) and (6.190) as f_1^L = 51.13 bar. Finally, the value of γ_1 is calculated as 1.285, which gives x_1 = 0.94 × 10/(1.285 × 51.13) = 0.144. Thus, K_1 = 1/0.144 = 6.9. To calculate K_1 from the Standing method, Eq. (6.202) should be used. From Table 6.13 for propane, b = 999.4 K and T_B = 231.1 K, and from Eq. (6.202) at 407 K and 10 bar, a = 0.1069, c = 0.8646, and K_1 = 5.3. ♦

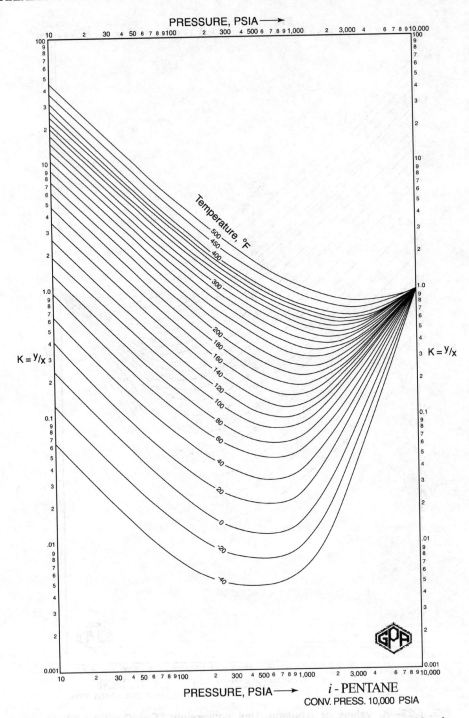

FIG. 6.24—K_i values of *i*-pentane. Unit conversion: °F = °C × 1.8 + 32 psia = 14.504 × bar. Taken with permission from Ref. [28].

To summarize methods of VLE calculations, recommended methods for some special cases are given in Table 6.14 [37].

6.8.3 Solid–Liquid Equilibria—Solid Solubility

Formulation of SLE is similar to that of VLE and it is made through Eq. (6.174) with equality of fugacity of i in solid and liquid phases, where the relations for calculation of $\hat{f}_i^S$ and $\hat{f}_i^L$ are given in Section 6.6. To formulate solubility of a solid in a liquid, the solid phase is assumed pure, $\hat{f}_i^S = f_i^S$, and the

above relation becomes

$$(6.205) \qquad f_i^S = x_i \gamma_i f_i^L$$

by substituting f_i^S from Eq. (6.155) we get

$$\ln \frac{1}{\gamma_i x_i} = -\frac{\Delta H_i^f}{R T_{Mi}} \left(1 - \frac{T_{Mi}}{T} \right) + \frac{\Delta C_{Pi}}{R} \left(1 - \frac{T_{Mi}}{T} \right)$$
$$(6.206) \qquad \qquad + \frac{\Delta C_{Pi}}{R} \ln \frac{T_{Mi}}{T}$$

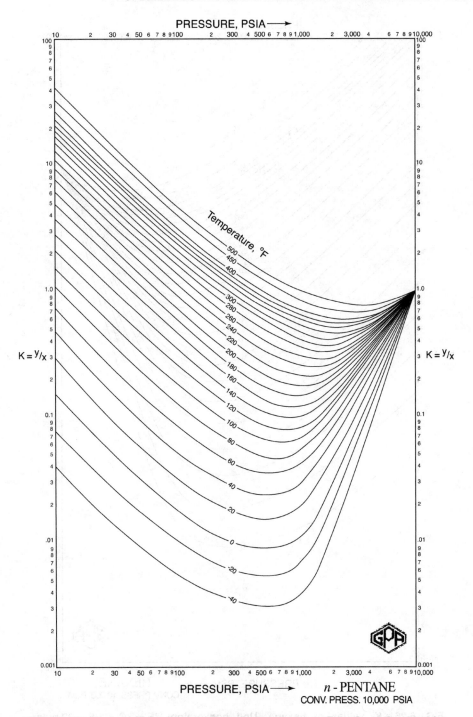

PRESSURE, PSIA ⟶

$K = y/x$

$K = y/x$

PRESSURE, PSIA ⟶   *n* - PENTANE
CONV. PRESS. 10,000 PSIA

FIG. 6.25—K_i values of *n*-pentane. Unit conversion: °F = °C × 1.8 + 32 psia = 14.504 × bar. Taken with permission from Ref. [28].

It should be noted that this equation can be used to calculate solubility of a pure solid into a solvent. γ_i (a function of x_i) can be calculated from methods given in Section 6.6.6 and x_i must be found by trial-and-error procedure with initial value of x_1 calculated at $\gamma_i = 1.0$. However, for ideal solutions where γ_i is equal to unity the above equation can be used to calculate solubility directly. Since actual values of $\Delta C_{Pi}/R$ are generally small (see Fig. 6.11) with a fair approximation the above relation for ideal solutions can be simplified as [17, 21, 43]

$$(6.207) \qquad x_i = \exp\left[\frac{\Delta H_i^{\mathrm{f}}}{RT_{\mathrm{M}i}}\left(1 - \frac{T_{\mathrm{M}i}}{T}\right)\right]$$

The above equation provides a quick way of calculating solubility of a solid into a solvent where chemical nature of solute is similar to that of solvent; therefore, only properties of solute are needed. Calculation of heat of fusion (ΔH_i^{f}) was discussed in Section 6.6.6 and calculation of freezing point ($T_{\mathrm{M}i}$) was discussed in Section 2.6.4. Solubility of naphthalene in several hydrocarbons is given in Ref. [43]. At 20°C mole fraction of solid naphthalene in solvents hexane, benzene, and toluene is 0.09, 0.241, and 0.224, respectively. Naphthalene (aromatics) has higher solubility in benzene and toluene (also aromatics) than in hexane (a paraffinic). Naphthalene

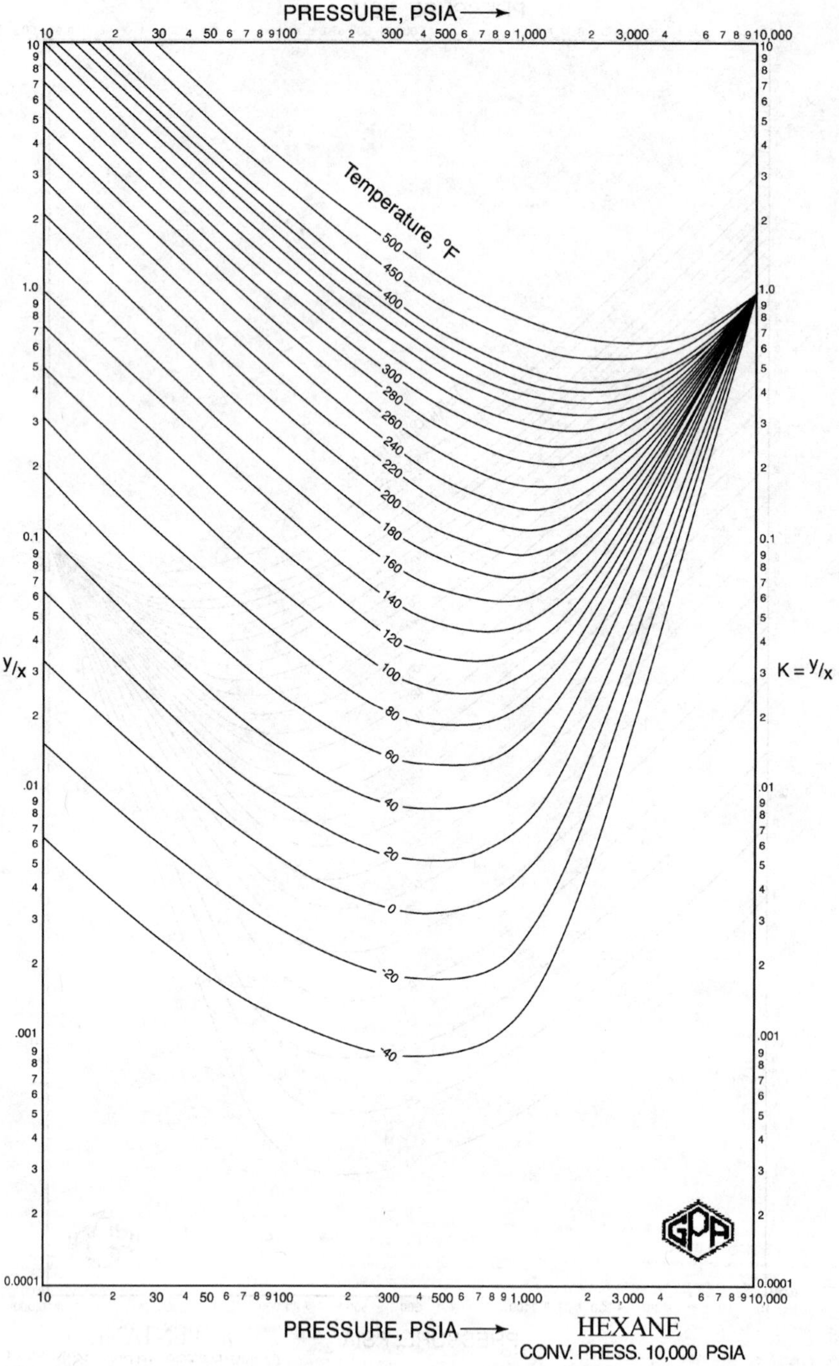

FIG. 6.26—K_i values of hexane. Unit conversion: °F = °C × 1.8 + 32 psia = 14.504 × bar. Taken with permission from Ref. [28].

has ΔH_i^f of 18.58 kJ/mol and its melting point is 80°C [43]. Solubility of naphthalene calculated through Eq. (6.207) is $x_1 = 0.27$. Better prediction can be obtained by calculating γ_1 for each system. For example, through regular solution theory for naphthalene (1)–toluene (2) system at 20°C γ_1 is calculated as $\gamma_1 = 1.17$ (method of calculation was shown in Example 6.10). The corrected solubility is 0.27/1.17 = 0.23, which is in good agreement with the experimental value of 0.224. As discussed before, compounds with similar structures have better solubility.

For solid precipitation such as wax precipitation, the solid phase is a mixture and the general relation for SLE is given by Eq. (6.174). A SLE ratio, K_i^{SL}, can be defined similar to VLE ratio as

$$(6.208) \qquad x_i^S = K_i^{SL} x_i^L$$

By combining Eqs. (6.114) and (6.161) with Eq. (6.174) and use of the above definition we get [17]

$$(6.209) \qquad K_i^{SL} = \frac{f_i^L(T, P)\gamma_i^L(T, P, x_i^L)}{f_i^S(T, P)\gamma_i^S(T, P, x_i^S)}$$

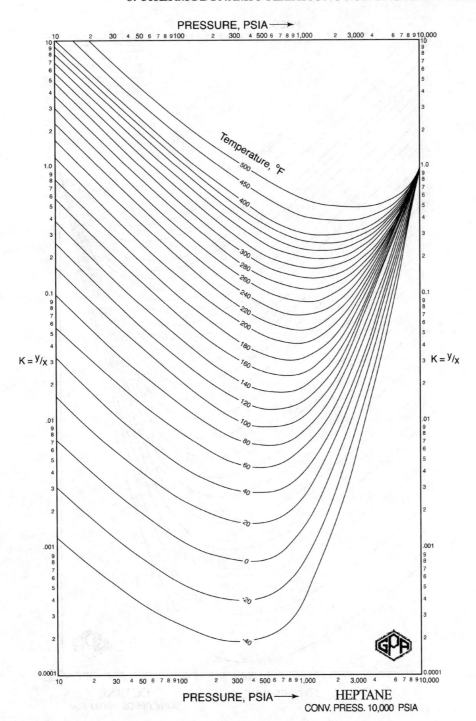

FIG. 6.27—K_i values of heptane. Unit conversion: °F = °C × 1.8 + 32 psia = 14.504 × bar. Taken with permission from Ref. [28].

where for ideal liquid and solid solutions γ_i^L and γ_i^S are unity and the pure component fugacity ratio f_i^L/f_i^S can be calculated from Eq. (6.155). For nonideal solutions, γ_i^S may be calculated from Eq. (6.153). However, δ^S calculated from Eq. (6.153) must be used for δ in Eq. (6.154). Equations (6.208) and (6.209) can be used to construct freezing/melting points or liquid–solid phase diagram ($Tx^S x^L$) based on SLE calculations as shown in Chapter 9. For an ideal binary system the $Tx^S x^L$ diagram is shown in Fig. 6.32. Such figures are useful to determine the temperature at which freezing

begins for a mixture (see Problem 6.29). Multicomponent SLE calculations become very easy once the stability analysis is made. From stability analysis consideration a component in a liquid mixture with mole fraction z_i may exist as a pure solid if the following inequality holds [17]:

$$(6.210) \qquad \hat{f}_i(T, P, z_i) - f_i^S(T, P) \geq 0$$

where f_i^S is fugacity of pure solid i. This equation is the basis of judgment to see if a component in a liquid mixture will precipitate as solid or not. The answer is yes if the above

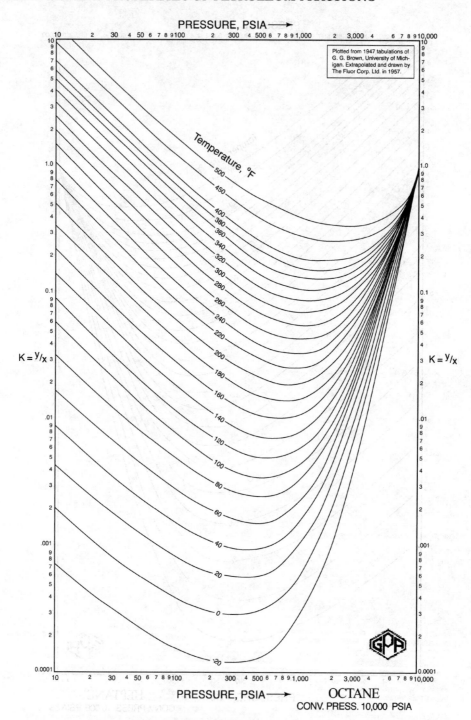

FIG. 6.28—K_i **values of octane. Unit conversion:** $°F = °C × 1.8 + 32$ psia $=$
14.504 × bar. Taken with permission from Ref. [28].

inequality holds for that component. Equality in the above equation is equivalent to Eq. (6.206). The same criteria apply to precipitation of a component from a gas mixture, where in the above inequality $\hat{f}_i(T, P, z_i)$ would refer to fugacity of i in the gas phase with mole fraction z_i. Similar principle applies in formation of liquid i from a gas mixture when the temperature decreases. One main application of this inequality is to determine the temperature at which solid begins to form from an oil. This temperature is equivalent to cloud point

of the oil. Applications of Eqs. (6.208) and (6.209) for calculation of cloud point and wax formation are demonstrated in Chapter 9. Full description of a thermodynamic model for wax precipitation is provided in Ref. [17]. Application of these relations to calculate cloud point of crude oils and reservoir fluids are given in Chapter 9.

Example 6.12—How much (in grams) n-hexacontane (n-C_{36}) can be dissolved in 100 g of n-heptane, so that when the

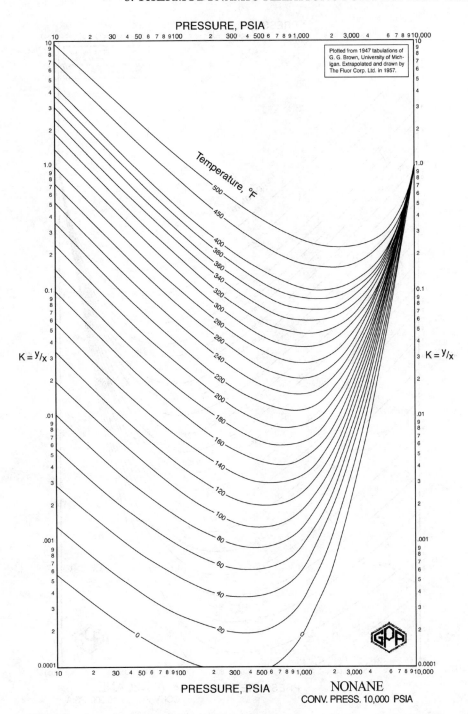

FIG. 6.29—K_i values of nonane. Unit conversion: °F = °C × 1.8 + 32 psia = 14.504 × bar. Taken with permission from Ref. [28].

temperature of the solution is reduced to 15°C, the solid phase begins to form.

Solution—We have a binary system of component 1 (C_{36}) and component 2 (n-C_7). Component 1 is solute and component 2 is considered as solvent. From Table 6.10, $M_2 = 100.2$, $T_{M_2} = 182.57$ K, $\Delta H_2^f / RT_{M_2} = 9.2557$, $M_1 = 506.95$, $T_{M_1} = 349$ K, and $\Delta H_1^f / RT_{M_1} = 30.6066$. Assuming ideal solution we use Eq. (6.207) for calculation of x_1 at $T = 288.2$ K: $x_1 = \exp[30.6066(1 - 349/288.2)] = 0.0016$. With respect to

$M_1 = 506.9$ and $M_2 = 100.2$, from Eq. (1.15) we get $x_{wi} = 0.08$, which is equivalent to 0.807 g of n-C_{36} in 100 g of n-C_7. ◆

6.8.4 Freezing Point Depression and Boiling Point Elevation

When a small amount of a pure solid (solute) is added to a solvent, the freezing point of solvent decreases while its boiling point increases. Upon addition of a solute (component 1) to a solvent (component 2) mole fraction of solvent

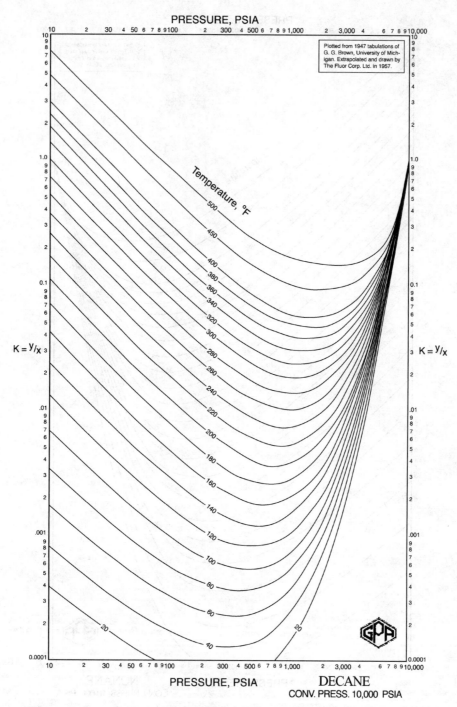

FIG. 6.30—K_i values of decane. Unit conversion: °F = °C × 1.8 + 32 psia = 14.504 × bar. Taken with permission from Ref. [28].

(x_2) reduces from unity. This slight reduction in mole fraction of x_2 causes slight reduction in chemical potential according to Eq. (6.120). Therefore, at the freezing point when liquid solvent is in equilibrium with its solid, the activity of pure solid must be lower than its value that corresponds to normal freezing point. This decrease in freezing point is called *freezing point depression*. At freezing point temperature, liquid and solid phases are in equilibrium and Eq. (6.206) applies. If the solution is assumed ideal, Eq. (6.206) can be written for the solvent (component 2) in the following form

neglecting ΔC_{Pi}:

$$(6.211) \qquad \ln\frac{1}{x_2} = -\frac{\Delta H_2^f}{RT_{M2}}\left(1 - \frac{T_{M2}}{T}\right)$$

where ΔH_2^f is the molar heat of fusion for pure solvent and T_{M2} is the solvent melting point. T is the temperature at which solid and liquid phases are in equilibrium and is the same as the freezing point of solution after addition of solute. The amount of decrease in freezing point is shown by ΔT_{M2}, which

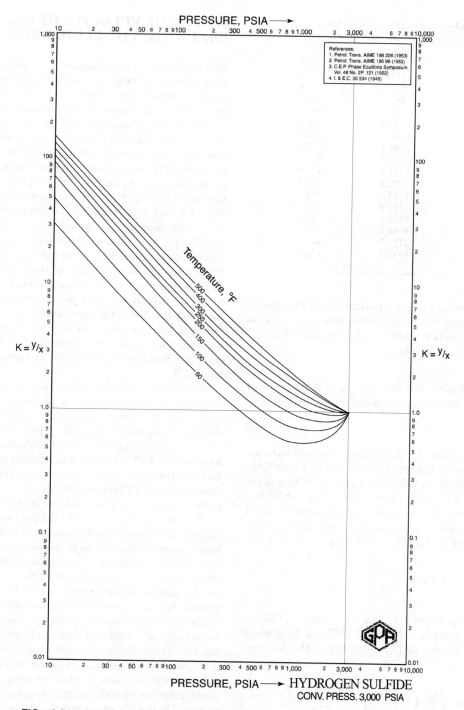

FIG. 6.31—K_i values of hydrogen sulfide. Unit conversion: °F = °C × 1.8 + 32 psia = 14.504 × bar. Taken with permission from Ref. [28].

is equal to $(T_{M2} - T)$. The amount of solute in the solution is $x_1 (= 1 - x_2)$, which is very small ($x_1 \ll 1$), and from a mathematical approximation we have

$$(6.212) \qquad \ln(1 - x_1) \approx -x_1$$

Therefore, Eq. (6.211) can be solved to find ΔT_{M2} [43]:

$$(6.213) \qquad \Delta T_{M2} \cong \frac{x_1 R T_{M2}^2}{\Delta H_2^f}$$

In deriving this equation since the change in the freezing point is small, $T T_{M2}$ is approximated by T_{M2}^2. Equation (6.213) is

approximate but it is quite useful for calculation of freezing point depression for hydrocarbon systems. For nonideal solutions, Eq. (6.211) must be used by replacing x_2 with $\gamma_2 x_2$, where γ_2 is a function of x_2.

By similar analysis from VLE relation for ideal solutions it can be shown that boiling point elevation may be estimated from the following simplified relation [43]:

$$(6.214) \qquad \Delta T_{b2} \cong \frac{x_1 R T_{B2}^2}{\Delta H_2^{vap}}$$

TABLE 6.13—Values of b and T_B for use in computing K_i values from Eq. (6.202) [Ref. 41].

Compound	b, K	T_B, K
N_2	261.1	60.6
CO_2	362.2	107.8
H_2S	631.1	183.9
C_1	166.7	52.2
C_2	636.1	168.3
C_3	999.4	231.1
i-C_4	1131.7	261.7
n-C_4	1196.1	272.8
i-C_5	1315.6	301.1
n-C_5	1377.8	309.4
i-C_6 (all)	1497.8	335.0
n-C_6	1544.4	342.2
n-C_7	1704.4	371.7
n-C_8	1852.8	398.9
n-C_9	1994.4	423.9
n-C_{10}	2126.7	447.2
C_6 (lumped)	1521.1	338.9
C_{7+}	Use Eq. (6.203)	

where ΔH_2^{vap} is the molar heat of vaporization for the solvent and ΔT_{b2} is the increase in boiling point when mole fraction of solute in the solution is x_1. Methods of estimation of ΔH_2^{vap} are discussed in the next chapter.

Example 6.13—Calculate the freezing point depression of toluene when 5 g of benzoic acid is dissolved in 100 g of benzene at 20°C.

Solution—For this system the solute is benzoic acid (component 1) and the solvent is benzene (component 2). From Table 6.10 for benzoic acid, $M_1 = 122.1$ and $T_{M1} = 395.5$ K, and for toluene, $M_2 = 78.1$, $T_{M2} = 278.6$ K, and $\Delta H_2^f/RT_{M2} = 4.26$. For 5 g benzoic acid and 100 g benzene from a reverse form of Eq. (1.15) we get $x_1 = 0.031$. To calculate freezing point depression we can use Eq. (6.213):

$$\Delta T_{M2} \cong \frac{x_1 R T_{M2}^2}{\Delta H_2^f}$$

where $x_1 = 0.031$, $T_{M2} = 278.6$ K, and $RT_{M2}/\Delta H_2^f = 1/4.26 = 0.2347$. Thus $\Delta T_{M2} = 0.031 \times 278.6 \times 0.2347 = 2$ K. A more accurate result can be obtained by use of Eq. (6.211) for nonideal systems as

$$\ln\frac{1}{\gamma_2 x_2} = -\frac{\Delta H_2^f}{RT_{M2}}\left(1 - \frac{T_{M2}}{T}\right) \approx \frac{\Delta H_2^f \Delta T_{M2}}{RT_{M2}^2}$$

For this system since x_2 is near unity, $\gamma_2 = 1.0$ and same value for ΔT_{M2} is obtained; however, for cases that x_2 is substantially lower than unity this equation gives different result. ♦

6.9 USE OF VELOCITY OF SOUND IN PREDICTION OF FLUID PROPERTIES

One application of fundamental relations discussed in this chapter is to develop an equation of state based on the velocity of sound. The importance of PVT relations and equations of state in estimation of physical and thermodynamic properties and phase equilibrium were shown in Chapter 5 as well as in this chapter. Cubic equations of state and generalized corresponding states correlations are powerful tools for predicting thermodynamic properties and phase equilibria calculations. In general most of these correlations provide reliable data if accurate input parameters are used (see Figs. 1.4 and 1.5). Accuracy of thermodynamic PVT models largely depends on the accuracy of their input parameters (T_c, P_c, and ω) particularly for mixtures where no measured data are available on the pseudocritical properties and acentric factor. While values of these parameters are available for pure and light hydrocarbons or they may be estimated accurately for light petroleum fractions (Chapter 2), for heavy fractions and heavy compounds found in reservoir fluids such data are not available. Various methods of predicting these parameters give significantly different values especially for high-molecular-weight compounds (see Figs. 2.18 and 2.20).

One way to tackle this difficulty is to use a measured property such as density or vapor pressure to calculate critical properties. It is impractical to do this for reservoir fluids under reservoir conditions, as it requires sampling and laboratory measurements. Since any thermodynamic property can be related to PVT relations, if accurately measured values of a thermodynamic property exist, they can be used to extract parameters in a PVT relation. In this way there is no need to use various mixing rules or predictive methods for calculation of T_c, P_c, and ω of mixtures and EOS parameters can be directly calculated from a set of thermodynamic data. One thermodynamic property that can be used to estimate EOS parameter is velocity of sound that may be measured directly in a reservoir fluid under reservoir conditions without sampling. Such data can be used to obtain an accurate PVT relation for the reservoir fluids. For this reason Riazi and Mansoori [44] used thermodynamic relations to develop an equation of state based on velocity of sound and then sonic velocity data have been used to obtain thermodynamic properties [8, 44, 45]. Colgate et al. [45, 46] used velocity of sound data to determine critical properties of substances. Most recently, Ball et al. [48] have constructed an ultrasonic apparatus for measuring the speed of sound in liquids and compressed gases. They also reported speed of sound data for an oil sample up to pressure of 700 bars (see Fig. 6.34) and discussed prospects for use of velocity of sound in determining bubble point, density, and viscosity of oils.

TABLE 6.14—Recommended methods for VLE calculations.

Pressure	Mixtures of similar substances	Mixtures of dissimilar substances
<3.45 bar (50 psia)	Raoult's law (Eq. 6.180)	Modified Raoult's law (Eq. 6.181)
<13.8 bar (200 psia)	Lewis rule (Eq. 6.198)	Activity coefficients (Eq. 6.179)
$P < 5$–10 bar,	Henry's law (Eq. 6.184) for dilute liquid systems ($x_i < \sim 0.03$)	
Any P, 255 $< T <$ 645 K	Chao–Seader (Eq. 6.200) for nonpolar systems and outside critical region	
> 13.8 bar (200 psia)	Eq. (6.197) with SRK or PR EOS using appropriate BIPs	
$P < 69$ bar (1000 psia)	Standing correlation (Eq. 6.202) for natural gases, gas condensate reservoir fluids and light hydrocarbon systems with little C_{7+}	

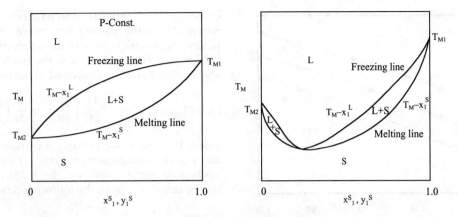

FIG. 6.32—Schematic of freezing–melting diagram for ideal and nonideal binary systems.

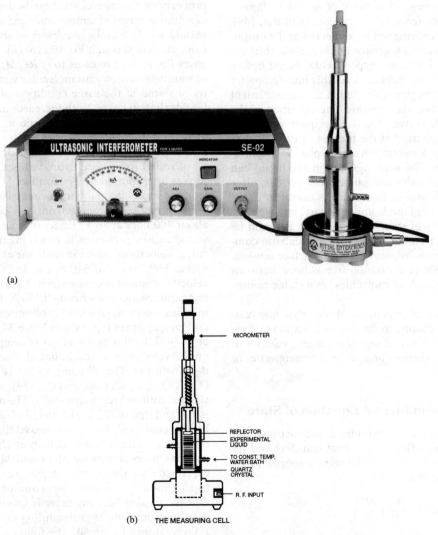

FIG. 6.33—Schematic of interferometer for measuring velocity of sound in liquids. (*a*) General view of ultrasonic interferometer and (*b*) cross section of ultrasonic cell. Taken with permission from Ref. [49].

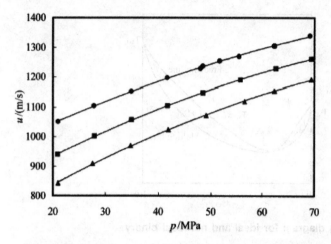

FIG. 6.34—Speed of sound in oil sample. ● 335.1 K, ■ 370.7 K, and ▲ 402.1 K. The lines are quadratic fits. Taken with permission from Ref. [48].

Method of measurement of velocity of sound in liquids through ultrasonic interferometer is presented in Ref. [49]. In this method the measuring cell is connected to the output terminal of a high-frequency generator through a shielded cable. The cell is filled with the experimental liquid before switching on the generator. Schematic of this interferometer is shown in Fig. 6.33. The principle used in the measurement of velocity (c) is based on the accurate determination of the wavelength. Ultrasonic waves of known frequency are produced by a quartz plate fixed at the bottom of the cell. The waves are reflected by a movable metallic plate kept parallel to the quartz plate. The sonic velocity in the liquid can be determined from the following relation: c = wavelength × frequency. This simple measuring device is useful to determine velocity of sound in liquids under normal atmospheric pressure. From velocity of sound measurement it would be possible to directly determine isothermal or adiabatic compressibilities, excess enthalpy, heat capacity, surface tension, miscibility, van der Waal's constants, free volume between molecules, mean free path of molecules, molecular radius, etc. [49].

The purpose of this section is to use thermodynamic relations discussed in this chapter to develop an equation of state based on the velocity of sound and then to use velocity of sound data to estimate thermal and volumetric properties of fluids and fluid mixtures.

6.9.1 Velocity of Sound Based Equation of State

Sound waves in a fluid are longitudinal contractions and rarefactions, which are adiabatic (no heat transfer) and reversible (no energy loss) and which travel at a speed c given by the thermodynamic quantity [10]:

$$(6.215) \qquad c^2 = -\frac{V^2}{M}\left(\frac{\partial P}{\partial V}\right)_S$$

where c is the velocity of sound, V is the molar volume, M is the molecular weight, and constant S refers to the fact the wave transmission is a constant entropy process (adiabatic and reversible). It should be noted that if in the above

equation V is the specific volume then parameter M must be removed from the relation. Equivalent forms of this equation in terms of specific volume or mass density are also commonly used in various sources. From thermodynamic relations the above relation can be converted to the following form:

$$(6.216) \qquad c^2 = -\frac{\gamma V^2}{M}\left(\frac{\partial P}{\partial V}\right)_T = \frac{\gamma}{M}\left(\frac{\partial P}{\partial \rho}\right)_T$$

where V is the molar volume, ρ is the molar density ($1/V$), and γ is the heat capacity ratio (C_P/C_V). Using definition of isothermal compressibility, κ, (Eq. 6.25), the velocity of sound can be calculated from the following relation:

$$(6.217) \qquad c^2 = \frac{\gamma}{M\rho\kappa}$$

From this relation it is apparent that the velocity of sound in a fluid depends on the fluid properties and it is somewhat less than mean velocity of molecules as shown from the kinetic theory of gases [10]. Since speed of sound is a state function property, an equation of state can be developed for the velocity of sound in terms of temperature and density as independent variables [43]. Similarly velocity of sound can be calculated from an EOS through Eq. (6.216) [8]. For example, for ideal gases Eq. (6.216) reduces to $(\gamma RT/M)^{1/2}$. In general velocity of sound decreases with molecular weight of the fluid. Velocity of sound at the same condition of T and P is higher in liquids than in gases. With increases in temperature, velocity of sound in gases increases while in liquids decreases. Velocity of sound increases with pressure for both gases and liquids. Some experimental and calculated data on velocity of sound for several hydrocarbons in gaseous and liquid phases are reported by Firoozabadi [17]. As an example, velocity of sound in methane gas increases from 450 to 750 m/s when pressure increases from low pressures (< 1 bar) to about 400 bars at 16°C. Effect of temperature on velocity of sound at low pressures is much greater than at high pressures. Velocity of sound in methane at 50 bar increases from 430 at 16°C to about 540 m/s at 167°C. For liquid n-hexane velocity of sound decreases from 1200 to about 860 m/s when temperature increases from −10 to 70°C [17]. Experimentally measured velocity of sound in oil sample at various pressures and temperatures is shown in Fig. 6.34 as determined by Ball et al. [48]. In this figure effect of temperature and pressure on the velocity of sound in liquid phase for a live oil is well demonstrated. The oil composition is given as follows: CO_2 (1), C_1 (34), C_2–C_6 (26), and C_{7+} (39), where the numbers inside parentheses represent mol%. The molecular weight of oil is 102 and that of C_{7+} is 212. Detail of oil composition is given by Ball et al. [48]. They also showed that velocity of sound in oils increases linearly with density at a fixed temperature [48].

It has been shown by Alem and Mansoori [50] that the expression for the entropy departure of a hard-sphere fluid can be used for entropy departure of a real fluid provided that the hard sphere diameter is taken as temperature- and density-dependent. By substituting Carnahan–Starling EOS, Eq. (5.93) into Eq. (6.50), the following relation is obtained for the entropy departure of hard-sphere fluids:

$$(6.218) \qquad (S - S^{ig})^{HS} = -\frac{R\zeta(4 - 3\zeta)}{(1 - \zeta)^2}$$

in which ξ is dimensionless packing factor defined by Eq. (5.86) in terms of hard-sphere diameter σ. In general for nonassociating fluids, σ can be taken as a linear function of $1/T$ and ρ [50], i.e., $\sigma = d_0 + d_1\rho + d_2/T + d_3\rho/T$. From Eq. (6.218) we have

$$(6.219) \qquad S = S(T, V)$$

Since c is a state function we can write it as a function of only two independent properties for a pure fluid or fluid mixtures of constant composition in a single phase (see the phase rule in Chapter 5):

$$(6.220) \qquad c = c(S, V)$$

Differentiating Eq. (6.220) with respect to S at constant V gives

$$(6.221) \quad 2cdc = -\frac{V^2}{M}\left\{\frac{\partial}{\partial S}\left[\left(\frac{\partial P}{\partial V}\right)_S\right]_V\right\}dS \quad \text{(at constant } V)$$

Applying Eqs. (6.10) to (6.219) gives

$$(6.222) \qquad dS = \left(\frac{\partial S}{\partial T}\right)_V dT + \left(\frac{\partial S}{\partial V}\right)_T dV$$

which at constant V becomes

$$(6.223) \qquad dS = \left(\frac{\partial S}{\partial T}\right)_T dT \quad \text{(at constant } V)$$

From mathematical relations we know

$$(6.224) \qquad \frac{\partial}{\partial S}\left[\left(\frac{\partial P}{\partial V}\right)_S\right]_V = \frac{\partial}{\partial V}\left[\left(\frac{\partial P}{\partial S}\right)_V\right]_S$$

The Maxwell's relation given by Eq. (6.10) gives $(\partial P/\partial S)_V = -(\partial T/\partial V)_S$, where $-(\partial T/\partial V)_S$ can be determined from dividing both sides of Eq. (6.222) by ∂V at constant S as

$$(6.225) \qquad \left(\frac{\partial T}{\partial V}\right)_S = -\frac{(\partial S/\partial V)_T}{(\partial S/\partial T)_V}$$

Substituting Eqs. (6.223)–(6.224) into Eq. (6.221) and integrating from T to $T \to \infty$, where $c \to c^{HS}$ gives the following relation for c in terms of T and V:

$$(6.226) \quad c^2 = \left(c^{HS}\right)^2 - \frac{V^2}{M}\int\limits_T^\infty \left(\frac{\partial^2 T}{\partial V^2}\right)_S\left(\frac{\partial S}{\partial T}\right)_V dT$$

c^{HS} can be calculated from Eq. (6.216) using the CS EOS, Eq. (5.93). Derivative $(\partial S/\partial T)_V$ can be determined from Eq. (6.218) or (6.219) as a function of T and V only. $(\partial^2 T/\partial V^2)_S$ can be determined from Eq. (6.225) as a function of T and V. Therefore, the RHS of above equation is in terms of only T and V, which can be written as

$$(6.227) \qquad c = c(T, V)$$

Equation (6.226) or (6.227) is a cVT relation and is called velocity of sound based equation of state [44]. One direct application of this equation is that when a set of experimental data on cVT or cPT for a fluid or a fluid mixture of constant composition are available they can be used with the above relations to obtain the PVT relation of the fluid. This is the essence of use of velocity of sound in obtaining PVT relations. This is demonstrated in the next section by use of velocity of sound data to obtain EOS parameters. Once the PVT relation for a fluid is determined all other thermodynamic properties can be calculated from various methods presented in this chapter.

6.9.2 Equation of State Parameters from Velocity of Sound Data

In this section the relations developed in the previous section for the velocity of sound are used to obtain EOS parameters. These parameters have been compared with those obtained from critical constants or other properties in the form of prediction of volumetric and thermodynamic properties. Truncated virial (Eq. 5.76), Carnahan–Starling–Lennard–Jones (Eq. 5.96), and common cubic equations (Eq. 5.40) have been used for the evaluations and testing of the suggested method. Although the idea of the proposed method is for heavy hydrocarbon mixtures and reservoir fluids, but because of lack of data on the sonic velocity of such mixtures applicability of the method is demonstrated with use of acoustic data on light and pure hydrocarbons [8, 44].

6.9.2.1 Virial Coefficients

Since any equation of state can be converted into virial form, in this stage second and third virial coefficients have been obtained from sonic velocity for a number of pure substances. Assuming that the entropy departure for a real fluid is the same as for a hard sphere and by rearranging Eq. (6.218) the packing fraction of hard sphere can be calculated from real fluid entropy departure:

$$(6.228) \quad \zeta = \left(\frac{\pi}{6}\right)\rho N_A\sigma^3 = \frac{2 - \frac{S-S^{ig}}{R} - \left(4 - \frac{S-S^{ig}}{R}\right)^{1/2}}{3 - \frac{S-S^{ig}}{R}}$$

Calculated values of σ from the above equation indicate that there is a simple relation between hard-sphere diameter as in the following form [44]:

$$(6.229) \qquad \sigma = d_0 + \frac{d_1}{T}$$

Application of the virial equation truncated after the third term, Eq. (5.76), to hard sphere fluids gives

$$(6.230) \qquad Z^{HS} = 1 + \frac{B^{HS}}{V} + \frac{C^{HS}}{V^2}$$

By converting the HS EOS, Eq. (5.93), into the above virial form one gets [51]

$$(6.231) \qquad \begin{aligned} B^{HS} &= \frac{2}{3}\pi N_A\sigma^3 \\ C^{HS} &= \frac{5}{18}\pi^2 N_A^2\sigma^6 \end{aligned}$$

Substituting Eqs. (5.76) and (6.230) for real and hard-sphere fluids virial equations into Eq. (6.39) one can calculate entropy departures for real and hard-sphere fluids as

$$(6.232) \quad \left[\frac{S-S^{ig}}{R}\right]_{real\,fluid} = -\left(\ln P + \frac{B+TB'}{V} + \frac{C+TC'}{2V^2}\right)$$

$$(6.233) \quad \left[\frac{S-S^{ig}}{R}\right]_{hard\,sphere} = -\left(\ln P + \frac{B^{HS}}{V} + \frac{C^{HS}}{2V^2}\right)$$

Since it is assumed that the left sides of the above two equations are equal, so the RHSs must also be equal, which result in the following relations:

$$(6.234) \qquad \begin{aligned} TB' + B &= B^{HS} \\ TC' + C &= C^{HS} \end{aligned}$$

where $B' = dB/dT$ and $C' = dC/dT$. Substituting for B^{HS} and C^{HS} from Eq. (6.231) into Eq. (6.234) and, combining with Eq. (6.229), gives two nonhomogeneous differential equations that after their solutions we get:

$$(6.235) \quad \begin{aligned} B(T) &= q_0 \frac{\ln T}{T} + \sum_{n=0}^{3} \frac{p_n}{T^n} \\ C(T) &= q_1 \frac{\ln T}{T} + \sum_{n=0}^{3} \frac{L_n}{T^n} \end{aligned}$$

Parameters p_1 and L_1 are constants of integration while all other constants are related to parameters d_0 and d_1 in Eq. (6.229) [44]. For example, parameters q_0 and L_0 are related to d_0 and d_1 as follows: $q_0 = 2\pi N_A d_0 d_1$ and $L_0 = (5/18)\pi^2 N_A^2 d_0^6$, where N_A is the Avogadro's number. Substitution of the truncated virial EOS, Eq. (5.76), into Eq. (6.216) gives the following relation for the velocity of sound in terms of virial coefficients:

$$(6.236) \quad c^2 = \frac{\gamma R T}{M}[1 + \rho(2B + 3C\rho)]$$

where γ is the heat capacity ratio (C_P/C_V) and ρ is the molar density $(1/V)$. Once B and C are determined from Eq. (6.235), C_P and C_V can be calculated from Eqs. (6.64) and (6.65) and upon substitution into Eq. (6.236) one can calculate velocity of sound. *Vice versa* the sonic velocity data can be used to obtain virial coefficients and consequently constants p_1 and L_1 in Eqs. (6.235) by minimizing the following objective function:

$$(6.237) \quad RC = \sum_{i=1}^{N} (c_{i,\text{calc.}} - c_{i,\text{exp.}})^2$$

where N is the number of data points on the velocity of sound.

Thermodynamic data, including velocity of sound for methane, ethane, and propane, are given by Goodwin et al. [52–54]. Entropy data on methane [52] were used to obtain constants d_0 and d_1 by substituting Eq. (6.235) into Eq. (6.232). Values of $d_0 = 2.516 \times 10^{-10}$ m and $d_1 = 554.15 \times 10^{-10}$ m · K have been obtained for methane from entropy data [44]. With knowledge of d_0 and d_1 all constants in Eq. (6.235) were determined except p_1 and L_1. For simplicity, truncated virial equation after the second term (Eq. 5.75) was used to obtain constant p_1 for the second virial coefficient, B, by minimizing RC in Eq. (6.237). For methane in the temperature range of 90–500 K and pressures up to 100 bar, it was found that $p_1 = -8.1 \times 10^3$ cm³ · K/mol. Using this value into constants for B in Eq. (6.235) the following relation was found [44]:

$$(6.238) \quad \begin{aligned} B(T) &= 13274 \frac{\ln T}{T} + 20.1 - \frac{81000}{T} \\ &\quad - \frac{2.924 \times 10^6}{T^2} - \frac{1.073 \times 10^{10}}{T^3} \end{aligned}$$

TABLE 6.15—*Constants in Eq. (6.239) for calculation of second virial coefficient.*

Compound	a	B	c	%AAD for Z
Methane	0.02854	19.4	1.6582	0.5
Ethane	0.16	250	0.88	1.1
Propane	0.22	230	1.29	1.4

Taken with permission from Ref. [44].
Number of data points for each compound: 150; pressure range: 0.1–200 bar; temperature range: 90–500 K for C₁, 90–600 K for C₂, and 90–700 K for C₃.

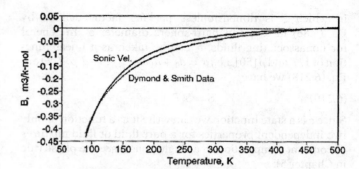

FIG. 6.35—Prediction of second virial coefficient of methane from velocity of sound data (Eq. 6.239). Taken with permission from Ref. [44].

where B is in cm³/mol and T is in kelvin. This equation can be fairly approximated by the following simpler form for the second virial coefficient:

$$(6.239) \quad B(T) = a\frac{10^6 \times \ln T}{T} - b - \frac{10^6 \times c}{T}$$

where B is in cm³/mol and T is in kelvin. All three constants a, b, and c have been directly determined from velocity of sound data for methane, ethane, and propane and are given in Table 6.15. When this equation is used to calculate c from Eq. (6.236) with $C = 0$, an error of 0.5% was obtained for 150 data points for methane [44]. If virial equation with coefficients B and C (Eq. 5.76) were used obviously lower error could be obtained. Errors for prediction of compressibility factor of each compound using Eq. (5.75) with coefficient B estimated from Eq. (6.239) are also given in Table 6.15. Graphical evaluation of predicted coefficient B for methane from Eq. (6.239) is shown in Fig. 6.35. Predicted compressibility factor (Z) for methane at 30 bar, using B determined from velocity of sound and truncated virial equation (Eq. 5.75), is shown in Fig. 6.36. Further development in relation between sonic velocity and virial coefficient is discussed in Ref. [55].

6.9.2.2 Lennard–Jones and van der Waals Parameters

In a similar way Lennard–Jones potential parameters, ε and σ have been determined from velocity of sound data using CSLJ EOS (Eq. 5.96). Calculated parameters have been compared with those determined for other methods and are given

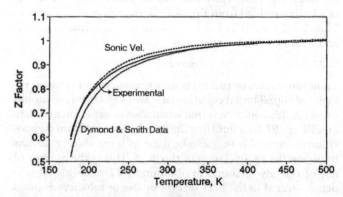

FIG. 6.36—Prediction of Z factor of methane at 30 bar from truncated virial EOS with second coefficient from velocity of sound data. Taken with permission from Ref. [44].

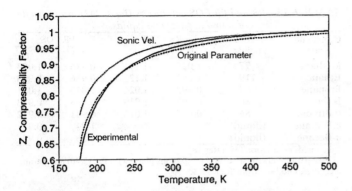

FIG. 6.37—Prediction of *Z* factor for methane at 30 bar from vdW EOS using parameters from velocity of sound data. Taken with permission from Ref. [44].

in Table 6.17. Errors for calculated *Z* values with use of LJ parameters from different methods are also given in this table. Van der Waals EOS parameters determined from velocity of sound are given in Table 6.18 and predicted *Z* values for methane and ethane are shown in Figs. 6.37 and 6.38, respectively. Predicted *Z* factor for propane from CSLJ EOS (Eq. 5.96) is shown in Fig. 6.39. Results presented in Tables 6.15–6.17 and Figs. 6.36–6.39 show that EOS parameters determined from velocity of sound provide reliable PVT data and may be used to calculate other thermodynamic properties.

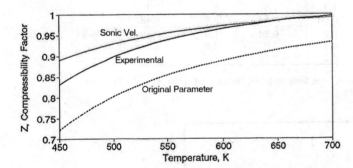

FIG. 6.38—Prediction of *Z* factor for ethane at 100 bar from vdW EOS using parameters from velocity of sound data. Taken with permission from Ref. [44].

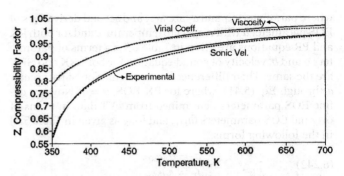

FIG. 6.39—Prediction of *Z* factor of propane at 30 bar from LJ EOS with parameters from different methods. Taken with permission from Ref. [44].

6.9.2.3 RK and PR EOS Parameters—Property Estimation

To further investigate the possibility of using velocity of sound for calculation of PVT and thermodynamic data, RK and PR EOS parameters were determined for both gases and liquids through velocity of sound data. Using parameters defined in Table 6.1 for calculation of γ, V, and $(\partial P/\partial V)_T$ and substituting them into Eq. (6.216), velocity of sound, c, can be estimated. For both RK and SRK equations the relation for c becomes

$$c^2_{\text{RK,SRK}} = -\frac{V^2}{M}\left[-\frac{RT}{(V-b)^2} + \frac{a(2V+b)}{V^2(V+b)^2} \right.$$
$$\left. -T\left(\frac{R}{V-b} - \frac{a_1}{V^2+bV}\right)^2 \times \left(C_P^{\text{ig}} - R - \frac{Ta_2}{b}\ln\frac{V}{V+b}\right)^{-1} \right]$$

(6.240)

and for PR EOS the relation for c becomes

$$c^2_{\text{PR}} = -\frac{V^2}{M}\left[-\frac{RT}{(V-b)^2} + \frac{2a(V+b)}{(V^2+2bV-b^2)^2} \right.$$
$$-T\left(\frac{R}{V-b} - \frac{a_1}{V^2+2bV-b^2}\right)^2$$

(6.241)
$$\left. \times \left(C_P^{\text{ig}} - R - \frac{Ta_2}{2\sqrt{2}b}\ln\frac{V+b-\sqrt{2}b}{V+b+\sqrt{2}b}\right)^{-1} \right]$$

TABLE 6.16—*The Lennard–Jones parameters from the velocity of sound data and other sources.*

Compound	Velocity of sound			Second virial coefficient[a]			Viscosity data[a]		
	ε/k_B, K	σ, Å	%AAD for Z	ε/k_B, K	σ, Å	%AAD for Z	ε/k_B, K	σ, Å	%AAD for Z
Methane	178.1	3.97	0.8	148.2	3.817	4.0	144.0	3.796	4.7
Ethane	300.0	4.25	0.5	243.0	3.594	3.0	230.0	4.418	3.4
Propane	350.0	5.0	1.1	242.0	5.637	11.5	254.0	5.061	8.0

Taken with permission from Ref. [44].
[a]The LJ parameters are used with Eq. (5.96) to calculate Z. The LJ parameters from the second virial coefficient and viscosity are taken from Hirschfelder et al. [56]. k_B is the Boltzman constant (1.381×10^{-23} J/K) and 1 Å = 10^{-10} m.

TABLE 6.17—*The van der Waals constants from the velocity of sound data.*

Compounds	Velocity of sound			Original constants[a]		
	$a \times 10^{-6}$	b	%AAD for Z	$a \times 10^{-6}$	b	%AAD for Z
Methane	1.88583	44.78	1.0	2.27209	43.05	0.8
Ethane	3.84613	57.18	1.8	5.49447	51.98	2.4
Propane	8.34060	90.51	1.4	9.26734	90.51	1.5

Taken with permission from Ref. [44]. *a* is in cm^6/mol$^2 \cdot$ bar and *b* is in cm^3/mol.
[a]From Table 5.1.

where parameters a_1 and a_2 are first and second derivatives of EOS parameter a with respect to temperature and for both RK and PR equations are given in Table 6.1. In terms of parameters a and b, velocity of sound equation for both RK and SRK are the same. Their difference lies in calculation of parameter a through Eq. (5.41), where for RK EOS, $\alpha = 1$. Now we define EOS parameters determined from cVT data in terms of original EOS parameters (a_{EOS} and b_{EOS} as given in Table 5.1) in the following forms:

$$(6.242) \qquad \begin{aligned} a_s &= \alpha_s a_{EOS} \\ b_s &= \beta_s b_{EOS} \end{aligned}$$

Parameters α_s and β_s can be determined for each compound or mixtures of constant composition from velocity of sound data. Parameters a_{EOS} and b_{EOS} can be calculated from their definition and use of critical constants. In fact values of the critical constants used in the calculations do not affect the outcome of results but they affect calculated values of α_s and β_s. For this reason α_s and β_s must be used with the same a_{EOS} and b_{EOS} that were used originally to determine these parameters. As an alternative approach and especially for petroleum

mixtures it would be appropriate to determine a_s and b_s from cVT data and directly use them in the corresponding EOS without calculation of a_{EOS} and b_{EOS} through critical properties. Therefore, for both RK and SRK we get same values of a_s and b_s since the original form of EOS is the same. For a number of light gases, parameters α_s and β_s have been determined from sonic velocity data for both RK and PR EOSs and they are given in Table 6.18. Once these parameters are used

TABLE 6.18—*RK and PR EOS parameters (Eq. 6.242) from velocity of sound in gases and liquids.*

Compound (gas)	No. of points	RK EOS[a]		PR EOS[a]	
		α_s	β_s	α_s	β_s
Methane	77	1.025	1.111	0.936	1.093
Ethane	119	1.043	1.123	0.956	0.993
Propane	63	0.992	1.026	1.013	1.031
Isobutane	80	0.993	1.019	0.983	0.983
n-Butane	86	0.987	1.015	0.941	0.912
n-Pentane	(liquid)			1.04	0.9
n-Decane	(liquid)			1.06	0.99

Taken with permission from Ref. [8].
[a] These parameters must be used for gaseous phase with Eq. (5.40) and parameters a_{EOS} and b_{EOS} from Tables 5.1.

TABLE 6.19—*Prediction of thermodynamic properties of light gases from RK and PR equations with use of velocity of sound and original parameters.[a]*

Gas system	No. of data points	Property	%AAD for RK EOS		%AAD for PR EOS	
			Sonic velocity	Original	Sonic velocity	Original
Pure gas	425	C	0.82	0.58	0.62	0.82
compounds	425	Z	0.76	0.92	0.5	0.77
C_1, C_2, C_3,	341	C_P	1.9	1.8	1.3	1.2
iC_4, nC_4	341	H	0.66	0.53	0.42	0.48
Gas mixture	61	C	9.2	0.84	1.47	0.89
69 Mol% C_1,	66	Z	4.1	2.0	4.0	1.9
31 Mol% C_2	66	C_V	8.2	2.85	6.5	7.0

Taken with permission from Ref. [8].
[a] For the velocity of sound parameters, values of α_s and β_s from Table 6.18 have been used. For the original parameters these corrections factors are taken as unity.

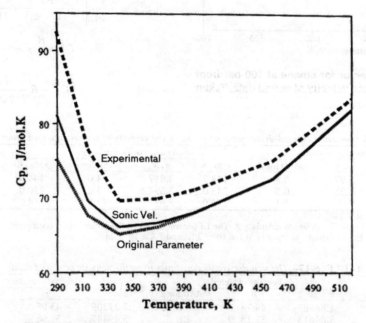

FIG. 6.40—**Prediction of constant pressure heat capacity of ethane gas at 30 bar from RK EOS using parameters from velocity of sound data.**

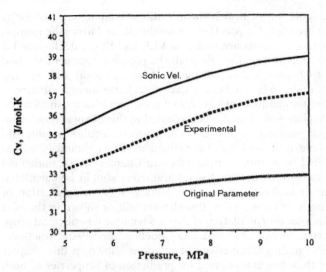

FIG. 6.41—Prediction of constant volume heat capacity of 69 mol% methane and 31 mol% ethane gas at 260 K from RK EOS using parameters from velocity of sound data.

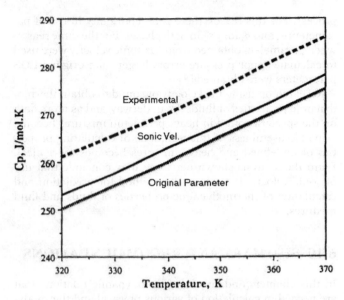

FIG. 6.43—Prediction of liquid heat capacity of *n*-octane at 100 bar from PR EOS using parameters from velocity of sound data.

to calculate various physical properties errors very similar to those obtained from original parameters are obtained as shown in Table 6.19 [8]. Predicted constant pressure heat capacity from RK EOS with parameters determined from velocity of sound for ethane a gas mixture of methane and ethane is shown in Figs. 6.40 and 6.41, respectively.

Similarly sonic velocity data for some liquids from C_5 to C_{10} were used to calculate EOS parameters. Calculated α_s and β_s parameters for use with PR EOS through Eq. (6.242) are also given in Table 6.18. When EOS parameters from velocity of sound are used to calculate C_P of liquids ranging from C_5 to C_{10} an average error of 6.4% is obtained in comparison with 7.6% error obtained from original parameters. Velocity

of sound for liquids can be estimated from original PR parameters with AAD of 9.7%; while using parameters calculated from sonic velocity, an error of 3.9% was obtained for 569 data points [8]. Graphical evaluations for prediction of liquid density of a mixture and constant volume heat capacity of *n*-octane are shown in Figs. 6.42 and 6.43. Results shown in these figures and Table 6.18 indicate that EOS parameters determined from velocity of sound are capable of predicting thermodynamic properties. It should be noted that data on velocity of sound were obtained either for compounds as gases or liquids but not for a single compound data on sonic velocity of both liquids and gases were available in this evaluation

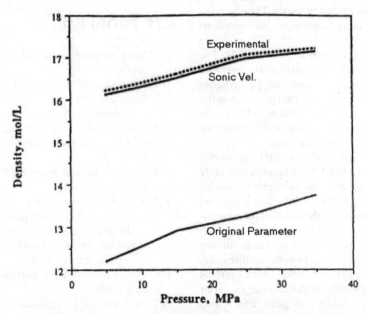

FIG. 6.42—Prediction of liquid density of 10 mol% *n*-hexadecane and 90 mol% carbon dioxide at 20°C from PR EOS using parameters from velocity of sound data.

process. For this reason there is no continuity in use of parameters α_s and β_s for use in both phases. For the same reason when parameters obtained from gas sonic velocity were used to calculate vapor pressure errors larger than original EOS parameters were obtained [8].

Research on using the velocity of sound to obtain thermodynamic properties of fluids are underway, and as more data on the speed of sound in heavy petroleum mixtures become available usefulness of this technique of calculating properties of undefined and heavy mixtures becomes more clear. From the analysis shown here, one may conclude that use of sonic velocity is a promising method for prediction and calculation of thermodynamic properties of fluids and fluid mixtures.

6.10 SUMMARY AND RECOMMENDATIONS

In this chapter fundamental thermodynamic relations that are needed in calculation of various physical and thermodynamic properties are presented. Through these relations various properties can be calculated from knowledge of a PVT relation or an equation of state. Methods of calculation of vapor pressure, enthalpy, heat capacity, entropy, fugacity, activity coefficient, and equilibrium ratios suitable for hydrocarbon systems and petroleum fractions are presented in this chapter. These methods should be used in conjunction with equations of states or generalized correlations presented in Chapter 5. In use of cubic equations of state for phase equilibrium calculations and calculation of K_i values, binary interaction parameters recommended in Chapter 5 should be used. Cubic equations are recommended for high-pressure phase equilibrium calculations while activity coefficient models are recommended for low-pressure systems. Methods of calculation of activity coefficient and Henry's law constants from a cubic EOS are presented. Recent studies show that cubic equations are not the best type of PVT relation for prediction of derivative properties such as enthalpy, Joule–Thomson coefficient, or heat capacity. For this reason noncubic equations such as statistical associating fluid theory (SAFT) are being investigated for prediction of such properties [38, 57]. The main purpose of this chapter was to demonstrate the role that theory plays in estimation of physical properties of petroleum fluids. However, among the methods presented in this chapter, the LK generalized correlations are the most suitable methods for calculation of enthalpy, heat capacity, and fugacity for both liquid and gas phases at elevated pressures.

While the cubic equations (i.e., SRK or PR) are useful for phase behavior calculations, the LK corresponding state correlations are recommended for calculation of density, enthalpy, entropy, and heat capacity of hydrocarbons and petroleum fractions. Partial molar properties and their methods of calculation have been presented for estimation of mixture properties. Calculation of volume change due to mixing or heat of mixing is shown. Fundamental phase equilibria relations especially for vapor–liquid and solid–liquid systems are developed. Through these relations calculation of vapor pressure of pure substances, solubility of gases and solids in liquids are demonstrated. Solubility parameters for pure compounds are given for calculation of activity coefficients without use of any VLE data. Correlations are presented for

calculation of heat of fusion, molar volume, and solubility parameters for paraffinic, naphthenic, and aromatic groups. These relations are useful in VLE and SLE calculations for petroleum fractions through the pseudocomponent method of Chapter 3. Data on the enthalpy of fusion and freezing pointd can be used to calculate freezing point of a mixture or the temperature at which first solid particles begin to form. Application of methods presented in this chapter require input parameters (critical properties, molecular weight, and acentric factor) that for defined mixtures should be calculated from mixing rules given in Chapter 5. For undefined petroleum fractions these parameters should be calculated from methods given in Chapters 2–4. Main application of methods presented in this chapter will be shown in the next chapter for calculation of thermodynamic and physical properties of hydrocarbons and undefined petroleum fractions. The main characteristic of relations shown in this chapter is that they can be used for prediction of properties of both gases and liquids through an equation of state. However, as it will be seen in the next chapter there are some empirically developed correlations that are mainly used for liquids with higher degree of accuracy. Generally properties of liquids are calculated with lesser accuracy than properties of gases.

With the help of fundamental relations presented in this chapter a generalized cVT relation based on the velocity of sound is developed. It has been shown that when EOS parameters are calculated through a measurable property such as velocity of sound, thermophysical properties such as density, enthalpy, heat capacity, and vapor pressure have been calculated with better accuracy for both liquid and vapor phases through the use of velocity of sound data. This technique is particularly useful for mixtures of unknown composition and reservoir fluids and it is a promising approach for estimation of thermodynamic properties of complex undefined mixtures.

6.11 PROBLEMS

6.1. Develop an equation of state in terms of parameters β and κ.
6.2. In storage of hydrocarbons in cylinders always a mixtures of both vapor and liquid (but not a single phase) are stored. Can you justify this?
6.3. Derive a relation for calculation of $(G - G^{ig})/RT$ in terms of PVT and then combine with Eq. (6.33) to derive Eq. (6.50).
6.4. Derive Edmister equation for acentric factor (Eq. 2.108) from Eq. (6.101).
6.5. a. Derive a relation for molar enthalpy from PR EOS.
   b. Use the result from part a to derive a relation for partial molar enthalpy from PR EOS.
   c. Repeat part a assuming parameter b is a temperature-dependent parameter.
6.6. Derive a relation for partial molar volume from PR EOS (Eq. 6.88).
6.7. Derive fugacity coefficient relation from SRK EOS for a pure substance and compare it with results from Eq. (6.126).
6.8. Derive Eq. (6.26) for the relation between C_P and C_V.

6.9. Show that

$$C_V - C_V^{ig} = \int_{V=\infty}^{V} \left[T \left(\frac{\partial^2 P}{\partial T^2} \right)_V \right]_T dV$$

Use this relation with truncated virial equation to derive Eq. (6.65).

6.10. The Joule–Thomson coefficient is defined as

$$\eta = \left(\frac{\partial T}{\partial P} \right)_H$$

a. Show that it can be related to PVT in the following form:

$$\eta = \frac{T \left(\frac{\partial V}{\partial T} \right)_P - V}{C_P}$$

b. Calculate η for methane at 320 K and 10 bar from the SRK EOS.

6.11. Similar to derivation of Eq. (6.38) for enthalpy departure at T and V, derive the following relation for the heat capacity departure and use it to calculate residual heat capacity from RK EOS. How do you judge validity of your result?

$$(C_P - C_P^{ig})_{T,V} = T \int_{V \to \infty}^{V} \left(\frac{\partial^2 P}{\partial T^2} \right)_V dV - \frac{T \left[\left(\frac{\partial P}{\partial T} \right)_V \right]^2}{\left(\frac{\partial P}{\partial V} \right)_T} - R$$

6.12. Show that Eqs. (6.50) and (6.51) for calculation of residual entropy are equivalent.

6.13. Prove Eq. (6.81) for the Gibbs–Duhem equation.

6.14. Derive Eq. (6.126) for fugacity coefficient of i in a mixture using SRK EOS.

6.15. Derive the following relation for calculation of fugacity of pure solids at $T < T_{tp}$.

$$f_i^S(T,P) = P_i^{sub} \exp \left[\frac{V_i^S(P - P_i^{sub})}{RT} \right]$$

where P_i^{sub} is the vapor pressure of pure solid i at temperature T.

6.16. Derive Eq. (6.216) for the velocity of sound.

6.17. A mixture of C_1 and C_5 exists at 311 K and 69.5 bar in both gas and liquid phases in equilibrium in a closed vessel. The mole fraction of C_1 in the mixture is $z_1 = 0.541$. In the gas $y_1 = 0.953$ and in the liquid $x_1 = 0.33$. Calculate K_1 and K_5 from the following methods:
a. Regular solution theory
b. Standing correlation
c. GPA/NIST graphs
d. PR EOS
In using PR EOS, use shift parameters of -0.2044 and -0.045 for C_1 and n-C_5, respectively. Also use BIP value of $k_{1-5} = 0.054$.

6.18. a. For a gas mixture that follows truncated virial EOS, show that

$$V^E = \frac{1}{2} \sum_i \sum_j y_i y_j \delta_{ij}$$

$$G^E = \frac{P}{2} \sum_i \sum_j y_i y_j \delta_{ij}$$

$$H^E = \frac{P}{2} \sum_i \sum_j y_i y_j \left(\delta_{ij} - T \frac{d\delta_{ij}}{dT} \right)$$

where δ_{ij} is defined in Eq. (5.70) in Chapter 5.
b. Derive a relation for heat of mixing of a binary gas that obeys truncated virial EOS.

6.19. In general for mixtures, equality of mixture fugacity between two phases is not valid in VLE calculations:

$$f_{mix}^V = f_{mix}^L$$

However, only under a certain condition this relation is true. What is that condition?

6.20. With the use of PR EOS and definition of solubility parameter (δ) by Eq. (6.147) one can derive the following relation for calculation of δ [17]:

$$\delta = \left[\frac{1}{2\sqrt{2}bV^L} \left(a - T \frac{da}{dT} \right) \ln \frac{V^L + (1+\sqrt{2})b}{V^L + (1-\sqrt{2})b} \right]$$

where da/dT for PR EOS can be obtained from Table 6.1. With use of volume translation for V^L estimate values of V^L and δ at 25°C for hydrocarbons C_5 and C_{10} and compare with values given in Table 6.10.

6.21. Calculate freezing point depression of toluene when it is saturated with solid naphthalene at 20°C.

6.22. Derive Eq. (6.240) for calculation of velocity of sound from RK/SRK EOS.

6.23. Consider the dry natural gas (fluid 1) and black oil (fluid 5) samples whose compositions are given in Table 1.2. Assume there are two reservoirs, one containing the natural gas and the other containing black oil, both at 400 K and 300 bar. Calculate velocity of sound in these two fluids using SRK EOS.

6.24. Calculate $(U - U^{ig})$, $(H - H^{ig})$, and $(S - S^{ig})$ for steam at 500°C and 100 bar from SRK and PR EOS. How do you evaluate your results?

6.25. Calculate the increase in enthalpy of n-pentane when its pressure increases from 600 to 2000 psia at 190.6°F using the following methods:
a. SRK EOS.
b. LK method.
c. Compare the results with the measured value of 188 Btu/lbmol [17].

6.26. Estimate C_P, C_V, and the speed sound in liquid hexane at atmospheric pressure and 269 and 300K from the following methods:
a. SRK EOS
b. PR EOS
c. Compare calculated sonic velocities with reported values of 1200 (at 269 K) and 1071 m/s (at 300 K) (Fig. 3.33, Ref. [17]).

6.27. Estimate γ_i^∞ for the system of n-C_4 and n-C_{32} at 100°C from PR EOS and compare with the value from Fig. 6.8.

6.28. Using results from Problem 6.27, estimate the Henry's law constant for the system of $n\text{-}C_4$ and $n\text{-}C_{32}$ at 100°C from PR EOS.

6.29. For the mixture of benzene and toluene construct the freezing/melting diagram similar to Fig. 6.32.

6.30. One of the advanced liquid solution theories is known as the quasichemical approximation, which is particularly useful for mixtures containing molecules quantitatively different in size and shape. According to this theory the molar excess Gibbs energy for a binary system is given by [21]:

$$\frac{G^E}{RT} = \left(\frac{w}{kT}\right) x_1 x_2 \left[1 - \frac{1}{2}\left(\frac{2w}{zkT}\right) x_1 x_2 + \cdots\right]$$

where the higher terms are neglected. x_1 and x_2 are mole fractions of 1 and 2 and k is the Boltzman's constant. W and z are model parameters that must be determined for each system. W/zkT is less than unity and z is called *co-ordination number* and varies from 6 to 12 [21]. Typical value of z is 10. Use Eq. (6.137) to derive the relations for γ_1 and γ_2.

6.31. Show that at constant T and P, the Gibbs–Duhem equation in a multicomponent mixture can be written in the following forms:

$$\sum_i x_i \, d\ln \gamma_i = 0$$

or

$$\sum_i x_i \, d\ln \hat{f}_i = 0$$

6.32. Consider a binary solution of components 1 and 2. Show that in the region that Raoult's law is valid for component 1, Henry's law must be valid for component 2.

6.33. In a binary liquid, mixtures of 1 and 2, fugacity of component 1 at 20°C can be approximately presented by the equation:

$$\hat{f}_1^L = 30x_1 - 20x_1^2$$

where $\hat{f}_1^L$ is the fugacity of 1 in the mixture in bar. At 20°C and 30 bar determine:
a. The fugacity of pure component 1, f_1^L.
b. The fugacity coefficient of pure component 1, ϕ_1.
c. The Henry's law constant for component 1, k_1.
d. Relation for the activity coefficient γ_1 in terms of x_1 (based on the standard state of Lewis rule).
e. Relation for $\hat{f}_2^L$.

6.34. Consider three hydrocarbon components benzene, cyclohexane, and n-hexane all having six carbon atoms. Both quantitatively and qualitatively state that solubility of benzene in methylcylopentane is higher or benzene in n-hexane.

6.35. A solution is made at the temperature of 298 K by adding 5 g of naphthalene to a mixture of 50 g benzene and 50 g n-heptane. The temperature is gradually lowered until the particles of solid are observed. What is the temperature at this point? What is the temperature if 10 g of naphthalene is added?

6.36. Estimate vapor pressure of isobutane at 50°C from the following methods:
a. SRK EOS.
b. PR EOS.

c. Clausius–Clapeyron equation
d. Compare the results with a reported value.

6.37. A natural gas is composed of 85% methane, 10% ethane, and 5% propane. What are the mole fractions of each of the components in water (gas solubility) at 300 bar and 298 K.

6.38. Ninety barrels of $n\text{-}C_{36}$ are diluted with addition of 10 bbl of $n\text{-}C_5$ at 25°C. Calculate volume of the solution at 1 bar from the following methods:
a. Using partial molar volume from PR EOS.
b. Using API procedure.

6.39. Show that if Edmister equation (Eq. 2.108) is used for acentric factor, Wilson correlation for K_i-values (Eq. 6.204) reduces to Hoffman type correlation (Eq. 6.202).

6.40. Solubility of water in a gasoline sample at 1 atm can be determined approximately by Eq. (6.195). However, accurate solubility of water can be estimated through a thermodynamic model with activity coefficient calculations. A gasoline from California oil has mid boiling point of 404°F and API gravity of 43.5 with PNA composition of 30.9, 64.3, and 4.8% as reported by Lenoir and Hipkin [12]. Estimate solubility of water in this gasoline sample at 100°F and 1 atm from appropriate thermodynamic model and compare the predicted value with the value estimated from Eq. (6.195).

REFERENCES

[1] Smith, J. M., Van Ness, H. C., and Abbott, M. M., *Introduction to Chemical Engineering Thermodynamics*, 5th ed., McGraw-Hill, New York, 1996.

[2] Elliott, J. R. and Lira, C. T., *Introductory Chemical Engineering Thermodynamics*, Prentice Hall, New Jersey, 1999 (www.phptr.com).

[3] Daubert, T. E., Danner, R. P., Sibul, H. M., and Stebbins, C. C., *Physical and Thermodynamic Properties of Pure Compounds: Data Compilation*, DIPPR-AIChE, Taylor & Francis, Bristol, PA, 1994 (extant) (www.aiche.org/dippr). Updated reference: Rowley, R. L., Wilding, W. V., Oscarson, J. L., Zundel, N. A., Marshall, T. L., Daubert, T. E., and Danner, R. P., *DIPPR Data Compilation of Pure Compound Properties*, Design Institute for Physical Properties (DIPPR), Taylor & Francis, New York, 2002 (http://dippr.byu.edu).

[4] Poling, B. E., Prausnitz, J. M., and O'Connell, J. P., *Properties of Gases and Liquids*, 5th ed., Mc-Graw Hill, New York, 2000.

[5] Daubert, T. E. and Danner, R. P., Eds., *API Technical Data Book—Petroleum Refining*, 6th ed., American Petroleum Institute (API), Washington, DC, 1997.

[6] Garvin, J., "Use the Correct Constant–Volume Specific Heat," *Chemical Engineering Progress*, Vol. 98, No. 7, 2002, pp. 64–65.

[7] Polt, A., Platzer, B., and Maurer, G., "Parameter der thermischen Zustandsgl eichung von Bender fuer 14 mehratomige reine Stoffe," *Chemische Technik (Leipzig)*, Vol. 44, No. 6, 1982, pp. 216–224.

[8] Shabani, M. R., Riazi, M. R., and Shaban, H. I., "Use of Velocity of Sound in Predicting Thermodynamic Properties from Cubic Equations of State," *Canadian Journal of Chemical Engineering*, Vol. 76, 1998, pp. 281–289.

[9] Lee, B. I. and Kesler, M. G., "A Generalized Thermodynamic Correlation Based on Three-Parameter Corresponding States," *American Institute of Chemical Engineers Journal*, Vol. 21, 1975, pp. 510–527.

[10] Alberty, R. A. and Silbey, R. J., *Physical Chemistry*, 2nd ed., Wiley, New York, 1999.

[11] Reid, R. C., Prausnitz, J. M., and Poling, B. E., *Properties of Gases and Liquids*, 4th ed., Mc-Graw Hill, New York, 1987.

[12] Lenoir, J. M. and Hipkin, H. G., "Measured Enthalpies of Eight Hydrocarbon Fractions," *Journal of Chemical and Engineering Data*, Vol. 18, No. 2, 1973, pp. 195–202.

[13] AIChE DIPPR® Database, *Design Institute for Physical Property Data (DIPPR)*, EPCON International, Houston, TX, 1996 (for more updated database from NIST see: http://www.nist.gov/ srd/fluids.htm and http://webbook.nist.gov/chemistry/fluid/).

[14] Wagner, W. and Pruss, A., "The IAPWS Formulation for the Thermodynamic Properties of Ordinary Water Substance for General and Scientific Use," *Journal of Physical and Chemical Reference Data*, Vol. 31, No. 2, 2002, pp. 387–535 (also see: http://www.iapws.org/).

[15] Kesler, M. G. and Lee, B. I., "Improve Prediction of Enthalpy of Fractions," *Hydrocarbon Processing*, Vol. 55, 1976, pp. 153–158.

[16] Tsonopoulos, C., Heidman, J. L., and Hwang, S.-C., *Thermodynamic and Transport Properties of Coal Liquids*, An Exxon Monograph, Wiley, New York, 1986.

[17] Firoozabadi, A., *Thermodynamics of Hydrocarbon Reservoirs*, Mc-Graw Hill, New York, 1999.

[18] Riazi, M. R. and Daubert, T. E., "Simplify Property Predictions," *Hydrocarbon Processing*, Vol. 59, No. 3, 1980, pp.115–116.

[19] Denbigh, K., *The Principles of Chemical Equilibrium*, Cambridge University Press, Cambridge, UK, 1992.

[20] Danesh, A., *PVT and Phase Behavior of Petroleum Reservoir Fluids*, Elsevier, Amsterdam, 1998.

[21] Prausnitz, J. M., Lichtenthaler, R. N., and de Azevedo, E. G., *Molecular Thermodynamics of Fluid-Phase Equilibria*, Prentice-Hall, New Jersey, 2nd ed., 1986, 3rd ed., 1999.

[22] Sandler, S. I., *Chemical and Engineering Thermodynamics*, 3rd ed., Wiley, New York, 1999.

[23] Riazi, M. R. and Vera, J. H., "Method to Calculate the Solubility of Light Gases in Petroleum and Coal Liquid Fractions Based on their P/N/A Composition," *Industrial and Engineering Chemistry Research*, Vol. 44, 2005.

[24] Won, K. W., "Thermodynamics for Solid Solution–Liquid– Vapor Equilibria: Wax Phase Formation from Heavy Hydrocarbon Mixtures," *Fluid Phase Equilibria*, Vol. 30, 1986, pp. 265–279.

[25] Pan, H., Firoozabadi, A., and Fotland, P., "Pressure and Composition Effect on Wax Precipitation: Experimental Data and Model Results," *Society of Petroleum Engineers (SPE) Production and Facilities*, Vol. 12, No. 4, 1993, pp. 250–258.

[26] Lira-Galeana, C. A., Firoozabadi, A., and Prausnitz, J. M., "Thermodynamics of Wax Precipitation in Petroleum Mixtures," *American Institute of Chemical Engineers Journal*, Vol. 42, 1996, pp. 239–248.

[27] Pedersen, K. S., Skovborg, P., and Ronningsen, H. P., "Wax Precipitation from North Sea Crude Oils, 4: Thermodynamic Modeling," *Energy and Fuels*, Vol. 5, No. 6, 1991, pp. 924–932.

[28] *Engineering Data Book*, 10th ed., Volume II, Sections 17–26, Gas Processors (Suppliers) Association, Tulsa, OK, 1987.

[29] National Institute of Standards and Technology (NIST), http://www.nist.gov/srd/fluids.htm and http://webbook.nist.gov/chemistry/fluid/

[30] Baker, L. E., Pierce, A. C., and Luks, K. D., "Gibbs Energy Analysis of Phase Equilibria," *Society of Petroleum Engineers Journal* (Trans. AIME), Vol. 273, 1982, pp. 731–742.

[31] Whitson, C. H. and Brule, M. R., *Phase Behavior*, Monograph Vol. 20, Society of Petroleum Engineers, Richardson, TX, 2000.

[32] Manafi, H., Mansoori, G. A., and Ghotbi, S., "Phase Behavior Prediction of Petroleum Fluids with Minimum

Characterization Data," *Journal of Petroleum Science and Engineering*, Vol. 22, 1999, pp. 67–93.

[33] Plöcker, U., Knapp, H., and Prausnitz, J. M., "Calculation of High-Pressure Vapor-Liquid Equilibria from a Corresponding States Correlation with Emphasis on Asymmetric Mixtures," *Industrial and Engineering Chemistry, Process Design and Development*, Vol. 17, No. 13, 1978, pp. 324–332.

[34] Chao, K. C. and Seader, J. D., "A General Correlation of Vapor–Liquid Equilibria in Hydrocarbon Mixtures," *American Institute of Chemical Engineers Journal*, Vol. 7, 1961, pp. 598–605.

[35] Grayson, H. G. and Streed, C. W., "Vapor–Liquid Equlibria for High Temperature, High Pressure Hydrogen–Hydrocarbon Systems," *Proceedings of the 6th World Petroleum Congress*, Frankfurt/Main III, Paper 20-PD7, 1963, pp. 233–245.

[36] PRO/II, *Keyword Manual*, Simulation Sciences Inc., Fullerton, CA, October 1992.

[37] Tsonopoulos, C. and Riazi, M. R., *Physical Properties for Engineers*, Industrial Course Note for CHE7, OCCD, Kuwait University, Kuwait, March 2002.

[38] Li, X.-S. and Englezos, P., "Vapor–Liquid Equilibrium of Systems Containing Alcohols Using the Statistical Associating Fluid Theory Equation of State," *Industrial and Engineering Chemistry Research*, Vol. 42, 2003, pp. 4953–4961.

[39] Hoffman, A. E., Crump, J. S., and Hocott, C. R., "Equilibrium Constants for a Gas Condensate System," *Transactions of AIME*, Vol. 198, 1953, pp. 1–10.

[40] Standing, M. B., "A Set of Equations for Computing Equilibrium Ratios of a Crude Oil/Natural Gas System at Pressures Below 1000 psia," *Journal of Petroleum Technology*, Vol. 31, 1979 , pp. 1193–1195.

[41] Katz, D. L. and Hachmuth, K. H., "Vaporization Equilibrium Constants in a Crude-Oil–Natural-Gas System," *Industrial and Engineering Chemistry Research*, Vol. 29, 1937, pp. 1072– 1077.

[42] Glaso, O. S., "The Accuracy of PVT Parameters Calculated from Computer Flash Separation at Pressures Less than 1000 psia," *Journal of Petroleum Technology*, Vol. 34, 1982, pp. 1811–1813.

[43] Winnick, J., *Chemical Engineering Thermodynamics*, Wiley, New York, 1997.

[44] Riazi, M. R. and Mansoori, G. A., "Use of the Velocity of Sound in Predicting the PVT Relations," *Fluid Phase Equilibria*, Vol. 90, 1993, pp. 251–264.

[45] Riazi, M. R. and Roomi, Y. A., "Use of Velocity of Sound in Estimating Thermodynamic Properties of Petroleum Fractions," *Preprints of Division of Petroleum Chemistry Symposia*, American Chemical Society (ACS), Vol. 45, No. 4, 2000, pp. 661–664.

[46] Colgate, S. O., Silvarman, A., and Desjupa, C., "Acoustic Resonance Determination of Sonic Speed and the Critical Point," Paper presented at the *AIChE Spring Meeting*, Paper No. 29c, New Orleans, 1992.

[47] Colgate, S. O., Silvarman, A., and Desjupa, C., "Sonic Speed and Critical Point Measurements in Ethane by the Acoustic Resonance Method," *Fluid Phase Equilibria*, Vol. 76, 1992, pp. 175–185.

[48] Ball, S. J., Goodwin, A. R. H., and Trusler, J. P. M., "Phase Behavior and Physical Properties of Petroleum Reservoir Fluids from Acoustic Measurements," *Journal of Petroleum Science and Engineering*, Vol. 34, No. 1–4, 2002, pp. 1–11.

[49] Mittal Enterprises, New Delhi, India, 2004 (http://www.mittalenterprises.com).

[50] Alem, A. H. and Mansoori, G. A., "The VIM Theory of Molecular Thermodynamics," *American Institute of Chemical Engineers Journal*, Vol. 30, 1984, pp. 468–480.

[51] Walas, S. M., *Phase Equilibria in Chemical Engineering*, Butterworth, London, 1985.

[52] Goodwin, R. D., *The Thermophysical Properties of Methane from 90 to 500 K at Pressures to 700 Bar*, NBS Technical Note 653, National Bureau of Standards, April 1974.

[53] Goodwin, R. D., Roder, H. M., and Starty, G. C., *The Thermophysical Properties of Ethane from 90 to 600 K at Pressures to 700 Bar*, NBS Technical Note 684, National Bureau of Standards, August 1976.

[54] Goodwin, R. D. and Hayens, W. M., *The Thermophysical Properties of Propane from 85 to 700 K at Pressures to 70 MPa*, NBS Monograph 170, National Bureau of Standards, April 1982.

[55] Mossaad, E. and Eubank, P. T., "New Method for Conversion of Gas Sonic Velocities to Density Second Virial Coefficients," *American Institute of Chemical Engineers Journal*, Vol. 47, No. 1, 2001, pp. 222–229.

[56] Hirschfelder, O. J., Curtiss, C. F., and Bird, R. B., *Molecular Theory of Gases and Liquids*, Wiley, New York, 1964.

[57] Konttorp, K., Peters, C. J., and O'Connell, J. P., "Derivative Properties from Model Equations of State," Paper presented at the *Ninth International Conference on Properties and Phase Equilibria for Product and Process Design (PPEPPD 2001)*, Kurashiki, Japan, May 20–25, 2001.

Applications: Estimation of Thermophysical Properties

7

NOMENCLATURE

API API gravity defined in Eq. (2.4)

A, B, C, D, E Coefficients in various equations

a, b Cubic EOS parameters given in Table 5.1

a, b, c, d, e Constants in various equations

C_P Heat capacity at constant pressure defined by Eq. (6.17), J/mol·K

C_V Heat capacity at constant volume defined by Eq. (6.18), J/mol·K

d_T Liquid density at temperature T and 1 atm, g/cm³

$f^{(0)}, f^{(1)}$ Dimensionless functions for vapor pressure generalized correlation (Eq. 7.17)

H Enthalpy defined in Eq. (6.1), molar unit: J/mol; specific unit: kJ/kg

H% Hydrogen wt% in a petroleum fraction

I Refractive index parameter at temperature T, defined in Eq. (2.36) $[=(n^2-1)/(n^2+2)]$, dimensionless

K_W Watson characterization factor defined by Eq. (2.13)

log_{10} Common logarithm (base 10)

ln Natural logarithm (base e)

M Molecular weight (molar mass), g/mol [kg/kmol]

N Number of components in a mixture

N_C Number of carbon atoms

N% Nitrogen wt% in a petroleum fraction

n Liquid refractive index at temperature T and 1 atm, dimensionless

O% Oxygen wt% in a petroleum fraction

P Pressure, bar

P_c Critical pressure, bar

P_r Reduced pressure defined by Eq. (5.100) $(=P/P_c)$, dimensionless

P_{tp} Triple point pressure, bar

P^{vap} Vapor pressure at a given temperature, bar

P^{sub} Sublimation pressure (vapor pressure of a solid) at a given temperature, bar

P_r^{vap} Reduced vapor pressure at a given temperature $(=P^{vap}/P_c)$, dimensionless

Q A parameter defined in Eq. (7.21)

R Gas constant = 8.314 J/mol·K (values in different units are given in Section 1.7.24)

RVP Reid vapor pressure, bar

S% Sulfur wt% in a petroleum fraction

SG Specific gravity of liquid substance at 15.5°C (60°F) defined by Eq. (2.2), dimensionless

T Absolute temperature, K

T_b Normal boiling point, K

T_c Critical temperature, K

T_r Reduced temperature defined by Eq. (5.100) $(=T/T_c)$, dimensionless

T_o A reference temperature for use in Eq. (7.5), K

T_M Freezing (melting) point for a pure component at 1.013 bar, K

T_{br} Reduced boiling point $(=T_b/T_c)$, dimensionless

T_{pc} Pseudocritical temperature, K

T_{tp} Triple point temperature, K

V Molar volume at T and P, cm³/mol

V_c Critical molar volume, cm³/mol

V^S Molar volume of solid, cm³/mol

x_i Mole fraction of component i in a mixture, dimensionless

x_{wi} Weight fraction of component i in a mixture (usually used for liquids), dimensionless

x_P, x_N, x_A Fractions (i.e., mole) of paraffins, naphthenes, and aromatics in a petroleum fraction, dimensionless

Z Compressibility factor defined by Eq. (5.15), dimensionless

Greek Letters

Δ Difference between two values of a parameter

θ_m A molar property (i.e., molar enthalpy, molar volume, etc...)

θ_s A specific property (i.e., specific enthalpy, etc...)

ρ Density at a given temperature and pressure, g/cm³ (molar density unit: cm³/mol)

ρ_m Molar density at a given temperature and pressure, mol/cm³

ρ_r Reduced density $(=\rho/\rho_c = V_c/V)$, dimensionless

ρ^w Water density at a given temperature, g/cm³

ω Acentric factor defined by Eq. (2.10), dimensionless

ΔH_{298}^f Heat of formation at 298 K, kJ/mol

ΔH^{vap} Heat of vaporization (or latent heat) at temperature T, J/mol

Superscript

g Value of a property for gas phase

ig Value of a property for component i as ideal gas at temperature T and $P \to 0$

L Value of a property at liquid phase

V Value of a property at vapor phase

vap Change in value of a property due to vaporization

S Value of a property at solid phase

sat Value of a property at saturation pressure

sub Value of a property at sublimation pressure

$[]^{(0)}$ A dimensionless term in a generalized correlation for a property of simple fluids

$[]^{(1)}$ A dimensionless term in a generalized correlation for a property of acentric fluids

o Value of a property at low pressure (ideal gas state) condition at a given temperature

Subscripts

A Value of a property for component A

B Value of a property for component B

b Value of a property at the normal boiling point

c Value of a property at the critical point

i, j Value of a property for component i or j in a mixture

L Value of a property for liquid phase

m Molar property (quantity per unit mole)

m Mixture property

mix Value of a property for a mixture

nbp Value of a liquid phase property at the normal boiling point of a substance

pc Pseudocritical property

r Reduced property

ref Value of a property at the reference state

S Value of a property at the solid phase

S Value of a property for solvent (LMP)

s Specific property (quantity per unit mass)

T Values of property at temperature T

tp Value of a property at the triple point

W Values of a property for water

20 Values of property at 20°C

7+ Values of a property for C_{7+} fraction of an oil

Acronyms

API-TDB American Petroleum Institute—Technical Data Book (see Ref. [9])

BIP Binary interaction parameter

COSTALD Corresponding State Liquid Density (given by Eq. 5.130)

DIPPR Design Institute for Physical Property Data (see Ref. [10])

EOS Equation of state

GC Generalized correlation

HHV Higher heating value

LHV Lower heating value

MB Maxwell and Bonnell (see Eqs. (3.29), (3.30), and (7.20)–(7.22))

RVP Reid vapor pressure

PR Peng–Robinson EOS (see Eq. 5.39)

PNA Paraffins, naphthenes, aromatics content of a petroleum fraction

PVT Pressure–volume–temperature

SRK Soave–Redlich–Kwong EOS given by Eq. (5.38) and parameters in Table 5.1

scf Standard cubic foot (unit for volume of gas at 1 atm and 60°F)

stb Stock tank barrel (unit for volume of liquid oil at 1 atm and 60°F)

TVP True vapor pressure

VABP Volume average boiling point defined by Eq. (3.3)

%AAD Average absolute deviation percentage defined by Eq. (2.135)

%AD Absolute deviation percentage defined by Eq. (2.134)

wt% Weight percent

THE LAST THREE CHAPTERS of this book deal with application of methods presented in previous chapters to estimate various thermodynamic, physical, and transport properties of petroleum fractions. In this chapter, various methods for prediction of physical and thermodynamic properties of pure hydrocarbons and their mixtures, petroleum fractions, crude oils, natural gases, and reservoir fluids are presented. As it was discussed in Chapters 5 and 6, properties of gases may be estimated more accurately than properties of liquids. Theoretical methods of Chapters 5 and 6 for estimation of thermophysical properties generally can be applied to both liquids and gases; however, more accurate properties can be predicted through empirical correlations particularly developed for liquids. When these correlations are developed with some theoretical basis, they are more accurate and have wider range of applications. In this chapter some of these semitheoretical correlations are presented. Methods presented in Chapters 5 and 6 can be used to estimate properties such as density, enthalpy, heat capacity, heat of vaporization, and vapor pressure.

Characterization methods of Chapters 2–4 are used to determine the input parameters needed for various predictive methods. One important part of this chapter is prediction of vapor pressure that is needed for vapor–liquid equilibrium calculations of Chapter 9.

7.1 GENERAL APPROACH FOR PREDICTION OF THERMOPHYSICAL PROPERTIES OF PETROLEUM FRACTIONS AND DEFINED HYDROCARBON MIXTURES

Finding reliable values for inadequate or missing physical properties is the key to a successful simulation, which depends on the selection of correct estimation method [1]. In Chapters 5 and 6 theoretically developed methods for calculation of physical and thermodynamic properties of hydrocarbon fluids were presented. Parameters involved in these methods were mainly based on properties of pure compounds. Methods developed based on corresponding states approaches or complex equations of state usually predict the properties more accurately than those based on cubic EOSs. For the purpose of property calculations, fluids can be divided into gases and liquids and each group is further divided into two categories of pure components and mixtures. Furthermore, fluid mixtures are divided into two categories of defined and undefined mixtures. Examples of defined mixtures are hydrocarbon mixtures with a known composition, reservoir fluids with known compositions up to C_6, and pseudocompounds of the C_{7+} fraction. Also petroleum fractions expressed in terms of several pseudocomponents can be considered as defined mixtures. Examples of undefined mixtures are petroleum fractions and reservoir fluids whose compositions are not known. For such mixtures, some bulk properties are usually known.

Theoretically developed methods are generally more accurate for gases than for liquids. Kinetic theory provides sound predictive methods for physical properties of ideal gases [2, 3]. For this reason, empirical correlations for calculation of physical properties of liquids have been proposed. Similarly, theoretical methods provide a more accurate estimation of physical properties of pure compounds than of their mixtures. This is mainly due to the complexity of interaction of components in the mixtures especially in the liquid phase. For undefined mixtures such as petroleum fractions, properties can be calculated in three ways. One method is to consider them as a single pseudocomponent and to use the methods developed for pure components. The second method is to develop empirical correlations for petroleum fractions. Such empirically developed methods usually have limited applications and should be used with caution. They are accurate for those data for which correlation coefficients have been obtained but may not provide reliable values for properties of other fractions. These two approaches cannot be applied to mixtures with wide boiling range, such as wide fractions, crude oils, or reservoir fluids. The third approach is used for available data on the mixture to express the mixture in terms of several pseudocomponents, such as those methods discussed in Chapters 3 and 4. Then, methods available for prediction of properties of defined mixtures can be used for such petroleum fluids. This approach should particularly be used for wide boiling range fractions and reservoir fluids.

Fluid properties generally depend on temperature (T), pressure (P), and composition (x_i). Temperature has a significant effect on properties of both gases and liquids. Effect of pressure on properties of gases is much larger than effect of pressure on properties of liquids. The magnitude of this effect decreases for fluids at higher pressures. For the liquid fluids, generally at low pressures, effect of pressure on properties is neglected in empirically developed correlations. As pressure increases, properties of gases approach properties of liquids. Effect of composition on the properties of liquid is stronger than the effect of composition on properties of gases. Moreover, when components vary in size and properties the role of composition on property estimation becomes more important. For gases, the effect of composition on properties increases with increase in pressure. At higher pressures molecules are closer to each other and the effect of interaction between dissimilar species increases. For gases at atmospheric or lower pressures where the gas may be considered ideal, composition has no role on molar density of the mixture as seen from Eq. (5.14).

There are two approaches to calculate properties of defined mixtures. The first and more commonly used approach is to apply the mixing rules introduced in Chapter 5 for the input parameters (T_c, P_c, ω) of an EOS or generalized correlations and then to calculate the properties for the entire mixture. The second approach is to calculate desired property for each component in the mixture and then to apply an appropriate mixing rule on the property. This second approach usually provides more accurate results; however, calculations are more tedious and time-consuming, especially when the

number of components in the mixture is large since each property must be calculated for each component in the mixture. In applying a mixing rule, the role of binary interaction parameters (BIPs) is important when the mixture contains components of different size and structure. For example, in a reservoir fluid containing C_1 and a heavy component such as C_{30} the role of BIP between these two components cannot be ignored. Similarly when nonhydrocarbon components such as H_2S, N_2, H_2O, and CO_2 exist in the mixture, the BIPs of these compounds with hydrocarbons must be considered. For some empirically developed correlations specific interaction parameters are recommended that should be used.

Theoretically developed thermodynamic relations of Chapters 5 and 6 give thermodynamic properties in molar quantities. They should be converted into specific properties by using Eq. (5.3) and molecular weight. In cases that no specific mixing rule is available for a specific property the simple Kay's mixing rule (Section 3.4.1) may be used to calculate mixture properties from pure component properties at the same conditions of T and P. If molar properties for all components (θ_{mi}) are known, the mixture molar property (θ_m) may be calculated as

$$(7.1) \qquad \theta_m = \sum_i x_{mi}\theta_{mi}$$

where x_{mi} is mole fraction of component i and the summation is on all components present in the mixture. Subscript m indicates that the property is a molar quantity (value of property per unit mole). For gases especially at low pressures (<1 bar), the volume fraction, x_{Vi} may be used instead of mole fraction. Similarly for specific properties this equation can be written as

$$(7.2) \qquad \theta_s = \sum_i x_{wi}\theta_{si}$$

where x_{wi} is weight fraction of i in the mixture and subscript s indicates that the property is a specific quantity (per unit mass). In the above two equations, θ is a thermodynamic property such as volume (V), internal energy (U), enthalpy (H), heat capacity (C_P), entropy (S), Helmholtz free energy (A), or Gibbs free energy (G). Usually Eq. (7.1) is used to calculate molar property of the mixture as well as its molecular weight and then Eq. (5.3) is used to calculate specific property wherever is required. In fact Eqs. (7.1) and (7.2) are equivalent and one may combine Eqs. (5.3) and (1.15) with Eq. (7.1) to derive Eq. (7.2). These equations provide a good estimate of mixture properties for ideal solutions and mixtures of similar compounds where the interaction between species may be ignored.

Empirically developed correlations for properties of undefined or defined mixtures are based on a certain group of data on mixtures. Correlations specifically developed based on data of petroleum fractions usually cannot be used for estimation of properties of pure hydrocarbons. However, if in development of correlations for properties of undefined petroleum fractions pure component data are also used, then the resulting correlation will be more general. Such correlations can be applied to both pure components and undefined mixtures and they can be used more safely to fractions that have not been used in development of the correlation.

However, one important limitation in use of such correlations is boiling point range, carbon number, or molecular weight of fractions and compounds used in the development of the correlations. For example, correlations that are based on properties of petroleum fractions and pure components with carbon number range of C_5–C_{20} cannot be used for estimation of properties of light gases (natural gases or LPG), heavy residues, or crude oils. Another limitation of empirically developed correlations is the method of calculation of input parameters. For example, a generalized correlation developed for properties of heavy fractions requires critical properties as input parameter. For such correlations the same method of estimation of input parameters as the one used in the development of correlation should be used. The most reliable correlations are those that have some theoretical background, but the coefficients have been determined empirically from data on petroleum fractions as well as pure compounds. One technique that is often used in recent years to develop correlations for physical properties of both pure compounds and complex mixtures is the artificial neural network method [4–6]. These methods are called neural networks methods because artificial neural networks mimic the behavior of biological neurons. Although neural nets can be used to correlate data accurately and to identify correlative patterns between input and target values and the impact of each input parameter on the correlation, they lack necessary theoretical basis needed in physical property predictions. The resulting correlations from neural nets are complex and involve a large number of coefficients. For this reason correlations are inconvenient for practical applications and they have very limited power of extrapolation outside the ranges of data used in their developments. However, the neural net model can be used to identify correlating parameters in order to simplify theoretically developed correlations.

7.2 DENSITY

Density is perhaps one of the most important physical properties of a fluid, since in addition to its direct use in size calculations it is needed to predict other thermodynamic properties as shown in Chapter 6. As seen in Section 7.5, methods to estimate transport properties of dense fluids also require reduced density. Therefore once an accurate value of density is used as an input parameter for a correlation to estimate a physical property, a more reliable value of that property can be calculated. Methods of calculation of density of fluids have been discussed in Chapter 5. Density may be expressed in the form of absolute density (ρ, g/cm^3), molar density (ρ_m, mol/cm^3), specific volume (V, cm^3/g), molar volume (V_m, cm^3/mol), reduced density ($\rho_r = \rho/\rho_c = V_c/V$, dimensionless), or compressibility factor ($Z = PV/RT = V/V^{ig}$, dimensionless). Equations of states or generalized correlations discussed in Chapter 5 predict V_m or Z at a given T and P. Once Z is known, the absolute density can be calculated from

$$(7.3) \qquad \rho = \frac{MP}{ZRT}$$

where M is the molecular weight, R is the gas constant, and T is the absolute temperature. If M is in g/mol, P in bars, T in kelvin, and $R = 83.14$ bar·cm^3/mol·K, then ρ is calculated in g/cm^3, which is the standard unit for density in this

book. Equation (7.3) is valid for both liquids and gases once their Z values are calculated from an equation of state or a generalized correlation. If Z is known for all components in a mixture, then Z_m can be calculated from Eq. (7.1) and ρ_m from Eq. (7.3). Specific methods and recommendations for calculation of density of gases and liquids are given in the following sections.

7.2.1 Density of Gases

Generally both equations of state and the Lee–Kesler generalized correlation (Section 5.7) provide reliable prediction of gaseous densities. For high-pressure gases, cubic EOS such as PR or SRK EOS give acceptable values of density for both pure and mixtures and no volume translation (Section 5.5.3) is needed. For practical calculations, properties of gases can be calculated from simple equations of state. For example, Press [7] has shown that the original simple two-parameter Redlich–Kowng equation of state (RK EOS) gives reasonably acceptable results for predicting gas compressibility factors needed for calculation of valve sizes. For moderate pressures truncated virial equation (Eq. 5.76) can be used with coefficients (B and C) calculated from Eqs. (5.71) and (5.78). For low-pressure gases (<5 bar), virial equation truncated after the second term (Eq. 5.75) with predicted second virial coefficient from Tsonopoulos correlation (Eq. 5.71) is sufficient to predict gas densities. For light hydrocarbons and natural gases, the Hall–Yarborough correlation (Eq. 5.102) gives a good estimate of density. For defined gas mixtures the mixing rule may be applied to the input parameters (T_c, P_c, and ω) and the mixture Z value can be directly calculated from an EOS. For undefined natural gases, the input parameters may be calculated from gas-specific gravity using correlations given in Chapter 4 (see Section 4.2).

7.2.2 Density of Liquids

For high-pressure liquids, density may be estimated from cubic EOS such as PR or SRK equations. However, these equations break at carbon number of about C_{10} for liquid density calculations. They provide reasonable values of liquid density when appropriate volume translation introduced in Section 5.5.3 is used. The error of liquid density calculations from cubic equations of states increases at low and atmospheric pressures. For saturated liquids, special care should be taken to take the right Z value (the lowest root of a cubic equation). Once a cubic equation is used to calculate various thermodynamic properties (i.e., fugacity coefficient) at high pressures, it is appropriate to use a cubic equation such as SRK or PR with volume translation for both liquid and gases. However, when density of a liquid alone is required, PR or SRK are not the most appropriate method for calculation of liquid density. For heavy hydrocarbons and petroleum fractions, the modified RK equation of state based on refractive index proposed in Section 5.9 is appropriate for calculation of liquid densities. The refractive index of heavy petroleum fractions can be estimated accurately with methods outlined in Chapter 2. One should be careful that this method is not applicable to nonhydrocarbons (i.e., water, alcohols, or acids) or highly polar aromatic compounds.

For the range that Lee–Kesler generalized correlation (Eq. 5.107) and Table 5.9) can be used for liquids, it gives density

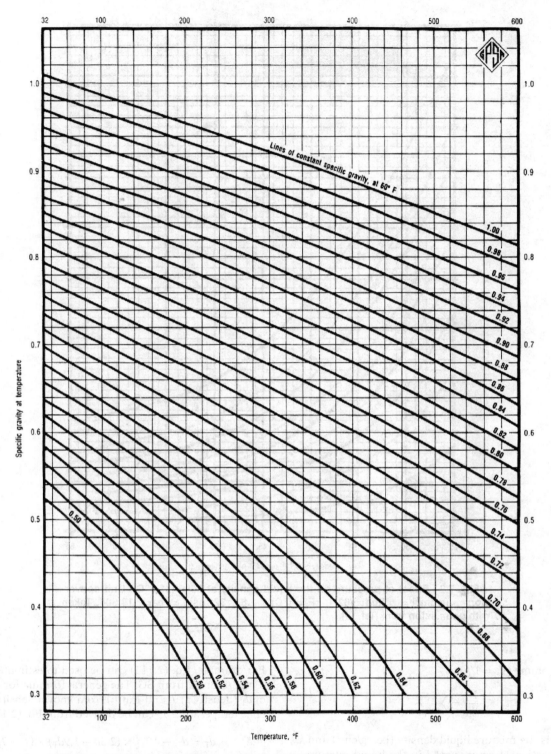

FIG. 7.1—Effect of temperature on the liquid specific gravity of hydrocarbons. Units conversion: °F = (°C) × 1.8 + 32. Taken with permission from Ref. [8].

values more accurate than SRK or PR equations without volume translation. The Lee–Kesler correlation is particularly useful for rapid-hand calculations for a single data point. The most accurate method for prediction of saturated liquid densities is through Rackett equation introduced in Sections 5.8. However, for high-pressure liquids the method of API (Eq. 5.129) or the COSTALD correlation (Eq. 5.130) may be used combined with the Rackett equation to provide very accurate density values for both pure components and petroleum fractions. These methods are also applicable to nonhydrocarbons as well. At low pressures or when the pressure is near saturation pressure, no correction on the effect of pressure is required and saturated liquid density calculated from Rackett equation may be directly used as the density of compressed (subcooled) liquid at pressure of interest.

For liquid mixtures with known composition, density can be accurately calculated from density of each component (or pseudocomponents) through Eq. (7.2) when it is applied

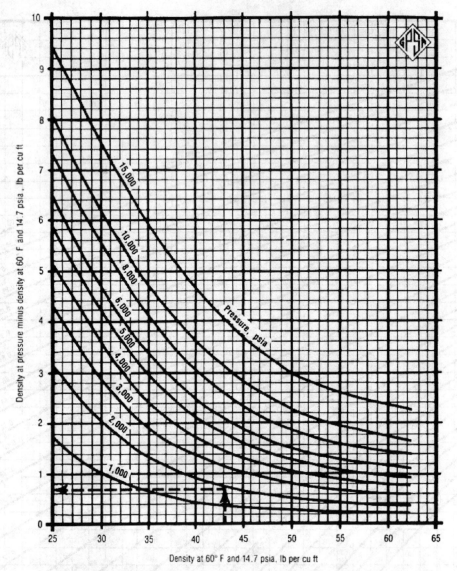

FIG. 7.2—Effect of pressure on the liquid density at 60°F (15.5°C). Unit conversion: ρ [g/cm³] = ρ [lb/ft³]/62.4; °F = (°C) × 1.8 + 32; psia = bar × 14.504. Taken with permission from Ref. [8].

to specific volume ($V_s = 1/\rho$) as

$$(7.4) \qquad \frac{1}{\rho_{mix}} = \sum_i \frac{x_{wi}}{\rho_i}$$

where ρ_{mix} is the mixture liquid density (i.e., g/cm³) and x_{wi} is weight fraction of component i in the liquid mixture. ρ_i should be known from database, experiment, or may be calculated from Rackett equation. Equation (7.4) can be applied to specific gravity but not to molar density. This equation is particularly useful for calculation of specific gravity and density of crude oils with known composition at atmospheric pressure, as it is shown later in this chapter.

When only a minimum of one data point for liquid density of a petroleum fraction at atmospheric pressure is known (i.e., SG, d_{20}, or d_{25}), then Eq. (2.110) may be used to calculate liquid density at atmospheric pressure and other temperatures.

For example, Eq. (2.111) can be used to estimate density at temperature T from SG. The general formula for calculation of d_T (density at T and 1 atm) from known density at a reference temperature T_o can be derived from Eq. (2.110) as

$$(7.5) \qquad d_T = d_{T_o} - 10^{-3} \times (2.34 - 1.9d_T) \times (T - T_o)$$

where both T and T_o are in kelvin or in °C and d_T and d_{T_o} are in g/cm³. This method provides reliable density values at temperatures near the reference temperature at which density is known (when T is near T_o). When the actual temperature is far from the reference temperature, this equation should be used with caution.

For quick density calculations several graphical methods have been developed, which are less accurate than the methods outlined above. Graphical methods for liquid density calculations recommended by GPA [8] are shown in Figs. 7.1–7.3.

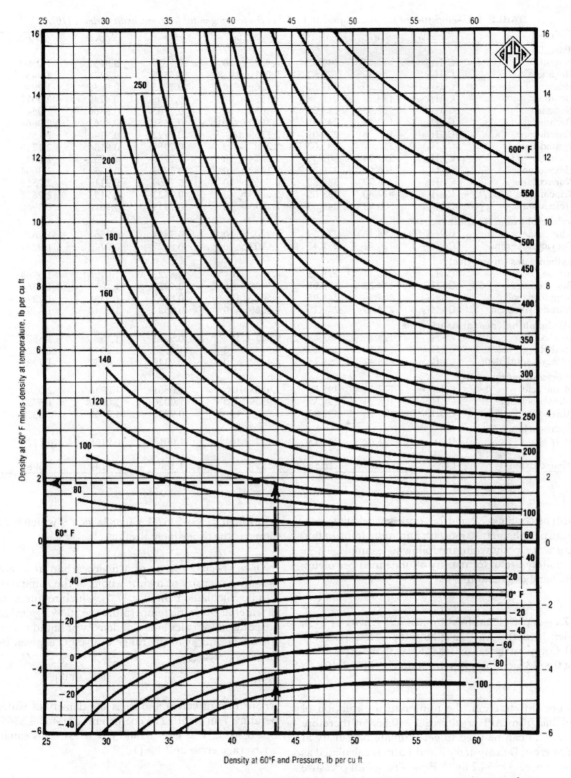

FIG. 7.3—Effect of temperature on liquid density at pressure P. Unit conversion: ρ [g/cm^3] = ρ[lb/ft^3]/62.4; °F = (°C) × 1.8 + 32. **Taken with permission from Ref. [8].**

Figure 7.1 gives effect of temperature on the liquid specific gravity (SG). Once SG at 60°F (15.5°C) is known, with the use of this figure specific gravity at temperature T (SG$_T$) can be determined. Then density at T and 1 atm can be determined through multiplying specific gravity by density of water. This figure may be used for calculation of density of liquids at low pressures where the effect of pressure on liquid density can be neglected (pressures less than ~50–70 bar). Figure 7.2 shows effect of pressure on density at 60°F and Fig. 7.3 shows effect of T on liquid density at any pressure P (P > 1000 psia or

TABLE 7.1—*Properties of saturated liquid and solid at the freezing point for some hydrocarbons [10].*

No.	Compound	Formula	M	T_M/T_{tp}, K	P_{tp}, bar	ρ^L, g/cm³	ρ^S, g/cm³	C_P^L, J/g·K	C_P^S, J/g·K
	n-Paraffins								
1	n-Pentane	C_5H_{12}	72.15	143.42	6.8642×10^{-7}	0.7557	0.9137	1.9509	1.4035
2	n-Hexane	C_6H_{14}	86.17	177.83	9.011×10^{-6}	0.7538	0.8471	1.9437	1.4386
3	n-Heptane	C_7H_{16}	100.20	182.57	1.8269×10^{-6}	0.7715	0.8636	1.9949	1.4628
4	n-Octane	C_8H_{18}	114.22	216.38	2.108×10^{-5}	0.7603	0.8749	2.0077	1.5699
5	n-Nonane	C_9H_{20}	128.25	219.66	4.3058×10^{-6}	0.7705	0.8860	2.0543	1.6276
6	n-Decane	$C_{10}H_{22}$	142.28	243.51	1.39297×10^{-5}	0.7656	0.8962	2.0669	1.6995
7	n-Tetradecane	$C_{14}H_{30}$	198.38	279.01	2.5269×10^{-6}	0.7722	0.9140	2.1589	1.8136
8	n-Pentadecane	$C_{15}H_{32}$	212.41	283.07	1.2887×10^{-6}	0.7752	0.9134	2.1713	...
9	n-Eicosane	$C_{20}H_{42}$	282.54	309.58	9.2574×10^{-8}	0.7769	0.8732	2.2049	2.2656
10	n-Hexacosane	$C_{26}H_{54}$	366.69	329.25	5.1582×10^{-9}	0.7803	0.9254	2.3094	2.2653
11	n-Nonacosane	$C_{29}H_{60}$	408.77	336.85	6.8462×10^{-10}	0.7804	0.9116	2.2553	1.8811
12	n-Triacontane	$C_{30}H_{62}$	422.80	338.65	2.0985×10^{-10}	0.7823	0.9133	2.2632	...
13	n-Hexacontane	$C_{36}H_{74}$	506.95	349.05	2.8975×10^{-12}	0.7819	0.9610	2.3960	2.4443
	n-Alkylcyclohexanes (naphthenes)								
14	Cyclohexane	C_6H_{12}	84.16	279.69	5.3802×10^{-2}	0.7894	0.8561	1.7627	1.6124
15	n-Decylcyclohexane	$C_{16}H_{32}$	224.42	271.42	4.5202×10^{-8}	0.8327	0.9740	1.9291	1.5398
	n-Alkylbenzenes (aromatics)								
16	Benzene	C_6H_6	78.11	278.65	4.764×10^{-4}	0.8922	1.0125	1.6964	1.6793
17	n-Butylbenzene	$C_{10}H_{14}$	134.22	185.25	1.5439×10^{-9}	0.9431	1.1033	1.5268	1.1309
18	n-Nonylbenzene	$C_{15}H_{24}$	204.36	248.95	6.603×10^{-9}	0.8857	1.0361	1.7270	1.6882
19	n-Tetradecylbenzene	$C_{20}H_{34}$	274.49	289.15	9.8069×10^{-9}	0.858	1.0046	1.8799	1.7305
	1-n-Alkylnaphthalenes (aromatics)								
20	Naphthalene	$C_{10}H_8$	128.16	353.43	9.913×10^{-3}	0.9783	1.157	1.687	1.6183
21	1-Methylnaphthalene	$C_{11}H_{10}$	142.19	242.67	4.3382×10^{-7}	1.0555	1.2343	1.4237	1.0796
22	1-n-Decylnaphthalene	$C_{20}H_{28}$	268.42	288.15	8.4212×10^{-9}	0.9348	1.0952	1.7289	1.5601
	Other organic compounds								
23	Benzoic acid	$C_7H_6O_2$	122.12	395.52	7.955×10^{-3}	1.0861	1.2946	2.0506	1.5684
24	Diphenylmethane	$C_{13}H_{12}$	168.24	298.39	1.9529×10^{-5}	1.0020	1.0900	1.5727	1.3816
25	Anthracene	$C_{14}H_{10}$	190.32	488.93	4.951×10^{-2}	0.9745	1.2167	2.0339	2.0182
	Nonhydrocarbons								
26	Water	H_2O	18.02	273.15/ 273.16	6.117×10^{-3}	1.0013	0.9168	4.227	2.1161
27	Carbon dioxide	CO_2	44.01	216.58	5.187	1.1807	1.5140	1.698	1.3844

70 bar). With use of Figs. 7.2 and 7.3, one may calculate density of a liquid petroleum fraction with minimum information on specific gravity as shown in the following example. These figures are mainly useful for density of undefined petroleum fractions by hand calculation.

Example 7.1—A petroleum fraction has API gravity of 31.4. Calculate density of this fraction at 20°C (68°F) and 372.3 bar (5400 psia). Compare the estimated value with the experimental value of 0.8838 g/cm³ as given in Chapter 6 of Ref. [9].

Solution—For this fraction, the minimum information of SG is available from API gravity (SG = 0.8686); therefore Figs. 7.2. and 7.3 can be used to get estimate of density at T and P of interest. Density at 60°F and 1 atm is calculated as $0.999 \times 0.8686 \times 62.4 = 54.2$ lb/ft³. From Fig. 7.2 for pressure of 5400 psia we read from the y axis the value of 1.2, which should be added to 54.2 to get density at 60°F and 5400 psia as $54.2 + 1.2 = 55.4$ lb/ft³. To consider the effect of temperature, use Fig. 7.3. For temperature of 68°F and at density of 55.4 lb/ft³ the difference between density at 60 and 68°F is read as 0.25 lb/ft³. This small value is due to small temperature difference of 8°F. Finally density at 68°F and 55.4 lb/ft³ is calculated as $55.4 - 0.25 = 55.15$ lb/ft³. This density is

equivalent to $55.15/62.4 = 0.8838$ g/cm³, which is exactly the same as the experimental value. ◆

Once specific gravity of a hydrocarbon at a temperature is known, density of hydrocarbons at the same temperature can be calculated using Eq. (2.1), which requires the density of water at the same temperature (i.e., 0.999 g/cm³ at 60°F). A correlation for calculation of density of liquid water at 1 atm for temperatures in the range of 0–60°C is given by DIPPR-EPCON [10] as

$$(7.6) \qquad d_T = A \times B^{-\left[1+(1-T/C)^D\right]}$$

where T is in kelvin and d_T is the density of water at temperature T in g/cm³. The coefficients are $A = 9.83455 \times 10^{-2}$, $B = 0.30542$, $C = 647.13$, and $D = 0.081$. This equation gives an average error of 0.1% [10].

7.2.3 Density of Solids

Although the subject of solid properties is outside of the discussion of this book, as shown in Chapter 6, such data are needed in solid–liquid equilibria (SLE) calculations. Densities of solids are less affected by pressure than are properties of liquids and can be assumed independent of pressure (see Fig. 5.2a). In addition to density, solid heat capacity and triple point temperature and pressure (T_{tp}, P_{tp}) are also needed in SLE calculations. Values of density and heat capacity of liquid and solid phases for some compounds at their melting points are given in Table 7.1, as obtained from DIPPR [10]. The triple point temperature (T_{tp}) is exactly the same as the melting or freezing point temperature (T_M). As seen from Fig. 5.2a and from calculations in Example 6.5, the effect of pressure on the melting point of a substance is very small and for a pressure change of a few bars no change in T_M is observed. Normal freezing point T_M represents melting point at pressure of 1 atm. P_{tp} for a pure substance is very small and the maximum difference between atmospheric pressure and P_{tp} is less than 1 atm. For this reason as it is seen in Table 7.1 values of T_M and T_{tp} are identical (except for water).

Effect of temperature on solid density in a limited temperature range can be expressed in the following linear form:

$$(7.7) \qquad \rho_m^S = A - (10^{-6} \times B)\, T$$

where ρ_m^S is the solid molar density at T in mol/cm^3. A and B are constants specific for each compound, and T is the absolute temperature in kelvin. Values of B for some compounds as given by DIPPR [10] are n-C$_5$: 6.0608; n-C$_{10}$: 2.46; n-C$_{20}$: 2.663; benzene: 0.3571; naphthalene: 2.276; benzoic acid: 2.32; and water (ice): 7.841. These values with Eq. (7.7) and values of solid density at the melting point given in Table 7.1 can be used to obtain density at any temperature as shown in the following example.

Example 7.2—Estimate density of ice at –50°C.

Solution—From Table 7.1 the values for water are obtained as $M = 18.02$, $T_M = 273.15$ K, $\rho^S = 0.9168$ g/cm^3 (at T_M). In Eq. (7.7) for water (ice) $B = 7.841$ and ρ_m^S is the molar density. At 273.15 K, $\rho_m^S = 0.050877$ mol/cm^3. Substituting in Eq. (7.7) we get $A = 0.053019$. With use of A and B in Eq. (7.7) at 223.15 K (−50°C) we get $\rho_m^S = 0.051269$ mol/cm^3 or $\rho^S = 0.051269 \times 18.02 = 0.9238$ g/cm^3. $\blacklozenge$

7.3 VAPOR PRESSURE

As shown in Chapters 2, 3, and 6, vapor pressure is required in many calculations related to safety as well as design and operation of various units. In Chapter 3, vapor pressure relations were introduced to convert distillation data at reduced pressures to normal boiling point at atmospheric pressure. In Chapter 2, vapor pressure was used for calculation of flammability potential of a fuel. Major applications of vapor pressure were shown in Chapter 6 for VLE and calculation of equilibrium ratios. As it was shown in Fig. 1.5, prediction of vapor pressure is very sensitive to the input data, particularly the critical temperature. Also it was shown in Fig. 1.7 that small errors in calculation of vapor pressure (or relative volatility) could lead to large errors in calculation of the height of absorption/distillation columns. Methods of calculation of vapor pressure of pure compounds and estimation methods using generalized correlations and calculation of vapor pressure of petroleum fractions are presented hereafter.

7.3.1 Pure Components

Experimental data for vapor pressure of pure hydrocarbons are given in the TRC Thermodynamic Tables [11]. Figures 7.4 and 7.5 show vapor pressure of some pure hydrocarbons from praffinic and aromatic groups as given in the API-TDB [9]. Further data on vapor pressure of pure compounds at 37.8°C (100°F) were given earlier in Table 2.2. For pure compounds the following dimensionless equation can be used to estimate vapor pressure [9]:

$$(7.8) \qquad \ln P_r^{vap} = (T_r^{-1}) \times (a\tau + b\tau^{1.5} + c\tau^{2.6} + d\tau^5)$$

where $\tau = 1 - T_r$ and P_r^{vap} is the reduced vapor pressure (P^{vap}/P_c), and T_r is the reduced temperature. Coefficients a–d with corresponding temperature ranges are given in Table 7.2 for a number of pure compounds. Equation (7.8) is a linearized form of Wagner equation. In the original Wagner equation, exponents 3 and 6 are used instead of 2.6 and 5 [12].

The primary correlation recommended in the API-TDB [9] for vapor pressure of pure compounds is given as

$$(7.9) \qquad \ln P^{vap} = A + \frac{B}{T} + C \ln T + DT^2 + \frac{E}{T^2}$$

where coefficients A–E are given in the API-TDB for some 300 compounds (hydrocarbons and nonhydrocarbons) with specified temperature range. This equation is a modified version of correlation originally developed by Abrams and Prausnitz based on the kinetic theory of gases. Note that performance of these correlations outside the temperature ranges specified is quite weak. In DIPPR [10], vapor pressure of pure hydrocarbons is correlated by the following equation:

$$(7.10) \qquad \ln P^{vap} = A + \frac{B}{T} + C \ln T + DT^E$$

where coefficients A–E are given for various compounds in Ref. [10]. In this equation, when $E = 6$, it reduces to the Riedel equation [12]. Another simple and commonly used relation to estimate vapor pressure of pure compounds is the Antoine equation given by Eq. (6.102). Antoine parameters for some 700 pure compounds are given by Yaws and Yang [13]. Antoine equation can be written as

$$(7.11) \qquad \ln P^{vap}(\text{bar}) = A - \frac{B}{T + C}$$

where T is in kelvin. Antoine proposed this simple modification of the Clasius–Clapeyron equation in 1888. The lower temperature range gives the higher accuracy. For some compounds, coefficients of Eq. (7.11) are given in Table 7.3. Equation (7.11) is convenient for hand calculations. Coefficients may vary from one source to another depending on the temperature range at which data have been used in the regression process. Antoine equation is reliable from about 10 to 1500 mm Hg (0.013–2 bars); however, the accuracy deteriorates rapidly beyond this range. It usually underpredicts vapor pressure at high pressures and overpredicts vapor pressures at low pressures. One of the convenient features of this equation is that either vapor pressure or the temperature can be directly calculated without iterative calculations. No generalized correlation has been reported on the Antoine constants and they should be determined from regression of experimental data.

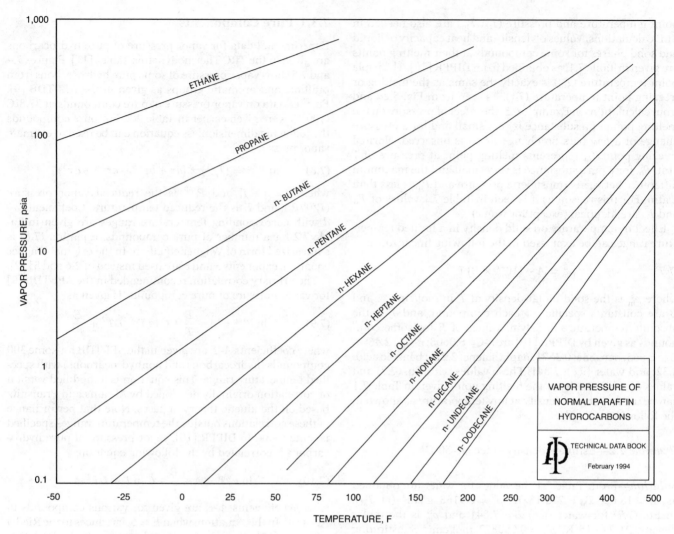

**FIG. 7.4—Vapor pressure of some *n*-alkane hydrocarbons. Unit conversion: °F = (°C) ×
1.8 + 32; psia = bar × 14.504. Taken with permission from Ref. [9].**

An expanded form of Antoine equation, which covers a
wider temperature range by including two additional terms
and a fourth parameter, is given in the following form as suggested by Cox [12]:

$$(7.12) \qquad \ln P^{\text{vap}} = A + \frac{B}{T} + CT + DT^2$$

Another correlation is the Miller equation, which has the following form [12]:

$$(7.13) \qquad \ln P_r^{\text{vap}} = -\frac{A}{T_r}\left[1 - T_r^2 + B(3 + T_r)(1 - T_r)^3\right]$$

where A and B are two constants specific for each compound.
These coefficients have been correlated to the reduced boiling
point T_{br} ($= T_b/T_c$) and P_c of pure hydrocarbon vapor pressure
in the following form:

$$A_1 = \frac{T_{\text{br}} \ln(P_c/1.01325)}{1 - T_{\text{br}}}$$

$$(7.14) \qquad A = 0.4835 + 0.4605 A_1$$

$$B = \frac{A/A_1 - (1 + T_{\text{br}})}{(3 + T_{\text{br}})(1 - T_{\text{br}})^2}$$

where P_c is in bar. Equations (7.13) and (7.14) work better
at superatmospheric pressures ($T > T_b$) rather than at subatmospheric pressures. The main advantage of this equation
is that it has only two constants. This was the reason that
it was used to develop Eq. (3.102) in Section (3.6.1.1) for
calculation of Reid vapor pressure (RVP) of petroleum fuels.
For RVP prediction, a vapor pressure correlation is applied at
a single temperature (100°F or 311 K) and a two-parameter
correlation should be sufficient. Some other forms of equations used to correlate vapor pressure data are given in
Ref. [12].

7.3.2 Predictive Methods—Generalized Correlations

In Section 6.5, estimation of vapor pressure from an equation of state (EOS) through Eq. (6.105) was shown. When an
appropriate EOS with accurate input parameters is used, accurate vapor pressure can be estimated through Eq. (6.105)
or Eq. (7.65) [see Problem 7.13]. As an example, prediction of
vapor pressure of *p*-xylene from a modified PR EOS is shown
in Fig. 7.6 [14].

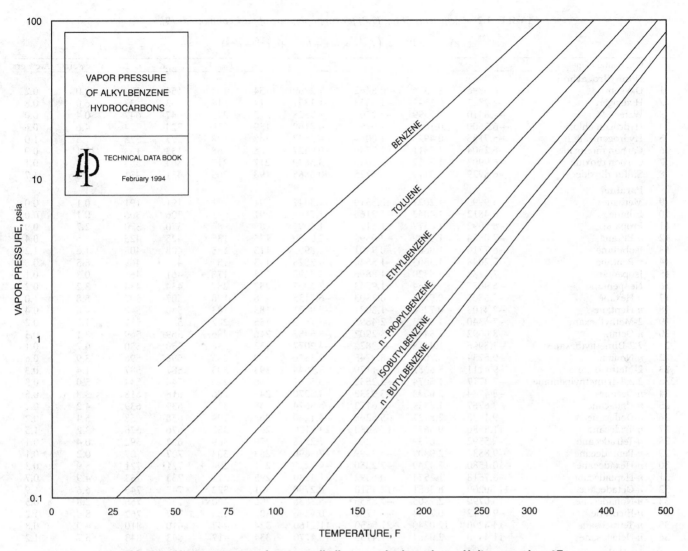

FIG. 7.5—Vapor pressure of some *n*-alkylbenzene hydrocarbons. Unit conversion: °F = (°C) × 1.8 + 32; psia = bar × 14.504. Taken with permission from Ref. [9].

Generally, vapor pressure is predicted through correlations similar to those presented in Section 7.3.2. These correlations require coefficients for individual components. A more useful correlation for vapor pressure is a generalized correlation for all compounds that use component basic properties (i.e., T_b) as an input parameter. A perfect relation for prediction of vapor pressure of compounds should be valid from triple point to the critical point of the substance. Generally no single correlation is valid for all compounds in this wide temperature range. As the number of coefficients in a correlation increases it is expected that it can be applied to a wider temperature range. However, a correct correlation for the vapor pressure in terms of reduced temperature and pressure is expected to satisfy the conditions that at $T = T_c$, $P^{vap} = P_c$ and at $T = T_b$, $P^{vap} = 1.0133$ bar. The temperature range $T_b \leq T \leq T_c$ is usually needed in practical engineering calculations. However, when a correlation is used for calculation of vapor pressure at $T \leq T_b$ ($P^{vap} \leq 1.0133$ bar), it is necessary to satisfy the following conditions: at $T = T_{tp}$, $P^{vap} = P_{tp}$ and at $T = T_b$, $P^{vap} = 1.0133$ bar, where T_{tp} and P_{tp} are the triple point temperature and pressure of the substance of interest.

The origin of most of predictive methods for vapor pressure calculations is the Clapeyron equation (Eq. 6.99). The simplest method of prediction of vapor pressure is through Eq. (6.101), which is derived from the Clapeyron equation. Two parameters of this equation can be determined from two data points on the vapor pressure. This equation is very approximate due to the assumptions made (ideal gas law, neglecting liquid volume, and constant heat of vaporization) in its derivation and is usually useful when two reference points on the vapor pressure curve are near each other. However, the two points that are usually known are the critical point (T_c, P_c) and normal boiling point (T_b and 1.013 bar) as demonstrated by Eq. (6.103). Equations (6.101) and (6.103) may be combined to yield the following relation in a dimensionless form:

$$(7.15) \quad \ln P_r^{vap} = \left[\ln\left(\frac{P_c}{1.01325}\right)\right] \times \left(\frac{T_{br}}{1 - T_{br}}\right) \times \left(1 - \frac{1}{T_r}\right)$$

where P_c is the critical pressure in bar and T_{br} is the reduced normal boiling point ($T_{br} = T_b/T_c$). The main advantage of Eq. (7.15) is simplicity and availability of input parameters

TABLE 7.2—*Coefficients of Eq. (7.8) for vapor pressure of pure compounds [9].*

$$(7.8) \qquad \ln P_r^{vap} = \left(T_r^{-1}\right) \times \left(a\tau + b\tau^{1.5} + c\tau^{2.6} + d\tau^5\right)$$

	Compound name	a	b	c	d	T_{tp}, K	T_{min}, K	T_{max}, K	T_c, K	Max% err	Ave% err
	Nonhydrocarbon										
1	Oxygen	−6.0896	1.3376	−0.8462	−1.2860	54	54	154	154	2.0	0.2
2	Hydrogen	−4.7322	0.5547	1.5353	−1.1391	14	14	33	33	6.0	0.8
3	Water	−7.8310	1.7399	−2.2505	−1.9828	273	273	647	647	0.4	0.0
4	Hydrogen chloride	−6.2600	0.1021	1.0793	−4.8162	159	159	324	324	4.6	0.6
5	Hydrogen sulfide	−5.7185	−0.4928	1.0044	−4.5547	188	188	373	373	7.3	1.0
6	Carbon monoxide	−6.2604	1.5811	−1.5740	−0.9427	68	68	133	133	10.1	0.9
7	Carbon dioxide	−6.9903	1.3912	−2.2046	−3.3649	217	217	304	304	0.8	0.1
8	Sulfur dioxlde	−6.8929	1.3119	−3.5225	0.6865	198	203	431	431	3.8	1.7
	Parafins										
9	Methane	−5.9999	1.2027	−0.5310	−1.3447	91	91	191	191	0.1	0.0
10	Ethane	−6.4812	1.4042	−1.2166	−1.7143	91	91	306	306	0.1	0.0
11	Propane	−6.8092	1.6377	−1.8173	−1.8094	86	86	370	370	2.7	0.9
12	n-Butane	−7.0524	1.6799	−2.0398	−2.0630	135	135	425	425	6.5	0.4
13	Isobutane	−6.7710	1.0669	−0.9201	−3.8903	113	125	408	408	1.6	1.2
14	n-Pentane	−7.2048	1.3503	−1.5540	−4.2828	143	157	469	469	6.5	0.8
15	Isopentane	−7.1383	1.5320	−1.8896	−2.7290	113	178	461	461	0.3	0.1
16	Neopentane	−6.9677	1.5464	−1.9563	−2.6057	257	257	434	434	3.2	0.2
17	n-Hexane	−7.3505	0.9275	−0.7303	−6.7135	178	178	507	507	5.8	1.0
18	n-Heptane	−7.4103	0.7296	−1.3081	−5.9021	183	183	540	540	4.6	0.4
19	2-Methylhexane	−7.6340	1.6113	−2.4895	−3.7538	155	222	531	531	1.9	0.2
20	n-Octane	−8.0092	1.8442	−3.2907	−3.5457	216	286	569	569	3.1	0.3
21	2,2-Dimethylhexane	−7.5996	1.4415	−2.3822	−4.2077	152	236	550	550	0.7	0.1
22	n-Nonane	−9.5734	5.7040	−8.9745	3.3386	219	233	596	596	3.9	0.6
23	2-Methyloctane	−9.4111	5.6082	−9.1179	3.9544	193	303	587	587	1.4	0.3
23a	2,2,4-Trimethylpentane	−7.4717	1.5074	−2.2532	−3.5291	166	225	544	544	5.0	0.3
24	n-Decane	−8.4734	2.0043	−3.9338	−4.5270	243	286	618	618	3.8	0.5
25	n-Undecane	−8.6767	1.8339	−3.6173	−6.5674	248	322	639	639	4.2	0.2
26	n-Dodecane	−9.1638	2.8127	−5.5268	−4.1240	263	294	658	658	5.4	0.2
27	n-Tridecane	−11.5580	9.5675	−17.8080	23.9100	268	333	676	676	3.2	1.5
28	n-Tetradecane	−9.5592	2.6739	−5.3261	−7.2218	279	369	692	692	0.4	0.1
29	n-Pentadecane	−9.8836	2.9809	−5.8999	−7.3690	283	383	707	707	0.2	0.1
30	n-Hexadecane	−10.1580	3.4349	−7.2350	−4.7220	291	294	721	721	5.9	0.3
31	n-Heptadecane	−8.7518	−1.2524	0.6392	−21.3230	295	311	733	733	4.9	0.7
32	n-Octadecane	−11.30200	6.3651	−12.4510	0.2790	301	322	745	745	6.6	1.9
33	n-Nonadecane	−10.0790	2.7305	−7.8556	−5.3836	306	400	756	756	2.8	0.5
34	n-Eicosane	−9.2912	0.7364	−8.1737	−0.4546	309	353	767	767	8.0	1.2
35	n-Tetracosane	−14.4290	12.0240	−21.5550	11.2160	324	447	810	810	4.0	0.8
36	n-Octacosane	−11.4490	2.0664	−7.4138	−15.4770	334	417	843	843	5.7	1.2
	Naphthenes										
37	Cyclopentane	−7.2042	2.2227	−2.8579	−1.2980	179	203	512	512	3.9	0.6
38	Methylcyclopentane	−7.1157	1.5063	−2.0252	−2.9670	131	178	533	533	2.0	0.1
39	Ethylcyclopentane	−7.2608	1.3487	−1.8800	−3.7286	134	183	569	569	1.5	0.1
40	n-Propylcyclopentane	−1.3961	0.2383	−5.7723	−6.0536	156	244	603	603	0.1	0.0
41	Cyclohexane	−7.0118	1.5792	−2.2610	−2.4077	279	279	553	553	4.3	0.7
42	Methylcyclohexane	−7.1204	1.4340	−1.9015	−3.3273	147	217	572	572	1.2	0.2
43	Ethylcyclohexane	−5.9783	−1.2708	0.2099	−5.3117	162	228	609	609	0.2	0.0
44	n-Propylcyclohexane	−5.6364	−2.1313	0.6054	−6.0405	178	228	639	639	0.4	0.0
45	Isopropylcyclohexane	−7.8041	2.0024	−2.8297	−3.4032	184	208	627	627	3.5	0.1
46	n-Butylcyclohexane	−4.9386	−3.9025	2.0300	−7.8420	198	286	667	667	0.2	0.0
47	n-Decylcyclohexane	−9.5188	2.4189	−4.5835	−7.7062	272	322	751	751	9.2	0.4
48	Cycloheptane	−7.3231	1.8407	−2.2637	−3.4498	265	265	604	604	0.7	0.1
	Olefins										
49	Ethylene	−6.3778	1.3298	−1.1667	−2.0209	104	104	282	282	8.9	0.6
50	Propylene	−6.7920	1.7836	−2.0451	−1.5370	88	88	366	366	5.2	0.3
51	1-Butene	−6.9041	1.3587	−1.3839	−3.7388	88	125	420	420	2.4	0.5
52	1-Pentene	−6.6117	0.0720	0.0003	−5.4313	108	167	465	465	6.5	1.5
53	1,3-Butadiene	−5.6060	−0.9772	−0.3358	−3.1876	127	156	484	484	0.3	0.1
	Diolefins and Acetylenes										
54	Acetylene	−7.3515	2.8334	−4.5075	6.8797	192	192	308	308	2.4	0.4
	Aromatics										
55	Benzene	−7.0200	1.5156	−1.9176	−3.5572	279	279	562	562	1.6	0.3
56	Toluene	−7.2827	1.5031	−2.0743	−3.1867	178	244	592	592	3.0	0.3
57	Ethylbenzene	−7.5640	1.7919	−2.7040	−2.8573	178	236	1	617	0.8	0.1
58	m-Xylene	−7.6212	1.6059	−2.4451	−3.0594	226	250	617	617	3.9	0.2
59	o-Xylene	−7.5579	1.5648	−2.1826	−3.7093	248	248	631	631	2.2	0.2
60	p-Xylene	−7.6935	1.8093	−2.5583	−3.0662	287	287	616	616	7.5	0.4

(Continued)

TABLE 7.2—(Continued).

	Compound name	a	b	c	d	T_{tp}, K	T_{min}, K	T_{max}, K	T_c, K	Max% err	Ave% err
61	i-Propylbenzene	−8.1015	2.6607	−3.8585	−2.2594	173	236	638	638	5.4	0.4
62	n-Butylbenzene	−7.8413	1.3055	−2.1437	−5.3415	186	233	661	661	5.6	0.7
63	n-Pentylbenzene	−8.7573	3.1808	−4.7169	−2.7442	198	311	680	680	2.8	0.2
64	n-Hexylbenzene	−8.0460	0.6792	−1.4190	−8.1068	212	333	698	698	1.8	0.2
65	n-Heptylbenzene	−9.1822	3.1454	−4.8927	−4.5218	225	356	714	714	2.0	0.2
66	n-Octylbenzene	−10.7760	7.0482	−10.5930	1.7304	237	311	729	729	8.0	0.8
67	Styrene	−6.3281	−1.2630	0.9920	−7.1282	243	243	636	636	0.6	0.1
68	n-Nonylbenzene	−10.7760	7.0038	−10.4060	1.1027	249	311	741	741	1.4	0.4
69	n-Decylbenzene	−10.5490	4.7502	−7.2424	−4.8469	259	333	753	753	0.1	0.0
70	n-Undecylbenzene	−11.8950	8.0001	−12.7000	4.6027	268	383	764	764	1.1	0.2
71	n-Dodecylbenzene	−10.6650	3.9860	−7.6855	−1.7721	276	333	659	774	9.6	1.8
72	n-Tridecylbenzene	−11.995	6.5968	−10.1880	−5.2923	283	417	783	783	2.1	0.4
73	Cumene	−7.4655	1.2449	−2.0897	−4.5973	177	228	631	631	2.6	0.3
	Diaromatics										
74	Naphthalene	−7.6159	1.8626	−2.6125	−3.1470	353	353	748	748	17.5	0.8
75	1-Methylnaphthalene	−7.4654	1.3322	−3.4401	−0.8854	243	261	772	772	7.1	1.7
76	2-Methylnaphthalene	−7.6745	1.0179	−1.3791	−5.6038	308	308	761	761	7.8	0.9
77	2,6-Dimethylnaphthalene	−7.8198	−2.5419	9.2934	−24.3130	383	383	777	777	0.1	0.0
78	i-Ethylnaphthalene	−6.7968	−0.5546	−1.2844	−5.4126	259	322	776	776	11.1	0.6
89	Anthracene	−8.4533	1.3409	−1.5302	−3.9310	489	489	873	873	5.6	0.5
80	Phenanthrene	−11.6620	9.2590	−10.0050	1.2110	372	372	869	869	1.0	0.2
	Oxygenated compounds										
81	Methanol	−8.6413	1.0671	−2.3184	−1.6780	176	176	513	513	5.9	0.7
82	Ethanol	−8.6857	1.0212	−4.9694	1.8866	159	194	514	514	4.9	0.4
83	Isopropanol	−7.9087	−0.6226	−4.8301	0.3828	186	200	508	508	8.4	1.6
84	Methyl-tert-butyl ether	−7.8925	3.3001	−4.9399	0.2242	164	172	497	497	8.0	1.3
85	tert-Butyl ethyl ether	−6.1886	−1.0802	−0.9282	−2.9318	179	179	514	514	8.7	4.8
86	Diisopropyl ether	−7.2695	0.4489	−0.9475	−5.2803	188	188	500	500	22.7	2.7
87	Methyl tert-pentyl ether	−7.8502	2.8081	−4.5318	−0.3252	158		534	534	1.3	0.4

T_{tp} is the triple point temperature and T_c is the critical temperature. T_{min} and T_{max} indicate the range at which Eq. (7.8) can be used with these coefficients. For quick and more convenient method use Antoine equation with coefficients given in Table 7.3.

(T_b, T_c, and P_c) for pure compounds. However, one should realize that since the base points in deriving the constants given by Eq. (6.103) are T_b and T_c, this equation should be used in the temperature range of $T_b \leq T \leq T_c$. Theoretically, a vapor pressure relation should be valid from triple point temperature to the critical temperature. But most vapor pressure correlations are very poor at temperatures near the triple point temperature. Using Eq. (7.15) at temperatures below T_b usually leads to unacceptable predicted values. For better prediction of vapor pressure near the triple point, the two base points should be normal boiling point ($T = T_b$, $P = 1.01325$ bar) and triple point (T_{tp}, P_{tp}). Values of T_{tp} and P_{tp} for some compounds are given in Table 7.1. Similarly if vapor pressure prediction near 37.8°C (100°F) is required the vapor pressure data given in Table 2 should be used as one of the reference points along with T_b, T_c, or T_{tp} to obtain the constants A and B in Eq. (6.101).

One of the latest developments for correlation of vapor pressure of pure hydrocarbons was proposed by Korsten [15]. He investigated modification of Eq. (6.101) with vapor pressure data of hydrocarbons and he found that $\ln P^{vap}$ varies linearly with $1/T^{1.3}$ for all hydrocarbons.

$$(7.16) \qquad \ln P^{vap} = A - \frac{B}{T^{1.3}}$$

where T is absolute temperature in kelvin and P^{vap} is the vapor pressure in bar. In fact the main difference between this equation and Eq. (6.101) is the exponent of T, which in this case is 1.3 (rather than 1 in the Clapeyron type equations). Parameters A and B can be determined from boiling and critical points as it was shown in Example 6.6. Parameters A and B in Eq. (7.16) can be determined from Eq. (6.103) with replacing T_b and T_c by $T_b^{1.3}$ and $T_c^{1.3}$. The linear relationship between $\ln P^{vap}$ and $1/T^{1.3}$ for large number of pure hydrocarbons is shown in Fig. 7.7.

Preliminary evaluation of Eq. (7.16) shows no major advantage over Eq. (7.15). A comparison of Eqs. (7.15) and (7.16) for n-hexane is shown in Fig. 7.8. Predicted vapor pressure from the method recommended in the API-TDB is also shown in Fig. 7.8. Clapeyron method refers to Eq. (7.15), while the Korsten method refers to Eq. (7.16), with parameters A and B determined from T_b, T_c, and P_c. Equation (7.15) agrees better than Eq. (7.16) with the API-TDB method. Substitution of Eq. (6.16) into Eq. (2.10) leads to Eq. (2.109) for prediction of acentric factor by Korsten method. Evaluation of methods of prediction of acentric factor presented in Section 2.9.4 also gives some idea on accuracy of vapor pressure correlations for pure hydrocarbons.

Korsten determined that all hydrocarbons exhibit a vapor pressure of 1867.68 bar at 1994.49 K as shown in Fig. 7.7. This data point for all hydrocarbons and the boiling point data can be used to determine parameters A and B in Eq. (7.16). In this way, the resulting equation requires only one input parameter (T_b) similar to Eq. (3.33), which is also shown in Section 7.3.3.1 (Eq. 7.25). Evaluation of Eqs. (7.25) and (7.16) with use of T_b as sole input parameter indicates that Eq. (7.25) is more accurate than Eq. (7.16) as shown in Fig. 7.8. However, note that Eq. (7.25) was developed for petroleum fractions and it may be used for pure hydrocarbons with $N_c \geq 5$.

Perhaps the most successful generalized correlation for prediction of vapor pressure was based on the theory of

TABLE 7.3—*Antione coefficients for calculation of vapor pressure from Eq. (7.11).*

$$\ln P^{\mathrm{vap}} = A - \frac{B}{T + C} \quad \text{Units: bar and K}$$

No.	Compound	T_b, K	A	B	C
	n-Alkanes				
1	Methane (C_1)	111.66	8.677752	911.2342	−6.340
2	Ethane (C_2)	184.55	9.104537	1528.272	−16.469
3	Propane (C_3)	231.02	9.045199	1851.272	−26.110
4	Butane (n-C_4)	272.66	9.055284	2154.697	−34.361
5	Isobutane (i-C_4)	261.34	9.216603	2181.791	−24.280
6	Pentane (n-C_5)	309.22	9.159361	2451.885	−41.136
7	Hexane (n-C_6)	341.88	9.213541	2696.039	−48.833
8	Heptane (n-C_7)	371.57	9.256922	2910.258	−56.718
9	Octane (n-C_8)	398.82	9.327197	3123.134	−63.515
10	Nonane (n-C_9)	423.97	9.379719	3311.186	−70.456
11	Decane (n-C_{10})	447.3	9.368137	3442.756	−79.292
12	Undecane (n-C_{11})	469.08	9.433921	3614.068	−85.450
13	Dodecane (n-C_{12})	489.48	9.493213	3774.559	−91.310
14	Tridecane (n-C_{13})	508.63	9.515341	3892.912	−98.930
15	Tetradecane (n-C_{14})	526.76	9.527867	4008.524	−105.430
16	Pentadecane (n-C_{15})	543.83	9.552251	4121.512	−111.770
17	Hexadecane (n-C_{16})	559.98	9.563948	4214.905	−118.700
18	Heptadecane (n-C_{17})	574.56	9.53086	4294.551	−123.950
19	Octadecane (n-C_{18})	588.3	9.502999	4361.787	−129.850
20	Nonadecane (n-C_{19})	602.34	9.533163	4450.436	−135.550
21	Eicosane (n-C_{20})	616.84	9.848387	4680.465	−141.050
	1-Alkenes				
22	Ethylene (C_2H_4)	169.42	9.011904	1373.561	−16.780
23	Propylene (C_3H_6)	225.46	9.109165	1818.176	−25.570
24	1-butane (C_4H_8)	266.92	9.021068	2092.589	−34.610
	Naphthenes				
25	Cyclopentane	322.38	9.366525	2653.900	−38.640
26	Methylcyclopentane	344.98	9.629388	2983.098	−34.760
27	Ethylcyclopentane	376.59	9.219735	2978.882	−53.030
28	Cyclohexane	353.93	9.049205	2723.438	−52.532
29	Methylcyclohexane	374.09	9.169631	2972.564	−49.449
	Aromatics				
30	Benzene (C_6H_6)	353.24	9.176331	2726.813	−55.578
31	Toluene (C_7H_8)	383.79	9.32646	3056.958	−55.525
32	Ethylbenzene	409.36	9.368321	3259.931	−60.850
33	Propylbenzene	432.35	9.38681	3434.996	−65.900
34	Butylbenzene	456.42	9.448543	3627.654	−71.950
35	o-Xylene (C_8H_{10})	417.59	9.43574	3358.795	−61.109
36	m-Xylene (C_8H_{10})	412.34	9.533877	3381.814	−57.030
37	p-Xylene (C_8H_{10})	411.53	9.451974	3331.454	−58.523
	Other hydrocarbons				
38	Isooctane	372.39	9.064034	2896.307	−52.383
39	Acetylene (C_2H_2)	188.40	8.459099	1217.308	−44.360
40	Naphthalene	491.16	9.522456	3992.015	−71.291
	Organics				
41	Acetone (C_3H_6O)	329.22	9.713225	2756.217	−45.090
42	Pyridine (C_5H_5N)	388.37	9.59600	3161.509	−58.460
43	Aniline (C_6H_7N)	457.17	10.15141	3897.747	−72.710
44	Methanol	337.69	11.97982	3638.269	−33.650
45	Ethanol	351.80	12.28832	3795.167	−42.232
46	Propanol	370.93	11.51272	3483.673	−67.343
	Nonhydrocarbons				
47	Hydrogen (H_2)	20.38	6.768541	153.8021	2.500
48	Oxygen (O_2)	90.17	8.787448	734.5546	−6.450
49	Nitrogen (N_2)	77.35	8.334138	588.7250	−6.600
50	Helium (He)	4.30	3.876632	18.77712	0.560
51	CO	81.66	8.793849	671.7631	−5.154
52	CO_2	194.65	10.77163	1956.250	−2.1117
53	Ammonia (NH_3)	239.82	10.32802	2132.498	−32.930
54	H_2S	212.84	9.737218	1858.032	−21.760
55	Sulfur (S)	717.75	9.137878	5756.739	−86.850
56	CCl_4	349.79	9.450845	2914.225	−41.002
57	Water (H_2O)	373.15	11.77920	3885.698	−42.980

The above coefficients may be used for pressure range of 0.02–2.0 bar except for water for which the pressure range is 0.01–16 bar as reported in Ref. [12]. These coefficents can generate vapor pressure near atmospheric pressure with error of less than 0.1%. There are other reported coefficients that give slightly more accurate results near the boiling point. For example, some other reported values for A, B, and C are given here. For water: 11.6568, 3799.89, and −46.8000; for acetone: 9.7864, 2795.82, and −43.15 or 10.11193, 2975.95, and −34.5228; for ethanol: 12.0706, 3674.49, and −46.70.

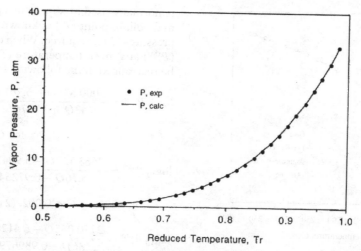

FIG. 7.6—Prediction of vapor pressure of *p*-xylene from modified PR EOS. Adopted with permission from Ref. [14].

corresponding states principle as described in Section 5.7 (see Eq. 5.107), which was proposed originally by Pitzer in the following form:

$$(7.17) \qquad \ln P_\mathrm{r}^{\mathrm{vap}} = f^{(0)}(T_\mathrm{r}) + \omega f^{(1)}(T_\mathrm{r})$$

where ω is the acentric factor. Lee and Kesler (1975) developed analytical correlation for $f^{(0)}$ and $f^{(1)}$ in the following

forms [16]:

$$\ln P_\mathrm{r}^{\mathrm{vap}} = 5.92714 - \frac{6.09648}{T_\mathrm{r}} - 1.28862 \ln T_\mathrm{r} + 0.169347 T_\mathrm{r}^6$$

$$(7.18) \qquad + \omega\left(15.2518 - \frac{15.6875}{T_\mathrm{r}} - 13.4721 \ln T_\mathrm{r} + 0.43577 T_\mathrm{r}^6\right)$$

where $P_\mathrm{r}^{\mathrm{vap}} = P^{\mathrm{vap}}/P_\mathrm{c}$ and $T_\mathrm{br} = T_\mathrm{b}/T_\mathrm{c}$. In 1989, Ambrose and Walton added a third term in Eq. (7.17) and proposed the following correlation for estimation of vapor pressure [12]:

$$T_\mathrm{r}(\ln P_\mathrm{r}^{\mathrm{vap}})$$
$$= -5.97616\tau + 1.29874\tau^{1.5} - 0.60394\tau^{2.5} - 1.06841\tau^5$$
$$+ \omega(-5.03365\tau + 1.11505\tau^{1.5} - 5.41217\tau^{2.5} - 7.46628\tau^5)$$
$$+ \omega^2(-0.64771\tau + 2.41539\tau^{1.5} - 4.26979\tau^{2.5} + 3.25259\tau^5)$$

$$(7.19)$$

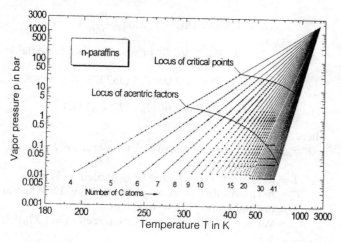

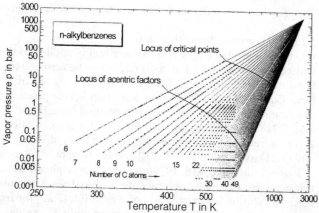

FIG. 7.7—Vapor pressure of pure hydrocarbons according to Eq. (7.16). Adopted with permission from Ref. [15].

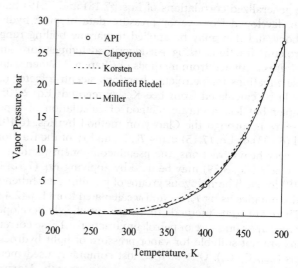

FIG. 7.8—Evaluation of various methods of calculation of vapor pressure of *n*-hexane. Methods: a. API: Eq. (7.8) with coefficients from Table 7.2; b. Clapeyron: Eq. (7.15); c. Korsten: Eq. (7.16); d. Modified Riedel: Eq. (7.24); e. Miller: Eqs. (7.13) and (7.14).

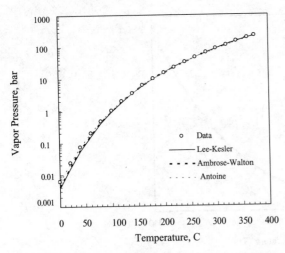

FIG. 7.9—Prediction of vapor pressure of water from Lee–Kesler (Eq. 7.18), Ambrose (Eq. 7.19), and Antoine (Eq. 7.11) correlations.

where $T_r = T/T_c$, $P_r^{vap} = P^{vap}/P_c$, and $\tau = 1 - T_r$. A graphical comparison between the Antoine equation (Eq. 7.11 with coefficients from Table 7.3), Lee–Kesler correlation, and Ambrose correlation for water from triple point to the critical point is shown in Fig. 7.9. Although Eq. (7.19) is more accurate than Eq. (7.18), the Lee–Kesler correlation (Eq. 7.18) generally provides reliable value for the vapor pressure and it is recommended by the API-TDB [9] for estimation of vapor pressure of pure hydrocarbons.

7.3.3 Vapor Pressure of Petroleum Fractions

Both analytical as well as graphical methods are presented here for calculation of vapor pressure of petroleum fractions, coal liquids, and crude oils.

7.3.3.1 Analytical Methods

The generalized correlations of Eqs. (7.18) and (7.19) have been developed from vapor pressure data of pure hydrocarbons and they may be applied to narrow boiling range petroleum fractions using pseudocritical temperature and pressure calculated from methods of Chapter 2. When using these equations for petroleum fractions, acentric factor (ω) should be calculated from Lee–Kesler method (Eq. 2.105). Simpler but less accurate method of calculation of vapor pressure is through the Clapeyron method by Eqs. (6.101) and (6.103) or Eq. (7.15) using T_b, T_c, and P_c of the fraction. For very heavy fractions, the pseudocomponent method of Chapter 3 (Eq. 3.39) may be used by applying Eq. (7.18) or (7.19) for each homologous groups of paraffins, naphthenes, and aromatics using T_c, P_c, and ω calculated from Eq. (2.42).

There are some methods that were specifically developed for the vapor pressure of petroleum fractions. These correlations are not suitable for vapor pressure of light hydrocarbons (i.e., C_1–C_4). One of the most commonly used methods for vapor pressure of petroleum fractions is the Maxwell and Bonnell (MB) correlation [17] presented by Eqs. (3.29)–(3.30). Usually Eq. (3.29) can be used at subatmospheric pressures ($P < 1$ atm.) for calculation of normal boiling point (T_b) from boiling points at low pressures (T). Equation (3.30)

is normally used at superatmospheric pressures where normal boiling point (T_b) is known and boiling point at higher pressures (T) is required. When calculation of vapor pressure (P^{vap}) at a given temperature (T) is required, Eq. (3.29) can be rearranged in the following form:

$$\log_{10} P^{vap} = \frac{3000.538Q - 6.761560}{43Q - 0.987672}$$

$$\text{for } Q > 0.0022 \ (P^{vap} < 2 \, \text{mm Hg})$$

$$\log_{10} P^{vap} = \frac{2663.129Q - 5.994296}{95.76Q - 0.972546}$$

$$\text{for } 0.0013 \le Q \le 0.0022 (2 \, \text{mm Hg} \le P^{vap} \le 760 \, \text{mm Hg})$$

$$\log_{10} P^{vap} = \frac{2770.085Q - 6.412631}{36Q - 0.989679}$$

(7.20) $\qquad\qquad$ for $Q < 0.0013 \ (P^{vap} > 760 \, \text{mm Hg})$

Parameter Q is defined as

(7.21) $$Q = \frac{\frac{T_b'}{T} - 0.00051606 T_b'}{748.1 - 0.3861 T_b'}$$

where T_b' can be calculated from the following relations:

$$T_b' = T_b - \Delta T_b$$

$$\Delta T_b = 1.3889 F(K_W - 12) \log_{10} \frac{P^{vap}}{760}$$

(7.22) $\quad F = 0 \quad (T_b < 367 \, \text{K})$ or when K_W is not available

$$F = -3.2985 + 0.009 T_b \quad (367 \, \text{K} \le T_b \le 478 \, \text{K})$$

$$F = -3.2985 + 0.009 T_b \quad (T_b > 478 \, \text{K})$$

where
$\quad P^{vap}$ = desired vapor pressure at temperature T, mm Hg ($= \text{bar} \times 750$)
$\qquad T$ = temperature at which P^{vap} is needed, in kelvin
$\qquad T_b'$ = normal boiling point corrected to $K_W = 12$, in kelvin
$\qquad T_b$ = normal boiling point, in kelvin
$\quad K_W$ = Watson (UOP) characterization factor [$=(1.8T_b)^{1/3}/$ SG]
$\qquad F$ = correction factor for the fractions with K_W different from 12
$\log_{10}$ = common logarithm (base 10)

It is recommended that when this method is applied to light hydrocarbons ($N_C < 5$), F in Eq. (7.22) must be zero and there is no need for value of K_W (i.e., $T_b' = T_b$). Calculation of P^{vap} from Eqs. (7.20)–(7.22) requires a trial-and-error procedure. The first initial value of P^{vap} can be obtained from Eqs. (7.20) and (7.21) by assuming $K_W = 12$ (or $T_b' = T_b$). If calculation of T is required at a certain pressure, reverse form of Eqs. (7.20) and (7.21) as given in Eqs. (3.29) and (3.30) should be used.

Tsonopoulos et al. [18, 19] stated that the original MB correlation is accurate for subatmospheric pressures. They modified the relation for calculation of ΔT_b in Eq. (7.22) for fractions with $K_W < 12$. Coal liquids have mainly K_W values of less than 12 and the modified MB correlation is suggested for vapor pressure of coal liquids. The relation for ΔT_b of coal

TABLE 7.4—*Prediction of vapor pressure of benzene at 400 K (260°F) from different methods in Example 7.3.*

	API [9] Fig. 7.5	Miller Eqs. (7.13) and (7.14)	Lee–Kesler Eq. (7.18)	Ambrose Eq. (7.19)	Riedel Eq. (7.24)	Eq. (7.25)	Clapeyron Eq. (7.15)	Korsten Eq. (7.16)	Maxwell Eqs. (7.20–7.22)	API Eq. (7.8)
P^{vap}, bar	3.45	3.74	3.48	3.44	3.50	3.53	3.43	3.11	3.44	3.53
%Error	...	8.4	0.9	−0.3	1.4	2.3	−0.6	−9.9	−0.3	2.3

liquids is [18, 19]:

$$T_b' = T_b - \Delta T_b$$

$$\Delta T_b = F_1 F_2 F_3$$

$$F_1 = \begin{cases} 0 & T_b \leq 366.5\,\text{K} \\ -1 + 0.009(T_b - 255.37) & T_b > 366.5\,\text{K} \end{cases}$$

$$F_2 = (K_W - 12) - 0.01304(K_W - 12)^2$$

$$F_3 = \begin{cases} 1.47422 \log_{10} P^{vap} & P^{vap} \leq 1\,\text{atm} \\ 1.47422 \log_{10} P^{vap} + 1.190833\,(\log_{10} P^{vap})^2 & P^{vap} > 1\,\text{atm} \end{cases}$$

(7.23)

where T_b' and T_b are in kelvin and P^{vap} is in atmospheres (=bar/1.01325). This equation was derived based on more than 900 data points for some model compounds in coal liquids including *n*-alkylbenzenes. Equation (7.23) may be used instead of Eq. (7.22) only for coal liquids and calculated T_b' should be used in Eq. (7.21).

Another relation that is proposed for estimation of vapor pressure of coal liquids is a modification of Riedel equation (Eq. 7.10) given in the following form by Tsonopoulos et al. [18, 19]:

$$\ln P_r^{vap} = A - \frac{B}{T_r} - C \ln T_r + D T_r^6$$

$$A = 5.671485 + 12.439604\omega$$

(7.24)

$$B = 5.809839 + 12.755971\omega$$

$$C = 0.867513 + 9.654169\omega$$

$$D = 0.1383536 + 0.316367\omega$$

This equation performs well for coal liquids if accurate input data on T_c, P_c, and ω are available. For coal fractions where these parameters cannot be determined accurately, modified MB (Eqs. 7.20–7.23) should be used. When evaluated with more than 200 data points for some 18 coal liquid fractions modified BR equations gives an average error of 4.6%, while the modified Riedel (Eq. 7.24) gives an error of 4.9% [18].

The simplest method for estimation of vapor pressure of petroleum fractions is given by Eq. (3.33) as

$$\log_{10} P^{vap} = 3.2041 \left(1 - 0.998 \times \frac{T_b - 41}{T - 41} \times \frac{1393 - T}{1393 - T_b}\right)$$

(7.25)

where T_b is the normal boiling point and T is the temperature at which vapor pressure P^{vap} is required. The corresponding units for T and P are kelvin and bar, respectively. Accuracy of this equation for vapor pressure of pure compounds is about 1%. Evaluation of this single parameter correlation is shown in Fig. 7.8. It is a useful relation for quick calculations or when only T_b is available as a sole parameter. This equation is highly accurate at temperatures near T_b.

Example 7.3—Estimate vapor pressure of benzene at 400 K from the following methods:

a. Miller (Eqs. (7.13) and (7.14))
b. Lee–Kesler (Eq. 7.18)
c. Ambrose (Eq. 7.19)
d. Modified Riedel (Eq. 7.24)
e. Equation (7.25)
f. Equations (6.101)–(6.103) or Eq. (7.15)
g. Korsten (Eq. 7.16)
h. Maxwell–Bonnell (Eqs. (7.20)–(7.22))
i. API method (Eq. 7.8)
j. Compare predicted values from different methods with the value from Fig. 7.5.

Solution—For benzene from Table 2.1 we have $T_b = 353.3$ K, SG = 0.8832, $T_c = 562.1$ K, $P_c = 48.95$ bar, and $\omega = 0.212$. $T = 400$ K, $T_r = 0.7116$, and $T_{br} = 0.6285$. The calculation methods are straightforward, and the results are summarized in Table 7.4. The highest error corresponds to the Korsten method. A preliminary evaluation with some other data also indicates that the simple Clapeyron equation (Eq. 7.15) is more accurate than the Korsten method (Eq. 7.16). The Antoine equation (Eq. 7.11) with coefficients given in Table 7.3 gives a value of 3.523 bar with accuracy nearly the same as Eq. (7.8). ◆

7.3.3.2 Graphical Methods for Vapor Pressure of Petroleum Products and Crude Oils

For petroleum fractions, especially gasolines and naphthas, laboratories usually report RVP as a characteristic related to quality of the fuel (see Table 4.3). As discussed in Section 3.6.1.1, the RVP is slightly less than true vapor pressure (TVP) at 100°F (37.8°C) and for this reason Eq. (7.25) or (3.33) was used to get an approximate value of RVP from a TVP correlation. However, once RVP is available from laboratory measurements, one may use this value as a basis for calculation of TVP at other temperatures. Two graphical methods for calculation of vapor pressure of petroleum finished products and crude oils from RVP are provided by the API-TDB [9]. These figures are presented in Figs. 7.10 and 7.11, for the finished products and crude oils, respectively. When using Fig. 7.10 the ASTM 10% slope is defined as $SL_{10} = (T_{15} - T_5)/10$, where T_5 and T_{15} are temperatures on the ASTM D 86 distillation curve at 5 and 15 vol% distilled both in degrees fahrenheit. In cases where ASTM temperatures at these points are not available, values of 3 (for motor gasoline), 2 (aviation gasoline), 3.5 (for light naphthas with RVP of 9–14 psi), and 2.5 (for naphthas with RVP of 2–8 psi) are recommended [9]. To use these figures, the first step is to locate a point on the RVP line and then a straight line is drawn between this point and the temperature of interest. The interception with the vertical line of TVP gives the reading. Values of TVP estimated from these figures are approximate especially at temperatures far from 100°F (37.8°C) but useful when only RVP is available from experimental measurements. Values of RVP for use in

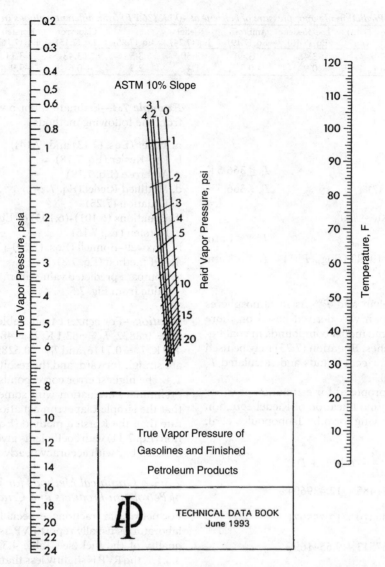

FIG. 7.10—True vapor pressure of petroleum products from RVP.
Unit conversion: °F = (°C) × 1.8 + 32; psia = bar × 14.504. Taken
with permission from Ref. [9].

Fig. 7.10 should be experimental rather than estimated from methods of Section 3.6.1.1. If no experimental data on RVP are available the TVP should be calculated directly from methods discussed in Sections 7.3.2 and 7.3.3.1.

For computer applications, analytical correlations have been developed from these two figures for calculation of vapor pressure of petroleum products and crude oils from RVP data [9]. For petroleum products, Fig. 7.10 has been presented by a complex correlation with 15 constants in terms of RVP and slope of ASTM D 86 curve at 10%. Similarly for crude oils the mathematical relation developed based on Fig. 7.11 is given as [9]

$$\ln P^{\mathrm{vap}} = A_1 + A_2 \ln(\mathrm{RVP}) + A_3(\mathrm{RVP}) + A_4 T$$

$$(7.26) \qquad + \frac{[B_1 + B_2 \ln(\mathrm{RVP}) + B_3(\mathrm{RVP})^4]}{T}$$

where P^{vap} and RVP are in psia, T is in °R. Ranges of application are $0F < T(°F) < 140F$ and 2 psi < RVP < 15 psi. The

coefficients are given as $A_1 = 7.78511307$, $A_2 = -1.08100387$, $A_3 = 0.05319502$, $A_4 = 0.00451316$, $B_1 = -5756.8562305$, $B_2 = 1104.41248797$, and $B_3 = -0.00068023$. There is no information on reliability of these methods. Figures 7.10 and 7.11 or Eq. (7.26) are particularly useful in obtaining values of vapor pressure of products and crude oils needed in estimation of hydrocarbon losses from storage tanks [20].

7.3.4 Vapor Pressure of Solids

Figure 5.2a shows the equilibrium curve between solid and vapor phases, which is known as a sublimation curve. In fact, at pressures below triple point pressure ($P < P_{\mathrm{tp}}$), a solid directly vaporizes without going through the liquid phase. This type of vaporization is called *sublimation* and the enthalpy change is called heat of sublimation (ΔH^{sub}). For ice, heat of sublimation is about 50.97 kJ/mol. Through phase equilibrium analysis similar to the analysis made for VLE of pure substances in Section 6.5 and beginning with Eq. (6.96) for

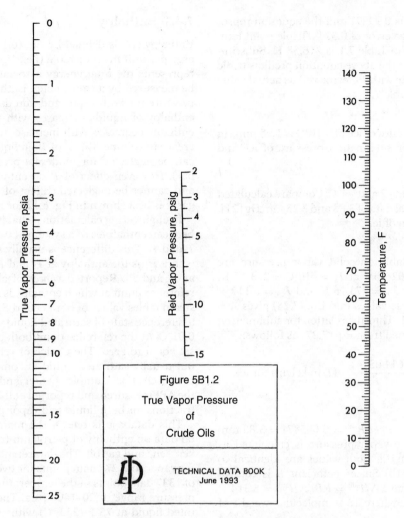

Figure 5B1.2

True Vapor Pressure

of

Crude Oils

TECHNICAL DATA BOOK
June 1993

FIG. 7.11—True vapor pressure of crude oils from RVP. Unit conversion:
°F = (°C) × 1.8 + 32; psia = bar × 14.504. Taken with permission from
Ref. [9].

vapor–solid equilibria (VSE) one can derive a relation similar to Eq. (6.101) for estimation of vapor pressure of solids:

$$\text{(7.27)} \qquad \ln P^{sub} = A - \frac{B}{T}$$

where P^{sub} is the vapor pressure of a pure solid also known as *sublimation pressure* and A and B are two constants specific for each compound. Values of P^{sub} are less than P_{tp} and one base point to obtain constant A is the triple point (T_{tp}, P_{tp}). Values of T_{tp} and P_{tp} for some selected compounds are given in Table 7.1. If a value of saturation pressure (P_1^{sub}) at a reference temperature of T_1 is known it can be used along with the triple point to obtain A and B in Eq. (7.27) as follows:

$$\text{(7.28)} \qquad B = \frac{\ln\left(\dfrac{P_{tp}}{P_1^{sub}}\right)}{\dfrac{1}{T_1} - \dfrac{1}{T_{tp}}}$$

$$A = \ln\left(P_1^{sub}\right) + \frac{B}{T_1}$$

where T_{tp} and T_1 are in kelvin. Parameter B is equivalent to $\Delta H^{sub}/R$. In deriving this equation, it is assumed that ΔH^{sub} is constant with temperature. This assumption can be justified

as the temperature variation along the sublimation curve is limited. In addition it is assumed that $\Delta V^{sub} = V^{vap} - V^s \cong RT/P^{sub}$. This assumption is reasonable as $V^s \ll V^v$ and P^{sub} is very small so that the vapor is considered as an ideal gas. In fact accuracy of Eq. (7.27) is more than Eq. (6.101) because the assumptions made in derivation of this equation are more realistic. The relations for sublimation pressure of naphthalene is given as [21]

$$\ln P^{sub}(\text{bar}) = 8.722 - \frac{3783}{T}$$

$$\text{(7.29)} \qquad (T \text{ in kelvin}) \text{ for solid naphthalene}$$

Vapor pressures of solid CO_2 are given at several temperatures as: 9.81 torr (at $-120°C$), 34.63 (-110), 104.81 (-100), 279.5 (-90) where the pressures are in mm Hg (torr) (1 bar = 750.06 mm Hg) and the numbers in the parentheses are the corresponding temperatures in °C as given by Levine [22]. Linear regression of $\ln P^{sub}$ versus $1/T$ gives constants A and B in Eq. (7.27) as

$$\ln P^{sub}(\text{bar}) = 16.117 - \frac{3131.97}{T} \quad (T \text{ in kelvin}) \text{ for solid } CO_2$$

$$\text{(7.30)}$$

The R^2 for this equation is 0.99991 and the equation reproduces data with an average error of 0.37%. Triple point temperature of CO_2 as given in Table 7.1 is 216.58 K. Substitution of this temperature in the above equation predicts triple point pressure of 5.238 bar with 1% error versus actual value of 5.189 bar as given in Table 7.1.

Example 7.4—Vapor pressure of ice at $-10°C$ is 1.95 mm Hg [21]. Derive a relation for sublimation pressure of ice and estimate the following:

a. Sublimation pressures at -2 and $-4°C$. Compare calculated values with experimental data of 3.88 and 3.28 mm Hg [21].
b. The heat of sublimation of ice.

Solution—(a) Data available on solid vapor pressure are $P_1^{sub} = 1.95$ mm Hg = 0.0026 bar at $T_1 = -10°C = 263.15$ K. From Table 7.1 for water $T_{tp} = 273.16$ K and $P_{tp} = 6.117 \times 10^{-3}$ bar. Substituting these values into Eq. (7.28) gives $A = 17.397$ and $B = 6144.3741$. Thus the relation for sublimation pressure of ice is determined from Eq. (7.27) as follows:

$$\ln P^{sub}(\text{bar}) = 17.397 - \frac{6144.3741}{T} \quad (T \text{ in kelvin}) \text{ for ice}$$

(7.31)

At $T_1 = -2°C = 271.15$ K we get $P^{sub} = 0.005871 = 3.88$ mm Hg. Similarly at $-4°C$ the vapor pressure is calculated as 0.004375 bar or 3.28 mm Hg. Both values are identical to the experimental values. (b) Since coefficient B is equivalent to $\Delta H^{sub}/R$ thus we have $\Delta H^{sub} = RB$. $\Delta H^{sub} = 8.314 \times 6144.3741/18 = 2.84$ kJ/g, where 18 is molecular weight of water. ♦

7.4 THERMAL PROPERTIES

In this section, methods of estimation of thermal properties such as enthalpy, heat capacity, heat of vaporization, and heating values for petroleum fractions are presented. These properties are required in calculations related to energy balances around various process units as well as design and operation of heat transfer related equipment. The fundamental equations for calculation of enthalpy and heat capacity were discussed in Chapter 6. In this section application of those methods and some empirical correlations developed for prediction of such properties are presented. Heat capacity, heats of vaporization, and combustion can be evaluated from enthalpy data, but independent methods are presented for convenience and better accuracy.

7.4.1 Enthalpy

Enthalpy (H) is defined by Eq. (6.1) and has the unit of energy per unit mass or mole (i.e., J/g or J/mol). This property represents the total energy associated with a fluid and can be measured by a calorimeter. Enthalpy increases with temperature for both vapor and liquids. According to Eq. (6.1), enthalpy of liquids increases with pressure, but for vapors enthalpy decreases with increase in pressure because of decrease in volume. Effect of P on liquid enthalpy is small and can be neglected for moderate pressure changes ($\Delta P \cong 10$ bar). However, effect of P on enthalpy of vapors is greater and cannot be neglected. Effect of T and P on enthalpy of gases is best shown in Fig. 6.12 for methane.

In engineering calculations what is needed is the difference between enthalpies of a system at two different conditions of T and P. This difference is usually shown by $\Delta H = H_2 - H_1$ where H_1 is the enthalpy at T_1 and P_1 and H_2 is the enthalpy at T_2 and P_2. Reported values of absolute enthalpy have a reference point at which enthalpy is zero. For example, in the steam tables values of both H and S are given with respect to a reference state of saturated liquid water at its triple point of $0.01°C$. At the reference point both enthalpy and entropy are set equal to zero. The choice of reference state is arbitrary but usually saturated liquid at some reference temperature is chosen. For example, Lenoir and Hipkin in a project for the API measured and reported enthalpies of eight petroleum fractions for both liquid and vapor phases [23].

This database is one of the main sources of experimental data on enthalpy of petroleum fractions from naphtha to kerosene and gas oil. The dataset includes 729 for liquid, 331 for vapor, and 277 data points for two-phase region with total of 1337 data points in the temperature range of $75–600°F$ and pressure range of 20–1400 psia. The reference state is saturated liquid at $75°F$ ($23.9°C$) with corresponding saturation pressure of about 20–40 psia. Some values of enthalpy from this database are given in Table 7.5. For all three fractions the reference state is saturated liquid at $75°F$ and 20 psia. One should be careful in reading absolute values of enthalpy, entropy, or internal energy since reported values depend on the choice of reference state. However, no matter what is choice of reference state calculation of ΔH is independent of reference state.

Heavy petroleum fractions possess lower enthalpy (per unit mass) than do light fractions at the same conditions of T and P. For example, for fractions with $K_W = 10$ and at 530 K, when the API gravity increases from 0 to 30, liquid enthalpy increases from 628 to 721 kJ/kg. Under the same conditions, for the vapor phase enthalpy increases from 884 to 988 kJ/kg as shown by Kesler and Lee [24]. Based on data measured by Lenoir and Hipkin [23], variation of enthalpy of two petroleum fractions (jet fuel and gas oil) versus temperature and two different pressures is shown in Fig. 7.12. Gas oil is

TABLE 7.5—*Enthalpies of some petroleum fractions from Lenoir–Hipkin dataset [23].*

Petroleum fraction	K_W	API	20 psi and 300°F, liquid	1400 psi and 500°F, liquid	600°F and (P, psi), vapor	20 psi and (T, °F), vapor
Jet fuel	11.48	44.4	117.6	245.4	401.9 (100)	311 (440)
Kerosene	11.80	43.5	120	250.9	404.1 (80)	358.6 (520)
Fuel oil	11.68	33.0	115.8	243.1	346.0 (25)	378.4 (600)

Reference enthalpy ($H = 0$): saturated liquid at $75°F$ and 20 psia for all samples. H values are in Btu/lb.

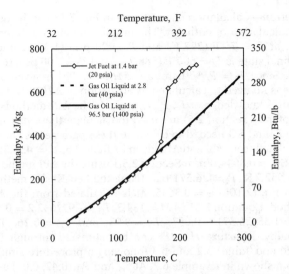

FIG. 7.12—Enthalpy of two petroleum fractions. Reference state: $H = 0$ for saturated liquid at 23.9°C (75°F) and 1.38 bar for jet fuel, and 23.9°C (75°F) and 2.76 bar (40 psia) for gas oil. Specifications: Jet fuel, $M = 144$, $T_b = 160.5$°C, SG = 0.804; gas oil, $M = 214$, $T_b = 279.4$°C, SG = 0.848. Gas oil is in liquid state for entire temperature range. Jet fuel has bubble point temperature of 166.8°C and dew point temperature of 183.1°C at 1.4 bar (20 psia). Data source Ref. [23].

heavier than jet fuel and its enthalpy as liquid is just slightly less than enthalpy of liquid jet fuel. However, there is a sharp increase in the enthalpy of jet fuel during vaporization. Pressure has little effect on liquid enthalpy of gas oil.

As it was discussed in Chapter 6, to calculate H one should first calculate enthalpy departure or the residual enthalpy from ideal gas state shown by $H^R = H - H^{ig}$. General methods for calculation of H^R were presented in Section 6.2. H^R is related to PVT relation through Eqs. (6.33) or (6.38). For gases that follow truncated virial equation of state ($T_r > 0.686 + 0.439P_r$ or $V_r > 2.0$), Eq. (6.63) can be used to calculate H^R. Calculation of H^R from cubic equations of state was shown in Table 6.1. However, the most accurate method of calculation of H^R is through generalized correlation of Lee–Kesler given by Eq. (6.56) in the form of dimensionless group H^R/RT_c. Then H may be calculated from the following relation:

$$(7.32) \qquad H = \frac{RT_c}{M}\left[\frac{H - H^{ig}}{RT_c}\right] + H^{ig}$$

where both H and H^{ig} are in kJ/kg, T_c in kelvin, $R = 8.314$ J/mol·K, and M is the molecular weight in g/mol. The ideal gas enthalpy H^{ig} is a function of only temperature and must be calculated at the same temperature at which H is to be calculated. For pure hydrocarbons H^{ig} may be calculated through Eq. (6.68). In this equation the constant A_H depends on the choice of reference state and in calculation of ΔH it will be eliminated. If the reference state is known, ΔH can be determined from $H = 0$, at the reference state of T and P. As it is seen shortly, it is the ΔH^{ig} that must be calculated in calculation of ΔH. This term can be calculated from the following

relation based on ideal gas heat capacity, C_P^{ig}.

$$(7.33) \qquad \Delta H^{ig} = \int_{T_1}^{T_2} C_P^{ig}(T)\,dT$$

where T_1 and T_2 are the same temperature points that ΔH^{ig} must be calculated. For pure compounds C_P^{ig} can be calculated from Eq. (6.66) and combining with the above equation ΔH^{ig} can be calculated. For petroleum fractions, Eq. (6.72) is recommended for calculation of C_P^{ig} and when it is combined with Eq. (7.33) the following equation is obtained for calculation of ΔH^{ig} from T_1 to T_2:

$$\Delta H^{ig} = M\left\{\left[A_0(T_2 - T_1) + \frac{A_1}{2}\left(T_2^2 - T_1^2\right) + \frac{A_2}{3}\left(T_2^3 - T_1^3\right)\right] \right.$$
$$\left. - C\left[B_0(T_2 - T_1) + \frac{B_1}{2}\left(T_2^2 - T_1^2\right) + \frac{B_2}{3}\left(T_2^3 - T_1^3\right)\right]\right\}$$

(7.34)

where T_1 and T_2 are in kelvin, ΔH^{ig} is in J/mol, M is the molecular weight, coefficients A, B, and C are given in Eq. (6.72) in terms of Watson K_W and ω. This equation should not be applied to light hydrocarbons ($N_C < 5$) as stated in the application of Eq. (6.72). H^{ig} or C_P^{ig} of a petroleum fraction may also be calculated from the pseudocompound approach discussed in Chapter 3 (Eq. 3.39). In this way H^{ig} or C_P^{ig} must be calculated from Eqs. (6.68) or (6.66) for three pseudocompounds from groups of n-alkane, n-alkylcyclopentane, and n-alkylbenzene having boiling points the same as that of the fraction. Then H^{ig} is calculated from the following equation:

$$(7.35) \qquad H^{ig} = x_P H_P^{ig} + x_N H_N^{ig} + x_A H_A^{ig}$$

where x_P, x_N, and x_A refer to the fractions of paraffins (P), naphthenes (N), and aromatics (A) in the mixture, which is known from PNA composition or may be determined from methods given in Section 3.5. C_P^{ig} of a petroleum fraction may be calculated from the same equation but Eq. (6.66) is used to calculate C_P^{ig} of the P, N, and A compounds having boiling points the same as that of the fraction.

A summary of the calculation procedure for ΔH from an initial state at T_1 and P_1 (state 1) to a final state at T_2 and P_2 (state 6), for a general case that the initial state is a subcooled (compressed) liquid and the final state is a superheated vapor, is shown in Fig. 7.13. The technique involves step-by-step calculation of ΔH in a way that in each step the calculation procedure is available. The subcooled liquid is transferred to a saturated liquid at T_1 and P_1^{sat} where P_1^{sat} is the vapor pressure of liquid at temperature T_1. For this step (1 to 2), ΔH_1 represents the change in enthalpy of liquid phase at constant temperature of T_1 from pressure P_1 to pressure P_1^{sat}. Methods of estimation of P_1^{sat} are discussed in Section 7.3. In most cases, the difference between P_1 and P_1^{sat} is not significant and the effect of pressure on liquid enthalpy can be neglected without serious error. This means that $\Delta H_1 \cong 0$. However, for cases that this difference is large it may be calculated through a cubic EOS or generalized correlation of Lee–Kesler as discussed in Chapter 6. However, a more convenient approach is to calculate T_1^{sat} at pressure P_1, where T_1^{sat} is the saturation temperature corresponding to pressure P_1 and it may be calculated from vapor pressure correlations presented in Section

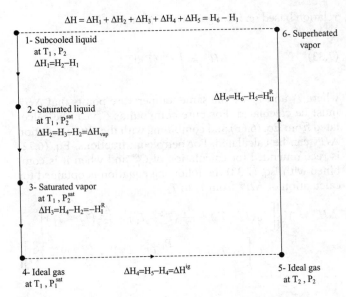

FIG. 7.13—Diagram of enthalpy calculation.

7.3. Then state 2 will be saturated liquid at T_1^{sat} and P_1 and ΔH_1 represents constant pressure enthalpy change of a liquid from temperature T_1 to T_1^{sat} and can be calculated from the following relation:

$$(7.36) \qquad \Delta H_1 = \int_{T_1}^{T_1^{sat}} C_P^L(T)dT$$

where C_P^L is the heat capacity of liquid and it may be calculated from methods presented in Section 7.4.2.

Since in most cases the initial state is low-pressure liquid, the approach presented in Fig. 7.13 to show the calculation procedures is used. Step 2 is vaporization of liquid at constant T and P. ΔH_2 represents heat of vaporization at T_1 and it can be calculated from the methods discussed in Section 7.4.3. Step 3 is transfer of saturated vapor to ideal gas vapor at constant T_1 and P_1^{sat} (or T_1^{sat} and P_1). $\Delta H_3 = -H_I^R$ in which H_I^R is the residual enthalpy at T_1 and P_1^{sat} and its calculation was discussed earlier. Step 4 is converting ideal gas at T_1 and P_1^{sat} to ideal gas at T_2 and P_2. Thus, $\Delta H_4 = \Delta H^{ig}$, where ΔH^{ig} can be calculated from Eq. (7.33). The final step is to convert ideal gas at T_2 and P_2 to a real gas at the same T_2 and P_2 and $\Delta H_5 = H_{II}^R$, where H_{II}^R is the residual enthalpy at T_2 and P_2. Once ΔH for each step is calculated, the overall ΔH can be calculated from sum of these ΔHs as shown in Fig. 7.13. Similar diagrams can be constructed for other cases. For example, if the initial state is a gas at atmospheric pressure, one may assume the initial state as an ideal gas and only steps 4 and 5 in Fig. 7.13 are necessary for calculation of ΔH. If the initial state is the chosen reference state, then calculated overall ΔH represents absolute enthalpy at T_2 and P_2. This is demonstrated in the following example.

Example 7.5—Calculate enthalpy of jet fuel of Table 7.5 at 600°F and 100 psia. Compare your result with the experimental value of 401.9 Btu/lb. The reference state is saturated liquid at 75°F and 20 psia.

Solution—Calculation chart shown in Fig. 7.13 can be used for calculation of enthalpy. The initial state is the reference state at $T_1 = 75°F$ (297 K) and $P_1 = 20$ psia (1.38 bar) and the final state is $T_2 = 600°F$ (588.7 K) and $P_2 = 100$ psia (6.89 bar). Since $P_1 = P_1^{sat}$, therefore, $\Delta H_1 = 0$. P_1^{sat} is given and there is no need to calculate it. Calculation of ΔH^{vap} and H^R requires knowledge of T_c, P_c, ω, and M. The API methods of Chapter 2 (Section 2.5) are used to calculate these parameters. T_b and SG needed to calculate these parameters can be calculated from K_W and API given in Table 7.5. $T_b = 437.55$ K and SG = 0.8044. From Section 2.5.1 using the API methods, $T_c = 632.2$ K, $P_c = 26.571$ bar. Using the Lee–Kesler method from Eq. (2.105) $\omega = 0.3645$. M is calculated from the API method, Equation 2.51 as $M = 134.3$. $T_{r1} = 297/632.2 = 0.47$, $P_{r1} = 1.38/26.571 = 0.052$, $T_{r2} = 0.93$, and $P_{r2} = 0.26$. The enthalpy departure $H - H^{ig}$ can be estimated through Eq. (6.56) and Tables 6.2 and 6.3 following a procedure similar to that shown in Example 6.2. At T_{r1} and P_{r1} (0.47, 0.05) as it is clear from Table 6.2, the system is in liquid region while the residual enthalpy for saturated vapor is needed. The reason for this difference is that the system is a petroleum mixture with estimated T_c and P_c different from true critical properties as needed for phase determination. For this reason, one should be careful to use extrapolated values for calculation of $[(H - H^{ig})/RT_c]^{(0)}$ and $[(H - H^{ig})/RT_c]^{(1)}$ at T_{r1} and P_{r1}. Therefore, with extrapolation of values at $T_r = 0.65$ and $T_r = 0.7$ to $T_r = 0.47$ for $P_r = 0.05$ we get $[(H - H^{ig})/RT_c]_I = -0.179 + 0.3645 \times (-0.83) = -0.4815$. At T_{r2} and P_{r2} (0.93, 0.26) the system is as superheated vapor: $[(H - H^{ig})/RT_c]_{II} = -0.3357 + 0.3645 \times (-0.3691) = -0.47$ or $(H - H^{ig})_I = -2530.8$ J/mol and $(H - H^{ig})_{II} = -2470.4$ J/mol. Thus, from Fig. 7.13 $\Delta H_3 = -(H - H^{ig})_I = +2530.8$ J/mol and $\Delta H_5 = +(H - H^{ig})_{II} = -2470.4$ J/mol. ΔH^{ig} can be calculated from Eq. (7.34) with coefficients given in Eq. (6.72). The input parameters are $K_W = 11.48$, $\omega = 0.3645$, $M = 134.3$, $T_1 = 297$, and $T_2 = 588.7$ K. The calculation result is $\Delta H_4 = \Delta H^{ig} = 78412$ J/mol. ΔH^{vap} can be calculated from methods of Section 7.4.3. (Eqs. 7.54 and 7.57), which gives $\Delta H_2 = \Delta H^{vap} = 46612$ J/mol. Thus, $\Delta H = \Delta H_1 + \Delta H_2 + \Delta H_3 + \Delta H_4 + \Delta H_5 = 0 + 46612 + 2530.8 + 78412 - 2470.4 = 125084.4$ J/mol = 125084.4/134.3 = 930.7 J/g = 930.7 kJ/kg. From Section 1.7.17 we get 1 J/g = 0.42993 Btu/lb. Therefore, $\Delta H = 930.7 \times 0.42993 = 400.1$ Btu/lb. Since the initial state is the chosen reference state, at the final T and P (600°F and 100 psia) the calculated absolute enthalpy is 400.1 Btu/lb, which differs by 1.8 Btu/lb or 0.4% from the experimental value of 401.9 Btu/lb. This is a good prediction of enthalpy considering the fact that minimum information on boiling point and specific gravity has been used for estimation of various basic parameters. ◆

In addition to the analytical methods for calculation of enthalpy of petroleum fractions, there are some graphical methods for quick estimation of this property. For example, Kesler and Lee [24] developed graphical correlations for calculation of enthalpy of vapor and liquid petroleum fractions. They proposed a series of graphs where K_W and API gravity were used as the input parameters for calculation of H at a given T and P. Further discussion on heat capacity and enthalpy is provided in the next section. Once H and V are calculated, the internal energy (U) can be calculated from Eq. (6.1).

7.4.2 Heat Capacity

Heat capacity is one of the most important thermal properties and is defined at both constant pressure (C_P) and constant volume (C_V) by Eqs. (6.17) and (6.18). It can be measured using a calorimeter. For constant pressure processes, C_P and in constant volume processes, C_V is needed. C_P can be obtained from enthalpy using Eq. (6.20).

Experimental data on liquid heat capacity of some pure hydrocarbons are given in Table 7.6 as reported by Poling et al. [12]. For defined mixtures where specific heat capacity for each compound in the mixture is known, the mixing rule given by Eq. (7.2) may be used to calculate mixture heat capacity of liquids $C_{P_i}^L$. Heat capacities of gases are lower than liquid heat capacities under the same conditions of T and P. For example, for propane at low pressures (ideal gas state) the value of C_P^{ig} is 1.677 J/g·K at 298 K and 3.52 J/g·K at 800 K. Values of C_P^{ig} of n-heptane are 1.658 J/g·K at 298 K and 3.403 J/g·K at 800 K. However, for liquid state and at 300 K, C_P^L of C_3 is 3.04 and that of n-C_5 is 2.71 J/g·K as reported by Reid et al. [12]. While molar heat capacity increases with M, specific heat capacity decreases with increase in M. Heat capacity increases with temperature.

The general approach to calculate C_P is to estimate heat capacity departure from ideal gas $[C_P - C_P^{ig}]$ and combine it with ideal gas heat capacity (C_P^{ig}). A similar approach can be used to calculate C_V. The relation for calculation of C_P^{ig} of petroleum fractions was given by Eq. (6.72), which requires K_W and ω as input parameters. C_V^{ig} can be calculated from C_P^{ig} through Eq. (6.23). Both C_P^{ig} and C_V^{ig} are functions of only temperature. For petroleum fractions, C_P^{ig} can also be calculated from the pseudocompound method of Chapter 3 (Eq. 3.39) by using Eq. (6.66) for pure hydrocarbons from different families similar to calculation of ideal gas enthalpy (Eq. 7.35). The most accurate method for calculation of $[C_P - C_P^{ig}]$ is through generalized correlation of Lee–Kesler (Eq. 6.57). Relations for calculation of $[C_P - C_P^{ig}]$ and $[C_V - C_V^{ig}]$ from cubic equations of state are given in Table 6.1. For gases at moderate pressures the departure functions for heat capacity can be estimated through virial equation of state (Eqs. 6.64 and 6.65). Once heat capacity departure and ideal gas properties are determined, C_P is calculated from the following relation:

$$(7.37) \qquad C_P = [C_P - C_P^{ig}] + C_P^{ig}$$

Relations given in Chapter 6 for the calculation of $[C_P - C_P^{ig}]$ and C_P^{ig} are in molar units. If specific unit of J/g·°C for heat capacity is needed, calculated values from Eq. (7.37) should be divided by molecular weight of the substance. Generalized correlation of Lee–Kesler normally provide reliable values of C_P for gases, but for liquids more specific correlations especially at low pressures have been proposed in the literature. Estimation of C_P and C_V from equations of state was demonstrated in Example 6.2.

For solids the effect of pressure on heat capacity is neglected and it varies only with temperature: $C_P^S \cong C_V^S = f(T)$. At moderate and low pressures the effect of pressure on liquids may also be neglected as $C_P^L \cong C_V^L = f(T)$. However, this assumption is not valid for liquids at high pressures. Some specific correlations are given in the literature for calculation of heat capacity of hydrocarbon liquids and solids at atmospheric pressures. At low pressures a generalized expression in a polynomial form of up to fourth orders is used to correlate C_P with temperature:

$$(7.38) \qquad \begin{aligned} C_P^L/R &= C_V^L/R = A_1 + A_2T + A_3T^2 + A_4T^3 + A_5T^4 \\ C_P^S/R &= C_V^S/R = B_1 + B_2T + B_3T^2 + B_4T^3 + B_5T^4 \end{aligned}$$

where T is in kelvin. Coefficients A_1–A_5 and B_1–B_5 for a number of compounds are given in Table 7.7 as given by DIPPR [10]. Some of the coefficients are zero for some compounds and for most solids the polynomial up to T^3 is needed. In fact Debye's statistical–mechanical theory of solids and experimental data show that specific heats of nonmetallic solids at very low temperatures obey the following [22]:

$$(7.39) \qquad C_P^S = aT^3$$

where T is the absolute temperature in kelvin. In this relation there is only one coefficient that can be determined from one data point on solid heat capacity. Values of heat capacity of solids at melting point given in Table 7.1 may be used as the reference point to find coefficient a in Eq. (7.39). Equation (7.39) can be used for a very narrow temperature range near the point where coefficient a is determined.

Cubic equations of states or the generalized correlation of Lee–Kesler for calculation of the residual heat capacity of liquids $[C_P^L - C_P^{ig}]$ do not provide very accurate values especially at low pressures. For this reason, attempts have been made to develop separate correlations for liquid heat capacity. Based on principle of corresponding states and using pure compounds' liquid heat capacity data, Bondi modified previous correlations into the following form [12]:

$$\begin{aligned} \frac{C_P^L - C_P^{ig}}{R} = {} & 1.586 + \frac{0.49}{1 - T_r} \\ (7.40) \qquad & + \omega \left[4.2775 + \frac{6.3(1 - T_r)^{1/3}}{T_r} + \frac{0.4355}{1 - T_r} \right] \end{aligned}$$

TABLE 7.6—*Some experimental values of liquid heat capacity of hydrocarbons, C_P^L [12].*

Compound	T, K	C_P^L, J/g·K	Compound	T, K	C_P^L, J/g·K	Compound	T, K	C_P^L, J/g·K
Methane	100	3.372	n-Pentane	250	2.129	n-Decane	460	2.905
Methane	180	6.769	n-Pentane	350	2.583	Cyclohexane	280	1.774
Propane	100	1.932	n-Heptane	190	2.014	Cyclohexane	400	2.410
Propane	200	2.120	n-Heptane	300	2.251	Cyclohexane	500	3.220
Propane	300	2.767	n-Heptane	400	2.703	Benzene	290	1.719
i-Butane	300	2.467	n-Heptane	480	3.236	Benzene	400	2.069
n-Pentane	150	1.963	n-Decane	250	2.091	Benzene	490	2.618

TABLE 7.7—*Coefficients of Eq. (7.38) for liquid (A_i, s) and solid (B_i, s) heat capacity for some selected compounds [10].*

$$C_P^L/R = C_V^L/R = A_1 + A_2T + A_3T^2 + A_4T^3 + A_5T^4$$

(7.38) $\quad C_P^S/R = C_V^S/R = B_1 + B_2T + B_3T^2 + B_4T^3 + B_5T^4$

Compound.	M	A_1	A_2	A_3	A_4	A_5	T_{min}, K	T_{max}, K
Liquid heat capacity, C_P^L								
n-Pentane	72.2	19.134	-3.254×10^{-2}	1.197×10^{-4}	0	0	143	390
n-Hexane	86.2	20.702	-2.210×10^{-2}	1.067×10^{-4}	0	0	178	460
n-Decane	142.3	33.512	-2.380×10^{-2}	1.291×10^{-4}	0	0	243	460
n-Pentadecane	212.4	41.726	2.641×10^{-2}	7.894×10^{-5}	0	0	283	544
n-Eicosane	282.5	42.425	9.710×10^{-2}	2.552×10^{-5}	0	0	309	617
n-Hexatriacontane (C_{36})	507.0	84.311	1.771×10^{-1}	0	0	0	353	770
Cyclohexane	84.2	−26.534	3.751×10^{-1}	-1.13×10^{-3}	1.285×10^{-6}	0	280	400
Methylcyclohexane	98.2	15.797	-7.590×10^{-3}	9.773×10^{-5}	0	0	146	320
Benzene	78.1	19.598	-4.149×10^{-2}	1.029×10^{-4}	0	0	279	500
Toluene	92.1	16.856	-1.832×10^{-2}	8.359×10^{-5}	0	0	178	500
Naphthalene	128.2	3.584	6.345×10^{-2}	0	0	0	353	491
Anthracene	178.2	9.203	7.325×10^{-2}	-5.93×10^{-6}	0	0	489	655
Carbon dioxide	44.0	−998.833	1.255×10	-5.21×10^{-2}	7.223×10^{-5}	0	220	290
Water	18.0	33.242	-2.514×10^{-1}	9.77×10^{-4}	-1.698×10^{-6}	1.127×10^{-9}	273	533

	B_1	B_2	B_3	B_4	B_5	T_{min}, K	T_{max}, K
Solid heat capacity, C_P^S							
n-Pentane	−1.209	0.1215	5.136×10^{-4}	-1.22×10^{-5}	5.08×10^{-8}	12	134
n-Hexane	−2.330	0.1992	-1.01×10^{-3}	2.43×10^{-6}	0	20	178
n-Decane	−4.198	0.3041	-1.52×10^{-3}	3.43×10^{-6}	0	20	240
n-Pentadecane	−311.823	1.3822	0	0	0	271	283
n-Eicosane	−0.650	0.3877	-1.57×10^{-3}	3.65×10^{-6}	0	93	268
n-Hexatriacontane (C_{36})	−200.000	1.0000	0	0	0	300	325
Cyclohexane	15.763	−0.0469	1.747×10^{-4}	0	0	191	271
Methylcyclohexane	−1.471	0.1597	-9.55×10^{-4}	3.06×10^{-6}	0	12	146
Benzene	0.890	0.0752	-3.23×10^{-4}	8.80×10^{-7}	0	40	279
Toluene	−0.433	0.1557	-1.05×10^{-3}	2.97×10^{-6}	0	40	274
Naphthalene	0.341	0.0949	-3.79×10^{-4}	1.34×10^{-6}	-1.34×10^{-9}	30	353
Anthracene	2.436	0.0531	1.04×10^{-4}	-8.82×10^{-8}	3.69×10^{-12}	40	489
Carbon dioxide	−2.199	0.1636	-1.46×10^{-3}	6.20×10^{-6}	-9.26×10^{-9}	25	216
Water	-3.157×10^{-2}	0.0169	0	0	0	3	273

where C_P^{ig} is the ideal gas molar heat capacity. Liquid heat capacity increases with temperature. This equation can also be applied to nonhydrocarbons as well. This equation is recommended for $T_r \leq 0.8$ and an average error of about 2.5% was obtained for estimation of C_P^L of some 200 compounds at 25°C [12]. For $0.8 < T_r < 0.99$ values obtained from Eq. (7.40) may be corrected if heat capacity of saturated liquid is required:

(7.41) $\quad \dfrac{C_P^L - C_{sat}^L}{R} = \exp(2.1 T_r - 17.9) + \exp(8.655 T_r - 8.385)$

where C_P^L should be calculated from Eq. (7.40). When $T_r \leq 0.8$, it can be assumed that $C_P^L \cong C_{sat}^L$ and the correction term may be neglected. C_{sat}^L represents the energy required while maintaining the liquid in a saturated state. Most often C_{sat}^L is measured experimentally while most predictive methods estimate C_P^L [12].

For petroleum fractions the pseudocomponent method similar to Eq. (7.35) can be used with M or T_b of the fraction as a characteristic parameter. However, there are some generalized correlations developed particularly for estimation of heat capacity of liquid petroleum fractions. Kesler and Lee [24] developed the following correlation for C_P^L of petroleum fractions at low pressures:

$$C_P^L = a(b + cT)$$

For liquid petroleum fractions in the temperature range:

$$145 \leq T \leq 0.8 T_c \text{ (T and T_c both in kelvin)}$$

$$a = 1.4651 + 0.2302 K_W$$

$$b = 0.306469 - 0.16734 \, SG$$

(7.42) $\quad c = 0.001467 - 0.000551 \, SG$

where K_W is the Watson characterization factor defined in Eq. (2.13). Preliminary calculations show that this equation overpredicts values of C_P^L of pure hydrocarbons and accuracy of this equation is about 5%. Equation (7.42) is recommended in the ASTM D 2890 test method for calculation of heat capacity of petroleum distillate fuels [25]. There are other forms similar to Eq. (7.42) correlating C_P^L of petroleum fractions to SG, K_W, and T using higher terms and orders for temperature but generally give similar results as that of Eq. (7.42). Simpler forms of relations for estimation of C_P^L of liquid petroleum fractions in terms of SG and T are also available in the literature [26]. But their ability to predict C_P^L is very poor and in some cases lack information on the units or involve with some errors in the coefficients reported. The corresponding states correlation of Eq. (7.40) may also be used for calculation of heat capacity of liquid petroleum fractions using T_c, ω, and C_P^{ig} of the fraction. The API method [9] for calculation of C_P^L of liquid petroleum fractions is given in the following form for $T_r \leq 0.85$:

$$C_P^L = A_1 + A_2T + A_3T^2$$

$$A_1 = -4.90383 + (0.099319 + 0.104281 SG)K_W$$

$$+ \left(\frac{4.81407 - 0.194833\,K}{SG} \right)$$

$$A_2 = (7.53624 + 6.214610 K_W) \times \left(1.12172 - \frac{0.27634}{SG} \right) \times 10^{-4}$$

$$A_3 = -(1.35652 + 1.11863 K_W) \times \left(2.9027 - \frac{0.70958}{SG} \right) \times 10^{-7}$$

(7.43)

where C_P^L is in kJ/kg·K and T is in kelvin. This equation was developed by Lee and Kesler of Mobile Oil Corporation in 1975. From this relation, the following equation for estimation of enthalpy of liquid petroleum fractions can be obtained.

$$
\begin{aligned}
H^L &= \int_{T_{ref}}^{T} C_P^L dT + H_{ref}^L = A_1(T - T_{ref}) + \frac{A_2}{2}(T - T_{ref})^2 \\
&\quad + \frac{A_3}{3}(T - T_{ref})^3 + H_{ref}^L
\end{aligned}
$$

(7.44)

where H_{ref}^L is usually zero at the reference temperature of T_{ref}. Equation (7.43) is not recommended for pure hydrocarbons. The following modified form of Watson and Nelson correlation is recommended by Tsonopoulos et al. [18] for calculation of liquid heat capacity of coal liquids and aromatics:

$$
\begin{aligned}
C_P^L &= (0.28299 + 0.23605 K_W) \\
&\quad \times \left[0.645 - 0.05959\,SG + (2.32056 - 0.94752\,SG) \right. \\
&\quad \left. \times \left(\frac{T}{1000} - 0.25537 \right) \right]
\end{aligned}
$$

(7.45)

where C_P^L is in kJ/kg·K and T is in kelvin. This equation predicts heat capacity of coal liquids with an average error of about 3.7% for about 400 data points [18]. The following example shows various methods of calculation of heat capacity of liquids.

Example 7.6—Calculate C_P^L of 1,4-pentadiene at 20°C using the following methods and compare with the value of 1.994 J/g·°C reported by Reid et al. [12].

a. SRK EOS
b. DIPPR correlation [10]
c. Lee–Kesler generalized corresponding states correlation (Eq. 6.57)
d. Bondi's correlation (Eq. 7.40)
e. Kesler–Lee correlation (Eq. 7.42)—ASTM D 2890 method
f. Tsonopoulos et al. correlation (Eq. 7.45)

Solution—Basic properties of 1,4-pentadiene are not given in Table 2.1. Its properties obtained from other sources such as DIPPR [10] are as follows: $M = 68.1185$, $T_b = 25.96°C$, SG = 0.6633, $T_c = 205.85°C$, $P_c = 37.4$ bar, $Z_c = 0.285$, $\omega = 0.08365$,

and $C_P^{ig} = 1.419$ J/g·°C (at 20°C). From T_b and SG, $K_W = 12.264$.

(a) To use SRK EOS use equations given in Table 6.1 and follow similar calculations as in Example 6.2: $A = 0.039685$, $B = 0.003835$, $Z^L = 0.00492$, $V^L = 118.3$ cm³/mol, the volume translation is $c = 13.5$ cm³/mol, V^L(corrected) = 104.8 cm³/mol, and Z^L(correc.) = 0.00436. From Table 6.1, $P_1 = 6.9868$, $P_2 = -103.976$, $P_3 = 0.5445$, and $[C_P - C_P^{ig}] = 21.41$ J/mol·K = 21.41/68.12 = 0.3144 J/g·K. $C_P^L = 0.3144 + 1.419 = 1.733$ J/g·K (error of −13%). (b) DIPPR [10] gives the value of $C_P^L = 2.138$ J/g·K (error of +7%). (c) From the Lee–Kesler correlation of Eq. (6.57), $T_r = 0.612$ and $P_r = 0.0271$. From Tables 6.4 and 6.5, using interpolation (for P_r) and extrapolation (for T_r, extrapolation from the liquid region) we get $[(C_P - C_P^{ig})/R]^{(0)} = 1.291$ and $[(C_P - C_P^{ig})/R]^{(1)} = 5.558$. In obtaining these values special care should be made not to use values in the gas regions. From Eqs. (6.57) and (7.38) using parameters R, M, and C_P^{ig} we get $C_P^L = 1.633$ (error of −18%). (d) From Eq. (7.40), $[(C_P - C_P^{ig})/R] = 3.9287$, $C_P^L = 1.899$ (error of −4.8%). (e) From Eq. (7.42), $a = 4.2884$, $b = 0.19547$, $c = 0.0011$, $C_P^L = 2.223$ J/g·K (error of +11.5%). This is the same as ASTM D 2890 test method. (f) From Tsonopoulos correlation, Eq. (7.45), $C_P^L = 2.127$ J/g·K (error of +6.6%). The generalized Lee–Kesler correlation (Eq. 6.57) gives very high error because this method is mainly accurate for gases. For liquids, Eq. (7.40) is more accurate than is Eq. (6.57). Equation (7.45) although recommended for coal liquids predicts liquid heat capacity of hydrocarbons relatively with relative good accuracy. ◆

There are some other methods developed for calculation of C_P^L. In general heat capacity of a substance is proportional to molar volume and can be related to the free space between molecules. As this space increases the heat capacity decreases. Since parameter I (defined by Eq. 2.36) also represents molar volume occupied by the molecules Riazi et al. [27] showed that C_P^L varies linearly with $I/(1 - I)$. They obtained the following relation for heat capacity of homologous hydrocarbon groups:

$$\frac{C_P^L}{R} = (a_1 M + b_1) \times \left(\frac{I}{1 - I} \right) + c_1 M + d_1$$

(7.46)

In the above relation M is molecular weight, R is the gas constant, and coefficients a_1–d_1 are specific for each hydrocarbon family. Parameters I is calculated throughout Eqs. (2.36) and (2.118) at the same temperature at which C_P^L is being calculated. Parameters a_1–d_1 for different hydrocarbon families and solid phase are given in Table 7.8.

7.4.3 Heats of Phase Changes—Heat of Vaporization

Generally there are three types of phase changes: solid to liquid known as fusion (or melting), liquid to vapor (vaporization), and solid to vapor (sublimation), which occurs at pressures below triple point pressure as shown in Fig. 5.2a. During phase change for a pure substance or mixtures of constant composition, the temperature and pressure remain constant. According to the first law of thermodynamics, the heat

TABLE 7.8—*Constants for estimation of heat capacity from refractive index (Eq. 7.46).*

$$C_P/R = (a_1 M + b_1)[I/(1 - I)] + c_1 M + d_1$$

Group	State	Carbon range	Temp. range, °C	a_1	b_1	c_1	d_1	No. of data points	AAD%	MAD%
n-Alkanes	Liquid	C_5–C_{20}	−15–344	−0.9861	−43.692	0.6509	5.457	225	0.89	1.36
1-Alkenes	Liquid	C_5–C_{20}	−60–330	−1.533	40.357	0.836	−21.683	210	1.5	5.93
n-Alkyl-cyclopentane	Liquid	C_5–C_{20}	−75–340	−1.815	56.671	0.941	−28.884	225	1.05	2.7
n-Alkyl-cyclohexane	Liquid	C_6–C_{20}	−100–290	−2.725	165.644	1.270	−68.186	225	1.93	2.3
n-Alkyl-benzene	Liquid	C_6–C_{20}	−250–354	−1.149	4.357	0.692	−3.065	225	1.06	4.71
n-Alkanes	Solid	C_5–C_{20}	−180–3	−1.288	−66.33	0.704	14.678	195	2.3	5.84

AAD%: Average absolute deviation percent. MAD%: Maximum absolute deviation percent. Coefficients are taken from Ref. [27]. Data source: DIPPR [10].

transferred to a system at constant pressure is the same as the enthalpy change. This amount of heat (Q) is called (latent) heat of phase change.

$$Q \text{ (latent heat)} = \Delta H \text{ (phase transition)}$$

(7.47)   at constant T and P

The term latent is normally not used. Since during phase transition, temperature is also constant, thus the entropy change is given as

$$\Delta S \text{ (phase change)} = \frac{\Delta H \text{ (phase change)}}{T \text{ (phase change)}}$$

(7.48)   at constant T and P

Heat of fusion was discussed in Section 6.6.5 (Eq. 6.157) and is usually needed in calculations related to cloud point and precipitation of solids in petroleum fluids (Section 9.3.3). In this section calculation methods for heat of vaporization of petroleum fractions are discussed.

Heat of vaporization (ΔH^{vap}) can be calculated in the temperature range from triple point to the critical point. Thermodynamically, ΔH^{vap} is defined by Eq. (6.98), which can be rearranged as

(7.49) $\Delta H^{vap} = (H^V - H^{ig})^{sat} - (H^L - H^{ig})^{sat}$

where $(H^V - H^{ig})^{sat}$ and $(H^L - H^{ig})^{sat}$ can be both calculated from a generalized correlation or a cubic equation of state at T and corresponding P^{sat} (i.e., see Example 7.7). At the critical point where H^V and H^L become identical, ΔH^{vap} becomes zero. For several compounds, variation of ΔH^{vap} versus temperature is shown in Fig. 7.14. The figure is constructed based

on data generated from correlations provided in Ref. [10]. Specific value of ΔH_{nbp}^{vap} (kJ/g) decreases as carbon number of hydrocarbon (or molecular weight) increases, while the molar values (kJ/mol) increases with increase in the carbon number or molecular weight. In the API-TDB [9], ΔH_T^{vap} for pure compounds is correlated to temperature in the following form:

(7.50) $\Delta H_T^{vap} = A (1 - T_r)^{B + CT_r}$

where coefficients A, B, and C for a large number of compounds are provided [9]. For most hydrocarbons coefficient C is zero [9]. For some compounds values of A, B, and C are given in Table 7.9 as provided in the API-TDB [9].

The most approximate and simple rule to calculate ΔH^{vap} is the Trouton's rule, which assumes ΔS^{vap} at the normal boiling point (T_b) is roughly $10.5R$ (~87.5 J/mol · K) [22]. In some references value of 87 or 88 is used instead of 87.5. Thus, from Eq. (7.48)

(7.51) $H_{nbp}^{vap} = 87.5 T_b$

where ΔH_{nbp}^{vap} is the heat of vaporization at the normal boiling point in J/mol and T_b is in K. This equation is not valid for certain compounds and temperature ranges. The accuracy of this equation can be improved substantially by taking ΔS_{nbp}^{vap} as a function of T_b, which gives the following relation for ΔH_{nbp}^{vap} [22]:

(7.52) $H_{nbp}^{vap} = RT_b (4.5 + \ln T_b)$

where R is 8.314 J/mol · K. This equation at $T_b = 400$ K reduces to Eq. (7.51). In general, ΔH^{vap} can be determined from a vapor pressure correlation through Eq. (6.99).

(7.53) $\dfrac{\Delta H^{vap}}{RT_c} = \Delta Z^{vap} \left[-\dfrac{d \ln P_r^{sat}}{d (1/T_r)} \right]$

where P_r^{sat} is the reduced vapor (saturation) pressure at reduced temperature of T_r. ΔZ^{vap} is the difference between Z^V and Z^L where at low pressures $Z^L \ll Z^V$ and ΔZ^{vap} can be approximated as Z^V. Furthermore, at low pressure if the gas is assumed ideal, then $\Delta Z^{vap} = Z^V = 1$. Under these conditions, use of Eq. (6.101) in the above equation would result in $\Delta H^{vap} = RB$, where B is the coefficient in Eq. (6.101). Obviously, because of the assumptions made to derive Eq. (6.101), this method of calculation of ΔH^{vap} is very approximate. More accurate predictive correlations for ΔH^{vap} can be obtained by using a more accurate relation for the vapor pressure such as Eqs. (7.17) and (7.18).

There are a number of generalized correlations for prediction of ΔH^{vap} based on the principle of corresponding states theory. Pitzer correlated $\Delta H^{vap}/RT_c$ to T_r through acentric factor ω similar to Eq. (7.17). In such correlations,

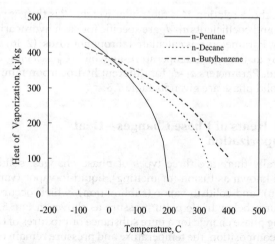

FIG. 7.14—Enthalpy of vaporization of several hydrocarbons versus temperature.

TABLE 7.9—*Coefficients of Eq. 7.50 for calculation of enthalpy of vaporization of pure compounds versus temperature [9].*

$$\Delta H_T^{vap}[kJ/kg] = A(1 - T_r)^{B+CT_r}$$

Compound	A	B	C	Compound	A	B	C
Water	2612.982	−0.0577	0.3870	Methylcyclopentane	527.6931	0.3967	0.0000
Ammonia	1644.157	−0.017	0.3739	Ethylcyclopentane	502.1246	0.3912	0.0000
H_2S	754.073	0.3736	0.0000	Pentylcyclopentane	442.0789	0.3800	0.0000
CO_2	346.1986	−0.6692	0.9386	Decylcyclopentane	397.8670	0.3800	0.0000
N_2	228.9177	−0.1137	0.4281	Pentadecylcyclopentane	372.6050	0.3800	0.0000
CH_4	570.8220	−0.1119	0.4127	Cyclohexane	534.5225	0.3974	0.0000
C_2H_6	588.1554	0.0045	0.3236	Methylcyclohexane	503.9656	0.4152	0.0000
C_3H_8	610.2175	0.3649	0.0000	Ethylene	679.2083	0.3746	0.0000
$n\text{-}C_4H_{10}$	568.6540	0.3769	0.0000	Propylene	539.9479	0.0169	0.0000
$n\text{-}C_5H_{12}$	540.6440	0.3838	0.0000	Benzene	651.8210	0.6775	−0.2695
$n\text{-}C_6H_{14}$	515.2685	0.3861	0.0000	Toluene	544.7929	0.3859	0.0000
$n\text{-}C_7H_{16}$	497.0039	0.3834	0.0000	Ethylbenzene	515.2839	0.3922	0.0000
$n\text{-}C_8H_{18}$	489.0450	0.4004	0.0000	o-Xylene	521.7788	0.3771	0.0000
$n\text{-}C_{10}H_{22}$	461.4396	0.3909	0.0000	Propylbenzene	500.4582	0.3967	0.0000
$n\text{-}C_{15}H_{32}$	431.6786	0.4185	0.0000	n-Butylbenzene	470.0009	0.3808	0.0000
$n\text{-}C_{20}H_{42}$	407.3617	0.4089	0.0000	n-Octylbenzene	456.0581	0.4281	0.0000
Cyclopentane	517.7318	0.1808	0.1706	Naphthalene	371.4852	−0.3910	0.0000

$f^{(0)}(T_r)$ and $f^{(1)}(T_r)$ are correlated to $(1 - T_r)$, where as $T_r \to 1$, $\Delta H^{vap}/RT_c \to 0$. However, more accurate predictive methods are developed in two steps. In the first step heat of vaporization at normal boiling point, ΔH_{nbp}^{vap}, is calculated and then corrected to the desired temperature by a second correlation. One of the most successful correlations for prediction of ΔH_{nbp}^{vap} was proposed by Riedel [12]:

$$(7.54) \qquad \Delta H_{nbp}^{vap} = 1.093 R T_c T_{br} \frac{\ln P_c - 1.013}{0.93 - T_{br}}$$

where T_{br} is the reduced boiling point (T_b/T_c) and P_c is the critical pressure in bars. The unit of ΔH_{nbp}^{vap} depends on the units of R and T_c. Later Chen and Vetere developed similar correlations for calculation of ΔH_{nbp}^{vap} in terms of P_c and T_{br} [12]. For example, Chen correlation is in the following form:

$$(7.55) \quad \Delta H_{nbp}^{vap} = R T_c T_{br} \frac{3.978 T_{br} - 3.958 + 1.555 \ln P_c}{1.07 - T_{br}}$$

Although for certain pure compounds the Chen correlation is slightly superior to the Riedel method, but for practical applications especially for petroleum fractions in which T_c and P_c are calculated values, the Riedel equation is reasonably accurate. A more direct calculation of ΔH_{nbp}^{vap} for petroleum fractions is use of fraction's bulk properties such as T_b and SG or other available parameters in an equation similar to Eq. (2.38) [28]:

$$(7.56) \qquad \Delta H_{nbp}^{vap} = a\theta_1^b\theta_2^c$$

where ΔH_{nbp}^{vap} is in J/mol (or kJ/kmol) and constants a, b, and c are given in Table 7.10 for a number of different input parameters. Once the value of ΔH_{nbp}^{vap} is calculated, it should be divided by M to convert its unit from kJ/kmol to kJ/kg.

Equation (7.56) with coefficients given in Table 7.10 can be used for fractions with molecular weight range of 70–300 ($\sim T_b$ of 300–600 K) with accuracy of about 2% when tested against 138 pure hydrocarbons. Application of the equation can be extended up to 700 K with reasonable accuracy. Once ΔH_{nbp}^{vap} is determined, the Watson relation can be used to calculate ΔH^{vap} at the desired temperature (T).

$$(7.57) \qquad \Delta H^{vap} = \Delta H_{nbp}^{vap} \left(\frac{1 - T_r}{1 - T_{br}} \right)^{0.38}$$

where T_r and T_{br} are the reduced temperature and reduced boiling point, respectively. The same equation can be used to calculate ΔH^{vap} at any temperature when its value at one temperature is available. As it was shown in Example 7.5, use of Eqs. (7.54) and (7.57) predicts ΔH^{vap} of petroleum fractions with good accuracy. Tsonopoulos et al. [18] modified the original Lee–Kesler correlation for calculation of heat of vaporization of coal liquids and aromatics in the following form:

$$(7.58) \qquad (\Delta H^{vap})_{T_r=0.8} = R T_c (4.0439 + 5.3826\omega)$$

where R is 8.314 J/mol · K, T_c is the critical temperature in kelvin, ω is the acentric factor, and ΔH^{vap} is the heat of vaporization at $T = 0.8 T_c$ in J/mol. Equation (7.57) can be used to calculate ΔH^{vap} at temperatures other than $T_r = 0.8$. An evaluation of various methods for estimation of ΔH_{nbp}^{vap} of several coal liquid samples is shown in Tables 7.11 and 7.12. Basic calculated parameters are given in Table 7.11, while estimated values of ΔH_{nbp}^{vap} from Riedel, Vetere, Riazi–Daubert, and Lee–Kesler are given in Table 7.12. In Table 7.11, M is calculated from Eq. (2.51), which is recommended for heavy fractions. If Eq. (2.50) were used to estimate M, the %AAD for the four methods increase to 4.5, 3.2, 4.9, and 2.3, respectively. Equation (2.50) is not applicable to heavy fractions $(M > 300)$, which shows the importance of the characterization method used to calculate molecular weight of hydrocarbon fractions. Evaluations shown in Table 7.12 indicate that both the Riedel method and Eq. (7.56) predict heats of vaporization with good accuracy despite their simplicity. For a coal liquid sample 5HC in Table 7.11, experimental data on

TABLE 7.10—*Coefficients of Eq. (7.56) for estimation of heat of vaporization of petroleum fractions at the normal boiling point [28].*

$$(7.56) \qquad \Delta H_{nbp}^{vap} = a\theta_1^b\theta_2^c$$

ΔH_{nbp}^{vap}, J/mol	θ_1	θ_2	A	b	c
$\Delta H_{1,nbp}^{vap}$	T_b, K	SG	37.32315	1.14086	9.77089×10^{-3}
$\Delta H_{2,nbp}^{vap}$	T_b, K	I	39.7655	1.13529	0.024139
$\Delta H_{3,nbp}^{vap}$	M	I	5238.3846	0.5379	0.48021

TABLE 7.11—*Experimental data on heat of vaporization of some coal liquid fractions with calculated basic parameters [28].*

Fraction (a)	T_b, K	SG	ΔH_{nbp}^{vap}, kJ/kg	M	T_c, K	P_c, bar	ω
5HC	433.2	0.8827	309.4	121.8	649.1	33.1	0.302
8HC	519.8	0.9718	281.4	162.7	748.1	27.1	0.394
11HC	612.6	1.0359	269.6	223.1	843.1	21.5	0.512
16HC	658.7	1.0910	245.4	247.9	896.4	20.5	0.552
17HC	692.6	1.1204	239.3	272.0	932.2	19.4	0.590

M from Eq. (2.51), T_c and P_c from Eqs. (2.63) and (2.64), ω from Eq. (2.108). Experimental value on T_b, SG, and ΔH_{nbp} are taken from J. A. Gray, Report DOE/ET/10104-7, April 1981; Department of Energy, Washington, DC and are also given in Ref. [28].

TABLE 7.12—*Evaluation of various methods of prediction of heat of vaporization of petroleum fractions with data of Table 7.11.*

Fraction	ΔH_{bp}^{vap} exp.	Riedel, Eq. (7.54)		Chen, Eq. (7.55)		RD, Eq. (7.56)		MLK, Eq. (7.58)	
		Calc.	%Dev.	Calc.	%Dev.	Calc.	%Dev.	Calc.	%Dev.
5HC	309.4	305.9	−1.1	303.9	−1.8	311.8	0.8	304.7	−1.5
8HC	281.4	282.5	0.4	278.9	−0.9	287.7	2.2	276.6	−1.7
11HC	269.6	252.2	−6.4	246.3	−8.6	253.2	−6.1	240.5	−10.8
16HC	245.4	248.5	1.3	241.5	−1.6	247.6	0.9	234.7	−4.4
17HC	239.3	241.8	1.0	233.8	−2.3	239.0	−0.1	226.2	−5.5
%AAD	...	...	2.0	...	3.0	...	2.0	...	4.8

Values of M, T_c, P_c, and ω from Table 7.10 have been used for the calculations. RD refers to Riazi–Daubert method or Eq. (7.56) in terms of T_b and SG as given in Table 7.10. MLK refers to modified Lee–Kesler correlation or Eq. (7.58). In use of Eq. (7.58), values of ΔH_{nbp}^{vap} have been obtained by correcting estimated values at $T_r = 0.8$ to $T_r = T_{rb}$, using Eq. (7.57).

ΔH_T^{vap} in the temperature range of 350–550 K are given in Ref. [28]. Predicted values from Eq. (7.57) with use of different methods for calculation of ΔH_{nbp}^{vap} as given in Table 7.12 are compared graphically in Fig. 7.15. The average deviations for the Riedel, Vetre, Riazi–Daubert, and Lee–Kesler are 1.5, 1.8, 1.9 and 1.7%, respectively. The data show that the Riedel method gives the best result for both ΔH_{nbp}^{vap} and ΔH_T^{vap} when the latter is calculated from the Watson method.

As a final method, ΔH^{vap} can be calculated from Eq. (7.49) by calculating residual enthalpy for both saturated vapor and liquid from an equation of state. This is demonstrated in the following example for calculation of ΔH^{vap} from SRK EOS.

Example 7.7—Derive a relation for the heat of vaporization from SRK equation of state.

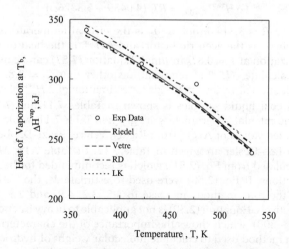

FIG. 7.15—**Evaluation of various methods for estimation of heat of vaporization of coal liquid 5HC. Characteristics of 5HC fraction and description of methods are given in Tables 7.11 and 7.12.**

Solution—The enthalpy departure from SRK is given in Table 6.1. If it is applied to both saturated vapor and saturated liquid at the same temperature and pressure and subtracted from each other based on Eq. (7.49) we get:

$$H^V - H^L = \Delta H^{vap} = RT\left(Z^V - Z^L\right)$$

$$(7.59) \qquad + \left(\frac{a}{b} - T\frac{a_1}{b}\right)\left[\ln\left(\frac{Z^V}{Z^V + B}\right) - \ln\left(\frac{Z^L}{Z^L + B}\right)\right]$$

where a_1 is da/dT as given in Table 6.1 for the SRK EOS. Replacing for $Z = PV/RT$ and $B = bP/RT$ and considering that the ratio of $V^V/(V^V + b)$ is nearly unity (since $b \ll V^V$):

$$(7.60) \quad \Delta H^{vap} = RT\left(Z^V - Z^L\right) + \left(\frac{a}{b} - T\frac{a_1}{b}\right)\left[\ln\left(1 + \frac{b}{V^L}\right)\right]$$

at low temperatures where $Z^L \ll Z^V$, the first term in the right-hand side can be replaced by RTZ^V. At higher temperatures where the difference between Z^V and Z^L decreases Z^L cannot be neglected in comparison with Z^V; however, the term $(Z^V - Z^L)$ becomes zero at the critical point. In calculation of ΔH^{vap} from the above equation one should be careful of the units of a, b and V. If a is in bar (cm⁶/mol²) and b is in cm³/mol, then the second term in the right-hand side of the above equation should be divided by factor 10 to have the unit of J/mol and R in the first term should have the value of 8.314 J/mol·K. Eubank and Wang [29] also developed a new identity to derive heat of vaporization from a cubic equation of state (see Eq. (7.65) in problem 7.13). ♦

7.4.4 Heat of Combustion—Heating Value

Combustion is a chemical reaction wherein the products of the reaction are $H_2O(g)$, $CO_2(g)$, $SO_2(g)$, and $N_2(g)$, where (g) refers to the gaseous state. The main reactants in the reaction are a fuel (i.e., hydrocarbon, H_2, SO, CO, C, ...) and oxygen (O_2). In case of combustion of H_2 or CO, the product is only one compound (i.e., H_2O or CO_2). However, when a hydrocarbon (C_xH_y) is burned the only products are H_2O and CO_2. Combustion is a reaction in which the enthalpy of products is less than enthalpy of reactants and as a result the heat of reaction (enthalpy of products − enthalpy of reactants) is always negative. This heat of reaction is called *heat of combustion* and is shown by ΔH^C. Heat of combustion depends on the temperature at which the combustion takes place. The standard temperature at which usually values of ΔH^C are reported is $25°C$ (298 K).

Amount of heat released by burning one unit mass (i.e., kg, g, or lb) of a fuel is called *heating value* or calorific value and has the unit of kJ/kg or Btu/lb (1 kJ/kg = 0.42993 Btu/lb). In some cases for liquid fuels the heating values are given per unit volume (i.e., kJ/L of fuel), which differs from specific (mass unit) heating values by liquid density. If in the combustion process produced H_2O is considered as liquid, then the heat produced is called gross heat of combustion or *higher heating value* (HHV). When produced H_2O is considered as vapor (as in the actual cases), then the heat produced is called *lower heating value* (LHV). The LHV is also known as the *net heating value* (NHV). The difference between HHV and LHV is due to the heat required to vaporize produced water from liquid to vapor form at the standard temperature (43.97 kJ/mol or 2.443 kJ/g of H_2O). The amount of H_2O formed depends on the hydrogen content of fuel. If the hydrogen wt% of fuel is H% then the relation between HHV and LHV is given as [30]:

$$(7.61) \qquad LHV = HHV - 0.22H\%$$

where both LHV and HHV are in kJ/g. The heating values can also be determined from standard heats of formation (ΔH^f_{298}). Values of ΔH^f_{298} for any element (i.e., H_2, O_2, C, S, etc.) is zero and for formed molecules such as H_2O are given in most thermodynamics references [12, 21, 31]. For example, for $H_2O(g)$, $CO_2(g)$, $CO(g)$, SO_2, $CH_4(g)$, $C_2H_6(g)$, $C_3H_8(g)$, and $n\text{-}C_{10}H_{22}$ the respective values of ΔH^f_{298} are −241.81, −393.51, −110.53, −296.81, −74.52, −83.82, −104.68, and −249.46 kJ/mol. The following example shows calculation of heating values from heats of formations.

Example 7.8—Calculate HHV and LHV of hydrogen, methane, propane, carbon, and sulfur from heats of formation.

Solution—Here the calculation of heating value of CH_4 is demonstrated and a similar approach can be used for other fuels. The chemical reaction of combustion of CH_4 is $CH_4(g) + 2O_2(g) \rightarrow 2H_2O(g) + CO_2(g) + \Delta H^C$, where $\Delta H^C = 2\Delta H^f_{298}(H_2O) + \Delta H^f_{298}(CO_2) - \Delta H^f_{298}(CH_4) - \Delta H^f_{298}(O_2) = 2 \times (-241.835) + (-393.51) - (-74.8936) - (0) = -802.286$ kJ/mol. Since the produced water is assumed to be in gas phase so the LHV is calculated as 802.286/16.04 = 50.01 kJ/g. This is equivalent to 11953 cal/g or 21500 Btu/lb. The HHV can be calculated by adding heat of vaporization of water (2 × 43.97 = 87.94 kJ/mol) to the molar LHV. HHV = 802.286 + 87.94 = 890.2 kJ/mol or 55.5 kJ/g of CH_4. Equation (7.61) to convert LHV to HHV or *vice versa* using H% of fuel may also be used. In this case, H% of CH_4 = (4/16) × 100 = 25 wt%. Thus HHV = 50 + 0.22 × 25 = 55.5 kJ/g. Similarly for H_2, LHV = 241.81/2.0 = 121 kJ/g. The

HHV = 121 + 0.22 × 100 = 143 kJ/g or 61000 Btu/lb. The heating values of other fuels are calculated as follows:

Fuel	Hydrogen, H_2	Methane, CH_4	Propane, C_3H_8	Carbon, C	Sulfur, S
LHV, kJ/g	121	50	46.4	32.8	9.3
HHV, kJ/g	143	55.5	50.4	32.8	9.3

As it can be seen from these calculations, hydrogen has the highest heating value and carbon has the lowest heating value. Thus hydrogen is the best, while carbon is considered as the worst fuel. Sulfur heating value is even less than carbon but sulfur is not really considered as a fuel. Some values of HHV for several other fuels as reported by Felder and Rousseau [32] are given in Table 7.13. The calculated value of HHV of C is near the HHV of hard coal (i.e., solid form) as given in Table 7.13. In natural gases since there are some hydrocarbons heavier than methane, its heating value is somewhat lower than that of pure methane. ◆

Example 7.8 shows that the heating value generally increases as the hydrogen content of fuel increases and carbon content decreases. In other words, as CH weight ratio increases the heating value decreases. Furthermore, presence of sulfur further reduces the heating value. For this reason, some researchers have correlated HHV to wt% of C, H, S, N, and O content of fuel. For example, Tsonopoulos et al. [18] proposed the following relation for estimation of HHV of coal liquids:

$$HHV\,[kJ/g] = 0.3506\,(C\%) + 1.1453\,(H\%) + 0.2054\,(S\%)$$
$$(7.62) \qquad + 0.0617\,(N\%) - 0.0873\,(O\%)$$

S, N, and O are usually found in heavy fuels and aromatic rich fuels such as coal liquids. This equation predicts HHV of coal liquids with %AAD of 0.55 for some 130 fuels. This equation predicts HHV of pure C as 35 kJ/g. However, this equation is not recommended for light fuels, petroleum fraction, or pure compounds. There is a simpler relation for calculation of LHV of heavy fuels and petroleum fractions [30]:

$$(7.63) \qquad LHV\,[kJ/g] = 55.5 - 14.4 \times SG - 0.32S\%$$

where S% is the sulfur wt% in the fuel. A very simple but approximate formula for calculation of HHV of crude oils is [26]:

$$(7.64) \qquad HHV = 51.9 - 8.8 \times SG^2$$

where HHV is in kJ/g (or MJ/kg) and SG is the specific gravity of crude and S% is the sulfur wt% of the crude. Accuracy of these equations is usually about 1%. A typical crude oil has heating value of about 10 500 cal/g (~44 kJ/g). Increase in hydrogen content of a fuel not only increases the heating value

TABLE 7.13—*Heating values of some fuels. Taken with permission from Ref. [32].*

Fuel	Higher heating value	
	kJ/g	Btu/lb
Wood	18	7700
Hard coal	35	15 000
Crude oil	44	19 000
Natural gas	54	23 000
Hydrogen	143	61 000

of the fuel but also decreases amount of unburned hydrocarbon and CO/CO_2 production. For these reasons, natural gas is considered to be a clean fuel but the cleanest and most valuable fuel is hydrogen. This is the reason for global acceleration of development of hydrogen fuel cells as a clean energy, although still production of energy from hydrogen is very costly [33].

HHV can be measured in the laboratory through combustion of the fuel in a bomb calorimeter surrounded by water. Heat produced can be calculated from the rise in the temperature of water. The experimental procedure to measure heating value is explained in ASTM D 240 test method. The heating value of a fuel is one of the characteristics that determines price of a fuel.

7.5 SUMMARY AND RECOMMENDATIONS

In this chapter, application of methods and procedures presented in the book for calculation and estimation of various thermophysical properties are shown for pure hydrocarbons and their defined mixtures, natural gases and nonhydrocarbon gases associated with them (i.e. H_2S, CO_2, N_2, H_2O), defined and undefined petroleum fractions, crude oils, coal liquids, and reservoir fluids. Characterization methods of Chapters 2–4 and thermodynamic relations of Chapters 5 and 6 are essential for such property calculations. Basically, thermophysical properties can be estimated through equations of state or generalized correlations. However, for some special cases empirical methods in the forms of graphical or analytical correlations have been presented for quick estimation of certain properties.

Methods of prediction of properties introduced in the previous chapters such as density, enthalpy, heats of vaporization and melting, heat capacity at constant pressure and volume, vapor pressure, and fuels' heating values are presented.

For calculation of properties of pure components when a correlation for a specific compound is available it must be used wherever applicable. Generalized correlations should be used for calculation of properties of pure hydrocarbons when specific correlation (analytical or graphical) for the given compound is not available. For defined mixtures the best way of calculation of mixture properties when experimental data on properties of individual components of the mixture are available is through appropriate mixing rules for a given property using pure components properties and mixture composition. For defined mixtures wherein properties of pure components are not available, the basic input parameters for equations of states or generalized correlations should be calculated from appropriate mixing rules given in Chapter 5. These basic properties are generally T_c, P_c, V_c, ω, M, and C_P^{ig}, which are known for pure components. For petroleum fractions these parameters should be estimated and the method of their estimations has a great impact on accuracy of predicted physical properties. In fact the impact of estimation of basic input properties is greater than the impact of selected thermodynamic method on the accuracy of property predictions.

For prediction of properties of petroleum fractions, special methods are provided for undefined mixtures. For both pure compounds and petroleum mixtures, properties of gases can be estimated with greater accuracy than for properties of liquids. This is mainly due to better understanding of intermolecular forces in gaseous systems. Similarly properties of gases at low pressures can be estimated with better accuracy in comparison with gases at high pressures. Effect of pressure on properties of gases is much greater than the effect of pressure on properties of liquids. At high pressures as we approach the critical region properties of gases and liquids approach each other and under such conditions a unique generalized correlation for both gases and liquids termed dense fluids may be used for prediction of properties of both gases and liquids. For wide boiling range fractions or crude oils the mixture should be split into a number of pseudocomponents and treat the fluid as a defined mixture. A more accurate approach would be to consider the fluid as a continuous mixture

When using a thermodynamic model, cubic equations of state (i.e., PR or SRK) should be used for calculation of PVT and equilibrium properties at pressures greater than about 13 bars (~200 psia). At low pressures and especially for liquids, properties calculated from a cubic EOS are not reliable. For liquid systems specific generalized correlations developed based on liquid properties are more accurate than other methods. It is on this basis that Rackett equation provides more accurate data on liquid density than any other correlation. In application of EOS to petroleum mixtures the BIPs especially for the key components have significant impact on accuracy of predicted results. Wherever possible BIP of key components (i.e., C_1–C_{7+} in a reservoir fluid) can be tuned with available experimental data (i.e., density or saturation pressure) to improve prediction by an EOS model [34].

TABLE 7.14—*Summary of recommended methods for various properties.*

Property	Methods of estimation for various fluids
Density	• Eq. (7.3) for gases with Lee–Kesler generalized correlation for calculation of Z (Ch 5), also see Section 7.2.1. • For pure liquid hydrocarbons, Table 2.1 and Eq. (7.5) or Rackett equation. • Eq. (7.4) for defined liquid mixture. • For petroleum fractions use Rackett equation, Eq. (7.5), or Figs. 7.1–7.3. • See Section 7.2.2 for other cases.
Vapor pressure	• Eq. (7.8) for pure compounds and if coefficients are not known use Eqs. (7.18) or (7.19). • Use Eqs. (7.20) and (7.22) for petroleum fractions and Eqs. (7.21) and (7.23) for coal liquids. • For crude oils use Eq. (7.26) or Fig. 7.11.
Enthalpy	• Use Eq (7.32) and Fig. 7.13 with Lee–Kesler correlations of Ch 6 for petroleum fractions. • Use Eq. (7.34) for ΔH^{ig}. • For special cases see Section 7.4.1.
Liquid heat capacity, C_P^L	• Eq. (7.40) for pure compounds. • Eq. (7.43) for petroleum fractions. • Eq. (7.45) for coal liquids.
ΔH^{vap}	• Eq. (7.50) or Eqs. (7.54) and (7.57) for pure compounds. • Eqs. (7.54) and (7.57) for petroleum fractions. • Eqs. (7.58) and (7.57) for coal liquids.
Heating value	See Section 7.4.4.

These recommendations are not general and for special cases one should see specific recommendations in each section.

In general when model parameters are tuned with available experimental data especially for complex mixtures and heavy fractions, accuracy of model prediction for the given systems can be greatly improved. A summary of some recommended methods for different physical and thermodynamic properties is given in Table 7.14.

7.6 PROBLEMS

7.1. For storage of light hydrocarbons and their mixtures in sealed tanks, always a mixture of liquid and vapor are stored. Why is this practiced, rather than storing 100% gas or 100% liquid phase?

7.2. Figure 7.16 shows reported laboratory data on variation of P with V for a fluid mixture at constant T and composition. Can you comment on the data?

7.3. Figure 7.17 shows reported laboratory data on variation of H with T for a fluid mixture at constant P and composition. Can you comment on the data?

7.4. A kerosene sample has specific gravity and molecular weight of 0.784 and 167, respectively. Methane is dissolved in this liquid at 333 K and 20.7 bar. The mole fraction of methane is 0.08. Use the graphical method suggested in this chapter to calculate molar density of the mixture and compare it with the value of 5.224 kmol/m³ as given in Ref. [35]. What is the predicted value from an EOS?

7.5. Derive Eq. (4.120) for vapor pressure.

7.6. A petroleum product has mid boiling point of 385 K and specific gravity of 0.746. Estimate its vapor pressure at 323 K from three most suitable methods.

7.7. For the petroleum product of Problem 7.6 calculate RVP from an appropriate method in Chapter 3 and then use Fig. 7.10 to calculate TVP. Compare the result with those estimated in Problem 7.6.

7.8. Sublimation pressure of benzene at –36.7°C is 1.333 Pa [21]. Derive a relation for sublimation pressure of benzene. Calculate sublimation pressure of benzene at –11.5 and –2.6°C. Compare estimated values with reported values of 13.33 and 26.67 Pa [21]. Also estimate heat of sublimation of benzene.

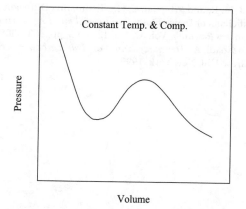

FIG. 7.16—Pressure–volume data for Problem 7.2.

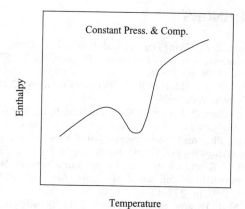

FIG. 7.17—Enthalpy–temperature data for Problem 7.3.

7.9. When n-pentane is heated from 190.6°F and 600 psia to 370.7°F and 2000 psia, the enthalpy increases by 8655 Btu/lbmol [36]. Calculate this enthalpy change from Lee–Kesler and PR EOSs and compare calculated values with the experimental value.

7.10. In the previous problem consider the initial pressure is 2000 psia. In this case the process becomes heating at constant pressure. Calculate the enthalpy change and compare with the experimental value of 8467 Btu/lbmol [36].

7.11. Calculate enthalpy of kerosene of Table 7.5 in liquid phase at 500°F and 1400 psia using Lee–Kesler generalized correlation and SRK EOS. Compare the results with experimental value given in Table 7.5. Use the API methods for prediction of M, T_c, P_c, and the Lee–Kesler correlation for calculation of acentric factor. Repeat the calculations using Lee–Kesler generalized correlation for the enthalpy departure and Twu correlations for M, T_c, and P_c.

7.12. The purpose of this problem is to show the impact of both the selected predictive method and the selected characterization method on the estimation of thermal properties of hydrocarbon fractions. A coal liquid has a boiling point of 476°F and specific gravity of 0.9718. Estimate its heat of vaporization at 600°F using the methods proposed by Riedel, Chen, and Riazi–Daubert. For each method use API, Riazi–Daubert (1980), Lee–Kesler, and Twu methods for estimation of input parameters. The experimental value is 110.9 Btu/lb [28].

7.13. For pure components, the Maxwell Equal Area Rule (MEAR) is a thermodynamic identity for vapor–liquid equilibria [29]:

$$(7.65) \qquad P^{\text{sat}}(V^{\text{V}} - V^{\text{L}}) = \int_{V^{\text{L}}}^{V^{\text{V}}} P^{\text{EOS}}(dV)_{\text{T}}$$

Use this equation to calculate vapor pressure of benzene at 25, 100, 140, and 220°C from SRK EOS and compare with actual data.

7.14. Derive relations for heat of vaporization based on RK and PR EOS.

REFERENCES

[1] Carlson, E. C., "Don't Gamble with Physical Properties for Simulations," *Chemical Engineering Progress*, Vol. 92, No. 10, October 1996, pp. 35–46.

[2] Alberty, R. A. and Silbey, R. J., *"Physical Chemistry,"* 2nd ed., Wiley, New York, 1999.

[3] Hirschfelder, O. J., Curtiss, C. F., and Bird, R. B., *Molecular Theory of Gases and Liquids*, Wiley, New York, 1964.

[4] Bhagat, Ph., "An Introduction to Neural Nets," *Chemical Engineering Progress*, Vol. 86, No. 8, August 1990, pp. 55–60.

[5] Riazi, M. R. and Elkamel, A., "Using Neural Network Models to Estimate Critical Constants," *Simulators International*, Vol. 29, No. 3, 1997, pp. 172–177.

[6] Lee, M.-J., Hwang,, S.-M., and Chen, J.-T., "Density and Viscosity Calculations for Polar Solutions via Neural Networks," *Journal of Chemical Engineering of Japan*, Vol. 27, No. 6, 1994, pp. 749–754.

[7] Press, J., "Working with Non-Ideal Gases," *Chemical Engineering Progress*, Vol. 99, No. 3, 2003, pp. 39–41.

[8] *Engineering Data Book, Vol. II*, Sections 17–26, 10th ed., Gas Processors (Suppliers) Association, Tulsa, OK, 1987.

[9] Daubert, T. E. and Danner, R. P., Eds, *API Technical Data Book—Petroleum Refining*, 6th ed., American Petroleum Institute (API), Washington, DC, 1997.

[10] AIChE DIPPR® Database, Design Institute for Physical Property Data (DIPPR), EPCON International, Houston, TX, 1996.

[11] Hall, K. R., Ed., *TRC Thermodynamic Tables—Hydrocarbons*, Thermodynamic Research Center, The Texas A&M University System, College Station, TX, 1993.

[12] Poling, B. E., Prausnitz, J. M., and O'Connell, J. P., *Properties of Gases and Liquids*, 5th ed., McGraw-Hill, New York, 2000. Reid, R. C., Prausnitz, J. M., Poling, B. E., Eds., *Properties of Gases and Liquids,* 4th ed., McGraw-Hill, New York, 1987.

[13] Yaws, C. L. and Yang, H.-C., "To Estimate Vapor Pressure Easily," *Hydrocarbon Processing*, Vol. 68, 1989, pp. 65–68.

[14] Feyzi, F., Riazi, M. R., Shaban, H. I., and Ghotbi, S., "Improving Cubic Equations of State for Heavy Reservoir Fluids and Critical Region," *Chemical Engineering Communications*, Vol. 167, 1998, pp. 147–166.

[15] Korsten, H., "Internally Consistent Prediction of Vapor Pressure and Related Properties," *Industrial and Engineering Chemistry Research*, Vol. 39, 2000, pp. 813–820.

[16] Lee, B. I. and Kesler, M. G., "A Generalized Thermodynamic Correlation Based on Three-Parameter Corresponding States," *American Institute of Chemical Engineers Journal*, Vol. 21, 1975, pp. 510–527.

[17] Maxwell, J. B. and Bonnell, L. S., *Vapor Pressure Charts for Petroleum Engineers*, Exxon Research and Engineering Company, Florham Park, NJ, 1955. Reprinted in 1974. "Deviation and Precision of a New Vapor Pressure Correlation for Petroleum Hydrocarbons," *Industrial and Engineering Chemistry*, Vol. 49, 1957, pp. 1187–1196.

[18] Tsonopoulos, C., Heidman, J. L., and Hwang, S.-C., *Thermodynamic and Transport Properties of Coal Liquids*, An Exxon Monograph, Wiley, New York, 1986.

[19] Wilson, G. M., Johnston, R. H., Hwang, S. C., and Tsonopoulos, C., "Volatility of Coal Liquids at High Temperatures and Pressures," *Industrial Engineering Chemistry, Process Design and Development*, Vol. 20, No. 1, 1981, pp. 94–104.

[20] Marks, A., *Petroleum Storage Principles*, PennWell, Tulsa, OK, 1983.

[21] Sandler, S. I., *Chemical and Engineering Thermodynamics*, 3rd ed., Wiley, New York, 1999.

[22] Levine, I. N., *Physical Chemistry*, 4th ed., McGraw-Hill, New York, 1995.

[23] Lenoir, J. M. and Hipkin, H. G., "Measured Enthalpies of Eight Hydrocarbon Fractions," *Journal of Chemical and Engineering Data*, Vol. 18, No. 2, 1973, pp. 195–202.

[24] Kesler, M. G. and Lee, B. I., "Improve Prediction of Enthalpy of Fractions," *Hydrocarbon Processing*, Vol. 55, No. 3, 1976, pp. 153–158.

[25] ASTM, *Annual Book of Standards*, Section Five, Petroleum Products, Lubricants, and Fossil Fuels (in 5 Vols.), ASTM International, West Conshohocken, PA, 2002.

[26] Speight, J. G., *The Chemistry and Technology of Petroleum*, 3rd ed., Marcel Dekker, New York, 1999.

[27] Riazi, M. R. and Roomi, Y., "Use of the Refractive Index in the Estimation of Thermophysical Properties of Hydrocarbons and Their Mixtures," *Industrial and Engineering Chemistry Research*, Vol. 40, No. 8, 2001, pp. 1975–1984.

[28] Riazi, M. R. and Daubert, T. E., "Characterization Parameters for Petroleum Fractions," *Industrial and Engineering Chemistry Research*, Vol. 26, 1987, pp. 755–759. Corrections, p. 1268.

[29] Eubank, P. T. and Wang, X., "Saturation Properties from Equations of State," *Industrial and Engineering Chemistry Research*, Vol. 42, No. 16, 2003, pp. 3838–3844.

[30] Wauquier, J.-P., *Petroleum Refining, Vol. 1: Crude Oil, Petroleum Products, Process Flowsheets*, Editions Technip, Paris, 1995.

[31] Smith, J. M., Van Ness, H. C., and Abbott, M. M., *Introduction to Chemical Engineering Thermodynamics*, 5th ed., McGraw-Hill, New York, 1996.

[32] Felder, R. M. and Rousseau, R. W., *Elementary Principles of Chemical Processes*, 2nd ed., Wiley, New York, 1986.

[33] Mang, R. A., "Clean Energy for the Hydrogen Planet," *The Globe and Mall*, Canada's National Newspaper, June 9, 2003, Page H1.

[34] Manafi, H., Mansoori, G. A., and Ghotbi, S., "Phase Behavior Prediction of Petroleum Fluids with Minimum Characterization Data," *Journal of Petroleum Science & Engineering*, 1999, Vol. 22, pp. 67–93.

[35] Riazi, M. R. and Whitson, C. H., "Estimating Diffusion Coefficients of Dense Fluids," *Industrial and Engineering Chemistry Research*, Vol. 32, No. 12, 1993, pp. 3081–3088.

[36] Firoozabadi, A., *Thermodynamics of Hydrocarbon Reservoirs*, McGraw-Hill, New York, 1999.

Applications: Estimation of Transport Properties

8

NOMENCLATURE

A Helmholtz free energy defined in Eq. (6.7), J/mol

A_T Adhesion tension (Eq. 8.85), N or dyne

API   API gravity defined in Eq. (2.4)

A, B, C, D, E Coefficients in various equations

a, b, c, d, e Constants in various equations

BI_{vis} Blending index for viscosity of liquid hydrocarbons (see Eq. 8.20), dimensionless

c_s Velocity of sound, m/s

D_A Self diffusion coefficient of component A, cm^2/s ($= 10^{-4}$ m^2/s)

D_{AB} Binary (mutual) diffusion coefficient (diffusivity) of component A in B, cm^2/s

$D_{A\text{-mix}}$ Effective diffusion coefficient (diffusivity) of component A in a mixture, cm^2/s

$D_{AB}^{\infty L}$ Liquid binary diffusion coefficient (diffusivity) of component A in B at infinite dilution ($x_A \to 0$), cm^2/s ($= 10^{-4}$ m^2/s)

d Molecular diameter, m (1 Å $= 10^{-10}$ m)

E Activation energy (see Eq. 8.55), kcal/mol

F Formation resistivity factor in a porous media (see Eq. 8.72), dimensionless

g Acceleration of gravity ($= 9.8$ m^2/s)

I Refractive index parameter defined in Eq. (2.36) $[= (n^2 - 1)/(n^2 + 2)]$, dimensionless

J_{Ay} Mass diffusion flux of component A in the y direction (i.e., $g/cm^2 \cdot s$)

K_W Watson characterization factor defined by Eq. (2.13)

k Thermal conductivity, $W/m \cdot K$

k_b^L Thermal conductivity of liquid at normal boiling point, $W/m \cdot K$

k_M^L Thermal conductivity of liquid at normal melting point, $W/m \cdot K$

k_B Boltzman constant ($= R/N_A = 1.381 \times 10^{-23}$ J/K)

$\log_{10}$ Common logarithm (base 10)

$\ln$ Natural logarithm (base e)

M Molecular weight (molar mass), g/mol (kg/kmol)

m Mass of one molecule ($= M/N_A$), kg

m Cementation factor, a parameter characteristic of a porous media, dimensionless (i.e., see Eqs. (8.73) and (8.74))

N Number of components in a mixture

N_A Avogadro number = number of molecules in one mole (6.022×10^{23} mol^{-1})

N_{Pr} Prandtl number defined in Eq. (8.29), dimensionless

n A parameter in various equations

n Liquid refractive index at temperature T and 1 atm, dimensionless

P Pressure, bar

P_a Parachor number (see Eq. 8.86)

P_c Critical pressure, bar

P_r Reduced pressure defined by Eq. (5.100) ($= P/P_c$), dimensionless

p A dimensionless parameter for use in Eq. (8.78)

Q A parameter for use in Eq. (8.88)

q_y Heat flux in the y direction, W/m^2

R Gas constant = 8.314 $J/mol \cdot K$ (values in different units are given in Section 1.7.24)

r Radius of capillary tube (see Eqs. (8.80)–(8.82))

SG   Specific gravity of liquid substance at 15.5°C (60°F) defined by Eq. (2.2), dimensionless

T Absolute temperature, K

t Temperature related parameter (i.e., see Eq. (8.37) or (8.44))

T_b Normal boiling point, K

T_c Critical temperature, K

T_r Reduced temperature defined by Eq. (5.100) ($= T/T_c$), dimensionless

T_M Freezing (melting) point for a pure component at 1.013 bar, K

T_{br} Reduced boiling point ($= T_b/T_c$), dimensionless

V Molar volume, cm^3/mol

V_A Liquid molar volume of pure component A at normal boiling point, cm^3/mol

V_c Critical volume (molar), cm^3/mol (or critical specific volume, cm^3/g)

V_i Molar volume of pure component i at T and P, cm^3/mol

V_r Reduced volume ($= V/V_c$), dimensionless

V_x Velocity of fluid in the x direction, m/s

V_m^L Molar volume of liquid mixture, cm^3/mol

x_i Mole fraction of component i in a mixture (usually used for liquids), dimensionless

x_{wi} Weight fraction of component i in a mixture (usually used for liquids), dimensionless

x_P, x_N, x_A Fractions (i.e., mole) of paraffins, naphthenes, and aromatics in a petroleum fraction, dimensionless

y_i Mole fraction of i in a mixture (usually used for gases), dimensionless

Z_c Critical compressibility factor [$Z = P_c V_c/RT_c$], dimensionless

Greek Letters

α Thermal diffusivity ($= k/\rho C_p$), m^2/s or cm^2/s

α_{AB} A thermodynamic parameter for nonideality of a liquid mixture defined by Eq. (8.63), dimensionless

ε_i Energy parameter for component i (see Eq. 8.57)

Φ_{ij} Dimensionless parameter defined in Eq. (8.7)

ϕ Porosity of a porous media (Eq. 8.73), dimensionless

θ_{wo} Oil–water contact angle, in degrees as used in Eq. (8.84)

ρ Density at a given temperature and pressure, g/cm^3 (molar density unit: cm^3/mol)

ρ_M Molar density at a given temperature and pressure, mol/cm^3

ρ_r Reduced density ($= \rho/\rho_c = V_c/V$), dimensionless

ρ_o Oil density at a given temperature, g/cm^3

ρ_w Water density at a given temperature, g/cm^3

σ Molecular size parameter, Å [1 Å = 10^{-10} m]

σ Surface tension of a liquid at a given temperature, dyn/cm

σ_H Surface tension of a hydrocarbon at a given temperature, dyn/cm

σ_{wo} Interfacial tension of oil and water at a given temperature, dyn/cm

σ_{so} Surface tension of water with rock surface, dyn/cm

ω Acentric factor defined by Eq. (2.10), dimensionless

ξ Viscosity parameter defined by Eq. (8.5), (cp)$^{-1}$

μ Absolute viscosity, mPa·s (cp)

μ_a Viscosity at atmospheric pressure, mPa·s (cp)

μ_c Critical viscosity, mPa·s (cp)

μ_P Viscosity at pressure P, mPa·s

μ_r Reduced viscosity ($= \mu/\mu_c$), dimensionless

ν Kinematic viscosity ($= \mu/\rho$), cSt (10^{-2} cm^2/s)

$\nu_{38(100)}$ Kinematic viscosity of a liquid at 37.8°C (100°F), cSt (10^{-2} cm^2/s)

Ω Molecular energy parameter (i.e., see Eqs. 8.31 or 8.57)

Γ Parameter defined in Eq. (8.38)

λ Parameter defined in Eq. (8.34), m·s·mol^{-1}.

γ_A Activity coefficient of component A in liquid solution defined by Eq. (6.112), dimensionless

Ψ_{AB} Association parameter defined in Eq. (8.60), dimensionless

τ Tortuosity, dimensionless parameter defined for pore connection structure in a porous media system (see Eq. 8.71)

τ_{yx} x component of momentum flux in the y direction, N/m^2

π A numerical constant = 3.14159265

Superscript

g Value of a property for gas phase

ig Value of a property for component "i" as ideal gas at temperature T and $P \rightarrow 0$

L Value of a property at liquid phase

V Value of a property at vapor phase

o Value of a property at low pressure (ideal gas state) condition at a given temperature

Subscripts

A Value of a property for component A

B Value of a property for component B

b Value of a property at the normal boiling point

c Value of a property at the critical point

i,j Value of a property for component i or j in a mixture

L Value of a property for liquid phase

m Mixture property

od Value of a property for dead oil (crude oil) at atmospheric pressure

r Reduced property

T Values of property at temperature T

w Values of a property for water

20 Values of property at 20°C

Acronyms

API-TDB American Petroleum Institute—Technical Data Book (see Ref. [5])

BIP Binary interaction parameter

bbl Barrel, unit of volume of liquid as given in Section 1.7.11.

cp Centipoise, unit of viscosity, (1 cp = 0.01 p = 0.01 g·cm·s = 1 mPa·s = 10^{-3} kg/m·s)

cSt Centistoke, unit of kinematic viscosity, (1 cSt = 0.01 St = 0.01 cm^2/s)

DIPPR Design Institute for Physical Property Data (see Ref. [10])

EOS Equation of state

GLR Gas-to-liquid ratio

IFT Interfacial tension

PNA Paraffins, naphthenes, aromatics content of a petroleum fraction

scf Standard cubic foot (unit for volume of gas at 1 atm and 60°F).

stb Stock tank barrel (unit for volume of lquid oil at 1 atm and 60°F).

IN THIS CHAPTER, application of various methods presented in previous chapters is extended to estimate another type of physical properties, namely, transport properties for various petroleum fractions and hydrocarbon mixtures. Transport properties generally include viscosity, thermal conductivity, and diffusion coefficient (diffusivity). These are molecular properties of a substance that indicate the rate at which specific (per unit volume) momentum, heat, or mass are transferred. Science of the study of these processes is called *transport phenomenon*. One good text that describes these processes was written by Bird et al. [1]. The first edition appeared in 1960 and remained a leading source for four decades until its second publication in 1999. A fourth property that also determines transport of a fluid is surface or interfacial tension (IFT), which is needed in calculations related to the rise of a liquid in capillary tubes or its rate of spreading over a surface. Among these properties, viscosity is considered as one of the most important physical properties for calculations related to fluid flow followed by thermal conductivity and diffusivity. Interfacial tension is important in reservoir engineering calculations to determine the rate of oil recovery and for process engineers it can be used to determine foaming characteristics of hydrocarbons in separation units.

As was discussed in Chapter 7, properties of gases may be estimated more accurately than can the properties of liquids. Kinetic theory provides a good approach for development of predictive methods for transport properties of gases. However, for liquids more empirically developed methods are used for accurate prediction of transport properties. Perhaps combination of both approaches provides most reliable and general methods for estimation of transport properties of fluids. For petroleum fractions and crude oils, characterization methods should be used to estimate the input parameters. It is shown that choice of characterization method may have a significant impact on the accuracy of predicted transport property. Use of methods given in Chapters 5 and 6 on the development of a new experimental technique for measurement of diffusion coefficients in high-pressure fluids is also demonstrated.

8.1 ESTIMATION OF VISCOSITY

Viscosity is defined according to the Newton's law of viscosity:

$$(8.1) \qquad \tau_{yx} = -\mu \frac{\partial V_x}{\partial y} = -\nu \frac{\partial (\rho V_x)}{\partial y} \quad \nu = \frac{\mu}{\rho}$$

where τ_{yx} is the x component of flux of momentum in the y direction in which y is perpendicular to the direction of flow x. Velocity component in the x direction is V_x. ρ is the density and ρV_x is the specific momentum (momentum per unit volume). $\partial V_x/\partial y$ is the velocity gradient or shear rate (with dimension of reciprocal of time, i.e., s^{-1}) and $\partial(\rho V_x)/\partial y$ is gradient of specific momentum. τ represents tangent force to the fluid layers and is called *shear stress* with the dimension of force per unit area (same as pressure). Velocity is a vector quantity, while shear stress is a tensor quantity. While pressure represents normal force per unit area, τ represents tangent (stress) force per unit area, which is in fact the same as momentum (mass × velocity) per unit area per unit time or momentum flux. Thus μ has the dimension of mass per unit length per unit time. For example, in the cgs unit system it has unit of g/cm · s, which is called poise. The most widely used unit for viscosity is centipoise (1 cp = 0.01 p = 10^{-4} micropoise). The ratio of μ/ρ is called kinematic viscosity and is usually shown by ν with the unit of stoke (cm²/s) in the cgs unit system. The common unit for ν is cSt (0.01 St), which is equivalent to mm²/s. Liquid water at 20°C exhibits a viscosity of about 1 cp, while its vapor at atmospheric pressure has viscosity of about 0.01 cp. More viscous fluids (i.e., oils) have viscosities higher than the viscosity of water at the same temperature. Fluids that follow linear relation between shear stress and shear rate (i.e., Eq. 8.1) are called Newtonian. Polymer solutions and many heavy oils with large amount of wax or asphaltene contents are considered non-Newtonian and follow other relations between shear stress and shear rate. Viscosity is in fact a measure of resistance to motion and the reciprocal of viscosity is called fluidity. Fluids with higher viscosity require more power for their transportation. Viscosity is undoubtedly the most important transport property and it has been studied both experimentally and theoretically more than other transport properties. In addition to its direct use

for fluid-flow calculations, it is needed in calculation of other properties such as diffusion coefficient. Experimental values of gas viscosity at 1 atm versus temperature for several hydrocarbon gases and liquids are shown in Fig. 8.1. As it is seen from this figure, viscosity of liquids increases with molecular weight of hydrocarbon while viscosity of gases decreases.

8.1.1 Viscosity of Gases

Viscosity of gases can be predicted more accurately than can the viscosity of liquids. At low pressures (ideal gas condition) viscosity can be well predicted from the kinetic theory of gases [1, 3, 4].

$$(8.2) \qquad \mu = \frac{2}{3\pi^{2/3}} \frac{\sqrt{mk_B T}}{d^2}$$

where m is the mass of one molecule in kg ($m = 0.001 M/N_A$), k_B is the Boltzmann constant ($= R/N_A$), and d is the molecular diameter. In this relation if m is in kg, k_B in J/K, T in K, and d in m, then μ would be in kg/m · s (1000 cp). This relation has been obtained for hard-sphere molecules. Similar relations can be derived for viscosity based on other relations for the intermolecular forces [1]. The well-known Chapman–Enskog equations for transport properties of gases at low densities (low pressure) are developed on this basis by using Lennard–Jones potential function (Eq. 5.11). The relation is very similar to Eq. (8.2), where μ is proportional to $(MT)^{1/2}/(\sigma^2\Omega)$ in which σ is the molecular collision diameter and Ω is a function of $k_B T/\epsilon$. Parameters σ and ϵ are the size and energy parameters in the Lennard–Jones potential (Eq. 5.11). From such relations, one may obtain molecular collision diameters or potential energy parameters from viscosity data. At low pressures, viscosity of gases changes with temperature. As shown by the above equation as T increases, gas viscosity also increases. This is mainly due to increase in the intermolecular collision that is caused by an increase in molecular friction. At high pressures, the behavior of the viscosity of gases and liquids approach each other.

For pure vapor compounds, the following correlation was developed by the API-TDB group at Penn State and is recommended in the API-TDB for temperature ranges specified for each compound [5]:

$$(8.3) \qquad \mu = \frac{1000 A T^B}{\left(1 + \frac{C}{T} + \frac{D}{T^2}\right)}$$

where correlation coefficients A–D are given for some selected compounds in Table 8.1. The average error over the entire temperature range is about 5% but usually errors are less than 2%. This equation should not be applied at pressures in which $P_r > 0.6$.

The following relation developed originally by Yoon and Thodos [6] is recommended in the previous editions of API-TDB and DIPPR manuals for estimation of viscosity of hydrocarbons as well as nonhydrocarbons and nonpolar gases at atmospheric pressures:

$$(8.4) \qquad \begin{aligned} \mu\xi \times 10^5 &= 1 + 46.1 T_r^{0.618} - 20.4 \exp(-0.449 T_r) \\ &\quad + 19.4 \exp(-4.058 T_r) \end{aligned}$$

$$(8.5) \qquad \xi = T_c^{\frac{1}{6}} M^{-\frac{1}{2}} (0.987 P_c)^{-\frac{2}{3}}$$

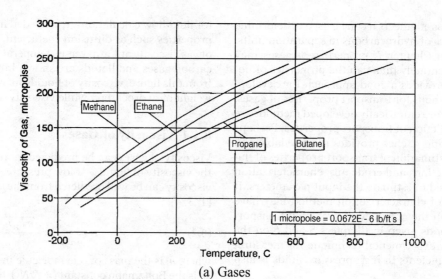

(a) Gases

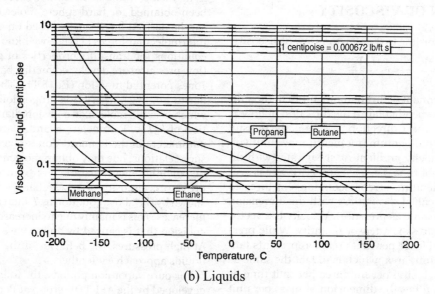

(b) Liquids

FIG. 8.1—Viscosity of several light hydrocarbons versus temperature at atmospheric pressure. Taken with permission from Ref. [2].

where μ is in cp, T_r is the reduced temperature, and ξ is a parameter that has a dimension of inverse viscosity and is obtained from kinetic theory of gases. The factor 0.987 comes from the original definition that unit of atm was used for P_c. In the above equation P_c, T_c, and M are in bar, kelvin, and g/mol, respectively. In cases where data on gas viscosity is available, it would be more appropriate to determine ξ from viscosity data rather than to calculate it from the above equation. Reliability of this equation is about 3–5%. For some specific compounds such as hydrogen, the numerical coefficients in Eq. (8.4) are slightly different and in the same order as given in the DIPPR manual are 47.65, −20.0, −0.858, +19.0, and −3.995. In the 1997 edition of the API-TDB [5], the more commonly used correlation developed earlier by Stiel and Thodos [7] is recommended:

$$(8.6) \quad \begin{aligned} \mu\xi &= 3.4 \times 10^{-4} T_r^{0.94} & \text{for } T_r \leq 1.5 \\ \mu\xi &= 1.778 \times 10^{-4}(4.58T_r - 1.67)^{0.625} & \text{for } T_r > 1.5 \end{aligned}$$

where units of μ and ξ are the same as in Eq. (8.4). For defined gas mixtures at low pressures, Eq. (8.6) may be used with T_c, P_c, and M calculated from Kay's mixing rule (Eq. 7.1). However, when viscosity of individual gases in a mixture are known, a more accurate method of estimation of mixture viscosity is provided by Wilke, which can be applied for pressures with $P_r < 0.6$ [1, 5]:

$$(8.7) \quad \begin{aligned} \mu_m &= \sum_{i=1}^{N} \frac{x_i \mu_i}{\sum_{j=1}^{N} x_j \phi_{ij}} \\ \phi_{ij} &= \frac{1}{\sqrt{8}} \left(1 + \frac{M_i}{M_j}\right)^{-\frac{1}{2}} \left[1 + \left(\frac{\mu_i}{\mu_j}\right)^{\frac{1}{2}} \left(\frac{M_j}{M_i}\right)^{\frac{1}{4}}\right]^2 \end{aligned}$$

This semiempirical method is recommended by both API-TDB and DIPPR for calculation of viscosity of gas mixtures of known composition at low pressures. Accuracy of this equation is about 3% [5, 8]. In the above relation $\phi_{ij} \neq \phi_{ji}$.

TABLE 8.1—*Coefficients of Eq. (8.3) for viscosity of pure vapor compounds. (Taken with permission from Ref. [5].)*

$$\mu = \frac{1000AT^{B}}{\left(1+\frac{C}{T}+\frac{D}{T^{2}}\right)} \quad \text{Units: cp and kelvin} \quad (8.3)$$

API No.	Compound	A	B	C	D	T_{min}, K	T_{max}, K
794	Oxygen	1.1010E−06	5.6340E−01	9.6278E+01	0.0000E+00	54	1500
781	Hydrogen	1.7964E−07	6.8500E−01	−5.8889E−01	1.4000E+02	14	3000
845	Water	6.1842E−07	6.7780E−01	8.4722E+02	−7.4074E+04	273	1073
771	Ammonia	4.1856E−08	9.8060E−01	3.0800E+01	0.0000E+00	196	1000
786	Hydrogen sulfide	5.8597E−08	1.0170E+00	3.7239E+02	−6.4198E+04	250	480
789	Nitrogen	6.5593E−07	6.0810E−01	5.4711E+01	0.0000E+00	63	1970
774	Carbon monoxide	1.1131E−06	5.3380E−01	9.4722E+01	0.0000E+00	68	1250
775	Carbon dioxide	2.1479E−06	4.6000E−01	2.9000E+02	0.0000E+00	194	1500
797	Sulfur trioxide	3.9062E−06	3.8450E−01	4.7011E+02	0.0000E+00	298	694
770	Air	1.4241E−06	5.0390E−01	1.0828E+02	0.0000E+00	80	2000
	Paraffins						
1	Methane	5.2553E−07	5.9010E−01	1.0572E+02	0.0000E+00	91	1000
2	Ethane	2.5904E−07	6.7990E−01	9.8889E+01	0.0000E+00	91	1000
3	Propane	2.4995E−07	6.8610E−01	1.7928E+02	−8.2407E+03	86	1000
4	*n*-Butane	2.2982E−07	6.9440E−01	2.2772E+02	−1.4599E+04	135	1000
5	Isobutane	6.9154E−07	5.2140E−01	2.2900E+02	0.0000E+00	150	1000
6	*n*-Pentane	6.3411E−08	8.4760E−01	4.1722E+01	0.0000E+00	143	1000
7	Isopentane	1.1490E−06	4.5720E−01	3.6261E+02	−4.9691E+03	113	1000
8	Neopentane	4.8643E−07	5.6780E−01	2.1289E+02	0.0000E+00	257	1000
9	*n*-Hexane	1.7505E−06	7.0740E−01	1.5711E+02	0.0000E+00	178	1000
10	2-Methylpentane	1.1160E−06	4.5370E−01	3.7472E+02	0.0000E+00	119	1000
14	*n*-Heptane	6.6719E−08	8.2840E−01	8.5778E+01	0.0000E+00	183	1000
15	2-Methylhexane	1.0130E−06	4.5610E−01	3.5978E+02	0.0000E+00	155	1000
23	*n*-Octane	3.1183E−08	9.2920E−01	5.5089E+01	0.0000E+00	216	1000
24	2-Methylheptane	4.4595E−07	5.5350E−01	2.2222E+02	0.0000E+00	164	1000
37	2,2,4-Trimethylpentane	1.1070E−07	7.4600E−01	7.2389E+01	0.0000E+00	166	1000
41	*n*-Nonane	1.0339E−07	7.7300E−01	2.2050E+02	0.0000E+00	219	1000
62	*n*-Decane	2.6408E−08	9.4870E−01	7.1000E+01	0.0000E+00	243	1000
73	*n*-Undecane	3.5939E−08	9.0520E−01	1.2500E+02	0.0000E+00	248	1000
74	*n*-Dodecane	6.3443E−08	8.2870E−01	2.1950E+02	0.0000E+00	263	1000
75	*n*-Tridecane	3.5581E−08	8.9870E−01	1.6528E+02	0.0000E+00	268	1000
76	*n*-Tetradecane	4.4566E−08	8.6840E−01	2.2822E+02	4.3519E+03	279	1000
77	*n*-Pentadecane	4.0830E−08	8.7660E−01	2.1272E+02	0.0000E+00	283	1000
78	*n*-Hexadecane	1.2460E−07	7.3220E−01	3.9500E+02	6.0000E+03	291	1000
79	*n*-Heptadecane	3.1340E−07	6.2380E−01	6.9222E+02	0.0000E+00	295	1000
80	*n*-Octadecane	3.2089E−07	6.1840E−01	7.0889E+02	0.0000E+00	301	1000
81	*n*-Nonadecane	3.0460E−07	6.2220E−01	7.0556E+02	0.0000E+00	305	1000
82	*n*-Eicosane	2.9247E−07	6.2460E−01	7.0278E+02	0.0000E+00	309	1000
86	*n*-Tetracosane	2.6674E−07	6.2530E−01	7.0000E+02	0.0000E+00	324	1000
90	*n*-Octacosane	2.5864E−07	6.1860E−01	6.9833E+02	0.0000E+00	334	1000
	Naphthenes						
101	Cyclopentane	2.3623E−07	6.7460E−01	1.3900E+02	0.0000E+00	179	1000
102	Methylcyclopentane	9.0803E−07	4.9500E−01	3.5589E+02	0.0000E+00	131	1000
103	Ethylcyclopentane	2.1695E−06	3.8120E−01	5.7778E+02	0.0000E+00	134	1000
109	*n*-Propylcyclopentane	2.6053E−06	3.4590E−01	5.8556E+02	0.0000E+00	156	1000
146	Cyclohexane	6.7700E−08	8.3670E−01	3.6700E+01	0.0000E+00	279	900
147	Methylcyclohexane	6.5276E−07	5.2940E−01	3.1061E+02	0.0000E+00	147	1000
148	Ethylcyclohexane	4.1065E−07	5.7140E−01	2.3011E+02	0.0000E+00	162	1000
156	*n*-Propylcyclohexane	9.7976E−07	4.5420E−01	3.8589E+02	0.0000E+00	178	1000
157	Isopropylcyclohexane	5.7125E−07	5.2610E−01	2.7989E+02	0.0000E+00	184	1000
158	*n*-Butylcyclohexane	5.3514E−07	5.2090E−01	2.7711E+02	0.0000E+00	198	1000
168	*n*-Oecylcyclohexane	3.3761E−07	5.4480E−01	2.0728E+02	0.0000E+00	272	1000
	Olefins						
192	Ethylene	2.0793E−06	4.1630E−01	3.5272E+02	0.0000E+00	169	1000
193	Propylene	8.3395E−07	5.2700E−01	2.8339E+02	0.0000E+00	88	1000
194	1-Butene	1.0320E−06	4.8960E−01	3.4739E+02	0.0000E+00	175	1000
198	1-Pentene	1.6706E−06	4.1110E−01	4.3028E+02	0.0000E+00	108	1000
204	1-Hexene	1.3137E−06	4.3220E−01	4.0211E+02	0.0000E+00	133	1000
	Diolefins and acetylene						
292	1,3-Butadiene	2.6963E−07	6.7150E−01	1.3472E+02	0.0000E+00	164	1000
322	Acetylene	1.2019E−06	4.9520E−01	2.9139E+02	0.0000E+00	192	600
	Aromatics						
335	Benzene	3.1347E−08	9.6760E−01	7.9000E+00	0.0000E+00	279	1000
336	Toluene	8.7274E−07	4.9400E−01	3.2378E+02	0.0000E+00	178	1000
337	Ethylbenzene	3.8777E−07	5.9270E−01	2.2772E+02	0.0000E+00	178	1000

(Continued)

TABLE 8.1—(Continued)

API No.	Compound	A	B	C	D	T_{min}, K	T_{max}, K
338	o-Xylene	3.8080E−06	3.1520E−01	7.7444E+02	0.0000E+00	248	1000
339	m-Xylene	4.3098E−07	5.7490E−01	2.3861E+02	0.0000E+00	226	1000
340	p-Xylene	5.7656E−07	5.3820E−01	2.8700E+02	0.0000E+00	287	1000
341	n-Propylbenzene	1.6304E−06	4.1170E−01	5.4722E+02	0.0000E+00	173	1000
349	n-Butylbenzene	9.9652E−07	4.6320E−01	4.3278E−02	0.0000E+00	186	1000
371	n-Pentylbenzene	4.2643E−07	5.5740E−01	2.5900E+02	0.0000E+00	198	1000
372	n-Hexylbenzene	5.5928E−07	5.1090E−01	2.8722E+02	0.0000E+00	212	1000
373	n-Heptylbenzene	4.3188E−07	5.3580E−01	2.4561E−02	0.0000E+00	225	1000
374	n-Octylbenzene	5.4301E−07	4.9890E−01	2.7711E−02	0.0000E+00	237	1000
375	n-Nonylbenzene	4.8731E−07	5.0900E−01	2.6178E−02	0.0000E+00	249	1000
376	n-Decylbenzene	4.6333E−07	5.1060E−01	2.5611E−02	0.0000E+00	259	1000
377	n-Undecylbenzene	4.3614E−07	5.1410E−01	2.4761E−02	0.0000E+00	268	1000
378	n-Dodecylbenzene	3.7485E−07	5.2390E−01	2.1878E−02	0.0000E+00	276	1000
379	n-Tridecylbenzene	3.5290E−07	5.2760E−01	2.1039E−02	0.0000E+00	283	1000
384	Styrene	6.3856E−07	5.2540E−01	2.9511E+02	0.0000E+00	243	1000
342	Cumene	4.1805E−06	3.0520E−01	8.8000E+02	0.0000E+00	177	1000
	Diaromatics						
427	Naphthalene	6.4323E−07	5.3890E−01	4.0022E+02	0.0000E+00	353	1000
428	1-Methylnaphthalene	2.6217E−07	6.4260E−01	2.3522E+02	0.0000E+00	243	1000
474	Anthracene	7.3176E−08	7.5320E−01	1.0000E+00	0.0000E+00	489	1000
475	Phenanthrene	4.3474E−07	5.2720E−01	2.3828E+02	0.0000E+00	372	1000
	Aromatics amines						
746	Pyridine	5.2402E−08	9.0080E−01	6.2722E+01	0.0000E+00	232	1000
749	Quinoline	1.3725E−06	4.8350E−01	9.2389E+02	−6.7901E+04	511	1000
	Sulfur						
776	Carbonyl sulfide	2.2405E−05	2.0430E−01	1.3728E+03	0.0000E+00	134	1000
828	Methyl mercaptan	1.6372E−07	7.6710E−01	1.0800E+02	0.0000E+00	150	1000
891	Thiophene	1.0300E−06	5.4970E−01	5.6944E+02	0.0000E+00	235	1000
892	Tetrahydrothiophene	1.6446E−07	7.4400E−01	1.4472E+02	0.0000E+00	394	1000
	Alcohols						
709	Methanol	3.07E−007	6.9650E−001	2.0500E+02	0.0000E+00	240	1000
710	Ethanol	1.06E−006	8.0660E−001	5.2700E+02	0.0000E+00	200	1000
712	Isopropanol	1.99E−007	7.2330E−001	1.7800E+02	0.0000E+00	186	1000
766	Methyl-tert-butyl ether	1.54E−007	7.3600E−001	1.0822E+02	0.0000E+00	164	1000

A simpler version of Eq. (8.7) for a gas mixture is given as [9]:

$$(8.8) \qquad \mu_{om} = \frac{\sum_{i=1}^{N} x_i \phi_i \mu_{oi}}{\sum_{i=1}^{N} x_i \phi_i}$$

where N is the total number of compounds in the mixture, $\phi_i = M_i^{1/2}$, and subscript o indicates low pressure (atmospheric and below) while subscript m indicates mixture property. By assuming $\phi_i = 1$ this equation reduces to Kay's mixing rule ($\mu_m = \sum x_i \mu_i$), which usually gives a reasonably acceptable result at very low pressure.

Pressure has a good effect on the viscosity of real gases and at a constant temperature with increase in pressure viscosity also increases. For simple gases at high pressures, reduced viscosity (μ_r) is usually correlated to T_r and P_r based on the theory of corresponding states [1]. μ_r is defined as the ratio of μ/μ_c, where μ_c is called critical viscosity and represents viscosity of a gas at its critical point (T_c and P_c).

$$(8.9) \qquad \mu_c = 6.16 \times 10^{-3}(MT_c)^{\frac{1}{2}}(V_c)^{-\frac{2}{3}}$$

$$(8.10) \qquad \mu_c = 7.7 \times 10^{-4}\xi^{-1}$$

In the above relations, μ_c is in cp, T_c in kelvin, V_c is in cm³/mol, and ξ is defined by Eq. (8.5). Equation (8.10) can be obtained by combining Eqs. (8.9) and (8.5) with Eq. (2.8) assuming $Z_c = 0.27$. In some predictive methods, reduced viscosity is defined with respect to viscosity at atmospheric pressure (i.e., $\mu_r = \mu/\mu_a$), where μ_a is the viscosity at 1 atm and temperature T at which μ must be calculated. Another reduced form of

viscosity is $(\mu - \mu_a)\xi$, which is also called as residual viscosity (similar to residual heat capacity) and is usually correlated to the reduced density ($\rho_r = \rho/\rho_c = V_c/V$). For pure hydrocarbon gases at high pressures the following method is recommended in the API-TDB [5]:

$$(\mu - \mu_a)\xi = 1.08 \times 10^{-4} \left[\exp(1.439\rho_r) - \exp(-1.11\rho_r^{1.858})\right]$$
$$(8.11)$$

The same equation can be applied to mixtures if T_c, P_c, M, and V_c of the mixture are calculated from Eq. (7.1). V or ρ can be estimated from methods of Chapter 5. For mixtures, in cases that there is at least one data point on μ, it can be used to obtain μ_a rather than to use its estimated value. Equation (8.11) may also be used for nonpolar nonhydrocarbons as recommended in the DIPPR manual [10]. However, in the API-TDB another generalized correlation for nonhydrocarbons is given in the form of μ/μ_a versus T_r and P_r with some 22 numerical constants. The advantage of this method is mainly simplicity in calculations since there is no need to calculate ρ_r and μ can be directly calculated through μ_a and T_r and P_r.

In the petroleum industry one of the most widely used correlations for estimation of viscosity of dense hydrocarbons is proposed by Jossi et al. [11]:

$$\left[(\mu - \mu_o)\xi + 10^{-4}\right]^{\frac{1}{4}} = 0.1023 + 0.023364\rho_r + 0.058533\rho_r^2$$
$$(8.12) \qquad\qquad - 0.040758\rho_r^3 + 0.0093324\rho_r^4$$

This equation is, in fact, a modification of Eq. (8.11) and was originally developed for nonpolar gases in the range of $0.1 < \rho_r < 3$. μ_o is the viscosity at low pressure and at the same temperature at which μ is to be calculated. μ_o may be calculated from Eqs. (8.6)–(8.8). However, this equation is also used by reservoir engineers for the calculation of the viscosity of reservoir fluids under reservoir conditions [9, 12]. Later Stiel and Thodos [13] proposed similar correlations for the residual viscosity of polar gases:

$$(\mu - \mu_o)\xi = 1.656 \times 10^{-4}\rho_r^{1.111} \qquad \text{for } \rho_r \leq 0.1$$

$$(\mu - \mu_o)\xi = 6.07 \times 10^{-6} \times (9.045\rho_r + 0.63)^{1.739}$$

$$(8.13) \qquad\qquad \text{for } 0.1 \leq \rho_r \leq 0.9$$

$$\log_{10}\left\{4 - \log_{10}\left[(\mu - \mu_o) \times 10^4 \xi\right]\right\} = 0.6439 - 0.1005\rho_r$$

$$\text{for } 0.9 \leq \rho_r \leq 2.2$$

These equations are mainly recommended for calculation of viscosity of dense polar and nonhydrocarbon gases. At higher reduced densities accuracy of Eqs. (8.11)–(8.13) reduces.

For undefined gas mixtures with known molecular weight M, the following relation can be used to estimate viscosity at temperature T [5]:

$$\mu_o^g = -0.0092696 + \sqrt{T}(0.001383 - 5.9712 \times 10^{-5}\sqrt{M})$$

$$(8.14) \qquad + 1.1249 \times 10^{-5}M$$

where T is in kelvin and μ_o^g is the viscosity of gas at low pressure in cp. Reliability of this equation is about 6% [5]. There are a number of empirical correlations for calculation of viscosity of natural gases at any T and P; one widely used correlation was proposed by Lee et al. [14]:

$$\mu^g = 10^{-4}A\left[\exp\left(B \times \rho^C\right)\right]$$

$$A = \left[(12.6 + 0.021M)\,T^{1.5}\right]/(116 + 10.6M + T)$$

$$(8.15)$$

$$B = 3.45 + 0.01M + \frac{548}{T}$$

$$C = 2.4 - 0.2B$$

where μ^g is the viscosity of natural gas in cp, M is the gas molecular weight, T is absolute temperature in kelvin, and ρ is the gas density in g/cm^3 at the same T and P that μ^g should be calculated. This equation may be used up to 550 bar and in the temperature range of 300–450 K. For cases where M is not known, it may be calculated from specific gravity of the gas as discussed in Chapter 3 ($M = 29\,SG_g$). For sour natural gases, correlations in terms of H_2S content of natural gas are available in handbooks of reservoir engineering [15, 16].

8.1.2 Viscosity of Liquids

Methods for the prediction of the viscosity of liquids are less accurate than the methods for gases, especially for the estimation of viscosity of undefined petroleum fractions and crude oils. Errors of 20–50% or even 100% in prediction of liquid viscosity are not unusual. Crude oil viscosity at room temperature varies from less than 10 cp (light oils) to many thousands of cp (very heavy oils). Usually conventional oils with API gravities from 35 to 20 have viscosities from 10 to 100 cp and heavy crude oils with API gravities from 20 to 10 have viscosities from 100 to 10000 cp [17]. Most of the methods developed for estimation of liquid viscosity are empirical in

nature. An approximate theory for liquid transport properties is the Eyring rate theory [1, 4]. Effect of pressure on the liquid viscosity is less than its effect on viscosity of gases. At low and moderate pressure, liquid viscosity may be considered as a function of temperature only. Viscosity of liquids decreases with increase in temperature. According to the Eyring rate model the following relation can be derived on a semitheoretical basis:

$$(8.16) \qquad \mu = \frac{N_A h}{V} \exp\left(\frac{3.8T_b}{T}\right)$$

where μ is the liquid viscosity in posie at temperature T, N_A is the Avogadro number (6.023×10^{23} gmol^{-1}), h is the Planck's constant (6.624×10^{-27} g$\cdot$cm^2/s), V is the molar volume at temperature T in cm^3/mol, and T_b is the normal boiling point. Both T_b and T are in kelvin. Equation (8.16) suggests that $\ln \mu$ versus $1/T$ is linear, which is very similar to the Clasius–Clapeyron equation (Eq. 7.27) for vapor pressure. More accurate correlations for temperature dependency of liquid viscosities can be obtained based on a more accurate relation for vapor pressure. In the API-TDB [5] liquid viscosity of pure compounds is correlated according to the following relation:

$$(8.17) \qquad \mu = 1000 \exp\left(A + B/T + C \ln T + DT^E\right)$$

where T is in kelvin and μ is in cp. Coefficients A–E for a number of compounds are given in Table 8.2 [5]. Liquid viscosity of some n-alkanes versus temperature calculated from Eq. (8.17) is shown in Fig. 8.2. Equation (8.17) has uncertainty of better than ±5% over the entire temperature ranges given in Table 8.2. In most cases the errors are less than 2% as shown in the API-TDB [5].

For defined liquid mixtures the following mixing rules are recommended in the API-TDB and DIPPR manuals [5, 10]:

$$\mu_m = \left(\sum_{i=1}^{N} x_i \mu_i^{1/3}\right)^3 \qquad \text{for liquid hydrocarbons}$$

$$(8.18)$$

$$\ln \mu_m = \sum_{i=1}^{N} x_i \ln \mu_i \qquad \text{for liquid nonhydrocarbons}$$

where μ_m is the mixture viscosity in cp and x_i is the mole fraction of component i with viscosity μ_i. There are some other mixing rules that are available in the literature for liquid viscosity of mixtures [18].

For liquid petroleum fractions (undefined mixtures), usually kinematic viscosity v is either available from experimental measurements or can be estimated from Eqs. (2.128)–(2.130), at low pressures and temperatures. The following equation developed by Singh may also be used to estimate v at any T as recommended in the API-TDB [5]:

$$\log_{10}(v_T) = A\left(\frac{311}{T}\right)^B - 0.8696$$

$$(8.19) \qquad A = \log_{10}(v_{38(100)}) + 0.8696$$

$$B = 0.28008 \times \log_{10}(v_{38(100)}) + 1.8616$$

where T is in kelvin and $v_{38(100)}$ is the kinematic viscosity at 100°F (37.8°C or 311 K) in cSt, which is usually known from experiment. The average error for this method is about 6%. For blending of petroleum fractions the simplest method is

TABLE 8.2—*Coefficients of Eq. (8.17) for viscosity of pure liquid compounds. (Taken with permission from Ref. [5].)*

$$\mu = 1000 \exp(A + B/T + C \ln T + DT^E) \quad (8.17)$$

API No.	Compound	A	B	C	D	E	T_{min}, K	T_{max}, K
794	Oxygen	−4.1480E+00	9.4039E+01	−1.2070E+00	0.0000E+00	0.0000E+00	54	150
781	Hydrogen	−1.1660E+01	2.4700E+01	−2.6100E−01	−4.1000E−16	1.0000E+01	14	33
845	Water	−5.2840E+01	3.7040E+03	5.8660E+00	−5.8791E−29	1.0000E+01	273	646
771	Ammonia	−6.7430E+00	5.9828E+02	−7.3410E−01	−3.6901E−27	1.0000E+01	196	393
786	Hydrogen sulfide	−1.0900E+01	7.6211E+02	−1.1860E−01	0.0000E+00	0.0000E+00	188	350
798	Nitrogen	1.6000E+01	−1.8160E+02	−5.1550E+00	0.0000E+00	0.0000E+00	63	124
775	Carbon dioxide	1.8770E+01	−4.0290E+02	−4.6850E+00	−6.9999E−26	1.0000E+01	219	304
	Paraffins							
1	Methane	−6.1570E+00	1.7810E+02	−9.5240E−01	−9.0611E−24	1.0000E+01	91	188
2	Ethane	−3.4130E+00	1.9700E+02	−1.2190E+00	−9.2022E−26	1.0000E+01	91	300
3	Propane	−6.9280E+00	4.2080E+02	−6.3280E−01	−1.7130E−26	1.0000E+01	86	360
4	n-Butane	−7.2470E+00	5.3480E+02	−5.7470E−01	−4.6620E−27	1.0000E+01	135	420
5	Isobutane	−1.8340E+01	1.0200E+03	1.0980E+00	−6.1001E−27	1.0000E+01	190	400
6	n-Pentane	−2.0380E+01	1.0500E+03	1.4870E+00	−2.0170E−27	1.0000E+01	143	465
7	Isopentane	−1.2600E+01	8.8911E+02	2.0470E−01	0.0000E+00	0.0000E+00	150	310
8	Neopentane	−5.6060E+01	3.0290E+03	6.5860E+00	0.0000E+00	0.0000E+00	257	304
9	n-Hexame	−2.0710E+01	1.2080E+03	1.4990E+00	0.0000E+00	0.0000E+00	178	343
10	2-Methylpentane	−1.2860E+01	9.4689E−04	2.6190E−01	0.0000E+00	0.0000E+00	119	333
14	n-Heptane	−2.4450E+01	1.5330E+03	2.0090E+00	0.0000E+00	0.0000E+00	183	373
15	2-Methylhexane	−1.2220E+01	1.0210E+03	1.5190E−01	0.0000E+00	0.0000E+00	155	363
23	n-Octane	−2.0460E+01	1.4970E+03	1.3790E+00	0.0000E+00	0.0000E+00	216	399
24	2-Methylheptane	−1.1340E+01	1.0740E+03	1.3050E−02	0.0000E+00	0.0000E+00	164	391
37	2,2,4-Trimethylpentane	−1.2770E+01	1.1300E+03	2.3460E−01	−3.7069E−28	1.0000E+01	166	541
41	n-Nonane	−2.1150E+01	1.6580E+03	1.4540E+00	0.0000E+00	0.0000E+00	219	424
62	n-Decane	−1.6470E+01	1.5340E+03	7.5110E−01	0.0000E+00	0.0000E+00	243	448
73	n-Undecane	−1.9320E+01	1.7930E+03	1.1430E+00	0.0000E+00	0.0000E+00	248	469
74	n-Dodecane	−2.1386E+05	1.9430E+03	1.3200E+00	0.0000E+00	0.0000E+00	263	489
75	n-Tridecane	−2.1010E+01	2.0430E+03	1.3690E+00	0.0000E+00	0.0000E+00	268	509
76	n-Tetradecane	−2.0490E+01	2.0880E+03	1.2850E+00	0.0000E+00	0.0000E+00	279	528
77	n-Pentadecane	−1.9300E+01	2.0890E+03	1.1090E+00	0.0000E+00	0.0000E+00	283	544
78	n-Hexadecane	−2.0180E+01	2.2040E+03	1.2290E+00	0.0000E+00	0.0000E+00	291	564
79	n-Heptadecane	−1.9990E+01	2.2450E+03	1.1980E+00	0.0000E+00	0.0000E+00	295	576
80	n-Octadecane	−2.2690E+01	2.4660E+03	1.5700E+00	0.0000E+00	0.0000E+00	301	590
81	n-Nonadecane	−1.63995E+01	2.1200E+03	6.8810E−01	0.0000E+00	0.0000E+00	305	603
82	n-Eicosane	−1.8310E+01	2.2840E+03	9.5480E−01	0.0000E+00	0.0000E+00	309	617
86	n-Tetracosane	−2.0610E+01	2.5360E+03	1.2940E+00	−7.0442E−30	1.0000E+01	324	793
	Naphthenes							
101	Cyclopentane	−3.2610E+00	6.1422E+02	−1.1560E+00	0.0000E+00	0.0000E+00	225	325
102	Methylcyclopentane	−1.8550E+00	6.1261E+02	−1.3770E+00	0.0000E+00	0.0000E+00	248	353
103	Ethylcyclopentane	−6.8940E+00	8.1861E+02	−5.9410E−01	0.0000E+00	0.0000E+00	253	378
109	n-Propylcyclopentane	−2.3300E+01	1.6180E+03	1.8470E+00	0.0000E+00	0.0000E+00	200	404
110	Isopropylcyclopentane	−1.0500E+01	1.0840E+03	−8.2650E−02	0.0000E+00	0.0000E+00	162	399
146	Cyclohexane	−6.9310E+01	4.0860E+03	8.5250E+00	0.0000E+00	0.0000E+00	285	354
147	Methylcyclohexane	−1.5920E+01	1.4440E+03	6.6120E−01	2.1830E−27	1.0000E+01	200	393
148	Ethylcyclohexane	−2.2110E+01	1.6730E+03	1.6410E+00	0.0000E+00	0.0000E+00	200	405
156	n-Propylcyclohexane	−3.1230E+01	2.1790E+03	2.9730E+00	0.0000E+00	0.0000E+00	248	430
158	n-Butylcyclohexane	−3.9820E+01	2.6870E+03	4.2270E+00	0.0000E+00	0.0000E+00	253	454
168	n-Decylcyclohexane	−2.7670E+01	2.9210E+03	2.1910E+00	0.0000E+00	0.0000E+00	272	420
	Olefins							
192	Ethylene	1.8880E+00	7.8861E+01	−2.1550E+00	0.0000E+00	0.0000E+00	104	250
193	Propylene	−9.1480E+00	5.0090E+02	−3.1740E−01	0.0000E+00	0.0000E+00	88	320
	Diolefins and acetylenes							
322	Acetylene	6.2240E+00	−1.5180E+02	−2.6550E+00	0.0000E+00	0.0000E+00	204	384
	Aromatics							
335	Benzene	−7.3700E+00	1.0380E+03	−6.1810E−01	−1.1020E−28	1.0000E+01	279	545
336	Toluene	−6.0670E+01	3.1490E+03	7.4820E+00	−5.7092E−27	1.0000E+01	178	384
337	Ethylbenzene	−1.0450E+01	1.0480E+03	−7.1500E−02	0.0000E+00	0.0000E+00	248	413
338	o-Xylene	−1.5680E+01	1.4040E+03	6.6410E−01	0.0000E+00	0.0000E+00	248	418
341	n-Propylbenzene	−1.8280E+01	1.5500E+03	1.0450E+00	0.0000E+00	0.0000E+00	200	432
349	n-Butylbenzene	−2.3800E+01	1.8870E+03	1.8480E+00	0.0000E+00	0.0000E+00	200	457
371	n-Pentylbenzene	−7.8290E+01	4.4840E+03	9.9270E+00	−2.3490E−27	1.0000E+01	220	478
372	n-Hexylbenzene	−8.8060E+01	5.0320E+03	1.1360E+01	−2.6390E−27	1.0000E+01	220	499
373	n-Heptylbenzene	−9.5724E+01	5.4770E+03	1.2480E+01	−2.8510E−27	1.0000E+01	336	519
374	n-Octylbenzene	−9.4614E+01	5.5678E+03	1.2260E+01	−1.8370E−27	1.0000E+01	237	538

(Continued)

TABLE 8.2—*(Continued)*

API No.	Compound	A	B	C	D	E	T_{min}, K	T_{max}, K
375	n-Nonylbenzene	−1.0510E+02	6.1272E+03	1.3820E+01	−2.8910E−27	1.0000E+01	360	555
376	n-Decylbenzene	−1.0710E+02	6.3311E+03	1.4080E+01	−2.7260E−27	1.0000E+01	253	571
377	n-Undecylbenzene	−1.0260E+02	6.2200E+03	1.3380E+01	−2.4450E−27	1.0000E+01	258	587
378	n-Dodecylbenzene	−8.8250E+01	5.6472E+03	1.1230E+01	−1.8200E−27	1.0000E+01	268	601
379	n-Tridecylbenzene	−4.5740E+01	3.6870E+03	4.9450E+00	−5.8391E−28	1.0000E+01	328	614
383	Cyclohexylbenzene	−4.3530E+00	1.4700E+03	−1.1600E+00	0.0000E+00	0.0000E+00	280	513
386	Styrene	−2.2670E+01	1.7580E+03	1.6700E+00	0.0000E+00	0.0000E+00	243	418
342	Cumene	−2.4962E+01	1.8079E+03	2.0556E+00	0.0000E+00	0.0000E+00	200	400
	Diaromatics and condensed rings							
427	Naphthalene	−1.9310E+01	1.8230E+03	1.2180E+00	0.0000E+00	0.0000E+00	353	633
472	Acenaphthene	2.0430E+01	1.0380E+02	−4.6070E+00	0.0000E+00	0.0000E+00	367	551
473	Fluorene	4.1850E+00	7.2328E+02	−2.1490E+00	0.0000E+00	0.0000E+00	388	571
474	Anthracene	−2.7430E+02	2.1060E+04	3.6180E+01	0.0000E+00	0.0000E+00	489	595
709	Methanol	1.2135E+04	1.7890E+03	2.0690E+04	0.0000E+00	0.0000E+00	176	338
710	Ethanol	7.8750E+00	7.8200E+02	−3.0420E+00	0.0000E+00	0.0000E+00	200	440

to use Eq. (3.105) by calculating blending index of the mixture. The viscosity-blending index can be calculated from the following relation proposed by Chevron Research Company [19]:

$$(8.20) \qquad BI_{vis} = \frac{\log_{10} \nu}{3 + \log_{10} \nu}$$

$$BI_{mix} = \sum x_{vi} BI_i$$

in which ν is the kinematic viscosity in cSt. Once ν is determined absolute viscosity of a petroleum fraction can be estimated from density ($\mu = \rho \times \nu$). It should be noted that Eqs. (2.128)–(2.130) or Eqs. (8.19) and (8.20) are not suitable for pure hydrocarbons.

To consider the effect of pressure on liquid viscosity of hydrocarbons, the three-parameter corresponding states correlations may be used for prediction of viscosity of high-pressure liquids [5]:

$$(8.21) \qquad \mu_r = \frac{\mu}{\mu_c} = [\mu_r]^{(0)} + \omega [\mu_r]^{(1)}$$

where $[\mu_r]^{(0)}$ and $[\mu_r]^{(1)}$ are functions of T_r and P_r. These functions are given in the API-TDB [5] in the form of polynomials in terms of T_r and P_r with more than 70 numerical constants.

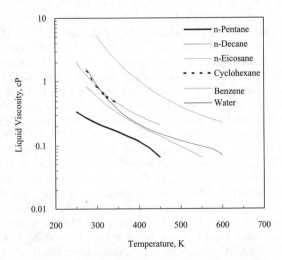

FIG. 8.2—Liquid viscosity of several compounds versus temperature at atmospheric pressure.

More recently a corresponding state correlation similar to this equation was proposed for estimation of viscosity of hydrocarbon fluids at elevated pressures in which the reduced molar refraction (parameter r defined by Eq. 5.129) was used instead of ω [20]. Parameters $[\mu_r]^{(0)}$ and $[\mu_r]^{(1)}$ have been correlated to T_r and P_r. Results show that for hydrocarbon systems, parameter ω can be replaced by r in the corresponding states correlations. Such correlations have higher power of extrapolation to heavier hydrocarbons. Moreover, parameter r can be accurately calculated for heavy petroleum fractions and undefined hydrocarbon mixtures as discussed in Section 5.9.

Equation (8.21) is recommended for low-molecular-weight hydrocarbons [5]. For such systems, Jossi's correlation (Eq. 8.12) can also be used for calculation of viscosity of high-pressure liquids. However, this approach is not appropriate for heavy or high-molecular-weight liquid hydrocarbons and their mixtures. For such liquids the Kouzel correlation is recommended in the API-TDB [5]:

$$(8.22) \qquad \log_{10}\left(\frac{\mu_P}{\mu_a}\right) = \frac{P - 1.0133}{10000}\left(-1.48 + 5.86\mu_a^{0.181}\right)$$

where P is pressure in bar and μ_a is low-pressure (1 atm) viscosity at a given temperature in cp. μ_P is the viscosity at pressure P and given temperature in cp. The maximum pressure for use in the above equation is about 1380 bar (~20000 psi) and average error is about 10% [5].

When a gas is dissolved in a pure or mixed liquid hydrocarbons viscosity of solution can be calculated from viscosity of gas-free hydrocarbon (μ_a) and gas-to-liquid ratio (GLR) using the following relation [5]:

$$\frac{\mu_m}{\mu_a} = \left\{\frac{1.651(GLR) + 137\mu_a^{1/3} + 538.4}{\mu_a^{1/3}[137 + 4.891(GLR)] + 538.4}\right\}^3$$

$$(8.23)$$

$$\log_{10}\mu_T = -1.209 + 132.8\left(\frac{1.209 + \log_{10}(\mu_m)}{T - 178}\right)$$

where both μ_m and μ_a are at 37.8°C (100°F) in cp and GLR is in m³/m³. μ_T is the viscosity of solution at temperature T, where T is in kelvin. This equation should not be used for pressures above 350 bar. If μ_a at 37.8°C (100°F) is not available, it may be estimated; however, if μ_a at the same temperature at which μ is to be calculated is available then μ may be estimated from

$\mu = A(\mu_a)^B$, where A and B are functions of GLR (see Problem 8.4). GLR were calculated from the following relation:

$$(8.24) \qquad \text{GLR} = \frac{379 x_A}{(1 - x_A) \times \left(\dfrac{M_B}{62.4 \text{SG}_B}\right)}$$

where x_A is the mole fraction of dissolved gas in liquid, M_B is molecular weight of liquid, and SG_B is the specific gravity of liquid. In this relation GLR is calculated as stm³ of gas/stm³ of liquid (1 m³/m³ = 1 scf/st·ft³ = 5.615 scf/bbl). Units of GLR are discussed in Section 1.7.23. Prediction of viscosity of crude oils (gas free dead oils at 1 atm) is quite difficult due to complexity of mixtures. However, there are many empirical correlations developed for calculation of crude oils [15, 16]. For example the Glaso's correlation for viscosity of crude oils is given as

$$\mu_{od} = (3.141 \times 10^{10}) \times [(1.8T - 460)^{-3.444}] \times [\log_{10}(\text{API})]^n$$

$$n = 10.313 \left[\log_{10}(1.8T - 460)\right] - 36.447$$

$$(8.25)$$

where μ_{od} is the viscosity of dead oil (gas free at 1 atm.), T is temperature in kelvin, and API is the oil gravity. This equation should be used for crude oils with API gravity in the range of 20–48 and in the temperature range of 283–422 K (50–300°F). More advanced and accurate methods of calculation of viscosity of crude oils is based on splitting the oil into several pseudocomponents and to use methods discussed in Chapter 4 for calculation of the mixture properties. Accurate prediction of viscosities of heavy crude oils is a difficult task and most correlations result in large errors and errors of 50–100% are quite common in such predictions.

As seen from Eqs. (8.11) and (8.25), viscosity of liquids and oils is mainly related to density. In general, heavier oils (lower API gravity) exhibit higher viscosity. Pure hydrocarbon paraffins have viscosity of about 0.35 cp (0.5 cSt.), naphthenes about 0.6 cp, n-alkylbenzenes (aromatics) about 0.8 cp (1.1 cSt.), gasoline about 0.6 cp, kerosene about 2 cp, and residual oils' viscosity is in the range of 10–100 000 cp [17]. The methods of measurement of viscosity of oils are given in ASTM D 445 and D 446. A graphical method for calculation of viscosity of the blend is given by ASTM D 341. For light oils capillary viscometers are suitable for measuring liquid viscosity in which viscosity is proportional to the pressure difference in two tubes.

Most recently Riazi et al. [21] developed a relation for estimation of viscosity of liquid petroleum fractions by using refractive index at 20°C as one of the input parameters in addition to molecular weight and boiling point (see Problem 8.3). Another development on the prediction of viscosity and other transport properties for liquid hydrocarbon systems was to use refractive index to estimate a transport property at the same temperature in which relative index is available. Theory of Hildebrand [22] suggests that fluidity ($1/\mu$) of a liquid is proportional to the free space between the molecules.

$$(8.26) \qquad \frac{1}{\mu} = E \left(\frac{V - V_0}{V_0}\right)$$

where E is a constant, V is the liquid volume (i.e., molar), and V_0 is the value of V at zero fluidity ($\mu \to 0$). Parameters E and V_0 may be determined from regression of experimental data.

The term $(V - V_0)$ represents the free space between molecules. As temperature increases V also increases and μ decreases. This theory is applicable to liquids at low pressures. In Chapter 2 it was shown that parameter I (defined by Eq. 2.36) is proportional with fraction of liquid occupied by molecules. Therefore parameter I is proportional to V_0/V and thus

$$(8.27) \qquad \mu^{-1} = C \left(I^{-1} - 1\right)$$

where μ and I are evaluated at given temperature. Methods of calculation of I were discussed in Chapter 2 (see Eqs. (2.36) and (2.118)). On this basis, one can see that $1/\mu$ varies linearly with $1/I$ for any substance. This relation has been also confirmed with experimental data [23]. Similar correlations for thermal conductivity and diffusivity were developed and the coefficients were related to hydrocarbon properties such as molecular weight [23, 24]. Equation (8.27) is applicable only to nonpolar and hydrocarbon liquid systems in which the intermolecular forces can be determined by London forces. Other developments in the calculation of liquid viscosity are reported by Chung et al. (generalized correlations for polar and nonpolar compounds) [25] and Quinones-Cisneros et al. (pure hydrocarbons and their mixtures) [26].

Example 8.1—Consider a liquid mixture of 74.2 mol% acetone and 25.8 mol% carbon tetrachloride (CCl_4) at 298.2 K and 1 atm. Estimate its viscosity assuming the only information known for this system are T_c, P_c, V_c, ω, M, and Z_{RA} of each compound. Compare estimated value with the experimental value of 0.395 mPa·s (cp) [10].

Solution—CCl_4 and acetone are nonhydrocarbons whose critical properties are not given in Table 2.1 and for this reason they are obtained from other sources such as DIPPR [10] or any chemical engineering thermodynamics text as [18,27]: for acetone, $T_c = 508.2$ K, $P_c = 47.01$ bar, $V_c = 209$ cm³/mol, $\omega = 0.3065$, $M = 58.08$ g/mol, and $Z_{RA} = 0.2477$; for CCl_4, $T_c = 556.4$ K, $P_c = 45.6$ bar, $V_c = 276$ cm³/mol, $\omega = 0.1926$, $M = 153.82$ g/mol, and $Z_{RA} = 0.2722$ [18]. Using the Kay's mixing rule (Eq. 7.1) with $x_1 = 0.742$ and $x_2 = 0.258$: $T_c = 520.6$ K, $P_c = 46.6$ bar, $V_c = 226.3$ cm³/mol, $\omega = 0.2274$, $M = 82.8$, and $Z_{RA} = 0.254$. Mixture liquid density at 298 K is calculated from Racket equation (Eq. 5.121): $V^s = 80.5$ cm³/mol ($\rho_{25} = 1.0286$ g/cm³). This gives $\rho_r = V_c/V = 226.3/80.5 = 2.8112$. For calculation of residual viscosity a generalized correlation in terms of ρ_r may be used. Although Eq. (8.12) is proposed for hydrocarbons and nonpolar fluids, for liquids ρ_r is quite high and the equation can be used up to ρ_r of 3.0. From Eq. (8.5), $\xi = 0.02428$ and $T_r = T/T_c = 0.5724 < 1.5$. From Eq. (8.6), $\mu_o = 0.00829$ cp. From Eq. (8.12), $\mu = 0.374$ cp, which in comparison with experimental value of 0.395 cp gives an error of only −5.3%. This is a good prediction considering the fact that the mixture contains a highly polar compound (acetone) and predicted density was used instead of a measured value. If actual values of ρ_{25} [18] for pure compounds were used ($\rho_{25} = 0.784$ for acetone and $\rho_{25} = 1.584$ g/cm³ for CCl_4) and density is calculated from Eq. (7.4) we get $\rho_{25} = 1.03446$ g/cm³ ($\rho_r = 2.828$), which predicts $\mu_{mix} = 0.392$ cp (error of only −0.8%). ♦

8.2 ESTIMATION OF THERMAL CONDUCTIVITY

Thermal conductivity is a molecular property that is required for calculations related to heat transfer and design and operation of heat exchangers. It is defined according to the Fourier's law:

$$(8.28) \qquad q_y = -k\frac{\partial T}{\partial y} = -\alpha\frac{\partial (\rho C_P T)}{\partial y}$$

$$\alpha = \frac{k}{\rho C_P}$$

where q_y is the heat flux (heat transferred per unit area per unit time, i.e., $J/m^2 \cdot s$ or W/m^2) in the y direction, $\partial T/\partial y$ is the temperature gradient, and the negative sign indicates that heat is being transferred in the direction of decreasing temperature. The proportionality constant is called thermal conductivity and is shown by k. This equation shows that in the SI unit systems, k has the unit of $W/m \cdot K$, where K may be replaced by °C since it represents a temperature difference. In English unit system it is usually expressed in terms of $Btu/ft \cdot h \cdot °F$ (= 1.7307 $W/m \cdot K$). The unit conversions are given in Section 1.7.19. In Eq. (8.28), $\rho C_P T$ represents heat per unit volume and coefficient $k/\rho C_P$ is called *thermal diffusivity* and is shown by α. A comparison between Eq. (8.28) and Eq. (8.1) shows that these two equations are very similar in nature as one represents flux of momentum and the other flux of heat. Coefficients ν and α have the same unit (i.e., cm^2/s) and their ratio is a dimensionless number called *Prandtl* number N_{Pr}, which is an important number in calculation of heat transfer by conduction in flow systems. In use of correlations for calculation of heat transfer coefficients, N_{Pr} is needed [28].

$$(8.29) \qquad N_{Pr} = \frac{\nu}{\alpha} = \frac{\mu C_P}{k}$$

At 15.5°C (60°F), values of N_{Pr} for *n*-heptane, *n*-octane, benzene, toluene, and water are 6.0, 5.0, 7.3, 6.5, and 7.7, respectively. These values at 100°C (212°F) are 4.2, 3.6, 3.8, 3.8, and 1.5, respectively [28]. Vapors have lower N_{Pr} numbers, i.e., for water vapor $N_{Pr} = 1.06$. Thermal conductivity is a molecular property that varies with both temperature and pressure. Vapors have k values less than those for liquids. Thermal conductivity of liquids decreases with an increase in temperature as the space between molecules increases, while for vapors thermal conductivity increases with temperature as molecular collision increases. Pressure increases thermal conductivity of both vapors and liquids. However, at low pressures k is independent of pressure. For some light hydrocarbons thermal conductivities of both gases and liquids versus temperature are shown Fig. 8.3.

Methods of prediction of thermal conductivity are very similar to those of viscosity. However, thermal conductivity of gases can generally be estimated more accurately than can liquid viscosity. For dense fluids, residual thermal conductivity is usually correlated to the reduced density similar to that of viscosity (i.e., see Eqs. (8.11)–(8.13)).

8.2.1 Thermal Conductivity of Gases

Kinetic theory provides the basis of prediction of thermal conductivity of gases. For example, based on the potential relation for hard-sphere molecules, the following equation is developed for monoatomic gases.

$$(8.30) \qquad k = \frac{1}{d^2}\sqrt{\frac{k_B^3 T}{\pi^3 m}}$$

where the parameters are defined in Eq. (8.2). This equation is independent of pressure and is valid up to pressure of 10 atm for most gases [1]. The Chapmman–Enskog theory discussed in Section 8.1.1 provides a more accurate relation in the following form:

$$(8.31) \qquad k = \frac{1.9\times 10^{-4}\left(\frac{T}{M}\right)^{1/2}}{\sigma^2\Omega}$$

where k is in $cal/cm \cdot s \cdot K$, σ is in Å, and Ω is a parameter that is a weak function of T as given for viscosity or diffusivity. This function is given later in Section 8.3.1 (Eq. 8.57). From Eq. (8.31) it is seen that thermal conductivity of gases decreases with increase in molecular weight. For polyatomic gases the Eucken formula for Prandtl number is [1]

$$(8.32) \qquad N_{Pr} = \frac{C_P}{C_P + 1.25R}$$

where C_P is the molar heat capacity in the same unit as for gas constant R. This relation is derived from theory and errors as high as 20% can be observed.

For pure hydrocarbon gases the following equation is given in the API-TDB for the estimation of thermal conductivity [5]:

$$(8.33) \qquad k = A + BT + CT^2$$

where k is in $W/m \cdot K$ and T is in kelvin. Coefficients A, B, and C for a number of hydrocarbons with corresponding temperature ranges are given in Table 8.3. This equation can be used for gases at pressures below 3.45 bar (50 psia) and has accuracy of ±5%. A generalized correlation for thermal conductivity of pure hydrocarbon gases for $P < 3.45$ bar is given as follows [5]:

$$k = 4.911\times 10^{-4}\frac{T_r C_P}{\lambda}$$

(a) only for methane and cyclic compounds at $T_r < 1$

$$k = \left[11.04\times 10^{-5}(14.52T_r - 5.14)^{2/3}\right]\frac{C_P}{\lambda}$$

(b) for all compounds at any T except (a)

$$\lambda = 1.11264\frac{T_c^{1/6}M^{1/2}}{P_c^{2/3}}$$

(8.34)

Equation (8.34) also applies to methane and cyclic compounds at $T_r > 1$, but for other compounds can be used at any temperature. The units are as follows: C_P in $J/mol \cdot K$, T_c in K, P_c in bar, and k in $W/m \cdot K$. This equation gives an average error of about 5%.

For gas mixtures the following mixing rule similar to Eq. (8.7) can be used [18]:

$$(8.35) \qquad k_m = \sum_{i=1}^{N}\frac{x_i k_i}{\sum_{j=1}^{N}x_j A_{ij}}$$

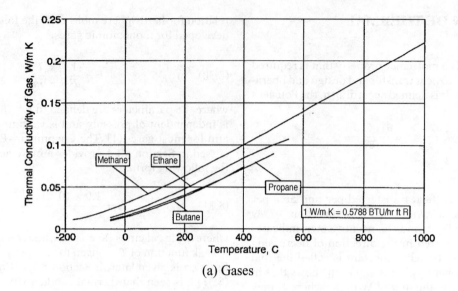

(a) Gases

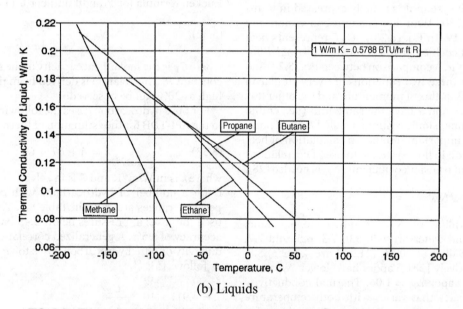

(b) Liquids

FIG. 8.3—Thermal conductivity of several light hydrocarbons versus temperature at atmospheric pressure. Taken with permission from Ref. [2].

where A_{ij} may be set equal to ϕ_{ij} given in Eq. (8.7). Another mixing rule that does not require viscosity of pure component is given by Poling et al. [18]. A more advanced mixing rule for calculation of mixture thermal conductivity of gases and liquids is provided by Mathias et al. [29]. For vapors from undefined petroleum faction, the following equation has been derived from regression of an old figure developed in the 1940s [5]:

$$k = A + B(T - 255.4)$$

(8.36)
$$A = 0.00231 + \frac{0.42624}{M} + \frac{1.9891}{M^2}$$

$$B = 1.0208 \times 10^{-4} + \frac{1.3047 \times 10^{-4}}{M} + \frac{0.00574}{M^2}$$

where k is in W/m·K and T is in kelvin. The equation should be used for pressure below 3.45 bar, for petroleum fractions with M between 50–150 and T in the range of 260–811 K. This

equation is oversimplified and should be used when other methods are not applicable. Riazi and Faghri [30] used the general relationship between k, T, and P at the critical point (T_c, P_c) to develop an equation similar to Eq. (2.38) for estimation of thermal conductivity of petroleum fractions and pure hydrocarbons.

(8.37)
$$k = 1.7307A(1.8T_b)^B SG^C$$
$$A = \exp\left(21.78 - 8.07986t + 1.12981t^2 - 0.05309t^3\right)$$
$$B = -4.13948 + 1.29924t - 0.17813t^2 + 0.00833t^3$$
$$C = 0.19876 - 0.0312t - 0.00567t^2$$
$$t = \frac{1.8T - 460}{100}$$

where k is in W/m·K, T_b and T are in kelvin. Factors 1.7307 and 1.8 come from the fact that the original units were in English. This equation can be applied to pure hydrocarbons

TABLE 8.3—*Coefficients of Eq. (8.33) for thermal conductivity of pure gases [5].*

$$k = A + BT + CT^2 \quad (8.33)$$

No.	Compound name	$A \times 10^{-1}$	$B \times 10^{-4}$	$C \times 10^{-7}$	Range, K
1	Methane	−0.0076	0.9753	0.7486	97–800
2	Ethane	−0.1444	0.9623	0.7649	273–728
3	Propane	−0.0649	0.4829	1.1050	233–811
4	*n*-Butane	0.0000	0.0614	1.5930	273–444
5	*n*-Pentane	0.0327	−0.0676	1.5580	273–444
6	*n*-Hexane	0.0147	0.0654	1.2220	273–683
7	*n*-Heptane	−0.0471	0.2788	0.9449	378–694
8	*n*-Octane	−0.1105	0.5077	0.6589	416–672
9	*n*-Nonane	−0.0876	0.4099	0.6937	450–678
10	*n*-Decane	−0.2249	0.8623	0.2636	450–678
11	*n*-Undecane	−0.1245	0.4485	0.6230	472–672
12	*n*-Dodecane	−0.2535	0.8778	0.2271	516–666
13	*n*-Pentadecane	−0.3972	1.3280	−0.2523	566–644
14	Ethene	−0.0174	0.3939	1.1990	178–589
15	Propene	−0.0844	0.6138	0.8086	294–644
16	Cyclohexane	−0.0201	0.0154	1.4420	372–633
17	Benzene	−0.2069	0.9620	0.0897	372–666
18	Toluene	−0.3124	1.3260	−0.1542	422–661
19	Ethylbenzene	−0.3383	1.3240	−0.1295	455–678
20	1,2-Dimethylbenzene (*o*-Xylene)	−0.1430	0.8962	0.0533	461–694
21	*n*-Propylbenzene	−0.3012	0.9695	0.7099	455–616

(C_5–C_{16}) or to petroleum fractions with $M > 70$ (boiling point range of 65–300°C) in the temperature range of 200–370°C ($\sim$400–700°F). Accuracy of this equation for pure compounds within the above ranges is about 3%.

The effect of pressure on the thermal conductivity of gases is usually considered through generalized correlations similar to those given for gas viscosity at high pressures. The following relation for calculation of thermal conductivity of dense gases and nonpolar fluids by Stiel and Thodos [31] is widely used with accuracy of about 5–6% as reported in various sources [10, 18]:

$$(8.38) \quad k = k^\circ + \frac{A}{\Gamma}\left[\exp(B\rho_r) + C\right]$$

$$\Gamma = 4.642 \times 10^4 \left(\frac{M^3 T_c}{P_c^4}\right)^{1/6} Z_c^5$$

For $\rho_r < 0.5$: $A = 2.702$, $B = 0.535$, $C = -1.000$

$0.5 < \rho_r < 2.0$: $A = 2.528$, $B = 0.670$, $C = -1.069$

$2.0 < \rho_r < 2.8$: $A = 0.574$, $B = 1.155$, $C = 2.016$

where k° is the thermal conductivity of low-pressure (atmospheric pressure) gas at given temperature and k is the corresponding thermal conductivity at given temperature and pressure of interest. ρ_r is the reduced density (V_c/V), T_c is in K, P_c is in bar, and Z_c is the critical compressibility factor. Both k and k° are in W/m·K. In the API-DTB [5] a generalized correlation developed by Crook and Daubert is recommended for calculation of k of dense hydrocarbon gases. However, this method requires calculation of isochoric (constant volume) heat capacity (C_V) at the T and P of interest. Another generalized correlation for estimation of thermal conductivity of gases at high pressure was developed by Riazi and Faghri [32]:

$$(8.39) \quad k_r = \frac{k}{k_c} = (0.5 - \omega)k_r^1 + \omega k_r^2$$

where k_r is the reduced thermal conductivity and k is the thermal conductivity at T and P of interest in W/m·K while k_c is the thermal conductivity at the critical point (T_c and P_c). Parameters $k_r^{(1)}$ and $k_r^{(2)}$ are determined as a function of T_r and

P_r. Values of k_c were determined from experimental data for a number of hydrocarbons and are given in Table 8.4 [32]. Values of $k_r^{(1)}$ and $k_r^{(2)}$ are given in Table 8.5. For those compounds for which values of k_c are not available they may be determined from Eq. (8.39) if only one data on k is available ($k_c = k/k_r$). In some cases, as shown in the following example, k_c can be obtained from inter/extrapolation of values given in Table 8.4.

Values of k_c reported in Table 8.4 are based on extrapolation from experimental data at subcritical conditions. It is believed that there is a great enlargement of thermal conductivity at the critical point for fluids. For mixtures, the critical enhancements are significant but the thermal conductivity remains finite [29]. Actual values of critical thermal conductivity may be substantially different from the values given in this table. For example, value of k_c from methane as shown by Mathias et al. [29] is 0.079 W/m·K, while the value given in Table 8.4 is 0.0312 W/m·K. However for ethane the value of k_c from this table is the same as obtained from method of Mathias et al. [29]. Equation (8.39) is mainly recommended for conditions different from the critical point and as long as values of k_c from Table 8.4 are used, predicted values from Eq. (8.39) are reliable.

Example 8.2—Consider *n*-pentane vapor at 300°C and 100 bar. Calculate its thermal conductivity from Stiel–Thodos and Riazi–Faghri methods.

Solution—From Table 2.1, $M = 72.2$, $T_c = 196.55°C = 469.7$ K, $P_c = 33.7$ bar, $V_c = 313.05$ cm³/mol, and $Z_c = 0.2702$. From Eq. (8.33) and coefficients for *n*-C_5 in Table 8.3 at

TABLE 8.4—*Critical thermal conductivity of some pure compounds [32].*

Compound	k_c, W/mK	Compound	k_c, W/mK
Methane	0.0312	Ethene	0.0379
Ethane	0.0319	Cyclohexane	0.0533
Propane	0.0433	Benzene	0.0472
n-Butane	0.0478	Toluene	0.0526
n-Heptane	0.0535	Ethylbenzne	0.0526

TABLE 8.5—*Values of $k_r^{(1)}$ and $k_r^{(2)}$ for Eq. (8.39). (Taken with permission from Ref. [32].)*

$$k_r = \frac{k}{k_c} = (0.5 - \omega)k_T^1 + \omega k_T^2 \quad (8.101)$$

T_r	P_r									
	0.2	0.5	1.0	1.5	2.0	3.0	4.0	6.0	8.0	10.0
Values of $k_r^{(1)}$ versus T_r and P_r										
1.00	1.1880	1.3307	2.0000	4.1517	4.4282	4.7900	5.2140	5.7989	6.2080	6.5132
1.05	1.3002	1.3640	1.8922	3.2806	3.7990	4.4915	4.7590	5.2817	5.7710	6.2040
1.10	1.4300	1.4810	1.8660	2.5989	3.3334	4.1068	4.4746	4.9502	5.3740	5.8812
1.15	1.5182	1.5365	1.8356	2.2978	2.9769	3.8583	4.4676	4.9404	5.3734	5.8760
1.20	1.8311	1.8956	2.1200	2.3983	2.8809	3.5626	4.2067	4.9285	5.3731	5.8699
1.40	2.1838	2.2520	2.3589	2.5291	2.7120	3.3000	4.0020	4.6327	5.2404	5.7656
1.60	2.5971	2.6589	2.7305	2.8572	3.0035	3.3760	3.8239	4.4385	4.8967	5.3031
2.00	3.6763	3.6984	3.7418	3.9161	3.9594	4.1370	4.3768	4.7138	5.0462	5.3614
3.00	6.9896	7.0010	7.0310	7.0617	7.1079	7.1452	7.2197	7.4077	7.5915	7.7685
Values of $k_r^{(2)}$ versus T_r and P_r										
1.00	1.6900	1.6990	2.0000	2.0619	2.3112	2.3140	2.3160	2.3180	2.3210	2.3212
1.05	1.7200	1.7290	1.8100	1.8170	2.1318	2.1912	2.3010	2.8380	2.3398	2.3400
1.10	1.8001	1.8211	1.8300	1.8310	1.9672	2.1384	2.1369	2.3614	2.3988	2.4105
1.15	2.0599	2.0601	2.0661	2.0700	2.0801	2.1269	2.2246	2.3780	2.4618	2.4622
1.20	2.1441	2.1539	2.1629	2.1681	2.1689	2.1901	2.2319	2.3981	2.4640	2.4701
1.40	2.6496	2.6772	2.6865	2.6889	2.6900	2.6911	2.7001	2.7119	2.8079	2.8810
1.60	3.2184	3.2448	3.2559	3.2886	3.3142	8.8292	3.3343	3.3352	3.8869	3.4525
2.00	4.5222	4.5330	4.5465	4.6871	4.6378	4.7108	4.8148	4.8119	4.8850	4.9885
3.00	8.4002	8.4158	8.4234	8.4503	8.4504	8.5038	8.6083	8.6204	8.6732	8.7454

$T = 573.2$ K (300°C) $k° = 0.048$ W/m·K. From Lee–Kesler correlation (Eq. 5.107), the molar volume at 573.2 K and 100 bar is calculated as $Z = 0.59$ or $V = 281$ cm³/mol. Thus $\rho_r = V_c/V = 313.05/281 = 1.114$. Since $0.5 < \rho_r < 2$, from Eq. (8.38) $\Gamma = 151.82$, $A = 2.702$, $B = 0.67$, $C = -1.069$, and $k = 0.048 + 0.017 = 0.065$ W/m·K.

To calculate k from Eq. (8.39), k_c is required. Since in Table 8.4 value of k_c for n-C_5 is not given, one can obtain it from interpolation of values given for C_4 and C_7 by assuming a linear relation between k_c and T_c. For C_4, $k_c = 0.0478$ and $T_c = 425.2$ K and for C_7, $k_c = 0.0535$ and $T_c = 540.2$ K. For C_5 with $T_c = 469.7$ by linear interpolation, $k_c = [(0.0535 - 0.0478)/(540.2 - 425.2)] \times (469.7 - 425.2) + 0.0478 = 0.05$ W/m·K. Extrapolation between values of k_c for C_3 and C_4 to k_c of C_5 gives a slightly different value. At T and P of interest, $T_r = 1.22$ and $P_r = 2.97$. From Table 8.5, $k_r^{(1)} = 3.5$ and $k_r^{(2)} = 2.2$. From Eq. (8.39), $k_r = 1.42$ and $k = 0.05 \times 1.42 = 0.071$ W/m·K. Stiel–Thodos method varies by 8.5% from Riazi–Faghri method, which represents a reasonable deviation. In this case the Stiel–Thodos method is more accurate since the value of $k°$ is calculated more accurately. ♦

8.2.2 Thermal Conductivity of Liquids

Theory of thermal conductivity of liquids was proposed by Bridgman [1]. In this theory, it is assumed that molecules are arranged as cubic lattice with center-to-center spacing of $(V/N_A)^{1/3}$, in which V is the molar volume and N_A is the Avogadro number. Furthermore, it is assumed that energy is transferred from one lattice to another at the speed of sound, c_s. This theory provides the basis of prediction of thermal conductivity of liquids. For monoatomic liquids the following relation can be obtained from this theory [1]:

$$ (8.40) \qquad k = 3\left(\frac{N_A}{V}\right)^{\frac{2}{3}} k_B c_s $$

where k_B is the Boltzman's constant and methods of calculation of c_s have been discussed in Section 6.9. For pure liquid hydrocarbons, thermal conductivity varies linearly with temperature:

$$ (8.41) \qquad k = A + BT $$

Coefficients A and B can be determined if at least two data points on thermal conductivity are available. Values of thermal conductivity of some compounds at melting and boiling points are given in Table 8.6, as given in the API-TDB [5]. Liquid thermal conductivity of several n-paraffins as calculated from Eq. (8.41) (or Eq. 8.42) is shown in Fig. 8.4.

If values of thermal conductivity at melting and boiling points are taken as reference points, then Eq. (8.41) can be used to obtain value of thermal conductivity at any other temperature:

$$ (8.42) \qquad k_T^L = k_M^L + (k_b^L - k_M^L)\frac{T - T_M}{T_b - T_M} $$

where T_M and T_b are normal melting (or triple) and boiling points, respectively. k_M^L and k_b^L are values of liquid thermal conductivity at T_M and T_b, respectively. k_T^L is value of liquid thermal conductivity at temperature T. According to API-TDB [5] this equation can predict values of liquid thermal conductivity of pure compounds up to pressure of 35 bar with an accuracy of about 5% [5]. There are a number of generalized correlations developed for prediction of thermal conductivity of pure hydrocarbon liquids. The Riedel method is included in the API-TDB [5]:

$$ (8.43) \qquad k^L = \frac{CM^n}{V_{25}^L}\left[\frac{3 + 20(1 - T_r)^{2/3}}{3 + 20\left(1 - \frac{298.15}{T_r}\right)^{2/3}}\right] $$

For unbranched, straight-chain hydrocarbons,

$n = 1.001$ and $C = 0.1811$

For branched and cyclic hydrocarbons,

$n = 0.7717$ and $C = 0.4407$

TABLE 8.6—*Liquid thermal conductivity of some pure compounds at their normal melting and boiling points [5].*

No.	Compound	T_M, K	k at T_M, W/mK	T_b, K	k at T_b, W/mK
1	Methane	90.69	0.2247	111.66	0.1883
2	Propane	85.47	0.2131	231.11	0.1289
3	n-Butane	134.86	0.1869	272.65	0.1176
4	n-Pentane	143.42	0.1783	309.22	0.1086
5	n-Hexane	177.83	0.1623	341.88	0.1042
6	2-Methylpentane	119.55	0.1600	333.41	0.1000
7	3-Methylpentane	110.25	0.1646	336.42	0.1010
8	n-Heptane	182.57	0.1599	371.58	0.1025
9	n-Octane	216.38	0.1520	398.82	0.0981
10	2.24-Trimethylpentane	165.78	0.1284	372.39	0.0815
11	n-Nonane	219.66	0.1512	423.97	0.0972
12	n-Decane	243.51	0.1456	447.31	0.0946
13	n-Undecane	247.57	0.1461	469.04	0.0930
14	n-Dodecane	263.57	0.1436	489.47	0.0909
15	n-Tridecane	267.76	0.1441	508.62	0.0896
16	n-Tetradecane	279.01	0.1423	526.73	0.0882
17	n-Pentadecane	283.07	0.1446	543.83	0.0874
18	n-Hexadecane	291.31	0.1438	560.02	0.0849
19	n-Heptadecane	295.13	0.1441	575.26	0.0819
20	n-Octadecane	301.31	0.1460	589.86	0.0810
21	n-Nonadecane	305.04	0.1453	603.05	0.0797
22	n-Eicosane	309.58	0.1488	616.94	0.0801
23	n-Heneicosane	313.35	0.1499	629.66	0.0799
24	n-Docosane	317.15	0.1513	641.75	0.0809
25	n-Tricosane	320.65	0.1516	653.35	0.0811
26	n-Tetracosane	323.75	0.1530	664.45	0.0819
27	Cyclopentane	179.31	0.1584	322.40	0.1198
28	Methylcyclopentane	130.73	0.1605	344.96	0.1071
29	Cyclohexane	279.69	0.1282	353.87	0.1096
30	Methylcyclohexane	146.58	0.1449	374.04	0.0935
31	Cyclohexane	169.67	0.1653	356.12	0.1167
32	Benzene	278.68	0.1494	353.24	0.1266
33	Methylbenzene (toluene)	178.18	0.1616	383.78	0.1117
34	Ethylbenzene	178.20	0.1576	409.35	0.1025
35	n-Propylbenzene	173.55	0.1528	432.39	0.1014
36	n-Butylbenzene	185.30	0.1501	456.46	0.0957

where V_{25}^L is the liquid molar volume at 25°C (298 K) in cm³/mol. For some compounds these values of V_{25}^L are given in Table 6.10. M is the molecular weight in g/mol and k^L is desired liquid thermal conductivity at T in W/m · K. Average error for this equation is about 5% as reported in the

API-TDB [5]. This equation can be used for temperatures at $T_r < 0.8$ and pressure below 35 bar. For estimation of k^L at temperatures above normal boiling point (compressed or saturated liquids), there are a number of methods that use reduced density ρ_r as a correlating parameter [5, 8]. Riazi and Faghri [30] also developed a method similar to Eq. (8.37) for prediction of thermal conductivity of liquid hydrocarbons for pentanes and heavier.

$$
\begin{aligned}
k &= 1.7307A(1.8T_b)^B SG^C \\
A &= \exp\left(-4.5093 - 0.6844t - 0.1305t^2\right) \\
B &= 0.3003 + 0.0918t + 0.01195t^2 \\
C &= 0.1029 + 0.0894t + 0.0292t^2 \\
t &= (1.8T - 460)/100
\end{aligned}
$$

(8.44)

where k is in W/m · K, while T_b and T are in kelvin. This equation can be applied to pure hydrocarbons (C_5–C_{22}) or to petroleum fractions with $70 < M < 300$ (boiling point range of 65–360°C) in the temperature range of −20–150°C (~0–300°F) and pressures below 30–35 bar. If Eq. (8.44) is applied to thermal conductivity data at two reference temperatures of 0 and 300°F (256 and 422 K) one can get

$$
\begin{aligned}
k_{256} &= 1.1594 \times 10^{-3} T_b^{0.7534} SG^{0.5478} \\
k_{422} &= 2.2989 \times 10^{-2} T_b^{0.2983} SG^{0.0094}
\end{aligned}
$$

(8.45)

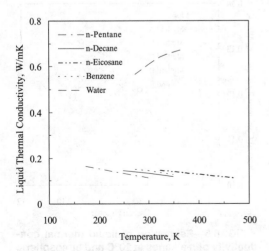

FIG. 8.4—Liquid thermal conductivity of n-alkanes versus temperature at atmospheric pressure.

TABLE 8.7—*Comparison of various methods of calculation of liquid thermal conductivity at 20°C (Example 8.3).*

N_C	k_{exp}	Linear,[a] Eq. (8.42)		API, Eq. (8.43)		Eq. (8.52)		RF,[b] Eq. (8.44)		RF, Eq. (8.46)	
		K	%Dev	k	%Dev	k	%Dev	k	%Dev	k	%Dev
5	0.114	0.114	0.4	0.113	−0.8	0.107	−6.6	0.107	−6.5	0.113	−0.6
6	0.121	0.121	0.2	0.119	−1.4	0.111	−8.4	0.112	−7.8	0.118	−2.4
7	0.1262	0.126	−0.1	0.124	−1.7	0.114	−9.7	0.116	−8.3	0.122	−3.3
8	0.1292	0.129	0.0	0.127	−1.5	0.116	−10.0	0.119	−7.6	0.126	−2.8
9	0.1316	0.132	0.0	0.130	−1.2	0.118	−10.2	0.123	−6.9	0.129	−2.2
10	0.133	0.133	0.0	0.132	−0.6	0.120	−9.9	0.125	−5.8	0.132	−1.1
Overall			0.1		1.2		9.1		7.1		2.1

[a]Linear refers to linear relation betweern k and T.
[b]RF referes to Riazi–Faghri methods.

where k_{256} refers to the value of k at 256 K (0°F) and k_{422} is the value of k at 422 K (300°F). Using Eq. (8.41) and on the basis of linear interpolation of thermal conductivity from the above equations, the following relation was also derived for the temperature range and molecular weight ranges specified for Eq. (8.44):

$$k = 10^{-2} \left(0.11594 T_b^{0.7534} SG^{0.5478} - 2.2989 T_b^{0.2983} SG^{0.0094} \right)$$
$$\times \left(\frac{1.8T - 460}{300} \right) + 2.2989 \times 10^{-2} T_b^{0.2983} SG^{0.0094}$$
(8.46)

where T_b and T are in kelvin and k is in W/m · K. Accuracy of this equation for pure compounds with the specified ranges is about 3.8% [30] and it is recommended instead of Eq. (8.44).

Example 8.3—Estimate values of thermal conductivity of liquid normal alkanes from C_5 to C_{10} at 20°C and 1 atm, using methods given in Eqs. (8.42)–(8.44) and (8.46). Compare calculated values with experimental data as given in the literature [8, 10].

Solution—Sample calculations are shown for n-C_5 and similar approach can be used to estimate values of k^L for other n-alkane compounds. From Table 2.1, for n-pentane $T_b = 36.1°C$ (309.3 K), SG = 0.6317, $T_M = -129.7°C$ (143.45 K), and $T_c = 196.55°C$ (469.8 K). From reference [10], $k_{20} = 0.114$ W/m · K. From Table 8.4, $k_m^L = 0.1758$ and $k_b^L = 0.1079$ W/m · K. Substituting in Eq. (8.42) $k_T^L = 0.1758 + (0.1079 - 0.1758) \times (298.15 - 143.45)/(309.3 - 143.45) = 0.1758 - 0.06334 = 0.1145$ W/m · K. This gives an error of + 0.43%. From Eq. (8.43), $n = 1.001$ and $C = 0.1811$ and it gives $k^L = 0.115$, with 0.7% error. From Eq. (8.44), $t = 0.68$, $A = 0.006524$, $B = 0.36787$, $C = 0.17677$, and $k^L = 0.107$ (error of − 6.5%). Equation (8.46) gives $k^L = 0.1134$, with error of − 0.57%. Later in this section several other empirical correlations for estimation of liquid thermal conductivity are presented. For example, Eq. (8.52) is proposed for thermal conductivty of coal liquids. This equation gives a value of 0.107 (−6.6%). Summary of results are given in Table 8.7 and also shown in Fig. 8.5. As expected, Eq. (8.44) because of its simplicity and Eq. (8.52) proposed for coal liquids give the highest errors in estimation of thermal conductivity of liquid hydrocarbons. ♦

For defined mixtures the following mixing rule proposed by Li is recommended in the API-TDB [5] for calculation of

liquid thermal conductivity of hydrocarbon systems:

$$k_m^L = \sum_i \sum_j \Phi_i \Phi_j k_{ij}^L$$

(8.47) $\quad k_{ij}^L = 2 \left(\frac{1}{k_i^L} + \frac{1}{k_j^L} \right) \quad$ where $k_{ij} = k_{ji}$ and $k_{ii} = k_i$

$$\Phi_i = \frac{x_i V_i^L}{\sum_i x_i V_i^L}$$

in which k_m^L is the thermal conductivity of liquid mixture, V_i^L is the liquid molar volume at a reference temperature (20 or 25°C), x_i is mole fraction, and ϕ_i is the volume fraction of component i in the mixture. Average error for this method is about 5% [5]. Li proposed a simpler mixing rule, which is recommended in the DIPPR manual [10] for nonhydrocarbon liquids:

(8.48) $$k_m^L = \left[\sum_i \frac{x_{wi}}{(k_i^L)^2} \right]^{-1/2}$$

where x_{wi} is the weight fraction of i in the mixture. This equation gives an average deviation of about 4–6% [10]. The Jamieson method for a binary liquid mixture is suggested by Poling et al. [18]:

(8.49) $\quad k_m^L = x_{w1} k_1^L + x_{w2} k_2^L - \alpha_{12} \left(k_2^L - k_1^L \right) \left(1 - \sqrt{x_{w2}} \right) x_{w2}$

Parameter α_{12} is an adjustable parameter that can be determined from an experimental data on mixture thermal

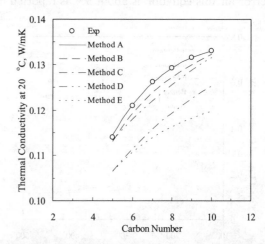

FIG. 8.5—Estimation of liquid thermal conductivity of n-alkanes at 20°C and atmospheric pressure (Example 8.3). Method A: Eq. (8.42); Method B: Eq. (8.43); Method C: Eq. (8.44); Method D: Eq. (8.46); Method E: Eq. (8.52).

conductivity and when no data are available it can be considered as unity [18].

For calculation of thermal conductivity of liquid petroleum fractions, if the PNA composition is available the pseudocomponent method using Eq. (8.42) and Table 8.4 may be applied. The simplest method of calculation of k_T^L for petroleum fractions when there is no information on a fraction is provided in the API-TDB [5]:

$$(8.50) \qquad k_T^L = 0.164 - 1.277 \times 10^{-4}T$$

where T is in kelvin and k^L in W/m·K. In other references this equation is reported with slight difference in the coefficients. For example, Wauquier [8] gives $k^L = 0.17 - 1.418 \times 10^{-4}T$. At 298 K (25°C), this relation gives a value of 0.128 W/m·K (near k of n-C_8), while Eq. (8.50) gives a value of 0.126, which is the same as the value of k for n-heptane. The error for this equation is high, especially for light and branched hydrocarbons. Average error of 10% is reported for this equation [5] and it may be used in absence of any information on a fraction. A more accurate relation uses average boiling point of the fraction as an input parameter and was developed by the API group at Penn State [5]:

$$(8.51) \qquad k_T^L = T_b^{0.2904} \times (2.551 \times 10^{-2} - 1.982 \times 10^{-5}T)$$

where both T_b and T are in kelvin. This equation gives an average error of about 6%. For n-C_5 of Example 8.3, this equation gives a value of 0.104 W/m·K with −9% error. However, this equation is not recommended for pure hydrocarbons. For petroleum fractions Eq. (8.46) can be used with better accuracy with the specified ranges of boiling point and temperature when both T_b and SG are available. For coal liquids and heavy fractions, Tsonopoulos et al. [33] developed the following relation based on the corresponding states method of Sato and Riedel:

$$(8.52) \qquad k^L = 0.05351 + 0.10177(1 - T_r)^{2/3}$$

where k^L is in W/m·K. This equation is not recommended for pure hydrocarbons. For some eight coal liquid samples and 74 data points this equation gives an average error of about 3% [33].

For liquid hydrocarbons and petroleum fractions when pressure exceeds 30–35 atm, effect of pressure on liquid thermal conductivity should be considered. However, this effect is not significant for pressures up to 70–100 atm. For the reduced temperature range of 0.4–0.8 and pressures above 35 atm, the following correction factor for the effect of pressure on liquid thermal conductivity is recommended in the API-TDB [5]:

$$(8.53) \qquad \begin{aligned} k_2^L &= k_1^L \frac{C_2}{C_1} \\ C &= 17.77 + 0.065P_r - 7.764T_r - \frac{2.054T_r^2}{\exp(0.2P_r)} \end{aligned}$$

To calculate value of k_2^L at T_2 and P_2, value of k_1^L at T_1 and P_1 must be known. In case of lack of an experimental value, the value of k_1^L at T_1 and P_1 can be calculated from Eqs. (8.41)–(8.43). There are some other generalized correlations based on the theory of corresponding states for prediction of both viscosity and thermal conductivity of dense fluids [25, 34]. However, these methods, although complex, are not widely

recommended for practical applications in the petroleum industry.

8.3 DIFFUSION COEFFICIENTS

Diffusion coefficient or diffusivity is the third transport property that is required in calculations related to molecular diffusion and mass transfer in processes such as mixing and dissolution. In the petroleum and chemical processing, diffusion coefficients of gases in liquids are needed in design and operation of gas absorption columns and gas–gas diffusion coefficients are required to determine rate of reactions in catalytic-gas-phase reactions, where mass transfer is a controlling step. In the petroleum production, knowledge of diffusion coefficient of a gas in oil is needed in the study of gas injection projects for improved oil recovery. If a binary system of components A and B is considered, where there is a gradient of concentration of A in the fluid, then it diffuses in the direction of decreasing concentration (or density)—a process similar to heat conduction due to temperature gradient. In this case the diffusion coefficient of component A in the system of A and B is called binary (or mutual) diffusion coefficient and is usually shown by D_{AB}, which is defined by the Fick's law [1]:

$$(8.54) \qquad J_{Ay} = -D_{AB}\frac{d\rho_A}{dy} = -\rho D_{AB}\frac{dx_{wA}}{dy}$$

where J_{Ay} is the mass flux of component A in the y direction (i.e., g/cm²·s) and $d\rho_A/dy$ is the gradient of mass density in the y direction. ρ is the mass density (g/cm³) of mixture and x_{wA} is the weight fraction of component A in the mixture. In the above relation, the second equality holds when ρ is constant with respect to y. J_A represents the rate of transport of mass in the direction of reducing density of A. It can be shown that in binary systems D_{AB} is the same as D_{BA} [1]. From the above equation, it can be seen that the unit of diffusivity in the cgs unit system is cm²/s (or 1 cm²/s = 10^{-4} m²/s). Diffusion of a component within its own molecules (D_{AA}) is called self-diffusion coefficient. From thermodynamic equilibrium point of view, the driving force behind molecular diffusion is gradient of chemical potential $\partial\mu_A/\partial y$. Since chemical potential is a function of T, P, and concentration, for systems with uniform temperature and pressure, μ_a is only a function of concentration (see Eq. 6.121) and Eq. (8.54) is justified. Various forms of Fick's law can be established in the forms of gradients of molar concentration, mole, weight, or mass fractions [1]. A comparison between Eqs. (8.1), (8.28), and (8.54) shows the similarity in momentum, heat, and mass transfer processes. The corresponding molecular properties (i.e., kinematic viscosity (v), thermal diffusivity (α), and diffusion coefficient (D)) that characterize the rate of these processes have the same unit (i.e., cm²/s or ft²/h). This is the reason that these physical properties are called *transport properties*. The diffusion process may also be termed mass transfer by conduction. The ratio of v/D or ($\mu/\rho D$) is a dimensionless number called *Schmidt number* (N_{Sc}) and is similar to the Prandtl number (N_{Pr}) in heat transfer (see Eq. 8.29). Schmidt number represents the ratio of mass transfer by convection to mass transfer by diffusion. Values of N_{Sc} of methane, propane, and n-octane in the air at 0°C and 1 atm are 0.69, 1.42, and 2.62,

TABLE 8.8—*Order of magnitude of binary diffusion coefficient and its concentration dependency for various systems [35].*

Type of system	Order of magnitude of D, cm^2/s	Activation energy (E), kcal/mol	Concentration dependence
Gas–gas (vapor–gas)	0.1–1.0	$E < 5$	Very weak
Gas–liquid	$\sim 10^{-5}$	$E \leq 5$	Weak
Normal liquids	10^{-5}–10^{-6}	5–10	$\pm 100\%$
Polymer solutions	10^{-5}–10^{-8}	10–20	$\pm 1000\%$
Gas or liquid in polymer or solids	$\sim 10^{-12}$–10^{-15}	$E \geq 40$	Factor of 1000%

respectively. The order of magnitude of N_{Sc} in liquids such as water is 10^3.

Diffusion coefficient like any other thermodynamic property is a function of the state of a system and depends on T, P, and concentration (i.e., x_i). One theory that describes molecular diffusion is based on the assumption that molecular diffusion requires a jump in their energy level. This energy is called *activation energy* and is shown by E_A. This activation energy, although not the same, is very similar to the activation energy required for a chemical reaction to occur. Heavier molecules have higher activation energy and as a result lower diffusion coefficients. Based on this theory, dependency of D with T can be expressed by Arrhenius-type equation in the following form:

$$(8.55) \qquad D = D_o \exp\left(-\frac{E_A}{RT}\right)$$

where D_o is a constant (with respect to T), E_A is the activation energy, R is the gas constant, and T is the absolute temperature. The order of magnitude of D and E_A in various systems and concentration dependency of D are shown in Table 8.8. Diffusion coefficients depend on the ability of molecules to move. Therefore, larger molecules have more difficulty to move and consequently their diffusivity is lower. Similarly in liquids where the space between molecules is small, diffusion coefficients are lower than in the gases. Increase in T would increase diffusion coefficients, while increase in P decreases diffusivity. The effect of P on diffusivity of liquids is less than its effect on the diffusivity of gases. At very high pressures, values of diffusion coefficients of liquids approach their values for the gas phase. At the critical point, both liquid and gas phases have the same diffusion coefficient called critical diffusion coefficient and it is represented by D_C.

In this section, methods of estimation of diffusion coefficients in gases and liquids as well as in multicomponent systems and the effect of porous media on diffusivity are presented. In the last part a new method different from conventional methods for experimental measurement of diffusion coefficients in dense hydrocarbon fluids (both gases and liquids) is presented.

8.3.1 Diffusivity of Gases at Low Pressures

Similar to viscosity and thermal conductivity, kinetic theory provides a relatively accurate relation for diffusivity of rigid (hard) molecules with different size. Based on this theory, for gases at low pressures (ideal gas conditions) the following relation is developed for gas–gas diffusivities [1, 3]:

$$D_{AB} = \frac{3 \times \pi}{8} \times 10^{-1} \times \left(\frac{k_B^3}{\pi^3}\right)^{1/2} \left(\frac{1}{2m_A} + \frac{1}{2m_B}\right)^{1/2} \frac{T^{3/2}}{P\left(\frac{d_A + d_B}{2}\right)^2}$$

$$(8.56)$$

where D_{AB} is in cm^2/s, k_B is the Boltzman's constant (1.381×10^{-23} J/K), T is temperature in kelvin, P is the pressure in bar, m is the molecular mass in kg [M/N_A, i.e., $m_A = M_A \times 10^{-3}/(6.022 \times 10^{23})$], and d is the hard sphere molecular diameter in m (1 nm = 1×10^{-9} m). Values of d may be determined from measured viscosity or thermal conductivity data by Eqs. (8.2) and (8.30), respectively. For example, for CH$_4$ value of d from viscosity is 0.414 nm while from thermal conductivity is 0.405 nm. For O$_2$, H$_2$ and CO$_2$, values of d are 0.36, 0.272, and 0.464 nm, respectively [3]. As an example, the self-diffusion coefficient of CH$_4$ at 1 bar and 298 K from the kinetic theory is calculated as $m_A = m_B = m = 2.66 \times 10^{-26}$ kg, $d_A = d_B = d = 0.414 \times 10^{-9}$ m and from Eq. (8.56) $D_{AB} = 0.194$ cm^2/s. Thus one can calculate diffusion coefficient from viscosity data through calculation of molecular diameter. For gases at low pressures D varies inversely with pressure, while it is proportional to $T^{3/2}$. Furthermore, D_A varies with $M_A^{-1/2}$, that is, heavier molecules have lower diffusivity under the same conditions of T and P. In practical cases molecular diameters can be estimated from liquid molar volumes in which actual data are available, as will be seen in Eq. (8.59). A more accurate equation for estimation of diffusivity of ideal gases was derived independently by Chapman and Enskog from the kinetic theory and is known as Chapman–Enskog equation, which may be written as [1, 9]

$$(\rho D_{AB})^\circ = \frac{2.2648 \times 10^{-5} T^{0.5} \left(\frac{1}{M_A} + \frac{1}{M_B}\right)^{0.5}}{\sigma_{AB}^2 \Omega_{AB}}$$

$$\sigma_{AB} = \frac{\sigma_A + \sigma_B}{2}$$

$$\sigma_i = 0.1866 V_{ci}^{1/3} Z_{ci}^{-6/5}$$

$$\Omega_{AB} = \frac{1.06036}{\left(T_{AB}^*\right)^{0.1561}} + 0.193 \exp\left(-0.47635 T_{AB}^*\right) + 1.76474$$

$$\times \exp\left(-3.89411 T_{AB}^*\right) + 1.03587 \exp\left(-1.52996 T_{AB}^*\right)$$

$$T_{AB}^* = T/\varepsilon_{AB}$$

$$\varepsilon_{AB} = \left(\varepsilon_A \varepsilon_B\right)^{1/2}$$

$$\varepsilon_i = 65.3 T_{ci} Z_{ci}^{18/5}$$

$$(8.57)$$

where $(\rho D_{AB})^\circ$ represents the product of density–diffusivity of ideal gas at low-pressure conditions according to the Chapman–Enskog theory and is in mol/cm·s. ε and σ are the energy and size parameters in the potential energy relation (i.e., Eq. 5.11). σ is in Å, T and T_c are in kelvin, and V_c is in cm^3/mol. The correlations for calculation of Lennard–Jones (LJ) parameters (ε and σ) from critical constants as given in Eq. (8.57) were developed by Stiel and Thodos [36]. There are some other correlations given in the literature for calculation of LJ parameters [18]. Typical values of ε and σ determined from various properties are given in Table 6.16. In the above relation low-pressure diffusivity can be calculated through dividing $(\rho D_{AB})^\circ$ by ρ° ($= 83.14T/P$) in which T is in kelvin and P is in bar. Calculated D_{AB} would be in cm^2/s.

For practical calculations, a more accurate estimation method is required. Most of these correlations are based on the modified version of Chapman–Enskog theory [18]. The empirical correlation of Chen–Othmer for estimation of D_{AB} of gases at low pressures is in the following form [28]:

$$(8.58) \qquad D_{AB}^\circ = \frac{1.518 \times 10^{-2} T^{1.81} \left(\frac{1}{M_A} + \frac{1}{M_B} \right)^{1/2}}{P \left(T_{cA} T_{cB} \right)^{0.1405} \left(V_{cA}^{0.4} + V_{cA}^{0.4} \right)^2}$$

where D_{AB}° is the diffusivity of A in B at low pressures in cm^2/s, T is in kelvin, P is in bar, M in g/mol, and T_{cA} and V_{cA} are the critical temperature and volume of A in kelvin and cm^3/mol, respectively. This method can be used safely up to pressure of about 5 bar. This equation predicts self-diffusion coefficient of methane at 298 K and 1 bar as 0.248 cm^2/s versus the value of 0.194 from the kinetic theory (Eq. 8.56). For hydrocarbon–hydrocarbon systems the API-TDB recommends the Gilliland method in the following form [5]:

$$(8.59) \qquad D_{AB}^\circ = \frac{4.36 \times 10^{-3} T^{1.5} \left(\frac{1}{M_A} + \frac{1}{M_B} \right)^{1/2}}{P \left(V_A^{1/3} + V_B^{1/3} \right)^2}$$

$$V_i = 0.285 V_{ci}^{1.048}$$

where V_i is the liquid molar volume of component i at its normal boiling point and V_{ci} is the molar critical volume and both are in cm^3/mol. Other units are the same as in Eq. (8.58). This equation can be used up to pressure of 35 bar with an accuracy of about 4% as reported in the API-TDB [5]. Several other methods for prediction of gas diffusivity at low pressures are given by Poling et al. [18].

8.3.2 Diffusivity of Liquids at Low Pressures

Calculation of diffusion coefficients for liquids is less accurate than gases as for any other physical property. This is mainly due to the lack of a perfect theory for liquids. Generally there are three theories for diffusivity in liquids: (1) hydrodynamic theory, which usually applies to systems of solids dissolved in liquids, (2) Eyring rate theory, and (3) the free-volume theory. In the hydrodynamic theory, it is assumed that fluid slides over a particle according to the Stoke's law of motion. The Eyring theory was presented earlier by Eq. (8.55) in which molecules require an energy jump before being able to diffuse. The free-volume theory says that for a molecule to jump to a higher energy level (activation energy), it needs a critical free-volume (V_A^*) and Eq. (8.55) can be modified by multiplying the right-hand side by factor $\exp\left(V_A^*/V \right)$, where V is the apparent molar volume of liquid. None of these theories is perfect; however, it can be shown by both the Eyring rate and

hydrodynamic theories that in liquid systems diffusion coefficient is inversely proportional to viscosity of solvent. For example, based on the hydrodynamic theory and the Stokes–Einstein equation, Wilke and Chang developed the following relation for estimation of diffusion coefficient at infinite dilution [18, 28]:

$$(8.60) \qquad D_{AB}^{\infty L} = 7.4 \times 10^{-8} \frac{(\Psi_B M_B)^{1/2} T}{\mu_B V_A^{0.6}}$$

where $D_{AB}^{\infty L}$ is the diffusion coefficient (in cm^2/s) of solute A in solvent B, when concentration of A is small (dilute solution). The superscript ∞ indicates the system is dilute in solute and for this reason concentration of solute is not included in this equation. M_B is the molecular weight of solvent (g/mol), T is absolute temperature in kelvin, and μ_B is the viscosity of solvent B (in cp). Because the solution is dilute, μ_B is almost the same as viscosity of solution. V_A is the molar volume of solute A at its normal boiling point in cm^3/mol and it may be calculated from V_c according to the relation given in Eq. (8.59). ψ_B is called *association parameter* for solvent where for water the value of 2.6 is recommended [28]. For methanol and ethanol, ψ_B is 1.9 and 1.5, respectively. For benzene, heptane, and unassociated solvents (most hydrocarbons) its value is 1.0 [18, 28]. The average error for this equation for some 250 systems is about 10% [18].

Another simple method derived from Tyn and Calus equation and is given as follows [18]:

$$(8.61) \qquad D_{AB}^{\infty L} = 8.93 \times 10^{-8} \left(\frac{V_B^{0.267}}{V_A^{0.433}} \right) \frac{T}{\mu_B}$$

where the parameters and units are the same as those given in Eq. (8.60). V_B is the molar volume of solvent at its boiling point and can be calculated from V_{cB} similar to V_A. Equation (8.61) is suitable for organic and hydrocarbon systems. Because of higher accuracy, the Wilke–Chang method (Eq. 8.60) is widely used for calculation of diffusion coefficient of liquids and it is also recommended in the API-TDB [5].

As shown in Table 8.8 diffusion coefficient of a binary liquid system depends on the concentration of solute. This is the reason that most experimental data on liquid diffusivity are reported for dilute solutions without concentration dependency and for the same reason predictive methods (Eqs. (8.60) and (8.61)) are developed for diffusion coefficients of dilute solutions. There are a number of relations that are proposed to calculate D_{AB}^L at different concentrations. The Vignes method suggests calculation of D_{AB} from $D_{AB}^{\infty L}$ and $D_{BA}^{\infty L}$ as follows [35]:

$$(8.62) \qquad D_{AB}^L = \left(D_{AB}^{\infty L} \right)^{x_B} \left(D_{BA}^{\infty L} \right)^{x_A} \alpha_{AB}$$

where x_A is the mole fraction of solute and x_B is equal to $1 - x_A$. Parameter α_{AB} is a dimensionless thermodynamic factor indicating nonideality of a solution defined as

$$(8.63) \qquad \alpha_{AB} = 1 + \left(\frac{\partial \ln \gamma_A}{\partial \ln x_A} \right)_{T,P} = 1 + \left(\frac{\partial \ln \gamma_B}{\partial \ln x_B} \right)_{T,P}$$

where γ_A is the activity coefficient of solute A and can be estimated from methods of Chapter 6. For ideal systems or dilute solutions ($x_A \cong 0$), $\alpha_{AB} = 1.0$. For simplicity in calculations for hydrocarbon–hydrocarbon systems this parameter is taken as unity.

Another simple relation is suggested by Caldwell and Babb and is also recommended in the API-TDB for hydrocarbon

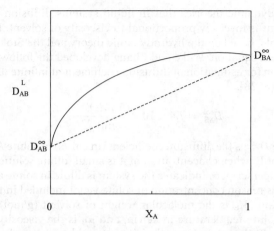

FIG. 8.6—Dependency of liquid diffusion coefficients with composition.

systems:

$$(8.64) \qquad D_{AB}^L = (1 - x_A) \times \left(D_{AB}^{\infty L}\right) + x_A \left(D_{BA}^{\infty L}\right)$$

where $D_{AB}^{\infty L}$ and $D_{BA}^{\infty L}$ are diffusivities at infinite dilutions and are known from experiments or may be calculated from Eq. (8.60) or (8.61). For nonideal systems D_{AB}^L calculated from Eq. (8.64) must be multiplied by factor α_{AB} defined in Eq. (8.63). This is demonstrated in Fig. 8.6 in which the binary diffusion coefficient of ideal systems is shown by a dotted line while the actual diffusivity of nonideal solutions is shown by a solid line. Riazi and Daubert [37] showed that corresponding state approach can also be used to correlate diffusion coefficient of liquids and developed a generalized chart for reduced diffusivity ($D_r = D/D_C$), in a form similar to Eq. (8.21), for calculation of liquid diffusivity at low pressures.

8.3.3 Diffusivity of Gases and Liquids at High Pressures

Pressure has significant effect on diffusivity of gases while it has lesser effect on liquid diffusivity. At very high pressures diffusion coefficients of gases approach those of liquids. For calculation of diffusion coefficients of gases at high pressures,

Slattery and Bird [38] developed a generalized chart in terms of $(PD)/(PD)^\circ$ versus T_r and P_r. The chart is in graphical form and is based on a very few data on self-diffusion coefficient of simple gases, which were available six decades ago. Later Takahashi [39] proposed a similar and identical chart but using more data on self- as well as some binary-diffusion coefficients. Obviously these methods cannot be used with computer tools and use of the charts is inconvenient to obtain an accurate value of diffusion coefficient. However, Slattery–Bird chart has been included in the API-TDB [5].

Sigmund [40] measured and reported binary diffusion coefficient of dense gases for C_1, C_2, C_3, n-C_4, and N_2 for the pressure range of 200–2500 psia (14–170 bars), temperature range of 38–105°C (100–220°F) and mole fraction range of 0.1–0.9 for methane. Sample of Sigmund's dataset for some binary systems are given in Table 8.9. Sigmund also reported experimental data on the density of mixtures and based on the original work of Dawson et al. [41] correlated reduced density–diffusivity product (ρD_{AB}) to the reduced density in a polynomial form as follows:

$$\frac{(\rho D_{AB})}{(\rho D_{AB})^\circ} = 0.99589 + 0.096016\rho_r - 0.22035\rho_r^2 + 0.032874\rho_r^3$$

(8.65)

where $(\rho D_{AB})^\circ$ is a value of (ρD_{AB}) at low pressure for an ideal gas and should be calculated from Eq. (8.57). For developing this correlation, Sigmund used liquid diffusivity data for binary systems of C_1–n-C_n (n varied from 6 to 16) in addition to diffusivity data of dense gases. The main advantage of this equation is that it can be used for both gases and liquids and for this reason reservoir engineers usually use this method for calculation of diffusion coefficients of reservoir fluids under reservoir conditions. However, the main disadvantage of this method is its sensitivity to reduced density for liquid systems where reduced density approaches 3. This is shown in Fig. 8.7 in which $(\rho D_{AB})/(\rho D_{AB})^\circ$ is plotted versus ρ_r according to Eq. (8.65). For gases where $\rho_r < 1$, reduced diffusivity $(\rho D_{AB})/(\rho D_{AB})^\circ$ is about unity; however, for liquids where $\rho_r > 2.5$ the curve is nearly vertical and small error in ρ would result in a much larger error in diffusivity calculation. For this reason, this equation generally gives higher errors for calculation of diffusion coefficient of liquids even

TABLE 8.9—*Diffusion coefficient of gases at high pressures [40].*

No.	Component A	Component B	x_A	T, K	P, bar	$10^5 D_{AB}$, cm²/s
1	Methane	Propane	0.896	311	14.0	883
2	Methane	Propane	0.472	311	137.9	22.5
3	Methane	Propane	0.091	311	206.8	16.9
4	Methane	Propane	0.886	344	13.9	1196
5	Methane	Propane	0.15	344	206.8	21.6
6	Methane	Propane	0.9	378	13.7	1267
7	Methane	Propane	0.116	378	168.9	36.5
8	Methane	n-Butane	0.946	311	137.2	55.79
9	Methane	n-Butane	0.973	344	13.8	1017
10	Methane	n-Butane	0.971	344	172.4	62.99
11	Methane	n-Butane	0.126	344	135.4	16.34
12	Methane	n-Butane	0.973	378	13.8	1275
13	Methane	n-Butane	0.124	378	135.1	26.82
14	Methane	Nitrogen	0.5	313	14.1	1870
15	Methane	Nitrogen	0.5	313	137.9	164
16	Methane	Nitrogen	0.5	366	137.8	232

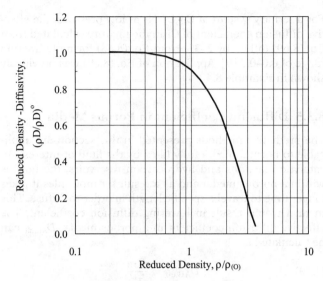

FIG. 8.7—Correlation of reduced diffusivity versus reduced density (Eq. 8.65).

reduced diffusivity and reduced viscosity [9]:

$$\frac{(\rho D_{AB})}{(\rho D_{AB})^\circ} = a \left(\frac{\mu}{\mu^\circ} \right)^{b+cP_r}$$

(8.67)

$$a = 1.07 \quad b = -0.27 - 0.38\omega$$

$$c = -0.05 + 0.1\omega \quad P_r = P/P_c$$

$$T_c = x_A T_{cA} + x_B T_{cB} \quad P_c = x_A P_{cA} + x_B P_{cB}$$

$$\omega = x_A \omega_A + x_B \omega_B$$

where $(\rho D_{AB})^\circ$ must be determined from Eq. (8.57). μ° must be calculated from Eqs. (8.6) and (8.8). If experimental data on μ are not available it should be calculated from Eq. (8.12) for both liquids and gases. Coefficients a, b, and c have been determined from data on diffusion coefficients of some 300 binary systems as shown in Table 8.10. Errors for both Eqs. (8.65) and (8.67) are also shown in this table. In evaluation of Eq. (8.65) the coefficients were reevaluated from the same data bank as given in Table 8.10. When Eq. (8.67) is evaluated against 17 diffusivity data points for binary systems that were not used in the development of this equation, an average error of 9% was observed [9]. Furthermore Eq. (8.67) was evaluated with $D_{AB}^{\infty L}$ of some dilute binary liquids at atmospheric pressure and results show that it is comparable with the Wilke–Chang equation (8.60) specifically developed for liquids [9].

The main objective of development of Eq. (8.67) was to have a unified predictive method for both gas and liquid diffusivities, which can be safely used for diffusivity prediction of heavy hydrocarbon fluids. The extrapolation ability of Eq. (8.67) can be seen from the linear relationship between $(\rho D_{AB})/(\rho D_{AB})^\circ$ and (μ/μ°) on a log–log scale. For this reason, this equation can be used with good accuracy for heavy oils up to molecular weight of 350. Equation (8.67) was developed for dense gases and for this reason data on diffusion coefficient of gases at atmospheric pressure were not used in determination of its coefficients. Theoretically, coefficient a in Eq. (8.67) must be unity, but value of 1.07 was obtained from regression of experimental data. This is mainly due to the fact that majority of data used were at high pressure (see Table 8.10). However, even at low pressure where $\mu/\mu^\circ = 1$, this equation gives average deviation of 7% from the Stokes–Einstein equation, which is within the range of errors for calculation of diffusivity at higher pressures. The Stokes–Einstein equation (Eq. 8.57) usually underpredicts diffusivity at atmospheric pressure and for this reason coefficient of 1.07 improves accuracy of prediction of diffusivity at low pressures. However, for low-pressure gases and liquids, methods proposed in previous sections may be used. Although this equation was developed for hydrocarbon systems, but when applied to some nonhydrocarbon systems, reasonably good results have been obtained as shown in the following example.

though some modifications have been proposed for $\rho_r > 3$. For example, it is suggested that the right-hand side of Eq. (8.65) be replaced by $0.18839 \exp(3 - \rho_r)$, when $\rho_r > 3$ [42]. There are some other empirical correlations for estimation of diffusion coefficient of light gases in reservoir fluids. For example, Renner proposed the following empirical correlation for calculation of $D_{i\text{-oil}}$ in gas injection projects [43]:

$$(8.66) \quad D_{A\text{-oil}} = 7.47 \times 10^{-8} \mu_{oil}^{-0.4562} M_A^{-0.6898} \rho_{MA}^{1.706} P^{-1.831} T^{4.524}$$

where $D_{A\text{-oil}}$ is the effective diffusivity of light gas A (C_1, C_2, C_3, CO_2) in an oil (reservoir fluid) in cm²/s. μ_{oil} is the viscosity of oil (free of gas A) at T and P in cp, M_A is molecular weight of gas A, ρ_{MA} is molar density of gas A at T and P in mol/cm³, P is pressure in bar, and T is absolute temperature in kelvin. Exponent 4.524 on T indicates that estimated value of $D_{A\text{-oil}}$ is quite sensitive to the value of T considering that the value of T is a large number. This equation was developed based on 140 data points for the ranges $1 < P < 176$ bar, $273 < T < 333$ K, and $16 < M_i < 44$. As mentioned earlier such empirical correlations are mainly accurate for the data used in their development.

Another generalized correlation for diffusion coefficient of dense fluids was developed by Riazi [9]. For liquids, according to the Stokes–Einstein and Eyring theories [44], diffusion coefficient is inversely proportional to viscosity ($D \propto 1/\mu$). If it is further assumed that the deviation of diffusivity of a gas from ideal gas diffusivity is proportional to the viscosity deviation the following correlation can be developed between

TABLE 8.10—Data used for development of Eq. (8.67).

Dense fluid	Binary systems	No. of data	M range of barrier[a]	P range, bar	T range, K	(μ/μ°) Range	$10^4 D_{AB}$ cm²/s	%AAD Eq. (8.67)	Eq. (8.65)
Gases	N_2, C_1, C_2, C_3, C_4	140	16–58	7–416	155–354	1–15	1.4–240	8.1	10.2
Liquids	C_1, C_3, C_6, C_{10}, Oil	143	44–340	2–310	274–411	4–20000	0.01–5	15.4	48.9

[a]Molecular weight range of heavier component in the binary systems. Ref. [9].

Example 8.4—Estimate the diffusivity of benzene in a binary mixture of 74.2 mol% acetone and 25.8 mol% carbon tetrachloride (CCl_4) at 298 K and 1 atm pressure.

Solution—The system is a ternary mixture of benzene, acetone, and CCl_4. Consider benzene, the solute, as component A and the mixture of acetone and CCl_4, the solvent, as component B. Because amount of benzene is small (dilute system), $x_A = 0.0$ and $x_B = 1.0$. $T_{cB} = 520$, $P_{cB} = 46.6$ bar, $V_{cB} = 226.3$ cm^3/mol, $\omega = 0.2274$, $M = 82.8$. These properties are calculated from properties of acetone and CCl_4 as given in Ref. [45]. Actually the liquid solvent is the same as the liquid in Example 8.1, calculated properties of which are $\rho = 0.012422$ mol/cm^3, $\mu^\circ = 0.00829$ and $\mu = 0.374$, thus $\mu/\mu^\circ = 45.1677$. From Eq. (8.67), $b = -0.356$, $c = -0.02726$, and $(\rho D_{AB})/(\rho D_{AB})^\circ = 0.2745$. From Eq. (8.57), $(\rho D_{AB})^\circ = 1.28 \times 10^{-6}$ $mol/cm \cdot s$. Therefore, $D_{AB} = 1.28 \times 10^{-6} \times 0.2745/0.012422 = 2.83 \times 10^{-5}$ cm^2/s. In comparison with the experimental value of 2.84×10^{-5} cm^2/s [10] an error of -0.4% is obtained. In this example both μ and ρ have been calculated, while in many cases these values may be known from experimental measurements. ♦

8.3.4 Diffusion Coefficients in Mutlicomponent Systems

In multicomponent systems, diffusion coefficient of a component (A) in the mixture of N components is called *effective diffusion coefficient* and is shown by $D_{A\text{-mix}}$. Based on the material balance and ideal gas law Wilke derived the following relation for calculation of $D_{A\text{-mix}}$ [46]:

$$(8.68) \qquad D_{A\text{-mix}} = \frac{1 - y_A}{\sum_{i \neq A}^{N} \frac{y_i}{D_{A-i}}}$$

where y_i is the mole fraction of i and D_{A-i} is the binary diffusion coefficient of A in i. This equation may be used for pressures up to 35 bar; however, because of lack of a reliable method, this is also used for high-pressure gases and liquids as well [9]. For calculation of $D_{A\text{-mix}}$ in liquids the method of Leffler and Cullinan is recommended in the API-TDB [5]. This method requires binary diffusion coefficients at infinite dilution $D_{A-i}^{\infty L}$, mole fraction of each component x_i, liquid viscosity of each component μ_i, and viscosity of liquid mixture μ_m. However, this method is not recommended in other sources and is not widely practiced by petroleum engineers.

Riazi has proposed calculation of $D_{A\text{-mix}}$ for both gases and liquids at low and high-pressure systems by assuming that the mixture can be considered as a binary solution of A and B where B is a pseudocomponent composed of all components in the mixture except A. $D_{A\text{-mix}}$ is assumed to be the same as binary diffusivity, D_{AB}, which can be calculated from Eq. (8.67). $D_{A\text{-mix}}$ is calculated from the following relations [9]:

$$(8.69) \qquad \begin{aligned} D_{A\text{-mix}} &= D_{AB} \\ \theta_B &= \frac{\sum_{\substack{i=1 \\ i \neq A}}^{N} x_i \theta_i}{\sum_{\substack{i=1 \\ i \neq A}}^{N} x_i} \end{aligned}$$

where θ_B is a property such as T_c, P_c, or ω for pseudocomponent B. This method is equivalent to the Wilke's method (Eq. 8.68) for low-pressure gases at infinite dilution (i.e., $x_A \to 0$).

For a ternary system of C_1–C_3–N_2 at low pressure, the effective diffusion coefficient of C_1 in the mixture calculated from Eq. (8.69) differs by 2–3% from Eq. (8.68) for mole fraction range of 0.0–0.5 [9]. Application of Eq. (8.69) was previously shown in Example 8.4.

8.3.5 Diffusion Coefficient in Porous Media

The predictive methods presented in this section are applicable to normal media fully filled by the fluid of interest. In catalytic reactions and hydrocarbon reservoirs, the fluid is within a porous media and as a result for molecules it takes longer time to travel a specific length in order to diffuse. This in turn would result in lowering diffusion coefficient. The effective diffusion coefficient in a porous media, $D_{AB,eff}$ can be calculated as

$$(8.70) \qquad D_{AB,eff} = \frac{D_{AB}}{\tau^n}$$

where D_{AB} is the diffusion coefficient in absence of porous media and exponent n is usually taken as one but other values of n are also recommended for some porous media systems [47]. τ is a dimensionless parameter called *tortuosity* defined to indicate degree of complexity in connection of free paths in a porous media. Its definition is demonstrated in Fig. 8.8 according to the following relation:

$$\tau = \frac{\text{Actual free distance between points } a \text{ and } b \text{ in porous media}}{\text{Distance of a straight line between } a \text{ and } b}$$

(8.71)

Since actual distance between a and b is always greater than a straight line connecting the two points, $\tau > 1.0$. For determination of τ in an ideal media, assuming all particles that form a porous media are spherical, then as shown in Fig. 8.9 the approximate value of tortuosity can be calculated as $\tau \cong 1.4$. In actual cases such as for petroleum reservoirs where the

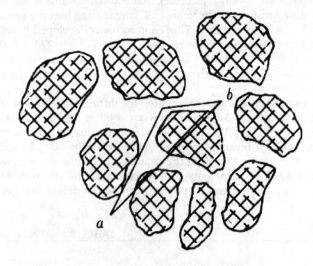

$$\tau = \frac{\text{Free distance between a and b}}{\text{Distance of straight line between a and b}}$$

FIG. 8.8—Distance for traveling a molecule from *a* to *b* in a porous media and concept of tortuosity.

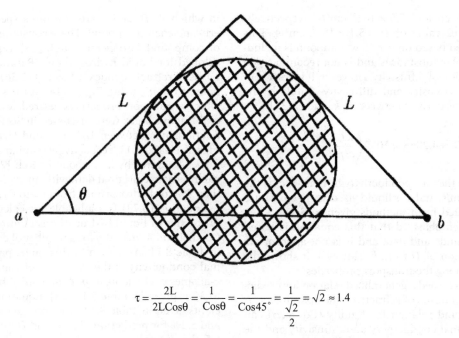

$$\tau = \frac{2L}{2L\cos\theta} = \frac{1}{\cos\theta} = \frac{1}{\cos 45°} = \frac{1}{\frac{\sqrt{2}}{2}} = \sqrt{2} \approx 1.4$$

FIG. 8.9—Approximate calculation of tortuosity (τ).

size and shape of particles are all different, value of τ varies from 3 to 5.

In a porous media τ is related to the *formation resistivity factor* and *porosity* as

$$(8.72) \qquad \tau = (F\phi)^{n_1}$$

where F is the resistivity and ϕ is the porosity, both are dimensionless parameters. ϕ is the fraction of connected empty space in a porous media and F is an indication of electrical resistance of materials that form the porous media and is always greater than unity. n_1 is a dimensionless empirical parameter that depends on the type of porous media. Theoretically, value of n_1 in Eq. (8.72) is one; however, in practice n_1 is taken as 1.2. Various relations between τ and ϕ are given by Amyx et al. [48] and Langness et al. [49]. One general relation is given as follows [48]:

$$(8.73) \qquad \tau = a\phi^{-m}$$

where parameters a and m are specific of a porous media. Parameter m is called *cementation factor* and it is specifically a characteristic of a porous media and it usually varies from 1.3 to 2.5. Some researchers have attempted to correlate parameter m with porosity and resistivity. For some reservoirs $a = 0.62$ and $m = 2.15$, while for some other reservoirs, when $\phi > 0.15$, $a = 0.75$ and $m = 2$ and for $\phi < 0.15$, $a = 1$ and $m = 2$. By combining Eqs. (8.72) and (8.73) with $n_1 = 1.2$ and $a = 1$:

$$(8.74) \qquad \tau = \phi^{1.2-1.2m}$$

Equation (8.74) can be combined with Eq. (8.70) to estimate effective diffusion coefficients in a porous media. Parameter m in Eq. (8.74) can be taken as an adjustable parameter, while for simplicity, parameter n in Eq. (8.70) can be taken as unity.

In practical applications, engineers use simpler relations between tortuosity and porosity. For example, Fontes et al.

[50] suggest that for calculation of diffusion coefficients of gases in porous solids (i.e., catalytic reactors) effective diffusion coefficients can be calculated from the following equation:

$$(8.75) \qquad D_{\text{eff}} = \phi^{-1.5}D$$

This equation can be obtained from Eq. (8.70) by assuming $\tau^n = \phi^{1.5}$.

8.4 INTERRELATIONSHIP AMONG TRANSPORT PROPERTIES

In previous sections three transport properties of μ, k, and D were introduced. In the predictive methods for these molecular properties, there exist some similarities among these properties. Most of the predictive methods for transport properties of dense fluids are developed through reduced density, ρ_r. In addition, diffusion coefficients of dense fluids and liquids are related to viscosity. Riazi and Daubert developed several relationships between μ, k, and D based on the principle of dimensional analysis [37]. For example, they found that for liquids $\ln(\mu^{2/3}D/T)$ versus $\ln(T/T_b)$ is linear and obtained the following relations:

$$\frac{\mu^{2/3}}{T}D = 6.3 \times 10^{-8}\left(\frac{T}{T_b}\right)^{0.7805} \qquad \text{for liquids except water}$$

$$\frac{\mu^{2/3}}{T}D = 10.03 \times 10^{-8}\left(\frac{T}{T_b}\right)^{1.0245} \qquad \text{for liquid water}$$

$$(8.76)$$

where μ is liquid viscosity in cp (mPa·s), T is temperature in kelvin, D is liquid self-diffusivity in cm²/s, and T_b is normal boiling point in kelvin. For example, for n-C_5 in which T_b is 309 K the viscosity and self-diffusion coefficient at 25°C

(298.2 K) are 0.215 cp and 5.5×10^{-4} cm²/s, respectively. Equation (8.76) gives value of $D = 5.1 \times 10^{-5}$ cm²/s. This equation is developed based on very few compounds including polar and nonpolar substances and is not recommended for accurate estimation of diffusivity. However, it gives a general trend between viscosity and diffusivity. Similarly the following relation was derived between μ, k, and D [37]:

$$(8.77) \qquad \left(\frac{k}{c_s}\right)^{1/2} D = 4.2868 \times 10^{-9} \left(\frac{kT}{c_s^2 \mu}\right)^{1.0128}$$

where k is the liquid thermal conductivity in W/m · s, c_s is the velocity of sound in m/s, and μ is liquid viscosity in cp. Values of c_s can be calculated from methods given in Section 6.9. Again it should be emphasized that this equation is based on very few compounds and data and it is not appropriate for accurate prediction of D from k. However, it shows the interrelationship among the transport properties.

Riazi et al. [23] developed a generalized relation for prediction of μ, k, and D in terms of refractive index parameter, I. In fact, the Hildebrand relation for fluidity (Eq. 8.26) can be extended to thermal conductivity and diffusivity and the following relation can be derived based on Eq. (8.27):

$$(8.78) \qquad \theta^p = A \left(\frac{1}{I} - 1\right)^p + B$$

$$\text{where } \theta = \frac{1}{\mu}, \frac{1}{k}, D$$

in which A, B, and p are constants specific for each property and each compound. These constants for a large number of compounds are given in Ref. [23]. Equation (8.78) is developed for liquid hydrocarbons. Parameter I is defined in terms of refractive index (n) by Eq. (2.36) and n must be evaluated from n_{20} using Eq. (2.114) at the same temperature at which a transport property is desired. Methods of estimation of refractive index were discussed in Section 2.6.2. The linear relationships between $1/\mu$ or D and $(1/I - 1)$ are shown in Figs. 8.10 and 8.11, respectively. Similar relations are shown for k of several hydrocarbons in Ref. [23]. Equation (8.78) can reproduce original data with an average deviation of less than 1% for hydrocarbons from C_5 to C_{20}.

Equation (8.78) is applicable for calculation of transport properties of liquid hydrocarbons at atmospheric pressures. Coefficients A, B, and p for a number of compounds are given in Table 8.11. As shown in this table, parameter p for thermal conductivity is the same for all compounds as 0.1. For n-alkanes coefficients of p, A, and $-B/A$ have been correlated to M as given in Table 8.12 [23]. Equation (8.78) with coefficients given in Table 8.12 give average deviations of 0.7, 2.1, and 5.2% for prediction of μ, k, and D, respectively. Example 8.5 shows application of this method of prediction of transport properties.

Example 8.5—Estimate the thermal conductivity of n-decane at 349 K using Eq. (8.78) with coefficients predicted from correlations of Table 9.11. The experimental value is 0.119 W/m · K as given by Reid et al. [18].

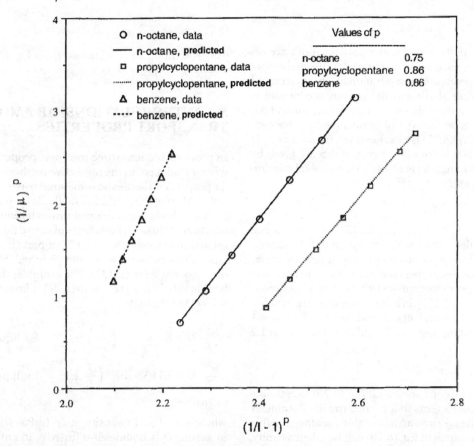

FIG. 8.10—Relationship between fluidity ($1/\mu$) (μ is in mPa · s (cp)) and refractive index parameter I from Eq. (8.78). Adopted from Ref. [23].

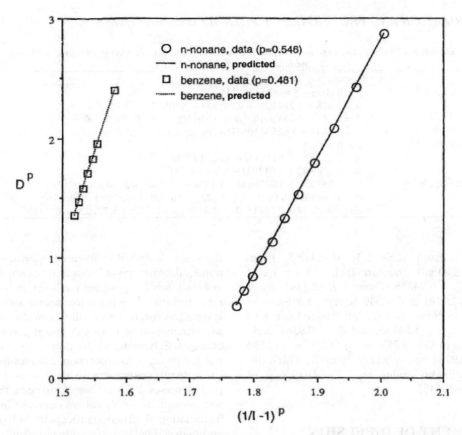

FIG. 8.11—Relationship between diffusivity (D) (D is in 10^5 cm²/s) and refractive index parameter I from Eq. (8.78). With permission from Ref. [23].

TABLE 8.11—Coefficients of Eq. (8.78) for some liquid hydrocarbons with permission from Ref. [23].

Compound	M	n_{20}	p	A	$-B/A$	T range, K
Coefficients for viscosity (mPa · s or cp)						
n-Pentane	72.2	1.3575	0.747	7.8802	2.2040	144–297
n-Decane	142.3	1.4119	0.709	6.3394	2.0226	256–436
n-Eicosane	282.6	1.4424	0.649	4.5250	1.8791	311–603
Cyclopentane	70.1	1.4065	0.525	8.3935	1.6169	250–322
Methylcyclopentane	84.2	1.4097	0.584	7.9856	1.7272	255–345
n-Decylcyclopentane	210.4	1.4487	0.349	8.5664	1.3444	255–378
Cyclohexane	84.2	1.4262	0.567	8.8898	1.7153	288–345
n-Pentylcyclohexane	154.3	1.4434	0.650	6.5114	1.8300	255–378
n-Decylcyclohexane	224.4	1.4534	0.443	7.7700	1.4899	255–378
Benzene	78.1	1.5011	0.863	11.2888	1.9936	278–344
Toluene	92.1	1.4969	0.777	9.9699	1.8321	233–389
n-Pentylbenzene	148.2	1.4882	0.740	7.6244	1.8472	255–411
n-Decylbenzene	218.4	1.4832	0.565	7.0362	1.6117	255–411
Water	18	1.3330	0.750	6.3827	2.5979	273–373
Methanol	32.0	1.3288	0.919	4.8375	3.1701	268–328
Ethanol	46.1	1.3610	0.440	8.2649	1.6273	280–338
Coefficients for thermal conductivity (W/mK)						
n-Pentane	72.2	1.3575	0.1	2.6357	0.6638	335–513
n-Decane	142.3	1.4119	0.1	2.0358	0.5152	256–436
n-Eicosane	282.6	1.4424	0.1	1.6308	0.3661	427–672
Cyclopentane	70.1	1.4065	0.1	2.5246	0.6335	328–551
Methylcyclopentane	84.2	1.4097	0.1	2.3954	0.6010	328–551
Cyclohexane	84.2	1.4262	0.1	1.9327	0.4755	411–544
Benzene	78.1	1.5011	0.1	2.6750	0.6384	410–566
Toluene	92.1	1.4969	0.1	2.1977	0.5358	354–577
Ethylbenzene	106.2	1.4959	0.1	2.0965	0.5072	354–577
Coefficients for self-diffusion coefficients ($10^5 \times$ cm²/s)						
n-Pentane	72.2	1.3575	0.270	10.4596	1.2595	195–309
n-Decane	142.3	1.4119	0.555	10.0126	1.7130	227–417
Benzene	78.1	1.5011	0.481	16.8022	1.4379	288–313
Water	18	1.3330	0.633	14.6030	2.2396	273–373
Methanol	32.0	1.3288	0.241	11.6705	1.2875	268–328
Ethanol	46.1	1.3610	0.220	15.1893	1.2548	280–338

TABLE 8.12—*Coefficients of Eq. (8.78) for estimation of transport properties of liquid n-alkanes with permission from Ref. [23].*

θ	Coefficients of Eq. (8.78) for n-alkanes
$1/\mu$ (cp)$^{-1}$	$p = 0.8036 - 5.8492 \times 10^{-4}M$
	$A = 2.638 + 5.214 \ln M + 0.0458M - 2.408M^{0.5}$
	$-B/A = 2.216 - 1.235 \times 10^{-3}M - 94(\ln M)^{-5} + 2.1809 \times 10^3 M^{-2.2}$, if $M < 185$
	$-B/A = 5.9644 - 3.625 \times 10^{-3}M + 788(\ln M)^{-3} - 71.441M - 0.4$, if $M > 185$
$1/k$ (W/mK)$^{-1}$	$p = 0.1$
	$A = 3.27857 - 0.01174M + 1.6 \times 10^{-5}M^2$
	$B = -2.50942 + 0.0139M - 2.0 \times 10^{-5}M^2$
$10^5 D$, cm^2/s	$p = -0.99259 + 0.02706M - 1.4936 \times 10^{-4}M^2 + 2.5383 \times 10^{-7}M^3$
	$A = 10.06464 + 0.02191M - 2.6223 \times 10^{-4}M^2 + 6.17943 \times 10^{-7}M^3$
	$-B = -9.80924 + 0.518156M - 3.31368 \times 10^{-3}M^2 + 5.70209 \times 10^{-6}M^3$

Solution—For n-C$_{10}$, from Table 2.1, $M = 142.3$. From Eq. (2.42) with coefficients given in Table 2.6 for I_{20} of n-alkanes we get $I_{20} = 0.24875$. From Eq. (2.114), $n_{20} = 1.41185$. From Eq. (2.118) at $T = 349$ K, $n_T = 1.38945$ and from Eq. (2.14) we calculate $I_T = 0.2368$. From Table 8.12 for $1/k$ we get $p = 0.1$, $A = 1.93196$, and $B = -0.9364$. Substituting these values in Eq. (8.78) we get $(1/k)^{0.1} = 1.93196$ $(1/0.2368 - 1)^{0.1} - 0.9364$ or $k = 0.1206$ W/m · K, which differs from the experimental value by 1.3%. DIPPR gives value of 0.1215 W/m · K [45]. ♦

8.5 MEASUREMENT OF DIFFUSION COEFFICIENTS IN RESERVOIR FLUIDS

Molecular diffusion is an important property needed in simulation and evaluation of several oil recovery processes. Examples are vertical miscible gas flooding, nonthermal recovery of heavy oil by solvent injection, and solution-gas-derived reservoirs. In these cases when pressure is reduced below bubble point of oil, gas bubbles are formed and the rate of their diffusion is the controlling step. Attempts in measurement of gas diffusivity in hydrocarbons under high-pressure conditions goes back to the early 1930s and has continued to the recent years [37, 51–57]. In general, methods of measuring diffusion coefficients in hydrocarbon systems can be divided into two categories. In the first category, during the experiment samples of the fluid are taken at various times and are analyzed by gas chromatography or other analytical tools [37, 55]. In the second category, samples are not analyzed but self-diffusion coefficients are measured by equipment such as NMR and then binary diffusion coefficients are calculated [41]. Other methods involve measuring volume of gas dissolved in oil versus time at constant pressure in order to determine gas diffusivity in reservoir fluids [43].

In the early 1990s a simple method to determine diffusion coefficients in both gas–gas and gas–liquid for binary and multicomponent systems at high pressures without compositional measurement was proposed by Riazi [56]. In this method, gas and oil are initially placed in a PVT under constant temperature condition. As the system approaches its equilibrium the pressure as well as gas–liquid interphase position in the cell vary and are measured versus time. Based on the rate of change of pressure or the liquid level, rate of diffusion in each phase can be determined [56]. The mechanism of diffusion process is based on the principle of thermodynamic equilibrium and the deriving force in molecular diffusion is

the system's deviation from equilibrium. Therefore, once a nonequilibrium gas is brought into contact with a liquid, the system tends to approach equilibrium so that the Gibbs energy, and therefore pressure, decreases with time. Once the system has reached an equilibrium state the pressure as well as composition of both gas and liquid phases remains unchanged. Schematic of the process is shown in Fig. 8.12. If the gas phase is hydrocarbon, dissolution of a hydrocarbon gas in an oil causes increase in oil volume and height of liquid (L_o) increases. For the case of nitrogen, the result is opposite and dissolution of N_2 causes decrease in the oil volume. In formulation of diffusion process in each phase, the Fick's law and material balance equations are applied for each component in the system. At the interphase, equilibrium criterion is imposed on each component ($\hat{f}_i^{oil} = \hat{f}_i^{gas}$). In addition, at the interphase the rates of diffusion in each phase are equal for each component. A semianalytical model for calculation of rates of diffusion process in both phases of gas and liquid is given by Riazi [56]. The model is a combination of material balance and vapor–liquid equilibrium calculations. When the diffusion processes come to an end the system will be at equilibrium. Diffusion coefficients needed in the model are calculated through a method such as Eq. (8.67). The model predicts composition of each phase, location of the liquid interface, and pressure of the system versus time.

To evaluate the proposed method, pure methane was placed on pure n-pentane at 311 K (100°F) and 102 bar in a PVT cell of 21.943 cm height and 2.56 cm diameter. The initial volume of liquid was 35% of the cell volume. Pressures were measured and recorded manually at selected times and continuously on a strip chart. The liquid level was measured manually with a precision of ±0.02 mm. Measurements were continued until there is no change in both pressure and liquid length at which the system reaches equilibrium. Diffusion coefficients were corrected so that predicted pressure curve versus time matches the experimental data as shown in Fig. 8.13. When diffusivities calculated by Eq. (8.67) are multiplied by 1.1 the model prediction perfectly matches experimental data. This technique measures diffusion coefficient of C$_1$–C$_5$ in liquid phase at 311 K and 71 bars as 1.51×10^{-4} cm^2/s, while the experimental data reported by Reamer et al. [52] is 1.43×10^{-4} cm^2/s. Diffusion coefficients of C$_1$–C$_5$, in both gas and liquid phases, versus pressure, and composition are shown in Figs. 8.14 and 8.15, respectively. Diffusivity of methane in heavy oils (bitumens) at 50 bar and 50°C is within the order of magnitude of 5×10^{-4} cm^2/s, while ethane diffusivity in such oils is about 2×10^{-4} cm^2/s [55].

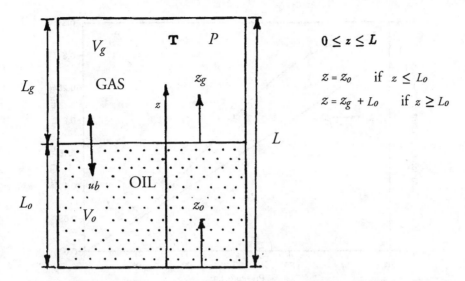

FIG. 8.12—Schematic and dimensions of a constant volume cell. Taken with permission from Ref. [23].

This method can be extended to multicomponent systems and it has been successfully used to measure gas diffusivity in heavy oils [57, 58]. In this method, pressure measurement is more accurate than measuring the interphase location. Furthermore, the initial measurements are more critical than measurements near the final equilibrium condition. The amount of initial liquid or gas determines diffusivity of which phase can be measured more accurately [56]. As it can be seen from Fig. 8.13 once a correct value of diffusion coefficient is used, the model prediction matches experimental data throughout the curve. This confirms the validity of the model and the assumptions made in its formulation. Methods that use unrealistic assumptions, i.e., neglecting natural convection terms when it exists, or oversimplified boundary conditions (i.e., semiinfinite assumption) lead to predictions that do not match the entire curve. In these cases, reported diffusion coefficients are based on a portion of experimental data and this is the reason that in such cases differences as large as ±100% are reported for diffusion coefficients in liquids at high pressures for the same systems under the same conditions. The technique can also be used to measure diffusivity in porous media by placing a reservoir core in the bottom of a PVT cell saturated initially with liquid oil. With such experiments and availability of more data, Eq. (8.67) can be further studied, modified, and improved.

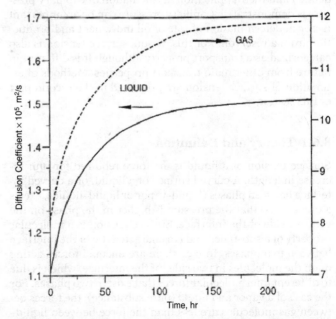

FIG. 8.13—Variation of pressure for the C_1–C_5 constant volume diffusion experiment at 311 K. — Diffusion coefficient from Eq. (8.67); ------ diffusion coefficient from Eq. (8.67) multiplied by 1.1; ····· diffusion coefficient from Eq. (8.65). Taken with permission from Ref. [56].

FIG. 8.14—Diffusion coefficient of the methane–n-pentane system at 311 K for the liquid and gas phases. Taken with permission from Ref. [56].

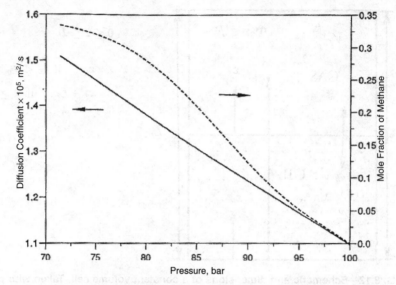

FIG. 8.15—Diffusion coefficient of liquid phase methane–*n*-pentane system at 311 K versus pressure and composition. Taken with permission from Ref. [56].

8.6 SURFACE/INTERFACIAL TENSION

Surface tension is an important molecular property in reservoir engineering calculations. In addition, surface tension is needed for the design and operation of gas–liquid separation units such as distillation and absorption columns. Based on the difference between surface tension of top and bottom products, one can determine whether or not foaming would occur in a distillation or absorption column. Foam formation is the cause of major problems in separation of gas and liquid phases. In this section surface and interfacial tensions are defined and their application in calculation of capillary pressure is demonstrated. Capillary pressure can be an important factor in determination of rate of oil movement and production from a reservoir. For this reason, surface tension is also categorized as a transport property although it is different in nature from other main transport properties. Methods of estimation of surface tension are presented in the second part of this section.

8.6.1 Theory and Definition

Surface tension of a liquid is the force required for unit increase in length. A curved surface of a liquid, or a curved interface between phases (liquid–vapor or liquid–liquid), exerts a pressure so that the pressure is higher in the phase on the concave side of the interface. Surface tension is a molecular property of a substance and is a characteristic of the interface between two phases. In fact, there are unequal forces acting upon the molecules in two sides of the interface, which is due to different intermolecular forces that exist in two phases. For the case of a vapor and liquid (pure substance), the forces between gas molecules are less than the force between liquid–liquid molecules, which cause the curvature on the liquid surface. It is due to this phenomenon, that liquid droplets form spherical shapes on a solid surface (i.e., droplet of liquid

mercury as seen in Fig. 8.16). Generally tension for vapor–liquid interface (pure substances) is referred to as *surface tension* and the tension between two different liquids (i.e., oil–water) is referred as *interfacial tension* (IFT). However, these two terms are used interchangeably. Surface tension is shown by σ and in the SI unit system it has the unit of N/m but usually the unit of dyne/cm (1 dyne/cm = 10^{-3} N/m = 1 mN/m) is used.

Based on the principle of phase equilibrium, one can show that for a droplet of pure liquids the difference between pressure in the liquid and vapor sides is proportional to the droplet radius. Consider a liquid droplet of radius r and that its surface is expanded in a closed container at constant temperature. Because of the extension of the surface droplet, radius changes by dr. Total volume (liquid and vapor) is constant ($V^{\text{total}} = V^{\text{V}} + V^{\text{L}}$ = constant or $dV^{\text{total}} = 0$) and as a result we have $dV^{\text{V}} = -dV^{\text{L}}$. The surface areas and volume of liquid droplet (V^{L}) are given as $S = 4\pi r^2$ and $V^{\text{L}} = (4/3)\pi r^3$. In this process (constant temperature and volume), the principle of equilibrium is formulated in terms of Helmholtz energy, A as follows:

$$(8.79) \qquad dA_{T,V} = 0$$

where $A = A^{\text{L}} + A^{\text{V}}$. With respect to definition of Helmholtz energy, one can have $dA^{\text{L}} = -P^{\text{L}}dV^{\text{L}} + \mu^{\text{L}}dn^{\text{L}} + \sigma dS$, where σdS represents the work required to expand liquid droplet by dr. Similarly for the vapor phase $dA^{\text{V}} = -P^{\text{V}}dV^{\text{V}} + \mu^{\text{V}}dn^{\text{V}}$ in which at equilibrium $\mu^{\text{V}} = \mu^{\text{L}}$ and $dn^{\text{V}} = -dn^{\text{L}}$. Substituting dA^{L} and dA^{V} into Eq. (8.79) the following relation is obtained:

$$(8.80) \qquad P^{\text{L}} - P^{\text{V}} = \frac{2\sigma}{r}$$

In the case of a bubble in the liquid, where pressure in the gas side is higher than that of liquid, the left side of the above relation becomes $P^{\text{V}} - P^{\text{L}}$. This can be formulated through contact angle θ, which is defined to determine degree of liquid

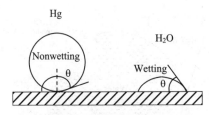

FIG. 8.16—The contact angle of a liquid surface and concept of wettability.

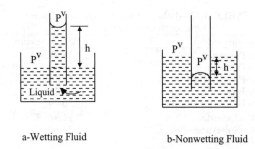

a-Wetting Fluid   b-Nonwetting Fluid

FIG. 8.17—Wetting and nonwetting liquids in capillary tubes.

wettability. Consider droplets of water and mercury on a solid surface as shown in Fig. 8.16. For mercury $\theta > 90°$ and it is called a *nonwetting* fluid, while for water with $\theta < 90°$ is an example of a *wetting liquid*. For other liquids θ is varying between 0 and 180° and have different degrees of wettability.

Equation (8.80) was derived on the assumption that the droplet is spherical. However, when a liquid is in contact with a solid surface where the liquid curvature is not fully spherical the above equation is corrected as

$$(8.81) \qquad P^V - P^L = \frac{2\sigma \, Cos\theta}{r}$$

For a fully nonwetting liquid $\theta = 180°$ (or $Cos\theta = -1$), Eq. (8.81) reduces to Eq. (8.80). If a wetting liquid (i.e., water) and a nonwetting liquid (i.e., mercury) are placed in two capillary tubes of radius r (diameter $2r$), the wetting liquid rises while nonwetting liquid depresses in the tube, as shown in Fig. 8.17. The height of liquid rise is determined from the pressure difference $P^V - P^L \, [= (\rho^L - \rho^V)gh]$ in which by substituting into Eq. (8.81) one can get:

$$(8.82) \qquad h = \frac{2\sigma \, Cos\theta}{(\rho^L - \rho^V)gr}$$

where ρ^L and ρ^V are the liquid and vapor density, respectively, and g is the acceleration of gravity (9.8 m/s^2). At low or atmospheric pressures where $\rho^V \ll \rho^L$, for simplicity ρ^V can be neglected. At high pressures, the pressure difference $(P^V - P^L)$ causes liquid rise and it is called *capillary pressure* shown by P_{cap}. For nonwetting liquids, such as mercury, where $\theta > 90°$, $Cos\theta < 0$ and according to Eq. (8.81) the liquid depresses in the tube as shown in Fig. 8.17. From this equation, when the

radius of tube decreases the height of liquid rise increases. In the case of oil and water, Eq. (8.82) becomes

$$(8.83) \qquad h = \frac{2\sigma_{wo} \, Cos\theta}{(\rho^w - \rho^o)gr}$$

where σ_{wo} is the interfacial tension between oil and water phases. ρ^W and ρ^o are density of water and oil, respectively. In this equation, if σ_{wo} is in N/m and ρ is in kg/m^3, then h and r must be in m.

The instrument that measures surface tension of a liquid is called *tensiometer*, which may be manual or digital. Most commonly used methods of measuring surface tension include classical ring method, capillary rise, pendant drop, and bubble pressure. The pendant method is most commonly used to measure surface tension of liquid oils. Schematic of apparatus to measure interfacial tension using the pendant drop method is shown in Fig. 8.18 [59]. Millette et al. [60] recommends maximum bubble pressure method to measure surface tension of hydrocarbons at high temperatures and pressures. Most advanced instruments can measure surface tension with an accuracy of ±0.001 mN/m.

Surface tension usually decreases with both pressure and temperature. Effect of temperature is greater than effect of pressure on surface tension. As pressure increases the difference between $(\rho^L - \rho^V)$ decreases and as a result surface tension also decreases, according to Eq. (8.82). The effect of pressure on IFT is discussed later. Surface tension increases with increase in molecular weight of a compound within a homologous hydrocarbon group. Some values of surface tension

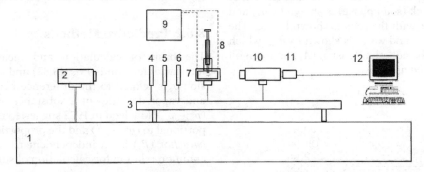

FIG. 8.18—Schematic of apparatus to measure interfacial tension using the pendant drop method. Taken with permission from Ref. [59].

(1) anti-vibration table; (2) light source; (3) optical rail; (4) light diffuser; (5) iris; (6) green filter; (7) thermostated interfacial tension cell with optical flats; (8) syringe to form pendant drops; (9) thermostat; (10) photomacrographic Tessovar zoom lens; (11) CCD camera; (12) computer with digitizing board.

TABLE 8.13—*Values of surface tension of some hydrocarbons at 25°C [45].*

Compound	σ at 25°C, dyne/cm
n-Pentane	15.47
n-Decane	23.39
n-Pentadecane	26.71
n-Eicosane	28.56
n-Hexatriacontane	30.44
Cyclopentane	21.78
Cyclohexane	24.64
Benzene	28.21
Decylbenzene	30.52
Pentadecylbenzene	31.97
Water (at 15°C)	74.83
Water (at 25°C)	72.82

for pure hydrocarbons and water are given in Table 8.13. Surface tension increases from paraffins to naphthenes and to aromatics for a same carbon number. Water has significantly higher surface tension than hydrocarbons. Surface tension of mercury is quite high and at 20°C it is 476 mN/m. Liquid metals have even higher surface tensions [18].

Example 8.6—Consider water at 15°C in a capillary tube open to atmosphere, as shown in Fig. 8.17. If the diameter of the tube is 10^{-4} cm, calculate the rise of water in the tube. What is the capillary pressure of water?

Solution—From Table 8.13 for water, σ at 15°C = 74.83 mN/m and liquid density of water at 15°C is 0.999 g/cm^3. Equation (8.82) must be used to calculate liquid rise. For water (assuming full wettability), $\theta = 0$ and $\text{Cos}(\theta) = 1$, $r = 5 \times 10^{-7}$ m, $\sigma = 74.83 \times 10^{-3}$ N/m, and $\rho^L = 999$ kg/m^3. Substituting this in Eq. (8.82) gives $h = (2 \times 74.83 \times 10^{-3} \times 1)/(999 \times 9.8 \times 5 \times 10^{-7}) = 30.57$ m. When r increases the rise in liquid height decreases. The capillary pressure is calculated from Eq. (8.80) as $P_{\text{cap}} = 2.99$ bar. ◆

One of the main applications of IFT between oil and water is to determine the type of rock wettability in a petroleum reservoir. Wettability may be defined as "the tendency of one fluid to spread on or adhere to a solid surface in the presence of other immiscible fluids" [15]. Consider oil and water in a reservoir as shown in Fig. 8.19. Assume the surface tension of oil with the reservoir rock (solid phase) is shown by σ_{so} and surface tension of water with the rock is shown by σ_{sw}. The contact angle between oil and water is shown by θ_{wo}, which varies from 0 to 180°. The adhesion tension (A_T) between oil and water A_T is calculated as follows:

$$(8.84) \qquad A_T = \sigma_{\text{so}} - \sigma_{\text{sw}} = \sigma_{\text{wo}} \text{Cos}(\theta_{\text{wo}})$$

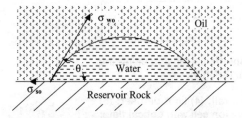

FIG. 8.19—Wettability of oil and water on a reservoir rock consisting mainly of calcium carbonate (CaCO₃).

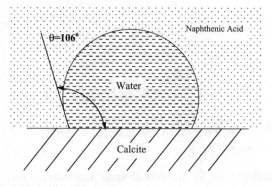

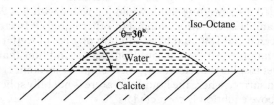

FIG. 8.20—Comparison of wettability of two different fluids on a calcite surface.

where σ_{wo} is the IFT between oil and water. If $A_T > 0$, the heavier liquid (in this case water), is the wettable fluid. The higher value of A_T indicates higher degree of wettability, which means the wetting fluid spreads better on the solid surface. If θ_{wo} is small (large A_T), the heavier fluid quickly spreads the solid surface. If $\theta_{\text{wo}} < 90°$, the solid surface is wettable with respect to water and if $\theta_{\text{wo}} > 90°$, the solid surface is wettable with respect to oil. Wettability of isooctane (*i*-C₈) and naphthenic acid on a calcite (a rock consisting mainly CaCO₃) is shown in Fig. 8.20. For the case of *i*-C₈ and water, the surface of calcite is wettable with water, while for the case of naphthenic acid, the calcite surface is wettable with respect to acid since $\theta > 90°$. Wettability of reservoir rocks has direct effect on the performance of miscible gas flooding in enhanced oil recovery (EOR) processes. For example, water flooding has better performance for reservoirs that are strongly water wet than those which are oil wet. For oil wet reservoirs water flooding must be followed by gas flooding to have effective improved oil recovery [61].

8.6.2 Predictive Methods

The basis of calculation and measurement of surface/interfacial tension is Eqs. (8.82) and (8.83). For surface tension σ is related to the difference between saturated liquid and vapor densities of a substance at a given temperature ($\rho^L - \rho^V$). Macleod in 1923 suggested that $\sigma^{1/4}$ is directly proportional to ($\rho^L - \rho^V$) and the proportionality constant called *parachor* (P_a) is an independent parameter [18]. The most common relation for calculation of surface tension is

$$(8.85) \qquad \sigma^{1/n} = P_a \frac{(\rho^L - \rho^V)}{M}$$

where M is molecular weight, ρ is density in g/cm^3, and σ is in mN/m (dyn/cm). This relation is usually referred to as Macleod–Sugden correlation. Parachor is a parameter that is defined to correlate surface tension and varies from one molecule to another. Different values for parameter n in

TABLE 8.14—*Values of parachor for some hydrocarbons for use in Eq. (8.85) with n = 3.88 [16].*

Compound	Parachor
Methane	74.05
n-Pentane	236.0
Isopentane	229.37
n-Hexane	276.71
n-Decane	440.69
n-Pentadecane	647.43
n-Eicosane	853.67
Cyclopentane	210.05
Cyclohexane	247.89
Methylcyclohexane	289.00
Benzene	210.96
Toluene	252.33
Ethylbenzene	292.27
Carbon dioxide	82.00
Hydrogen sulfide	85.50

Eq. (8.85) are suggested, the most commonly used values are 4, 11/3 (= 3.67), and 3.88. For example, values of parachors reported in the API-TDB [5] are given for $n = 4$, while in Ref. [16] parameters are given for the value of $n = 3.88$. Parachor number of pure compounds may be estimated from group contribution methods [5, 18]. For example, for n-alkanes the following equation can be obtained based on a group contribution method suggested by Poling et al. [18]:

$$(8.86) \quad P_a = 111 + a(N_C - 2) \quad \text{for } n = 4 \text{ in Eq. (8.85)}$$

where N_C is the carbon number of n-alkane hydrocarbon and $a = 40$ if $2 \le N_C \le 14$ or $a = 40.3$ if $N_C > 14$. Calculated values of surface tension by Eq. (8.85) are quite sensitive to the value of parachor. Values of parachor for some compounds as given in Ref. [16] for use in Eq. (8.85) with $n = 3.88$ are given in Table 8.14. For defined mixtures the Kay's mixing rule (Eq. 7.1) can be used as $\sigma_{mix} = \sum x_i \sigma_i$ for quick calculations. For more accurate calculations, the following equation is suggested in the API-TDB to calculate surface tension of defined mixtures [5]:

$$(8.87) \quad \sigma_{mix} = \left\{ \sum_{i=1}^{N} \left[P_{a,i} \left(\frac{\rho^L}{M^L} x_i - \frac{\rho^V}{M^V} y_i \right) \right] \right\}^n$$

where M^L and M^V are molecular weight of liquid and vapor mixtures, respectively. x_i and y_i are mole fractions of liquid and vapor phases. ρ^L and ρ^V are densities of saturated liquid and vapor mixtures at given temperature in g/cm³. Some attempts to correlate surface tension to liquid viscosity have been made in the form of $\sigma = A \exp(-B\mu)$ in which A is related to PNA composition and parameter B is correlated to M as well as PNA distribution [34]. At higher pressures where the difference between liquid and vapor properties reduces, μ could be replaced by $\Delta\mu = (\mu_L^{0.5} - \mu_V^{0.5})^2$. Such correlations, however, are not widely used in the industry.

Temperature dependency of surface tension can be observed from the effect of temperature on density as shown in Eq. (8.85). At the critical point, $\rho^L - \rho^V = 0$ and surface tension reduces to zero ($\sigma = 0$). In fact, there is a direct correlation between $(\rho^L - \rho^V)$ and $(T_c - T)$, and one can assume $(\rho^L - \rho^V) = K(1 - T_r)^m$ where K and m are constants that depend on the fluid where n is approximately equal to 0.3. Combination of this relation with Eq. (8.85) gives a correlation between σ and $(1 - T_r)^{0.3n}$ in which n is close to 4.0.

Generally, corresponding state correlation in terms of reduced surface tension versus $(1 - T_r)$ are proposed [18]. The group $\sigma/P_c^{2/3} T_c^{1/3}$ is a dimensionless parameter except for the numerical constant that depends on the units of σ, P_c, and T_c. There are a number of generalized correlations for calculation of σ. For example, Block and Bird correlation is given as follows [18]:

$$(8.88) \quad \sigma = P_c^{2/3} T_c^{1/3} Q (1 - T_r)^{11/9}$$
$$Q = 0.1196 \left[1 + \frac{T_{br} \ln(P_c/1.01325)}{1 - T_{br}} \right] - 0.279$$

where σ is in dyn/cm, P_c in bar, T_c in kelvin, and T_{br} is the reduced boiling point (T_b/T_c). This equation is relatively accurate for hydrocarbons; however, for nonhydrocarbons errors as high as 40–50% are observed. In general, the accuracy of this equation is about 5%. Another generalized correlation was developed by Miqueu et al. [62] based on an earlier correlation proposed by Schmidt and it is given in the following form:

$$\sigma = k_B T_c \left(\frac{N_A}{V_c} \right)^{2/3} \times (4.35 + 4.14\omega) \times \left(1 + 0.19\tau^{0.5} - 0.25\tau \right) \tau^{1.26}$$
$$(8.89)$$

where $\tau = 1 - T_r$, σ is in dyn/cm, $k_B(= 1.381 \times 10^{-16}$ dyn·cm/K), N_A, T_c, T_r, V_c, and ω are the Boltzmann constant, Avogadro number, the critical temperature in kelvin, reduced temperature, the critical volume in cm³/mol, and acentric factor, respectively. This equation was developed based on experimental data for surface tensions of N_2, O_2, Kr, hydrocarbons from C_1 to n-C_8 (including i-C_4 and i-C_5) and 16 halogenated hydrocarbons (refrigerants) with an average reported error of 3.5%.

For undefined petroleum fractions the following relation suggested in the API-TDB [5] can be used for calculation of surface tension:

$$(8.90) \quad \sigma = \frac{673.7 (1 - T_r)^{1.232}}{K_W}$$

where T_r is the reduced temperature and K_W is the Watson characterization factor. Tsonopoulos et al. [33] have correlated parachor of hydrocarbons, petroleum fractions, and coal liquids to boiling point and specific gravity in a form similar to that of Eq. (2.38):

$$(8.91) \quad \sigma^{1/4} = \frac{P_a}{M} (\rho^L - \rho^V)$$
$$\frac{P_a}{M} = 1.7237 T_b^{0.05873} SG^{-0.64927}$$

where T_b is the boiling point in kelvin and SG is the specific gravity. Units for the other parameters are the same as those in Eq. (9.85). This equation can predict surface tension of pure hydrocarbons with an average deviation of about 1% [33].

Recently, Miqueu et al. [59] reported some experimental data on IFT of petroleum fractions and evaluated various predictive methods. They recommended the following method

TABLE 8.15—*Effect of characterization method on prediction of interfacial tension of some petroleum fractions through Eq. (8.91).*

						% Error on prediction of IFT[a]		
Fraction	T_b, K	SG	ρ_{25}, g/cm³	M	σ at 25°C, mN/m	Method 1	Method 2	Method 3
1	429	0.769	0.761	130.9	22.3	26.5	2.7	14.8
2	499	0.870	0.863	167.7	30.7	−29.3	−7.5	−2.0
3	433	0.865	0.858	120.2	29.2	−15.4	3.4	22.9
4	505	0.764	0.756	184.4	25.6	−4.7	7.8	−10.9
Overall						19.0	5.4	12.7

[a]Experimental data are taken from Miqueu et al. [59]. Method 1: T_c and P_c from Kesler–Lee (Eqs. (2.69) and (2.70)) and ω from Lee–Kesler (Eq. 2.105). Method 2: T_c and P_c from API-TDB (Eqs. (2.65) and (2.66)) [5] and ω from Lee–Kesler (Eq. 2.105). Method 3: T_c and P_c from Twu (Eqs. (2.80) and (2.86)) and ω from Lee–Kesler (Eq. 2.105).

for calculation of surface tension of undefined petroleum fractions:

$$\sigma = \left[\frac{P_a}{M} \left(\rho^L - \rho^V \right) \right]^{11/3}$$

$$(8.92) \qquad P_a = \frac{(0.85 - 0.19\omega) \, T_c^{12/11}}{(P_c/10)^{9/11}}$$

In this method, n in Eq. (8.85) is equal to 11/3 or 3.6667. In the above equation, T_c and P_c are in kelvin and bar, respectively, σ is in mN/m (dyn/cm), and ρ is in g/cm³. Predicted values of surface tension by this method strongly depend on the characterization method used to calculate T_c, P_c, and M. For four petroleum fractions predicted values of surface tension by three different characterization methods described in Chapter 2 are given in Table 8.15. As it is seen from this table, the API method of calculating T_c, P_c, ω, and M (Section 2.5) yields the lowest error for estimation of surface tension. Miqueu et al. [59] used the pseudocomponent method (Section 3.3.4, Eq. 3.39) to develop the following equation for estimation of parachor and surface tension of defined petroleum fractions with known PNA composition.

$$\sigma = \left[\frac{P_a}{M} \left(\rho^L - \rho^V \right) \right]^{11/3}$$

$$P_a = x_P P_{a,P} + x_N P_{a,N} + x_A P_{a,A}$$

$$(8.93) \qquad P_{a,P} = 27.503 + 2.9963M$$

$$P_{a,N} = 18.384 + 2.7367M$$

$$P_{a,A} = 25.511 + 2.8332M$$

where x_P, x_N, and x_A are mole fractions of paraffins, naphthenes, and aromatics in the fraction. Units are the same as in Eq. (8.92). Experimental data of Darwish et al. [63] on surface tension consist PNA distribution of some petroleum fractions. For undefined fractions, the PNA composition may be estimated from methods of Chapter 3. For cases where accurate PNA composition data are not available the parachor number of an undefined petroleum fraction may be directly calculated from molecular weight of the fraction (M), using the following correlation originally provided by Fawcett and recommended by Miqueu et al. [59]:

$$(8.94) \qquad P_a = 81.2 + 2.448M \quad \text{value of } n \text{ in Eq. (8.85)} = 11/3$$

In this method, only M and liquid density are needed to calculate surface tension at atmospheric pressure. Firoozabadi [64] also provided a similar correlation ($P_a = 11.4 + 3.23M - 0.0022M^2$), which is reliable up to C_{10}, but for heavier hydrocarbons it seriously underpredicts values of surface tension.

An evaluation of various methods for prediction of surface tension of n-alkanes is shown in Fig. 8.21. Data are taken from DIPPR [45]. The most accurate method for calculation of surface tension of pure hydrocarbons is through Eq. (8.85) with values of parachor from Table 8.14 or Eq. (8.86). Method of Block and Bird (Eq. 8.88) or Eq. (8.90) for petroleum fractions also provide reliable values for surface tension of pure hydrocarbons with average errors of about 3%. Equation (8.90) is perhaps the most accurate method as it gives the lowest error for surface tension of n-alkanes (error of $\sim$2%), while it is proposed for petroleum fractions. Equations (8.92)–(8.94) give generally very large errors, especially for hydrocarbons heavier than C_{10}. Equation (8.93) is developed for petroleum fractions ranging from C_5 to C_{10} and Eq. (8.94) is not suitable for heavy hydrocarbons as shown in Fig. 8.21.

Interfacial tension (IFT) between hydrocarbon and water is important in understanding the calculations related to oil recovery processes. The following simple relation is suggested in the API-TDB [5] to calculate σ_{HW} from surface tension of hydrocarbon σ_H and that of water σ_W:

$$(8.95) \qquad \sigma_{HW} = \sigma_H + \sigma_W - 1.10 \, (\sigma_H \sigma_W)^{1/2}$$

Use of this method is also demonstrated in Example 8.7. Another relation for IFT of hydrocarbon–water systems under reservoir conditions was proposed by Firoozabadi and Ramey [16, 65] in the following form:

$$(8.96) \qquad \sigma_{HW} = 111 \, (\rho_W - \rho_H)^{1.024} \, (T/T_{cH})^{-1.25}$$

where σ_{HW} is the hydrocarbon–water IFT in dyn/cm (mN/m), ρ_W and ρ_H are water and hydrocarbon densities in g/cm³, T is temperature in kelvin, and T_{cH} is the pure hydrocarbon critical temperature in kelvon. Errors as high as 30% are reported for this correlation [16]. IFT similar to surface tension decreases with increase in temperature. For liquid–liquids, such as oil–water systems, IFT usually increases slightly with pressure; however, for gas–liquid systems, such as methane–water, the IFT slightly decreases with increase in pressure.

Example 8.7—A kerosene sample has boiling point and specific gravity of 499 K and 0.87, respectively. Calculate the IFT of this oil with water at 25°C. Liquid density of the fraction at this temperature is 0.863 g/cm³.

Solution—$T_b = 499$ K and SG = 0.87. From Eq. (2.51), $M = 167.7$. Parachor can be calculated from the Fawcett method as given in Eq. (8.94): $P_a = 491.73$. From data $\rho_{25} = 0.863$ g/cm³. Substituting values of M, P_a, and ρ_{25} (for ρ^L) in Eq. (8.85) with $n = 11/3$ gives $\sigma_{25} = 30.1$ mN/m, where in comparison with the experimental value of 30.74 mN/m [59] the error is −2.1%. When using Eq. (8.85), the value of ρ^V is neglected

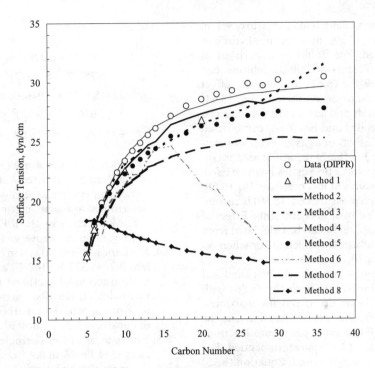

FIG. 8.21—Prediction of surface tension of *n*-alkanes from various methods. Method 1: Eq. (8.85) and Table 8.14; Method 2: Eqs. (8.85) and (8.86); Method 3: Eq. (8.88); Method 4: Eq. (8.90); Method 5: Eq. (8.91); Method 6: Eq. (8.92); Method 7: Eq. (8.93); Method 8: Eq. (8.94).

with respect to ρ^L at atmospheric pressure. To calculate IFT of water–oil, Eq. (8.95) can be used. From Table 8.13 for water at 25°C, $\sigma_W = 72.8$ mN/m. From Eq. (8.95), $\sigma_{W-oil} = 72.8 + 30.1 - 1.1(72.8 \times 30.1)^{1/2} = 51.4$ mN/m. To calculate σ_{W-oil} from Eq. (8.96), T_{cH} is calculated from the API method (Eq. 2.65) as 705 K and at 25°C, $\rho_W = 0.995$ g/cm^3. From Eq. (8.96), $\sigma_{W-oil} = 40.9$ mN/m. This is about 20% less than the value calculated from Eq. (8.95). As mentioned before large error may be observed from Eq. (8.96) for calculation of IFT. ◆

8.7 SUMMARY AND RECOMMENDATIONS

In this chapter, methods and procedures presented in the previous chapters are used for estimation of four transport properties: viscosity, thermal conductivity, diffusion coefficient, and surface tension. In general semitheoretical methods for estimation of transport properties have wider range of applications than do pure empirical correlations and their development and applications are discussed in this chapter. A summary of recommended methods is given below.

For calculation of viscosity of pure gases at atmospheric pressure, Eq. (8.3) should be used and for compounds for which the coefficients are not available, Eq. (8.6) may be used. For defined gas mixture when viscosity of components are known Eq. (8.7) or (8.8) can be used. For hydrocarbon gases at high pressure, viscosity can be calculated from Eq. (8.12) and for nonhydrocarbons Eq. (8.13) can be used. For estimation of viscosity of natural gas at atmospheric pressure, Eq. (8.14) and at higher pressure Eq. (8.15) are recommended.

To estimate viscosity of pure liquids, Eq. (8.17) is recommended and for a defined hydrocarbon mixture Eq. (8.18) can be used. For petroleum fractions when kinematic viscosity at 100°F (37.8°C) is available, Eq. (8.19) can be used. When two petroleum fractions are mixed, Eq. (8.20) is useful. Viscosity of liquid hydrocarbons at high pressure can be calculated from Eq. (8.22). For crude oil at atmospheric pressure Eq. (8.25) is useful; however, for reservoir fluids Eq. (8.12) can be used for both gases and liquids or their mixtures.

Thermal conductivity of pure hydrocarbon gases at low pressures should be calculated from Eq. (8.33) and for those for which the coefficients are not available, Eq. (8.34) should be used. For defined hydrocarbon gas mixtures Eq. (8.35) and for undefined petroleum vapor fractions Eq. (8.37) is recommended. For vapor fractions at temperatures in which Eq. (8.37) is not applicable, Eq. (8.36) is recommended. For hydrocarbon gases at high pressures Eq. (8.39) may be used and if not possible Eq. (8.38) can be used for both pure gases and undefined gas mixtures.

For pure hydrocarbon liquids at low pressures, Eq. (8.42) is recommended and for those compounds whose thermal conductivity at two reference points are not known, Eq. (8.43) is recommended. For undefined liquid petroleum fractions, Eq. (8.46) and for defined liquid mixtures Eq. (8.48) can be used. For fractions without any characterization data, Eq. (8.50) can be used for determination of approximate value of thermal conductivity. For fractions with only boiling point available, Eq. (8.51) should be used and for coal liquid fractions Eq. (8.52) is recommended. For liquid fractions at high pressures, Eq. (8.53) is recommended.

To estimate binary diffusion coefficients for hydrocarbon gases at low pressures, Eq. (8.59) and for nonhydrocarbons Eq. (8.58) can be used. For liquid hydrocarbons at low pressure, diffusion coefficients at infinite dilution can be estimated from Eq. (8.60) or (8.61) and for the effect of concentration on binary diffusion coefficients Eq. (8.64) should be used. For both liquids and gases at high pressures, Eq. (8.67) is highly recommended and Eq. (8.66) can be used as alternative method for diffusivity of a gas in oil under reservoir conditions. When using Eq. (8.67) recommended methods for calculation of low-pressures properties must be used. For multicomponent gas mixtures at low pressure, Eq. (8.68) and for liquids or gases at high pressures Eq. (8.69) is recommended to calculate effective diffusion coefficients. Effect of porous media on diffusion coefficient can be calculated from Eqs. (8.70) and (8.74). Self-diffusion coefficients or when refractive index is available, Eq. (8.78) can be used.

Surface tension of pure compounds should be calculated from Eq. (8.85) and defined mixtures from Eq. (8.86) with parachors given in Table 8.14 or Eq. (8.86) for *n*-alkanes. For undefined petroleum fraction surface tension can be calculated from Eq. (8.90). For defined petroleum fractions (known PNA composition), Eq. (8.93) is recommended. For coal liquid fractions Eq. (8.91) may be used. Equation (8.95) is recommended for calculation of IFT of water–hydrocarbon systems. For specific cases, recommended methods are discussed in Section 8.6.2.

In addition to predictive methods, two methods for experimental measurement of diffusion coefficient and surface tension are presented in Sections 8.5 and 8.6.1. Furthermore, the interrelationship among various transport properties, effects of porous media and concept of wettability, calculation of capillary pressure and the role, and importance of interfacial tension in enhanced oil recovery processes are discussed. It is also shown that choice of characterization method could have a significant impact on calculation of transport properties of petroleum fractions.

8.8 PROBLEMS

8.1. Pure methane gas is being displaced in a fluid mixture of C_1, n-C_4, and n-C_{10} with composition of 41, 27, and 32 mol%, respectively. Reported measured diffusion coefficient of pure methane in the fluid mixture under the conditions of 344 K and 300 bar is 1.01×10^{-4} cm^2/s [9].
   a. Calculate density and viscosity of fluid.
   b. Estimate diffusion coefficient of methane from Sigmund method (Eq. 8.65).
   c. Estimate diffusion coefficient of methane from Eq. 8.67.

8.2. Hill and Lacy measured viscosity of a kerosene sample at 333 K and 1 atm as 1.245 mPa · s [51]. For this petroleum fraction, $M = 167$ and SG = 0.7837. Estimate the viscosity from two most suitable methods and compare with given experimental value.

8.3. Riazi and Otaibi [21] developed the following relation for estimation of viscosity of liquid petroleum fractions based on Eq. (8.78):

$$1/\mu = A + B/I$$

where

$$A = 37.34745 - 0.20611M + 141.1265\text{SG} - 637.727I_{20}$$
$$- 6.757T_b^* + 6.98(T_b^*)^2 - 0.81(T_b^*)^3,$$
$$B = -15.5437 + 0.046603M - 42.8873\text{SG} + 211.6542I_{20}$$
$$+ 1.676T_b^* - 1.8(T_b^*)^2 + 0.212(T_b^*)^3,$$
$$T_b^* = (1.8T_b - 459.67)/1.8$$

in which T_b is the average boiling point in Kelvin, μ is in cP, and parameter I should be determined at the same temperature as μ is desired. (Parameter I can be determined as discussed for its use in Eq. (8.78).)
   For kerosene sample of Problem 8.2, calculate viscosity based on the above method and obtain the error.

8.4. Methane gas is dissolved in the kerosene sample of Problem 8.2, at 333 K (140°F) and 20.7 bar (300 psia). The mole fraction of methane is 0.08. For this fluid mixture calculate density, viscosity, and thermal conductivity from appropriate methods. The experimental value of density is 5.224 kmol/m^3.

8.5. Estimate diffusion coefficient of methane in kerosene sample of Problem 8.4 from Eqs. (8.65)–(8.67).

8.6. Estimate thermal conductivity of N_2 at 600°F and 3750 and 10 000 psia. Compare the result with values of 0.029 and 0.0365 Btu/ft · h · °F as reported in the API-TDB [5].

8.7. Consider an equimolar mixture of C_1, C_3, and N_2 at 14 bar and 311 K. The binary diffusion coefficient of D_{C1-C3} and D_{C1-N2} are 88.3×10^{-4} and 187×10^{-4} cm^2/s, respectively. The mixture density is 0.551 kmol/m^3. Estimate the effective diffusion coefficient of methane in the mixture from Eq. (8.68) and compare it with the value calculated from Eqs. (8.67) and (8.69).

8.8. A petroleum fraction has boiling point and specific gravity of 429 K and 0.761, respectively. The experimental value of surface tension at 25°C is 22.3 mN/m [59]. Calculate the surface tension at this temperature from the following methods and compare them against the experimental value.
   a. Five different methods presented by Eqs. (8.88)–(8.92) with estimated input parameters from the API-TDB methods.
   b. Equation (8.93) with predicted PNA distribution.
   c. Fawcett's method for parachor (Eq. 8.94).
   d. Firoozabadi's method for parachor.

REFERENCES

[1] Bird, R. B., Stewart, W. E., and Lightfoot, E. N., *Transport Phenomena Processes*, Wiley, New York, 1960; 2nd ed., 1999.

[2] Gallant, R. W. and Yaws, C. L., *Physical Properties of Hydrocarbons, Vol. 1*, 2nd ed., Gulf Publishing, Houston, TX, 1992.

[3] Alberty, R. A. and Silbey, R. J., *Physical Chemistry*, 2nd ed., Wiley, New York, 1999.

[4] Hirschfelder, O. J., Curtiss, C. F., and Bird, R. B., *Molecular Theory of Gases and Liquids*, Wiley, New York, 1964.

[5] Daubert, T. E. and Danner, R. P., Eds., *API Technical Data Book—Petroleum Refining*, 6th ed., American Petroleum Institute (API), Washington, DC, 1997.

[6] Yoon, P. and Thodos, G., "Viscosity of Nonpolar Gaseous Mixtures at Normal Pressures," *American Institute of Chemical Engineers Journal*, Vol. 16, No. 2, 1970, pp. 300–304.

[7] Stiel, L. I. and Thodos, G., "The Viscosity of Nonpolar Gases at Normal Pressures," *American Institute of Chemical Engineers Journal*, Vol. 7, No. 4, 1961, pp. 611–615.

[8] Wauquier, J.-P., *Petroleum Refining, Vol. 1: Crude Oil, Petroleum Products, Process Flowsheets*, Editions Technip, Paris, 1995.

[9] Riazi, M. R. and Whitson, C. H., "Estimating Diffusion Coefficients of Dense Fluids," *Industrial and Engineering Chemistry Research*, Vol. 32, No. 12, 1993, pp. 3081–3088.

[10] Danner, R. P. and Daubert, T. E., *Manual for Predicting Chemical Process Design Data*, Design Institute for Physical Property Data (DIPPR), AIChE, New York, 1986. Rowley, R. L., Wilding, W. V., Oscarson, J. L., Zundel, N. A., Marshall, T. L., Daubert, T. E., and Danner, R. P., *DIPPR Data Compilation of Pure Compound Properties*, Design Institute for Physical Properties, Taylor & Francis, New York, 2002 (http://dippr.byu.edu/).

[11] Jossi, J. A., Stiel, L. I., and Thodos, G., "The Viscosity of Pure Substances in the Dense Gaseous and Liquid Phases," *American Institute of Chemical Engineers Journal*, Vol. 8, 1962, pp. 59–63.

[12] Lohrenz, J., Bray, B. G., and Clark, C. R., "Calculating Viscosity of Reservoir Fluids from Their Compositions," *Journal of Petroleum Technology*, October 1964, pp. 1171–1176.

[13] Stiel, L. I. and Thodos, G., "The Viscosity of Polar Substances in the Dense Gaseous and Liquid Regions," *American Institute of Chemical Engineers Journal*, Vol. 10, No. 2, 1964, pp. 275–277.

[14] Lee, A., Gonzalez, M. H., and Eakin, B. E., "The Viscosity of Natural Gases," *Journal of Petroleum Technology*, August 1966, pp. 997–1000.

[15] Ahmed, T., *Reservoir Engineering Handbook*, Gulf Publishing, Houston, TX, 2000.

[16] Danesh, A., *PVT and Phase Behavior of Petroleum Reservoir Fluids*, Elsevier, Amsterdam, 1998.

[17] Speight, J. G., *The Chemistry and Technology of Petroleum*, 3rd ed., Marcel Dekker, New York, 1999.

[18] Poling, B. E., Prausnitz, J. M., and O'Connell, J. P., *Properties of Gases and Liquids*, 5th ed., Mc-Graw Hill, New York, 2000. Reid, R. C., Prausnitz, J. M., and Poling, B. E., *Properties of Gases and Liquids*, 4th ed., Mc-Graw Hill, New York, 1987.

[19] *Crude Oil Yields and Product Properties*, Cud Thomas Baird IV, 1222 Vezenaz, Geneva, Switzerland, June 1981.

[20] Riazi, M. R., Mahdi, K., and Alqallaf, M., "A Generalized Correlation for Viscosity of Hydrocarbons Based on Corresponding States Principles and Molar Refraction," *Journal of Chemical and Engineering Data*, American Chemical Society, Vol. 50, 2005.

[21] Riazi, M. R. and Al-Otaibi, G. N., "Estimation of Viscosity of Petroleum Fractions," *Fuel*, Vol. 80, 2001, pp. 27–32.

[22] Hildebrand, J. H., "Motions of Molecules in Liquids: Viscosity and Diffusivity," *Science*, Vol. 174, 1971, pp. 490–493.

[23] Riazi, M. R., Al-Enezi, G., and Soleimani, S., "Estimation of Transport Properties of Liquids," *Chemical Engineering Communications*, Vol. 176, 1999, pp. 175–193.

[24] Riazi, M. R. and Roomi, Y., "Use of the Refractive Index in the Estimation of Thermophysical Properties of Hydrocarbons and Their Mixtures," *Industrial and Engineering Chemistry Research*, Vol. 40, No. 8, 2001, pp. 1975–1984.

[25] Chung, T. H., Ajlan, M., Lee, L. L., and Starling, K. E., "Generalized Multiparameter Correlation for Nonpolar and Polar Fluid Transport Properties," *Industrial and Engineering Chemistry Research*, Vol. 27, 1988, pp. 671–679.

[26] Quinones-Cisneros, S. E., Zeberg-Mikkelsen, C. K., and Stenby, E. H., "The Friction Theory (f-Theory) for Viscosity Modeling," *Fluid Phase Equilibria*, Vol. 169, 2000, pp. 249–276.

[27] Smith, J. M., Van Ness, H. C., and Abbott, M. M., *Introduction to Chemical Engineering Thermodynamics*, 5th ed., McGraw-Hill, New York, 1996.

[28] McCabe, W. L., Smith, J. C., and Harriot, P., *Unit Operations of Chemical Engineering*, 5th ed., McGraw-Hill, New York, 2001.

[29] Mathias, P. M., Parekh, V. S., and Miller, E. J., "Prediction and Correlation of the Thermal Conductivity of Pure Fluids and Mixtures, Including the Critical Region," *Industrial and Engineering Chemistry Research*, Vol. 41, 2002, pp. 989–999.

[30] Riazi, M. R. and Faghri, A., "Thermal Conductivity of Liquid and Vapor Hydrocarbon Systems: Pentanes and Heavier at Low Pressures," *Industrial and Engineering Chemistry, Process Design and Development*, Vol. 24, No. 2, 1985, pp. 398–401.

[31] Stiel, L. I. and Thodos, G., "The Thermal Conductivity of Nonpolar Substances in the Dense Gaseous and Liquid Regions," *American Institute of Chemical Engineers Journal*, Vol. 10, No. 1, 1964, pp. 26–30.

[32] Riazi, M. R. and Faghri, A., "Prediction of Thermal Conductivity of Gases at High Pressures," *American Institute of Chemical Engineers Journal*, Vol. 31, No. 1, 1985, pp. 164–167.

[33] Tsonopoulos, C., Heidman, J. L., and Hwang, S.-C., *Thermodynamic and Transport Properties of Coal Liquids*, An Exxon Monograph, Wiley, New York, 1986.

[34] Pedersen, K. S. and Fredenslund, Aa., "An Improved Corresponding States Model for the Prediction of Oil and Gas Viscosities and Thermal Conductivities," *Chemical Engineering Science*, Vol. 42, 1987, pp. 182–186.

[35] Duda, J. L., "Personal Notes in Transport Phenomena," Instructor's note for ChE 545 Course, Chemical Engineering Department, Pennsylvania State University, University Park, PA, 1977.

[36] Stiel, L. I. and Thodos, G., "Lennard-Jones Constants Predicted from Critical Properties," *Journal of Chemical and Engineering Data*, Vol. 7, No. 2, 1962, pp. 234–236.

[37] Riazi, M. R. and Daubert, T. E., "Application of Corresponding States Principles for Prediction of Self-Diffusion Coefficients in Liquids," *American Institute of Chemical Engineers Journal*, Vol. 26, No. 3, 1980, pp. 386–391.

[38] Slattery, J. C. and Bird, R. B., "Calculation of the Diffusion Coefficients of Dilute Gases and of the Self-Diffusion Coefficients of Dense Gases," *American Institute of Chemical Engineers Journal*, Vol. 4, 1958, pp. 137–142.

[39] Takahashi, S., "Preparation of a Generalized Chart for the Diffusion Coefficients of Gases at High Pressures," *Journal of Chemical Engineering of Japan*, Vol. 7, 1974, pp. 417–421.

[40] Sigmund, P. M., "Prediction of Molecular Diffusion at Reservoir Conditions. Part I—Measurement and Prediction of Binary Dense Gas Diffusion Coefficients," *Canadian Journal of Petroleum Technology*, Vol. 15, No. 2, 1976, pp. 48–57.

[41] Dawson R., Khoury, F., and Kobayashi, R., "Self-Diffusion Measurements in Methane by Pulsed Nuclear Magnetic Resonance," *American Institute of Chemical Engineers Journal*, Vol. 16, No. 5, 1970, pp. 725–729.

[42] Whitson, C. H. and Brule, M. R., *Phase Behavior*, Monograph Vol. 20, Society of Petroleum Engineers, Richardson, TX, 2000.

[43] Renner, T. A., "Measurement and Correlation of Diffusion Coefficients for CO_2 and Rich Gas Applications," *SPE Reservoir Engineering*, No. 3 (May), 1988, pp. 517–523.

[44] Eyring, H., *Significant Liquid Structure*, Wiley, New York, 1969.

[45] AIChE DIPPR® Database, Design Institute for Physical Property Data (DIPPR), EPCON International, Houston, TX, 1996.

[46] Wilke, C. R., "Diffusional Properties of Multicomponent Gases," *Chemical Engineering Progress*, Vol. 46, No. 2, 1950, pp. 95–104.

[47] Saidi, A. M., *Reservoir Engineering of Fracture Reservoirs*, TOTAL Edition Presse, Paris, 1987, Ch. 8.

[48] Amyx, J. W., Bass, D. M., and Whiting, R. L., *Petroleum Reservoir Engineering, Physical Properties*, McGraw-Hill, New York, pp. 129 (total of 600 pp.), 1960.

[49] Langness, G. L., Robertson, J. O., Jr., and Chilingar, G. V., *Secondary Recovery and Carbonate Reservoirs*, Elsevier, New York, 304 pp. 1972.

[50] Fontes, E. D., Byrne, P., and Hernell, O., "Put More Punch Into Catalytic Reactors," *Chemical Engineering Progress*, Vol. 99, No. 3, 2003, pp. 48–53.

[51] Hill, E. S. and Lacy, W. N., "Rate of Solution of Methane in Quiescent Liquid Hydrocarbons," *Industrial and Engineering Chemistry*, Vol. 26, 1934, pp. 1324–1327.

[52] Reamer, H. H., Duffy, C. H., and Sage, B. H., "Diffusion Coefficients in Hydrocarbon Systems: Methane-Pentane in Liquid Phase," *Industrial and Engineering Chemistry*, Vol. 48, 1956, pp. 275–282.

[53] Lo, H. Y., "Diffusion Coefficients in Binary Liquid *n*-Alkanes Systems," *Journal of Chemical and Engineering Data*, Vol. 19, No. 3, 1974, pp. 239–241.

[54] McKay, W. N., "Experiments Concerning Diffusion of Multicomponent Systems at Reservoir Conditions," *Journal of Canadian Petroleum Technology*, Vol. 10 (April–June), 1971, pp. 25–32.

[55] Nguyen, T. A. and Farouq Ali, S. M., "Effect of Nitrogen on the Solubility and Diffusivity of Carbon Dioxide into Oil and Oil Recovery by the Immiscible WAG Process," *Journal of Canadian Petroleum Technology*, Vol. 37, No. 2, 1998, pp. 24–31.

[56] Riazi, M. R., "A New Method for Experimental Measurement of Diffusion Coefficient in Reservoir Fluids," *Journal of Petroleum Science and Engineering*, Vol. 14, 1996, pp. 235–250.

[57] Zhang, Y. P., Hyndman, C. L., and Maini, B. B., "Measurement of Gas Diffusivity in Heavy Oils," *Journal of Petroleum Science and Engineering*, Vol. 25, 2000, pp. 37–47.

[58] Upreti, S. R. and Mehrotra, A. K., "Diffusivity of CO_2, CH_4, C_2H_6 and N_2 in Athabasca Bitumen," *The Canadian Journal of Chemical Engineering*, Vol. 80, 2002, pp. 117–125.

[59] Miqueu, C., Satherley, J., Mendiboure, B., Lachiase, J., and Graciaa, A., "The Effect of P/N/A Distribution on the Parachors of Petroleum Fractions," *Fluid Phase Equilibria*, Vol. 180, 2001, pp. 327–344.

[60] Millette, J. P., Scott, D. S., Reilly, I. G., Majerski, P., Piskorz, J., Radlein, D., and de Bruijin, T. J. W., "An Apparatus for the Measurement of Surface Tensions at High Pressures and Temperatures," *The Canadian Journal of Chemical Engineering*, Vol. 80, 2002, pp. 126–134.

[61] Rao, D. N., Girard, M., and Sayegh, S. G., "The Influence of Reservoir Wettability on Waterflood and Miscible Flood Performance," *Journal of Canadian Petroleum Technology*, Vol. 31, No. 6, 1992, pp. 47–55.

[62] Miqueu, C., Broseta, D., Satherley, J., Mendiboure, B., Lachiase, J., and Graciaa, A., "An Extended Scaled Equation for the Temperature Dependence of the Surface Tension of Pure Compounds Inferred From an Analysis of Experimental Data," *Fluid Phase Equilibria*, Vol. 172, 2000, pp. 169–182.

[63] Darwish, E., Al-Sahhaf, T. A., and Fahim, M. A., "Prediction and Correlation of Surface Tension of Naphtha Reformate and Crude Oil," *Fuel*, Vol. 74, No. 4, 1995, pp. 575–581.

[64] Firoozabadi, A., Katz, D. L., and Soroosh, H., *SPE Reservoir Engineering*, Paper No. 13826, February 1988, pp. 265–272.

[65] Firoozabadi, A. and Ramey, Jr., H. J., "Surface Tension of Water-Hydrocarbon Systems at Reservoir Conditions," *Journal of Canadian Petroleum Technology*, Vol. 27, No. 3, May-June 1988, pp. 41–48.

Applications: Phase Equilibrium Calculations

NOMENCLATURE

API	API gravity defined in Eq. (2.4)
a, b, c, d, e	Constants in various equations
b	A parameter defined in the Standing correlation, Eq. (6.202), K
C_P	Heat capacity at constant pressure defined by Eq. (6.17), J/mol·K
F	Number of moles for the feed in VLSE unit, mol (feed rate in mol/s)
$F(V_F)$	Objective function defined in Eq. (9.4) to find value of V_F
F^{SL}	Objective function defined in Eq. (9.19) to find value of S_F
f_i	Fugacity of component i in a mixture defined by Eq. (6.109), bar
$f_i^L(T, P, x_i^L)$	Fugacity of component i in a liquid mixture of composition x_i^L at T and P, bar
K_i	Equilibrium ratio in vapor–liquid equilibria ($K_i = y_i/x_i$) defined in Eq. (6.196), dimensionless
K_i^{VS}	Equilibrium ratio in vapor–solid equilibria ($K_i^{SL} = y_i/x_i^S$), dimensionless
k_{AB}	Binary interaction coefficient of asphaltene and asphaltene-free crude oil, dimensionless
L	Number of moles of liquid formed in VLE process, mol (rate in mol/s)
L_F	Mole of liquid formed in VLSE process for each mole of feed ($F = 1$), dimensionless
M	Molecular weight (molar mass), g/mol [kg/kmol]
M_B	Molecular weight (molar mass) of asphaltene-free crude oil, g/mol
N	Number of components in a mixture
n_j^S	Number of moles of component j in the solid phase, mol
P	Pressure, bar
P_b	Bubble point pressure, bar
P_c	Critical pressure, bar
P_{tp}	Triple point pressure, bar
R	Gas constant = 8.314 J/mol·K (values in different units are given in Section 1.7.24)
R_i	Refractivity intercept [$= n_{20} - d_{20}/2$] defined in Eq. (2.14)
R_S	Dilution ratio of LMP solvent to oil (cm³ of solvent added to 1 g of oil), cm³/g
S	Number of moles of solid formed in VLSE separation process, mol (rate in mol/s)
S_F	Moles of solid formed in VLSE separation process for each mole of initial feed ($F = 1$), dimensionless
SG	Specific gravity of liquid substance at 15.5°C (60°F) defined by Eq. (2.2), dimensionless
T	Absolute temperature, K
T_b	Normal boiling point, K
T_c	Critical temperature, K
T_M	Freezing (melting) point for a pure component at 1.013 bar, K
T_{pc}	Pseudocritical temperature, K
T_{tc}	True-critical temperature, K
T_{tp}	Triple point temperature, K
V	Molar volume, cm³/mol
V	Number of moles of vapor formed in VLSE separation process, mol (rate in mol/s)
V_A	Liquid molar volume of pure component A at normal boiling point, cm³/mol
V_F	Mole of vapor formed in VLSE separation process for each mole of feed ($F = 1$), dimensionless
V_c	Critical molar volume, cm³/mol (or critical specific volume, cm³/g)
V_i	Molar volume of pure component i at T and P, cm³/mol
V_m^L	Molar volume of liquid mixture, cm³/mol
x_i	Mole fraction of component i in a mixture (usually used for liquids), dimensionless
x_i^S	Mole fraction of component i in a solid mixture, dimensionless
y_i	Mole fraction of i in a mixture (usually used for gases), dimensionless
Z	Compressibility factor defined by Eq. (5.15), dimensionless
Z_c	Critical compressibility factor [$Z = P_c V_c/RT_c$], dimensionless
z_i	Mole fraction of i in the feed mixture (in VLE or VLSE separation process), dimensionless

Greek Letters

Δ	Difference between two values of a parameter
ε	Convergence tolerance (e.g., 10^{-5})
ϕ_i	Volume fraction of component i in a mixture defined by Eq. (9.11), dimensionless
Φ_i	Volume fraction of component i in a mixture defined by Eq. (9.33), dimensionless
$\hat{\phi}_i$	Fugacity coefficient of component i in a mixture at T and P defined by Eq. (6.110)

ρ Density at a given temperature and pressure, g/cm^3 (molar density unit: cm^3/mol)

ρ_M Molar density at a given temperature and pressure, mol/cm^3

ω Acentric factor defined by Eq. (2.10), dimensionless

$\hat{\mu}_i$ Chemical potential of component i in a mixture defined by Eq. (6.115)

δ_i Solubility parameter for i defined in Eq. (6.147), (J/cm^3)$^{1/2}$ or (cal/cm^3)$^{1/2}$

γ_i Activity coefficient of component i in liquid solution defined by Eq. (6.112), dimensionless

ΔC_{Pi} Difference between heat capacity of liquid and solid for pure component i at its melting (freezing) point ($= C_{Pi}^{L} - C_{Pi}^{S}$), J/mol·K

ΔH_i^{f} Heat of fusion (or latent heat of melting) for pure component i at the freezing point and 1.013 bar, J/mol

Superscript

L   Value of a property at liquid phase
V   Value of a property at vapor phase
S   Value of a property at solid phase

Subscripts

A   Value of a property for component A
A   Value of a property for asphaltenes
c   Value of a property at the critical point
i,j Value of a property for component i or j in a mixture
L   Value of a property for liquid phase
M   Value of a property at the melting point of a substance
pc   Pseudocritical property
S   Value of a property at the solid phase
S   Value of a property for solvent (LMP)
s   Specific property (quantity per unit mass)
T Values of property at temperature T
tc   True critical property
tr   Value of a property at the triple point
20   Values of property at 20°C
7+   Values of a property for C$_{7+}$ fraction of an oil

Acronyms

ABSA   Alkyl benzene sulfonic acid
API-TDB   American Petroleum Institute—Technical Data Book (see Ref. [12])
BIP   Binary interaction parameter
bbl   Barrel, unit of volume of liquid as given in Section 1.7.11
CPT   Cloud-point temperature
cp   Centipoise, unit of viscosity, (1 cp = 0.01 p = 0.01 g·cm·s = 1 mPa·s = 10^{-3} kg/m·s)
cSt   Centistoke, unit of kinematic viscosity, (1 cSt = 0.01 St = 0.01 cm^2/s)
EOR   Enhanced oil recovery
EOS   Equation of state
FH   Flory–Huggins

GC   Gas condensate (a type of reservoir fluid defined in Chapter 1)
GOR   Gas-to-oil ratio, scf/bbl
HFT   Hydrate formation temperature
IFT   Interfacial tension
LLE   Liquid–liquid equilibria
LMP   Low molecular weight n-paraffins (i.e., C$_3$, n-C$_5$, n-C$_7$)
LVS   liquid–vapor–solid
LS   Liquid–solid
MeOH   Methanol
PR   Peng–Robinson EOS (see Eq. 5.39)
SRK   Soave–Redlich–Kwong EOS given by Eq. (5.38) and parameters in Table 5.1
SAFT   Statistical associating fluid theory (see Eq. 5.98)
SLE   Solid–liquid equilibrium
scf   Standard cubic foot (unit for volume of gas at 1 atm and 60°F)
stb   Stock tank barrel (unit for volume of liquid oil at 1 atm and 60°F)
VABP   Volume average boiling point defined by Eq. (3.3).
VLE   Vapor–liquid equilibrium
VLSE   Vapor–liquid–solid equilibrium
VS   Vapor–solid
VSE   Vapor–solid equilibrium
WAT   Wax appearance temperature
WPT   Wax precipitation temperature
%AAD   Average absolute deviation percentage defined by Eq. (2.135)
%AD   Absolute deviation percentage defined by Eq. (2.134)
wt%   Weight percent

ONE OF THE MAIN APPLICATIONS of science of thermodynamics in the petroleum industry is for the prediction of phase behavior of petroleum fluids. In this chapter calculations related to vapor–liquid and solid–liquid equilibrium in petroleum fluids are presented. Their application to calculate gas–oil ratio, crude oil composition, and the amount of wax or asphaltene precipitation in oils under certain conditions of temperature, pressure, and composition is presented. Methods of calculation of wax formation temperature, cloud point temperature of crude oils, determination of onset of asphaltene, hydrate formation temperature, and methods of prevention of solid formation are also discussed. Finally application of characterization techniques, methods of prediction of transport properties, equations of state, and phase equilibrium calculations are demonstrated in modeling and evaluation of gas injection projects.

9.1 TYPES OF PHASE EQUILIBRIUM CALCULATIONS

Three types of phase equilibrium, namely, vapor–liquid (VLE), solid–liquid (SLE), and liquid–liquid (LLE), are of particular interest in the petroleum industry. Furthermore, vapor–solid (VSE), vapor–liquid–solid (VLSE), and vapor–liquid–liquid (VLLE) equilibrium are also of importance in

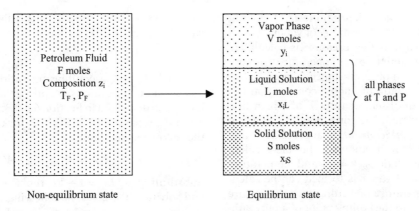

FIG. 9.1—Typical vapor–liquid–solid equilibrium for solid precipitation.

calculations related to petroleum and natural gas production. VLE calculations are needed in design and operation of separation units such as multistage surface separators at the surface facilities of production fields, distillation, and gas absorption columns in petroleum and natural gas processing as well as phase determination of reservoir fluids. LLE calculations are useful in determination of amount of water dissolved in oil or amount of oil dissolved in water under reservoir conditions. SLE calculations can be used to determine amount and the conditions at which a solid (wax or asphaltene) may be formed from a petroleum fluid. Cloud-point temperature (CPT) can be accurately calculated through SLE calculations. VSE calculation is used to calculate hydrate formation and the conditions at which it can be prevented.

Schematic of a system at vapor–liquid–solid equilibrium (VLSE) is shown in Fig. 9.1. The system at its initial conditions of T_F and P_F is in a nonequilibrium state. When it reaches to equilibrium state, the conditions change to T and P and new phases may be formed. The initial composition of the fluid mixture is z_i; however, at the final equilibrium conditions, compositions of vapor, liquid, and solid in terms of mole fractions are specified as y_i, x_i^L, and x_i^S, respectively. The amount of feed, vapor, liquid, and solid in terms of number of moles is specified by F, V, L, and S, respectively. Under VLE conditions, no solid is formed ($S = 0$) and at VSE state no liquid exists at the final equilibrium state ($L = 0$). The system variables are F, z_i, T, P, V, y_i, L, x_i^L, S, and x_i^S, where in a typical equilibrium calculation, F, z_i, T, and P are known, and V, L, S, y_i, x_i^L, and x_i^S are to be calculated. In some calculations such as bubble point calculations, T or P may be unknown and must be calculated from given information on P or T and the amount of V, L, or S. Calculations are formulated through both equilibrium relations and material balance for all components in the system. Two-phase equilibrium such as VLE or SLE calculations are somewhat simpler than three-phase equilibrium such as VLSE calculations.

In this chapter various types VLE and SLE calculations are formulated and applied to various petroleum fluids. Principles of phase equilibria were discussed in Section 6.8 through Eqs. (6.171)–(6.174). VLE calculations are formulated through equilibrium ratios (K_i) and Eq. (6.201), while SLE calculations can be formulated through Eq. (6.208). In addition there are five types of VLE calculations that are discussed in the next section. Flash and bubble point pressure

calculations are the most widely used VLE calculations by both chemical and reservoir engineers in the petroleum processing and production.

9.2 VAPOR–LIQUID EQUILIBRIUM CALCULATIONS

VLE calculations are perhaps the most important types of phase behavior calculations in the petroleum industry. They involve calculations related to equilibrium between two phases of liquid and vapor in a multicomponent system. Consider a fluid mixture with mole fraction of each component shown by z_i is available in a sealed vessel at T and P. Under these conditions assume the fluid can exist as both vapor and liquid in equilibrium. Furthermore, assume there are total of F mol of fluid in the vessel at initial temperature and pressure of T_F and P_F as shown in Fig. 9.1. The conditions of the vessel change to temperature T and pressure P at which both vapor and liquid can coexist in equilibrium. Assume V mol of vapor with composition y_i and $L(= F - V)$ mol of liquid with composition x_i are produced as a result of phase separation due to equilibrium conditions. No solid exists at the equilibrium state and $S = 0$ and for this reason composition of liquid phase is simply shown by x_i. The amount of vapor may be expressed by the ratio of V/F or V_F for each mole of the mixture. The parameters involved in this equilibrium problem are T, P, z_i, x_i, y_i, and V_F (for the case of $F = 1$). The VLE calculations involve calculation of three of these parameters from three other known parameters.

Generally there are five types of VLE calculations: (i) Flash, (ii) bubble-P, (iii) bubble-T, (iv) dew-P, and (v) dew-T. (i) In flash calculations, usually z_i, T, and P are known while x_i, y_i, and V are the unknown parameters. Obviously calculations can be performed so that P or T can be found for a known value of V. Flash separation is also referred as flash distillation. (ii) In the bubble-P calculations, pressure of a liquid of known composition is reduced at constant T until the first vapor molecules are formed. The corresponding pressure is called *bubble point pressure* (P_b) at temperature T and estimation of this pressure is known as bubble-P calculations. For analysis of VLE properties, consider the system in Fig. 9.1 without solid phase ($S = 0$). Also assume the feed is a liquid with composition ($x_i = z_i$) at $T = T_F$ and P_F. Now at constant

T, pressure is reduced to P at which infinitesimal amount of vapor is produced ($\sim V = 0$ or beginning of vaporization). Through bubble-P calculations this pressure is calculated. Bubble point pressure for a mixture at temperature T is similar to the vapor pressure of a pure substance at given T. (iii) In bubble-T calculations, liquid of known composition (x_i) at pressure P is heated until temperature T at which first molecules of vapor are formed. The corresponding temperature is known as *bubble point temperature* at pressure P and estimation of this temperature is known as bubble-T calculations. In this type of calculations, $P = P_F$ and temperature T at which small amount of vapor is formed can be calculated. Bubble point temperature or saturation temperature for a mixture is equivalent to the boiling point of a pure substance at pressure P. (iv) In dew-P calculations a vapor of known composition ($y_i = z_i$) at temperature $T = T_F$ is compressed to pressure P at which infinitesimal amount of liquid is produced ($\sim L = 0$ or beginning of condensation). Through dew-P calculations this pressure known as *dew point pressure* (P_d) is calculated. For a pure substance the dew point pressure at temperature T is equivalent to its vapor pressure at T. (v) In dew-T calculations, a vapor of known composition is cooled at constant P until temperature T at which first molecules of liquid are formed. The corresponding temperature is known as *dew point temperature* at pressure P and estimation of this temperature is known as dew-T calculations. In these calculations, $P = P_F$ and temperature T at which condensation begins is calculated. Flash, bubble, and dew points calculations are widely used in the petroleum industry and are discussed in the following sections.

9.2.1 Flash Calculations—Gas-to-Oil Ratio

In typical flash calculations a feed fluid mixture of composition z_i enters a separator at T and P. Products of a flash separator for F mol of feed are V mol of vapor with composition y_i and L mol of liquid with composition x_i. Calculations can be performed for each mole of the feed ($F = 1$). By calculating vapor-to-feed mole ratio ($V_F = V/F$), one can calculate the gas-to-oil ratio (GOR) or gas-to-liquid ratio (GLR). This parameter is particularly important in operation of surface separators at the oil production fields in which production of maximum liquid (oil) is desired by having low value of GOR. Schematic of a continuous flash separator unit is shown in Fig. 9.2.

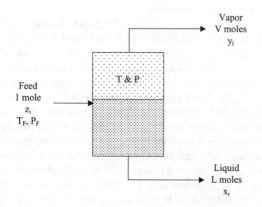

FIG. 9.2—A continuous flash separator.

Since vapor and liquid leaving a flash unit are in equilibrium from Eq. (6.201) we have

$$(9.1) \qquad y_i = K_i x_i$$

in which K_i is the equilibrium ratio of component i at T and P and compositions x_i and y_i. Calculations of K_i values have been discussed in Section 6.8.2.3. Mole balance equation around a separator unit (Fig. 9.2) for component i is given by the following equation:

$$(9.2) \qquad 1 \times z_i = L_F \times x_i + V_F \times y_i$$

Substituting for $L_F = 1 - V_F$, replacing for y_i from Eq. (9.1), and solving for x_i gives the following:

$$(9.3) \qquad x_i = \frac{z_i}{1 + V_F(K_i - 1)}$$

Substituting Eq. (9.3) into Eq. (9.1) gives a relation for calculation of y_i. Since for both vapor and liquid products we must have $\sum x_i = \sum y_i = 1$ or $\sum (y_i - x_i) = 0$. Substituting x_i and y_i from the above equations gives the following objective function for calculation of V_F:

$$(9.4) \qquad F(V_F) = \sum_{i=1}^{N} \frac{z_i(K_i - 1)}{1 + V_F(K_i - 1)} = 0$$

Reservoir engineers usually refer to this equation as Rachford–Rice method [1]. When $V_F = 0$, the fluid is a liquid at its bubble point (saturated liquid) and if $V_F = 1$, the system is a vapor at its dew point (saturated vapor). Correct solution of Eq. (9.4) should give positive values for all x_i and y_i, which match the conditions $\sum x_i = \sum y_i = 1$. The following step-by-step procedure can be used to calculate V_F:

1. Consider the case that values of z_i (feed composition), T, and P (flash condition) are known.
2. Calculate all K_i values assuming ideal solution (i.e., using Eqs. 6.198, 6.202, or 6.204). In this way knowledge of x_i and y_i are not required.
3. Guess an estimate of V_F value. A good initial guess may be calculated from the following relationship [2]: $V_F = A/(A - B)$, where $A = \sum [z_i (K_i - 1)]$ and $B = \sum [z_i(K_i - 1)/K_i]$.
4. Calculate $F(V)$ from Eq. (9.4) using assumed value of V_F in Step 3.
5. If calculated $F(V_F)$ is smaller than a preset tolerance, ε (e.g., 10^{-15}), then assumed value of V_F is the desired answer. If $F(V_F) > \varepsilon$, then a new value of V_F must be calculated from the following relation:

$$(9.5) \qquad V_F^{\text{new}} = V_F - \frac{F(V_F)}{\frac{dF(V_F)}{dV_F}}$$

In which $dF(V_F)/dV_F$ is the first-order derivative of $F(V_F)$ with respect to V_F.

$$(9.6) \qquad \frac{dF(V_F)}{dV_F} = -\sum_{i=1}^{N} \left\{ \frac{z_i (K_i - 1)^2}{\left[V_F(K_i - 1) + 1\right]^2} \right\}$$

The procedure is repeated until the correct value of V_F is obtained. Generally, if $F(V_F) > 0$, V_F must be reduced and if $F(V_F) < 0$, V_F must be increased to approach the solution.

6. Calculate liquid composition, x_i, from Eq. (9.3) and the vapor phase composition, y_i, from Eq. (9.1).

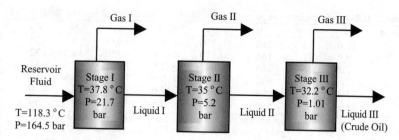

FIG. 9.3—Schematic of a three-stage separator test in a Middle East production field.

7. Calculate K_i values from a more accurate method using x_i and y_i calculated in Step 6. For example, K_i can be calculated from Eq. (6.197) by a cubic equation of state (i.e., SRK EOS) through calculating $\hat{\phi}_i^L$ and $\hat{\phi}_i^V$ using Eq. (6.126). Subsequently $\hat{f}_i^L$ and $\hat{f}_i^V$ can be calculated from Eq. (6.113). For isothermal flash we must have

$$
(9.7) \qquad \sum_{i=1}^{N} \left(\frac{\hat{f}_i^L}{\hat{f}_i^V} - 1 \right)^2 < \varepsilon
$$

where ε is a convergence tolerance, (e.g., 1×10^{-13}).

8. Repeat a new round of calculations from Step 4 with calculated V_F from the previous round until there is no change in values of V_F, x_i, and y_i and inequality (9.7) is satisfied.

Various other methods of flash calculations for fast convergence are given in different references [1–4]. For example, Whitson [1] suggests that the initial guess for V_F must be between two values of $V_{F,min}$ and $V_{F,max}$ to obtain fast convergence. Michelsen also gives a stability test for flash calculations [5, 6]. Accuracy of results of VLE calculations largely depends on the method used for estimation of K_i values and for this reason recommended methods in Table 6.15 can be used as a guide for selection of an appropriate method for VLE calculation. Another important factor for the accuracy of VLE calculations is the method of characterization of C_{7+} fraction of the petroleum fluid. Application of continuous functions, as it was shown in Section 4.5, can improve results of calculations. The impact of characterization on phase behavior of reservoir fluids is also demonstrated in Section 9.2.3.

The above procedure can be easily extended to LLE or vapor–liquid–liquid equilibrium (VLLE) in which two immiscible liquids are in equilibrium with themselves and their vapor phase (see Problem 9.1).

Once value of V_F is calculated in a VLE flash calculation, the gas-to-liquid ratio (GLR) or gas-to-oil ratio (GOR) can be calculated from the following relation [7]:

$$
(9.8) \qquad \text{GOR} \left[\text{scf/stb} \right] = \frac{1.33 \times 10^5 \rho_L V_F}{(1 - V_F) M_L}
$$

where ρ_L (in g/cm^3) and M_L (in g/mol) are the density and molecular weight of a liquid product, respectively (see Problem 9.2). The best method of calculation of ρ_L for a liquid mixture is to calculate it through Eq. (7.4), using pure component liquid densities. If the liquid is at atmospheric pressure and temperature, then ρ_L can be replaced by liquid specific gravity, SG_L, which may also be calculated from Eq. (7.4) and components SG values. The method of calculations is demonstrated in Example 9.1.

Example 9.1 (Three-stage surface separator)—Schematic of a three-stage separator for analysis of a reservoir fluid to produce crude oil is shown in Fig. 9.3. The composition of reservoir fluid and products as well as GOR in each stage and the overall GOR are given in Table 9.1. Calculate final crude composition and the overall GOR from an appropriate model.

Solution—The first step in calculation is to express the C_{7+} fraction into a number of pseudocomponents with known

TABLE 9.1—*Experimental data for a Middle East reservoir fluid in a three-stage separator test. Taken with permission from Ref. [7].*

No.	Component	Feed	1st-Stage gas	2nd-Stage gas	3rd-Stage gas	3rd-Stage liquid
1	N_2	0.09	0.77	0.16	0.15	0.00
2	CO_2	2.09	4.02	3.92	1.41	0.00
3	H_2S	1.89	1.35	4.42	5.29	0.00
4	H_2O	0.00	0.00	0.00	0.00	0.00
5	C_1	29.18	63.27	31.78	5.10	0.00
6	C_2	13.60	20.15	33.17	26.33	0.19
7	C_3	9.20	7.56	18.84	36.02	1.88
8	$n\text{-}C_4$	4.30	1.5	4.14	13.6	3.92
9	$i\text{-}C_4$	0.95	0.43	1.24	3.62	0.62
10	$n\text{-}C_5$	2.60	0.36	0.92	3.50	4.46
11	$i\text{-}C_6$	1.38	0.24	0.63	2.46	2.11
12	C_6	4.32	0.24	0.57	2.09	8.59
13	C_{7+}	30.40	0.11	0.21	0.43	78.23
SG at 60°F						0.8150
Temp,°F		245	105	100	90	90
Pressure, psia		2387	315	75	15	15
GOR, scf/stb		850	601	142	107	

TABLE 9.2—*Characterization parameters of the C_{7+} fraction of sample of Table 9.1 [7].*

Pseudocomponent	mol%	wt%	M	SG	T_b, K	n_{20}	N_C	P%	N%	A%
C_{7+} (1)	10.0	12.5	110	0.750	391.8	1.419	8	58	22	20
C_{7+} (2)	9.0	17.1	168	0.810	487.9	1.450	12.3	32	35	33
C_{7+} (3)	7.7	23.1	263	0.862	602.1	1.478	19.3	17	37	46
C_{7+} (4)	2.5	11.6	402	0.903	709.0	1.501	28.9	6	34	60
C_{7+} (5)	1.2	8.2	608	0.949	777.6	1.538	44	0	45	55
Total C_{7+}	30.4	72.5	209.8	0.843	576.7	1.469	15.3	25	34	41

Experimental values on M_{7+} and SG_{7+}. Distribution parameters (for Eq. 4.56) and calculated values: $M_{7+} = 209.8$; $M_o = 86.8$; $S_o = 0.65$; $S_{7+} = 0.844$; $B_M = 1$; $A_s = 0.119$; $n_{7+} = 1.4698$; $A_M = 1.417$; $B_S = 3$; $M_{av} = 209.8$; $S_{av} = 0.847$.

characterization parameters (i.e., M, T_b, SG, n_{20}, N_C, and PNA composition). This is done using the distribution model described in Section 4.5.4 with M_{7+} and SG_{7+} as the input parameters. The basic parameters (T_b, n_{20}) are calculated from the methods described in Chapter 2, while the PNA composition for each pseudocomponent is calculated from methods given in Section 3.5.1.2 (Eqs. 3.74–3.81). The calculation results with distribution parameters for Eq. (4.56) are given in Table 9.2. Molar and specific gravity distributions of the C_{7+} fraction are shown in Fig. 9.4. The PNA composition is needed for calculation of properties through pseudocomponent approach (Section 3.3.4). Such information is also needed when a simulator (i.e., EOR software) is used for phase behavior calculations [9].

To generate the composition of gases and liquids in separators, see Fig. 9.3, the feed to the first stage is considered as a mixture of 17 components (12 components listed in Table 9.1 and 5 components listed in Table 9.2). For pure components (first 11 components of Table 9.1), T_c, P_c, V_c, and ω are taken from Table 2.1. For C_6 fraction (SCN) and C_{7+} fractions (Table 9.3) critical properties can be obtained from methods of Chapter 2 (Section 2.5) or from Table 4.6. For this example, Lee–Kesler correlations for calculation of T_c, P_c, and ω and Riazi–Daubert correlations (the API methods) for calculation of V_c and M (or T_b) have been used. The binary interaction parameters (BIPs) for nonhydrocarbon–hydrocarbon are taken from Table 5.3 and for hydrocarbon–hydrocarbon pairs are calculated from Eq. (5.63). Parameter A in this equation has been used as an adjustable parameter so that at least one predicted property matches the experimental data. This property can be saturation pressure or a liquid density data. For this calculation, parameter A was determined so that predicted liquid specific gravity from last stage matches experimental value of 0.815. Liquid SG is calculated from Eq. (7.4) using SG of all components in the mixture. It was found that when $A = 0.18$, a good match is obtained. Another adjustable parameter can be the BIP of methane and the first pseudocomponent of heptane-plus, $C_7(1)$. The value of BIP of this pair exhibits a major impact in the calculation results. K_i values are calculated from SRK EOS and flash calculations are performed for three stages shown in Fig. 9.3. The liquid product from the first stage is used as the feed for the second stage separator and flash calculation for this stage is performed to calculate composition of feed for the last stage. Similarly, the final crude oil is produced from the third stage at atmospheric pressure. Composition of C_{7+} in each stream can be calculated from sum of mole fractions of the five pseudocomponents of C_{7i}. GOR for each stage is calculated from Eq. (9.8). Summary of results are given in Table 9.3. Overall GOR is calculated as 853 compared with actual value of 850 scf/stb. This is a very good prediction mainly due to adjusting BIPs with liquid density of produced crude oil. The calculated compositions in Table 9.3 are also in good agreement with actual data of Table 9.1.

The method of characterization selected for treatment of C_{7+} has a major impact on the results of calculations as shown by Riazi et al. [7]. Table 9.4 shows results of GOR calculations for the three stages from different characterization methods. In the Standing method, Eqs. (6.204) and (6.205) have been used to estimate K_i values, assuming ideal solution mixture. As shown in this table, as the number of pseudocomponents for the C_{7+} fraction increases better results can be obtained. ◆

9.2.2 Bubble and Dew Points Calculations

Bubble point pressure calculation is performed through the following steps:

1. Assume a liquid mixture of known x_i and T is available.
2. Calculate P_i^{sat} (vapor pressure) of all components at T from methods described in Section 7.3.
3. Calculate initial values of y_i and P_{bub} from Raoult's law as $P = \sum x_i P_i^{sat}$ and $y_i = x_i P_i^{sat}/P$.
4. Calculate K_i from Eq. (6.197) using T, P, x_i, and y_i.
5. Check if $|\sum x_i K_i - 1| < \varepsilon$, where ε is a convergence tolerance, (e.g., 1×10^{-12}) and then go to Step 6. If not, repeat calculations from Step 4 by guessing a new value for pressure P and $y_i = K_i x_i$. If $\sum x_i K_i - 1 < 0$, reduce P and if $\sum x_i K_i - 1 > 0$, increase value of P.
6. Write P as the bubble point pressure and y_i as the composition of vapor phase. Bubble P can also be calculated through flash calculations by finding a pressure at which $V_F \cong 0$. In bubble T calculation x_i and P are known. The calculation procedure is similar to bubble P calculation method except that T must be guessed instead of guessing P.

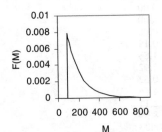

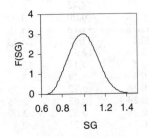

FIG. 9.4—Probability density functions for molecular weight and specific gravity of the C_{7+} fraction given in Table 9.2 [8].

TABLE 9.3—*Calculated values for the data given in Table 9.1 using proposed characterization method. Taken with permission from Ref. [7].*

No.	Component	Feed	1st-Stage gas	2nd-Stage gas	3rd-Stage gas	3rd-Stage liquid
1	N_2	0.09	0.54	0.12	0.05	0.00
2	CO_2	2.09	3.91	4.09	1.44	0.02
3	H_2S	1.89	1.47	4.38	5.06	0.14
4	H_2O	0.00	0.00	0.00	0.00	0.00
5	C_1	29.18	64.10	32.12	5.68	0.03
6	C_2	13.60	19.62	32.65	25.41	0.38
7	C_3	9.20	7.41	18.24	35.47	3.05
8	$n\text{-}C_4$	4.30	1.48	4.56	13.92	4.38
9	$i\text{-}C_4$	0.95	0.41	1.23	3.47	0.78
10	$n\text{-}C_5$	2.60	0.36	1.01	3.98	4.81
11	$i\text{-}C_6$	1.38	0.24	0.68	2.61	2.37
12	C_6	4.32	0.27	0.61	2.22	9.01
13	C_{7+}	30.40	0.19	0.31	0.69	75.03
	SG at 60°F					0.8105
	Temp,°F	245	105	100	90	90
	Pressure, psia	2197	315	75	15	15
	GOR, scf/stb	853	580	156	117	

For vapors of known composition dew P or dew T can be calculated as outlined below:

1. Assume a vapor mixture of known y_i and T is available.
2. Calculate P_i^{sat} (vapor pressure) of all components at T from methods of Section 7.3.
3. Calculate initial values of x_i and P_{dew} from Raoult's law as $1/P = \sum y_i / P_i^{sat}$ and $x_i = y_i P / P_i^{sat}$.
4. Calculate K_i from Eq. (6.197), using T, P, x_i, and y_i.
5. Check if $\left|\sum y_i / K_i - 1\right| < \varepsilon$, where ε is a convergence tolerance, (e.g., 1×10^{-12}) go to Step 6. If not, repeat calculations from Step 4 by guessing a new value for pressure P and $x_i = y_i / K_i$. If $\sum y_i / K_i - 1 < 0$, increase P and if $\sum y_i / K_i - 1 > 0$, decrease value of P.
6. Write P as the dew point pressure and x_i as the composition of formed liquid phase.

Dew P can also be calculated through flash calculations by finding a pressure at which $V_F = 1$. In dew T calculation y_i and P are known. The calculation procedure is similar to dew P calculation method except that T must be guessed instead of guessing P. In this case if $\sum y_i / K_i - 1 < 0$, decrease T and if $\sum y_i / K_i - 1 > 0$, increase T. Bubble and dew point calculations are used to calculate PT diagrams as shown in the next section.

Reservoir engineers usually use empirically developed correlations to estimate bubble and dew points for reservoir fluid mixtures. For example, Standing, Glaso, and Vazquez and Beggs correlations for prediction of bubble point pressure of reservoir fluids are given in terms of temperature, GOR, gas specific gravity, and stock tank oil specific gravity (or API

gravity). These correlations are widely used by reservoir engineers for quick and convenient calculation of bubble point pressures [1, 3, 10]. The Standing correlation for prediction of bubble point pressure is [1, 3]

$$P_b(\text{psia}) = 18.2(a \times 10^b - 1.4)$$

(9.9)
$$a = (\text{GOR}/\text{SG}_{gas})^{0.83}$$

$$b = 0.00091T - 0.0125\,(\text{API}_{oil})$$

$$T = \text{Temperature, °F}$$

where P_b is the bubble point pressure, SG_{gas} is the gas specific gravity ($= M^g/29$), API_{oil} is the API gravity of produced liquid crude oil at stock tank condition, and GOR is the solution gas-to-oil ratio in scf/stb. Use of this correlation is shown in the following example. A deviation of about 15% is expected from the above correlation [3]. Marhoun developed the following relation for calculation of P_b based on PVT data of 69 oil samples from the Middle East [10]:

$$P_b(\text{psia}) = a\,(\text{GOR})^b\,(\text{SG}_{gas})^c\,(\text{SG}_{oil})^d\,(T)^e$$

(9.10)
$$a = 5.38088 \times 10^{-3} \quad b = 0.715082 \quad c = -1.87784$$

$$d = 3.1437 \quad e = 1.32657 \quad T = \text{temperature, °R}$$

where SG_{oil} is the specific gravity of stock tank oil and GOR is in scf/stb. The average error for this equation is about $\pm 4\%$.

Example 9.2—Calculate bubble point pressure of reservoir fluid of Table 9.1 at 245°F from the following methods and compare the results with an experimental value of 2387 psia.

TABLE 9.4—*Calculated GOR from different C_{7+} characterization methods. Taken with permission from Ref. [7].*

Method	Input for C_{7+}	No. of C_{7+} fractions	Overall GOR, scf/stb	Stage 1	Stage 2	Stage 3
Lab data			850	601	142	107
Proposed	M_{7+} and SG_{7+}	5	853	580	156	117
Standing (Eqs. 6.202 and 6.203)	M_{7+} and SG_{7+}	1	799	534	134	131
Simulation 1[a]	N_C & T_b	1	699	472	141	86
Simulation 2	N_C & T_b	5	750	516	142	92
Simulation 3	M & PNA	1	779	542	142	95
Simulation 4	M & PNA	5	797	559	143	95

[a]Calculations have been performed through PR EOS using a PVT simulator [9].

a. Thermodynamic model with use of SRK EOS similar to the one used in Example 9.1.
b. Standing correlation, Eq. (9.9).
c. Mahroun's correlation, Eq. (9.10).

Solution—(a) The saturation pressure of the reservoir fluid (Feed in Table 9.1) at 245°F can be calculated along flash calculations, using the method outlined above. Through flash calculations (see Example 9.1) one can find a pressure at 245°F and that the amount of vapor produced is nearly zero ($V_F \cong 0$). The pressure is equivalent to bubble (or saturation) pressure. This is a single-stage flash calculation that gives $P^{sat} = 2197$ psia, which differs by -8% from the experimental value of 2387 psia. (b) A simpler method is given by Eq. (9.9). This equation requires GOR, API_{oil}, and SG_{gas}. GOR is given in Table 9.1 as 850 scf/stb. API_{oil} is calculated from the specific gravity of liquid from the third stage (SG = 0.815), which gives $API_{oil} = 42.12$. SG_{gas} is calculated from gas molecular weight, M_{gas}, and definition of gas specific gravity by Eq. (2.6). Since gases are produced in three stages, M_{gas} for these stages are calculated from the gas composition and molecular weights of components as 23.92, 31.74, and 44.00, respectively. M_{gas} for the whole gas produced from the feed may be calculated from GOR of each stage as $M_{gas} = (601 \times 23.92 + 142 \times 31.74 + 107 \times 44.00)/(601 + 142 + 107) = 27.76$. $SG_{gas} = 27.76/29 = 0.957$. From Eq. (9.9), $A = 139.18$ and $P_b = 2507.6$ psia, which differs by $+5.1\%$ from the experimental value. (c) Using Marhoun's correlation (Eq. 9.10) with $T = 705\,°R$, $SG_{oil} = 0.815$, $SG_{gas} = 0.957$, and GOR = 850 we get $P_b = 2292$ psia (error of -4%). In this example, Marhoun's correlation gives the best result since it was mainly developed from PVT data of oils from the Middle East. ♦

9.2.3 Generation of P–T Diagrams—True Critical Properties

A typical temperature–pressure (TP) diagram of a reservoir fluid was shown in Fig. 5.3. The critical temperature and pressure (critical point) in a PT diagram are true critical properties and not the pseudocritical. For pure substances, both the true and pseudocritical properties are identical. The main application of a PT diagram is to determine the phase (liquid, vapor or solid) of a fluid mixture. For a mixture of known composition, pseudocritical temperature and pressure (T_{pc}, P_{pc}) may be calculated from the Kay's mixing rule (Eq. 7.1) or other mixing rules presented in Chapter 5 (i.e., Table 5.17). Methods of calculation of critical properties of undefined petroleum fractions presented in Section 2.5 all give pseudocritical properties. While pseudocritical properties are useful for generalized correlations and EOS calculations, they do not represent the true critical point of a mixture, which indicates phase behavior of fluids. Calculated true critical temperature and pressure for the reservoir fluid of Table 9.1 by simulations 1 and 2 in Table 9.4 are given in Table 9.5. Generated PT diagrams by these two simulations are shown in Fig. 9.5. The bubble point curves are shown by solid lines while the dew point curves are shown by a broken line. This figure shows the effect of number of pseudocomponents for the C_{7+} on the PT diagram. Critical properties given in Table 9.5 are true critical properties and values calculated with five pseudocomponents for the C_{7+} are more accurate. Obviously as discussed

TABLE 9.5—*Effect of C_{7+} characterization methods on calculated mixture critical properties [7].*

Charac. scheme	Input for C_{7+} of Table 9.3	No. of C_{7+} Fractions	T_c, K	P_c, bars	Z_c
Simulation 1	N_C & T_b	1	634	98	0.738
Simulation 2	N_C & T_b	5	651	141	0.831

Calculations have been performed through PR EOS using a PVT simulator [9].

in Chapter 4, for lighter reservoir fluids such as gas condensate samples detailed treatment of C_{7+} has less effect on the phase equilibrium calculations of the fluid.

The true critical temperature (T_{tc}) of a defined mixture may also be calculated from the following simple mixing rule proposed by Li [11]:

$$T_{tc} = \sum_i \phi_i T_{ci}$$

(9.11)

$$\phi_i = \frac{x_i V_{ci}}{\sum_i x_i V_{ci}}$$

where x_i, T_{ci}, and V_{ci} are mole fraction, critical temperature, and volume of component i in the mixture, respectively. The average error for this method is about 0.6% ($\sim$3 K) with maximum deviation of about 1.6% ($\sim$8 K) [12]. The Kreglewski–Kay correlation for calculation of true critical pressure, P_{tc}, is given as [13] follows:

$$P_{tc} = P_{pc}\left[1 + (5.808 + 4.93\omega)\left(\frac{T_{tc}}{T_{pc}} - 1\right)\right]$$

(9.12)

$$T_{pc} = \sum_i x_i T_{ci} \quad P_{pc} = \sum_i x_i P_{ci} \quad \text{and} \quad \omega = \sum_i x_i \omega_i$$

where T_{pc} and P_{pc} are pseudocritical temperature and pressure calculated through Kay's mixing rule (Eq. 7.1). The average deviation for this method is reported as 3.8% ($\sim$2 bar) for nonmethane systems and average deviation of 50% ($\sim$48 bar) may be observed for methane–hydrocarbon systems [12]. These methods are recommended in the API-TDB [12] as well as other sources [3].

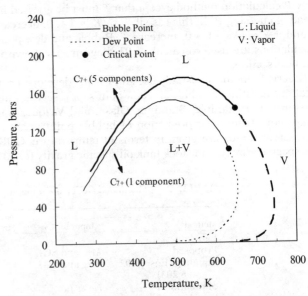

FIG. 9.5—**The PT diagram for simulations 1 and 2 given in Table 9.5 with use of N_C and T_b. Taken with permission from Ref. [7].**

For undefined petroleum fractions the following correlation may also be used to estimate true critical temperature and pressure from specific gravity and volume average boiling point (VABP) of the fraction [12]:

$$T_{tc} = 358.79 + 1.6667\Delta - 1.2827(10^{-3})\Delta^2$$

(9.13) $$\Delta = SG\,(VABP - 199.82)$$

$$\log_{10}(P_{tc}/P_{pc}) = 0.05 + 5.656 \times \log_{10}(T_{tc}/T_{pc})$$

where T_{tc}, T_{pc}, and VABP are in kelvin and P_{pc} and P_{tc} are in bars. It is important to note that both T_{pc} and P_{pc} must be calculated from the methods given in Section 2.5 for critical properties of undefined petroleum fractions. The average error for calculation of T_{tc} from the above method is about 0.7% (~3.3 K) with maximum error of 2.6% (~12 K). Reliability of the above method for prediction of true critical pressure of undefined petroleum fractions is about 5% as reported in the API-TDB [12]. The above equation for calculation of P_{tc} is slightly modified from the correlation suggested in the API-TDB. This correlation is developed based on an empirical graph of Smith and Watson proposed in the 1930s. For this reason it should be used with special caution. The following method is recommended for calculation of true critical volume in some petroleum-related references [3]:

$$V_{tc} = \frac{Z_{tc}RT_{tc}}{P_{tc}}$$

(9.14)

$$Z_{tc} = \sum_i x_i Z_{ci}$$

Method of calculation of true critical points (T_{tc}, P_{tc}, and V_{tc}) of defined mixtures through an equation of state (i.e., SRK) requires rigorous vapor–liquid thermodynamic relationships as presented in Procedure 4.B4.1 in Chapter 4 of the API-TDB [12]. At the true critical point, a correct VLE calculation should show that $x_i = y_i$. Most cubic EOSs fail to perform properly at the critical point and for this reason attempts have been made and are still continuing to improve EOS phase behavior predictions at this point.

9.3 VAPOR–LIQUID–SOLID EQUILIBRIUM— SOLID PRECIPITATION

In this section, practical application of three-phase equilibrium in the petroleum industry is demonstrated. Upon reducing the temperature, heavy hydrocarbons present in a petroleum fluid may precipitate as a solid phase and the liquid becomes in equilibrium with both the solid and the vapor phase. In such cases, the solid is at the bottom, liquid is in the middle, and the vapor phase is on top of the liquid phase. A general schematic of typical vapor–liquid–solid equilibrium (VLSE) during solid precipitation in a petroleum fluid is shown in Fig. 9.1. Solid precipitation is a serious problem in the petroleum industry and the basic question is: what is the temperature at which precipitation starts and under certain temperature, pressure, and composition how much solid can be precipitated from a petroleum fluid? These two questions are answered in this section. Since solids are formed at low temperatures, under these conditions the amount of vapor produced is low and the problem reduces to SLE such as the case for asphaltene precipitation. Initially, this section

discusses the nature of heavy compounds that are present in petroleum residua and heavy oils. Precipitation of these heavy compounds under certain conditions of temperature and pressure or composition follow general principles of SLE, which were discussed in Section 6.8.3. In this section, the problems associated with such heavy compounds as well as methods that can be used to predict the certain conditions at which they precipitate will be discussed. Based on the principle of phase equilibrium discussed in Section 6.8.1, a thermodynamic model is presented for accurate calculation of cloud point of crude oils under various conditions. Methods for calculating the amount of solid precipitation from sophisticated thermodynamic models as well as readily available parameters for a petroleum fluid are also discussed in this section.

9.3.1 Nature of Heavy Compounds, Mechanism of their Precipitation, and Prevention Methods

Petroleum fluids, especially heavy oils and residues, contain heavy hydrocarbons from paraffinic, naphthenic, and aromatic groups. Generally, there are three types of heavy hydrocarbons that may exist in a heavy petroleum fluid: (1) waxes, (2) resins, and (3) asphaltenes. As discussed in Section 1.1.3, the main type of waxes in petroleum fluids are paraffinic waxes. They are mainly n-paraffins with carbon number range of C_{16}–C_{36} and average molecular weight of about 350. Waxes that exist in petroleum distillates usually have freezing points between 30 and 70°C. Another group of waxes called crystalline waxes are primarily isoparaffins and cycloparaffins (with long-chain alkyl groups) with carbon number range of 30–60 and molecular weight range of 500–800. The melting points of commercial grade waxes are in the 70–90°C range. Solvent de-oiling of petroleum or heavy residue results in dark-colored waxes or a sticky, plastic to hard nature material [14]. Waxes present in a petroleum fluid may precipitate when the conditions of temperature and pressure change. When the temperature falls, heavy hydrocarbons in a crude or even a gas condensate may precipitate as wax crystals. The temperature at which a wax begins to precipitate is directly related to the cloud point of the oil [15, 16]. Effects of pressure and composition on wax precipitation are discussed by Pan et al. [17].

Wax formation is undesirable and for this reason, different additives usually polymer-based materials are used to lower pour points of crude oils. Wax inhibitor materials include polyalkyl acrylates and methacrylates, low-molecular-weight polyethylene waxes, and ethyl-vinyl acetate (EVA) copolymers. The EVA copolymers are probably the most commonly used wax inhibitors [14]. These inhibitors usually contain 20–40 wt% EVA. Molecular weight of such materials is usually greater than 10 000. The amount of EVA added to an oil is important in its effect on lowering pour point. For example, when 100 ppm of EVA is added to an oil it reduces pour point from 30 to 9°C, while if 200 ppm of same inhibitor is added to another oil, it causes an increase in the pour point from 21°C to 25°C [14].

Asphaltenes are multiring aromatics (see Fig. 1.2) that are insoluble in low-molecular-weight n-paraffins (LMP) such as C_3, n-C_4, n-C_5, or even n-C_7 but soluble in benzene, carbon disulfide (CS_2), chloroform, or other chlorinated hydrocarbon solvents [15]. They exist in reservoir fluids and heavy petroleum fractions as pellets of 34–40 microns and are maintained in suspension by resins [16, 18, 19]. Petroleum fluids with low-resin contents or under specific conditions of temperature, pressure, and LMP concentration may demonstrate asphaltene deposition in oil-producing wells. Asphaltene deposition may also be attributed to the reduction of pressure in the reservoirs or due to addition of solvents as in the case of CO_2 injection in enhanced oil recovery (EOR) processes.

Resins play a critical role in the solubility of the asphaltenes and must be present for the asphaltenes to remain in the solution. Although the exact mechanism is unknown, current theory states that resins act as mutual solvent or form stability peptide bonds with asphaltenes [16]. Both oils and asphaltenes are soluble in resins. Structure of resins is not well known, but it contains molecules with aromatic as well as naphthenic rings. Resins can be separated from oil by ASTM D 2006 method. Resins are soluble in *n*-pentane or *n*-heptane (while asphaltenes are not) and can be adsorbed on surface-active material such as alumina. Resins when separated are red to brown semisolids and can be desorbed by a solvent such as pyridine or a benzene/methanol mixed solvent [15]. The amount of sulfur in asphaltenes is more than that of resins and sulfur content of resins is more than that of oils [15]. Oils with higher sulfur contents have higher asphaltene content. Approximate values of molecular weight, H/C weight ratio, molar volume, and molecular diameter of asphaltenes, resins and oils are given in Table 9.6. In the absence of actual data typical values of M, d_{25}, ΔH_i^f, and T_M are also given for monomeric asphaltene separated by *n*-heptane as suggested by Pan and Firoozabadi [20]. In general $M_{asph.} > M_{res.} > M_{wax}$ and $(H/C)_{wax} > (H/C)_{resi} > (H/C)_{asph}$. Waxes have H/C atomic ratio of 2–2.1 greater than those of resins and asphaltenes because they are mainly paraffinic.

In general, crude oil asphaltene content increases with decrease in the API gravity (or increase in its density) and for the residues the asphaltene content increases with increase in carbon residue. Approximately, when Conradson carbon residue increases from 3 to 20%, asphaltene content increases from 5 to 20% by weight [15]. For crude oils when the carbon residue increases from 0 to 40 wt%, asphaltene, sulfur, and nitrogen contents increase from 0 to 40, 10, and 1.0, respectively [15]. Oils with asphaltene contents of about 20 and 40 wt% exhibit viscosities of about 5×10^6 and 10×10^6 poises, respectively.

As discussed in Section 6.8.2.2, generally two substances with different structures are not very soluble in each other. For this reason, when a low-molecular-weight *n*-paraffin

compound such as *n*-C_7 is added to a petroleum mixture, the asphaltene components (heavy aromatics) begin to precipitate. If propane is added to the same oil more asphaltenes precipitate as the difference in solubilities of C_3-asphaltene is greater than that of nC_7-asphaltene. Addition of an aromatic hydrocarbon such as benzene will not cause precipitation of asphaltic compounds as both are aromatics and similar in structure; therefore they are more soluble in each other in comparison with LMP hydrocarbons. When three parameters for a petroleum fluid change, heavy deposition may occur. These parameters are temperature, pressure, and fluid composition that determine location of state of a system on the PT phase diagram of the fluid mixture. Precipitation of a solid from liquid phase is a matter of solid–liquid equilibrium (SLE) with fundamental relations introduced in Sections 6.6.6 and 6.8.3.

Estimation of the amount of asphaltene and resins in crude oils and derived fractions is very important in design and operation of petroleum-related industries. As experimental determination of asphaltene or resin content of various oils is time-consuming and costly, reliable methods to estimate asphaltene and resin contents from easily measurable or available parameters are useful. Waxes are insoluble in 1:2 mixture of acetone and methylene chloride. Resins are insoluble in 80:20 mixture of isobutyl alcohol–cyclohexane and asphaltenes are insoluble in hexane [15]. ASTM D 4124 method uses *n*-heptane to separate asphaltenes from oils. Other ASTM test methods for separation of asphaltenes include D 893 for separation of insolubles in lubricating oils [21]. The most widely used test method for determination of asphaltene content of crude oils is IP 143 [22]. Asphaltene proportions in a typical petroleum residua is shown in Fig. 9.6. Since these are basically polar compounds with very large molecules, most of correlations developed for typical petroleum fractions and hydrocarbons fail when applied to such materials. Methods developed for polymeric solutions are more applicable to asphaltic oils as shown in Section 7.6.5.4.

Complexity and significance of asphaltenes and resins in petroleum residua is clearly shown in Fig. 9.6. Speight [15] as well as Goual and Firoozabadi [23] considered a petroleum fluid as a mixture of primarily three species: asphaltenes, resins, and oils. They assumed that while the oil component is nonpolar, resins and asphaltene components are polar. The degrees of polarities of asphaltenes and resins for several oils were determined by measuring dipole moment. They reported that while dipole moment of oil component of various crudes is usually less than 0.7 debye (D) and for many oils zero, the dipole moment of resins is within 2–3 D and for asphaltenes (separated by *n*-C_7) is within the range of 4–8 D. Dipole moment of waxy oils is zero, while for asphaltic crudes is about 0.7 D. Therefore, one may determine degree of asphaltene content of oil through measuring dipole moment. Values of dipole moments of some pure compounds are given in Table 9.7. *n*-Paraffins have dipole moment of zero, while hydrocarbons with double bonds or branched hydrocarbons have higher degree of polarity. Presence of heteroatoms such as N or O significantly increases degrees of polarity.

The problems associated with asphaltene deposition are even more severe than those associated with wax deposition. Asphaltene also affects the wettability of reservoir fluid on solid surface of reservoir. Asphaltene may cause wettability

TABLE 9.6—*Properties of typical asphaltenes, resins and oils.*

Hydrocarbons	M	H%	H/C	V	d, Å	D
Asphaltene	1000–5000	9.2–10.5	1.0–1.4	900	14.2	4–8
Resin	800–1000	10.5–12.5	1.4–1.7	700	13	2–3
Oil	200–600	12.5–13.1	1.7–1.8	200–500	8–12	0–0.7

M is molecular weight in g/mol. H% is the hydrogen content in wt%. H/C is the hydrogen-to-carbon atomic ratio. V is the liquid molar volume at 25°C. d is molecular diameter calculated from average molar volume in which for methane molecules is about 4 Å (1 Å = 10^{-10} m). D is the dipole moment in Debye. These values are approximate and represent properties of typical asphaltenes and oils. For practical calculations for resins one can assume M = 800 g/mol and for a typical monomeric asphaltene separated by *n*-heptane approximate values of some properties are as follows: M = 1000 g/mol. Density of liquid ≈ density of solid ≈ 1.1 g/cm³. Enthalpy of fusion at the melting point: ΔH_M = 7300 cal/mol, melting point: T_M = 583 K. Data source: Pan and Firoozabadi [20].

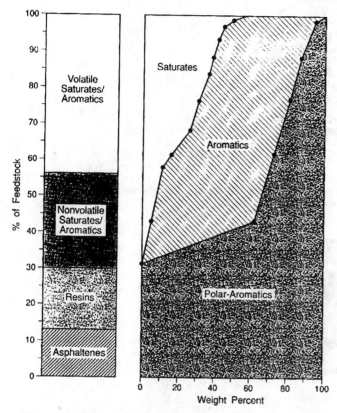

FIG. 9.6—Representation of proportions of resins and asphaltenes in a petroleum residua. Taken with permission from Ref. [15].

TABLE 9.7—*Dipole moments of some compounds and oil mixtures.*

No.	Compound	Dipole, debye
1	Methane (C_1)	0.0
2	Eicosane (C_{20})	0.0
3	Tetracosane (C_{24})	0.0
4	2-methylpentane	0.1
5	2,3-dimethylbutane	0.2
6	Propene	0.4
7	1-butene	0.3
8	Cyclopentane	0.0
9	Methylcyclopentane	0.3
10	Cyclopentene	0.9
11	Benzene	0.0
12	Toluene	0.4
13	Ethylbenzene	0.2
14	o-Xylene	0.5
15	Acetone (C_3H_6O)	2.9
16	Pyridine	2.3
17	Aniline	1.6
18	NH_3	1.5
19	H_2S	0.9
20	CO_2	0.0
21	CCl_4	0.0
22	Methanol	1.7
23	Ethanol	1.7
24	Water	1.8
	Oil mixtures	
25	Crude Oils	<0.7
26	Resins	2–3
27	Asphaltenes[a]	4–8

Data source for pure compounds: Poling, B. E., Prausnitz, J. M., O'Connell, J. P., *Properties of Gases and Liquids*, 5th ed., McGraw-Hill, New York, 2000.
[a]Asphaltenes separated by *n*-heptane. Data source for oil mixtures: Goul and Firoozabadi [23].

reversal and its understanding will help plan for more efficient oil recovery processes [24]. Similarly, asphaltene deposition negatively affects the EOR gas flooding projects [25]. Asphaltene precipitation may also occur during oil processing in refineries or transportation in pipelines and causes major problems by plugging pipes and catalysts pores [26, 27]. The problem is more severe for heavy oils. Further information on problems associated with asphaltene precipitation during production, especially in the Middle East fields, is given by Riazi et al. [28].

There are a number of models and theories that are proposed to describe mechanism of asphaltene formation [29]. Understanding of kinetics of asphaltene formation is much more difficult than wax formation. There is no universally accepted model for asphaltene formation; however, most researchers agree on two models: (1) colloidal and (2) micellar. Schematic of colloidal model is shown in Fig. 9.7. The nature and shape of the resulting aggregates will determine their effect on the behavior of the petroleum fluid [30, 31]. In this model, asphaltene particles come together to form larger molecules (irreversible aggregation), which grow in size. According to this model the surface of asphaltene molecules must be fully covered by resin molecules. For this reason when concentration of resin exceeds from a certain level, rate of asphaltene deposition decreases even if its concentration is high. Because of this it is often possible that an oil with higher asphaltene content results in less precipitation due to high resin content in comparison with an oil with lower asphaltene and resin contents. Knowledge of the concentration of resin in oil is crucial in determination of the amount of asphaltene precipitation.

In the micellar model, it is assumed that asphaltene molecules exist as micelles in crude and micellar formation is a reversible process. Furthermore, it is assumed that the micellar shape is spherical, the micellar sizes are monodispersed (i.e., all having the same size), and the asphaltene micellar core is surrounded by a solvated shell as shown in Fig. 9.8. In this model too, resins may cover asphaltene cores and prevent precipitation. Thermodynamic models to describe phase behavior of asphaltic oils depend on such models to describe nature of asphaltene molecules.

Asphaltenes precipitate when conditions of temperature, pressure, or composition change. The condition under which precipitation begins is called the *onset* of asphaltene precipitation. In general to select a right method for determination of asphaltene onset, asphaltene content or asphaltene prevention one must know the mechanism of asphaltene precipitation, which as mentioned earlier very much depends on the oil composition. Asphaltenes flocculate due to excess amounts of paraffins in the solution and micellization (self-association) of asphaltene is mainly due to increase in aromaticity (polarity) of its medium [32].

During the past decade, various techniques have been developed to determine asphaltene onset from easily measurable properties. These methods include measuring refractive index to obtain the onset [33]. Fotland et al. [34] proposed measuring electric conductivity to determine the asphaltene onset. Escobedo and Mansoori [35] proposed a method to determine the onset of asphaltene by measuring viscosity of crude oil diluted with a solvent (n-C_5, n-C_7, n-C_9). They showed that with a decrease in deposition rates with increasing crude

(a) Colloidal Phenomenon Due to Increase in Concentration of Polar Miscible Solvent (such as polar aromatic hydrocarbons shown by solid ellipses) in crude oil

(b) Asphaltene Flocculation and Precipitation

(c) Steric Colloid Formation

FIG. 9.7—Schematic of colloidal model for asphaltene formation. Taken with permission from Ref. [29].

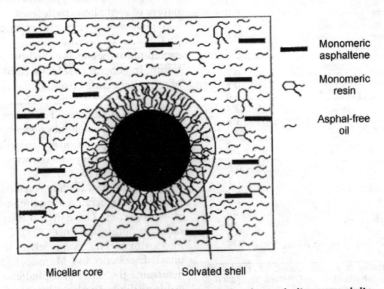

FIG. 9.8—Schematic of micellar formation in asphaltene precipitation. Taken with permission from Ref. [20].

376

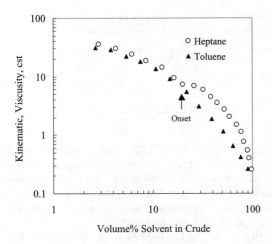

FIG. 9.9—Determination of asphaltene onset from viscosity. Taken with permission from Ref. [35].

oil kinematic viscosity and with increase in production rate, deposition also increases [35]. Determination of asphaltene onset through viscosity measurement is shown in Fig. 9.9 [35]. When a solvent such as toluene is added to a crude oil the viscosity of crude–solvent solution decreases as concentration of solvent increases. For this solvent, asphaltene does not precipitate and the curve of viscosity versus solvent concentration is smooth. However, when a solvent such as n-heptane is added before asphaltene onset, viscosity decreases smoothly with solvent concentration similar to the case of toluene solvent, but as asphaltene molecules begin to aggregate and form larger particles viscosity does not fall for a short time. This is due to the fact that the increase in viscosity is due to particle formation that will offset a decrease in viscosity due to

dilution of the crude. However, as soon as particles become large enough to precipitate, viscosity of the crude begins to drop again but more rapidly than before the onset. Therefore, the onset of asphaltene precipitation is at the concentration level where viscosity curve shows a change in its trend. For the example that is shown in Fig. 9.9, this point is at 20 vol% solvent addition. Another technique to determine the onset of asphaltene is through measuring interfacial tension (IFT) in which when precipitation occurs there is a sudden change in IFT. Various methods of determination of the onset of asphaltene are discussed by Mansoori [32].

To remove precipitated asphaltenes, special chemicals known as inhibitors are used. Asphaltenes can be precipitated when a solvent is added to a crude oil, but once the asphaltenes are precipitated they are difficult to redissolve by a diluent. Some aromatics are used to inhibit asphaltene precipitation in crude oils. Because aromatics are similar in nature with asphaltene (also an aromatic compound) they are more soluble in each other than in other types of hydrocarbons and as a result precipitation is reduced. Benzene and toluene are not commonly used as an asphaltene inhibitor because a large concentration is required [16]. The effect of toluene in reducing amount of asphaltene precipitation for a reservoir fluid with different level of CO_2 concentration is shown in Fig. 9.10 [36]. Other types of asphaltene inhibitors include n-dodecyl-benzenesulfonic acid (DBSA), which has stronger effect than benzene in reducing asphaltene precipitation. In petroleum reservoirs, the main problem associated with asphaltene deposition is its adsorption on formation rocks. Adsorbed asphaltene negatively affects well performance and removal of the asphaltene is desired.

Piro et al. [37] made a good study on the evaluation of several chemicals for asphaltene removal and related test methods. Toluene is a typical solvent for asphaltenes and shows a very high uptake (several tens of a wt%) when the asphaltenes

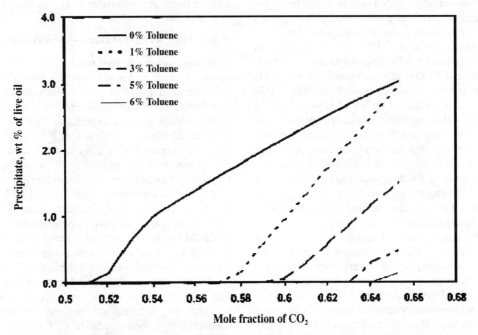

FIG. 9.10—Effect of toluene in reducing amount of asphaltene precipitation for a reservoir fluid. Taken with permission from Ref. [36].

TABLE 9.8—*Evaluation of three types of additives for asphaltene removal on a rock surface [37].*

Additive wt% in solvent	Time, h	Removal efficiency (wt% of adsorbed asphaltene) for different additives		
		Additive A	Additive B	Additive C
0.1	1	40.5	14.0	2.7
0.1	18	48.9	21.4	6.0
2.0	1	49.9	32.9	8.9
2.0	18	51.5	46.0	10.1

are in the bulk state; on the contrary, the asphaltene up-take by toluene is very low (10–20 wt%) when the same material is adsorbed on a rock surface (clays, dolomia, quartz, etc.), as experienced by Piro et al. [37]. For this reason, they used additives dissolved in toluene for asphaltenes' removal when they are adsorbed on rock surface. Three types of additives were evaluated: additive A was based on alkyl benzene sulfonic acid (ABSA); additive B was based on complex polymer; additive C was based on another complex polymer. Asphaltic materials were obtained from a crude oil of 42 API gravity by precipitation with *n*-heptane. The rock on which asphaltenes were adsorbed was powdered dolomite (average particle size of 60 μm and surface area of 10 m^2/g) and toluene was employed as solvent for the additives. Experiments were conducted to study the effect of different types of additives, concentration level, and time on the amount of asphaltenes up-take. A summary of results of experiments is given in Table 9.8. The results show that addition of additive A (0.1 wt% in toluene solution) can remove up to 41% after 1 h and up to 49% of asphaltene after 18 h [37]. Therefore, at higher additive concentrations the contact time can be reduced.

Deasphalted oils may also be used as asphaltene inhibitor since they contain resins that are effective in keeping asphaltene molecules soluble in the oil in addition to their potential for greater solvency. There are some synthetic resins such 2-hexadecyl naphthalene that can also be used as asphaltene inhibitor. Most of these inhibitors are expensive and research on manufacturing of commercially feasible asphaltene inhibitors is continuing. Asphaltenes or other heavy organics are precipitated under certain conditions that can be determined through phase diagram (i.e., PT or Px diagrams). An example of such diagrams is the Px diagram at constant temperature of 24°C for an oil–CO$_2$ system as shown in Fig. 9.11. Some specifications for this oil are given in Table 9.9. In this figure, the solid phase is indicated by S and regions of LVS and LS are the regions that asphaltenes may precipitate and should be avoided. The best way to prevent asphaltene precipitation is to avoid the region in the phase diagram where asphaltene precipitation can occur. It is for this reason that phase behavior of petroleum fluids containing heavy organics is important in determining the conditions in which precipitation can be avoided. Construction of such phase diagrams is extremely useful to determine the conditions where precipitation occurs. Unfortunately such diagrams for various oils and solvents are not cited in the open literature. Figure 9.11 shows that the solid phase is formed at very high concentration of CO$_2$, that is, the region that is not of practical application and should be considered with caution. Thermodynamic models, along with appropriate characterization schemes can be applied to waxy or asphaltic oils to determine possibility and amount of precipitation under certain conditions. For

example, Kawanaka et al. [30] used a thermodynamic approach to study the phase behavior and deposition region in CO$_2$–crude mixtures at different pressures, temperatures, and compositions. In the next few sections, thermodynamic models for solid formation are presented to calculate the onset and amount of solid precipitation.

For the same tank oil shown in Table 9.9, Pan and Firoozabai [20] used their thermodynamic model based on micellar theory of asphaltene formation to calculate asphaltene precipitation for various solvents. Their data are shown in Fig. 9.12, where amount of precipitation is shown versus dilution ratio. The dilution ratio (shown by R_S) represents volume (in cm^3) of solvent added to each gram of crude oil. The amount of precipitated resin under the same conditions is also shown in this figure. The onset of asphaltene formation is clearly shown at the point where amount of precipitation does not change with a further increase in solvent-to-oil ratio. Lighter solvents cause higher precipitation. Generally value of R_S at the onset for a given oil is a function of solvent molecular weight (M_S) and it increases with increase in M_S [38]. Effect of temperature on asphaltene precipitation depends on the type of solvent as shown in Fig. 9.13 [39]. The amount of solid deposition increases with temperature for propane, while for *n*-heptane the effect of temperature is opposite. Effect of pressure on asphaltene precipitation is shown in Fig. 9.14. Above the bubble point of oil, increase in pressure decreases the amount of precipitation, while below bubble point precipitation increases with pressure.

9.3.2 Wax Precipitation—Solid Solution Model

There are generally two models for wax formation calculations. The first and more commonly used model is the solid-solution model. In this model, the solid phase is treated as a homogenous solution similar to liquid solutions. Formulation of SLE calculations according to this model is very similar to VLE calculations with use of Eq. (6.205) and equilibrium ratio, K_i^{SL}, from Eq. (6.209) instead of K_i for the VLE. This model was first introduced by Won [41] and later was used to predict wax precipitation from North Sea oils by Pedersen et al. [14, 42]. The second model called *multisolid-phase model* was proposed by Lira-Galeana et al. in 1996 [43], which has also found some industrial applications [16]. In this model, the solid mixture is not considered as a solution but it is described as a mixture of pure components; each solid phase does not mix with other solid phases. The multisolid-phase model is particularly useful for calculation of CPT of oils. The temperature at which wax appears is known as *wax appearance (or precipitation) temperature* (WAT or WPT), which theoretically is the same as the CPT. Both models are based on the following relation expressing equilibrium between vapor,

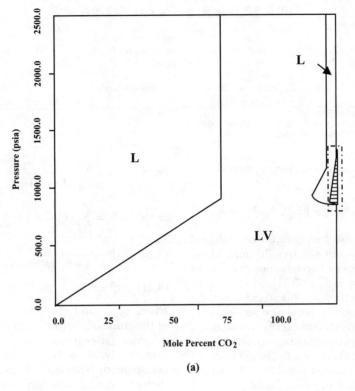

(a)

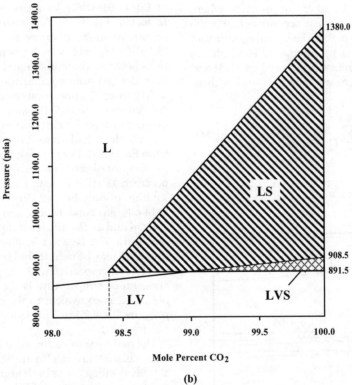

(b)

FIG. 9.11—Phase diagram for oil–CO₂ mixtures at 24°C. Asphaltene precipitation occurs in the LS and LVS regions. (a) Entire composition range. (b) Enlarged LS section. Oil properties are given in Table 9.9. Taken with permission from Ref. [30].

TABLE 9.9—*Data for shell tank oil of Fig. 9.12 [30].*

Oil specifications		Asphaltene specifications	
mol% $C_1 + C_2$	0.6	wt% (resin) in oil	14.1[a]
mol% C_3–C_5	10.6	wt% (asph.) in oil	4.02
mol% C_6	4.3	density, g/cm³	1.2
mol% C_{7+}	84.5		
M	221.5 ($M_{7+} = 250$)	M (precipitated)[b]	4500
SG	0.873 (SG$_{7+}$ = 0.96)[a]	δ_T^c	$12.66(1 - 8.28 \times 10^{-4}T)$

[a] Data taken from Refs. [20, 49].
[b] M for asphaltene in oil (monomer) is about 1000 [20].
[c] δ in (cal/cm³)$^{0.5}$ and T in kelvin.

liquid, and solid phases for a multicomponent system shown in Fig. 9.15.

$$(9.15) \quad \hat{f}_i^V(T, P, y_i) = \hat{f}_i^L(T, P, x_i^L) = \hat{f}_i^S(T, P, x_i^S)$$

This equation can be split into two parts, one for vapor–liquid equilibrium and the other for liquid–solid equilibrium. These two equations can be expressed by two relations in terms of equilibrium ratios as given by Eqs. (6.201) and (6.208). In this section the solid-solution model is discussed while the multisolid-phase model is presented in the next section.

In the solid-solution model the solid phase (S) is treated as a homogeneous solution that is in equilibrium with liquid solution (L) and its vapor. In Fig. 9.15, assume the initial moles of nonequilibrium fluid mixture (feed) is 1 mol ($F = 1$) and the molar fraction of feed converted to vapor, liquid, and solid phases are indicated by V_F, L_F, and S_F, respectively, where $L_F = 1 - V_F - S_F$. Following the same procedure as that in the VLE calculations and using the mass balance and equilibrium relations that exist between vapor, liquid, and solid phases yields the following set of equations similar to Eqs. (9.3) and (9.4) for calculation of V_F and S_F and compositions of three phases:

$$(9.16) \quad F^{VL} = \sum_{i=1}^{N} (y_i - x_i^L) = \sum_{i=1}^{N} \frac{z_i (K_i^{VL} - 1)}{(1 - S_F) + V_F (K_i^{VL} - 1)} = 0$$

$$(9.17) \quad x_i^L = \sum_{i=1}^{N} \frac{z_i}{(1 - S_F) + V_F (K_i^{VL} - 1)}$$

$$(9.18) \quad y_i = \sum_{i=1}^{N} \frac{z_i K_i^{VL}}{(1 - S_F) + V_F (K_i^{VL} - 1)}$$

$$(9.19) \quad F^{SL} = \sum_{i=1}^{N} (x_i^S - x_i^L) = \sum_{i=1}^{N} \frac{z_i (K_i^{SL} - 1)}{1 + S_F (K_i^{SL} - 1)} = 0$$

$$(9.20) \quad x_i^L = \frac{z_i}{1 + S_F (K_i^{SL} - 1)}$$

$$(9.21) \quad x_i^S = x_i^L K_i^{SL}$$

where z_i, x_i^L, and x_i^S are the compositions in mole fractions of the crude oil (before precipitation), the equilibrium liquid oil phase (after precipitation), and precipitated solid phase, respectively. S_F is number of moles of solid formed (wax precipitated) from each 1 mol of crude oil or initial fluid (before precipitation) and must be calculated from solution of Eq. (9.9), while V_F must be calculated from Eq. (9.16). In fact in Fig. 9.15, F is assumed to be 1 mol and 100 S_F represents mol% of crude that has precipitated. Equations (9.16)–(9.18) have been developed based on equilibrium relations between vapor and liquid, while Eqs. (9.19)–(9.21) have been derived from equilibrium relations between liquid and solid phases. Compositions of vapor and solid phases are calculated from Eqs. (9.18) and (9.21). Equation (9.20) is the prime equation for calculation of liquid composition, x_i^L. To validate the calculations it must be the same as x_i^L calculated from Eq. (9.17). For the case of crude oils and heavy residues, the amount of vapor produced is small (especially at low temperatures) so that $V_F = 0$. This simplifies the calculations and solution of only Eqs. (9.19)–(9.21) is required. However, for light oils, gas condensates, and natural gases V_F must be calculated and all the above six equations must be solved simultaneously. The Newton–Raphson method described in Section 9.2.1 may be used to find both V_F and S_F from Eqs. (9.16) and (9.19), respectively. The onset of solid formation or wax appearance temperature is the temperature at which $S \to 0$ [44]. This is equivalent to the calculation of dew point temperature (dew T) in VLE calculations that was discussed in Section 9.2.1.

The main parameter needed in this model is K_i^{SL} that may be calculated through Eq. (6.209). In the original Won model, activity coefficients of both liquid and solids become close to unity and Hansen et al. [45] recommended use of polymer-solution theory for calculation of activity coefficients through Eq. (6.150). On this basis the calculation of K_i^{SL} can be summarized as in the following steps:

a. Assume T, P, and compositions x_i^L and x_i^S for each i in the mixture are all known.

b. Calculate the ratio of f_i^L/f_i^S for each pure i at T and P from Eq. (6.155).

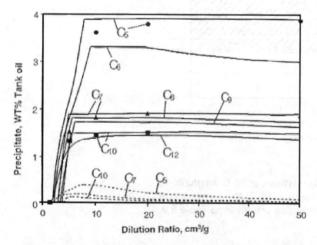

FIG. 9.12—**Precipitated amount of asphaltene (—) and resin (----) for the crude oil given in Table 9.9 at 1 bar and 295 K. Taken with permission from Ref. [20].**

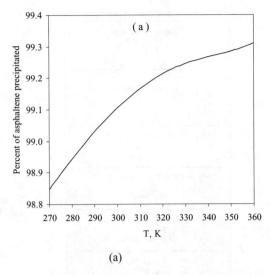

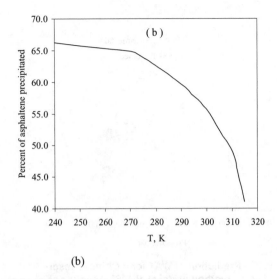

(a)   (b)

FIG. 9.13—Effect of temperature on asphaltene precipitation. (a) Propane diluent; (b) *n*-heptane diluent. Taken with permission from Ref. [39].

c. In calculation of f_i^L/f_i^S parameters T_{Mi}, ΔH_i^f, and ΔC_{Pi} must be calculated for each component i.

d. Calculate T_{Mi} from Eq. (6.156), ΔH_i^f from Eq. (6.157), and ΔC_{Pi} from Eq. (6.161).

e. Calculate both γ_i^L and γ_i^S from Eq. (6.154). In calculation of γ_i^S, calculate δ_i^S from Eq. (6.155). V_i^S and V_i^L can be obtained from Table 7.1.

f. Once f_i^L/f_i^S, γ_i^L, and γ_i^S have been determined, calculate K_i^{SL} from Eq. (6.209).

This is a typical solid–solution model for calculation of wax formation without the use of any adjustable parameter. All parameters can be calculated from the molecular weight of components or pseudocomponents as described in Sections 6.6.6. Using PNA composition for calculation of properties of C_{7+} pseudocomponents through Eqs. (6.149), (6.156), and (6.157) improves model predictions.

Pedersen et al. [42], based on their data for North Sea oils, showed that both Won and Hansen procedures significantly overestimate both the amount of wax precipitation and CPT. For this reason, they suggested a number of adjustable parameters to be used for calculation of various parameters. Chung [44] has used the following empirical set of correlations for calculation of properties of C_{7+} fractions for the wax formation prediction:

$$\Delta H_i^f = 0.9 T_{Mi} M_i^{0.55}$$

$$V_i^L = 3.8 M_i^{0.786}$$

(9.22)

$$\delta_i^L = 6.743 + 0.938 (\ln M_i) - 0.0395 (\ln M_i)^2$$

$$- 13.039 (\ln M_i)^{-1}$$

where T_{Mi} is the melting point in kelvin, ΔH_i^f is the molar heat of fusion in cal/mol, V_i^L is the molar liquid volume in

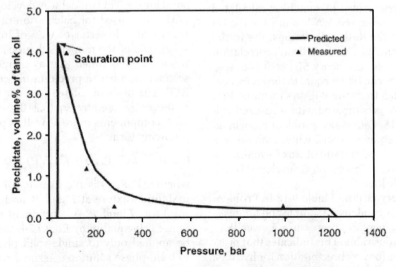

FIG. 9.14—Effect of pressure on asphaltene precipitation. Taken with permission from Ref. [40].

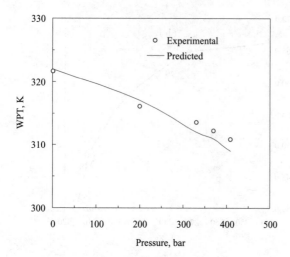

FIG. 9.15—Prediction of WAT for a Chinese reservoir fluid using method of Mei et al. [46]. Absolute error between calculated and experimental data is 1.6 K. Composition of reservoir fluid is given in Problem 9.10 (Table 9.18).

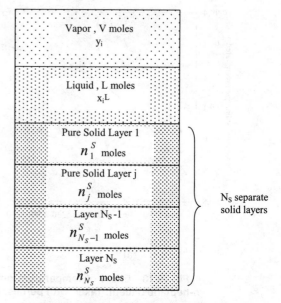

FIG. 9.16—Schematic of multiphase-solid model for wax precipitation. Courtesy of Lira-Galeana et al. [43].

cm^3/mol, and δ_i^L is the solubility parameter in $(cal/cm^3)^{0.5}$. Various researchers have used similar correlations but with different numerical coefficients. Most recently, Mei et al. [46] have applied the Pedersen et al. model to calculate wax precipitation in a live oil (oil under reservoir conditions) from Pubei Oil field located in the western part of China where gas injection is used in EOR processes. Composition of this oil is given in Problem 9.10 (Table 9.18). Basically, they used Won's correlations [41] for ΔC_{Pi}, T_{Mi}, and ΔH_i^f while the Thomas et al.'s correlation [47] was used for calculation of δ_i^S:

$$(9.23) \quad \delta_i^S = \begin{cases} 7.62 + 2.8a\,\{1 - \exp[-9.51 \\ \qquad \times 10^{-4}\,(M_i - 48.2)]\} & \text{for } M_i < 450 \\ 10.30 + 1.78 \times 10^{-3}a\,(M_i - 394.8) & \text{for } M_i \geq 450 \end{cases}$$

where a is an adjustable parameter. They used six adjustable parameters for calculation of ΔC_{Pi}, T_{Mi}, ΔH_i^f, and δ_i^S in terms of T and M, which were determined by matching calculated and experimental data on measured WAT values for the oil [46]. For the beginning of the flash calculations, the initial values of K_i^{VL} may be estimated from the Wilson's correlation (Eq. 6.204) assuming ideal solution theory. Mei et al. [46] suggested that initial K_i^{SL} values can be set equal to the reciprocal of K_i^{VL} values also calculated from the Wilson's formula. Predicted WAT versus pressure is compared with measured values and is given in Fig. 9.15. One major problem associated with this model is that it requires experimental data on wax precipitation temperature or the amount of wax formation to find the adjustable parameters. This graph is developed based on data reported in Ref. [46].

Composition of this reservoir fluid (Table 9.18 in Problem 9.10) indicates that it is a gas condensate sample and for gases usually WAT declines with increase in pressure. Lower WAT values for an oil are always desirable. This indicates that pressure behaves as an inhibitor for wax precipitation for live oils, gas condensate, or natural gas samples. However, this is not

the case for heavy liquid oils such as crude oils or dead oils. Pressure causes slight increase in the amount of wax precipitation as will be discussed in the next section where the exact method of calculation of CPT of crude oils is presented.

9.3.3 Wax Precipitation: Multisolid-Phase Model—Calculation of Cloud Point

One of the problems with the solid-solution model in prediction of wax formation is that without the use of adjustable parameters it usually overestimates amount of wax precipitation and cloud point of crude oils. In this section, another model that is particularly accurate for calculation of CPT of crude oils will be presented. In this model, the solid is considered as multilayer, each layer represents a pure component (or pseudocomponent) as a solid that is insoluble in other solid layers. This model was developed by Lira-Galeana et al. [43] and is used for calculation of both the amount of wax precipitated in terms of wt% of initial oil as well as CPT. A schematic of the model is demonstrated in Fig. 9.16. In this model, it is assumed that as temperature is reduced only a selected number of precipitating components will coexist in SLE. The basis of calculations for this model is the stability criteria expressed by Eq. (6.210), which should be applied to all N components (pure as well as pseudocomponents) in the following form:

$$(9.24) \quad \hat{f}_i(T, P, z_i) - f_i^S(T, P) \geq 0 \quad i = 1, 2, \ldots, N$$

where $\hat{f}_i(T, P, z_i)$ is the fugacity of component i in the original fluid mixture at T and P, and f_i^S is the fugacity of pure solid i at T and P. A component may exist as a pure solid phase if inequality by Eq. (9.24) is valid. This inequality can be applied only to single-solid phase and is not applicable to solid-phase solutions. Assume component 1 is the lightest (i.e., C_1 in a reservoir fluid) and N is the heaviest component (i.e., the last pseudocomponent of a C_{7+} fraction). If Eq. (9.24) is applied to all N components in the mixture the number of components that satisfy this equation is designated as N_S ($<N$). If $N_S = N$ it means that the mixture at T and P is initially in a solid phase (100% solid). All precipitating components must satisfy the following isofugacity equations:

$$(9.25) \quad \hat{f}_i^L(T, P, x_i^L) = f_i^S(T, P) \quad i = (N - N_S + 1), \ldots, N$$

The material balance equation for the nonprecipitating components is

$$z_i - x_i^L \left[1 - \sum_{j=(N-N_S+1)}^{N} \frac{n_j^S}{F} - \frac{V}{F} \right] - K_i^{VL} x_i^L \frac{V}{F} = 0$$

$$(9.26) \qquad\qquad\qquad i = 1, \ldots, (N - N_S)$$

where n_j^S is the moles of solid phase j and F is the number of moles of feed (initial fluid mixture). For the precipitating components where all solid phases are pure

$$z_i - x_i^L \left[1 - \sum_{j=(N-N_S+1)}^{N} \frac{n_j^S}{F} - \frac{V}{F} \right] - \frac{n_j^S}{F} - K_i^{VL} x_i^L \frac{V}{F} = 0$$

$$(9.27) \qquad [i = (N - N_S + 1), \ldots, N - 1], (N_S > 1)$$

In addition, all components must satisfy the following VLE isofugacity:

$$(9.28) \qquad \hat{f}_i^V(T, P, y_i) = \hat{f}_i^L(T, P, x_i^L) \quad i = 1, \ldots, N$$

There are two constraint equations for component i in the liquid and vapor phases:

$$(9.29) \qquad\qquad \sum_{i=1}^{N} x_i^L = \sum_{i=1}^{N} y_i = 1$$

Equation (9.28) is equivalent to Eq. (6.201) in terms of VLE ratios (K_i^{VL}). There are N_S equations through Eq. (9.24), ($N - N_S$) equations through Eq. (9.26), ($N_S - 1$) equations through Eq. (9.27), N equations through Eq. (9.28), and two equations through Eq. (9.29). Thus the total number of equations are $2N + N_S + 1$. The unknowns are x_i^L (N unknowns), y_i (N unknowns), n_j^S (N_S unknowns), and V/F (one unknown), with the sum of unknown same as the number of equations ($2N + N_S + 1$). Usually for crude oils and heavy residues, where under the conditions at which solid is formed, the amount of vapor is small and V/F can be ignored in the above equations. For such cases Eq. (9.28) and $\sum y_i = 0$ in Eq. (9.29) can be removed from the set of equations. On this assumption, the number of equations and unknowns reduces by $N + 1$ and y_i and V/F are omitted from the list of unknowns. Total number of moles of solid formed (S) is calculated as

$$(9.30) \qquad\qquad S = \sum_{j=(N-N_S+1)}^{N} n_j^S$$

The amount of wax precipitated in terms of percent of oil is calculated as

$$(9.31) \qquad \text{wax wt\% in oil} = 100 \times \frac{\sum_{i=1}^{N} M_i n_j^S}{F \sum_{i=1}^{N} z_i M_i}$$

where F is the total number of moles of initial oil and n_j^S is the moles of component i precipitated as solid. M_i is the molecular weight of component i and z_i is its mole fraction

in the initial fluid. The ratio of S/F is the same as S_F used in Eq. (9.17). The ratio of V/F in Eqs. (9.26) and (9.27) is the same as V_F in Eqs. (9.17) and (9.18).

The above set of equations can be solved by converting them into equations similar to Eqs. (9.17)–(9.19). For precipitating components, x_i^L can be calculated directly from Eq. (9.26), while for nonprecipitating components they must be calculated from Eq. (9.27) after finding V/F and S/F. Moles of solid formed for each component, n_j^S, must be calculated from Eq. (9.27). Values of V/F and S/F must be found by trial-and-error procedure so that Eq. (9.29) is satisfied.

The CPT of a crude can be calculated directly from Eq. (9.24) using trial-and-error procedure as follows:

a. Define the mixture and break C_{7+} into appropriate number of pseudocomponents as discussed in Chapter 4.
b. P and z_i are known for all component/pseudocomponents.
c. Guess a temperature that is higher than melting point of the heaviest components in the mixture so that no component in the mixture satisfies Eq. (9.24).
d. Reduce the temperature stepwise until at least one component (it must be the heaviest component) satisfies the equality in Eq. (9.24).
e. Record the temperature as calculated CPT of the crude oil.

A schematic of CPT and wax precipitation calculation using this model is illustrated in Fig. 9.17. To simplify and reduce the size of the calculations, Lira-Galeana et al. [43] suggest that solid phases can be combined into three or four groups where each group can be considered as one pesudocomponent. As the temperature decreases, the amount of precipitation increases. Compositions of six crude oils as well as their experimental and calculated values of CPT according to this model are given in Table 9.10. Calculated values of CPT very much depend on the properties (especially molecular weight) of the heaviest component in the mixture. For oils the C_{7+} fractions should be divided into several pseudocomponents according to the methods discussed in Chapter 4. In such cases, the heaviest component in the mixture is the last pseudocomponent of the C_{7+} and the value of its molecular weight significantly affects the calculated CPT. In such cases, the molecular weight of last pseudocomponent C_{7+} may be used as one of the adjustable parameters to match calculated amount of wax precipitation with the experimental values. Prediction of the amount of wax precipitation for oils 1 and 6 in Table 9.10 are shown in Fig. 9.18 as generated from the data provided in Ref. [43].

In the calculation of solid fugacity through Eq. (6.155), ΔC_P is required. In many calculations it is usually considered as zero; however, Lira et al. [43] show that without this term, considerable error may arise in calculation of solute composition in liquid phase for some oils as shown in Fig. 9.19. Effects of temperature and pressure according to the multisolid-phase model are clearly discussed by Pan et al. [17] and for several oils they have compared predicted CPT with experimental data at various pressures. They conclude that for heavy oils at low pressure or live oils (where light gases are dissolved in oil) the increase in pressure will decrease CPT as shown in Fig. 9.20. However, for heavy liquid oils (dead oils)

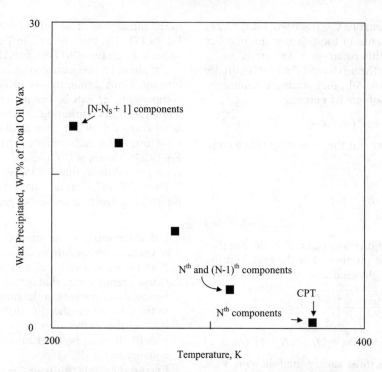

FIG. 9.17—Demonstration of multisolid-phase model for calculation of cloud-point temperature and wax precipitation for a typical crude oil. Courtesy of Lira-Galeana et al. [43].

and at high pressures, the effect of pressure is opposite and an increase in pressure will cause an increase in the CPT as shown in Fig. 9.21.

Values of WAT calculated from the solid-solution model are usually closer to pour points of an oil, while values from

multisolid-phase model are closer to the cloud point of the crude. As it was shown in Table 3.30, the pour point temperature is usually greater than the cloud point for most crude oils. As discussed earlier the method of characterization of oil and C_{7+} greatly affects result of calculations using this method.

TABLE 9.10—*Compositions and cloud point temperature of some oils. Taken with permission from Ref. [43].*

							Oil No.					
	1		2		3		4		5		6	
Comp	mol%	M	Mol%	M	mol%	M	mol%	M	mol%	M	mol%	M
C_1	0.056		0.000		0.016		0.000		0.000		0.021	
C_2	0.368		0.113		0.144		0.100		0.173		0.254	
C_3	1.171		1.224		1.385		0.118		1.605		1.236	
i-C_4	0.466		0.645		1.174		0.106		1.150		0.588	
n-C_4	1.486		2.832		3.073		0.099		3.596		2.512	
i-C_5	0.961		1.959		2.965		0.162		3.086		1.955	
n-C_5	1.396		3.335		3.783		0.038		4.171		3.486	
C_6	2.251		5.633		7.171		0.458		7.841		6.842	
C_7	6.536	88.8	9.933	92.8	11.27	94.1	2.194	90.8	11.11	94.1	12.86	92.2
C_8	8.607	101.0	10.75	106.3	12.41	107.0	2.847	106.5	13.43	105.4	13.99	105.4
C_9	4.882	116.0	7.179	120.0	7.745	122.0	1.932	122.3	9.420	119.0	9.195	119.0
C_{10}	2.830	133.0	6.561	134.0	5.288	136.0	5.750	135.0	5.583	135.0	6.438	134.0
C_{11}	3.019	143.0	5.494	148.0	5.008	147.0	4.874	149.0	4.890	148.0	5.119	148.0
C_{12}	3.119	154.0	4.547	161.0	3.969	161.0	5.660	162.0	3.864	162.0	4.111	161.0
C_{13}	3.687	167.0	4.837	175.0	3.850	175.0	6.607	176.0	4.300	175.0	4.231	175.0
C_{14}	3.687	181.0	3.700	189.0	3.609	189.0	6.149	189.0	3.272	188.0	3.682	188.0
C_{15}	3.687	195.0	3.520	203.0	3.149	203.0	5.551	202.0	2.274	203.0	3.044	202.0
C_{16}	3.079	207.0	2.922	216.0	2.300	214.0	5.321	213.0	2.791	216.0	2.255	214.0
C_{17}	3.657	225.0	3.072	233.0	2.460	230.0	5.022	230.0	2.311	232.0	2.405	230.0
C_{18}	3.289	242.0	2.214	248.0	2.801	244.0	4.016	244.0	1.960	246.0	2.006	245.0
C_{19}	3.109	253.0	2.493	260.0	2.100	258.0	4.176	256.0	1.821	256.0	1.770	257.0
C_{20+}	38.4	423.0	17.0	544.0	14.33	418.0	38.80	473.0	11.33	388.0	12.00	399.0
SG_{20+}	0.893		0.934		0.880		0.963		0.872		0.887	
CPT, K	313.15		311.15		314.15		295.15		305.15		308.15	
CPT, Calc.	312.4		308.2		316.0		299.3		301.2		309.5	
Error, K	0.75		2.95		−1.85		−4.15		3.95		−1.35	

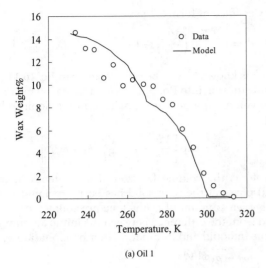

(a) Oil 1

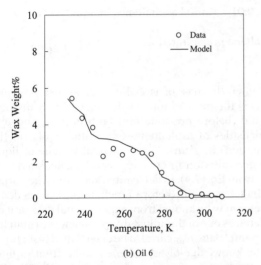

(b) Oil 6

FIG. 9.18—Prediction of wax precipitation and cloud-point temperature for oils 1 and 6 given in Table 9.10. Taken with permission from Ref. [43].

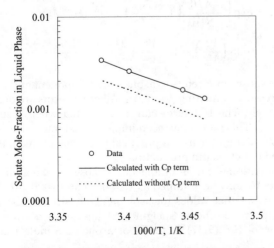

FIG. 9.19—Effect of ΔC_P term on calculation of solute solubilities at 1 bar [43]. Drawn based on data from Ref. [43].

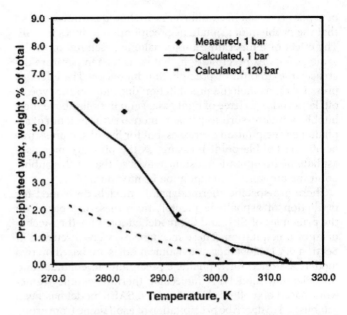

FIG. 9.20—Effect of pressure on cloud-point temperature and wax formation in a synthetic crude oil at low pressures. Taken with permission from Ref. [17].

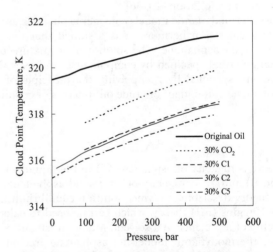

FIG. 9.21—Effect of pressure on cloud-point temperature at high pressures for a crude oil diluted by various light hydrocarbons. Taken with permission from Ref. [17].

9.4 ASPHALTENE PRECIPITATION: SOLID–LIQUID EQUILIBRIUM

Prediction of asphaltene precipitation is more difficult and complex than prediction of wax precipitation. The reason for this complexity is the complex nature of asphaltenes and the mechanism of their precipitation. The presence of resins further complicates modeling of asphaltene precipitation. In addition asphaltene molecules are polar and when aggregated they behave similar to polymer molecules. Asphaltenes are heavier than wax and they precipitate at a higher temperature than WAP. Asphaltenes usually exist in heavier oils and for the case of crude oils at atmospheric pressures the amount of vapor produced in equilibrium with liquid is quite small so that the problem of asphaltene precipitation reduces to LSE. The effect of temperature on asphaltene precipitation is the same as for wax precipitation, that is, as the temperature decreases the amount of precipitation increases. The effect of pressure on asphaltene precipitation depends on the type of oil. For crude oils (free of light gases) and live oils above their bubble point pressure, as pressure increases the amount of asphaltene precipitation decreases, but for live oils at pressures below the bubble point pressure, as the pressure increases asphaltene precipitation also increases so that at the bubble point the amount of precipitation is maximum [48].

There are specific thermodynamic models developed for prediction of asphaltene precipitation; these are based on the principles of SLE and the model adopted for the mechanism of precipitation. These mechanisms were discussed in Section 9.3.1. Most thermodynamic models are based on two models assumed for asphaltene precipitation: colloidal and micellization models. A molecular thermodynamic framework based on colloid theory and the SAFT model has been established to describe precipitation of asphaltene from crude oil by Wu–Prausnitz–Firoozabadi [39, 49]. Mansoori [29] also discusses various colloidal models and proposed some thermodynamic models. Pan and Firoozabadi have also developed a successful thermodynamic micellization models for asphaltene precipitation [20, 36].

Most thermodynamic models consider asphaltene as polymer molecules. Furthermore, it is assumed that the solid phase is pure asphaltene. The solution is a mixture of oil (asphaltene-free) specified by component B and asphaltene component specified by A. Applying the principle of SLE to asphaltene component of crude oil in terms of equality of fugacity

$$(9.32) \qquad \hat{f}_A^L = f_A^S$$

where $\hat{f}_A^L$ is fugacity of asphaltene (A) in the liquid solution (A and B) and f_A^S is fugacity of pure solid asphaltene. One good theory describing polymer–solution equilibrium is the Flory–Huggins (FH) theory, which can be used to calculate solubility of a polymer in a solvent. Many investigators who studied thermodynamic models for asphaltene precipitation have used the FH theory of polymer solutions for calculation of chemical potential of asphaltenes dissolved in oil [38, 50]. Nor-Azian and Adewumi [48] also used FH theory for the asphaltic oils. Moreover, they also considered the vapor phase in their model with VLE calculations between liquid oil and its vapor. According to the FH theory the chemical potential of component i (polymer) in the solution is given as

$$(9.33) \qquad \frac{\hat{\mu}_i^L - \mu_i^{\circ L}}{RT} = \ln(\Phi_i) + 1 - \frac{V_i^L}{V_m^L} + \frac{V_i^L}{RT}(\delta_i - \delta_m)^2$$

where $\Phi_i = \frac{x_i V_i^L}{V_m^L}$, $\delta_m = \sum_{i=1}^N \Phi_i \delta_i$, $\hat{\mu}_i^L$ is the chemical potential of component i in the liquid phase, and $\mu_i^{\circ L}$ is the chemical potential at reference state, which is normally taken as pure liquid i. Φ_i is the volume fraction of i, V_i^L is the liquid molar volume of pure i at T and P of the solution, R is the gas constant, V_m^L is the liquid molar volume of mixture, and δ_i is the solubility parameter for component i at T of the solution. The above equation can be conveniently converted into an

activity coefficient form (γ_i^L) as

$$(9.34) \quad \gamma_i^L = \exp\left[\ln\left(\frac{V_i^L}{V_m^L}\right) + 1 - \frac{V_i^L}{V_m^L} + \frac{V_i^L}{RT}(\delta_i - \delta_m)^2\right]$$

Once γ_i^L is known $\hat{f}_i^L$ can be calculated from Eq. (6.114) and after substituting into Eq. (9.32) we get the following relation for the volume fraction of asphaltene in the liquid solution [38]:

$$(9.35) \qquad \Phi_A^L = \exp\left[\frac{V_A^L}{V_m^L} - 1 - \frac{V_A^L}{RT}(\delta_A - \delta_m)^2\right]$$

where Φ_A^L is the volume fraction of asphaltenes in the oil (liquid) phase at the time solid has been precipitated. Once Φ_A^L is known, amount of asphaltene precipitated can be calculated from the difference between the initial amount of asphaltene in liquid and its amount after precipitation as

$$m_{AD} = m_{AT} - \rho_A \Phi_A^L V_T^L$$

$$m_{AT} = 0.01 \times (\text{initial asphaltenes in liquid, wt\%}) \times \rho_{mix}^L V_T^L$$

$$\text{asphaltene precipitated wt\%} = 100 \times \left(\frac{m_{AD}}{\rho_{mix}^L V_T^L}\right)$$

(9.36)

where m_{AD} is the mass of asphaltenes deposited (precipitated) and m_{AT} is the mass of total asphaltene initially dissolved in the liquid (before precipitation) both in g. ρ_A and ρ_{mix}^L are mass densities of asphaltenes and initial liquid oil (before precipitation) in g/cm³. V_T^L is the total volume of liquid oil before precipitation in cm³. ρ_{mix}^L can be calculated from an EOS or from Eq. (7.4). In determination of m_{AT}, the initial wt% of asphaltenes in oil (before precipitation) is needed. This parameter may be known from experimental data or it can be considered as one of the adjustable parameters to match other experimental data. m_{AT} can also be determined from Eq. (9.35) from the knowledge of asphaltene composition in liquid at the onset when m_{AD} is zero and $m_{AT} = \rho_A (\Phi_A^L V_T^L)^{onset}$. A more accurate model for calculation of asphaltene precipitation is based on Chung's model for SLE [44]. This model gives the following relation for asphaltene content of oil at temperature T and pressure P [51]:

$$x_A = \exp\left[-\frac{\Delta H_A^f}{RT_{MA}}\left(1 - \frac{T_{MA}}{T}\right) - \frac{V_A^S}{RT}(\delta_A - \delta_m)^2\right.$$

$$(9.37) \quad \left. - \ln\left(\frac{V_A^L}{V_m^L}\right) - 1 + \frac{V_A^L}{V_m^L} + \frac{(V_A^L - V_A^S) \times (P - P_{MA})}{RT}\right]$$

where subscript A refers to asphaltene component and P_{MA} is the pressure at melting point T_{MA}. All other terms are defined previously. The last term can be neglected when assumed $V_A^L \cong V_A^S$. This model has been implemented into some reservoir simulators for use in practical engineering calculations related to petroleum production [51].

As mentioned earlier (Table 9.6), in absence of actual data, ρ_A and M_A may be assumed as 1.1 cm³/g and 1000 g/mol, respectively. Other values for asphaltene density are also used by some researchers. Speight [15] has given a simplified version of Eq. (9.37) in terms of asphaltene mole fraction (x_A) as

$$(9.38) \qquad \ln x_A = -\frac{M_A (\delta_L - \delta_A)^2}{RT \rho_A}$$

where he assumed $\rho_A = 1.28$ cm^3/g and $M_A = 1000$ g/mol. δ_A and δ_L are the solubilities of asphaltene and liquid solvent (i.e., oil), respectively. If δ_A and δ_L are in (cal/cm^3)$^{1/2}$ and T is in kelvin then $R = 1.987$ cal/mol·K. This equation provides only a very approximate value of asphaltene solubility in oils. In fact one may obtain Eq. (9.38) from Eq. (9.35) by assuming molar volumes of both oil and asphaltene in liquid phase are equal: $V_A^L = V_m^L$. As this assumption can hardly be justified, one may realize the approximate nature of Eq. (9.38).

A mixture of asphaltenes and oil may be considered homogenous or heterogeneous. Kawanaka et al. [30] have developed a thermodynamic model for asphaltene precipitation based on the assumption that the oil is a heterogeneous solution of a polymer (asphaltenes) and oil. The asphaltenes and the C$_{7+}$ part of the oil are presented by a continuous model (as discussed in Chapter 4) and for each asphaltenes component the equilibrium relation has been applied as

$$(9.39) \quad \hat{\mu}_{Ai}^S(T, P, x_{Ai}^S) = \hat{\mu}_{Ai}^L(T, P, x_{Ai}^L) \quad i = 1, \ldots, N_A$$

where $\hat{\mu}_{Ai}^S$ and $\hat{\mu}_{Ai}^L$ are chemical potentials of ith component of asphaltene in the solid and liquid phase, respectively. Similarly x_{Ai}^S and x_{Ai}^L are the composition of asphaltene components in the solid and liquid phases. The sum $\sum x_{Ai}^S$ is unity but the sum $\sum x_{Ai}^L$ is equal to x_A^L the mole fraction of asphaltenes in the liquid phase after precipitation. N_A is the number of asphaltenes components determined from distribution model as it was discussed in Chapter 4. In this model, the solid phase is a mixture of N_A pseudocomponents for asphaltenes.

In this thermodynamic model, several parameters for asphaltenes are needed that include molecular weight (M_A), mass density (ρ_A), binary interaction coefficient between asphaltene and asphaltene-free crude (k_{AB}), and the asphaltene solubility parameter in liquid phase (δ_A^L). As discussed in Section 9.3.1, in lieu of experimental data on M_A and ρ_A they can be assumed as 1000 and 1.1 g/cm^3, respectively. Kawanaka et al. [30] recommends the following relations for calculation of δ_A^L and k_{AB} as a function of temperature:

$$(9.40) \quad \delta_A^L = 12.66(1 - 8.28 \times 10^{-4}T)$$

$$(9.41) \quad k_{AB} = -7.8109 \times 10^{-3} + 3.8852 \times 10^{-5}M_B$$

where δ_A^L is the asphaltene solubility parameter in (cal/cm^3)$^{0.5}$ and T is temperature in kelvin. M_B is the molecular weight of asphaltene-free crude oil. To calculate $\hat{\mu}_{Ai}^S$, values of δ_A^S and ρ_A^S are needed. δ_A^S can be calculated from Eq. (6.154) and ρ_A^S is assumed the same as ρ_A^L. It should be noted that in these relations asphaltene-free crude refers to the liquid phase in equilibrium with precipitated solid phase, which include added solvent (i.e., C$_3$, n-C$_5$, or n-C$_7$) and the original crude. The phase diagram shown in Fig. 9.11 was developed based on this compositional model [30]. Equation (9.40) gives value of 9.5 (cal/cm^3)$^{0.5}$ at 25°C, which is consistent with the value reported by other investigators. Equation (9.40) is named after Hirschberg who originally proposed the relation [52].

Most of the thermodynamic models discussed in this section predict data with good accuracy when the adjustable parameters in the model are determined from experimental data on asphaltene precipitation. Results of a thermodynamic model based on the colloidal model and SAFT theory for Suffield crude oil are shown in Fig. 9.22. The crude has

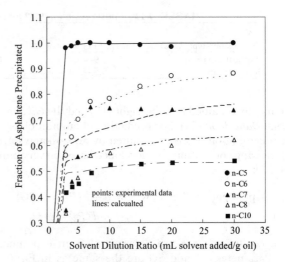

FIG. 9.22—Calculated versus experimental amount of asphaltene precipitated by various *n*-alkanes solvents added to Suffield crude oil. Taken with permission from Ref. [39].

specific gravity of 0.952 and average molecular weight of 360 with resin and asphaltene contents of 8 and 13 wt%, respectively. Effects of temperature and pressure on asphaltene and solid precipitation were discussed in Section 9.3.1.

To avoid complex calculations for quick and simple estimation of asphaltene and resin contents of crude oils, attempts were made to develop empirical correlations in terms of readily available parameters similar to those presented in Section 3.5.1.2 for composition of petroleum fractions. Because of the complex nature of asphaltenes and wide range of compounds available in a crude, such attempts were not as successful as those developed for narrow-boiling range petroleum fractions. However, Ghuraiba [53] developed the following simple correlation based on limited data collected from the literature for prediction of asphaltene and resin contents of crude oils:

wt% of asphaltene or resin in crude oil $= a + bR_i + cSG$

(9.42)

where R_i is the refractivity intercept defined in Eq. (2.14) as $R_i = n_{20} - d_{20}/2$. Amounts of asphaltenes and resin in a crude mainly depend on the composition of the crude. In Section 3.5.1.2, parameters R_i and SG were used to predict the composition of petroleum fractions. Calculation of n_{20} and d_{20} for a crude is not as accurate as for a fraction since the crude has a very wide boiling point range. For this reason, the above equation gives only an approximate value of asphaltene and resin contents. Coefficients a, b, and c in Eq. (9.42) are given in Table 9.11. These coefficients have been determined based on the calculation of n_{20} and d_{20} from Eq. (4.7) and Table 4.5. Only M is required for calculation of these two properties. If M is not available it may be estimated from other properties such as viscosity and SG (i.e., Eq. (2.52) or reversed form of Eq. (4.7) and Table 4.5. The above correlation generally predicts amount of asphaltene and resin contents with absolute deviation of 1.5–2 wt%. Experimental data points for resin contents were very limited and for this reason predicted values must be taken with caution. Data to develop these correlations

TABLE 9.11—*Constants in Eq. 9.42 for estimation of asphaltene and resin contents of crude oils.*

Composition type (wt%)	Range of API gravity of oil	Range of wt%	Constants in Eq. (9.42)			No. of oils	Absolute Dev.%	
			a	b	c		Avg	Max
Asphaltene in oil	5.9–43.4	0.1–20	−731	674	31	122	1.4	−5.2
Resin in oil	5.5–43.4	5.6–40	−2511.5	2467	−76	41	1.9	−4.9

were mostly obtained from Speight [15] and the *Oil and Gas Journal Data Book* [54]. For prediction of amount of asphaltene precipitation when it is diluted by an *n*-alkane solvent, the following correlation was developed based on very limited data [53]:

$$\text{Asphaltene predicted, wt\%} = a + b\,(R_i) + c\,(\text{SG})$$
$$(9.43) \qquad\qquad + d\,(R_\text{S}) + e\,(M_\text{S})$$

where coefficients *a–e* are determined from experimental data. Parameters R_i and SG are the same as in Eq. (9.42) and should be calculated in the same way. M_S is the solvent molecular weight (*n*-alkane) and R_S is the solvent-to-oil ratio in cm³/g. This correlation was developed based on the data available for three different Kuwaiti oils and 45 data points, and for this limited database the coefficients were determined as $a = -2332$, $b = 2325$, $c = -112.6$, $d = 0.0737$, and $e = -0.0265$. With these coefficients the above equation predicts asphaltene precipitation of Kuwait oils with AD of 0.5%. The correlation is not appropriate for other crude oils and to have a generalized correlation for various oils, the coefficients in Eq. (9.43) must be reevaluated with more data points for crude oils from around the world. The following example shows application of these equations.

Example 9.3—For Suffield crude oil the asphaltene precipitation by various solvents is shown in Fig. 9.22. Calculate

a. asphaltene content.
b. resin content.
c. amount of asphaltene (wt%) precipitated by adding 10 cm³/g *n*-decane.

Solution—For this oil, $M = 360$ and SG $= 0.952$. n_{20} and d_{20} should be calculated through M using Eq. (4.7) with coefficients in Table 4.5. The results are $n_{20} = 1.4954$, $d_{20} = 0.888$, and $R_i = 1.05115$. (a) From Eq. (9.42), asphaltene wt% = 7%. (b) From Eq. (9.42), resin wt% = 9.3%. (c) For calculation of asphaltene precipitation from Eq. (9.43) we have $M_\text{S} = 142$ and $R_\text{S} = 10$ cm³/g, thus wt% of asphaltene precipitated is calculated as 1.3%. The experimental value as shown in Fig. 9.22 is 0.5%. The experimental values for asphaltene and resin contents are 13 and 8%, respectively [39]. For resin content the calculated value is in error by 1.3% from the experimental data. This is considered as a good prediction. For the amount of asphaltene precipitated, Eq. (9.43) gives %AD of 0.8. The biggest error is for asphaltene content with %AD of 6. As mentioned these correlations are very approximate and based on limited data mainly from Middle East. However, the coefficients may be reevaluated for other oils when experimental data are available. In this example predicted values are relatively in good agreement with experimental data; however, this is very rare. For accurate calculations of asphaltene

precipitation appropriate thermodynamic models as introduced in this section should be used. ◆

9.5 VAPOR–SOLID EQUILIBRIUM— HYDRATE FORMATION

In this section, another application of phase equilibrium in the petroleum industry is demonstrated for prediction of hydrate formation from vapor–solid equilibrium (VSE) calculations. Hydrates are molecules of gas (C_1, C_2, C_3, iC_4, nC_4, N_2, CO_2, or H_2S) dissolved in solid crystals of water. Gas molecules, in fact, occupy the void spaces in water crystal lattice and the form resembles wet snow. In the oil fields hydrates look like grayish snow cone [1]. Gas hydrates are solid, semistable compounds that can cause plugging in natural gas transmission pipelines, gas handling equipments, nozzles, and gas separation units. Gas hydrates may be formed at temperatures below 35°C when a gas is in contact with water. However, at high pressures (>1000 bar), hydrate formation has been observed at temperatures above 35°C. Figure 9.23 shows temperature and pressure conditions that hydrates are formed for natural gases. As pressure increases hydrate can be formed at higher temperatures. Severe conditions in arctic and deep drilling have encouraged the development of predictive and preventive methods. It is generally believed that large amounts of energy is buried in hydrates, which upon their dissociation can be released.

Hydrates are the best example of the application of VSE calculations. Whitson [1] discusses various methods of calculation of the temperature at which a hydrate may form

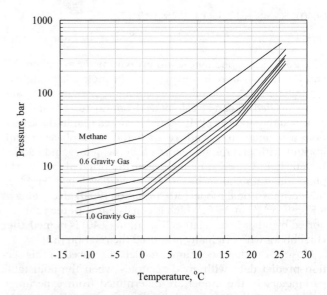

FIG. 9.23—Hydrate formation for methane and natural gases. Drawn based on data provided in Ref. [1].

TABLE 9.12—*Coefficients (C_i) for Eq. (9.46) for estimation of HFT at very high pressures. Taken with permission from Ref. [1].*

Pressure, bar (psia)	Methane	Ethane	Propane	i-Butane	n-Butane
414 (6000)	18933	20806	28382	30696	17340
483 (7000)	19096	20848	28709	30913	17358
552 (8000)	19246	20932	28764	30935	17491
620 (9000)	19367	21094	29182	31109	17868
690 (10000)	19489	21105	29200	30935	17868

at given pressure. Calculation of hydrate-formation temperature (HFT) is very similar to dewpoint temperature calculation in VLE. The equilibrium ratio for component i between vapor and solid phase is defined as $K_i^{VS} = y_i/x_i^S$, where x_i^S is the mole fraction of i in the solid hydrate phase. Hydrate is formed if at given T and P we have

$$(9.44) \qquad \sum_{i=1}^{N} \frac{y_i}{K_i^{VS}} \geq 1$$

where equality holds at temperature where hydrate formation begins. In the vapor phase the amount of water is very small (<0.001 mol%) thus its presence in the vapor phase can be neglected in the calculations ($y_i^W \cong 0$). To find the temperature at which a hydrate dissociates and hydrocarbons are released, a calculation similar to bubble point calculations can be performed so that $\sum x_i^S K_i^{VS} \geq 1$. Katz provided charts for calculation of K_i^{VS}, which later Sloan converted into empirical correlations in terms of T and P and they are used in the petroleum industry [1]. It should be noted that these K_i values are not true VSE ratios as the above calculations are based on water-free phases. This method can be applied to pressures below 70 bar ($\sim$1000 psia). For methane, ethane, propane, n-butane, and H_2S the correlations for calculation of K_i^{VS} are given as follows [1]:

$$\ln K_{C_1}^{VS} = 0.00173 + \frac{17.59}{T} - \frac{3.403}{P} + 1.3863 \times 10^{-4} PT$$
$$+ \frac{1.0356 P}{T} - 0.78338 \ln\left(\frac{P}{T}\right) - 23.9804\left(\frac{P}{T^2}\right)$$
$$- 1.34136 \times 10^{-6} T^3 - 1.8834 \times 10^{-5}\left(\frac{P^3}{T^2}\right)$$

$$\ln K_{C_2}^{VS} = 3.92157 - \frac{161.268}{T} + \frac{181.267}{P} + 1.8933 \times 10^{-5} P^2$$
$$+ \frac{1.04557 P}{T} - 1.19703 \ln\left(\frac{P}{T}\right) - \frac{402.16}{P^2}$$
$$- 8.8157\left(\frac{T}{P}\right) + 0.133231\left(\frac{T^2}{P}\right) - 21.2354\left(\frac{P}{T^2}\right)$$
$$+ 46.13339\left(\frac{T}{P^3}\right)$$

$$\ln K_{C_3}^{VS} = -7.59224 + \frac{26.1422}{T} - 3.0545 \times 10^{-5} PT + 2.315$$
$$\times 10^{-3} T^2 + 0.12348 \ln\left(\frac{P}{T}\right) + \frac{79.3379}{P^2}$$
$$+ 0.05209\left(\frac{T^2}{P}\right) - 26.4294\left(\frac{T}{P^3}\right) + 3.2076 \times 10^{-5} T^3$$

$$\ln K_{n-C_4}^{VS} = -37.211 + 1.5582 T + \frac{406.78}{T} + 1.9711 \times 10^{-3} T^2$$
$$- 8.6748\left(\frac{P}{T}\right) - 8.2183\left(\frac{T}{P}\right) + 540.976\left(\frac{T}{P^3}\right)$$
$$+ 4.6897 \times 10^{-3}\left(\frac{P^3}{T^2}\right) - 1.3227 \times 10^{-5} T^4$$

$$\ln K_{H_2S}^{VS} = -6.051 + 0.11146 T + \frac{45.9039}{T} - 1.9293 \times 10^{-4} PT$$
$$+ 1.94087\left(\frac{P}{T}\right) - 0.64405 \ln\left(\frac{P}{T}\right) - 56.87\left(\frac{P}{T^2}\right)$$
$$- 7.5816 \times 10^{-6} T^3$$

where $T =$ given temperature in kelvin $- 255.4$ and

$\qquad P =$ given pressure in bar

(9.45)

For pressures between 400 and 700 bars ($\sim$6000–10000 psia), a simple empirical method is proposed by McLeod and Campbell in the following form as given in Ref. [1]:

$$(9.46) \qquad T = 2.16\left(\sum_{i=1}^{N} y_i C_i\right)^{1/2}$$

where values of C_i for C_1–C_4 are given in Table 9.12 at several pressures encountered in deep-gaswell drilling.

This method can be used for quick estimation of HFT or to check the validity of estimated temperatures from other methods. More sophisticated methods using chemical potential and equations of state are discussed in other references [1].

Because of the problems associated with hydrate formation, hydrate inhibitors are used to reduce HFT. Commonly used hydrate inhibitors are methanol, ethanol, glycols, sodium chloride, and calcium chloride. These are nearly the same materials that are used as water antifreeze inhibitors. Effect of methanol (CH_3OH) on the depression of HFT of methane reservoir fluid is shown in Fig. 9.24 [55]. The composition of this condensate sample in terms of mol% is as follows: 0.64 N_2, 3.11 CO_2, 73.03 C_1, 8.04 C_2, 4.28 C_3, 0.73 i-C_4, 1.5 n-C_4, 0.54 i-C_5, 0.6 n-C_5, and 7.53 C_{6+} with mixture molecular weight of 32.4. The most commonly used equation to calculate the degree of decrease in HFT (ΔT) is given by Hammerschmidt, which is in the following form [1, 14]:

$$(9.47) \qquad \Delta T = \frac{A wt\%}{M(100 - wt\%)}$$

where ΔT is the decrease in HFT in °C (or in kelvin), wt% is the weight percent of inhibitor in the aqueous phase, and M is the molecular weight of the inhibitor. Values of M and A

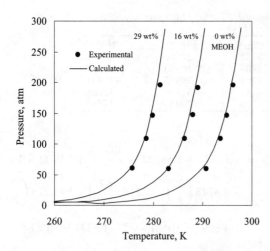

FIG. 9.24—Depression of hydrate formation temperature in methane by methanol-calculated versus measured values. Lines represent coexistence curves for methane, hydrate, and aqueous solutions of MeOH. Taken with permission from Ref. [55].

for some common inhibitors are given in Table 9.13. Values of A are corrected values as given in Ref. [14].

Equation (9.47) is recommended for sweet natural gases (H_2S content of less than 4 ppm on volume basis, also see Section 1.7.15) with inhibitor concentrations of less than 20 mol%. For concentrated methanol solutions, like those used to free a plugged-up tubing string in a high-pressure well, Whitson [1] suggests a modified form of Hammerschmidt equation:

$$(9.48) \qquad \Delta T = -72 \ln (1 - x_{MeOH})$$

where ΔT is the decrease in HFT in °C (or in kelvin) and x_{MeOH} is the mole fraction of methanol in the aqueous solution.

Example 9.4—Composition of a natural gas in terms of mol% is as follows: 85% C_1, 10% C_2, and 5% C_3. Calculate

a. HFT at 30 bars and composition of hydrate formed.
b. HFT at 414 bars.
c. wt% of methanol solution needed to decrease HFT to 5°C for each case.

Solution—(a) At 30 bar pressure (<70 bar) the HFT can be calculated from Eqs. (9.44) and (9.45) by trial-and-error method. Assuming HFT of 280 K, the sum in Eq. (9.44) is $\sum y_i/K_i^{VS} = 2.848$ since it is greater than 1, temperature should be increased in order to decrease K_i^{VS} values. At $T = 300$ K, $\sum y_i/K_i^{VS} = 0.308$;, at $T = 290$ K, $\sum y_i/K_i^{VS} = 0.504$;

TABLE 9.13—*Constants in Eq. (9.47) for hydrate formation inhibitors.*

Hydrate formation inhibitor	Formula	M	A
Methanol	CH_3OH	32	1297.2
Ethanol	C_2H_5OH	46	1297.2
Ethylene glycol	$C_2H_6O_2$	62	1500
Diethylene glycol	$C_4H_{10}O_3$	106	2222.2
Triethylene glycol	$C_6H_{14}O_4$	150	3000

and at 285 K, $\sum y_i/K_i^{VS} = 1.073$. Finally at $T = 285.3877$ K, $\sum y_i/K_i^{VS} = 1.00000$, which is the correct answer. Thus the HFT for this gas at 30 bar is 285.4 K or 12.2°C. At this temperature from Eq. (9.45), $K_1 = 2.222$, $K_2 = 0.7603$, and $K_3 = 0.113$. Composition of hydrocarbons in a water-free base hydrate is calculated as $x_i^S = y_i/K_i^{VS}$, which gives $x_1^S = 0.36, x_2^S = 0.197$, and $x_3^S = 0.443$. (b) At 414 bars Eq. (9.46) with coefficients in Table 9.12 should be used. At this pressure HFT is calculated as 303.2 K or HFT = 30°C. (c) To decrease HFT at 30 bars an inhibitor solution that can cause depression of $\Delta T = 12.2 - 5 = 7.2$°C is needed. Rearranging Eq. (9.47): wt% = $100[M\Delta T/(A + M\Delta T)]$, where wt% is the weight percent of inhibitor in aqueous solution. From Table 9.13 for methanol, $A = 1297.2$ and $M = 32$. Thus with $\Delta T = 7.2$, wt% = 15.1. Since calculated wt% of methanol is less than 20% use of Eq. (9.47) is justified. For pressure of 414 bars Eq. (9.48) should be used for methanol where upon rearrangement one can get $x_{MeOH} = 1 - \exp(-\Delta T/72)$. At $\Delta T = 30 - 5 = 25$°C we get $x_{MeOH} = 0.293$. For an aqueous solution ($M_{H_2O} = 18$) and from Eq. (1.15), the wt% of methanol ($M = 32$) can be calculated as: wt% = 42.4. ♦

9.6 APPLICATIONS: ENHANCED OIL RECOVERY—EVALUATION OF GAS INJECTION PROJECTS

In this section another application of some of the methods presented in this book is shown for the evaluation of gas injection projects. Gas is injected into oil reservoirs for different purposes: storage of gas, maintenance of reservoir pressure, and enhanced recovery of hydrocarbons. In the last case, understanding and modeling of the diffusion process is of importance to the planning and evaluation of gas injection projects. Gases such as natural gas, methane, ethane, liquefied petroleum gas (LPG), or carbon dioxide are used as miscible gas flooding in EOR techniques. Upon injection of a gas, it is dissolved into oil under reservoir conditions and increases the mobility of oil due to decrease in its viscosity. To reach a certain mobility limit a certain gas concentration is required. For planning and evaluation of such projects, it is desired to predict the amount of gas and duration of its injection in an oil reservoir. In such calculations, properties such as density, viscosity, diffusivity and phase behavior of oil and gas are needed. The purpose of this section is to show how to apply methods presented in this book to obtain desired information for such projects. This application is shown through modeling of fractured reservoirs for a North Sea reservoir for the study of nitrogen injection. Laboratory experimental data are used to evaluate model predictions as discussed by Riazi et al. [56].

An idealized matrix–fracture system is shown in Fig. 9.25, where matrix blocks are assumed to be rectangular cubes. Dimensions of matrix blocks may vary from 30 to 300 cm, and the thickness of fractures is about 10^{-2}–10^{-4} cm. When a gas is injected into a fractured reservoir, the gas flows through the fracture channels in horizontal and vertical directions. Therefore, all surfaces of a matrix block come into contact with the surrounding gas in the fracture. The injected gas comes into contact with oil in the matrix block at the matrix–fracture interface. The gas begins to diffuse into oil and light

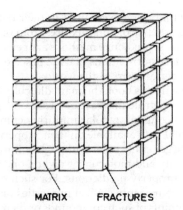

FIG. 9.25—Idealized fractured reservoirs (after Warren and Root [57]).

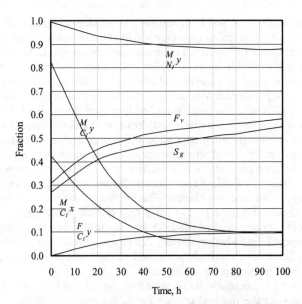

FIG. 9.27—Calculated compositions for oil in matrix and gas in fracture versus time. Taken from Ref. [56].

components in the oil diffuse in the opposite direction from matrix to the fracture. This process continues until the gas in the fracture reaches in equilibrium with the oil in the matrix block when no longer gas diffuses into oil. In such cases, it is assumed that the oil and gas inside the matrix blocks are in thermodynamic equilibrium at all times. Moreover, it is assumed that at the matrix–fracture interface, oil and gas are in equilibrium at all the times and there is no diffusion across the interface. To analyze the diffusion process, a laboratory experiment was conducted with a cell containing a porous core (from Ekofisk field) as shown in Fig. 9.26. The free volume in the cell can be considered as the fracture in real reservoirs. For simplicity in formulation of diffusion process and mathematical solutions, the matrix–fracture system was converted into a one-dimensional model. Details of the model and mathematical formulation are given in Ref. [56].

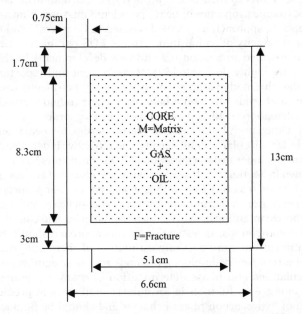

FIG. 9.26—Schematic of experimental cell for diffusion of gas in a matrix block. Taken with permission from Ref. [56].

In a particular experiment, the core was saturated with a live oil at its bubble point pressure of 382.8 bar and temperature of 403 K. The free volume around the core was filled with pure nitrogen. As nitrogen diffuses to the matrix block and light gas diffuses in the opposite direction to the free volume (fracture), composition of the gas in the fracture was measured versus time. Composition of oil was expressed by 15 components, including five pseudocomponents generated by methods of Chapter 4. Critical properties and acentric factor were estimated through methods of Chapter 2. Diffusion coefficients were calculated through methods presented in Chapter 8. Cubic equation of state (PR EOS) of Chapter 5 was used for calculation of PVT properties and flash calculations inside the matrix blocks. Through solution of diffusion equations concentration of all components in both the matrix and the fracture were determined. This composition in terms of mole fraction of key components (C_1 and N_2) in the matrix and fracture versus time is shown in Fig. 9.27. The system reaches final equilibrium conditions after 100 h. As dimension of matrix blocks increases, the time required to reach final state increases as well. Applying this model to real reservoirs one can determine how long the gas must be injected in order to reach the desired degree of oil mobility.

9.7 SUMMARY AND RECOMMENDATIONS

In this chapter, applications of methods and procedures presented in the book were shown in phase equilibria calculations of petroleum fluid mixtures. Five types of VLE calculations, namely, flash, bubble T, bubble P, dew P, and dew T, as well as construction of phase diagrams (i.e., PT or Px) are presented and their applications to petroleum reservoir fluids have been demonstrated. Furthermore, the principles of phase equilibria introduced in Chapter 6 is applied to VLE, SLE, VLSE, and VSE calculations for prediction of the onset and amount of solid formation in petroleum fluids. Some guidelines for quick convergence in the calculation and determination of key interaction parameters are given. Various models available in most recent publications for calculation of amounts of wax and asphaltene precipitations and their onsets are presented. Mechanism of solid formation, their negative effects in the petroleum industry and methods of their prevention are also discussed. Results of calculations from various models when applied to different petroleum mixtures are given. Effects of temperature and pressure on the wax and asphaltene precipitation for different oils are demonstrated. Methods of calculation of the conditions at which hydrates may be formed are shown. The impact of characterization methods of Chapters 2–4 on property and phase behavior predictions as well as methods of calculation of true critical properties are also presented. The chapter ends with another application of methods presented in the book in evaluation of gas injection projects for EOR.

In VLE calculations, accuracy of the results basically depends on the method chosen for calculation of equilibrium ratios. In this regards suggestions given in Table 6.14 should be used as a guide. For calculation of CPT and WFT the multisolid-phase model provides a reliable method without the need for adjustable parameters. A good prediction of onset of asphaltene precipitation is possible through measurement of kinematic viscosity.

9.8 FINAL WORDS

Variety of methods for prediction and calculation of various thermophysical properties for petroleum and related fluids is much wider than the methods presented in this book. However, attempts were made to include the most accurate and widely used methods by the people from industry and researchers. Limitations of application of methods, points of strength and weaknesses, and their degrees of accuracy have been discussed for different systems. Furthermore, the basis of development of nearly all methods discussed in this book have been discussed so the students and new researchers in this area can understand the basic concepts and fundamentals of property calculations. In addition, the approaches presented in the book should help researchers in expansion of the existing methods and be used as a guide in the development of new predictive methods. The methods presented in the book should also help users of various simulators (process, PVT, phase behavior, etc.) to be able to select the most appropriate method for their property prediction purposes.

Empirical correlations should be used with caution and as a last option in absence of experimental data or accurate fundamentally based thermodynamic models. In use of these correlations their limitations and sensitivity to the input parameters must be considered. Some of these methods are reliable when the input parameters are determined through recommended methods. Perhaps the most accurate methods are those based on fundamental theoretical approach combined with empirically determined coefficients and parameters. In development of such relations availability of input parameters and accuracy of their measurements should be considered. Furthermore, predictive methods can have general application for a wide range of petroleum fluids if properties

of pure compounds have been used in their development in addition to data on petroleum mixtures from oils around the world. The weakest predictive methods are perhaps those empirically developed correlations that are based on a set of data for oils from a certain part of the world.

As it is shown in this book the main difficulty in prediction of properties of petroleum fractions relies on properties of heavy fractions containing polar multiring compounds with few experimental data available on their properties. As heavy compounds are generally polar with high boiling points, data on specific gravity and molecular weight alone are not sufficient for their property predictions. For such compounds it is not possible to measure critical properties or even boiling point. Boiling points of such compounds or their mixtures are not measurable and estimated boiling points based on distillation data at low pressures have little practical applications as they do not represent true boiling points. For such compounds one has to look at other properties that are directly measurable and represent their characteristics.

Reported values of critical properties of heavy compounds are usually predicted from methods developed for lighter hydrocarbons. For example, in the API-TDB [12] reported values of critical constants for heavy compounds are calculated from group contribution methods. Kesler–Lee method for calculation of critical properties of heavy hydrocarbons are based on calculated values from vapor pressure data [58]. Predicted values of critical constants and boiling point from different methods for heavy compounds differ significantly from each other, especially as carbon number increases. This leads to an even greater difference in predicted thermodynamic properties. Presence of very heavy compounds in a mixture requires a rigorous mixing rule for calculation of mixture properties. Attempts in this area should be focused on standardization of values of critical constants for heavy hydrocarbons and characterization of heavy oils.

Use of directly measurable properties in calculation of thermodynamic properties of heavy petroleum mixtures is an appropriate approach as it was discussed in Chapters 5 and 6. Use of velocity of sound to determine EOS parameters was demonstrated in Section 6.9 and new developments in this area are highly desirable [59]. Measurement and reporting of this thermodynamic property on heavy petroleum fractions and crude oils would help researchers to find methods of calculation of EOS parameters from measurable properties. Other useful and measurable properties for heavy oils include molecular weight, density, and refractive index. Use of refractive index in determination of EOS parameters has been shown in Section 5.9. It seems that more advanced equations of state such as SAFT equations would be more appropriate for prediction of thermodynamic properties of heavy oils such as those containing heavy residues, asphaltenes, and complex polar compounds. Investigation of this approach should be continued for more accurate estimation of thermophysical properties. Newly developed methods for phase equilibrium calculations and phase determination of many-component systems are useful tools in formulation and efficient prediction of hydrocarbon phase behavior and should be pursued [60].

Another appropriate approach in characterization of heavy oils was taken by Goual and Firoozabadi [23] to measure dipole moments of such complex systems. Attempts in measurement and reporting of such data should be continued to enable us in our understanding of properties of heavy petroleum fluids. Upon availability of such data it would be possible to develop more accurate and physically sound methods for characterization of heavy petroleum fractions and crude oils based on their degrees of polarity. Use of dipole moment in correlation of transport properties of polar fluids was shown by Chung et al. [61]. Measurement and effects of heteroatoms in such complex compounds on physical properties should also be considered with great emphasis. Presence of heteroatoms such as S, N, or O in a hydrocarbon compound can have appreciable impact on the properties of the compound.

The market for heavy oils and residues are limited; however, production of light oil in the world is in decline. Therefore, heavy oil conversion becomes increasingly important. Theoretically, the resources for heavy oils are infinite, as it is near to impossible to produce the last barrels of oils from heavy oil reservoirs. Considering limited information available on properties of heavy compounds, the focus of future studies must be on characterization of heavy hydrocarbons and petroleum fractions. In the area of solid formation and prevention methods generation and development of phase envelope diagrams for different reservoir fluids would be of importance for designers and operating engineers. In this book attempts were made to address some of the difficulties associated with property prediction of heavy and complex petroleum mixtures and with limited data available appropriate approaches are recommended; however, the challenge in this area of petroleum research continues.

9.9 PROBLEMS

9.1. **Three-Phase Flash**—Consider three phases of water, hydrocarbon, and vapor in equilibrium under reservoir conditions. Water (L1) and hydrocarbons (L2) in the liquid phase form two immiscible phases. Develop appropriate equations for three-phase flash calculations and derive relations for calculation of x_i^{L1}, x_i^{L2}, and y_i. Measurement and prediction of VLLE in water–hydrocarbon systems by PR EOS has been presented by Eubank et al. [62].

9.2. Derive Eq. (9.8) for calculation of GOR.

9.3. Calculate composition of liquid and gas streams from the third stage in Table 9.1 (also see Fig. 9.3) using Standing correlations for calculation of K_i.

9.4. Consider the PVT cell and the core sample shown in Fig. 9.26. The free volume is 268 cm³ and is filled initially with pure N_2. The core (porous media) has porosity of 0.31 and is filled with saturated oil with the following composition in terms of mole fraction (Table 9.14). The C_{7+} has molecular weight (M_{7+}) and specific gravity (SG_{7+}) of 228 and 0.853, respectively. Nitrogen diffuses into the core and light gases from matrix into the free

volume. The system reaches to final equilibrium state at pressure of 270 bar when temperature is kept constant at 403 K. Determine the bubble point pressure of oil at 403 K. Also determine the final equilibrium composition of gas in terms of mole fractions of N_2, CO_2, C_1, C_2, ($C_3 + C_4$), and C_{5+} in the free volume.

9.5. Consider a constant volume–temperature cylinder as shown in Fig. 8.13. The volume of cylinder is 96.64 cm³ and its length is 20.5 cm. Initially the cell is filled with 30 vol% liquid *n*-pentane at 311.1 K and 100 bar. The rest of the cylinder is filled with pure methane at the same initial temperature and pressure. Since the system is not in equilibrium it approaches to a final equilibrium state at a lower pressure keeping temperature of the cell constant. Through constant volume isothermal flash calculations using PR EOS and information given in the problem complete Table 9.15.

9.6. Composition of a reservoir fluid (gas condensate) separated in a separator at 300 psig and 62°F is given in Table 9.16. The C_{7+} properties are given as follows: $SG_{7+} = 0.795$ and $M_{7+} = 143$. Laboratory measured value of produced stock tank liquid-to-well stream ratio is 133.9 bbl/MMscf and the gas-to-feed ratio is 801.66 Mscf/MMscf. Associated gas (separator product) specific gravity is $SG_{gas} = 0.735$ and the primary stage

TABLE 9.15—*Properties of gas and liquid phases in a constant volume cell.*

Specification	Initial state	Final state
Temperature, K		
Pressure, bar		
Volume of the cell, cm³		
Volume of the liquid phase, cm³		
Volume of the gas phase, cm³		
Moles of liquid, mol		
Moles of gas, mol		
Molecular weight of liquid phase		
Molecular weight of gas phase		
Mass of liquid, g		
Mass of gas phase, g		
Density of liquid phase, g/cm³		
Density of gas phase, g/cm³		
Molar density of liquid, mol/cm³		
Molar density of gas, mol/cm³		
Length of the cell, cm		
Length of the liquid phase, cm		
Length of the gas phase, cm		
Volume fraction of the liquid		
Mole fraction of the gas phase in the cell		
Equilibrium ratio of methane		
Mole fraction of methane in the liquid		
Mole fraction of methane in the gas		

TABLE 9.14—*Composition of oil for Problem 9.4.*

N_2	CO_2	C_1	C_2	C_3	$i\text{-}C_4$	$n\text{-}C_4$	$i\text{-}C_5$	$n\text{-}C_5$	C_6	C_7
0.00114	0.02623	0.58783	0.06534	0.03560	0.00494	0.01558	0.00500	0.00872	0.01442	0.23519

TABLE 9.16—*Composition of reservoir fluid of Problem 9.6.*

Component	Well stream, mol%	Separator liquid, mol%	Separator gas, mol%
CO_2	0.18	Trace	0.22
N_2	0.13	Trace	0.16
C_1	61.92	7.78	75.31
C_2	14.08	10.02	15.08
C_3	8.35	15.08	6.68
i-C_4	0.97	2.77	0.52
n-C_4	3.41	11.39	1.44
i-C_5	0.84	3.52	0.18
n-C_5	1.48	6.50	0.24
C_6	1.79	8.61	0.11
C_{7+}	6.85	34.33	0.06
Total	100	100	100

GOR is 4428 scf/bbl at 60 °F. The API gravity of produced crude oil is 58.5. Calculate the following:

a. Composition of separator gas and liquid using Standing correlation for K_i.

b. SG_{gas} for separator gas.

c. API gravity of separator liquid.

d. GOR in scf/bbl.

e. Stock tank liquid to well stream ratio in barrels/MMscf.

f. Gas-to-feed ratio in Mscf/MMscf.

g. Compare predicted values with available laboratory values.

9.7. For the gas condensate sample of Problem 9.6 calculate Z factor at the reservoir conditions of 186°F and 5713 psia and compare it with the reported value of 1.107. What is the value of gas condensate expansion factor in Mscf for each bbl at reservoir conditions? The measured value is 1.591 Mscf/bbl.

9.8. For the gas condensate sample of Problem 9.6 calculate dew point pressure (P_d) at 186°F and compare it with the measured value of 4000 psia.

9.9. The following data (Table 9.17) on two types of Chinese recombined crude oils are given by Hu et al. [63]:
The reservoir temperature is at 339 K and measured bubble point pressures for oils 1 and 2 are 102.8 and 74.2 bar, respectively. Densities of C_{11+} fraction at 20°C for oils 1 and 2 are 0.91 and 0.921 g/cm^3, respectively. M_{11+} for oils 1 and 2 are 428 and 443, respectively. At the reservoir pressure of 150 bar, viscosities of oils 1 and 2 are 5.8 and 6.1 cP, respectively. Estimate the bubble point pressures from an EOS for these two oils and compare with available data.

9.10. Mei et al. [46] have reported experimental data on composition of a well stream fluid from West China oil field with compositions of separator gas and produced oil as given in Table 9.18. Reservoir conditions (T and P), saturation pressure of fluid at reservoir T, and the GOR of reservoir fluid are also given in this table. Density of the reservoir fluid (well stream under reservoir T and P) has been measured and reported. From analysis of data it is observed that there is an error in the composition of well stream as the sum of all numbers is 90.96 rather than 100. In addition reservoir temperature of 94 K is not correct (too low). Perform the following calculations to get correct values for the well stream composition and reservoir temperature.

a. Recombine separator gas and oil tank to get the original well stream. Make appropriate material balance calculations, using molecular weight, to generate well stream composition. Also determine if given GOR is in stm^3/m^3 or m^3/m^3 at separator conditions.

b. Use bubble-T calculations to calculate reservoir temperature at which corresponding saturation pressure is 311.5 bar.

c. Use trial-and-error procedure to find a temperature at which calculated density of reservoir fluid matches measured reported value at reservoir pressure. This

TABLE 9.17—*Data on two Chinese crudes for Problem 9.9 [63].*

Compound	N_2	CO_2	C_1	C_2	C_3	i-C_4	n-C_4	i-C_5	n-C_5	C_6	C_7	C_8	C_9	C_{10}	C_{11+}
Oil 1	1.20	0.20	30.90	3.50	2.87	0.33	1.41	0.40	1.02	1.69	2.46	2.98	2.53	2.15	46.36
Oil 2	0.96	0.16	24.06	0.76	3.26	0.64	2.70	0.52	1.06	0.70	0.580	1.86	2.30	0.82	59.62

TABLE 9.18—*Composition of an oil sample from Western China field [46].*

Component	Gas in separator, mol%	Oil^a in tank, mol%	Well stream^a, mol%
CO_2	0.62		0.52
N_2	5.94		4.97
C_1	67.35		56.36
C_2	11.51	0.08	0.64
C_3	7.22	0.47	6.12
i-C_4	2.31	0.55	2.02
n-C_4	2.41	1.01	2.18
i-C_5	0.89	1.19	0.94
n-C_5	0.72	1.38	0.83
C_6	0.59	4.06	1.16
C_7	0.31	5.65	1.14
C_8	0.13	13.50	2.31
C_9		8.53	1.39
C_{10}		6.26	1.02
C_{11+}^b		57.32	9.36

Initial reservoir pressure, bar	Reservoir temp, K	GOR, m^3/m^3	Saturation pressure, bar	Density of reservoir fluid, g/cm^3
410	94^a	440	311.5	0.5364

^{a}Well stream composition and reservoir temperature are not correct. Find the correct values.
^{b}C_{11+} fraction: $M_{11+} = 311$ and $SG_{11+} = 0.838$.

temperature must be near the temperature calculated in part b.

9.11. For the reservoir fluid of Problem 9.10 calculate the amount of wax precipitated (in mol%) at 280, 300, and 320 K and 410 bar. Also estimate WAT at 410 bar using solid solution model.

9.12. Calculate the CPT for crude oil 6 in Table 9.10 using multisolid-phase model. Also calculate the amount of wax precipitation in wt% at 240 K.

9.13. A natural gas has the composition of 70 mol% methane, 15 mol% ethane, 7 mol% propane, 5 mol% n-butane, and 3 mol% H_2S. What is the hydrate formation temperature (HFT) for this gas at pressure of 15 bars? What methanol solution (in terms of wt%) is needed to reduce HFT of the gas to 0°C?

9.14. A gas mixture of 75 mol% C_1, 10 mol% C_2, 10 mol% C_3, and 5 mol% n-C_4 exists at 690 bars. Calculate hydrate formation temperature and the concentration of methanol solution required to reduce it to 10°C.

REFERENCES

[1] Whitson, C. H. and Brule, M. R., *Phase Behavior,* Monograph Vol. 20, Society of Petroleum Engineers, Richardson, TX, 2000.

[2] Ahmed, T., *Hydrocarbon Phase Behavior,* Gulf Publishing, Houston, TX, 1989.

[3] Danesh, A., *PVT and Phase Behavior of Petroleum Reservoir Fluids,* Elsevier, Amsterdam, 1998.

[4] Elliott, J. R. and Lira, C. T., *Introductory Chemical Engineering Thermodynamics,* Prentice Hall, New Jersey, 1999 (www.phptr.com).

[5] Michelsen, M. L., "The Isothermal Flash Problem. Part I: Stability," *Fluid Phase Equilibria,* Vol. 9, No. 1, 1982, pp. 1–19.

[6] Michelsen, M. L., "The Isothermal Flash Problem. Part II: Phase Split Calculation," *Fluid Phase Equilibria,* Vol. 9, No. 1, 1982, pp. 21–40.

[7] Riazi, M. R., Aladwani, H. A., and Bishara, A., "Characterization of Reservoir Fluids and Crude Oils: Options and Restrictions," *Journal of Petroleum Science and Engineering,* Vol. 44, No. 2–4, 2004, pp. 195–207.

[8] Riazi, M. R., "Improved Phase Behavior Calculations of Reservoir Fluids Using a Distribution Model," *International Journal of Science and Technology (Scientia),* Vol. 10, No. 2, 2003, pp. 341–345.

[9] "EOR V90.2 Simulator from DB Robinson for Phase Behavior Calculations," D. B. Robinson and Associates, Edmonton, Alberta, Canada, 1995 (www.dbra.com).

[10] Ahmed, T., *Reservoir Engineering Handbook,* Gulf Publishing, Houston, TX, 2000.

[11] Li, C. C., "Critical Temperature Estimation for Simple Mixtures," *Canadian Journal of Chemical Engineering,* Vol. 49, No. 5, 1971, pp. 709–710.

[12] Daubert, T. E. and Danner, R. P., Eds., *API Technical Data Book—Petroleum Refining,* 6th ed., American Petroleum Institute (API), Washington, DC, 1997.

[13] Kreglewski, A. and Kay, W. B., "The Critical Constants of Conformal Mixtures," *Journal of Physical Chemistry,* Vol. 73, No. 10, 1969, pp. 3359–3366.

[14] Pedersen, K. S., Fredenslund, Aa., and Thomassen, P., *Properties of Oils and Natural Gases,* Gulf Publishing, Houston, TX, 1989.

[15] Speight, J. G., *The Chemistry and Technology of Petroleum,* 3rd ed., Marcel Dekker, New York, 1999.

[16] Firoozabadi, A., *Thermodynamics of Hydrocarbon Reservoirs,* McGraw-Hill, New York, 1999.

[17] Pan, H., Firoozabadi, A., and Fotland, P., "Pressure and Composition Effect on Wax Precipitation: Experimental Data and Model Results," *Society of Petroleum Engineers (SPE) Production and Facilities,* Vol. 12, No. 4, 1997, pp. 250–258.

[18] Dubey, S. T. and Waxman, M. H., "Asphaltene Adsorption and Desorption from Mineral Surface," SPE, Paper No. 18462, 1989, pp. 51–62.

[19] Yen, T. F. and Chilingarian, G. V., *Asphaltenes and Asphalts,* Development in Petroleum Sciences, 40A, Elsevier Science, Amsterdam, 1994.

[20] Pan, H. and Firoozabadi, A., "Thermodynamic Micellization Model for Asphaltene Aggregation and Precipitation in Petroleum Fluids," *Society of Petroleum Engineers (SPE) Production and Facilities,* Vol. 13, May 1998, pp. 118–127.

[21] ASTM, *Annual Book of Standards, Section Five: Petroleum Products, Lubricants, and Fossil Fuels* (in 5 Vols.), ASTM International, West Conshohocken, PA, 2002.

[22] *Standard Methods for Analysis and Testing of Petroleum and Related Products and British Standard 2000 Parts,* IP 1–324 Vol. 1, IP 325–448, Vol. 2, The Institute of Petroleum, London, 1999.

[23] Goual, L. and Firoozabadi, A., "Measuring Asphaltenes and Resins, and Dipole Moment in Petroleum Fluids," *American Institute of Chemical Engineers Journal,* Vol. 48, No. 11, 2002, pp. 2646–2662.

[24] Kim, S. T., Boudth-Hir, M. E., and Mansoori, G. A., "The Role of Asphaltene in Wettability Reversal," Paper No. 20700, Paper presented at the *65th Annual Technical Conference and Exhibition of Society of Petroleum Engineers,* New Orleans, LA, September 23–26, 1990.

[25] Kawanaka, S., Park, S. J., and Mansoori, G. A., "The Role of Asphaltene Deposition in EOR Gas Flooding: A Predictive Technique," Paper No. 17376, Paper presented at the *SPE/DOE Enhanced Oil Recovery Symposium,* Tulsa, OK, April 17–20, 1988.

[26] Zou, R. and Liu, L., "The Role of Asphaltene in Petroleum Cracking and Refining," in *Asphlatenes and Asphalts, I: Developments in Petroleum Science,* Vol. 40, Yen, T. F. and Chilingarian, G. V., Eds., Elsevier, Amsterdam, 1994, Ch. 14, pp. 339–363.

[27] Escobedo, J. and Mansoori, G. A., "Asphaltene and Other Heavy Organic Particle Deposition During Transfer and Production Operation," Paper No. 30672, Paper presented at the *SPE Annual Technical Conference and Exhibition,* Dallas, TX, October 22–25, 1995.

[28] Riazi, M. R. and Merrill, R. C., Eds., "Petroleum Exploration and Production Research in the Middle East," *Journal of Petroleum Science and Engineering,* Vol. 42, Nos. 2–4, April 2004, pp. 73–272 (http://www.elsevier.com/locate/petrol/).

[29] Mansoori, G. A., "Modeling of Asphaltene and Other Heavy Organic Depositions," *Journal of Petroleum Science and Engineering,* Vol. 17, 1997, pp. 101–111.

[30] Kawanaka, S., Park, S. J., and Mansoori, G. A., "Organic Deposition from Reservoir Fluids: A Thermodynamic Predictive Technique," *SPE Reservoir Engineering,* Vol. 6, May 1991, pp. 185–192.

[31] Janradhan, A. S. and Mansoori, G. A., "Fractal Nature of Asphaltene Aggregation," *Journal of Petroleum Science and Engineering,* Vol. 9, 1993, pp. 17–27.

[32] Mousavi-Dehghani, S. A., Riazi, M. R., Vafaie-Sefti, M., and Mansoori, G. A., "An Analysis of Methods for Determination of Onsets of Asphaltene Phase Separations," *Journal of Petroleum Science and Engineering,* Vol. 42, pp. 145–156.

[33] Buckley, J. S., "Microscopic Investigation of the Onset of Asphaltene Precipitation," *Fuel Science and Technology International*, Vol. 14, No. 1/2, 1996, pp. 55–74.

[34] Fotland, P., Anfindsen, H., and Fadnes, F. H., "Detection of Asphaltene Precipitation and Amounts Precipitated by Measurement of Electrical Conductivity," *Fluid Phase Equilibria*, Vol. 82, 1993, pp. 157–164.

[35] Escobedo, J. and Mansoori, G. A., "Viscometric Determination of the Onset of Asphaltene Flocculation: A Novel Method," Paper No. 28018, *Society of Petroleum Engngineers, Production and Facilities*, Vol. 10, May 1995, pp. 115–118.

[36] Pan, H. and Firoozabadi, A., "Thermodynamic Micellization Model for Asphaltene Precipitation Inhibition," *American Institute of Chemical Engineers Journal*, Vol. 46, No. 2, 2000, pp. 416–426.

[37] Piro, G., Rabaioli, M. R., Canonico, L. B., and Mazzolini, E. I., "Evaluation of Asphaltene Removal Chemicals: A New Testing Method," Paper No. 27386, Paper presented at the *SPE International Symposium on Formation Damage Control*, Lafayette, LA, February 7–10, 1994.

[38] Rassamdana, H. B., Dabir, B., Nematy, M., Farhani, M., and Sahimi, M., "Asphaltene Flocculation and Deposition: I. The Onset of Precipitation," *American Institute of Chemical Engineers Journal*, Vol. 42, No. 1, 1996, pp. 10–22.

[39] Wu, J., Prausnitz, J. M., and Firoozabadi, A., "Molecular-Thermodynamic Framework for Asphaltene-Oil Equilibria," *American Institute of Chemical Engineers Journal*, Vol. 44, No., 5, 1998, pp. 1188–1199.

[40] Pan, H. and Firoozabadi, A., "Thermodynamic Micellization Model for Asphaltene Precipitation from Reservoir Crudes at High Pressures and Temperatures," *Society of Petroleum Engineers (SPE) Production & Facilities*, Vol. 15, No. 1, February 2000, pp. 58–65.

[41] Won, K. W., Thermodynamics for Solid Solution–Liquid–Vapor Equilibria: Wax Formation from Heavy Hydrocarbon Mixtures," *Fluid Phase Equilibria*, Vol. 30, 1986, pp. 265–279.

[42] Pedersen, K., Skovborg, P., and Ronningsen, H. P., "Wax Presipitation from North Sea Crude Oils, 4: Thermodynamic Modeling," *Energy and Fuels*, Vol. 5, No. 6, 1991, pp. 924–932.

[43] Lira-Galeana, C. A., Firoozabadi, A., and Prausnitz, J. M., "Thermodynamics of Wax Precipitation in Petroleum Mixtures," *American Institute of Chemical Engineers Journal*, Vol. 42, 1996, pp. 239–248.

[44] Chung, T.-H., "Thermodynamic Modeling for Organic Solid Precipitation," Paper no. 24851, Paper presented at the *67th Annual Technical Conference and Exhibition*, Washington, DC, October 4–7, 1992.

[45] Hansen, J. H., Fredenslund, Aa., Pedersen, K. S., and Ronningsen, H. P., "A Thermodynamic Model for Predicting Wax Formation in Crude Oils," *American Institute of Chemical Engineers Journal*, Vol. 34, No. 12, 1988, pp. 1937–1942.

[46] Mei, H., Zhang, M., Li, L., Sun, L., and Li, S., "Research of Experiment and Modeling on Wax Precipitation for Pubei Oilfield," Paper (No. 2002–006) presented at the *International Petroleum Conference 2002, Canadian Petroleum Society*, Calgary, Alberta, Canada, June 11–13, 2002.

[47] Thomas, F. B., Bennion, D. B., and Bennion, D. W., "Experimental and Theoretical Studies of Solids Precipitation from Reservoir Fluid," *Journal of Canadian Petroleum Technology*, Vol. 31, No. 1, 1992, pp. 22–31.

[48] Nor-Azian, N. and Adewumi, M. A., "Development of Asphaltene Phase Equilibria Predictive Model," SPE Paper No. 26905, Paper presented at the *Eastern Regional Conference and Exhibition*, Pittsburgh, PA, November 2–4, 1993.

[49] Wu, J., Prausnitz, J. M., and Firoozabadi, A., "Molecular Thermodynamics of Asphaltene Precipitation in Reservoir Fluids," *American Institute of Chemical Engineers Journal*, Vol. 46, No. 1, 2000, pp. 197–209.

[50] Burke, N. E., Hobbs, R. E., and Kashou, S. F., "Measurement and Modeling of Asphaltene Precipitation," *Journal of Petroleum Technology*, Vol. 42, November 1990, pp. 1440–1447.

[51] Qin, X., "Asphaltene Precipitation and Implementation of Group Contribution Equation of State Into UTCOMP," M.Sc. Thesis, Department of Petroleum Engineering, University of Texas, Austin, TX, August 1998.

[52] Hirschberg, A., Delong, L. N. J., Schipper, B. A., and Meijers, J. G., "Influence of Temperature and Pressure on Asphaltene Flocculation," *Society of Petroleum Engineers Journal*, Vol. 24, No. 3, 1984, pp. 283–293.

[53] Ghuraiba, M. M. A., "Characterization of Asphaltenes in Crude Oil," M.Sc. Thesis, Chemical Engineering Department, Kuwait University, Kuwait, 2000.

[54] *Oil and Gas Journal Data Book*, 2000 Edition, PennWell, Tulsa, OK, 2000.

[55] Munck, J., Skjold-Jorgensen, S., and Rsmussen, P., "Compositions of the Formation of Gas Hydrates," *Chemical Engineering Science*, Vol. 43, No. 10, 1988, pp. 2661–2672.

[56] Riazi, M. R., Whitson, C. H., and da Silva, F., "Modeling of Diffusional Mass Transfer in Naturally Fractured Reservoirs," *Journal of Petroleum Science and Engineering*, Vol. 10, 1994, pp. 239–253.

[57] Warren, J. E. and Root, P. J., "The Behavior of Naturally Fractured Reservoirs," *Society of Petroleum Engineers Journal*, Vol. 3, 1963, pp. 245–251.

[58] Kesler, M. G. and Lee, B. I., "Improve Prediction of Enthalpy of Fractions," *Hydrocarbon Processing*, March 1976, pp. 153–158.

[59] Mossaad, E. and Eubank, P. T., "New Method for Conversion of Gas Sonic Velocities to Density Second Virial Coefficients," *American Institute of Chemical Engineers' Journal*, Vol. 47, No. 1, 2001, pp. 222–229.

[60] Iglesias-Silva, G. A., Bonilla-Petriciolet, A., Eubank, P. T., Holste, J. C., and Hall, K. R., "An Algebraic Method that Includes Gibbs Minimization for Performing Phase Equilibrium Calculations for Any Number of Components and Phases," *Fluid Phase Equilibria*, Vol. 210, No. 2, 2003, pp. 229–245.

[61] Chung, T. H., Ajlan, M., Lee, L. L., and Starling, K. E., "Generalized Multiparameter Correlation for Nonpolar and Polar Fluid Transport Properties," *Industrial Engineering and Chemical Research*, Vol. 27, 1988, pp. 671–679.

[62] Eubank, P. T., Wu, C. H., Alvarado, F. J., Forero, A., and Beladi, M. K., "Measurements and Prediction of Three-Phase Water/ Hydrocarbon Equilibria," *Fluid Phase Equilibria*, Vol. 102, No. 2, 1994, pp. 181–203.

[63] Hu, Y.-F., Li, S., Liu, N., Chu, Y.-P., Park, S. J., Mansoori, G. A., and Guo, T.-M., "Measurement and Corresponding States Modeling of Asphaltene Precipitation in Jilin Reservoir Oils," *Journal of Petroleum Science and Engineering*, Vol. 41, 2004, pp. 169–182.

Appendix

ASTM DEFINITIONS OF TERMS

ASTM DICTIONARY OF SCIENCE AND TECHNOLOGY[1] defines various engineering terms in standard terminology. ASTM provides several definitions for most properties by its different committees. The closest definitions to the properties used in the book are given below. The identifier provided includes the standard designation in which the term appears followed by the committee having jurisdiction of that standard. For example, D02 represents the ASTM Committee on Petroleum Products and Lubricants.

Additive—Any substance added in small quantities to another substance, usually to improve properties; sometimes called a modifier. **D 16, D01**

Aniline point—The minimum equilibrium solution temperature for equal volumes of aniline (aminobenzene) and sample. **D 4175, D02**

API gravity—An arbitrary scale developed by the American Petroleum Institute and frequently used in reference to petroleum insulating oil. The relationship between API gravity and specific gravity 60/60°F is defined by the following: Degree API gravity at 60°F = 141.5/(SG 60/60°F)−131.5. [*Note*: For definition see Eq. (2.4) in this book.] **D 2864, D27**

Ash—Residue after the combustion of a substance under specified conditions. **D 2652, D28**

Assay—Analysis of a mixture to determine the presence or concentration of a particular component. **F 1494, F23**

Autoignition—The ignition of material caused by the application of pressure, heat, or radiation, rather than by an external ignition source, such as a spark, flame, or incandescent surface. **D 4175, D02**

Autoignition temperature—The minimum temperature at which autoignition occurs. **D 4175, D02**

Average (for a series of observations)—The total divided by the number of observations. **D123, D13**

Bar—Unit of pressure; 14.5 lb/in^2, 1.020 kg/cm^2, 0.987 atm, 0.1 MPa. **D 6161, D19**

Bitumen—A class of black or dark-colored (solid, semisolid, or viscous) cementitious substances, natural or manufactured, composed principally of high-molecular-weight hydrocarbons, of which asphalts, tars, pitches, and asphaltites are typical. **D 8, D04**

Boiling point—The temperature at which the vapor pressure of an engine coolant reaches atmospheric pressure under equilibrium boiling conditions. [*Note*: This definition is applicable to all types of liquids.] **D 4725, D15**

Boiling pressure—At a specified temperature, the pressure at which a liquid and its vapor are in equilibrium. **E 7, E04**

[1] *ASTM Dictionary of Engineering Science and Technology*, 9th ed., ASTM International, West Conshohocken, PA, 2000.

BTU—One British thermal unit is the amount of heat required to raise 1 lb of water 1°F. **E 1705, E48**

Carbon black—A material consisting essentially of elemental carbon in the form of near-spherical colloidal particles and coalesced particle aggregates of colloidal size, obtained by partial combustion or thermal decomposition of hydrocarbons. **D 1566, D11**

Carbon residue—The residue formed by evaporation and thermal degradation of a carbon-containing material. **D 4175, D02**

Catalyst—A substance whose presence initiates or changes the rate of a chemical reaction, but does not itself enter into the reaction. **C 904, C03**

Cetane number (cn)—A measure of the ignition performance of a diesel fuel obtained by comparing it to reference fuels in a standardized engine test. **D 4175, D02**

Chemical potential (μ_i or $\bar{G}_i$)—The partial molar free energy of component i, that is, the change in the free energy of a solution upon adding 1 mol of component i to an infinite amount of solution of given composition, $(\delta G/\delta n_i)_{T,P,n_i} = \bar{G}_i = \mu_i$, where G = Gibbs free energy and n_i = number of moles of the ith component. **E 7, E04**

Cloud point—The temperature at which a defined liquid mixture, under controlled cooling, produces perceptible haze or cloudiness due to the formation of fine particles of an incompatible material. **D 6440, D01**

Coal—A brown to black combustible sedimentary rock (in the geological sense) composed principally of consolidated and chemically altered plant remains. **D 121, D05**

Coke—A carbonaceous solid produced from coal, petroleum, or other materials by thermal decomposition with passage through a plastic state. **C 709, D02**

Combustion—A chemical process of oxidation that occurs at a rate fast enough to produce heat and usually light either as glow or flames. **D 123, D13**

Compressed natural gas (CNG)—Natural gas that is typically pressurized to 3600 psi. CNG is primarily used as a vehicular fuel. **D 4150, D03**

Concentration—Quantity of substance in a unit quantity of sample. **E 1605, E06**

Critical point—In a phase diagram, that specific value of composition, temperature, pressure, or combinations thereof at which the phases of a heterogeneous equilibrium become identical. **E 7, E04**

Critical pressure—Pressure at the critical point. **E 1142, E37**

Critical temperature—(1) Temperature above which the vapor phase cannot be condensed to liquid by an increase in pressure. **E 7, E04**
(2)Temperature at the critical point. **E 1142, E37**

Degradation—Damage by weakening or loss of some property, quality, or capability. **E 1749, E 06**

Degree Celsius (°C)—Derived unit of temperature in the International System of Units (SI). **E 344, E20**

Density—The mass per unit volume of a substrate at a specified temperature and pressure; usually expressed in g/mL, kg/L, g/cm^3, g/L, kg/m^3, or lb/gal. **D 16, D01**

Deposition—The chemical, mechanical, or biological processes through which sediments accumulate in a resting place. **D 4410, D19**

Dew point—The temperature at any given pressure at which liquid initially condenses from a gas or vapor. It is specifically applied to the temperature at which water vapor starts to condense from a gas mixture (water dew point) or at which hydrocarbons start to condense (hydrocarbon dew point). **D 4150, D03**

Diffusion—(1) Spreading of a constituent in a gas, liquid, or solid tending to make the composition of all parts uniform. (2) The spontaneous movement of atoms or molecules to new sites within a material. **B 374, B08**

Distillation—The act of vaporizing and condensing a liquid in sequential steps to effect separation from a liquid mixture. **E 1705, E 48**

Distillation temperature (in a column distillation)—The temperature of the saturated vapor measured just above the top of the fractionating column. **D 4175, D02**

Endothermic reaction—A chemical reaction in which heat is absorbed. **C 1145, C 28**

Enthalpy—A thermodynamic *function* defined by the equation $H = U + PV$, where H is the enthalpy, U is the *internal* energy, P is the pressure, and V the volume of the system. [*Note*: Also see Eq. (6.1) of this book.] **E 1142, E37**

Equilibrium—A state of dynamic balance between the opposing actions, reactions, or velocities of a reversible process. **E 7, E04**

Evaporation—Process where a liquid (water) passes from a liquid to a gaseous state. **D 6161, D19**

Fire point—The lowest temperature at which a liquid or solid specimen will sustain burning for 5 s. **D 4175, D02**

Flammable liquid—A liquid having a flash point below 37.8°C (100°F) and having a vapor pressure not exceeding 40 psi (absolute) at 37.8°C and known as a Class I liquid. **E 772, E44**

Flash point—The lowest temperature of a specimen corrected to a pressure of 760 mm Hg (101.3 kPa), at which application of an ignition source causes any vapor from the specimen to ignite under specified conditions of test. **D 1711, D09**

Fluidity—The reciprocal of viscosity. **D 1695, D01**

Freezing point—The temperature at which the liquid and solid states of a substance are in equilibrium at a given pressure (usually atmospheric). For pure substances it is identical with the melting point of the solid form. **D 4790, D16**

Gas—One of the states of matter, having neither independent shape nor volume and tending to expand indefinitely. **D 1356, D22**

Gasification—Any chemical or heat process used to convert a feedstock to a gaseous fuel. **E 1126, E 48**

Gasoline—A volatile mixture of liquid hydrocarbons, normally containing small amounts of additives, suitable for use as a fuel in spark-ignition internal combustion engines. **D 4175, D02**

Gibbs free energy—The maximum useful work that can be obtained from a chemical system without net change in temperature or pressure, $\Delta F = \Delta H - T \Delta S$. [*Note*: For definition see Eq. (6.6) in this book; the author has used G for Gibbs free energy.] **E 7, E04**

Grain—Unit of weight; 0.648 g, 0.000143 lb. **D 6161, D19**

Gross calorific value (synonym: higher heating value, HHV)—The energy released by combustion of a unit quantity of refuse-derived fuel at constant volume or constant pressure in a suitable calorimeter under specified conditions such that all water in the products is in liquid form. This the measure of calorific value is predominately used in the United States. **E 856, D34**

Heat capacity—The quantity of heat required to raise a system 1° in temperature either at constant volume or constant pressure. **D 5681, D34**

Heat flux (q)—The heat flow rate through a surface of unit area perpendicular to the direction of heat flow (q in SI units: W/m^2; q in inch-pound units: Btu/h/ft^2 = Btu/h · ft^2) **C 168, C16**

Henry's law—The principle that the mass of a gas dissolved in a liquid is proportional to the pressure of the gas above the liquid. **D 4175, D02**

Higher heating value (HHV)—A synonym for gross calorific value. **D 5681, D34**

Inert components—Those elements or components of natural gas (fuel gas) that do not contribute to the heating value. **D 4150, D03**

Inhibitor—A substance added to a material to retard or prevent deterioration. **D 4790, D16**

Initial boiling point—The temperature observed immediately after the first drop of distillate falls into the receiving cylinder during a distillation test. **D 4790, D 16**

Interface—A boundary between two phases with different chemical or physical properties. **E 673, E 42**

Interfacial tension (IFT)—The force existing in a liquid–liquid phase interface that tends to diminish the area of the interface. This force, which is analogous to the surface tension of liquid–vapor interfaces, acts at each point on the interface in the plane tangent at that point. **D 459, D12**

International System of Units, SI—A complete coherent system of units whose base units are the meter, kilogram, second, ampere, kelvin, mole, and candela. Other units are derived as combinations of the base units or are supplementary units. **A 340, A06**

Interphase—The region between two distinct phases over which there is a variation of a property. **E 673, E42**

ISO—Abbreviation for International Organization for Standards: An organization that develops and publishes international standards for a variety of technical applications, including data processing and communications. **E 1457, F05**

Jet fuel—Any liquid suitable for the generation of power by combustion in aircraft gas turbine engines. **D 4175, D02**

Joule (J)—The unit of energy in the SI system of units. One joule is 1 W···. **A 340, A06**

Kelvin (K)—The unit of thermodynamic temperature; the SI unit of temperature for which an interval of 1 kelvin (K) equals exactly an interval of 1°C and for which a level of 273.15 K equals exactly 0°C. **D 123, D13**

Liquefied petroleum gas (LPG)—A mixture of normally gaseous hydrocarbons, predominantly propane or butane or both, that has been liquefied by compression or cooling, or both, to facilitate storage, transport, and handling.
D 4175, D02

Liquid—A substance that has a definite volume but no definite form, except such given by its container. It has a viscosity of 1×10^{-3} to 1×10^3 St (1×10^{-7} to 1×10^{-1} m$^2 \cdot$ s^{-1}) at 104°F (40°C) or an equivalent viscosity at agreed upon temperature. (This does not include powders and granular materials.) Liquids are divided into two classes:

(1) Class A, low viscosity—A liquid having a viscosity of 1×10^{-3} to 25.00 St (1×10^{-7} to 25.00×10^{-4} m$^2 \cdot$ s^{-1}) at 104°F (40°C) or an equivalent viscosity at agreed upon temperature.
(2) Class B, high viscosity—A liquid having a viscosity of 25.01 to 1×10^3 St (25.01×10^{-4} to 1×10^{-1} m$^2 \cdot$ s^{-1}) at 104°F (40°C) or an equivalent viscosity at agreed upon temperature.
D 16, D01

Lower heating value (LHV)—A synonym for net calorific value.
D 5681, D34

Lubricant—Any material interposed between two surfaces that reduces the friction or wear between them.
D 4175, D02

Mass—The quantity of matter in a body (also see weight).
D 123, D13

Melting point—In a phase diagram, the temperature at which the liquids and solids coincide at an invariant point.
E 7, E04

Micron (μm, micrometer)—A metric unit of measurement equivalent to 10^{-6} m, 10^{-4} cm.
D 6161, D19

Molality—Moles (gram molecular weight) of solute per 1000 g of solvent.
D 6161, D19

Molarity—Moles (gram molecular weight) of solute per liter of total solution
D 6161, D19

Molecular diffusion—A process of spontaneous intermixing of different substances, attributable to molecular motion, and tending to produce uniformity of concentration.
D1356, D22

Mole fraction—The ratio of the number of molecules (or moles) of a compound or element to the total number of molecules (or moles) present.
D 4023, D22

Naphtha, aromatic solvent—A concentrate of aromatic hydrocarbons including C_8, C_9, and C_{10} homologs.
D 4790, D 16

Napthenic oil—An hydrocarbon process oil containing more than 30%, by mass, of naphthenic hydrocarbons.
D 1566, D11

Natural gas—A naturally occurring mixture of hydrocarbon and nonhydrocarbon gases found in porous geological formations (reservoirs) beneath the earth's surface, often in association with petroleum. The principal constituent of natural gas is methane.
D 4150, D03

Net calorific value (Net heat of combustion at constant pressure)—The heat produced by combustion of unit quantity of a solid or liquid fuel when burned, at constant pressure of 1 atm (0.1 MPa), under the conditions such that all the water in the products remains in the form of vapor.
D 121, D05

Net heat of combustion—The oxygen bomb (see Test Method D 3286) value for the heat of combustion, corrected for gaseous state of product water.
E 176, E05

Octane number (for spark ignition engine fuel)—Any one of several numerical indicators of resistance to knock obtained by comparison with reference fuels in standardized engine or vehicle tests.
D 4175, D02

Oxygenate—An oxygen-containing ashless organic compound, such as an alcohol or ether, which may be used as a fuel or fuel supplement.
D 4175, D02

Paraffinic oil—A petroleum oil (derived from paraffin crude oil) whose paraffinic carbon type content is typically greater than 60%.
E 1519, E35

Partial pressure—The contribution of one component of a system to the total pressure of its vapor at a specified temperature and gross composition.
E 7, E04

Porosity—The percentage of the total volume of a material occupied by both open and closed pores. [*Note*: In this book porosity represented by ϕ (see Eq. 8.72) is the fraction of total volume of a material occupied by *open pores* and is not identical to this definition.]
C 709, D02

Pour point—The lowest temperature at which a liquid can be observed to flow under specified conditions.
D 2864, D27

Precipitation—Separation of new phase from solid, liquid, or gaseous solutions, usually with changing conditions or temperature or pressure, or both.
E 7, E04

Pressure—The internal force per unit area exerted by any material. Since the pressure is directly dependent on the temperature, the latter must be specified.
D 3064, D10

Pressure, saturation—The pressure, for a pure substance at any given temperature, at which vapor and liquid, or vapor and solid, coexist in stable equilibrium. [*Note*: This is the definition of vapor pressure used in this book.]
E 41, G03

Quality—Collection of features and characteristics of a product, process, or service that confers its ability to satisfy stated or implied needs.
E 253, E18

Range—The region between the limits within which a quantity is measured and is expressed by stating the lower and upper range values.
E 344, E20

Refractive index—The ratio of the velocity of light (of specified wavelength) in air to its velocity in the substance under examination. This is relative refractive index of refraction. If absolute refractive index (that is, referred to vacuum) is desired, this value should be multiplied by the factor 1.00027, the absolute refractive index of air. [*Note*: In this book absolute refractive index is used.]
D 4175, D02

Saturation—The condition of coexistence in stable equilibrium of a vapor and a liquid or a vapor and solid phase of the same substance at the same temperature.
E 41, G03

Smoke point—The maximum height of a smokeless flame of fuel burned in a wick-fed lamp.
D 4175, D02

Solid—A state of matter in which the relative motion of molecules is restricted and in which molecules tend to retain a definite fixed position relative to each other. A solid may be said to have a definite shape and volume.
E 1547, E 15

Solubility—The extent that one material will dissolve in another, generally expressed as mass percent, or as volume percent, or parts per 100 parts of solvent by mass or volume. The temperature should be specified. **D 3064, D10**

Solubility parameter (of liquids)—The square root of the heat of vaporization minus work of vaporization (cohesive energy density) per unit volume of liquid at 298 K. **D 4175, D02**

Solutes—Matter dissolved in a solvent. **D 6161, D19**

Specific gravity (deprecated term of liquids)—The ratio of density of a substance to that of a reference substance such as water (for solids and liquids) or hydrogen (for gases) under specified conditions. Also called relative density. [*Note*: In this book the reference substance for definition of gas specific gravity is air]. **D 4175, D02**

Surface tension—Property that exists due to molecular forces in the surface film of all liquids and tends to prevent the liquid from spreading. **B 374, B08**

Temperature—The thermal state of matter as measured on a definite scale. **B 713, B01**

Thermal conductivity (λ)—Time rate of heat flow, under steady conditions, through unit area, per unit temperature gradient in the direction perpendicular to the area. **E 1142, E37**

Thermal diffusivity—Ratio of thermal conductivity of a substance to the product of its density and specific heat capacity. **E1142, E37**

Vapor—The gaseous phase of matter that normally exists in a liquid or solid state. **D 1356, D22**

Vapor pressure—The pressure exerted by the vapor of a liquid when in equilibrium with the liquid. **D 4175, D02**

Viscosity, absolute (η)—The ratio of shear stress to shear rate. It is the property of internal resistance of a fluid that opposes the relative motion of adjacent layers [*Note*: See Eq. (8.1) in this book.] The unit most commonly used for insulating fluids is centipoise. **D 2864, D27**

Viscosity, kinematic—The quotient of the absolute (dynamic) viscosity divided by the density, η/ρ both at the same temperature. For insulating liquids, the unit most commonly unit is the centistokes (100 cSt = 1 St). [*Note*: See Eq. (8.1) in this book.] **D 2864, D27**

Viscosity, Saybolt Universal—The efflux time in seconds of 60 mL of sample flowing through a calibrated Saybolt Universal orifice under specified conditions. **D 2864, D27**

Wax appearance point—The temperature at which wax or other solid substances first begin to separate from the liquid oil when it is cooled under prescribed conditions (refer to D 3117, Test Method for Wax Appearance Point of Distillate Fuels). **D 2864, D27**

Weight (synonymous with mass)—The mass of a body is a measure of its inertia, or resistance to change in motion. **E 867, E17**

Greek Alphabet

α	Alpha
β	Beta
Γ	Gamma (Uppercase)
γ	Gamma
Δ	Delta (Uppercase)
δ	Delta
ε	Epsilon
ζ	Zeta
η	Eta
Θ	Theta (Uppercase)
θ	Theta
K	Kappa (Uppercase)
κ	Kappa
Λ	Lambda (Uppercase)
λ	Lambda
μ	Mu
ν	Nu
ξ	Xi
Π	Pi (Uppercase)
π	Pi
ρ	Rho
Σ	Sigma (Uppercase)
σ	Sigma
τ	Tau
υ	Upsilon
Φ	Phi (Uppercase)
ϕ	Phi
φ	Phi
χ	Chi
Ψ	Psi (Uppercase)
Ω	Omega (Upper case)
ω	Omega

Index